Optics for Engineers

This accessible textbook introduces the fundamentals of geometric and physical optics as they relate to practical problems encountered by engineers and researchers in designing and analysing optical systems.

In this updated edition, the author focuses on topics that are critical to understanding how the basic principles of optics affect design decisions. In addition to information on breadboarding experiments and prototypes, the new edition also expands its coverage of diffraction and includes numerous complete examples, and practical reminders.

Professor Charles A. DiMarzio is an associate professor in the Department of Electrical and Computer Engineering, the Department of Mechanical and Industrial Engineering, and the Department of Bioengineering at Northeastern University in Boston, Massachusetts. He spent 14 years at Raytheon Company's Electro-Optics Systems Laboratory in coherent laser radar for air safety and meteorology. Among other projects there, he worked on an airborne laser radar, flown on the Galileo-II, to monitor airflow related to severe storms, pollution, and wind energy, and another laser radar to characterize the wake vortices of landing aircraft. His current research in biomedical optics focuses on microscopy including coherent imaging, structured illumination, and multimodal. He is also a founding member of Gordon-CenSSIS—the Gordon Center for Subsurface Sensing and Imaging Systems.

Optics for Engineers

Second Edition

Charles A. DiMarzio

CRC Press is an imprint of the
Taylor & Francis Group, an **informa** business

Designed cover image: shutterstock_ 9781032650401

Second edition published 2024
by CRC Press
2385 NW Executive Center Drive, Suite 320, Boca Raton FL 33431

and by CRC Press
4 Park Square, Milton Park, Abingdon, Oxon, OX14 4RN

CRC Press is an imprint of Taylor & Francis Group, LLC

ISBN: 978-1-032-65040-1 (hbk)
ISBN: 978-1-482-26323-7 (pbk)
ISBN: 978-1-315-15704-7 (ebk)

DOI: 10.1201/9781315157047

Typeset in CMR10 font
by KnowledgeWorks Global Ltd.

To my late Aunt, F. Esther DiMarzio, retired principal of Kingston (Massachusetts) Elementary School, who, by her example, encouraged me to walk the path of learning and teaching, and to my wife Sheila, who walks it with me.

Contents

List of Figures

List of Tables

Acknowledgments

Throughout my career in optics, I have been fortunate to work among exceptionally talented people who have contributed to my understanding and appreciation of the field. I chose the field as a result of my freshman physics course, taught by Clarence E. Bennett at the University of Maine. My serious interest in optics began at Worcester Polytechnic Institute, guided by Lorenzo Narducci, who so influenced me that during my time there, when I became engrossed in reading a research paper, I heard myself saying the words in his Italian-accented English. After WPI, it was time to start an industry career, and it was my good fortune to find myself in Raytheon Company's Electro-Optics Systems Laboratory. I began to write this preface on the evening of the funeral of Albert V. Jelalian, who led that laboratory, and I found myself reflecting with several friends and colleagues on the influence that "Big Al" had on our careers. I was indeed fortunate to start my career surrounded by the outstanding people and resources in his group. After 14 years, it was time for a new adventure, and Professor Michael B. Silevitch gave me the opportunity to work in his Center for Electromagnetics Research, taking over the laboratory of Richard E. Grojean. Dick was the guiding force behind the course, which is now called *Optics for Engineers.* The door was now open for me to complete a doctorate and become a member of the faculty. Michael's next venture, CenSSIS, the Gordon Center for Subsurface Sensing and Imaging Systems, allowed me to apply my experience in the growing field of biomedical optics and provided a strong community of mentors, colleagues, and students with a common interest in trying to solve the hardest imaging problems. My research would have not been possible without my CenSSIS collaborators, including Carol M. Warner, Dana H. Brooks, Ronald Roy, Todd Murray, Carey M. Rappaport, Stephen W. McKnight, Anthony J. Devaney, and many others. Ali Abur, my first department chair in the Department of Electrical and Computer Engineering, encouraged my writing and provided an exciting environment in which to teach and perform research. My long-time colleague and collaborator, Gregory J. Kowalski, provided a wealth of information about optics in mechanical engineering, introduced me to a number of other colleagues, and encouraged my secondary appointment in the Department of Mechanical and Industrial Engineering, where I continue to enjoy fruitful collaborations with Greg, Jeffrey W. Ruberti, Andrew Gouldstone, and others.

I am indebted to Dr. Milind Rajadhyaksha, who worked with me at Northeastern for several years and continues to collaborate from his new location at Memorial Sloan Kettering Cancer Center. Milind provided all my knowledge of confocal microscopy, along with professional and personal encouragement. The infrastructure of my research includes the W. M. Keck Three-Dimensional Fusion Microscope. This facility has been managed by Gary S. Laevsky and Josef Kerimo. Their efforts have been instrumental in my own continuing optics education and that of all my students.

I am grateful to my editor for the first edition, Lou Han, formerly at Taylor & Francis Group. Lou found the course material I had prepared for *Optics for Engineers*, suggested that I write this book, and guided me through the process of creating the first edition. Emma Morley and Betsy Byers provided valuable guidance on the second edition and I could not have completed it without their help, and their patience with many missed deadlines.

Thomas J. Gaudette was a student in my group for several years, and while he was still an undergraduate, introduced me to MATLAB®. Now working at MathWorks, he has continued to improve my programming skills, while providing encouragement and friendship. As I developed scripts to generate the figures, he provided answers to all my questions, and in many cases, rewrote my code to do what I wanted it to do.

I would like to acknowledge the many students who have taken this course over the years and provided continuous improvement to the notes that led to this book. Since 1987, nearly 300 students have passed through our Optical Science Laboratory, some for just a few weeks and some for several years. Every one of them has contributed to the body of knowledge that is contained here. A number of colleagues and students have helped greatly by reading drafts of chapters and making significant suggestions that have improved the final result. Josef Kerimo, Heidy Sierra, Yogesh Patel, Zhenhua Lai, Jason M. Kellicker, and Michael Iannucci all read and corrected chapters. I wish to acknowledge the special editing efforts of four people whose thoughtful and complete comments led to substantial revisions: Andrew Bongo, Joseph L. Hollmann, Matthew J. Bouchard, and William C. Warger II. Matt and Bill read almost every chapter of the initial draft, managed our microscope facility during the time between permanent managers, and made innumerable contributions to my research group and to this book. Bill and Joseph read large portions of the second edition and provided comments to improve the clarity and completeness.

Finally, I thank my wife, Sheila, for her unquestioning support, encouragement, and understanding during my nearly two years of writing and through all of my four-decade career.

Preface

Over the last few decades, the academic study of optics, once exclusively in the domain of physics departments, has expanded into a variety of engineering departments. Graduate students and advanced undergraduate students performing research in such diverse fields as electrical, mechanical, biomedical, and chemical engineering often need to assemble optical experimental equipment to complete their research. In industry, optics is now found in products such as laser printers, barcode scanners, and even mobile phones. This book is intended first as a textbook for an introductory optics course for graduate and advanced undergraduate students. It is also intended as a resource for researchers developing optical experimental systems for their work and for engineers faced with optical design problems in industry.

This book is based on a graduate course taught at Northeastern University to students in the Department of Electrical and Computer Engineering as well as to those from physics and other engineering departments. In over three decades as an advisor here, I have found that almost every time a student asks a question, I have been able to refer to the material presented in this book for an answer. The examples are mostly drawn from my own experience in laser radar and biomedical optics, but the results can almost always be applied in other fields. I have emphasized the broad fundamentals of optics as they apply to problems the intended readers will encounter.

My goal has been to provide a rigorous introduction to the field of optics with all the equations and assumptions used at each step. Each topic is addressed with some theoretical background, leading to the everyday equations in the field.

The background material is presented so that the reader may develop familiarity with it and so that the underlying assumptions are made clear, but a complete understanding of each step along the way is not required. In this second edition, these paragraphs are shown with a gray background like the current paragraph. These paragraphs may often be omitted by the user searching for a particular result. The reader interested in understanding the derivations of equations and their limitations is encouraged to read these details.

"Key equations," enclosed in boxes, include the most important and frequently used equations.

$$y = mx + b. \tag{0.1}$$

Most chapters and many sections of chapters include a summary and some examples of applications.

Take-Away

Take-Away Message:
In particular, key points are highlighted in boxes like this one, entitled "Take-Away." Usually these highlight

the main points to remember, and are often shown at the end of a section.

Example

Boxes entitled, "Example," show completed examples illustrating the application of the equations. Many of these are derived from the experience of the author and his students. These can often show a path for the reader to complete a calculation related to the design or analysis of a system.

In Practice

Boxes entitled, "In Practice," show how the material is related to helpful practical hints that may be useful in a computation or in an experiment.

Specific topics such as lasers, detectors, optical design, radiometry, and nonlinear optics are briefly introduced, but a complete study of each of these topics is best undertaken in a separate course, with a specialized textbook. Throughout this book, I provide pointers to a few of the many excellent texts in these subjects.

Organizing the material in a book of this scope was a challenge. I chose an approach that I believe makes it easiest for the student to assimilate the material in sequential order. It also allows the working engineer or graduate student in the laboratory to read a particular chapter or section that will help complete a particular task. For example, one may wish to design a common-aperture transmitter and receiver using a polarizing beam splitter, as in Section 6.6.5.2. Chapter 6 provides the necessary information to understand the design and to analyze its performance using components of given specifications. On some occasions, I will need to refer forward for details of a particular topic. For example, in discussing aberrations in Chapter 5, it is important to discuss the diffraction limit, which is not introduced until Chapter 8. In such cases, I have attempted to provide a short discussion of the phenomenon with a forward reference to the details. I have also repeated some material with the goal of making reading simpler for the reader who wishes to study a particular topic without reading the entire book sequentially. The index should be helpful to such a reader as well.

Chapter 1 provides an overview of the history of optics and a short discussion of Maxwell's equations, the wave equation, and the eikonal equation, which form the mathematical basis of the field of optics. Some readers will find in this background a review of familiar material, while others may struggle to understand it. The latter group need not be intimidated. Although we can address any topic in this book beginning with Maxwell's equations, in most of our everyday work we use equations and approximations that are derived from these fundamentals to describe imaging, interference, diffraction, and other phenomena. A level of comfort with Maxwell's equations is helpful, but not essential, to applying these derived equations.

After the introduction, Chapters 2 through 5 discuss geometric optics. Chapter 2, specifically, develops the concepts of real and virtual images, the lens equation to find the paraxial image location, and transverse, angular, and axial magnifications, all from Snell's law. Chapter 3 discusses the ABCD matrices, providing a simplified bookkeeping scheme for the equations of Chapter 2, but also providing the concept of principal planes and some good rules of thumb for lenses, and an important "invariant" relating transverse and angular magnification. Chapter 4 treats the issue of finite apertures, including aperture stops and field stops, and Chapter 5 gives a brief introduction to aberrations. While most readers will be eager to get beyond these four chapters, the material is essential to most optics projects, and many projects have failed because of a lack of attention to geometric optics.

Chapters 6 through 8 present the fundamentals of physical optics: polarization, interference, and diffraction. The material is presented in sufficient depth to enable the reader to solve many realistic problems. For example, Chapter 6 discusses polarization in progressively more detailed terms, ending with Jones calculus, coherency matrices, and Mueller calculus, which include partial polarization. Chapter 7 discusses basic concepts of interference, interferometers, and laser cavities, ending with a brief study of thin-film interference coatings for windows, beam splitters, and filters. Chapter 8 presents equations for Fresnel and Fraunhofer diffraction with a number of examples. Chapter 9 continues the discussion of diffraction with some closed-form expressions for the important case of Gaussian beams. The important concept of coherence looms large in physical optics, and is discussed in Chapter 10, which will help the reader understand the applicability of the results in Chapters 6 through 9 and will set the stage for a short discussion of Fourier optics in Chapter 11, which continues on the subject of diffraction.

The measurement and quantification of light are important in determining the performance limits of optical systems. Chapter 12 discusses the subjects of radiometry and photometry, giving the reader a set of quantities that are frequently used in the measurement of light, and Chapter 13 discusses the detection and measurement of light. Chapter 14 provides a brief introduction to nonlinear optics. Finally Chapter 15 in this second edition is based on another course entitled "Designing and prototyping optical systems," which introduces the optics laboratory environment and includes some experiments that expose the reader to techniques that will be useful in many applications of optics.

The material in this book was designed for an ambitious one-semester graduate course, although with additional detail, it can easily be extended to two semesters. Advanced undergraduates can understand most of the material, but for an undergraduate course I have found it best to skim quickly over most of Chapter 5 on aberrations and to leave out Section 6.7 on partial polarization and all of Chapters 10, 11, 13, and 14 on coherence, Fourier optics, detectors, and nonlinear optics, respectively.

At the end of most chapters are a few problems to challenge the reader. In contrast to most texts, I have chosen to offer a small number of more difficult but realistic problems, mostly drawn from my experience in laser radar, biomedical optics, and other areas. Although some solutions are provided in Appendix G, I recommend that the reader make a serious effort to formulate a solution before looking at the solution. There is much to be learned even through unsuccessful attempts. Particularly difficult problems are labeled with a (G), to denote that they are specifically for graduate students. A number of problems have been added to the second edition, mostly without solutions, and labeled (NS). To summarize the homework labels;

0.1 A problem. This homework problem is intended for all readers. Answers are included in Appendix G.

0.2 Another problem. (G) This one is recommended for graduate students or more advanced researchers. Answers are included in Appendix G.

0.3 Another problem. (NS) This one does **not** have a solution included.

The field of optics is so broad that it is difficult to achieve a uniform notation that is accepted by most practitioners, easy to understand, and avoids the use of the same symbol with different meanings. Although I have tried to reach these goals in this book, it has not always been possible. In geometric optics, I used capital letters to represent points and lowercase letters to represent corresponding distance. For example, the front focal point of a lens is called F, and the focal length is called f. In many cases, I had to make a choice among the goals above, and I usually chose to use the notation that

I believe to be most common among users. In some cases, this choice has led to some conflict. For example, in the early chapters, I use E for a scalar electric field and I for irradiance, while in Chapter 12, I use E for irradiance and I for intensity, in keeping with the standards in the field of radiometry. In such cases, footnotes will help the reader through the ambiguities that might arise. All the major symbols that are used more than once in the book are included in the index, to further help the reader. The field of optics contains a rich and complicated vocabulary of terms such as "field plane," "exit pupil," and many others that may be confusing at first. Such terms are highlighted in **bold font** where they are defined, and the page number in the index where the definition is given is also in **bold**.

The interested reader will undoubtedly search for other sources beyond this book. In each chapter, I have tried to provide a sampling of resources including papers and specific texts with more details on the subjects covered. The one resource that has dramatically changed the way we do research is the Internet, which I now use extensively in my own research and have used in the preparation of this book. This resource is far from perfect. Although a search may return hundreds of thousands of results, I have been disappointed at the number of times material is repeated, and it often seems that the exact equation that I need remains elusive. Nevertheless, Internet search engines often at least identify books and journal articles that can help the reader find information faster than was possible when such research was conducted by walking through library stacks. Valuable information is also to be found in many of the catalogs published by vendors of optical equipment. In recent years, most of these catalogs have migrated from paper to electronic form, and the information is freely available on the companies' websites.

The modern study and application of optics would be nearly impossible without the use of computers. In fact, the computer may be partly responsible for the growth of the field in recent decades. Lens designers today have the ability to perform in seconds calculations that would have taken months using computational tools such as slide rules, calculators, tables of logarithms, and plotting of data by hand. A number of freely available and commercial lens-design programs have user-friendly interfaces to simplify the design process. Computers have also enabled signal processing and image processing algorithms that are a part of many modern optical systems. Finally, every engineer now finds it useful to use a computer to evaluate equations and explore the effect of varying parameters on system performance. Over the years, I have developed and saved a number of computer programs and scripts that I have used repeatedly over different projects. Although FORTRAN® was a common computer language when I started, I have converted most of the programs to MATLAB® scripts and functions. Many of the figures in this book were generated using these scripts and functions. The ability to plot the results of an equation with these tools contributes to our understanding of the field in a way that would have been impossible when I first started my career. The reader is invited to download and use this code by searching for the book's web page at http://www.crcpress.com. Many of the functions are very general and may often be modified by the user for particular applications.

For MATLAB® and Simulink® product information, please contact:

The MathWorks, Inc.
3 Apple Hill Drive
Natick, MA, 01760-2098 USA
Tel: 508-647-7000
Fax: 508-647-7001
E-mail: info@mathworks.com
Web: www.mathworks.com

Second Edition

Roadmap: Roadmaps show the layout of a part of the book.

The Second Edition includes numerous corrections and revisions, along with the new Chapter 15 focused on experimental work. Along with the addition of take-away messages, in-practice hints, and examples mentioned above, there are some other features intended to make the book easier to follow.

FURTHER READING: Books and Papers.

Pointer to a topic elsewhere in the book.

Fig. 2.6:

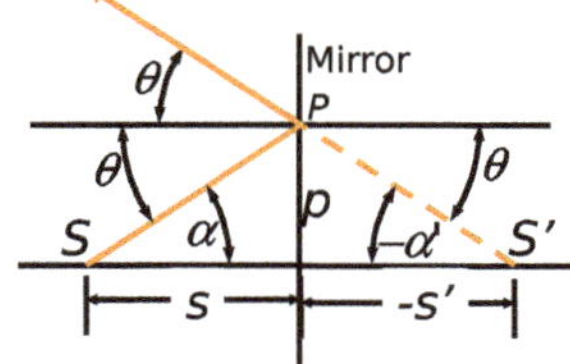

Eq. 0.1:
$y = mx + b.$

General Comments are shown in the margin to help the reader.

Roadmaps are provided in places to help the user follow the layout of the book, to find a particular section, or to put the current section into the context of the "bigger picture." These are shown in slanted type in the margins.

Suggestions for further reading are provided througout the book, with the eading "Further Reading," in the margins.

When a topic is discussed in detail elsewhere in the book, pointers are provided in the margins.

Figures and equations from elsewhere in the book are often reproduced in the margins to minimize the need for flipping pages when reading the text.

Helpful comments are provided in the margins so that the reader can more easily find specific points.

The author hopes that these additions will make the second edition of *Optics for Engineers* more useful to students, researchers, and developers in the optics industry.

Author

Professor Charles A. DiMarzio is an associate professor in the Department of Electrical and Computer Engineering, affiliated with the Department of Mechanical and Industrial Engineering, and the Department of Bioengineering at Northeastern University in Boston, Massachusetts.

He holds a BS in engineering physics from the University of Maine, an MS in physics from Worcester Polytechnic Institute, and a PhD in electrical and computer engineering from Northeastern. He spent 14 years at Raytheon Company's Electro-Optics Systems Laboratory in coherent laser radar for air safety and meteorology. Among other projects there, he worked on an airborne laser radar flown on the Galileo-II to monitor airflow related to severe storms, pollution, and wind energy, and another laser radar to characterize the wake vortices of landing aircraft.

At Northeastern, he extended his interest in coherent detection to optical quadrature microscopy—a method of quantitative phase imaging with several applications, most notably assessment of embryo viability. His current interests include coherent imaging, confocal microscopy for dermatology and other applications, structured illumination microscopy, multimodal microscopy, spectroscopy and imaging in turbid media, and the interaction of light and sound in tissue. His research ranges from computational models, through system development and testing, to signal processing. He is also a founding member of Gordon-CenSSIS – the Gordon Center for Subsurface Sensing and Imaging Systems. Prof. DiMarzio was a topical editor for *Optics Letters* from 2014 to 2020 and served as associate chair of his department from 2014 to 2016. He is active in providing international experiences for students, especially in Chile.

1 Introduction

Over recent decades, the academic study of optics has migrated from physics departments to a variety of engineering departments. In electrical engineering, the interest in optics followed the pattern of interest in electromagnetic waves at lower frequencies that occurred during and after World War II, with the growth of practical applications such as radar. More recently, optics has found a place in other departments such as mechanical and chemical engineering, driven in part by the use in these fields of optical techniques such as interferometry to measure small displacements, velocimetry to measure velocity of moving objects, and various types of spectroscopy to measure chemical composition. An additional motivational factor is the contribution that these fields make to optics, for example, in developing new stable platforms for interferometry and new light sources. In disciplines such as biomedical engineering and bioengineering, light finds application in both imaging and therapy, and the field is rich with exciting opportunities for new instruments and techniques. In civil and environmental engineering, optical techniques such as spectroscopy are used in the measurement of contaminants in water and air. Beyond engineering, optics finds applications in remote sensing, geology, oil exploration, medicine, art, agriculture, and more. In 2014, the Nobel Prize in Chemistry was awarded to Eric Betzig, Stefan W. Hell, and William E. Moerner for techniques that improve the resolution of optical microscopy. With optics so pervasive in so many disciplines even non-engineers may find it useful to develop some expertise in optics.

Our goal here is to develop an understanding of optics that is sufficiently rigorous in its mathematics to allow an engineer to work from a statement of a problem through a conceptual design, rigorous analysis, fabrication, and testing of an optical system. In this chapter, we discuss motivation and history, followed by some background on the design of both breadboard and custom optical systems. With that background, our study of optics for engineers will begin in earnest starting with Maxwell's equations, and then defining some of the fundamental quantities we will use throughout the book, their notation, and equations. We will endeavor to establish a consistent notation as much as possible without making changes to well-established definitions. After that, we will discuss the basic concepts of imaging. Finally, we will discuss the organization of the remainder of the book.

1.1 Why Optics?

Why do we study optics as an independent subject? The reader is likely familiar with the concept, to be introduced in Section 1.4, that light is an electromagnetic wave, just like a radio wave, but lying within a rather narrow band of wavelengths. Could we not just learn about electromagnetic waves in general as a radar engineer might, and then simply substitute the correct numbers into the equations? To some extent we could do so, but this particular wavelength band requires different approximations, uses different materials, and has a unique history, and is thus usually taught as a separate subject.

Despite the fact that mankind has found uses for electromagnetic waves from kilohertz (10^3 Hz) through exahertz (10^{18} Hz) and beyond, the single **octave** (band with the starting and ending frequency related by a factor of two) that is called the visible spectrum occupies a very special niche in our lives on Earth. The absorption spectrum of water is shown in

DOI: 10.1201/9781315157047-1

Water and air transmit visible light. They also transmit some ultraviolet and infrared wavelengths near the visible.

Other ultraviolet and infrared wavelengths are also addressed in the discipline of optics.

Optical wavelengths can resolve usefully small objects.

Figure 1.1A. The water that makes up most of our bodies, and in particular the vitreous humor that fills our eyes, is opaque to all but this narrow band of wavelengths; vision, as we know it, could not have developed in any other band. Figure 1.1B shows the optical absorption spectrum of the atmosphere. Here again, the visible spectrum and a small range of wavelengths around it are transmitted with very little loss, while most others are absorbed. This transmission band includes what we call the near ultraviolet (near UV), visible, and near-infrared (NIR) wavelengths. There are a few additional important transmission bands in the atmosphere. Among these are the middle infrared (mid-IR) roughly from 3 to 5 μm, and the far infrared (far-IR or FIR) roughly from 8 to 14 μm. We will claim all these wavelengths as the province of optics. We might also note that the longer wavelengths used in radar and radio are also, of course, transmitted by the atmosphere. However, we shall see in Chapter 8 that objects smaller than a wavelength are generally not resolvable, so these wavelengths would afford us the vision of very limited utility. Figure 1.2 illustrates how pervasive light is in our lives. The image is a mosaic of hundreds of photographs taken from satellites at night by the Defense Meteorological Satellite Program. The locations of high densities of humans, particularly affluent humans, the progress of civilization along coastal areas and rivers, and the distinctions between rich and poor countries are readily apparent. Light and civilization are tightly knitted together.

The study of optics began at least 3000 years ago, and it developed, largely on an empirical basis, well before we had any understanding of light as an electromagnetic wave. Rectilinear propagation, reflection and refraction, polarization, and the effect of curved mirrors were all analyzed and applied long before the first hints of wave theory. Even after wave theory was considered, well over a century elapsed before Maxwell developed his famous equations. The study of optics as a separate discipline is thus rooted partly in history.

However, there are fundamental technical reasons why optics is usually studied as a separate discipline. The propagation of longer wavelengths used in radar and radio is typically dominated by diffraction, while the propagation of shorter wavelengths such as x-rays is often adequately described by tracing rays in straight lines. Optical wavelengths lie in between the two, and often require the use of both techniques. Furthermore, this wavelength band uses some materials and concepts not normally used at other wavelengths. The concept of refractive optics (lenses) is most commonly found in optics. Lenses tend to be less practical with longer wavelengths, and in fact, focusing by either refractive or reflective elements is less useful because diffraction plays a bigger role in many applications of these wavelengths. Most of the material in the next four chapters concerns focusing components such as lenses and curved mirrors and their effect on imaging systems.

Take-Away

Why Optics?

- Light waves are electromagnetic waves.
- Water and air transmit visible light.
- They also transmit near-ultraviolet and near-infrared light.
- Optical wavelengths are short enough to resolve usefully small objects.
- Air also transmits mid- and far-infrared, which are useful in thermal imaging.

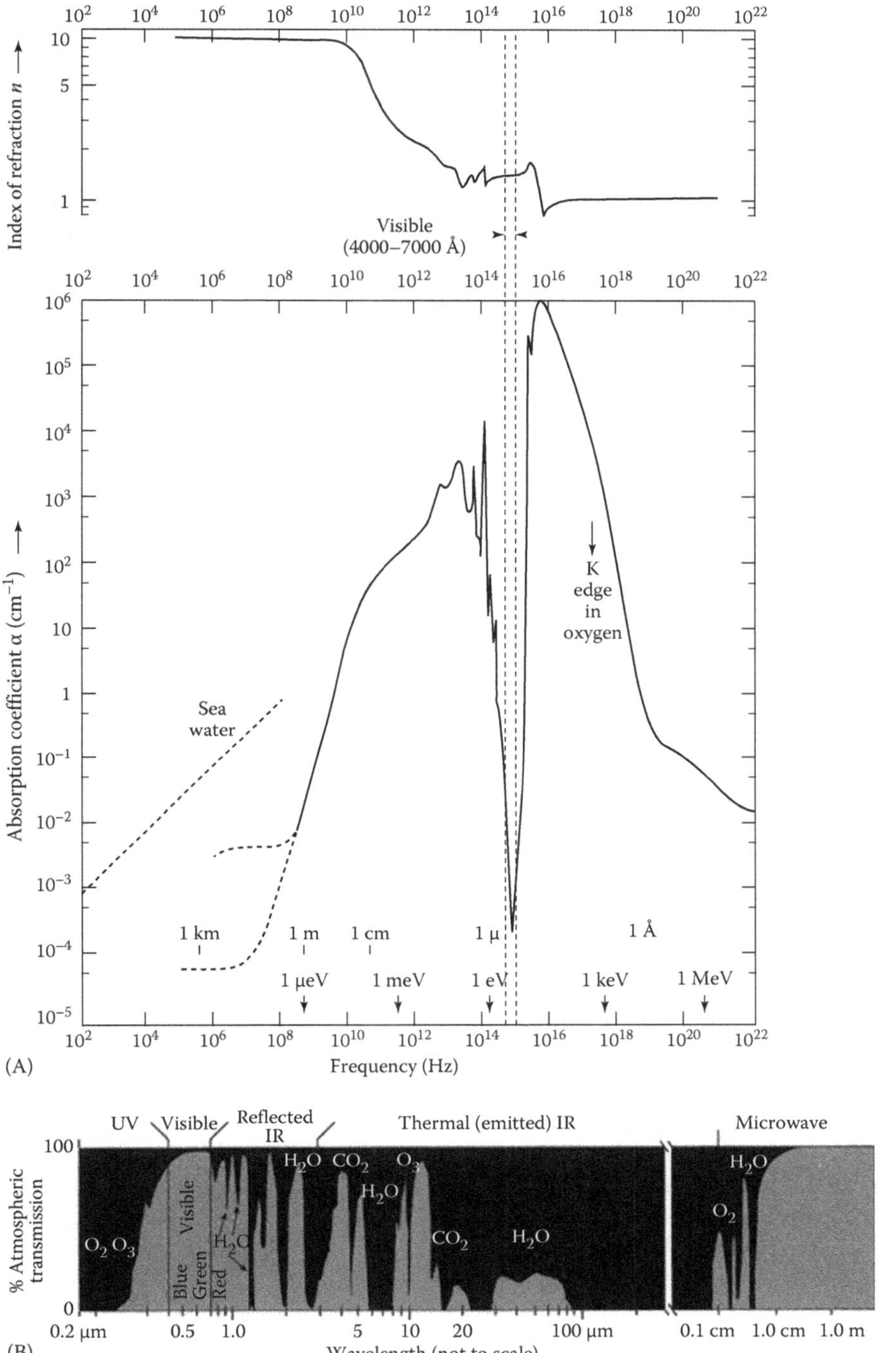

FIGURE 1.1
Electromagnetic transmission. (A) Water strongly absorbs most electromagnetic waves, with the exception of wavelengths near the visible spectrum. (Jackson, J.D., *Classical Electrodynamics*, 1975. Copyright Wiley-VCH Verlag GmbH & Co. KGaA. Reprinted with permission.) (B) The atmosphere also absorbs most wavelengths, except for very long wavelengths and a few transmission bands. (NASA's Earth Observatory.)

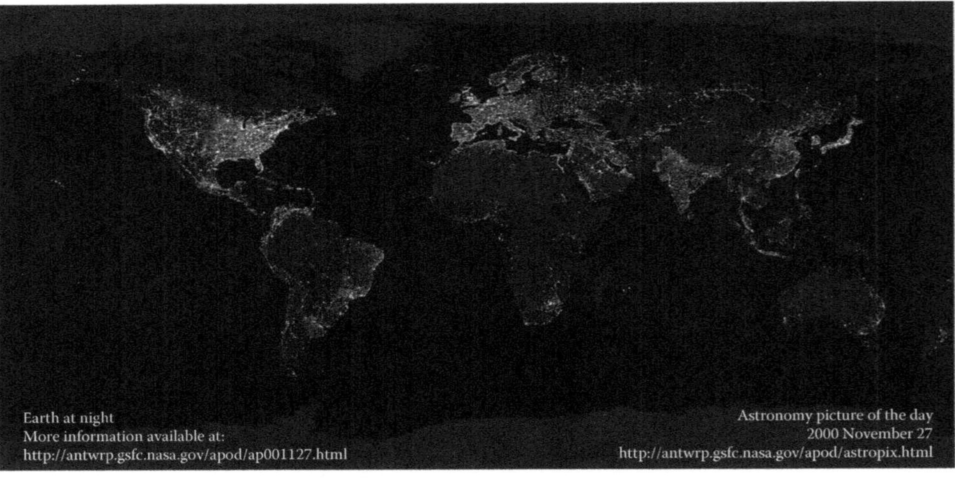

FIGURE 1.2
Earthlight. This mosaic of photographs of Earth at night from satellites shows the importance of light to human activity. (Courtesy of C. Mayhew & R. Simmon (NASA/GSFC), NOAA/NGDC, DMSP Digital Archive.)

1.2 History

FURTHER READING: Further Reading on History of Optics: Two books by Hecht[93;92]

Although we can derive the theory mostly from Maxwell's equations, with occasional use of quantum mechanics, many of the important laws that we use in everyday optical research and development were well-known as empirical results long before this firm foundation was understood. For this reason, it may be useful to discuss the history of the field. Jeff Hecht has written a number of books, articles, and web pages about the history of the field, particularly lasers[93] and fiber optics[92]. The interested reader can find numerous compilations of optics history in the literature for a more complete history than is possible here. Ronchi provides some overviews of the history[192;194] and a number of other sources describe the history of particular periods or individuals[142;141;246;47;37;252;48;244;53]. Various Internet search engines also can provide a wealth of information. The reader is encouraged to search for "optics history" as well as the names of individuals whose names are attached to many of the equations and principles we use.

1.2.1 Earliest Years

Perhaps the first mention of optics in the literature is a mention of brass mirrors or "looking glasses" in Exodus: "And he made the laver of brass, and the foot of it of brass, of the looking glasses of the women assembling, which assembled at the door of the tabernacle of the congregation[2]." Generally, the earliest discussion of optics in the context of what we would consider scientific methods dates to around 300 BCE, when Euclid[64] stated the theory of rectilinear propagation, the law of reflection, and relationships between size and angle. He postulated that light rays travel from the eye to the object, which now seems unnecessarily complicated and counterintuitive but can still provide a useful point of view in today's study of imaging. Hero (10–70 CE)[14] developed a theory that light travels from a source to a receiver along the shortest path, an assumption that is close to correct and was later refined by Fermat in the 1600s.

1.2.2 Al Hazan

Ibn al-Haitham Al Hazan (965–1040 CE) developed the concept of the plane of incidence and studied the interaction of light with curved mirrors, beginning the study of geometric optics that will be recognizable to the reader of Chapter 2. He also refuted Euclid's idea of rays traveling from the eye to the object and reported a relationship between the angle of refraction and the angle of incidence, although he fell short of obtaining what would later be called the law of refraction or Snell's law. He published his work in *Opticae Thesaurus*, the title being given by Risnero, who provided the Latin translation[189], which, according to Lindberg[142], is the only existing printed version of this work. In view of Al Hazan's rigorous scientific approach, he may well deserve the title of the "father of modern optics"[48]. During the following centuries, Roger Bacon and others built on Al Hazan's work[141].

Al Hazan, the father of modern optics.

1.2.3 1600–1800

The seventeenth century included a few decades of extraordinary growth in the knowledge of optics, providing at least some understanding of almost all of the concepts that we will discuss in this book. Perhaps this exciting time was fostered by Galileo's invention of the telescope in 1609[244], and a subsequent interest in developing theories of optics, but in any case several scientists of the time made important contributions that are fundamental to optics today. Willebrord van Royen Snel (or Snell) (1580–1626) developed the law of refraction that bears his name in 1621, from purely empirical considerations. Descartes (1596–1650) may or may not have been aware of Snell's work when he derived the same law[51] in 1637[116]. Descartes is also credited with the first description of light as a wave, although he suggested a pressure wave. Even with an erroneous understanding of the physical nature of the oscillation, wave theory provides a description of diffraction, which is in fact as correct as the frequently used scalar wave theory of optics. The second-order differential equation describing wave behavior is universal. In a letter in 1662, Pierre de Fermat described the idea of the principle of least time, which was initially controversial[146]. According to this remarkably simple and profound principle, light travels from one point to another along the path that requires the least time. On the face of it, such a principle requires that the light at the source must possess some property akin to "knowledge" of the paths open to it and their relative length. The principle becomes more palatable if we think of it as the result of some more basic physical laws rather than the fundamental source of those laws. It provides some exceptional insights into otherwise complicated optical problems. In 1678, Christian Huygens[102] (1629–1695) discussed the nature of refraction, the concept that light travels more slowly in media than it does in vacuum, and introduced the idea of "two kinds of light," an idea which was later placed on a firm foundation as the theory of polarization. Isaac Newton (1642–1727) is generally recognized for other scientific work but made significant contributions to optics. A considerable body of his work is now available thanks to the Newton Project[105]. Newton's contributions to the field, published in 1704[165], including the concept of corpuscles of light and the notion of the ether, rounded out this extraordinarily productive century. The corpuscle idea may be considered as an early precursor to the concept of photons, and the ether provided a convenient, if incorrect, foundation for growth of the wave theory of light.

Snell: Refraction

Descartes: Light is a wave.

Fermat: principle of least time.

Huygens: refraction, propagation, and polarization

Newton: "corpuscles" and the ether

1.2.4 1800–1900

The theoretical foundations of optics muddled along for another 100 years, until the early nineteenth century, when another group of researchers developed what is now our modern wave theory. First with what may have been a modern electromagnetic wave theory was Augustin Fresnel (1788–1827), although he proposed a longitudinal wave, and there is some uncertainty about the nature of the wave that he envisioned. Such a wave would

Fresnel: modern wave theory

have been inconsistent with Huygens's two kinds of light. Evidently, he recognized this weakness and considered some sort of "transverse vibrations" to explain it[252;37;47]. Young (1773–1829)[258] proposed a transverse wave theory[47], which allowed for our current notion of polarization (Chapter 6), in that there are two potential directions for the electric field of a wave propagating in the z direction – x and y. This time marked the beginning of the modern study of electromagnetism. Ampere and Faraday developed differential equations relating electric and magnetic fields. These equations were elegantly brought together by James Clerk Maxwell (1831–1879)[152], forming the foundation of the theory we use today. Maxwell's main contribution was to add the displacement current to Ampere's equation, although the compact set of equations that underpin the field of electromagnetic waves is generally given the title, "Maxwell's equations." The history of this revolutionary concept is recounted in a paper by Shapiro[205]. With the understanding provided by these equations, the existence of the ether was no longer required. Michelson and Morley, in perhaps the most famous failed experiment in optics, were unable to measure the drift of the ether caused by the Earth's rotation[156]. The experimental apparatus used in the experiments, now called the Michelson interferometer (Section 7.4), finds frequent use in modern applications.

Young: transverse waves

Maxwell's equations bring it all together.

Michelson and Morley: demise of the ether

1.2.5 Quantum Mechanics and Einstein's Miraculous Year

The fundamental wave theories of the day were shaken by the early development of quantum mechanics in the late nineteenth and early twentieth centuries. The radiation from an ideal black body, which we will discuss in Section 12.5, was a topic of much debate. The prevailing theories led to the prediction that the amount of light from a hot object increased with decreasing wavelength in such a way that the total amount of energy would be infinite. This so-called ultraviolet catastrophe and its obvious contradiction of experimental results contributed in part to the studies that resulted in the new field of quantum mechanics. Among the many people who developed the theory of quantum mechanics, Einstein stands out for his "miraculous year" (1905) in which he produced five papers that changed physics[217]. Among his many contributions were the notion of quantization of light[60], which helped resolve the ultraviolet catastrophe, and the concepts of spontaneous and stimulated emission[61;63], which ultimately led to the development of the laser and other devices. Our understanding of optics relies first on Maxwell's equations and second on the quantum theory. In the 1920s, Michelson measured the speed of light for the first time[154].

FURTHER READING: Einstein's Miraculous Year[217]

Einstein: Quantum nature of light.

1.2.6 Middle 1900s

The remainder of the twentieth century saw prolific development of new devices based on the theory that had grown during the early part of the century. It is impossible to provide a definitive list of the most important inventions, and we will simply make note of a few that find frequent use in our lives. Because fiber-optical communication and imaging are among the best-known current applications of optics, it is fitting to begin this list with the first optical fiber bundle by a German medical student, Heinrich Lamm[132]. Although the light-guiding properties of high-index materials had been known for decades (e.g., Colladon's illuminated water jets[42] in the 1840s), Lamm's was the first attempt at imaging. Edwin Land, a prolific researcher and inventor, developed low-cost polarizing sheets[133;134] that are now used in many optical applications. In 1935, Fritz Zernike[260] developed the technique of phase microscopy, for which he was awarded the Nobel Prize in Physics in 1953. In 1948, Dennis Gabor invented the field of holography[73], which now finds applications in research, art, and security. Holograms are normally produced with lasers, but this invention preceded the laser by over a decade. Gabor was awarded the Nobel Prize in Physics "for his investigation and development of holography" in 1971. Because the confocal microscope is featured in many examples in this text, it is worthwhile to point

Gabor: Holography.

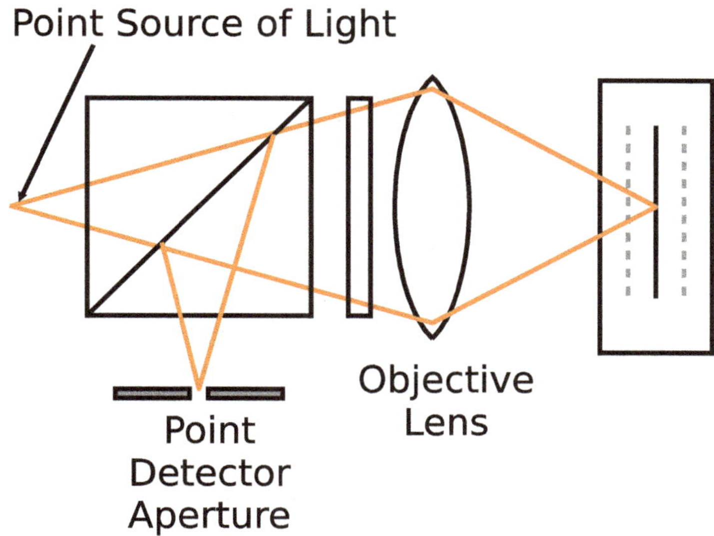

FIGURE 1.3
Confocal microscope. Light from a source is focused to a small focal region. In a reflectance confocal microscope, some of the light is scattered back toward the receiver, and in a fluorescence confocal microscope, some of the light is absorbed and then re-emitted at a longer wavelength. In either case, the collected light is imaged through a pinhole. Light scattered or emitted from regions other than the focal region is rejected by the pinhole. Thus, optical sectioning is achieved. The confocal microscope only images one point, so it must be scanned to produce an image.

out that this instrument, now almost always used with a laser source, was also invented before the laser[160] by Marvin Minsky using an arc lamp as a source. Best known for his work in artificial intelligence, Minsky might never have documented his invention, but his brother-in-law, a patent attorney, convinced him to do so[159]. A drawing of a confocal microscope is shown in Figure 1.3.

1.2.7 The Laser Arrives

The so-called cold war era in the middle of the twentieth century was a time of intense competition among many countries in science and engineering. Shortly after World War II, interest in microwave devices led to the invention of the MASER[85] (microwave amplification by stimulated emission of radiation), which in turn led to the theory of the "optical maser"[200] by Schawlow and Townes in 1958. Despite the fact that Schawlow coined the term "LASER" (light amplification by stimulated emission of radiation), the first working laser was invented by Theodore Maiman in 1960. The citation for this invention is not a journal publication but a news report[170] based on a press conference because *Physical Review Letters* rejected Maiman's paper[93]. The importance of the laser is highlighted by the number of Nobel Prizes related to it. The 1964 Nobel Prize in Physics went to Townes, Basov, and Prokhorov "for fundamental work in the field of quantum electronics, which has led to the construction of oscillators and amplifiers based on the maser–laser principle." In 1981, the prize went to Bloembergen, Schawlow, and Siegbahn, "for their contribution to the development of laser spectroscopy." Only a few months after Maiman's invention, the helium-neon laser was invented by Javan[110] in December 1960. This laser, because of its low cost, compact size, and visible light output, is ubiquitous in today's optics laboratories, and in a multitude of commercial applications.

Maiman's paper on the first laser is rejected.

1.2.8 The Space Decade

Perhaps spurred by U.S. President Kennedy's 1962 goal[117] to "go to the moon in this decade ... because that goal will serve to organize and measure the best of our energies and skills ...," a remarkable decade of research, development, and discovery began. Kennedy's bold proposal was probably a response to the 1957 launch of the Soviet satellite, Sputnik, and the intense international competition in science and engineering. In optics, the enthusiasm and excitement of this decade of science and Maiman's invention of the laser led to the further invention of countless lasers using solids, liquids, and gasses. The semiconductor laser or laser diode was invented by Alferov[15] and Kroemer[128;129]. In 2000, Alferov and Kroemer shared half the Nobel Prize in Physics "for developing semiconductor heterostructures used in high-speed- and opto-electronics." Their work is particularly important as it led to the development of small, inexpensive lasers that are now used in compact disk players, laser pointers, and laser printers. The carbon dioxide laser, with wavelengths around 10 μm, invented in 1964 by Patel[172] is notable for its high power and scalability. The argon laser, notable for producing moderate power in the green and blue (notably at 514 and 488 nm, but also other wavelengths) was invented by Bridges[34;35]. It continues to be used in many applications, although the more recent frequency-doubled Nd:YAG laser has provided a more convenient source of green light at a similar wavelength (532 nm). The neodymium-doped yttrium aluminum garnet (Nd:YAG)[76] invented by Geusic in 1964 is a solid-state laser, now frequently excited by semiconductor lasers, and can provide high power levels in CW or pulsed operation at 1064 nm. Frequency doubling of light waves was already known[71], and as laser technology developed progressively, more efficient frequency doubling and generation of higher harmonics greatly increased the number of wavelengths available. The green laser pointer consists of an Nd:YAG laser at 1064 nm pumped by a near-infrared diode laser around 808 nm, and frequency doubled to 532 nm.

Alferov and Kroemer invent the semiconductor laser.

Although light sources, and particularly lasers, figure prominently in most histories of optics, we must also be mindful of the extent to which advances in optical detectors have changed our lives. Photodiodes have been among the most sensitive detectors since Einstein's work on the photoelectric effect[60], but the early ones were vacuum photodiodes. Because semiconductor materials are sensitive to light, with the absorption of a photon transferring an electron from the valence band to the conduction band, the development of semiconductor technology through the middle of the twentieth century naturally led to a variety of semiconductor optical detectors. However, the early ones were single devices. Willard S. Boyle and George E. Smith invented the charge-coupled device or CCD camera at Bell Laboratories in 1969[32;33]. The CCD, as its name implies, transfers charge from one element to another, and the original intent was to make a better computer memory, although the most common application now is to acquire massively parallel arrays of image data, in what we call a CCD camera. Boyle and Smith shared the 2009 Nobel Prize in Physics with Charles Kuen Kao for his work in fiber-optics technology. The Nobel Prize committee, in announcing the prize, called them "the masters of light."

"The Masters of Light," Boyle and Smith develop the charge-coupled device, which changes the world of cameras.

1.2.9 The Late 1900s and Beyond

The post-Sputnik spirit affected technology growth in all areas, and many of the inventions in one field reinforced those in another. Optics benefited from developments in chemistry, physics, and engineering, but probably no field contributed to new applications so much as that of computers. Spectroscopy and imaging generate large volumes of data, and computers provide the ability to go far beyond displaying the data, allowing us to develop measurements that are only indirectly related to the phenomenon of interest, and then reconstruct the important information through signal processing. The combination of computing and communication has revolutionized optics from imaging at the smallest microscopic scale to remote sensing of the atmosphere and measurements in space.

Computers and communication make it possible to use all the data.

As a result of this recent period of development, we have seen such large-scale science projects as the Hubble Telescope in 1990, earth-imaging satellites, and airborne imagers, smaller laboratory-based projects such as development of novel microscopes and spectroscopic imaging systems, and such everyday applications as laser printers, compact disk players, light-emitting diode traffic lights, and liquid crystal displays. Optics also plays a role in our everyday lives "behind the scenes" in such diverse areas as optical communication, medical diagnosis and treatment, and the lithographic techniques used to make the chips for our computers.

The growth continues in the twenty-first century with new remote sensing techniques and components; new imaging techniques in medicine, biology, and chemistry[237;24]; and such advances in the fundamental science as laser cooling and negative-index materials. Steven Chu, Claude Cohen-Tannoudji, and William D. Phillips won the 1997 Nobel Prize in Physics, "for their developments of methods to cool and trap atoms with laser light." For a discussion of the early work in this field, there is a feature issue of the *Journal of the Optical Society of America*[40]. Negative-index materials were predicted in 1968[235] and are a current subject of research, with experimental results in the RF spectrum[213;173;206;171] moving toward shorter wavelengths. A recent review shows progress into the infrared[202].

FURTHER READING: Steven Chu discusses the early work in laser cooling[40]

FURTHER READING: Shalaev discusses negative-index materials in the infrared

As we look at recent trends in optics, it is tempting to try to predict which of the inventions of the past decade or two will be counted as milestones. On one hand, scientific discoveries will "push" new ideas that may have world-changing impact. Negative-index materials are still under study and may find unexpected applications in the near future. The advances in super-resolution microscopy may become common techniques in the pathology laboratory. Big science projects such as the James Webb Space Telescope, launched in 2021, may provide unexpected discoveries in astronomy. On the other hand, applications may "pull" some ideas into the marketplace by reducing costs and simplifying manufacturing. Such is the case already with light-emitting diodes that were used initially for low-power signaling applications. Now LED illumination, with reduced energy consumption and longer lifetimes, is a familiar sight in our homes and neighborhoods (Figure 1.4). Driven by application demands, some formerly expensive research instruments, thermal cameras, lidars, and hyperspectral imagers are now available at such low prices that they can be considered as consumer electronics.

1.3 Optical Engineering

As a result of the growth in the number of optical products, there is a need for engineers who understand the fundamental principles of the field but who can also apply these principles in the laboratory, shop, or factory. The results of our engineering work will look quite different, depending on whether we are building a temporary experimental system in the research laboratory, a few custom instruments in a small shop, or a high-volume commercial product. Nevertheless, the designs all require an understanding of certain basic principles of optics.

Breadboard optical systems.

Figure 1.5 shows a typical laboratory setup. Generally, we use a breadboard with an array of threaded holes to which various optical mounts may be attached. The size of the breadboard may be anything from a few centimeters square to several square meters on a large table filling most of a room. For interferometry, an air suspension is often used to maintain stability and isolate the experiment from vibrations of the building. We often call a large breadboard on such a suspension a "floating table." A number of vendors provide mounting hardware, such as that shown in Figure 1.5, to hold lenses, mirrors, and other components. Because precise alignment is required in many applications, some mounting hardware uses micro-positioning in position, angle, or both. Chapter 15 will discuss how to use this equipment.

FIGURE 1.4
Light-emitting diodes. Lighting is a significant source of energy consumption and environmental damage. LED devices may reduce this impact.

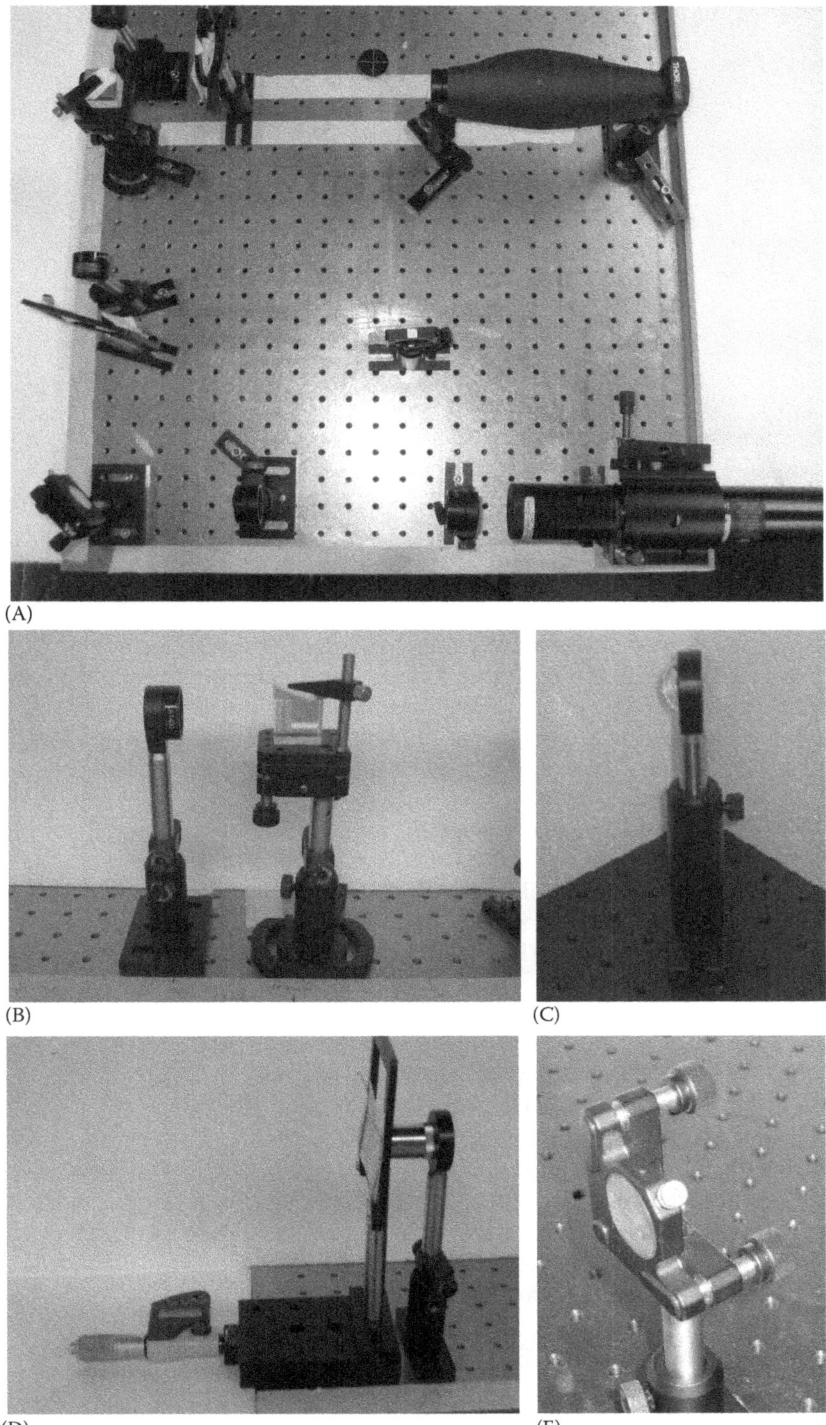

(A) (B) (C) (D) (E)

FIGURE 1.5
Laboratory system. The optical components are arranged on a breadboard (A) and held in place using general-purpose mounting hardware (B and C). Micro-positioning is often used for position (D) and angle (E).

FIGURE 1.6
Reflectance confocal microscope. This 405-nm, line-scanning reflectance confocal microscope includes a mixture of custom-designed components, commercial off-the-shelf mounting hardware, and a commercial microscope objective. (Photo by Gary Peterson of Memorial Sloan Kettering Cancer Center, New York.)

Some of the advantages of the breadboard systems are that they can be modified easily, the components can be used many times for different experiments, and adjustment is generally easy with the micro-positioning equipment carefully designed by the vendor. Some of the disadvantages are the large sizes of the mounting components, the potential for accidentally moving a component, the difficulty of controlling stray light, and the possible vibration of the posts supporting the mounts. The next alternative is a custom-designed device built in a machine shop in small quantities. An example is shown in Figure 1.6, where we see a custom-fabricated plate with several custom components and commercial components such as the microscope objective on the left and the adjustable mirror on the right. This 405 nm, line-scanning reflectance confocal microscope is a significant advance on the path toward commercialization. The custom design results in a much smaller and practical instrument than would be possible with the breadboard equipment used in Figure 1.5.

Mixed commercial and custom components.

Many examples of large-volume commercial products are readily recognizable to the reader. Examples include the LED devices mentioned earlier in Figure 1.4. Other examples are compact disk and DVD players, laser printers, microscopes, cameras, and projectors. Many optical designs will contain elements of all three categories. For example, a laboratory experiment will probably be built on an optical breadboard with posts, lens holders, micro-positioners, and other equipment from a variety of vendors. It may also contain a few custom-designed components where higher tolerances are required, space constraints are severe, or mounting hardware is not available or appropriate, and it may also contain commercial components such as cameras. The principles of optics needed to select and place the components of such systems is a major subject of this book.

Large-volume commercial products.

1.4 Electromagnetics Background

In this section, we develop some practical equations for our study of optics, beginning with the fundamental equations of Maxwell. We will develop a vector wave equation and postulate a harmonic solution. We will consider a scalar wave equation, which is often satisfactory for many applications. We will then develop the eikonal equation and use it in Fermat's principle. This approach will serve us well through Chapter 5. Then, we return to the wave theory to study what is commonly called "physical optics."

1.4.1 Maxwell's Equations

We begin our discussion of optics with **Maxwell's equations.** We shall use these equations to develop a wave theory of light and then will develop various approximations to the wave theory. The student with minimal electromagnetics background will likely find this section somewhat confusing but will usually be able to use the simpler approximations that we develop from it. On the other hand, the interested student may wish to delve deeper into this area using any electromagnetics textbook[108;208;18;220;197]. Maxwell's equations are usually presented in one of four forms, differential or integral, in MKS or CGS units. As engineers, we will choose to use the differential form and engineering units. Although CGS units were largely preferred for years in electromagnetics, Stratton[220] makes a case for MKS units saying, "there are few to whom the terms of the c.g.s. system occur more readily than volts, ohms, and amperes. All in all, it would seem that the sooner the old favorite is discarded, the sooner will an end be made to a wholly unnecessary source of confusion."

Maxwell's Equations lead to a wave equation, which in turn leads to other simplifications such as ray tracing.

FURTHER READING: Electromagnetics[108;208;18;220;197]

Faraday's law states that the curl, a spatial derivative, of the electric field vector, **E**, is proportional to the time derivative of the magnetic flux density vector, **B** (in this text, we will denote a vector by a letter in bold font, **V**)*:

Vectors are denoted by bold letters; **B**.

$$\nabla \times \mathbf{E} = -\frac{\partial \mathbf{B}}{\partial t}. \tag{1.1}$$

Likewise, **Ampere's law** relates a spatial derivative of the magnetic field strength, **H** to the time derivative of the electric displacement vector, **D**:

$$\nabla \times \mathbf{H} = \mathbf{J} + \frac{\partial \mathbf{D}}{\partial t},$$

where **J** is the source-current density. Because we are assuming $\mathbf{J} = 0$, we shall write

$$\nabla \times \mathbf{H} = \frac{\partial \mathbf{D}}{\partial t}. \tag{1.2}$$

The first of two **equations of Gauss** relates the divergence of the electric displacement vector to the charge density, which we will assume to be zero:

$$\nabla \cdot \mathbf{D} = \rho = 0. \tag{1.3}$$

The second equation of Gauss, by analogy, would relate the divergence of the magnetic flux density to the density of the magnetic analog of charge, or magnetic monopoles. Because these monopoles do not exist,

$$\nabla \cdot \mathbf{B} = 0. \tag{1.4}$$

*Equations shown in boxes like this one are noted as "key equations" either because they are considered fundamental, as is true here, or because they are frequently used.

We have introduced four field vectors, two electric (**E** and **D**) and two magnetic (**B** and **H**). We shall see that Faraday's and Ampere's laws can be combined to produce a wave equation in any one of these vectors, once we write two **constitutive relations**, one connecting the two electric vectors and another connecting the two magnetic vectors.

First, we note that a variable name in the bold calligraphic font, $\mathcal{M}$, will represent a matrix. Then, we define the dielectric tensor $\boldsymbol{\mathcal{E}}$ and the magnetic tensor $\boldsymbol{\mu}$ using

$$\mathbf{D} = \boldsymbol{\mathcal{E}}\mathbf{E} \quad \mathbf{B} = \boldsymbol{\mu}\mathbf{H}. \tag{1.5}$$

We may also define electric and magnetic **susceptibilities**, χ and χ_m.

$$\mathbf{D} = \epsilon_0 (1 + \chi) \mathbf{E} \quad \mathbf{B} = \mu_0 (1 + \chi_m) \mathbf{H} \tag{1.6}$$

Here, ϵ_0 is the free-space value of ϵ and μ_0 is the free-space value of μ, both of which are scalars. We can then separate the contributions to **D** and **B** into free-space parts and parts related to the material properties. Thus, we can write

$$\mathbf{D} = \epsilon_0\mathbf{E} + \mathbf{P} \quad \mathbf{B} = \mu_0\mathbf{H} + \mathbf{M}, \tag{1.7}$$

where we have defined the **polarization**, **P**, and **magnetization**, **M**, by

$$\mathbf{P} = \epsilon_0\chi\mathbf{E} \quad \mathbf{M} = \mu_0\chi_m\mathbf{H}. \tag{1.8}$$

Now, at optical frequencies, we can generally assume that

$$\chi_m = 0 \quad \text{so} \quad \boldsymbol{\mu} = \mu_0; \tag{1.9}$$

the magnetic properties of materials that are often important at microwave and lower frequencies are not significant in optics.

1.4.2 Wave Equation

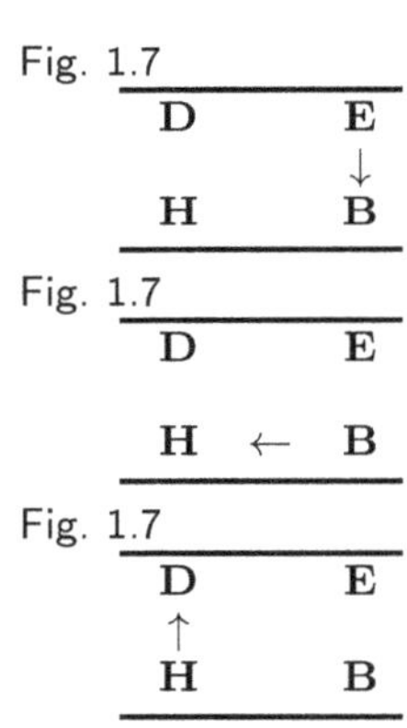

We may begin our derivation of the wave equation with any of the four vectors, **E**, **D**, **B**, and **H**, as shown in Figure 1.7. Let us begin with **E** and work our way around the loop clockwise. We repeat Equation 1.1:

$$\nabla \times \mathbf{E} = -\frac{\partial \mathbf{B}}{\partial t},$$

and the constitutive relation

$$\mathbf{H} = \mu\mathbf{B}$$

so that

$$\nabla \times \mathbf{E} = -\mu\frac{\partial \mathbf{H}}{\partial t}. \tag{1.10}$$

Next, we apply Ampere's law, Equation 1.2,

$$\nabla \times \mathbf{H} = \frac{\partial \mathbf{D}}{\partial t}$$

Taking the curl of both sides of Equation 1.10 and using Ampere's law,

$$\nabla \times (\nabla \times \mathbf{E}) = -\mu\nabla \times \left(\frac{\partial \mathbf{H}}{\partial t}\right),$$

$$\nabla \times (\nabla \times \mathbf{E}) = -\mu\frac{\partial^2 \mathbf{D}}{\partial t^2}. \tag{1.11}$$

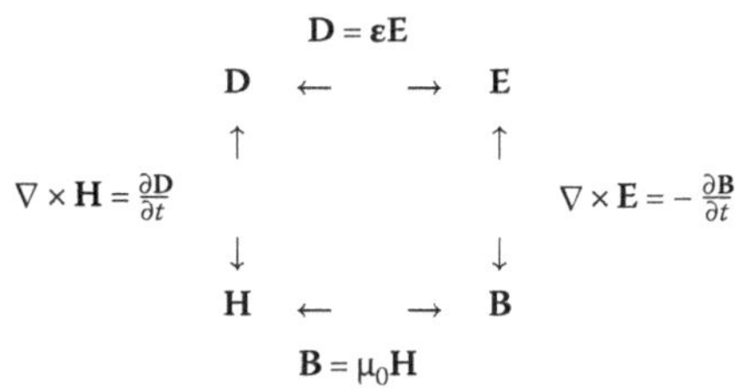

$$
\begin{array}{ccccc}
 & & \mathbf{D} = \varepsilon \mathbf{E} & & \\
 & \mathbf{D} & \leftarrow \quad \rightarrow & \mathbf{E} & \\
 & \uparrow & & \uparrow & \\
\nabla \times \mathbf{H} = \frac{\partial \mathbf{D}}{\partial t} & & & & \nabla \times \mathbf{E} = -\frac{\partial \mathbf{B}}{\partial t} \\
 & \downarrow & & \downarrow & \\
 & \mathbf{H} & \leftarrow \quad \rightarrow & \mathbf{B} & \\
 & & \mathbf{B} = \mu_0 \mathbf{H} & &
\end{array}
$$

FIGURE 1.7
The wave equation. The four vectors, $\mathbf{E}$, $\mathbf{D}$, $\mathbf{B}$, and $\mathbf{H}$ are related by Maxwell's equations and the constitutive equations. We may start with any of the vectors and derive a wave equation. At optical frequencies, $\mu = \mu_0$, so $\mathbf{B}$ is proportional to $\mathbf{H}$ and in the same direction. However, ε depends on the material properties and may be a tensor.

Finally, applying the constitutive relation for electric field from Equation eq:epsmu,

$$\mathbf{D} = \boldsymbol{\varepsilon}\mathbf{E}$$

we relate $\mathbf{D}$ to $\mathbf{E}$, with the result

$$\nabla \times (\nabla \times \mathbf{E}) = -\mu \frac{\partial^2 (\boldsymbol{\varepsilon}\mathbf{E})}{\partial t^2}. \tag{1.12}$$

Fig. 1.7

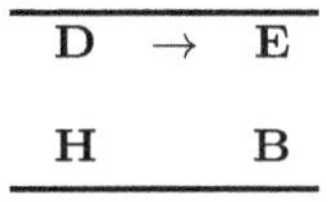

$\mathbf{D}$	$\rightarrow$	$\mathbf{E}$
$\mathbf{H}$		$\mathbf{B}$

We now have a differential equation that is second order in space and time, for $\mathbf{E}$. Using the vector identity,

$$\nabla \times (\nabla \times \mathbf{E}) = \nabla (\nabla \cdot \mathbf{E}) - \nabla^2 \mathbf{E} \tag{1.13}$$

Gray shaded regions in the text show details of derivations that may be useful for understanding but can be omitted by many readers.

and noting that the first term after the equality sign is zero by Gauss's law, Equation 1.3,

$$\nabla^2 \mathbf{E} = \mu \frac{\partial^2 (\boldsymbol{\varepsilon}\mathbf{E})}{\partial t^2}. \tag{1.14}$$

In the case that $\boldsymbol{\varepsilon}$ is a scalar constant, ϵ,

$$\nabla^2 \mathbf{E} = \mu \epsilon \frac{\partial^2 \mathbf{E}}{\partial t^2}. \tag{1.15}$$

Vector wave equation in isotropic medium.

The reader familiar with differential equations will recognize this equation immediately as the wave equation. We may guess that the solution will be a sinusoidal wave. One example is

$$\mathbf{E} = \hat{x} E_0 e^{j(\omega t - nkz)}, \tag{1.16}$$

Representative plane-wave solution.

which represents a complex electric field vector in the x direction, having an amplitude described by a wave traveling in the z direction at a speed

$$v = \frac{\omega}{nk}. \tag{1.17}$$

$e^{j\omega t}$, Harmonic time dependence is usually used. Time derivatives are replaced with multiplication by $j\omega$.

We will usually be justified in assuming this harmonic time dependence, $e^{j\omega t}$, in our fields, except when dealing with extremely short pulses. We have arbitrarily defined some constants in this equation. First, ω is the **angular frequency**. Second, we use the expression nk as the magnitude of the wave vector in the medium. The latter assumption may seem complicated but will prove useful later.

We may see that this vector satisfies Equation 1.15, by substitution and we obtain a relationship between ω and nk.

$$\frac{\partial^2 \mathbf{E}}{\partial z^2} = \mu\epsilon \frac{\partial^2 \mathbf{E}}{\partial t^2}.$$
$$-n^2 k^2 \mathbf{E} = -\mu\epsilon\omega^2 \mathbf{E}, \quad (1.18)$$

which is true provided that

$$n^2 k^2 = \mu\epsilon\omega^2,$$

so the wave speed, v, is given by

$$v^2 = \frac{\omega^2}{n^2 k^2} = \frac{1}{\mu\epsilon},$$
$$v = \frac{1}{\sqrt{\epsilon\mu}}. \quad (1.19)$$

In the case that $\epsilon = \epsilon_0$, the speed is

$$c = \frac{1}{\sqrt{\epsilon_0 \mu_0}} = 2.99792458 \times 10^8 \text{ m/s} \quad (1.20)$$

Universal Constant: c, the speed of light in vacuum.

The speed of light in a vacuum is one of the fundamental constants of nature.

The **frequency** of a wave, ν, is given by the equation

$$\omega = 2\pi\nu, \quad (1.21)$$

and is typically in the hundreds of terahertz. We usually use ν as the symbol for frequency, following the lead of Einstein. The repetition **period** is

$$T = \frac{1}{\nu} = \frac{2\pi}{\omega} \quad (1.22)$$

The **wavelength** is the distance over which the wave repeats itself in space and is given by

$$\lambda_{material} = \frac{2\pi}{nk}, \quad (1.23)$$

which has a form analogous to Equation 1.22.

We find it useful in optics to define the ratio of the speed of light in vacuum to that in the material:

$$n = \frac{c}{v} = \sqrt{\frac{\epsilon\mu}{\epsilon_0\mu_0}}. \quad (1.24)$$

n, the index of refraction is a characteristic of a material.

n values at http://refractiveindex.info

Some common approximate values that can be used in preliminary designs are shown in Table 1.1. The exact values depend on exact material composition, temperature, pressure, and wavelength. A more complete table is shown in Table F.5. As of 2014, the website http://refractiveindex.info has an excellent and easily accessible database on indices of refraction of various materials over different wavelengths. We will see that the **refraction**, or bending of light rays as they enter a medium, depends on this quantity, so we call it

TABLE 1.1
Approximate Indices of Refraction

Material	Approximate Index
Vacuum (also close for air)	1.00
Water (visible to NIR)	1.33
Glass	1.5
ZnSe (infrared)	2.4
Germanium (infrared)	4.0

These values are approximations useful at least for a preliminary design. The actual index of refraction of a material depends on many parameters.

the **index of refraction**. Radio frequency engineers often characterize a material by its dielectric constant, ϵ, and its permeability, μ. Because geometric optics, largely based on refraction, is an important tool at the relatively short wavelengths of light, and because $\mu = \mu_0$ at these wavelengths, we more frequently characterize an optical material by its index of refraction:

$$n = \sqrt{\frac{\epsilon}{\epsilon_0}} \quad \text{(Assuming } \mu = \mu_0\text{)}. \tag{1.25}$$

In any material, light travels at a speed slower than c, so the index of refraction is greater than, or equal to, one.

The index of refraction usually varies at least somewhat with wavelength. We say that the material is **dispersive**. In fact, it can be shown that any material that absorbs light must be dispersive. Because almost every material other than vacuum absorbs at least some light, dispersion is nearly universal, although it is very small in some materials and very large in others. Dispersion may be characterized by the derivative of the index of refraction with respect to wavelength, $dn/d\lambda$, but in the visible spectrum, it is often characterized by the **Abbe number**,

Dispersion: Index of refraction varies with wavelength.

$$V_d = \frac{n_d - 1}{n_F - n_c}, \tag{1.26}$$

where the indices of refraction are

$$\begin{aligned} n_d &\quad \text{at} \quad \lambda_d = 587.6\,\text{nm}, \\ n_F &\quad \text{at} \quad \lambda_F = 486.1\,\text{nm}, \\ n_c &\quad \text{at} \quad \lambda_c = 656.3\,\text{nm}. \end{aligned} \tag{1.27}$$

In this equation, the numerator is the difference between the index of refraction near the center of the visible and that of air, while the denominator is the change in index of refraction across much of the visible spectrum. These three wavelengths were chosen because they are prominent in the absorption spectrum of gasses in the atmosphere and were easily identified in solar spectra. Because we often want highly refractive materials with low dispersion (a large numerator and a small denominator), the Abbe number may be considered a figure of merit for optical glass.

V: A high Abbe number means low dispersion.

Optical materials are often shown on a **glass map** where each material is represented as a dot on a plot of index of refraction as a function of the Abbe number, as shown in Figure 1.8. Low dispersion glasses with an Abbe number above 50 are called **crown glasses** and those with higher dispersion are called **flint glasses**.

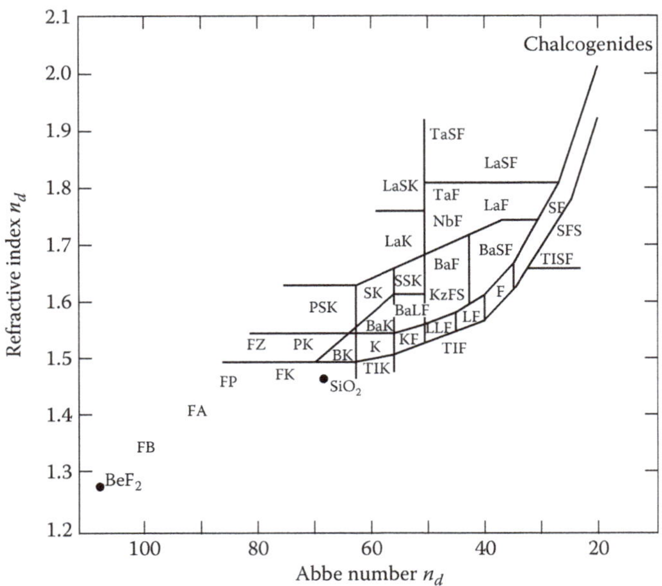

FIGURE 1.8
A glass map. Different optical materials are plotted on the glass map to indicate dispersion. A high Abbe number indicates low dispersion. Glasses with Abbe number above 50 are called crown glasses and those with lower Abbe numbers are called flints. (Reprinted from Weber, J., *CRC Handbook of Laser Science and Technology*, Supplement 2, CRC Press, Boca Raton, FL, 1995. With permission.)

At optical frequencies, the index of refraction is usually lower than it is in the RF portion of the spectrum. Materials with indices of refraction above 2 in the visible spectrum are relatively rare. Even in the infrared spectrum, $n = 4$ is considered a high index of refraction.

Example

How much does the index of refraction of glass vary across the visible spectrum?

First, let's pick a crown glass, BK7. A convenient source of indices of refraction is the website https://refractiveindex.info. We note that the Abbe number is 64.17 For BK7, the index of refraction is $n = 1.5308$ at $\lambda = 400\,\text{nm}$, and $n = 1.5112$ at $\lambda = 780\,\text{nm}$, for a difference of $\delta n = 0.0196$. For comparison, let's pick a flint, SF10, with an Abbe number of 28.53. Then $n = 1.7783$ at $\lambda = 400\,\text{nm}$, and $n = 1.7123$ at $\lambda = 780\,\text{nm}$, for a difference of $\delta n = 0.066$, or more than 3 times that of the BK7.

While RF engineers often talk in terms of frequency, in optics we generally prefer to use wavelength, and we thus find it useful to define the vacuum wavelength. A radar engineer can talk about a 10 GHz wave that retains its frequency as it passes through materials with different wavespeeds and changes its wavelength. It would be quite awkward for us to specify the wavelength of a helium-neon laser as 476 nm in water and 422 nm in glass. To avoid this confusion, we define the **vacuum wavelength** as

Vacuum wavelength, λ defines the frequency ν. The actual wavelength in a material will be shorter (Equation 1.30).

$$\lambda = \frac{2\pi c}{\omega} = \frac{c}{\nu}, \tag{1.28}$$

and the magnitude of the **vacuum wave vector** as

$$k = \frac{2\pi}{\lambda}. \tag{1.29}$$

Often, k is called the wavenumber, which may cause some confusion because spectroscopists often call the wavenumber $1/\lambda$, as we will see in Equation 1.63. We say that the helium-neon laser has a wavelength of 633 nm with the understanding that we are specifying the vacuum wavelength. The wavelength in a material with index of refraction, n, is

$$\lambda_{material} = \frac{\lambda}{n} \tag{1.30}$$

and the wave vector's magnitude is

$$k_{material} = nk. \tag{1.31}$$

Other authors may use a notation such as λ_0 for the vacuum wavelength and use $\lambda = \lambda_0/n$ for the wavelength in the medium, with corresponding definitions for k. Provided that the reader is aware of the convention used by each author, no confusion should occur.

We have somewhat arbitrarily chosen to develop our wave equation in terms of $\mathbf{E}$. We conclude this section by writing the remaining vectors in terms of $\mathbf{E}$. For each vector, there is an equation analogous to Equation 1.16:

$$\mathbf{D} = \hat{x} D_0 e^{j(\omega t - nkz)}, \tag{1.32}$$

$$\mathbf{H} = \hat{y} H_0 e^{j(\omega t - nkz)}, \tag{1.33}$$

and

$$\mathbf{B} = \hat{y} B_0 e^{j(\omega t - nkz)}. \tag{1.34}$$

In general, we can obtain $\mathbf{D}$ using Equation 1.5:

$$\mathbf{D} = \boldsymbol{\epsilon}\mathbf{E},$$

and in this specific case,

$$D_0 = \epsilon E_0.$$

For $\mathbf{H}$, we can use the expected harmonic behavior to obtain

$$\frac{\partial \mathbf{H}}{\partial t} = j\omega \mathbf{H},$$

which we can then substitute into Equation 1.10 to obtain

$$-\mu j\omega \mathbf{H} = \nabla \times \mathbf{E},$$

or

$$\mathbf{H} = \frac{j}{\mu\omega}\nabla \times \mathbf{E}. \qquad (1.35)$$

Eq. 1.16, $\mathbf{E} = \hat{x}E_0 e^{j(\omega t - nkz)}$.

In our example from Equation 1.16,

$$\nabla \times \mathbf{E} = jnkE_0\hat{y}, \qquad (1.36)$$

and

$$H_0 = -\frac{j}{\mu\omega}jnkE_0 = \frac{nk}{\mu\omega}E_0 \ldots$$

$$= \frac{2\pi n}{\lambda}\frac{1}{\mu 2\pi\nu}E_0 = \frac{n}{\mu\lambda\nu}E_0 = \frac{n}{\mu c}E_0.$$

Eq. 1.20, $c = \frac{1}{\sqrt{\epsilon_0\mu_0}}$.

Using Equation 1.20 and $\mu = \mu_0$,

$$H_0 = n\sqrt{\frac{\epsilon_0}{\mu_0}}E_0. \qquad (1.37)$$

For any harmonic waves, we will always find that this relationship between H_0 and E_0 is valid.

1.4.3 Vector and Scalar Waves

We may now write the **vector wave equation**, Equation 1.15, as

$$\nabla^2\mathbf{E} = -\omega^2\frac{n^2}{c^2}\mathbf{E} \qquad (1.38)$$

Vector Wave Equation

We will find it convenient to use a **scalar wave equation** frequently, where we are dealing exclusively with light having a specific polarization, and the vector direction of the field is not important to the calculation. The scalar wave can be defined in terms of $E = |\mathbf{E}|$. If the system does not change the polarization, we can convert the vector wave equation to a **scalar wave equation** or **Helmholtz equation**,

Scalar Wave Equation: Useful when we can ignore polarization

$$\nabla^2 E = -\omega^2\frac{n^2}{c^2}E. \qquad (1.39)$$

1.4.4 Impedance, Poynting Vector, and Irradiance

The **impedance** of a medium is the ratio of the electric field to the magnetic field. This relationship is analogous to that for impedance in an electrical circuit, the ratio of the voltage to the current. In a circuit, voltage is in volts, current is in amperes, and impedance is in ohms. Here the electric field is in volts per meter, the magnetic field in amperes per meter, and the impedance is still in ohms. It is denoted in some texts by Z and in others by η. From Equation 1.37,

$$Z = \eta = \frac{|\mathbf{E}|}{|\mathbf{H}|} = \frac{1}{n}\sqrt{\frac{\mu_0}{\epsilon_0}} = \sqrt{\frac{\mu_0}{\epsilon}}, \qquad (1.40)$$

or

$$Z = \frac{Z_0}{n}, \tag{1.41}$$

where

$$Z_0 = \sqrt{\frac{\mu_0}{\epsilon_0}} = 376.7303\,\text{Ohms}. \tag{1.42}$$

Z relates E and H fields.

Just as we can obtain current from voltage and vice versa in circuits, we can obtain electric and magnetic fields from each other. For a plane wave propagating in the z direction in an isotropic medium,

$$H_x = \frac{-E_y}{Z} \quad H_y = \frac{E_x}{Z}. \tag{1.43}$$

The **Poynting vector** determines the irradiance or power per unit area and its direction of propagation:

$$\mathbf{S} = \frac{1}{\mu_0}\mathbf{E} \times \mathbf{B}. \tag{1.44}$$

The direction of the Poynting vector is in the direction of energy propagation, which is not always the same as the direction of the wave vector, **k**. Because in optics, $\mu = \mu_0$, we can write

S, The Poynting vector describes the irradiance and the direction of energy transfer.

$$\mathbf{S} = \mathbf{E} \times \mathbf{H}. \tag{1.45}$$

The magnitude of the Poynting vector is the **irradiance**

$$\text{Irradiance} = |\mathbf{S}| = \frac{|\mathbf{E}|^2}{Z} = |\mathbf{H}|^2 Z. \tag{1.46}$$

I, Irradiance is power per projected area, typically in W/m^2.

Warning about irradiance variable names: Throughout the history of optics, many different names and notations have been used for irradiance and other radiometric quantities. In recent years, as there has been an effort to establish standards, the irradiance has been associated with the notation, E, in the field of radiometry. This is somewhat unfortunate as the vector **E** and the scalar field E have been associated with the electric field, and I has been associated with the power per unit area, often called "intensity." In Chapter 12, we will discuss the current standards extensively, using E for irradiance, and defining the intensity, I, as a power per unit solid angle. Fortunately, in the study of radiometry we seldom discuss fields, so we reluctantly adopt the convention of using I for irradiance, except in Chapter 12, and reminding ourselves of the notation whenever any confusion might arise.

Warning about irradiance units: In the literature of optics, it is common to define the irradiance as

$$I = |\mathbf{E}|^2 \quad \text{or} \quad I = |\mathbf{H}|^2 \quad \text{(Incorrect but often used)} \tag{1.47}$$

or in the case of the scalar field

$$I = |E|^2 = EE^* \quad \text{(Incorrect but often used)}, \tag{1.48}$$

instead of Equation 1.46,

$$I = |\mathbf{S}| = \frac{|\mathbf{E}|^2}{Z},$$

or

$$I = \frac{EE^*}{Z}.$$

While leaving out the impedance is strictly incorrect, doing so will not lead to errors provided that we begin and end our calculations with irradiance and the starting and ending media are the same.

In Practice

When we work with fields, for example in polarization (Chapter 6) or interference (Chapter 7) we often leave out the impedance, Z. At the beginning of problem, we calculate the fields from the irradiance. Then we manipulate the fields. Finally we determine the irradiance. If the starting and ending media are the same, the impedance cancels.

Example

For example, if we start a problem with an irradiance of $1\,\mathrm{W/m^2}$ in air, the correct electric field is $|\mathbf{E}_{in}| = \sqrt{E_{in}Z} = 19.4\,\mathrm{V/m}$. If we then perform a calculation that determines that the output field is $|\mathbf{E}_{out}| = |\mathbf{E}_{in}|/10 = 1.94\,\mathrm{V/m}$, then the output irradiance is $E_{out} = |\mathbf{E}_{out}|^2/Z = 0.01\,\mathrm{W/m^2}$. If we instead say that $|\mathbf{E}_{in}| = \sqrt{E_{in}} = 1$, then we will calculate $|\mathbf{E}_{out}| = 0.1$ and $E_{out} = |\mathbf{E}_{out}|^2 = 0.01\,\mathrm{W}$.

In this example, we have chosen not to identify the units of the electric field. The field is incorrect, but the final answer is correct. Many authors use this type of calculation, and we will do so when it is convenient and there is no ambiguity.

1.4.5 Wave Notation Conventions

Throughout the literature, there are various conventions regarding the use of exponential notation to represent sine waves. A plane wave propagating in the $\hat{z}$ direction is a function of $\omega t - kz$, or a (different) function of $kz - \omega t$.

A real monochromatic plane wave (one consisting of a sinusoidal wave at a single frequency*) has components given by

$$\cos(\omega t - kz) = \frac{1}{2}e^{\sqrt{-1}(\omega t - kz)} + \frac{1}{2}e^{-\sqrt{-1}(\omega t - kz)}, \tag{1.49}$$

and

$$\sin(\omega t - kz) = \frac{1}{2\sqrt{-1}}e^{\sqrt{-1}(\omega t - kz)} - \frac{1}{2\sqrt{-1}}e^{-\sqrt{-1}(\omega t - kz)}. \tag{1.50}$$

Because most of our study of optics involves linear behavior, it is sufficient to discuss the positive-frequency components and recall that the negative ones are assumed. However, the concept of negative frequency is artificially introduced here with the understanding that each positive-frequency component is matched by a corresponding negative one:

$$\mathbf{E}_r = \hat{\mathbf{x}}E_0 e^{\sqrt{-1}(\omega t - kz)} + \hat{\mathbf{x}}E_0^* e^{-\sqrt{-1}(\omega t - kz)}$$
$$\text{(Complex representation of real wave)}, \tag{1.51}$$

*A monochromatic wave would have to exist for all time. Any wave that exists for a time T will have a bandwidth of $1/T$. However, if we measure a sine wave for a time, $T_m \ll T$, the bandwidth of our signal will be less than the bandwidth we can resolve, $1/T_m$, and we can call the wave quasi-monochromatic.

where * denotes the complex conjugate. We use the notation $\mathbf{E}_r$ to denote the **time-domain variable** and the notation $\mathbf{E}$ for the frequency-domain variable. More compactly, we can write

$$\mathbf{E}_r = \mathbf{E}e^{\sqrt{-1}\omega t} + \mathbf{E}^* e^{-\sqrt{-1}\omega t}, \tag{1.52}$$

which is more generally valid, and is related to the plane wave by

$$\mathbf{E} = \hat{\mathbf{x}} E_0 e^{-jkz}. \tag{1.53}$$

Had we chosen to replace $\mathbf{E}$ by $\mathbf{E}^*$ and vice versa, we would have obtained the same result with

$$\mathbf{E}_r = \mathbf{E}e^{-\sqrt{-1}\omega t} + \mathbf{E}^* e^{\sqrt{-1}\omega t} \quad \text{(Alternative form not used)}. \tag{1.54}$$

Thus the choice of what we call "positive" frequency is arbitrary. Furthermore, the expression $\sqrt{-1}$ is represented in various disciplines as either i or j. Therefore, there are four choices for the exponential notation. Two of these are in common use. Physicists generally prefer to define the positive frequency term as

$$\mathbf{E}e^{-i\omega t} \tag{1.55}$$

and engineers prefer

$$\mathbf{E}e^{+j\omega t}. \tag{1.56}$$

The letter j is chosen to prevent confusion because of electrical engineers' frequent use of i to denote currents in electric circuits. This leads to the confusing statement, often offered humorously, that $i = -j$. The remaining two conventions are also used by some authors, and in most, but not all, cases, little is lost. In this book, we will adopt the engineers' convention of $e^{+j\omega t}$.

We define positive frequency as being associated with $e^{+j\omega t}$

Furthermore, the complex notation can be related to the actual field in two ways. We refer to the "actual" field as the **time-domain** field. It represents an actual electric field in units such as volts per meter, and is used in time-domain analysis. We refer to the complex fields as **frequency-domain** fields. Some authors prefer to say that the actual field is given by

$$\mathbf{E}_r = \Re(\mathbf{E}e^{j\omega t}), \tag{1.57}$$

while others choose the form we will use in this text:

We define the real field as the sum of positive- and negative-frequency components.

$$\boxed{\mathbf{E}_r = \mathbf{E}e^{j\omega t} + \mathbf{E}^* e^{-j\omega t}.} \tag{1.58}$$

The discrepancy between the two forms is a factor of two. In the former case, the irradiance is

Alternative form for S used by some others.

$$|\mathbf{S}| = \left|Re\left(\mathbf{E}e^{j\omega t}\right)\right|^2 = \frac{|\mathbf{E}|^2}{Z}, \tag{1.59}$$

while in the latter

This one is consistent with our notation.

$$|\mathbf{S}| = \left(\mathbf{E}e^{j\omega t} + \mathbf{E}^* e^{-j\omega t}\right)^2 = 2\frac{|\mathbf{E}|^2}{Z}. \tag{1.60}$$

The average irradiance over a cycle is half this value, so Equation 1.59 proves quite convenient:

$$\langle|\mathbf{S}|\rangle = \frac{|\mathbf{E}|^2}{Z}. \tag{1.61}$$

1.4.6 Summary of Electromagnetics

Section 1.4 has presented the fundamentals of electromagnetics which form the basis for our study of optics. Some practitioners of optics use Maxwell's equations as a starting point for most of their work. However, many readers will be able to use derived results with only a minimal need to return to these principles. For these readers it will usually be sufficient to remember the following points.

Take-Away

Summary of Electromagnetics:

- Light is an electromagnetic wave.
- Most of our study of optics is rooted in Maxwell's equations.
- Maxwell's equations lead to a wave equation for the field amplitudes.
- Optical materials are characterized by their indices of refraction.
- Optical waves are characterized by wavelength, phase, and strength of the oscillation.
- Normally, we characterize the strength of the oscillation not by the field amplitudes, but by the irradiance, $|\mathbf{E}\times\mathbf{H}|$, or by the power.
- Irradiance is $|\mathbf{E}|^2/Z = |\mathbf{H}|^2 Z$.
- We can often ignore the Z in these equations.

1.5 Wavelength, Frequency, Power, and Photons

Before we continue with an introduction to the principles of imaging, let us examine the parameters that characterize an optical wave, and their typical values.

1.5.1 Wavelength, Wavenumber, Frequency, and Period

We characterize the frequency of light waves with the vacuum wavelength,

$$\lambda = \frac{2\pi c}{\omega} = \frac{c}{\nu}$$

(Equation 1.28) .

Although it is sufficient to specify either the vacuum wavelength or the frequency, in optics we generally prefer to specify wavelength usually in nanometers (nm) or micrometers (μm), often called microns. The optical portion of the spectrum includes **visible** wavelengths from about 380–710 nm and the **ultraviolet** and **infrared** regions to either side. The ultraviolet region is further subdivided into bands labeled **UVA**, **UVB**, and **UVC**. Various definitions are used to subdivide the infrared spectrum, but we shall use **near-infrared** or **NIR**, **mid-infrared**, and **far-infrared** as defined in Table 1.2. The reader will undoubtedly find differences in these band definitions among other authors. For example, because the visible spectrum is defined in terms of the response of the human eye, which approaches zero rather slowly at the edges, light at wavelengths even shorter

TABLE 1.2
Spectral Regions

Band	Low λ	High λ	Characteristics
Vacuum ultraviolet	100 nm	300 nm	Requires vacuum for propagation
Ultraviolet C (UVA)	100 nm	280 nm	
Oxygen absorption		280 nm	
Ultraviolet B (UVB)	280 nm	320 nm	Causes sunburn. Is partially blocked by glass
Glass transmission	350 nm	2.5 μm	
Ultraviolet A (UVA)	320 nm	400 nm	Is used in a black light. Transmits through glass
Visible light	400 nm	750 nm	Is visible to humans, transmitted through glass, detected by silicon
Near-infrared (NIR)	750 nm	1.1 μm	Is transmitted through glass and biological tissue. Is detected by silicon
Si band edge	1.2 μm		Is not a sharp edge
Water absorption	1.4 μm		Is not a sharp edge
Mid-infrared	3 μm	5 μm	Is used for thermal imaging. Is transmitted by zinc selenide and germanium
Far-infrared (FIR)	8 μm	≈14 μm	Is used for thermal imaging, being near the peak for 300 K. Is transmitted through zinc selenide and germanium. Is detected by HgCdTe and other materials

The optical spectrum is divided into different regions, based on the ability of humans to detect light, on physical behavior, and on technological issues. Note that the definitions are somewhat arbitrary, and other sources may have slightly different definitions. For example, 390 nm, may be considered visible or UVA.

than 385 nm and longer than 760 nm may be visible at sufficient brightness. The FIR band is often called 8–10, 8–12, or 8–14 μm.

Although we shall seldom use frequency, it may be instructive to examine the numerical values at this time. **Frequency**, f or ν, is given by Equation 1.28:

$$f = \nu = \frac{c}{\lambda}.$$

Thus, the visible spectrum extends from about 790 THz (790 THz $= 790 \times 10^{12}$ Hz at $\lambda = 380$ nm) to 420 THz (710 nm). The strongest emission of the carbon dioxide laser is at $\lambda = 10.59$ μm or $\nu = 28.31$ THz, and the shortest UVA wavelength of 100 nm corresponds to 3 PHz (3 PHz $= 3 \times 10^{15}$ Hz). The **period** of the wave is the reciprocal of the frequency,

$$T = \frac{1}{\nu}, \tag{1.62}$$

and is usually expressed in femtoseconds (fs).

It is worth noting that some researchers, particularly those working in far-infrared and Raman spectroscopy (Section 1.6), use **wavenumber** instead of either wavelength or frequency. The wavenumber is the inverse of the wavelength,

Spectroscopists often use the wavenumber, $\tilde{\nu}$, as the inverse of the vacuum wavelength. This is convenient when adding and subtracting frequencies.

$$\tilde{\nu} = \frac{1}{\lambda}, \tag{1.63}$$

usually expressed in inverse centimeters (cm^{-1}). We note that the others use the term wavenumber for the magnitude of the wave vector as in Equation 1.29. Spectroscopists sometimes refer to the wavenumber as the "frequency" but express it in cm^{-1}. Unfortunately, many authors use the variable name, k for wavenumber, ($k = 1/\lambda$) contradicting our definition of the magnitude of the wave vector that $k = 2\pi/\lambda$. For this reason, we have chosen the notation $\tilde{\nu}$. The wavenumber at $\lambda = 10\,\mu\text{m}$ is $\tilde{\nu} = 1000\,\text{cm}^{-1}$, and at the green wavelength of 500 nm, the wavenumber is $\tilde{\nu} = 20,000\,\text{cm}^{-1}$. A spectroscopist might say that "the frequency of green light is 20,000 wavenumbers."

In Practice

The wavenumber $\tilde{\nu}$, is useful when we wish to subtract frequencies. For example, in Raman spectroscopy, we excite a material with light at some illumination wavelength, λ_{ex} and detect emission at a shorter wavelength, λ_{em}. The difference wavenumber is the quantity of interest, the wavenumber of the Raman spectrum.

$$\tilde{\nu}_{ex} - \tilde{\nu}_{em} = \tilde{\nu}_{Raman}.$$

Example

Carotinoids have a Raman spectrum with a peak at $1520\,\text{cm}^{-1}$, which is indicative of its chemical composition. When excited with an argon ion laser at 514.5 nm, what is the emission wavelength?

In inverse centimeters, the excitation wavenumber is

$$\tilde{\nu}_{ex} = \frac{0.01\,\text{m/cm}}{514 \times 10^{-9}\,\text{m}} = 19436\,\text{cm}^{-1}$$

Then

$$\tilde{\nu}_{em} = \tilde{\nu}_{ex} - \tilde{\nu}_{Raman} = 17916\,\text{cm}^{-1}$$

The emission wavelength is

$$\lambda_{em} = \frac{1}{17940\,\text{cm}^{-1} \times 10^{7}\,\text{nm/m}} = 558.2\,\text{nm}$$

The spectroscopist will know $\tilde{\nu}_{ex}$ and will display the emission spectrum from a Raman spectrometer as a function of ν_{Raman}, so that the peaks associated with different materials can readily be determined. The plot would look the same if a different excitation source were used. While those accustomed to working with wavelength units will find this confusing, a spectroscopist learns to "think in wavenumbers."

1.5.2 Field Strength, Irradiance, and Power

The irradiance is given by the magnitude of the Poynting vector in Equation 1.46:

$$|\mathbf{S}| = \frac{|\mathbf{E}|^2}{Z} = |\mathbf{H}|^2 Z.$$

Example

In optics, we seldom think of the field amplitudes, but it is useful to have an idea of their magnitudes. Let us consider the example of a 1 mW laser with a beam 1 mm in diameter. For present purposes, let us assume that the irradiance is uniform, so the irradiance is

$$\frac{P}{\pi\,(d/2)^2} = 1270\,\mathrm{W/m^2}. \tag{1.64}$$

The electric field is then

$$|\mathbf{E}| = \sqrt{1270\,\mathrm{W/m^2} Z_0} = 692\,\mathrm{V/m} \tag{1.65}$$

using Equation 1.42 for Z_0, and the magnetic field strength is

$$|\mathbf{H}| = \sqrt{\frac{1270\,\mathrm{W/m^2}}{Z_0}} = 1.83\,\mathrm{A/m}. \tag{1.66}$$

Example

It is interesting to note that one would expect to be able to ionize air using a laser beam of sufficient power that the electric field exceeds the dielectric breakdown strength of air, which is approximately $3 \times 10^6\,\mathrm{V/m}$. In fact ionization is more complicated, but this is still an instructive example. We can achieve this irradiance by focusing a beam of sufficiently high power to a sufficiently small spot. This field requires an irradiance of

$$\frac{|\mathbf{E}|^2}{Z_0} = 2.4 \times 10^{10}\,\mathrm{W/m^2}. \tag{1.67}$$

With a beam diameter of $d = 20\,\mu\mathrm{m}$, twice the wavelength of a CO_2 laser,

$$P = 2.4 \times 10^{10}\,\mathrm{W/m^2} \times \pi \left(\frac{d}{2}\right)^2 = 7.5\,\mathrm{W}. \tag{1.68}$$

1.5.3 Energy and Photons

Let us return to the milliwatt laser beam in the first part of Section 1.5.2, and ask how much **energy**, J is produced by the laser in one second:

$$J = Pt = 1\,\mathrm{mJ}. \tag{1.69}$$

If we attenuate the laser beam so that the power is only a nanowatt, and we measure for a microsecond, then the energy is reduced to a picojoule. Is there a lower limit? Einstein's study of the photoelectric effect[60] (Section 13.1) demonstrated that there is indeed a minimal quantum of energy which we call the energy of a photon:

$$J_1 = h\nu \tag{1.70}$$

Example

For a helium-neon laser, $\lambda = 633\,\text{nm}$, and

$$J_1 = \frac{hc}{\lambda} = 3.14 \times 10^{-19}\,\text{J}. \tag{1.71}$$

Thus, at this wavelength, a picojoule consists of over 3 million photons, and a millijoule consists of over 3×10^{15} photons.

Photon Energy

It is sometimes useful to think of light in the wave picture with an oscillating electric and magnetic field, and in other contexts, it is more useful to think of light as composed of particles (photons) having this fundamental unit of energy. The quantum theory developed in the early twentieth century resolved this apparent paradox, but the details are beyond the scope of this book and are seldom needed in most or our work. However, we should note that if a photon is a particle, we would expect it to have momentum. In fact, the momentum of a photon is

$$\mathbf{p}_1 = \frac{h\mathbf{k}_1}{2\pi}. \tag{1.72}$$

The momentum of our helium-neon-laser photon is

$$|p_1| = \frac{h}{\lambda} = 1.05 \times 10^{-27}\,\text{kg m/s}. \tag{1.73}$$

In a "collision" with an object, momentum must be conserved, and a photon is thus capable of exerting a force, as first observed in 1900[139]. This force finds application today in the laser-cooling work mentioned at the end of the historical overview in Section 1.2.

Take-Away

Wavelength, Frequency, Power, and Photons:

- Frequency is $\nu = c/\lambda$.
- Wavenumber is $\tilde{\nu} = 1/\lambda$.
- There is a fundamental unit of optical energy, called the photon, with energy $h\nu$.
- A photon also has momentum, hk, and can exert a force.

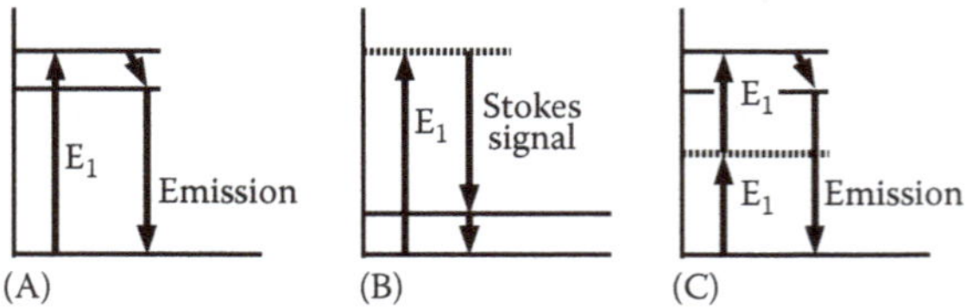

FIGURE 1.9
Energy-level diagrams. Three different types of interaction are shown. (A) Fluorescence. (B) Raman scattering. (C) Two-photon-excited fluorescence.

1.6 Energy Levels and Transitions

The optical absorption spectra of materials are best explained with the aid of energy-level diagrams, samples of which are shown in Figure 1.9. Many variations on these diagrams are used in the literature. The Jablonski diagram groups energy levels vertically using a scale proportional to energy, and spin multiplicity on the horizontal. Various authors have developed diagrams with different line styles and often colors to represent different types of levels and transitions. Another useful diagram is the energy-population diagram, where energies are again grouped vertically, but the horizontal scale is used to plot the population of each level. Such diagrams are particularly useful in developing laser rate equations[210]. Here, we use a simple diagram where the energies are plotted vertically, and the horizontal direction shows temporal order. The reader is likely familiar with Niels Bohr's description of the hydrogen atom, with electrons revolving around a nucleus in discrete orbits, each orbit associated with a specific energy level, with larger orbits having larger energies. A photon passing the atom can be **absorbed** if its energy, $h\nu$, matches the energy difference between the orbit occupied by the electron and a higher orbit. After the photon is absorbed, the electron moves to the higher-energy orbit. If the photon energy matches the difference between the energy levels of the occupied orbit and a lower one, **stimulated emission** may occur; the electron moves to the lower orbit and a photon is emitted and travels in the same direction as the one which stimulated the emission. In the absence of an incident photon, an electron may move to a lower level while a photon is emitted, in the process of **spontaneous emission**. Bohr's basic theory of the "energy ladder" was validated experimentally by James Franck and Gustav Ludwig Hertz who were awarded the Nobel Prize in Physics in 1925, "for their discovery of the laws governing the impact of an electron upon an atom." Of course, the Bohr model proved overly simplistic and modern theory is more complicated, but the basic concept of the "energy ladder" is valid and remains useful for discussing absorption and emission of light. The reader may find a book by Hoffmann[96] of interest, both for its very readable account of the development of quantum mechanics and for the titles of chapters describing Bohr's theory, "The Atom of Niels Bohr," and its subsequent replacement by more modern theories, "The Atom of Bohr Kneels."

Jablonski diagram
Ladder diagram

Absorption

Stimulated emission

Spontaneous emission

FURTHER READING: Quantum Mechanics and Bohr[96]

Although we are almost always interested in the interaction of light with more complicated materials than hydrogen atoms, every material can be described by an energy ladder. Figure 1.9A shows the process of absorption, followed by decay to an intermediate state and emission of a photon. The incident photon must have an energy matching the energy difference between the lowest, or "ground" state, and the upper level. The next decay may be radiative, emitting a photon, or non-radiative, for example, producing heat in the material. The non-radiative decay is usually faster than the radiative one. The material is then left for a time in a **metastable state**, which may last for several nanoseconds. The final, radiative, decay results in the spontaneous emission of a photon. We call this process **fluorescence**. The emitted photon must have an energy (and frequency) less than that of the incident photon. Thus, the wavelength of the emitted photon is always longer than that of the incident or photon.

Fluorescence

Some processes involve "virtual" states, shown on our diagrams by dotted lines. The material cannot remain in one of these states, but must find its way to a "real" state. One example of a virtual state is in Raman scattering[185] where a photon is "absorbed" and a longer-wavelength photon is emitted, leaving the material in a state with energy which is low, but still above the ground state (Figure 1.9B). Spectral analysis of these signals, or **Raman spectroscopy**, is an alternative to infrared absorption spectroscopy, which may allow better resolution and depth of penetration in biological media. Sir Chandrasekhara Venkata Raman won the Nobel Prize in Physics in 1930, "for his work on the scattering of light and for the discovery of the effect named after him."

Raman spectroscopy offers molecular specificity.

As a final example of our energy level diagrams, we consider two-photon-excited fluorescence in Figure 1.9C. The first prediction of two-photon excitation was by Maria

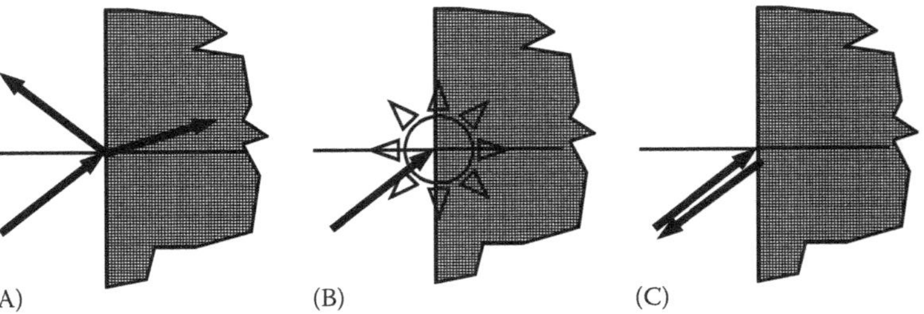

FIGURE 1.10
Light at an interface. (A) Light may undergo specular reflection and transmission at well-defined angles, or it may be scattered in all directions. (B) Scatter from a large number of particles leads to what we call diffuse transmission and reflection. (C) Many surfaces also produce retroreflection.

Göppert Mayer in 1931[84]. She shared half the Nobel Prize in Physics in 1963 with J. Hans D. Jensen, "for their discoveries concerning nuclear shell structure." The other half of the prize went to Eugene Paul Wigner "for his contributions to the theory of the atomic nucleus and the elementary particles. . . ." The first experimental realization of two-photon-excited fluorescence did not occur until the 1960s[175]. An incident photon excites a virtual state, from which a second photon excites a real state. After the material reaches the upper level, the fluorescence process is the same as in Figure 1.9A. Because the lifetime of the virtual state is very short, the two incident photons must arrive closely spaced in time. Extremely high irradiance is required, as will be discussed in Chapter 14. In that chapter, we will revisit the energy level diagrams in Section 14.4.1.

Two-photon excitation of fluorescence normally requires short pulses with high peak power.

This section has provided a very cursory description of some very complicated processes of light interacting with atoms and molecules. The concepts will be discussed in connection with specific processes later in the book, especially in Chapter 14. The interested reader will find many sources of additional material. An excellent discussion of energy level diagrams and their application to fluorescence spectroscopy is given in a book on the subject[131].

FURTHER READING: Fluorescence Spectroscopy[131].

1.7 Macroscopic Effects

Let us look briefly at how light interacts with materials at the macroscopic scale. The reader familiar with electromagnetic principles is aware of the concept of Fresnel reflection and transmission, by which light is reflected and refracted at a **dielectric interface**, where the index of refraction has different values on either side of the interface. The directions of propagation of the reflected and refracted light are determined exactly by the direction of propagation of the input wave, as we shall discuss in more detail in Chapter 2. The quantity of light reflected and refracted will be discussed in Section 6.4. We refer to these processes as specular reflection and transmission. Figure 1.10 shows the possible behaviors. Specular reflection and transmission occur only in directions determined by the direction of the incident light and the surface orientation. Therefore, we only see the specular light if we look in exactly the right direction.

Specular reflection and transmission

Light can be **scattered** from small particles into all directions. The details of the scattering are beyond the scope of this book and require calculations such as Mie scattering developed by Gustav Mie[157] for spheres, Rayleigh scattering developed by Lord Rayleigh (John William Strutt) [186;188] for small particles, and a few other special cases. More complicated scattering may be addressed using techniques such as finite-difference time-domain calculations[256]. Recent advances in computational power have made it possible to compute the scattering properties of structures as complex as somatic cells[58]. For further details on the subject of light scattering several texts are available[232;107;25].

Scattering: Diffuse reflection and transmission

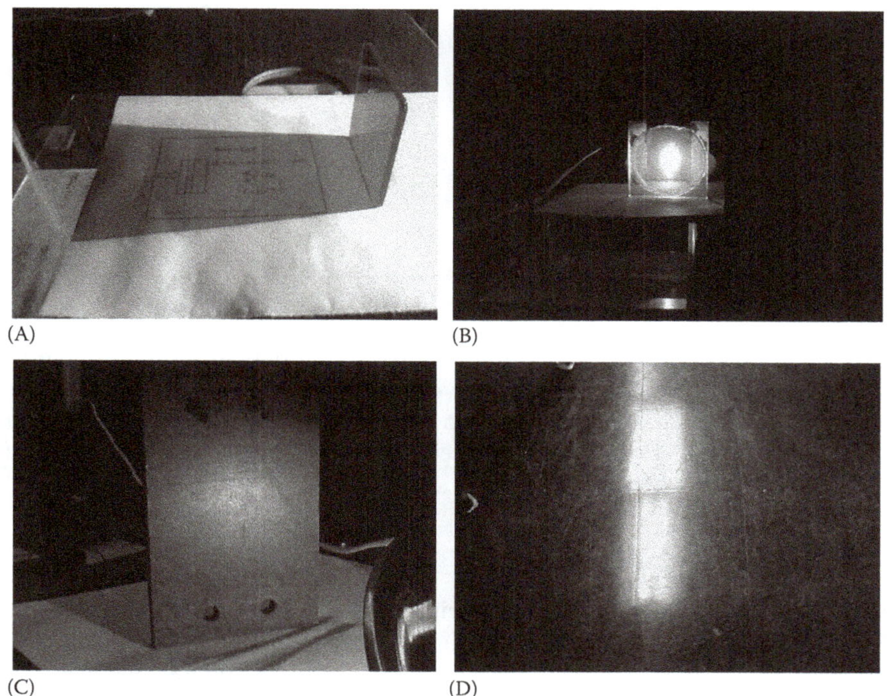

FIGURE 1.11
Examples of light in materials. Depending on the nature of the material, reflection and transmission may be more specular or more diffuse. A colored glass plate demonstrates mostly specular transmission, absorbing light of some colors more than others (A). A solution of milk in water may produce specular transmission at low concentrations or mostly diffuse transmission at higher concentrations (B). A rusty slab of iron (C) produces mostly diffuse reflection, although the bright spot in the middle indicates some specular reflection. The reflection in a highly polished floor is strongly specular, but still has some diffuse component (D).

Diffuse transmission and reflection occur when the surface is rough, or the material consists of many scatterers. An extreme example of diffuse reflection is the scattering of light by the titanium dioxide particles in white paint. If the paint layer is thin, there will be scattering both forward and backward. The scattered light is visible in all directions. A minimal example of diffuse behavior is from dirt or scratches on a glass surface. Many surfaces also exhibit a retroreflection, where the light is reflected directly back toward the source. We will see retroreflectors or corner cubes in Chapter 2, which are designed for this type of reflection. However, in many materials, there is an enhanced retroreflection, either by design or by the nature of the material. In most materials, some combination of these behaviors is observed, as shown in Figure 1.11. Furthermore, some light is usually absorbed by the material.

Take-Away

Light in Materials:

- Light may be specularly or diffusely reflected or retroreflected from a surface.
- Light may be specularly or diffusely transmitted through a surface.
- Light may be absorbed by a material.
- For most surfaces, all of these effects occur to some degree.

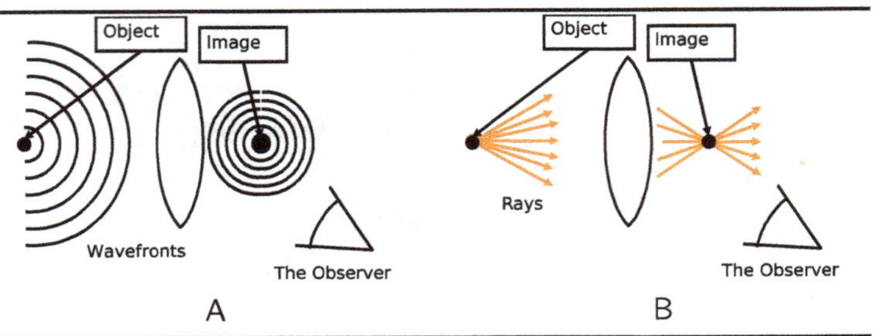

FIGURE 1.12
Imaging. We can sense the wavefronts from an object, and determine its location. If we can construct an optical system that makes the wavefronts appear to originate from a new location, we perceive an image at that location (A). Rays of light are everywhere perpendicular to the wavefronts. The origin of the rays is the object location, and the apparent origin after the optical system is the image location (B).

1.8 Basic Concepts of Imaging

Before we begin developing a rigorous description of image formation, let us think about what we mean by imaging. Suppose that we have a very small object that emits light, or scatters light from another source, toward our eye. The object emits spherical waves, and we detect the direction of the wavefronts which reach our eye to know where the object is in angle. If the wavefronts are sufficiently curved that they change direction over the distance between our eyes, we can also determine the distance. Figure 1.12 shows that if we can somehow modify the wavefronts so that they appear to come from a different point, we call that point an image of the object point. An alternative picture describes the process in terms of rays. Later we will make the definition of the rays more precise, but for now, they are everywhere perpendicular to the wavefronts and they show the direction of propagation of the light. Rays that emanate from a point in the object are changed in direction so that they appear to emanate from a point in the image.

Roadmap: Section 1.8

Our goal in the next few sections and in Chapter 2 is to determine how to transform the waves or rays to create a desired image. Recalling Fermat's principle, we will define an "optical path length" proportional to the time light takes to travel from one point to another in Section 1.8.1, and then minimize this path length in Section 1.8.2. Our analysis is rooted in Maxwell's equations through the vector wave equation, Equation 1.38, and its approximation, the scalar wave equation, Equation 1.39. We will postulate a general harmonic solution and derive the eikonal equation, which is an approximation as wavelength approaches zero.

1.8.1 Eikonal Equation and Optical Path Length

Beginning with Equation 1.39, we propose the general solution

$$E = e^{a(\mathbf{r})} e^{j[k\ell(\mathbf{r})-\omega t]}$$

or

$$E = e^{a(\mathbf{r})} e^{jk[\ell(\mathbf{r})-ct]}. \tag{1.74}$$

This is a completely general harmonic solution, where a describes the amplitude and ℓ is related to the phase. We shall shortly call ℓ the **optical path length**. The vector, $\mathbf{r}$, defines the position in three-dimensional space. Substituting into Equation 1.39,

$$\nabla^2 E = -\omega^2 \frac{n^2}{c^2} E,$$

and using $\omega = kc$,

$$\left\{\nabla^2 (a + jk\ell) + [\nabla (a + jk\ell)]^2\right\} E = -n^2k^2E$$

$$\nabla^2 a + jk\nabla^2 \ell + (\nabla a)^2 + 2jk\nabla a \nabla \ell - k^2 (\nabla \ell)^2 = -n^2k^2.$$

Dividing by k^2,

$$\frac{\nabla^2 a}{k^2} + j\frac{\nabla^2 \ell}{k} + \frac{(\nabla a)^2}{k^2} + 2j\frac{\nabla a}{k}\nabla \ell - (\nabla \ell)^2 = -n^2. \qquad (1.75)$$

Now we let the wavelength become small, or, equivalently, let $k = 2\pi/\lambda$ become large. We shall see that the small-wavelength approximation is the fundamental assumption of geometric optics. In contrast, if the wavelength is very large, then circuit theory can be used. In the case where the wavelength is moderate, we need diffraction theory, to which we shall return in Chapter 8.

How small is small? The first and second derivatives of a and ℓ must all be small enough that every colored term in Equation 1.75 is small compared to the terms printed in black. That is, the values of these parameters should not vary over a distance of one wavelength. For more details see Appendix C1A1 These conditions will often be violated, for example, at the edge of an aperture where a is discontinuous and thus has an infinite derivative. If such regions are a small fraction of the total, we can justify neglecting them. Therefore, we can use the theory we are about to develop to describe light propagation through a window provided that we do not look too closely at the edge of the shadow. However if the "window" size is only a few wavelengths, most of the light violates the small-wavelength approximation; so we cannot use this theory, and we must consider diffraction which we shall do in Chapter 8.

Finally then, from the non-zero terms in Equation 1.75 we have the **eikonal equation**

$$(\nabla \ell)^2 = n^2$$

$$|\nabla \ell| = n. \qquad (1.76)$$

Knowing that the gradient of ℓ is n, we can almost obtain $\ell(\mathbf{r})$, by integration along a path. We qualify this statement, because our answer can only be determined within a constant of integration. The **optical path length** from a point A to a point B is

ℓ, Optical path length (OPL)

$$\ell = OPL = \int_A^B n d\ell_p, \qquad (1.77)$$

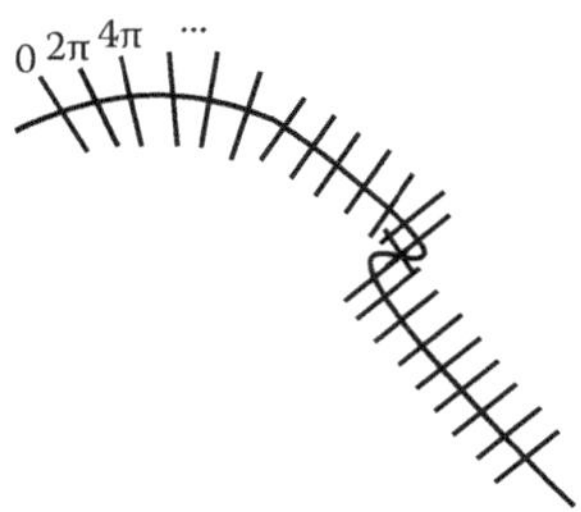

FIGURE 1.13
Optical path length. The OPL between any two points is obtained by integrating along a path between those two points. The eikonal equation ensures that the integral does not depend on the path chosen. The phase difference between the two points can be computed from the OPL.

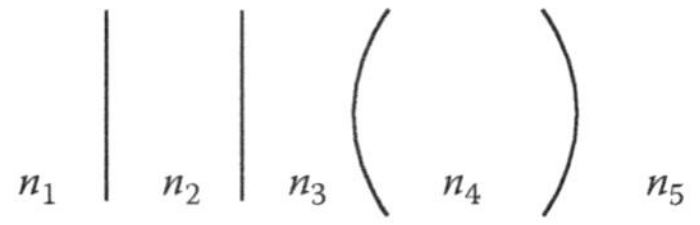

n_1 n_2 n_3 n_4 n_5

FIGURE 1.14
Optical path in discrete materials. The integrals become a sum.

where ℓ_p is the physical path length. The eikonal equation, Equation 1.76, ensures that the integral will be unique, regardless of the path taken from A to B.

If the index of refraction is piecewise constant, as shown Figure 1.14, then the integral becomes a sum,

$$OPL = \sum_m n_m \ell_m, \tag{1.78}$$

where

- n_m is the index of layer m and
- ℓ_m is its thickness.

Such a situation is common in systems of lenses, windows, and other optical materials.

The phase difference between the two points will be

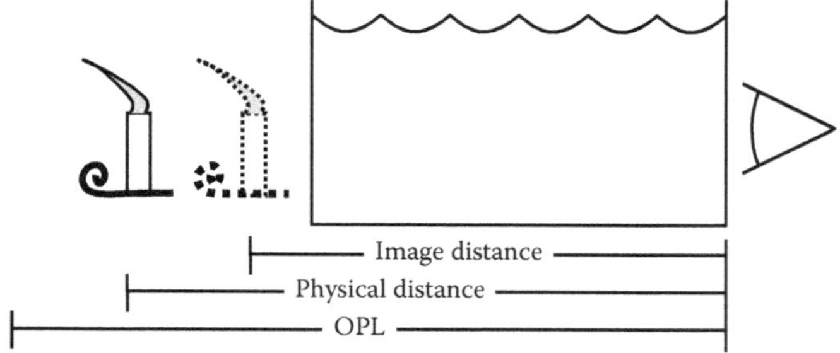

FIGURE 1.15
Optical path in water. The optical path is longer than the physical path, but the geometric path is shorter.

$$\Delta\phi = k\Delta\ell = 2\pi\frac{OPL}{\lambda}, \qquad (1.79)$$

as shown in Figure 1.13.

Example

In Figure 1.14, we place a source of light 1.00 cm to the left of the n_1–to–n_2 interface. $n_1 = 1.00$ for air. Then $n_2 = 1.5$ for a glass wall, 1 cm thick $n_3 = 1.33$ for a 2 cm thick region of water, and $n_4 = 1.50$ for a glass lens 0.5 cm thick. Our viewer is 3 cm after the lens in air with $n_5 = 1.00$. What is the optical path length?

$$\ell = 1\,\text{cm} \times 1.00 \ + \ 1\,\text{cm} \times 1.50 \ + \ 2\,\text{cm} \times 1.33 \ +$$

$$0.5\,\text{cm} \times 1.50 \ + \ 3\,\text{cm} \times 1.00 = 8.91\,\text{cm}.$$

By comparison, the physical path length is

$$\ell = 1\,\text{cm} \ + \ \ell = 1\,\text{cm} \ + \ \ell = 2\,\text{cm} \ +$$

$$\ell = 0.5\,\text{cm} \ + \ \ell = 3\,\text{cm} = 7.5\,\text{cm}.$$

The optical path length is not to be confused with the image distance. For example, the OPL of a fish tank filled with water is longer than its physical thickness by a factor of 1.33, the index of refraction of water. However, we will see in Chapter 2 that an object behind the tank (or inside it) appears to be closer than the physical distance, as shown in Figure 1.15.

Take-Away

Optical Path Length (OPL):
The optical path length is really a measure of transit time. if light travels along a physical path of length (PPL) with possibly different indices of refraction along the way in a given time, then light would travel a distance OPL in vacuum in the same time.

The OPL is often useful for computing phase. The phase change is given by Equation 1.79,

$$\Delta\phi = 2\pi * OPL/\lambda.$$

1.8.2 Fermat's Principle

We can now apply **Fermat's principle**, which we discussed in the historical overview of Section 1.2. The statement that "light travels from one point to another along the path which takes the least time" is equivalent to the statement that "light travels from one point to another along the path with the shortest OPL."

In the next chapter, we will use Fermat's principle with variational techniques to derive equations for reflection and refraction at surfaces. However, we can deduce one simple and important rule without further effort. If the index of refraction is uniform, then minimizing the OPL is the same as minimizing the physical path; light will travel along straight lines. These lines, or **rays**, must be perpendicular to the wavefronts by Equation 1.76.

A second useful result can be obtained in cases where the index of refraction is a smoothly varying function. Then Fermat's principle requires that a ray follow a path given by the **ray equation**[151]

$$\frac{d}{d\ell}\left[n(\mathbf{r})\frac{d\mathbf{r}}{d\ell}\right] = \nabla n(\mathbf{r}). \tag{1.80}$$

Using the product rule for differentiation, we can obtain an equation that can be solved numerically, from a given starting point,

$$n(\mathbf{r})\frac{d^2\mathbf{r}}{d\ell^2} = \nabla n(\mathbf{r}) - \frac{d\mathbf{r}}{d\ell}\frac{dn(\mathbf{r})}{d\ell}. \tag{1.81}$$

These equations and others derived from them are used to trace rays through **gradient index** materials, including some fibers, the atmosphere, and any other materials for which the index of refraction varies continuously in space. In some cases, a closed-form solution is possible[151]. In other cases, we can integrate Equation 1.81 directly, given an initial starting point, $\mathbf{r}$, and direction $d\mathbf{r}/d\ell$. A number of better approaches to the solution are possible using cylindrical symmetry[150;161], spherical symmetry[149], and other special cases.

Most of the work in this book will involve discrete optical components, where the index of refraction changes discontinuously at interfaces. Without knowing anything more, let us ask ourselves which path in Figure 1.16 the light will follow from the object to the image, both of which are in air, passing through the glass lens along the way. The curved path is obviously not going to be minimal. However paths that are piecewise linear all seem plausible. Let's think more about this. If we make the path too close to the center of the lens, the added thickness of glass will increase the optical path length contribution of ℓ_2 through the lens in

$$\sum_{m=1}^{3} n_m \ell_m = \ell_1 + n_{glass}\ell_2 + \ell_3, \tag{1.82}$$

but if we make the path too close to the edge, the paths in air, ℓ_1 nd ℓ_3, will increase.

If we choose the thickness of the glass carefully, maybe we can make these two effects cancel each other, and have all the paths equal, as in Figure 1.17. In this case, any light from the object point will arrive at the image point. We can use this simple result as a basis for developing a theory of imaging, but we will take a slightly different approach in Chapter 2.

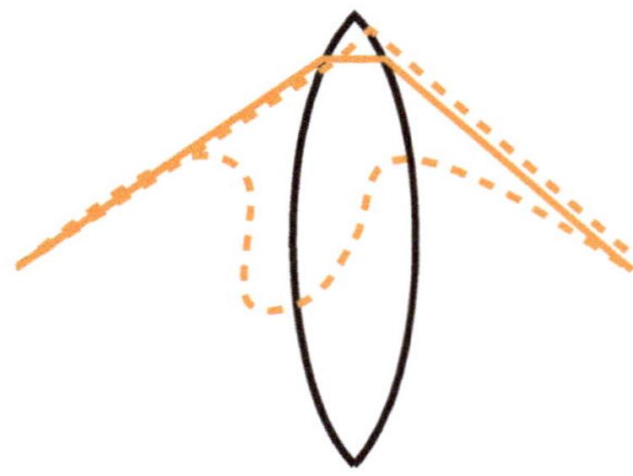

FIGURE 1.16
Fermat's principle. Light travels from one point to another along the shortest optical path.

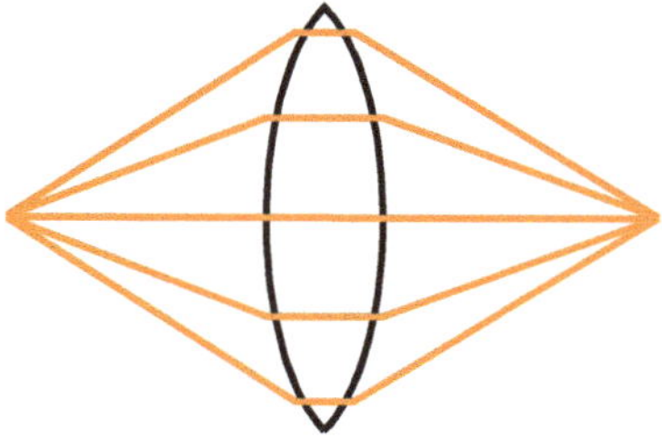

FIGURE 1.17
Imaging. If all the paths from one point to another are minimal, the points are said to be conjugate; one is the image of the other.

1.8.3 Summary

Take-Away

Imaging:

- An image is formed at a point from which wavefronts that originated at the object appear to be centered.
- Equivalently, an image is a point from which rays from a point in the image appear to diverge.
- The optical path length is the integral, $\int n d\ell$, between two points.
- Fermat's principle states that light will travel from one point to another along the shortest optical path length.
- Geometric optics can be derived from Fermat's principle and the concept of optical path length.
- In a homogeneous medium, the shortest optical path is the shortest physical path, a straight line.
- In a gradient index medium, the shortest optical path can be found by integrating Equation 1.81.

1.9 Overview of the Book

Roadmap: Ch. 2 to End

Chapters 2 through 5 all deal with various aspects of geometric optics. In Chapter 2, we will investigate the fundamental principles and develop equations for simple lenses, consisting of one glass element. Although the concepts are simple, the bookkeeping can easily become difficult if many elements are involved. Therefore, in Chapter 3 we develop a matrix formulation to simplify our work. We will find that any arbitrary combination of simple lenses can be reduced to a single effective lens, provided that we use certain definitions about the distances. These two chapters address so-called Gaussian or paraxial optics. Angles are assumed to be small, so that $\sin\theta \approx \tan\theta \approx \theta$ and $\cos\theta \approx 1$. We also give no consideration to the diameters of the optical elements. We assume they are all infinite. Then in Chapter 4, we consider the effect of finite diameters on the light-gathering ability and the field of view of imaging systems. We will define the "window" that limits

Ch. 2 and 3 on Gaussian, or paraxial optics.

Ch. 4 on finite apertures.

what can be seen and the "pupil" that limits the light-gathering ability. Larger diameters of lenses provide larger fields of view and allow the collection of more light. However, larger diameters also lead to more severe violations of the small-angle approximation. These violations lead to aberrations, discussed in Chapter 5. Implicit in these chapters is the idea that irradiances add and that the small-wavelength approximation is valid.

Ch. 5 on aberrations

Next, we turn our attention to wave phenomena. We will discuss polarization in Chapter 6. We investigate the basic principles, devices, and mathematical formulations that allow us to understand common systems using polarization. In Chapter 7, we develop the ideas of interferometers, laser cavities, and dielectric coated mirrors and beamsplitters. Chapter 8 addresses the limits imposed on light propagation by diffraction and determines the fundamental resolution limits of optical systems in terms of the size of the pupil, as defined in Chapter 4. Chapter 9 continues the discussion of diffraction, with particular attention to Gaussian beams, which are important in lasers and are a good first approximation to beams of other profiles.

Ch. 6 on polarization

Ch. 7 on interference

Ch. 8 on diffraction

Ch. 9 on Gaussian beams

In Chapter 10, we briefly discuss the concept of coherence, including its measurement and control. Chapter 11 offers a brief introduction to the field of Fourier optics, which is an application of the diffraction theory presented in Chapter 8. This chapter also brings together some of the ideas originally raised in the discussion of windows and pupils in Chapter 4. In Chapter 12, we study radiometry and photometry to understand how to quantify the amount of light passing through an optical system. A brief discussion of color will also be included.

Ch. 10 on coherence

Ch. 11 on Fourier optics

Ch. 12 on radiometry and photometry

Next are brief discussions of optical detection in Chapter 13 and nonlinear optics in Chapter 14. Each of these chapters is a brief introduction to a field for which the details must be left to other texts. We conclude with Chapter 15 on techniques for building prototype systems on optical breadboards.

Ch. 13 on detectors

Ch. 14 on non-linear optics

Ch. 15 on breadboarding

These chapters provide a solid foundation on which the reader can develop expertise in the field of optics. Whatever the reader's application area, these fundamentals will play a role in the design, fabrication, and testing of optical systems and devices.

Problems

1.1 Wavelength, Frequency, and Energy

Here are some "warmup" questions to build familiarity with the typical values of wavelength, frequency, and energy encountered in optics.

1.1a What is the frequency of an argon ion laser with a vacuum wavelength of 514 nm (green)?

1.1b In a nonlinear crystal, we can generate sum and difference frequencies. If the inputs to such a crystal are the argon ion laser with a vacuum wavelength of 514 nm (green) and another argon ion laser with a vacuum wavelength of 488 nm (blue), what are the wavelengths of the sum and difference frequencies that are generated? Can we use glass optics in this experiment?

1.1c A nonlinear crystal can also generate higher harmonics (multiples of the input laser's frequency). What range of input laser wavelengths can be used so that both the second and third harmonics will be in the visible spectrum?

1.1d How many photons are emitted per second by a helium-neon laser (633 nm wavelength) with a power of 1 mW?

1.1e It takes 4.18 J of energy to heat a cubic centimeter of water by 1 K. Suppose 100 mW of laser light is absorbed in a volume with a diameter of 2 μm and a length of 4 μm. Neglecting thermal diffusion, how long does it take to heat the water by 5 K?

1.1f A fluorescent material absorbs light at 488 nm and emits at 530 nm. What is the maximum possible efficiency, where efficiency means power out divided by power in.

1.2 Optical Path

1.2a We begin with an empty fish tank, 30 cm thick. Neglect the glass sides. We paint a scene on the back wall of the tank. How long does light take to travel from the scene on the back of the tank to the front wall?

1.2b Now we fill the tank with water and continue to neglect the glass. How far does the scene appear behind the front wall of the tank? Now how long does the light take to travel from back to front?

1.2c What is the optical-path length difference experienced by light traveling through a glass coverslip 170 μm thick in comparison to an equivalent length of air?

1.3 Raman Spectrometer

Repeat the example of Raman spectroscopy following Equation 1.63, assuming the excitation wavelength is 633 nm from a helium neon laser.

1.3a What is the emission wavelength?

1.3b What is the frequency associated with the Raman peak at $1520\,\text{cm}^{-1}$?

1.4 Photon Momentum (NS)

A photon has a momentum h/λ where λ is the wavelength, and h is Plancks constant. When a photon impinges on a particle, it produces an impulse, $\int f dt$, where f is the force. Thus, for N_t photons per unit time, the average force is $f = \frac{h}{\lambda} N_t$. How much optical power is needed to levitate a glass bead 50 μm in diameter (density 2.5 grams/cc)?

1.5 Fields (NS)

Consider a 1 −Watt laser beam focused to an area with radius of 5 microns. What are the magnitudes of the fields, E in Volts per meter, B, in Teslas, and H, in Amperes/meter?

1.6 Fluorescence (NS)

A molecule absorbs light at a wavelength of 488 nm and for each absorbed photon, it emits a photon at 532 nm. What is the photon energy at each wavelength and what fraction of the incident energy is absorbed by the molecule?

1.7 Laser Radar (NS)

What is the distance to a target if light makes the round trip in 1 μs? If it does so in 1 ms?

1.8 Wavelength (NS)

A helium–neon laser has a vacuum wavelength of $\lambda = 632.8$ nm. What is the frequency? What is the wavelength in water? In glass?

1.9 Energy (NS)

Someone suggested using a laser ice scraper to clear a car windshield of ice. Let's take a very quick look at feasibility. Suppose we use a laser with 100 Watts (very high power), and suppose the ice absorbs all the power. How long will it take to melt a layer of ice 2 mm thick?

Let's neglect any reflection and cooling effects so that all the energy goes into melting. Let's also assume that we need to melt all the ice. The former is overly optimistic and the latter is overly pessimistic so perhaps our number will be close to reality.

1.10 Pulsed Laser (NS)

A certain pulsed laser has an average power of 10 W, pulse width of 100 fs and pulse repetition frequency of 80 MHz. What is the pulse energy? What is the peak power assuming a rectangular pulse?

2 Basic Geometric Optics

The concepts of geometric optics were well understood long before the wave nature of light was fully appreciated. Although a scientific basis was lacking, reflection and refraction were understood by at least the 1600s. Nevertheless, we will follow the modern approach of basing geometric optics on the underlying wave theory. We will begin by deriving Snell's law from Fermat's principle in Section 2.1. From there, we will develop equations to characterize the images produced by mirrors (Section 2.3) and refractive interfaces (Section 2.4). Combining two refractive surfaces, we will develop equations for lenses in Section 2.5, and the particularly simple approximation of the thin lens in Section 2.5.1. We conclude with a discussion of prisms in Section 2.6 and reflective optics in Section 2.7.

Before we turn our attention to the main subject matter of this chapter, it may be helpful to take a look at a road map of the next few chapters.

At the end of this chapter, we will have a basic understanding of the location, size, and orientation of images. We will reach this goal by assuming that the diameters of lenses and mirrors are all infinite, and that angles are so small that their sines, tangents, and values in radians can all be considered equal and their cosines considered unity. We will develop a "thin-lens equation" and expressions for magnification that can be used for a single thin lens or a single refractive surface. These formulas can be represented by geometric constructions such as the example shown in Figure 2.1A. In early optics courses, students are often taught to trace the three rays shown here. The object is represented by a vertical arrow. We first define the **optical axis** as a straight line through the system. Normally this will be the axis of rotation. The ray from the tip of the arrow that is parallel to the optical axis is refracted through a point on the optical axis called the back focal point. The ray from the tip of the arrow that goes through the front focal point is refracted to be parallel to the optical axis. At the intersection of these rays, we place another arrow tip representing the image. We can check our work knowing that the ray through the center of the lens is undeviated. We can solve more complicated problems by consecutively treating the image from one surface as an object for the next.

Ch. 3 on matrix optics (bookkeeping for geometric optics)

This result is powerful, but incomplete. Some of the failures are listed here:

1. For multiple refractive elements, the bookkeeping becomes extremely complicated and the equations that result from long chains of lens equations yield little physical insight.

2. There is nothing in these equations to help us understand the amount of light passed through a lens system, or how large an object it can image.

3. The small-angle assumptions lead to the formation of perfect images. As we make the object smaller without limit, this approach predicts that the image also becomes smaller without limit. The resolution of the lens system is always perfect, contrary to observations. The magnification is the same at all points in the field of view, so there is no distortion.

Roadmap: Geometric Optics

In Chapter 3, we will develop matrix optics, a bookkeeping system that allows for arbitrarily complicated compound lenses. We will use matrix optics to reduce almost any complicated compound lens to a mathematical formulation identical to that of the thin lens, but with a different physical interpretation. We will learn that the lens equation is

DOI: 10.1201/9781315157047-2

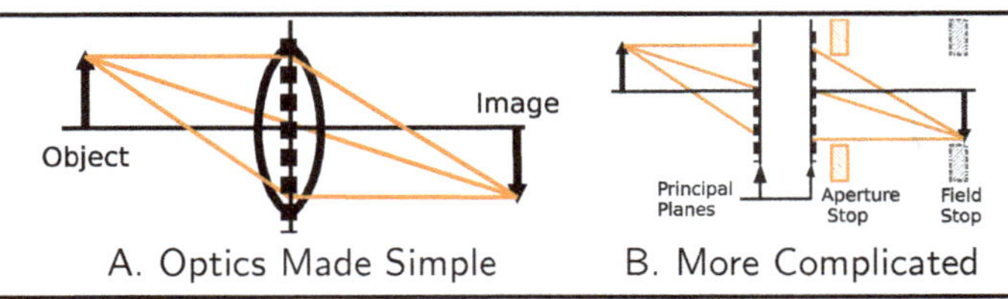

FIGURE 2.1
Geometric optics. To locate an image, we can use the simple construction (A). We can reduce a more complicated optical system to a simple one (B).

still valid if we measure distances from hypothetical planes called principal planes. This result will allow us to use the geometric constructions with only a slight modification as shown in Figure 2.1B. We will learn some general equations and simple approximations to locate the principal planes (thin vertical black lines with thick dashes in Figure 2.1B).

Ch. 4 on stops

In Chapter 4, we will explore the effect of finite apertures, defining the aperture stop, which limits the light-gathering ability of the system, and the field stop, which limits the field of view.

Ch. 5 on aberrations

In Chapter 5, we will remove the small-angle approximation from our equation and find some corrections to the simpler theory. The use of a better approximation for the angles may lead to shifts in image location, distortion, and limited resolution.

Appendix A for notation on geometric optics

It may be helpful from time to time to refer to Appendix A to better understand the figures in this and subsequent chapters. We will define a new notation the first time it is presented, and will try to keep a consistent notation in drawings and equations throughout the text.

2.1 Snell's Law

Angle of incidence is relative to the normal

The "law of sines" for refraction at a dielectric interface was derived by Descartes[51] in 1637. Descartes may or may not have been aware[116] of the work of Willebrord van Royen Snel, in 1621, and today this fundamental law of optics is almost universally called **Snell's law**, although with an extra "l" attached to his name. Snell's law can be derived from Fermat's principle from Section 1.8.2 that light travels from one point to another along the path that requires the shortest time. In Figure 2.2A, light travels from point A to point B. Our goal is to find the angles, θ and θ', for which the optical path length is shortest, thus satisfying Fermat's principle. We define angles from the normal to the surface. In the figure, the **angle of incidence**, θ, is the angle between the ray and the normal to the surface. Thus, an angle of zero corresponds to normal incidence and 90° corresponds to grazing incidence. The incident ray and the normal define the **plane of incidence**, which also contains the refracted and reflected rays.

Refraction derived from Fermat's principle

Feynman's Lectures in Physics[67] provides a very intuitive discussion of this topic; a lifeguard on land at A needs to rescue a person in the water at B, and wants to run and swim there in the shortest possible time. The straight line would be the shortest path, but because swimming is slower than running, it would not result in the shortest time. Running to the point on the shore closest to the swimmer would minimize the swimming time, but would lengthen the running time. The optimal point at which to enter the water is P.

We will take a variational approach: suppose that we already know the path goes through the point, P, as in Figure 2.2A, and displace the intersection point slightly to Q, as in Figure 2.2B. If we were correct in our choice of shortest path, then this deviation, in either direction, must increase the path length. Therefore, the differential change in

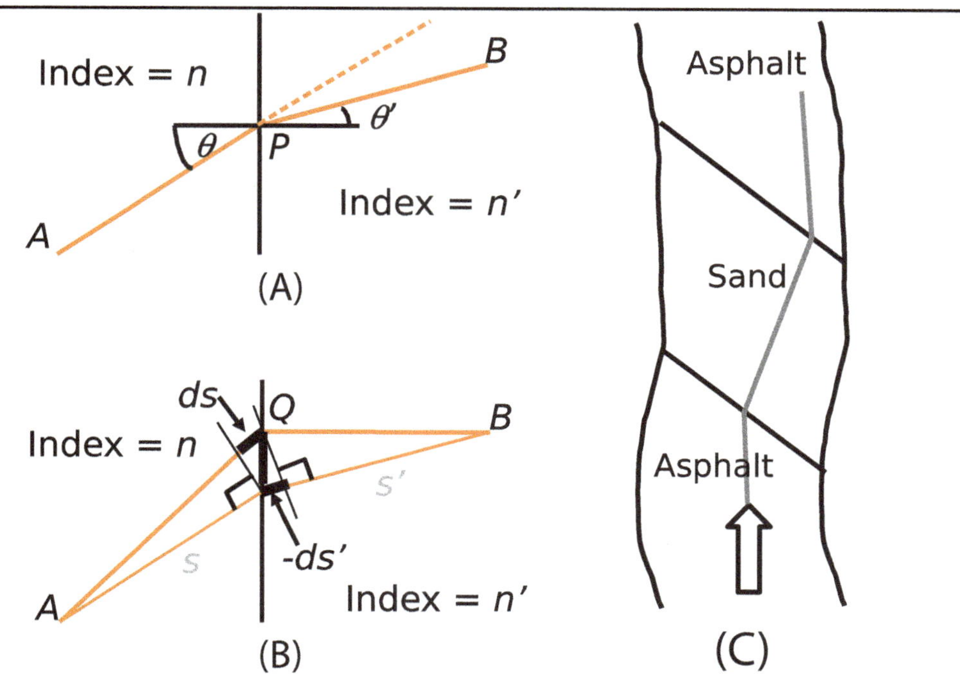

FIGURE 2.2
Snell's law. If P is along the shortest optical path, then the differential path length change in moving to a new point Q, must be zero. Note that the angle of incidence, θ, is always measured from the normal. An analogy is shown in C.

optical path length as defined in Equation 1.77 must be zero. Thus, in Figure 2.2B,

$$n\ ds = n' ds', \tag{2.1}$$

or

$$n \sin \theta = n' \sin \theta'. \tag{2.2}$$

This equation is perhaps the most important in geometric optics. If the index of refraction of the second medium is greater than that of the first, then a ray of light will always bend toward the normal, as shown in Figure 2.3A for glass and add trees and clouds to Figure 2.3D, Figure 2.3B for interfaces from air to various other materials including water in the visible spectrum and zinc selenide and germanium in the infrared region of the spectrum. We often perform a first approximation to a refraction problem assuming that the index of refraction of glass is 1.5. Although different types of glass have different indices of refraction, the differences are generally rather small, and this approximation is good for a first step.

Example

As a ray of light enters glass from air, how much does it refract?

- At normal incidence ($\theta = 0$), there is no refraction ($\theta' = 0$).

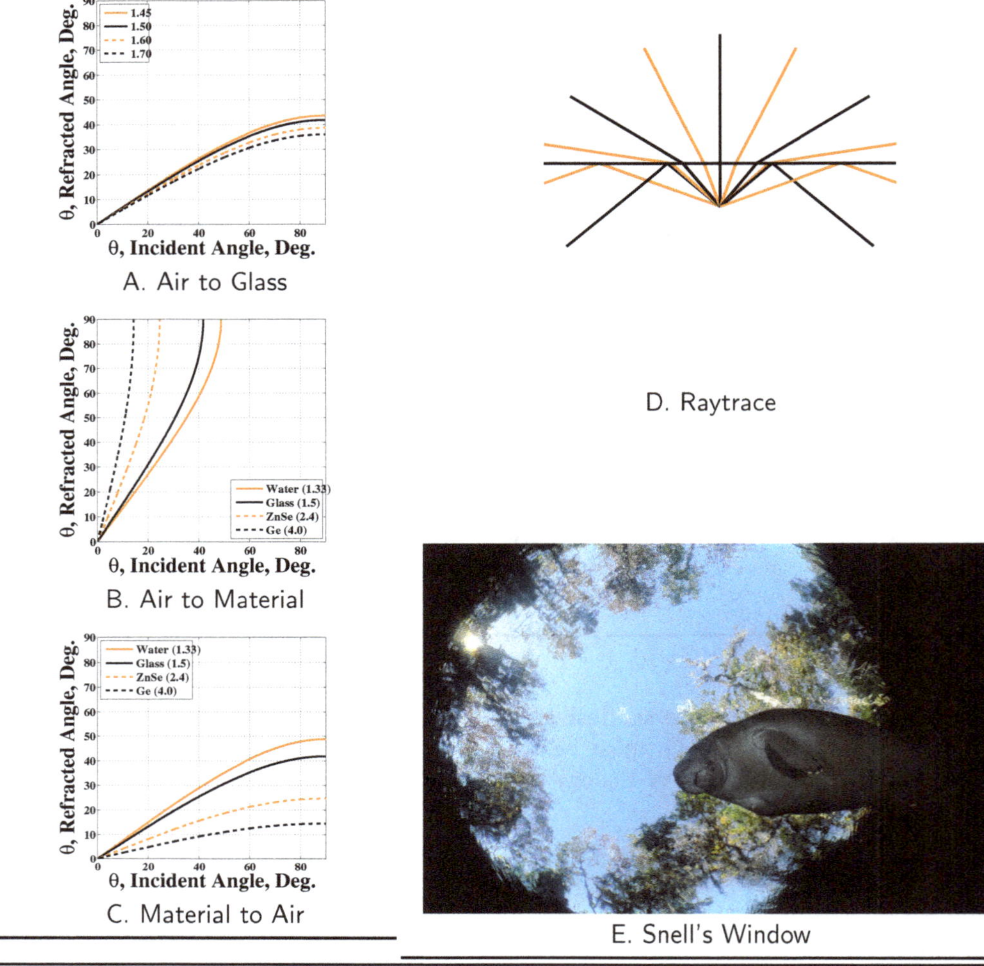

FIGURE 2.3
Snell's law. In refraction at an air-to-glass interface (A), the ray is refracted toward the normal, and the angle in the material is always smaller than in air. Differences among different glasses are small. Some materials have a much higher index of refraction than glass (B). At a material-air interface (C), the ray is refracted away from the normal. Note the critical angle. Rays from an underwater observer's eye are traced backward to show what the observer will see at each angle (D). The photograph (E, reprinted with permission of Carol Grant) shows Snell's window. The complete hemisphere above the surface is visible in a cone having a half-angle equal to the critical angle. The area beyond this cone acts as a 100% mirror. Because there is little light below the surface, this region appears dark.

- At ($\theta = 30^{\circ}$) from the normal, the refracted ray direction is $\sin\theta' = sin\theta/1.5$, or $\theta = 19.5^{\circ}$. The change in angle is 10.5°.
- At ($\theta = 60^{\circ}$) from the normal, the refracted ray direction is $\sin\theta' = sin\theta/1.5$, or $\theta = 35.3^{\circ}$, for a change of 27.4°.
- At a "grazing" angle of ($\theta = 89^{\circ}$) from the normal, the refracted ray direction is $\sin\theta' = sin\theta/1.5$, or $\theta = 41.8^{\circ}$, for a change of 47.2°.

We could have derived Snell's law from a wave picture of light, noting that the wave must slow as it enters the medium, picturing how the wavefronts must bend to accommodate this change of speed. We can remember the direction of refraction with a simple analogy. If we drive a car obliquely from a paved surface onto sand as in Figure 2.2C, such that the right wheel enters the sand first, then that wheel will be slowed and the car will pull toward the right, or toward the normal. This analogy is only tenuously connected to the wave propagation problem, but provides a convenient way to remember the direction of the effect.

Refraction analogy predicts direction of bending.

If the index of refraction of the second medium is lower, then the light refracts away from the normal.

Example

How does light refract going from glass to air?

- For an incident angle of $\theta = 15^\circ$, $\sin\theta' = 1.5\sin\theta$, so $\theta' = 22.8^\circ$.
- For $\theta = 30^\circ$, $\sin\theta' = 1.5\sin\theta$, so $\theta' = 48.6^\circ$.
- For $\theta = 45^\circ$, $\sin\theta' = 1.5\sin\theta$; The sine of θ' is greater than one. No refraction is possible.

Figure 2.3C shows a plot of Snell's law at a material-to-air interface. Now, the angle of refraction is always greater than the angle of incidence. There is an upper limit on angle of incidence, beyond which no light can escape the medium. We call this the **critical angle**, θ_c, and it is given by Snell's law when the angle of refraction is 90°:

Total internal reflection and the critical angle

$$n\sin\theta_c = 1. \tag{2.3}$$

Viewed from inside the material, the interface appears as a mirror at angles greater than the critical angle, as shown in Figure 2.3D. This effect is most vividly seen in the so-called Snell's window in Figure 2.3E. In the center of the photograph, one can see through Snell's window the complete hemisphere of sky above. Beyond the boundary of Snell's window, at $\theta_c = \sin^{-1}(1/1.33) \approx 48.7°$, the mirror-like interface reflects the dark ocean bottom.

When light is incident on a surface, we expect both reflection and refraction as shown in Figure 2.4. The **angle of reflection**, θ_r, is also measured from the normal, and is always the same as the angle of incidence:

Angle of Reflection

$$\theta_r = \theta. \tag{2.4}$$

Normally, the angle is defined to be positive for the direction opposite the angle of incidence. Equations 2.2 and 2.4 provide the only tools needed for the single surface imaging problems to be discussed in Section 2.2.

Take-Away

Snell's Law:

- At an interface, light is reflected and refracted.

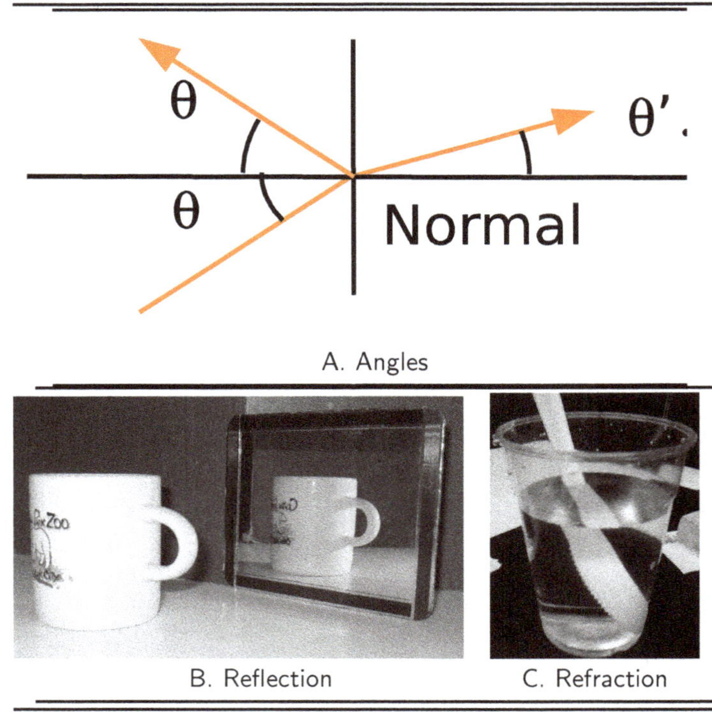

FIGURE 2.4
Reflection and refraction. The angle of reflection is equal to the angle of incidence, θ. The angle of refraction, θ′, is given by Snell's law (A). The plane of incidence contains the normal and the incident ray, and thus the reflected and refracted rays. Examples of reflection (B) and refraction (C) are shown.

- Snell's Law, Equation 2.2, is probably the most important equation in geometric optics. Although we can derive useful equations from it such as the lens equation, Snell's law by itself is sufficient to describe ray propagation.
- If the second index of refraction is less than the first, there will be a critical angle beyond which no light will be refracted into the medium given by Equation 2.3.
- The angle of reflection is equal to the angle of incidence according to Equation 2.4.

2.2 Imaging with a Single Interface

We are now ready to begin our discussion of imaging, starting with a single interface. We will proceed through the planar mirror, curved mirror, planar dielectric interface and curved dielectric interface, and then to simple lenses composed of a single glass element with two curved dielectric interfaces, and finally to compound lenses consisting of multiple simple lenses.

Among the most complicated issues are the sign conventions we must employ. Some authors use different choices, but we will use the most common. We also introduce a precise notation, which will hopefully help minimize confusion about the many different parameters. The conventions will be presented as they arise, but they are all summarized in Appendix A. The first conventions are shown in Figure 2.5. We label points with capital letters and distances with the corresponding lowercase letters. Thus, S is the object

Notation: Points: capital letters
Distances: lower-case letters

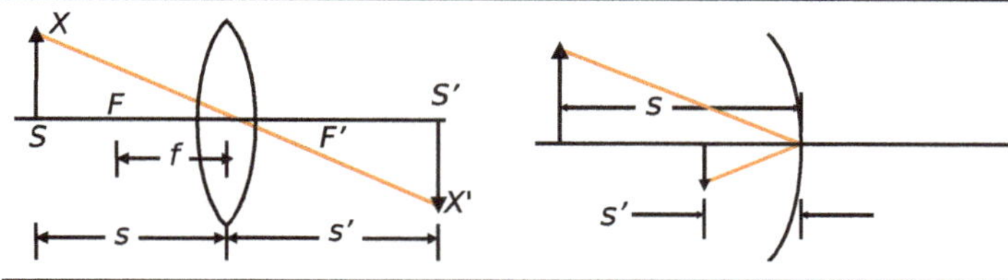

FIGURE 2.5
Sign conventions. All distances in these figures are positive. The object distance is positive if the object is to the left of the interface. The image distance for a refracting interface or a lens is positive if the image is to the right, and for a mirror, it is positive to the left. In general, the object distance is positive if light passes the object before reaching the interface, and the image distance is positive if the light reaches the interface before reaching the image. Complete discussion of sign conventions will be found in Appendix A.

TABLE 2.1
Imaging Equations

Quantity	Definition	Equation	Notes
Object distance	s		Positive to the left
Image distance	s'	$\frac{1}{s}+\frac{1}{s'}=\frac{1}{f}$	Positive to the right for interface or lens. Positive to the left for mirror
Magnification	$m=\frac{x'}{x}$	$m=-\frac{x'}{x}$	
Angular magnification	$m_\alpha=\frac{\partial\alpha'}{\partial\alpha}$	$\lvert m_\alpha\rvert=\frac{1}{\lvert m\rvert}$	
Axial magnification	$m_z=\frac{\partial s'}{\partial s}$	$\lvert m_z\rvert=\lvert m\rvert^2$	

The general definitions of the important imaging parameters are valid if the initial and final media are the same.

location and s is the **object distance**. The image is at S', and our imaging equation will establish a relationship between s' and s. We say that the planes perpendicular to the optical axis through points S and S' are **conjugate planes**. Any two planes such that an object in one is imaged in the other are described as conjugate. We consider object, s, and image, s', distances positive if light travels from the object to the interface to the image. Thus, for the lens in Figure 2.5, s is positive to the left, and s' is positive to the right. For the mirror in Figure 2.5, the image distance is positive to the left.

Conjugate planes defined

Sign convention for distances

The height of the object is x, the distance from S to X, and the height of the image, x', is from S' to X'. We define the magnification, m, as the ratio of x' to x. For each of our interfaces and lenses, we will compute the imaging equation, relating s' and s, the magnification, m, and various other magnifications, as shown in Table 2.1.

Magnification defined

Looking ahead to Table 2.2, we are going to calculate the relationship between object and image distance, as well as these magnifications, for each type of surface, plane and spherical reflecting surfaces in Section 2.3, plane and spherical refracting surfaces in Section 2.4, and then for a lens composed of two spherical refracting surfaces in Section 2.5. We'll also investigate the refraction through two planar refracting surfaces in Section 2.6.

2.3 Reflection

Reflection from a planar surface was probably one of the earliest applications of optics. In Greek mythology, Narcissus viewed his image reflected in a natural mirror, the surface of a pond, and today in optics, the word narcissus is used to describe unwanted reflections from components of an optical system. Biblical references to mirrors ("... the looking glasses of

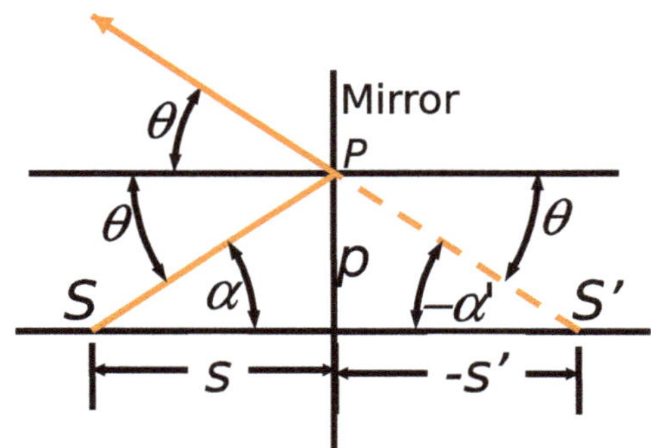

FIGURE 2.6
Imaging with a planar mirror. The image is behind the mirror by the same distance that the object is in front of it; $-s' = s$. The image is virtual.

the women..." Exodus 38:8) suggest the early fabrication of mirrors from polished brass and other metals. Here we begin our analysis with a planar surface, and then proceed to a curved surface. The analysis of angles is appropriate to either a high-reflectivity mirrored surface, or to reflection from a dielectric interface. In the former case, most of the light will be reflected, while in the latter, only a part of it will be. Nevertheless, the angles are the same in both cases.

2.3.1 Planar Surface

We will solve the problem of reflection at a planar surface with the aid of Figure 2.6. Light from a source point S on the optical axis is reflected from a planar mirror at a point P, a height p above the optical axis at an angle, θ. Because the angle of reflection is $\theta_r = \theta$ from Equation 2.4, the angle of the extended ray (which we show with a dotted line) from the image at S' must also be θ. There are two identical triangles, one with base and height being s and p, and the other $-s'$ and p. Thus, the imaging equation is

$$s' = -s \quad \text{(Planar reflector)}. \tag{2.5}$$

Reflection equation for a planar reflector

Virtual image in a mirror

The image is as far behind the mirror as the object is in front of it and is considered a **virtual image**, because the rays of light do not converge to it; they just appear to diverge from it. Using our drawing notation, the virtual image is formed by dotted ray extensions rather than the solid lines representing actual rays.

In Practice

In this book we will use dotted lines to represent virtual rays and images. This approach may aid in understanding the concept of real and virtual images; A virtual image is one formed by the intersection of dotted rays.

If a light source is placed at S, rays appear to diverge from the image at S', rather than converging to form an image, as would be the case for a **real image**.

Having found the image location, we next consider magnification. The **magnification** is the ratio of the image height to the object height, $m = x'/x$, which can be obtained using Figure 2.7. We place an object of height x at the point S. Once again, Equation 2.4 and the use of similar triangles provides the magnification,

$$m = \frac{x'}{x} = \frac{-s'}{s} = 1 \quad \text{(Planar reflector)}. \tag{2.6}$$

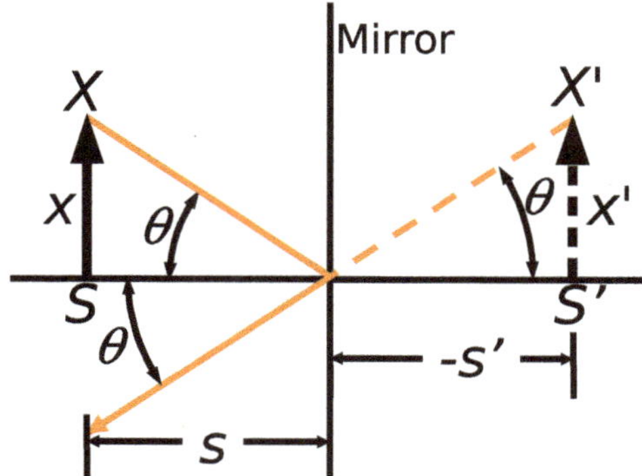

FIGURE 2.7
Magnification with a planar mirror. The magnification is $+1$, so the image is upright.

The magnification is positive, so the image is upright. With a planar mirror, as with most imaging problems, we have cylindrical symmetry, so we only work with x and z, and the results for x also hold for y.

$m = 1$ for planar reflector: Upright image.

The next quantity of interest is the **angular magnification**, defined by $m_\alpha = d\alpha'/d\alpha$, where α is the angle of a ray from the object relative to the axis, as shown in Figure 2.6. Noting that $\alpha = \theta$, and using alternate interior angles for the s', p triangle, $-\alpha' = \theta$:

$$m_\alpha = \frac{d\alpha'}{d\alpha} = -1 \quad \text{(Planar reflector).} \tag{2.7}$$

Angular magnification will become particularly important when we begin the discussion of telescopes.

Angular magnification, $m_\alpha = -1$ for planar reflector

Finally, we compute the **axial magnification**, $m_z = ds'/ds$. For our planar mirror, the solution is

$$m_z = \frac{ds'}{ds} = \frac{s'}{s} = -1 \quad \text{(Planar reflector).} \tag{2.8}$$

Axial magnification, $m_z = -1$ for planar reflector

At this point, we can enter the first line in Table 2.2 summarizing the imaging equations. We have already noted that the image is virtual and upright and because all magnifications have unit magnitude, $|m_z| = |m|$, we say the image is not distorted. We will find that the planar mirror is an exceptional case in this regard. The sign difference between m and m_z means that the coordinate system is "perverted"; it is not possible to rotate the image to the same orientation as the object. As shown in Figure 2.8, a right-handed coordinate system (dx, dy, ds) images to a left-handed (dx', dy', ds'). It is frequently said that a mirror reverses left and right, but this statement is incorrect, as becomes apparent when one realizes that there is no mathematical difference between the x and y directions and the mirror does not reverse the directions up and down. The directions that are reversed are front and back, in the direction of observation.

No distortion in a planar mirror.

Perversion of coordinates in a mirror

Take-Away

Plane Mirrors:
For plane mirrors, $s' = -s$, $m = 1$, and $m_z = -1$.

Before continuing to our next imaging problem, let us look at some applications of planar mirrors. Suppose that we have a **monostatic** laser radar for measuring distance or velocity. By "monostatic" we mean that the transmitted and received light pass along

TABLE 2.2
Imaging Equations

Surface	s'	m	m_α	m_z	Image	O*	D*	P*
Planar mirror	$s' = -s$	1	-1	-1	Virtual	Upright	No	Yes
Concave mirror								
$s > f$	$\frac{1}{s'} + \frac{1}{s} = \frac{1}{f}$	$-s'/s$	$-m^2$	$-1/m$	Real	Inverted	Yes	No
Convex mirror	$\frac{1}{s'} + \frac{1}{s} = \frac{1}{f}$	$-s'/s$	$-m^2$	$-1/m$	Virtual	Upright	Yes	Yes
Planar refractor	$\frac{s}{n} = \frac{s'}{n'}$	1	$\frac{n}{n'}$	$\frac{n'}{n}$	Virtual	Upright	Yes	No
Curved refractor								
$s > f$	$\frac{n}{s} + \frac{n'}{s'} = \frac{n'-n}{r}$	$-\frac{ns'}{n's}$	$-\frac{n}{n'}\frac{1}{m}$	$-\frac{n}{n'}m^2$	Real	Inverted	Yes	Yes
Lens in air								
$s > f$	$\frac{1}{s} + \frac{1}{s'} = \frac{1}{f}$	$-s'/s$	$-s/s'$	$-m^2$	Real	Inverted	Yes	Yes

These equations are valid in general. The image descriptions in the last columns assume a real object with positive s. In the last three columns, "O" mean orientation, "D" means distortion, and "P" means perversion of coordinates.

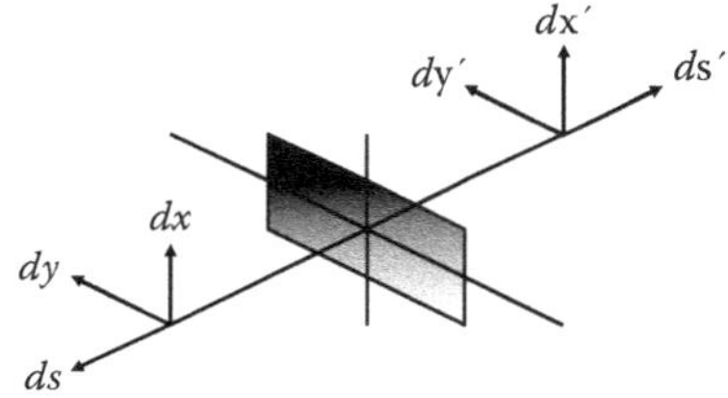

FIGURE 2.8
Axial magnification with a planar mirror. The axial magnification is -1, while the transverse magnification is $+1$. The image cannot be rotated to match the object. The front of the card, viewed directly, is out of focus because it is close to the camera. The image of the back is further away and properly focused, but inverted.

a common path. One application of a monostatic laser radar is in surveying, where the instrument can determine distance by measuring the time of flight and knowing the speed of light. When we test such an instrument, we often want to reflect a known (and preferably large) fraction of the transmitted light back to the receiver over a long path. We place the laser radar on the roof of one building and a mirror on a distant building. Now we have two difficult challenges. First, we must point the narrow laser beam at the mirror, and second we must align the mirror so that it reflects the beam back to the laser radar. We must do this in the presence of atmospheric turbulence, which changes the index of refraction of the air, causing the beam to wander. A retroreflector can be designed to make the second task easier. First, thinking in two dimensions, the pair of mirrors in

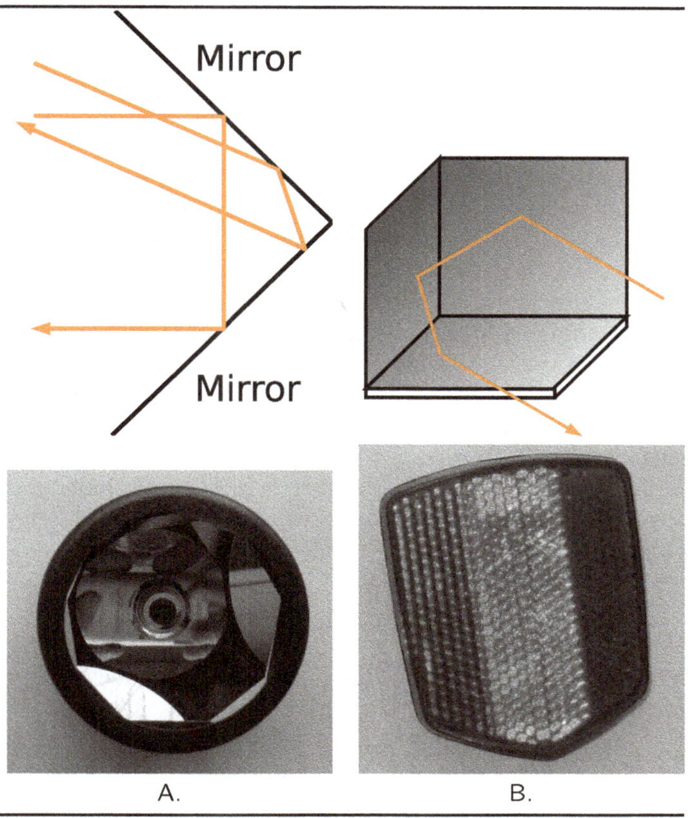

FIGURE 2.9
The retroreflector. Drawings show a two-dimensional retroreflector (A) and a three-dimensional one (B). A hollow (or open) retroreflector (C) is composed of three front-surfaced mirrors, and may be used over a wide range of wavelengths. Lines at the junctions of the three mirrors can be seen. An array of solid retroreflectors (D) is commonly used for safety applications. These arrays take advantage of total-internal reflection, so no metal mirrors are needed.

Figure 2.9A can help. If part of the transmitted beam is incident upon one mirror at 45°, then the angle of reflection is also 45° for a total change of direction of 90°. Then, the reflected light is incident upon the second mirror at the same angle and deviated another 90° resulting in a total direction change of 180° from the original direction. It can be shown that if the first angle increases, the second one decreases, so that the total change in direction is always 180°. The next drawing in Figure 2.9 extends this idea to three dimensions. The operation of this **retroreflector** or **corner-cube** will be familiar to any racquet-ball player. A racquet ball hit into a lower front corner of the court will reflect from (in any order) the front wall, side wall, and floor, and return along the same trajectory. Thus, in a configuration of three mirrors all orthogonal to each other, one will reflect the x component, k_x, of the propagation vector, $\mathbf{k}$, one the component k_y, and one k_z, so that the vector $\mathbf{k}$ itself is reflected, as shown in Figure 2.9B.

Retroreflector

For reflecting light that is outside the visible spectrum, "open" or "hollow" retroreflectors are usually used, consisting of three **front-surface mirrors**. These mirrors have the disadvantage that the reflective surface is not protected, and every cleaning causes some damage, so their useful lifetime may be shorter than mirrors with the reflective coating (metal or other) on the back, where it is protected by a layer of glass. The disadvantage of these **back-surface mirrors** is that there is a small reflection from the air–glass interface which may have an undesirable effect on the intended reflection from the back surface. Refraction at the front surface may also make alignment of a system more difficult. We will discuss these different types of mirrors in detail in Section 7.7. An open retroreflector is shown in Figure 2.9C, where the image of the camera can be seen.

Eq. 2.3, $n \sin \theta_c = 1$.

Low-cost retroreflectors use corner cubes with total internal reflection.

In the visible spectrum, it is easy to fabricate small retroreflectors by impressing the corner-cube pattern into the back of a sheet of plastic. Normally, a repetitive pattern is used for an array of reflectors. The diagonal of a cube is $\cos^{-1}\sqrt{1/3} = 54.7°$, which is considerably greater than the critical angle (Equation 2.3) $\sin^{-1} 1/1.5 = 41.8°$, so these plastic retroreflectors require no reflective coating and yet will work over a wide range of acceptance angles. Such low-cost retroreflectors (Figure 2.9D) are in common use on vehicles to reflect light from approaching headlights. In fact, it happens that a well-made retroreflector would reflect light in such a tight pattern that it would return to the headlights of the approaching vehicle, and not be particularly visible to the driver. To make the reflections more useful, reflectors for this application are made less than perfect, such that the divergence of the reflected light is increased. The use of total internal reflection makes these retroreflectors inexpensive to manufacture and easy to keep clean because the back surface need not be exposed to the air. In the infrared region of the spectrum, highly transmissive and inexpensive materials are not available, and open retroreflectors are generally made from front-surface mirrors.

In Practice

A retroreflector is useful for reflecting light along the incident path with little concern for alignment. It is useful in applications such as laser radar. It finds frequent application as reflective material for safety. The retroreflector may be deliberately made imperfect if the source and observer are not closely co-located.

2.3.2 Curved Surface

The next imaging problem is the curved reflector. For now, we will consider the reflector to be a **spherical mirror** which is a section of a sphere, as shown in Figure 2.10. Starting from an object of height, x, located at S with an object distance, s, we trace two rays from X and find their intersection at X' in the image. We know from Fermat's principle that any other ray from X will also go through X', so only these two are required to define the image location and magnification. The first ray we choose is the one reflected at the vertex, V. The **vertex** is defined as the intersection of the surface with the optical axis. The axis is normal to the mirror at this point, so the angle of reflection is equal to the angle of incidence, θ. The second ray passes through the center of curvature, R, and thus is incident normally upon the surface, and reflects back onto itself. Using the two similar triangles, S, X, R and S', X', R,

$$\frac{x}{s-r} = \frac{-x'}{r-s'} \tag{2.9}$$

and using two more similar triangles, S, X, V and S', X', V,

$$\frac{x}{s} = \frac{-x'}{s'}. \tag{2.10}$$

Combining these two equations,

$$\frac{1}{s'} + \frac{1}{s} = \frac{2}{r}. \tag{2.11}$$

Focal length of a concave mirror

We can write this equation in a form that will match the lens equation to be developed later by defining the **focal length** as half the radius of curvature:

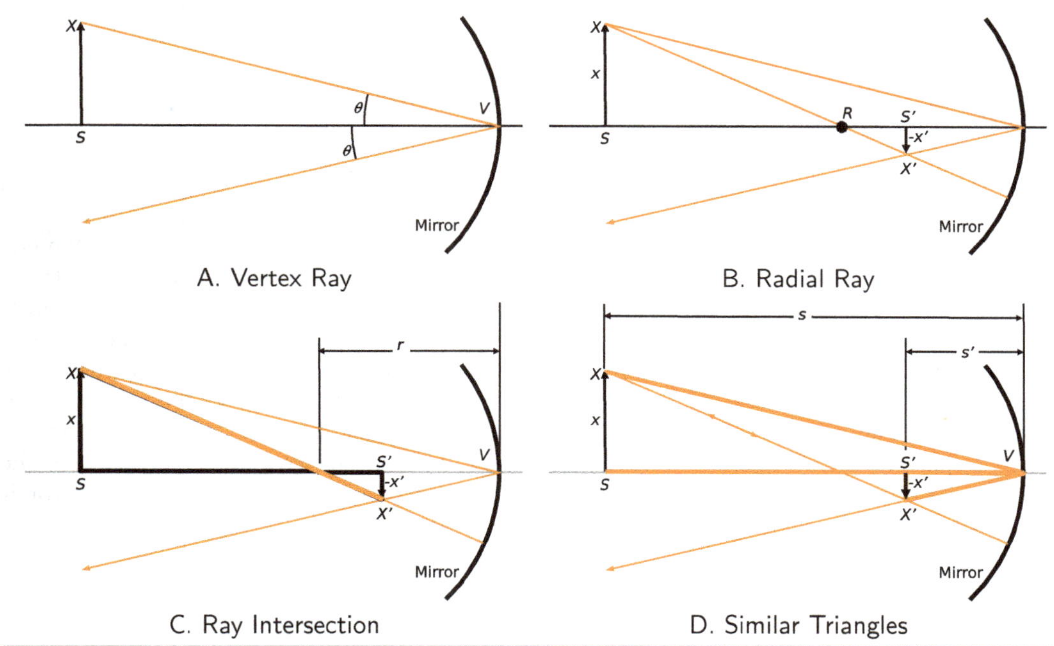

FIGURE 2.10
The curved mirror. The angle of reflection is the same as the angle of incidence for a ray reflected at the vertex (A). A ray through the center of curvature is reflected normally (B). These rays intersect to form the image (C). From this construction, the image location and magnification can be determined (D).

$$\frac{1}{s'} + \frac{1}{s} = \frac{1}{f} \quad f = \frac{r}{2} \quad \text{(Spherical reflector).} \tag{2.12}$$

We have used the definition here that r is positive for a concave mirror, and thus f is also positive. We call this a **converging mirror**. Does the definition of f have any physical significance? Let us consider Equation 2.12 when s or s' becomes infinite. Then

r is positive for a concave mirror (converging the rays)

$$s' \rightarrow f \quad s \rightarrow \infty \quad \text{or} \quad s \rightarrow f \quad s' \rightarrow \infty.$$

For example, light from a distant source such as a star will converge to a point at $s' = f$, or a point source such as a small filament placed at $s = f$ will be imaged at infinity.

Our next steps are to complete the entry in Table 2.2 for the spherical mirror. Using similar triangles as in Equation 2.10,

$$m = \frac{x'}{x} = -\frac{s'}{s} \quad \text{(Spherical reflector).} \tag{2.13}$$

We know that a ray from S to some point, P, on the mirror must pass through S'. Tracing such a ray, and examining triangles S, P, V and S', P, V, it can be shown that within the **small-angle approximation** ($\sin\theta \approx \tan\theta \approx \theta$ and $\cos\theta \approx 1$),

Magnification of a spherical mirror

$$m_\alpha = \frac{s}{s'} \quad |m_\alpha| = |1/m| \quad \text{(Spherical reflector).} \tag{2.14}$$

Angular magnification of a spherical mirror

Taking the derivative of Equation 2.12,

$$-\frac{ds}{s^2} - \frac{ds'}{(s')^2} = 0$$

$$m_z = \frac{ds'}{ds} = -\left(\frac{s'}{s}\right)^2 \qquad m_z = -m^2$$

$$|m_z| = |m|^2 \quad \text{(Spherical reflector).} \tag{2.15}$$

Axial magnification of a spherical mirror

Real image

The image is real, because the rays from the source do converge to the image. One can image sunlight and burn a piece of paper at the focus if the mirror is sufficiently large. This ability is one of the distinguishing features of a real image. The magnification is negative, so the image is inverted. The relationship $m_z = -m^2$ ensures that the image is distorted except for the case where $|m| = 1$, when the object is at the center of curvature: $s = s' = 2f = r$. However, both m and m_z have the same sign, so the image is not perverted. We can imagine rotating the image such that it has the same orientation as the object. A similar analysis holds for a convex surface (negative focal length) with the same equations, but with different conclusions, as shown in Table 2.2. Such a mirror is called a **diverging mirror**.

Eq. 2.13, $m = \frac{x'}{x} = -\frac{s'}{s}$.

Eq. 2.12, $\frac{1}{s'} + \frac{1}{s} = \frac{1}{f}$
$f = \frac{r}{2}$.

Example

Consider a diverging mirror used as the "side-view" mirror on a car or truck. The mirror is convex with a radius of curvature of 2 meters ($r = -2\,\text{m}$). A truck is approaching at a distance of $s = 10\,\text{m}$. Where is the image? Using Equation 2.12,

$$\frac{1}{s'} = \frac{1}{2\,\text{m}/2} - \frac{1}{10\,\text{m}} \qquad s' = -0.476\,\text{m}.$$

The image is a bit less than half a meter behind the mirror. Figure 2.11 shows the geometry.

How large is the image if the truck is 3 meters tall? Using Equation 2.13

$$m = -\frac{-0.476\,\text{m}}{10\,\text{m}} = 0.0476,$$

and

$$x' = 3\,\text{m} * m = 0.14\,\text{m}.$$

Now if the driver is 1 meter behind the mirror, she sees the truck a distance of 1.476 meters in front of her, with a height of $x' = 14\,\text{cm}$. The truck appears quite small, subtending an angle of $0.14\,\text{m}/1.476\,\text{m}$ or about 95 mrad, compared to $3\,\text{m}/11\,\text{m}$ or about 273 mrad in a flat mirror, and thus it appears much further away than it is. Note that the image is actually closer to the driver than the truck. The message, "Objects in this mirror are closer than they appear" is not true in the sense of Equation 2.12; the warning is nevertheless true because of our visual interpretation.

In Practice

It seems that the legality and acceptance of convex mirrors on vehicles varies from country to country. The advantage of a wider field of view that reduces the chance

FIGURE 2.11
Image of a truck in a side-view mirror. The image is actually closer than the object, but it is small, so it appears to be quite far away.

of not seeing something must be weighed against the image distortion that can give the impression that the approaching vehicle is further away than it is. In the United States, plane mirror seem to be required on the driver's side but curved ones are allowed on the passenger side.

Large telescopes are often made from multiple curved reflecting components or mixtures of reflecting and refracting components. Large reflectors are easier to fabricate than large refracting elements, and a metal reflector can operate over a wider band of wavelengths than can refractive materials. From Table 1.2, glass is transmissive in the visible and near-infrared, while germanium is transmissive in the mid- and far-infrared. A good metal coating can be highly reflecting across all these bands. Even if an application is restricted to the visible spectrum, most refractive materials have some dispersion (see Figure 1.8), which can result in chromatic aberrations (Section 5.5). Chromatic aberration can be avoided by using reflective optics.

Take-Away

Curved Mirrors:

- Reflective optical components are often used when large diameters are required.
- They are also often used when a wide range of wavelength is needed.
- The focal length of a spherical mirror of radius r is $f = r/2$.

- The focal length is positive (converging the light) for a concave mirror and negative for a convex one.

2.4 Refraction

Probably the first uses of refraction in human history involved water, perhaps in spheres used as magnifiers by gem cutters[212]. The optical power of a curved surface was recognized sufficiently for Alessandro di Spina to fabricate eyeglasses in the 1200s[196]. The modern descendents of these early optical devices include powerful telescopes, microscopes, and everyday instruments such as magnifiers. We will develop our understanding of the basic building blocks of refractive optics in the same way we did reflective optics.

2.4.1 Planar Surface

2 rays for image equation and magnification

Magnification, $m = 1$ for planar refractive surface

The planar refractive surface consists of an interface between two dielectric materials having indices of refraction, n and n', as shown in Figure 2.12A. Recall that for refraction, we define s' as positive to the right. We consider two rays from the object at X. The first is parallel to the axis and intersects the surface along the normal. Thus, $\theta = 0$ in Snell's law, and the ray does not refract. The second ray passes through the vertex, V. In the example, $n' < n$, and the ray refracts away from the normal. Extending these rays back, they intersect at X', which is therefore the image. The rays are drawn as dashed extensions of the real rays, and so this image is virtual. The ray parallel to the axis has already given us the magnification

$$m = \frac{x'}{x} = 1 \quad \text{(Refraction at a plane).} \tag{2.16}$$

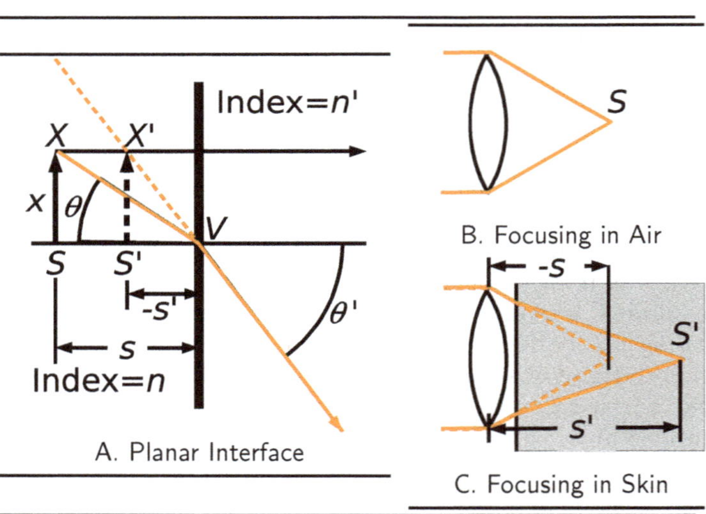

FIGURE 2.12
Refraction at a planar interface. The final index, n', is lower than the initial index, n, and rays refract away from the normal (A). A confocal microscope focused to a location S in water (B) will focus to a deeper location, S', in skin (C) because skin has a slightly higher index of refraction. The object that is actually at S' will appear to be at the shallower depth, S.

The image location can be computed by the triangles X, S, V with angle θ at V, and X', S', V with angle θ'. Using the small-angle approximation and Snell's law,

$$\tan\theta = \frac{x}{s} \qquad \tan\theta' = \frac{x'}{s'} = \frac{x}{s'}$$

$$n\sin\theta \approx n\frac{x}{s} \qquad n'\sin\theta' \approx n'\frac{x}{s'}$$

with the result,

Imaging equation for planar refractive surface

$$\frac{n}{s} = \frac{n'}{s'} \quad \text{(Refraction at a plane).} \tag{2.17}$$

In the example, with $n' < n$, we see that $s' < s$, as is well-known to anyone who has looked into a fish tank (Figure 2.12A). The **geometric thickness** or the apparent thickness of the fish tank as viewed by an observer in air is the actual thickness divided by the index of refraction of water. Recall that this definition is in contrast to the optical path length, which is the product of the physical path and the index of refraction. For a material of physical thickness ℓ,

$$\ell_g = \frac{\ell}{n} \quad \text{(Geometric thickness)} \tag{2.18}$$

$$OPL = \ell n \quad \text{(Optical path length)} \tag{2.19}$$

Finishing our analysis, we have directly from Snell's law

Angular magnification for planar refractive surface

$$m_\alpha = \frac{n}{n'}, \tag{2.20}$$

and from differentiating Equation 2.17,

$$m_z = \frac{ds'}{ds} = \frac{n'}{n} \quad \text{(Refraction at a plane).} \tag{2.21}$$

Axial magnification for planar refractive surface

The image is virtual, upright, and distorted but not perverted.

We consider one example where knowledge of refracting surfaces can be applied. In Figure 2.12B, laser light is focused through a microscope objective as part of a confocal microscope (Figure 1.3), and comes to a point at some distance f from the lens. Now, if we use this microscope to focus into skin* (Figure 2.12C), with an index of refraction slightly above that of water, then it will focus to a longer distance, s'. When confocal microscopes are used to create three-dimensional images, this correction is important.

Example

As we focus the microscope on image planes at 50, 100 and 150 micrometers below the skin surface, where is the actual object plane in each case?

$$s = s'\frac{n}{n'} = s'\frac{1.33}{1}$$

The object planes are actually at 66.5, 133, and 200 micrometers.

If we do not know the exact index of refraction of the specimen, then we cannot be sure of the exact depth at which we are imaging. In fact, we can use a combination of

*Optical properties of tissue are briefly discussed in Appendix D.

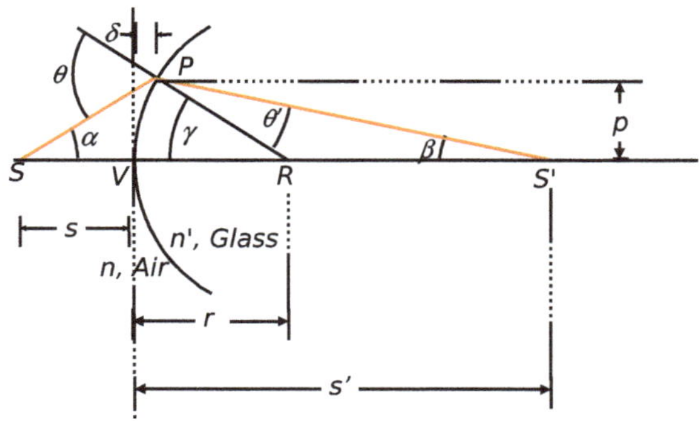

FIGURE 2.13
Refraction at a curved surface. Relationships among the angles can be used to determine the equation relating the image distance, s' to the object distance, s.

optical coherence tomography (Section 10.3.4), which measures optical path length and confocal microscopy in a concept called focus tracking[122] to determine both depth and index of refraction.

Take-Away

Flat Refracting Interface:
For a flat refracting interface, $s = s'\frac{n}{n'}$, $m = 1$, and $m_z = \frac{n'}{n}$.

2.4.2 Curved Surface

Finally, we turn our attention to perhaps the most common optical element, and likely the most interesting, the curved refracting surface. The geometry is shown in Figure 2.13. As in the reflective case, s and s' are object and image distances, and r is the radius of curvature, which here is positive when the surface is convex toward the source. We begin by deriving the equation for the image location.

First step toward a lens

Because we are dealing with refraction, we know that we need to employ Snell's law, relating θ to θ', but relationships among s, s', and r appear to involve α, β, and γ. Therefore, it will be helpful to note relationships between the exterior angles of two triangles:

$$\theta = \alpha + \gamma \quad \text{(From triangle } S, P, R\text{)}, \tag{2.22}$$

and

$$\gamma = \theta' + \beta \quad \text{(From triangle } S', P, R\text{)}. \tag{2.23}$$

Next, we find the tangents of three angles:

Assumptions: All angles are small and $\delta = 0$.

$$\tan\alpha = \frac{p}{s+\delta} \qquad \tan\beta = \frac{p}{s'-\delta} \qquad \tan\gamma = \frac{p}{r-\delta}. \tag{2.24}$$

With the small angle approximation, we represent these tangents by the angles in radians, and we let $\delta \to 0$, which is equivalent to letting cosines of angles go to unity. Then,

$$\alpha = \frac{p}{s} \qquad \beta = \frac{p}{s'} \qquad \gamma = \frac{p}{r}. \tag{2.25}$$

Substituting these results into Equations 2.22

$$\theta = \frac{p}{s} + \frac{p}{r}, \tag{2.26}$$

and into (2.23),

$$\frac{p}{r} = \theta' + \frac{p}{s'} \qquad \theta' = \frac{p}{r} - \frac{p}{s'}. \tag{2.27}$$

Applying Snell's law, also with the small angle approximation,

$$n\theta = n'\theta' \qquad \frac{np}{s} + \frac{np}{r} = \frac{n'p}{r} - \frac{n'p}{s'}, \tag{2.28}$$

and finally,

Imaging equation for spherical refracting surface

$$\frac{n}{s} + \frac{n'}{s'} = \frac{n'-n}{r} \qquad \text{(Refraction at a curved surface).} \tag{2.29}$$

This equation bears some similarity to Equation 2.12 for the curved mirror but with a bit of added complexity. We cannot define a focal length so easily as we did there. The reason is that the object and image are in media of different indices of refraction, while in the case of the mirror, the medium was the same for object and image. Setting the object distance to infinity,

$$s \to \infty,$$

$$BFL = f' = s' = \frac{n'r}{n'-n} \qquad \text{(Refraction at a curved surface)} \tag{2.30}$$

which we must now define as the **back focal length**, in contrast to the result

Back focal length defined

$$s' \to \infty$$

$$FFL = f = s = \frac{nr}{n'-n} \qquad \text{(Refraction at a curved surface),} \tag{2.31}$$

which defines the **front focal length**. Examples of the focal lengths are shown in Figure 2.14

Front focal length defined

The ratio of these focal lengths is

$$\frac{f'}{f} = \frac{n'}{n} \tag{2.32}$$

for a single surface, but we will see that this result is much more general. When we combine multiple surfaces in Section 2.5, we will find it useful to define the **refracting power** or sometimes just the **power** of a surface:

BFL/FFL ratio. This is always true as we will see in Equation 3.55).

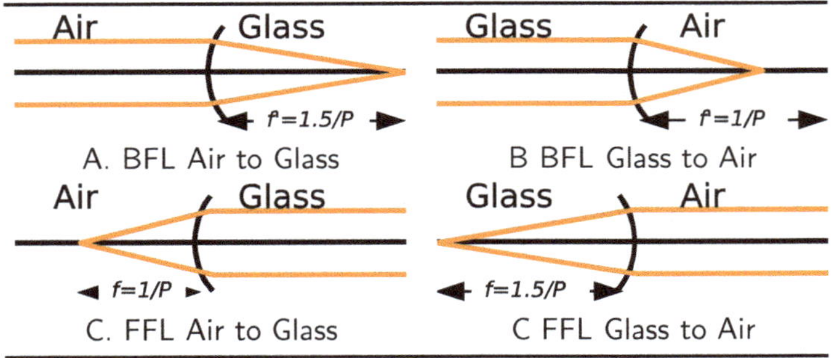

FIGURE 2.14
Front and back focal lengths. The surfaces shown all have positive refracting power. The front and back focal lengths depend on the medium.

$$P = \frac{n' - n}{r}. \tag{2.33}$$

By custom, we use the same letter, P, and the same name, "power," for both refracting power and power in the sense of energy per unit time. In most cases, there will be no ambiguity, because the distinction will be obvious in terms of context and certainly in terms of units. Refracting power has units of inverse length, and is generally given in diopters $= \text{m}^{-1}$.

Refracting power defined
Diopters

In Practice

Optical Power or **Refracting Power** is given by Equation 2.33

$$P = \frac{n' - n}{r}.$$

It is expressed in inverse meters, or diopters.

Example

What is the refracting power of a convex surface going from air to glass with a radius of curvature of 100 mm?

$$P = \frac{n' - n}{r}\frac{n' - n}{r}. = 5\,\text{diopters}.$$

What is the equivalent reflecting surface?

$$f = \frac{1}{P} = \frac{1}{5\,\text{diopters}} = 0.2\,\text{m}$$

$$r = 2f = 0.4\,\text{m}.$$

The surface is concave with a radius of curvature of 400 mm.

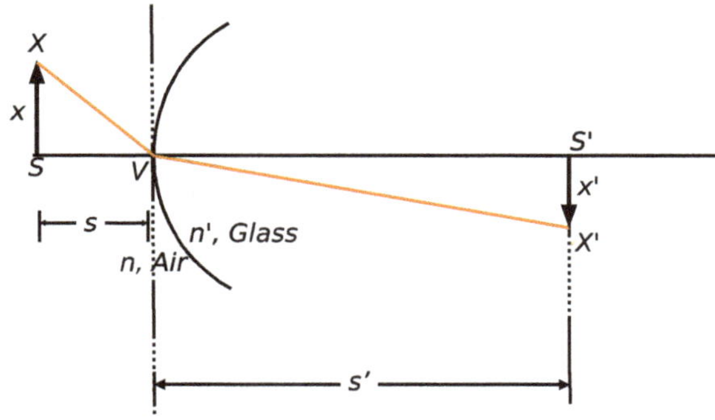

FIGURE 2.15
Magnification with a curved refractive surface. We compute the magnification using Snell's law at V.

Continuing our work on Table 2.2, to compute magnification, we trace a ray from an object at X, a height x above S, to V and then to the image, a height $-x'$ below S', as shown in Figure 2.15. Applying Snell's law at V, and using the small-angle approximation,

Magnification for a spherical refracting surface

$$m = -\frac{ns'}{n's}. \tag{2.34}$$

Noting that $p \approx s\alpha = s'\beta$ from Equation 2.25, the angular magnification is

Angular magnification for a spherical refracting surface

$$m_\alpha = \frac{-d\beta}{d\alpha} = -\frac{s'}{s} = -\frac{n}{n'}\frac{1}{m}, \tag{2.35}$$

the negative sign arising because an increase in α represents a ray moving up from the axis and an increase in β corresponds to the ray going down from the axis in the direction of propagation. The axial magnification is obtained by differentiating the imaging equation, Equation 2.29,

$$\frac{n}{s} + \frac{n'}{s'} = \frac{n'-n}{r}$$

$$-\frac{n\,ds}{s^2} - \frac{n'\,ds'}{(s')^2} = 0$$

$$\frac{ds'}{ds} = -\frac{n}{n'}\left(\frac{s'}{s^2}\right)^2$$

$$m_z = -\frac{n}{n'}m^2. \tag{2.36}$$

Some care is required in interpreting the results here. Both m and m_z are negative for positive s and s', but we must remember that s is defined positive to the left, and s' positive to the right. Thus, a right-handed coordinate system at the object will be imaged as a left-handed coordinate system at the image, and the image is perverted.

Axial magnification for a spherical refracting surface

Take-Away

Curved Refracting Interface:
A refracting surface has optical power or refracting power given by Equation 2.33;

$$P = \frac{n'-n}{r}.$$

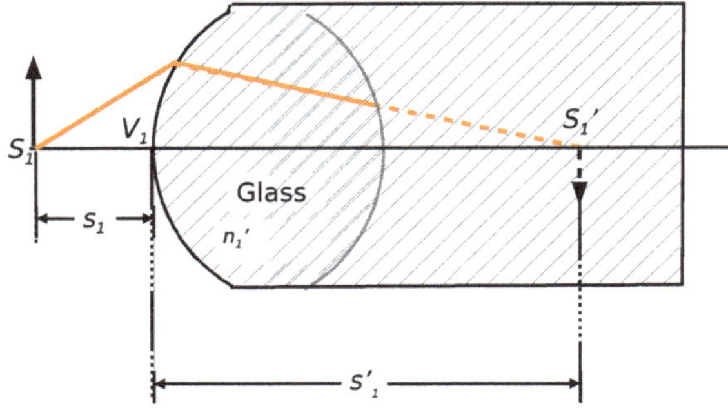

FIGURE 2.16
First surface of the lens. We begin our analysis of a lens by finding the image of the object as viewed through this surface as though the remainder of the path were in glass (shaded region).

> The two focal lengths (front and back) are related by Equation 2.32;
>
> $$\frac{BFL}{FFL} = \frac{f'}{f} = \frac{n'}{n}.$$

2.5 Simple Lens

We now have in place all the building blocks to determine the location, size, and shape of the image in a complicated optical system. The only new step is to treat the image from each surface as the object for the next, and work our way through the system. In fact, modern ray-tracing computer programs perform exactly that task, with the ability to omit the small-angle approximation for exact solutions and vary the design parameters for optimization. However, the use of a ray-tracing program generally begins with a simple system developed by hand. Further mathematical development will provide us with an understanding of lenses that will allow us to solve simple problems by hand, and will provide the intuition that is necessary when using the ray-tracing programs. We begin with the **simple lens**. The word "simple" in this context means "consisting of only two refracting surfaces." As we proceed through the section, we will examine some special cases, such as the thin lens and the thin lens in air, which are the starting points for most optical designs.

A simple lens consists of one glass element (2 surfaces)

We will build the lens using the equations of Section 2.4 and geometric considerations. It may be helpful to look briefly at Figures 2.16 through 2.19 to see the path. In Figure 2.16, we have the same problem we just solved in Section 2.4, with the result from Equation 2.29,

$$\frac{n_1}{s_1} + \frac{n_1'}{s_1'} = \frac{n_1' - n_1}{r_1}, \tag{2.37}$$

with a subscript, 1, on each variable to indicate the first surface of the lens.

Solve the imaging problem for the first surface.

Image from the first surface is object for the second.

The next step is to use this image as the object for the second surface. In Figure 2.16, the image would be a real image in the glass. However, when we consider the second surface in Figure 2.17 we see that the light intersects the second surface before the rays actually converge to an image. Nevertheless, we can still obtain the correct answers using the negative image distance. We say that we are starting this problem with a **virtual object**. This situation, with a virtual object, is quite common in practical lenses. In

Virtual object defined.

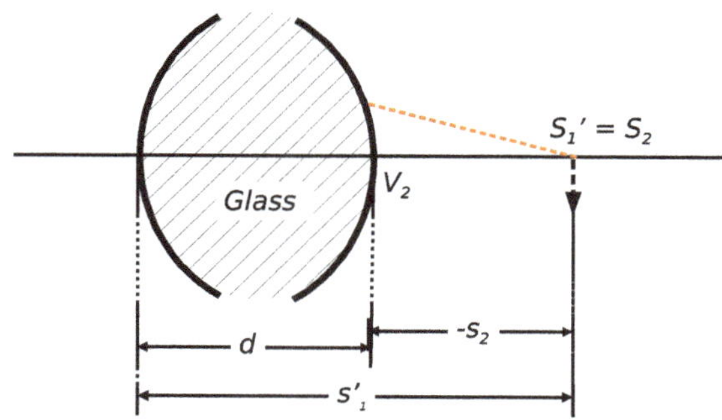

FIGURE 2.17
Distance relationships. The image from the first surface becomes the object for the second.

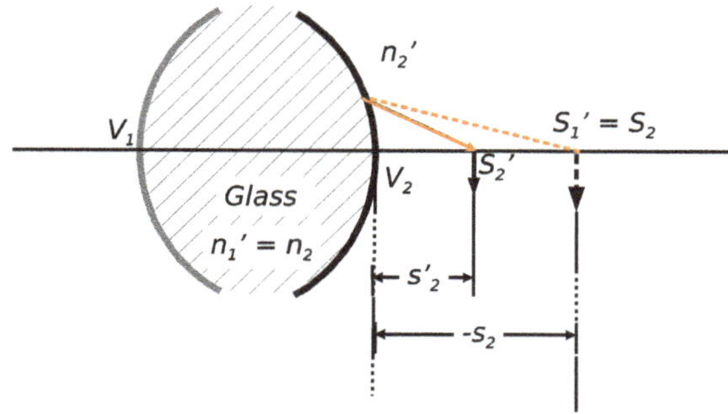

FIGURE 2.18
Second surface of the lens. The object is virtual, and $s_2 < 0$.

our drawing notation, the image is constructed with dashed rays. We can find the object distance from Figure 2.17 as

$$s_2 = -\left(s_1' - d\right), \tag{2.38}$$

and the index of refraction is

$$n_2 = n_1'. \tag{2.39}$$

Applying the imaging equation with new subscripts,

$$\frac{n_2}{s_2} + \frac{n_2'}{s_2'} = \frac{n_2' - n_2}{r_2} \qquad \frac{n_1'}{d - s_1'} + \frac{n_2'}{s_2'} = \frac{n_2' - n_1'}{r_2}, \tag{2.40}$$

as seen in Figure 2.18

At this point, it is possible to put the equations together and eliminate the intermediate distances, s_1' and s_2, to obtain

$$\frac{n_2'}{s_2'} = \frac{n_2' - n_1'}{r_2} - n_1' \frac{n_1' - n_1 - r_1 \frac{n_1}{s_1}}{d\left(n_1' - n_1\right) - n_1' r_1 - d n_1' r_1 \frac{n_1}{s_1}}. \tag{2.41}$$

Figure 2.19 shows the complete lens. We would like to eliminate the subscripts, so we will use some new notation,

$$w = s_1 \quad \text{and} \quad w' = s_2'. \tag{2.42}$$

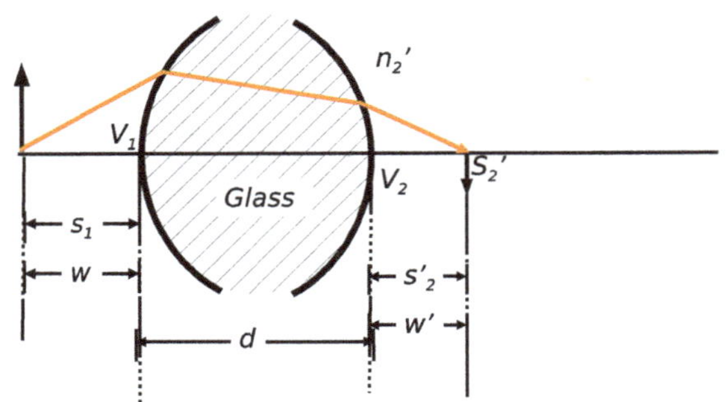

FIGURE 2.19
The complete lens. The object and image distance are shown.

With a thick lens, we will see that it is important to save the notation s and s' for another purpose (see Chapter 3). We can call w and w' the **working distances** .

Working distance defined

Simplify and remove redundant subscripts.

Eliminating subscripts from the indices of refraction, we are left with the initial and final indices of refraction and the index of refraction of the lens:

$$n = n_1 \qquad n' = n_2' \qquad n_\ell = n_1' = n_2. \tag{2.43}$$

We can now write our imaging equation as a relationship between w and w' that includes n, n', r_1, r_2, d, and n_ℓ:

$$\frac{n'}{w'} = \frac{n' - n_\ell}{r_2} - n_\ell \frac{n_\ell - n - r_1 \frac{n}{w}}{d\,(n_\ell - n) - n_\ell r_1 - d n_\ell r_1 \frac{n}{w}}. \tag{2.44}$$

We can simplify the equation slightly if we consider the initial and final materials to be air ($n = 1$):

$$\frac{1}{w'} = \frac{1 - n_\ell}{r_2} - n_\ell \frac{n_\ell - 1 - r_1 \frac{1}{w}}{d\,(n_\ell - 1) - n_\ell r_1 - d n_\ell \frac{r_1}{w}}. \tag{2.45}$$

These equations are usable, but rather complicated. We will develop a simpler formulation for this problem that can be extended to even more complicated systems when we discuss matrix optics in Chapter 3.

Example

Consider a plano-convex lens, 5 mm thick with water in front and air in back. A plano-convex lens has a flat front surface and a curved back surface that is convex toward the air. Then r_1 is infinite and by our sign convention r_2 is negative. Let's use $r_2 = -8\,\text{cm}$ in this example. We place an object $w = 10\,\text{cm}$ in front of the flat surface. Where is the image?

Using Equation 2.45, and letting $r_1 \rightarrow \infty$,

$$\frac{1}{w'} = \frac{1 - n_\ell}{r_2} - n_\ell \frac{\frac{1}{w}}{-n_\ell - d n_\ell \frac{1}{w}} \text{ or}$$

$$\frac{1}{w'} = \frac{1 - 1.5}{-8\,\text{cm}} - 1.5 \frac{\frac{1}{10\,\text{cm}}}{-1.5 - 0.5\,\text{cm} \times 1.5 \frac{1}{10\,\text{cm}}} = 0.1577\,\text{cm}^{-1}.$$

$$w = 6.34\,\text{cm}.$$

The image is 6.34 cm to the right of the curved vertex.

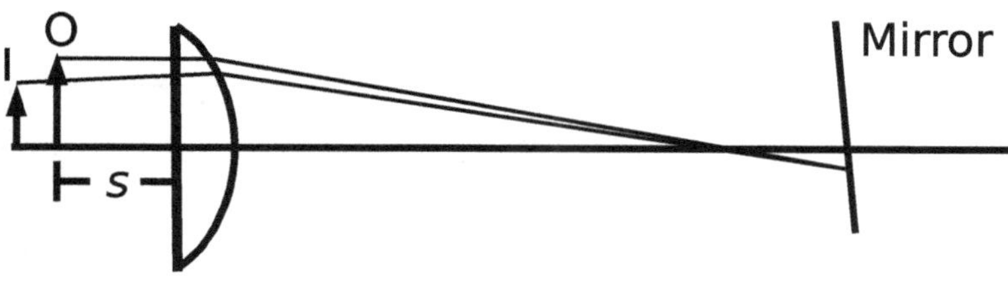

FIGURE 2.20
Autocollimation. An object at the front focal plane images to infinity. The image, reflected in a flat mirror, will focus back to the same plane.

In Practice

One can estimate the back focal plane of a lens experimentally by placing an object "far away." In this context, "far" means a distance long compared to the focal length. Outdoors the sun is a good object. Indoors, for focal lengths measured in millimeters, ceiling lights may be used. The result is only approximate (from what point on the lens do you make the measurement?). In Chapter 3, we will learn a more exact way to find focal length.

A more precise way to find a focal point is **autocollimation**. As we saw in the case of the thick lens, the distance to the focal plane is not measured from the vertex. In Chapter 3, we will learn about compound lenses for which locating the focal planes can be even more complicated, so being able to precisely locate the focal planes experimentally can be quite important. As seen in Figure 2.20, a sparse object, "O" placed near the front focal plane of a lens, $s \approx f$, will focus to an image distance, $s' \rightarrow$, close to infinity. If this image is reflected by a flat mirror, it becomes an object near infinity for the lens and the resulting image, "I" will be near the focal plane. If the mirror is not perfectly aligned, we can see the image overlaying the object, but probably slightly out of focus. Now if we move the lens or the object, so that the object distance becomes closer to the focal plane, the focused image will also move closer to the focal plane, and on the object it will be better focused. If we move in the wrong direction, the image will become more blurred. Thus we can quickly find the focal plane by adjusting for the best focus of the reflected image on the object. We need to record the distance from some reference surface such as the first vertex.

Autocollimation is the best way to find the focal planes of a lens experimentally.

After this test, we can reverse the lens and find the back focal plane in the same way. Sometimes we refer to the distance between the focal plane and the corresponding vertex as the **working distance**. In Chapter 3, we will learn how to compute focal lengths and working distances from the locations and radii of curvature of the surfaces in arbitrarily complicated compound lenses. The working distance can be quite important as the physical situation often requires some space between an object and the optics. For example, in a microscope we may need to view an object through a glass slide.

2.5.1 Thin Lens

Setting $d = 0$ in Equation 2.44,

$$\frac{n'}{w'} = \frac{n' - n_\ell}{r_2} - \frac{n_\ell - n - r_1\frac{n}{w}}{-r_1},$$

$$\frac{n}{w} + \frac{n'}{w'} = \frac{n' - n_\ell}{r_2} + \frac{n_\ell - n}{r_1}.$$

Because the lens is thin, the distinction between w and s is unnecessary, and we will write the more usual equation:

$$\frac{n}{s} + \frac{n'}{s'} = \frac{n' - n_\ell}{r_2} + \frac{n_\ell - n}{r_1}. \tag{2.46}$$

This equation is called the **thin lens equation**. In terms of refracting power,

$$\frac{n}{s} + \frac{n'}{s'} = P_1 + P_2 = P. \tag{2.47}$$

If we set the object distance s to infinity, s' gives us the back focal length,

$$f' = \frac{n'}{P_1 + P_2}, \tag{2.48}$$

and, instead, if we set s' to infinity, s gives us the front focal length,

$$f = \frac{n}{P_1 + P_2}. \tag{2.49}$$

These equations are deceptively simple, because the powers contain the radii of curvature and indices of refraction:

$$P_1 = \frac{n_\ell - n}{r_1} \quad \text{and} \quad P_2 = \frac{n' - n_\ell}{r_2}. \tag{2.50}$$

It is important to note that, just as for a single surface,

$$\frac{f'}{f} = \frac{n'}{n},$$

and that when $n = n'$, the front and back focal lengths are the same: $f = f'$.

Thin Lens in Air As a special case of the thin lens, we consider the thin lens in air. Then, $n = n' = 1$ and we have the usual form of the **lens equation**:

$$\frac{1}{s} + \frac{1}{s'} = \frac{1}{f}. \tag{2.51}$$

Take-Away

Lens Equation:

The lens equation is probably the most used equation in geometric optics. Although it is only an approximation to results that could be obtained with Snell's law, it is extremely useful. Coupled with the magnification equation, discussed below, it provides a mapping from object points, x, s to image points, x', s'.

Although this equation is presented here for this very special case, we will find that we can arrange more complicated problems so that it is applicable more generally, and we will make extensive use of it. The other important equation is the definition of f, the thin-lens form of the **lensmaker's equation**, so called because a lensmaker, asked to produce a lens of a specified focal length, can compute values of r_1 and r_2 for a given lens material with index n_ℓ:

The lensmaker's equation for a thin lens is used for designing the lens.

$$\frac{1}{f} = \frac{1}{f'} = P_1 + P_2 = (n_\ell - 1)\left(\frac{1}{r_1} - \frac{1}{r_2}\right). \tag{2.52}$$

Take-Away

Lensmaker's Equation:
The lensmaker's equation relates the focal length, f, to the radii of curvature and thus gives information about how to make the lens.

Of course, the lensmaker still has some freedom; although n_ℓ is usually limited by the range of materials that transmit at the desired wavelength, the lensmaker's equation provides only one equation that involves two variables, r_1 and r_2. The freedom to choose one of these radii independently can be used to reduce aberrations as we will discuss in Chapter 5.

The two contributions to the optical power,

$$P_1 = \frac{n_\ell - 1}{r_1} \qquad P_2 = \frac{n_\ell - 1}{-r_2}, \tag{2.53}$$

appear to have opposite signs. However, for a biconvex lens with both surfaces convex toward the air, our sign convention results in a negative value for r_2, which leads to a positive value for P_2. One significant result of these conventions is that the thin lens functions the same way if the light passes through it in the opposite direction, in which case r_1 and r_2 are exchanged and their signs are reversed. The result is that P_1 and P_2 are exchanged, and the total refracting power, $P_1 + P_2$, remains the same.

In Practice

Reversing a lens does not change its optical power. If the front and back media are different, it will exchange the front and back focal lengths.

Reversing a lens with the same medium on each side does not change the focal length.

Some books use a different sign convention, with r_1 and r_2 being positive if they are convex toward the lower index (usually the air). Such a convention is convenient for the lensmaker. In a biconvex lens, the same tool is used for both sides, so it could be convenient for the signs to be the same. However, we will continue with the convention that a surface with a positive radius of curvature is convex toward the source. The sign convention has been the source of many errors in specifying lenses. A diagram is often useful in communicating with vendors whenever any ambiguity is possible.

Be careful about signs on radii of curvature.

Example

Consider a thin biconvex germanium lens with radii of curvature 30 and −30 cm. This lens is symmetric; both surfaces are convex toward the air. The lens is to be used for an infrared camera at a wavelength around 10 μm.

What is the focal length? Where is the image of an object place 20 cm in front of this lens?

The index of refraction of germanium is 4 in the infrared. Using Equation 2.52,

$$\frac{1}{f} = \frac{1}{f'} = (4-1)\left(\frac{1}{30\,\text{cm}} - \frac{1}{-30\,\text{cm}}\right),$$

the focal length is

$$f = 5\,\text{cm}.$$

Now, using Equation 2.51 with the object distance at 20 cm.,

$$\frac{1}{20\,\text{cm}} + \frac{1}{s'} = \frac{1}{5\,\text{cm}},$$

we find the image distance,

$$s' = 6.667\,\text{cm}.$$

The object is 6.667 cm beyond the lens.

We still need to determine the magnification. For the thin lens, a ray from the tip of the object arrow in Figure 2.19 to the tip of the image arrow will be a straight line. By similar triangles, the ratio of heights is

$$\frac{x'}{x} = \frac{-s'}{s}$$

Magnification of a lens in air

so the magnification is

$$m = \frac{-s'}{s} \quad \text{(Lens in air).} \tag{2.54}$$

Magnification of a lens, generally

More generally, we can show that

$$m = \frac{-ns'}{n's}, \tag{2.55}$$

for any thin lens with index of refraction n in front of it and n' behind. We will later extend this equation to any lens.

Axial magnification of a lens

The axial magnification is obtained by differentiating Equation 2.46:

$$m_z = \frac{ds'}{ds} = -\frac{n}{n'}\left(\frac{s'}{s}\right)^2 = -\frac{n'}{n}m^2. \tag{2.56}$$

Example

An object 6 mm tall is place at the object location in the previous example $s = 20$ cm. What is the height of the image?

Using Equation 2.54

$$m = \frac{-6.667\,\text{cm}}{20\,\text{cm}} = -1/3..$$

The image is 6 mm/3 = 2 mm tall and inverted (because of the negative sign).

Example

A second object is place 1.8 mm before the one describe above. What is is the spacing between the two images? That is, what is the difference in image distances?

We could solve the problem using = 20.18 cm, with the answer $s' = 6.647\,\text{cm}$ and $6.667\,\text{cm} - 6.647\,\text{cm} = 0.0198\,\text{cm}$, but let's use the axial magnification. From Equation 2.56. With $n' = n$ the equation is simply

$$m_z = -\frac{n'}{n}m^2 = -m^2 = -1/9.$$

The object spacing was 1.8 mm so the image spacing is $-1.8\,\text{mm}/9 = -0.2\,\text{mm}$ using the axial magnification, as compared to 198 μm using the exact calculation.

With the simple lens, we are now near the end of our task of completing Table 2.2. We could extend this analysis to compound lenses by using multiple simple lenses. However, as we have seen, even with the simple lens, the bookkeeping becomes complicated quickly, it is easy to introduce errors into calculations, and physical insight begins to suffer. We will turn our attention in the next chapter to matrix optics which will provide a convenient mechanism for bookkeeping, and allow us to understand complex systems in the terms that we have developed here: object and image distances, optical power, and focal lengths.

Take-Away

Lenses in General:

- Lenses consist of multiple dielectric interfaces.
- A simple lens consists of one glass element and just two surfaces.
- The image location and size can always be found by treating one surface at a time, using the resulting image as the object for the next surface.
- The process is tedious and prone to errors, so some book-keeping scheme is important.

In Practice

In designing a new optical system, thin lenses often provide a good starting point. Once the basic design is established, they can be replaced by more complicated lenses with better performance.

A thin lens can be specified by just its optical power.

Most often, a lens is used in air and is specified by its focal length, f.

2.6 Prisms

Prisms find a wide diversity of applications in optics. Dispersive prisms are used to distribute different wavelengths of light across different angles in spectroscopy. Prisms are also used in scanning devices[195;16;242;241;123;222]. As shown in Figure 2.21A, the prism can be analyzed using Snell's law. For simplicity, let us assume that the medium surrounding

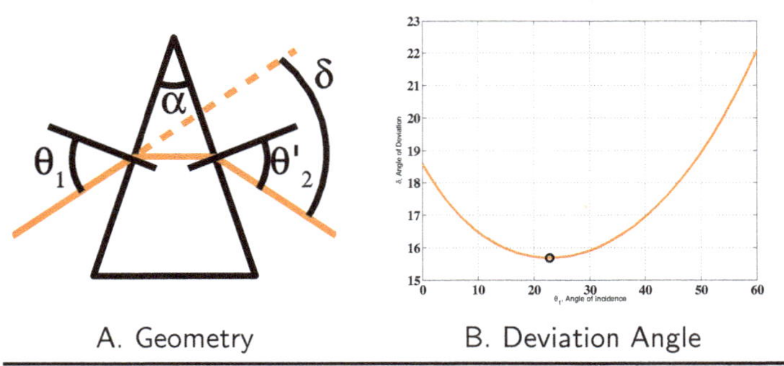

FIGURE 2.21
Prism. (A) A prism can be used to deflect the direction of light. (B) If the material is dispersive, different wavelengths will be refracted through different angles.

the prism is air, and that the prism has an index of refraction, n. The angle of refraction after the first surface is

$$\theta'_1 = \frac{\theta_1}{n}. \tag{2.57}$$

The angle of incidence on the next surface can be obtained by summing the angles in the triangle formed by the apex (angle α) and the points at which the light enters and exits the prism:

$$(90^\circ - \theta_2) + (90^\circ - \theta'_1) + \alpha = 180^\circ$$
$$\theta_2 + \theta'_1 = \alpha. \tag{2.58}$$

Applying Snell's law,

$$\sin\theta'_2 = n\sin\theta_2 = n\sin\alpha - \theta'_1$$
$$\sin\theta'_2 = n\left(\cos\theta'_1 \sin\alpha - n\sin\theta'_1 \cos\alpha\right)$$
$$\sin\theta'_2 = \sqrt{n^2 - \sin^2\theta_1}\sin\alpha - \sin\theta'_1\cos\alpha. \tag{2.59}$$

Beam deviation angle in general.

The deviation from the original path is

$$\delta = \theta_1 + \theta'_2 - \alpha. \tag{2.60}$$

Minimum deviation angle.

This equation is plotted against θ_1 in Figure 2.21B for $\alpha = 30^\circ$ and $n = 1.5$. It is clear that there is a minimum deviation of

$$\delta_{min} = 2\sin^{-1}\left(n\sin\frac{\alpha}{2}\right) - \alpha \quad \text{at} \quad \theta_1 = \sin^{-1}\left(n\sin\frac{\alpha}{2}\right) \tag{2.61}$$
$$\delta_{min} = 15.7^\circ \quad \text{at} \quad \theta_1 = 22.8^\circ,$$

and the minimum is quite broad. For small prism angles,

$$\delta_{min} \approx (n-1)\alpha \quad \text{at} \quad \theta_1 = \frac{n\alpha}{2} \quad \text{for} \quad (\alpha \to 0). \tag{2.62}$$

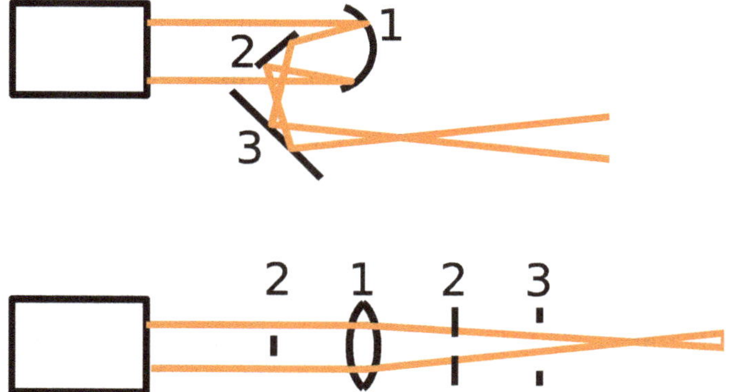

FIGURE 2.22
Reflective systems. Often it is conceptually easier to visualize a reflective system as though it were "unfolded." In this example, a light source is focused with a concave mirror with a radius, R, and then reflected by two flat mirrors (top). We may find it easier to imagine the light following a straight path in which we have replaced the mirror with a lens of focal length, $f = r/2$, and removed the folding mirrors. Mirror 2 appears twice, once as an obscuration when the light passes around it enroute to the curved mirror and once as an aperture when the light reflects from it.

Take-Away

Prisms:
A prism refracts light to a new direction.
Often it is used for dispersion, so that each wavelength is refracted to a different new direction.
There is a minimum deviation angle, δ, for any apex angle, α. This minimum deviation is, to a first approximation, $\delta = (n-1)\alpha$.

Example

We want to use a prism to separate different wavelengths. We would like to be able to tell the difference between the green argon ion laser line at 514.5 nm and the frequency-doubled Nd:YAG line at 532 nm. For a prism with an apex angle of 30 degrees, what will be the angular dispersion between these two lines?

Let's choose a flint glass (small Abbe number) so the dispersion will be large. See Figure 1.8. Turning to https://refractiveindex.info, we choose the SF10 glass, with $n = 1.7401$ at $\lambda = 514.5\,\text{nm}$ and $n = 1.7367$ at $\lambda = 532\,\text{nm}$. Using the approximation above, the nominal angle is

$$\delta = n\alpha \approx 1.7 \times 60\,\text{deg} = 51\,\text{deg}$$

and

$$\Delta\delta = \Delta n\alpha = 0.0034 \times 30\,\text{deg} = 0.102\,\text{deg} =$$

The difference is small but probably measurable.

2.7 Reflective Systems

We end this introduction to geometric optics with a short discussion of reflective systems. An example of a reflective system is shown in the top of Figure 2.22. The system consists of a large concave mirror (1), a folding mirror (2), which reflects the light focused by the concave mirror, but also obscures a portion of the incoming light, and (3) a scanning mirror (3) which directs the light to some target. After working in optics for a period of time, one becomes quite accustomed to refractive systems in which light travels from left to right through lenses. However, as we have indicated, sometimes reflective components can be useful, particularly when a large diameter is required, a large range of wavelengths is to be used, or when inexpensive refractive materials are not available at the desired wavelength. An optical system with multiple lenses can become quite long, and flat mirrors are often used to fold the path. Furthermore, scanning mirrors are often required and result in a folded path.

It is convenient to think of a "folded" system as if it were laid out in a straight line.

It is often convenient to "unfold" the optical path as shown in the bottom of Figure 2.22. The distances between components remain as they were in the original system. Curved mirrors are replaced by lenses with the equivalent focal length, $f = r/2$. It is now easier to visualize the behavior of rays in this system than in the actual folded system and the equations we have developed for imaging still hold. The dot shows that the beam "pivots" about the center of mirror 3. Such an analysis is useful in the early design phase of a system.

In Practice

A straight-line layout can make it easier to visualize the focusing and imaging behavior of an optical system. It provides a way of visualizing the scanning and alignment issues that may arise when the system is built.

Show the scanner and alignment locations with dots on the diagram, representing locations where the beam can "pivot."

It may be useful to draw multiple diagrams (*e.g.* for transmitter and receive paths or for the two transverse directions if the system is not cylindrically symmetric).

Problems

2.1 Dogleg

A beam of laser light is incident on a beamsplitter of thickness, t, at an angle, θ, as shown in Figure P2.1. The index of refraction of the medium on each side is the same, n, and that of the beamsplitter is n'. When I remove the beamsplitter, the light falls on a detector.

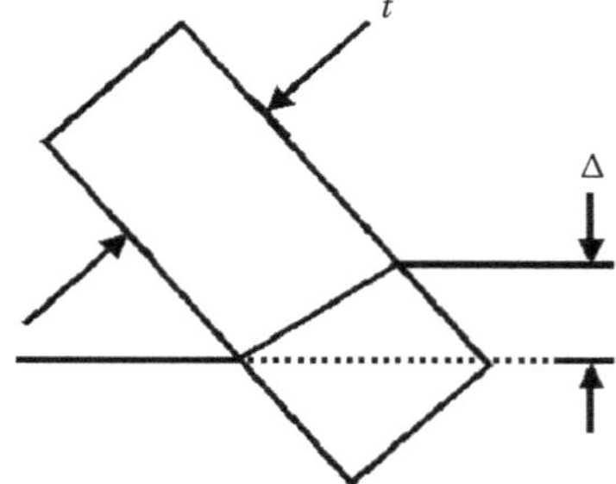

FIGURE P2.1
Layout for Problem 2.1.

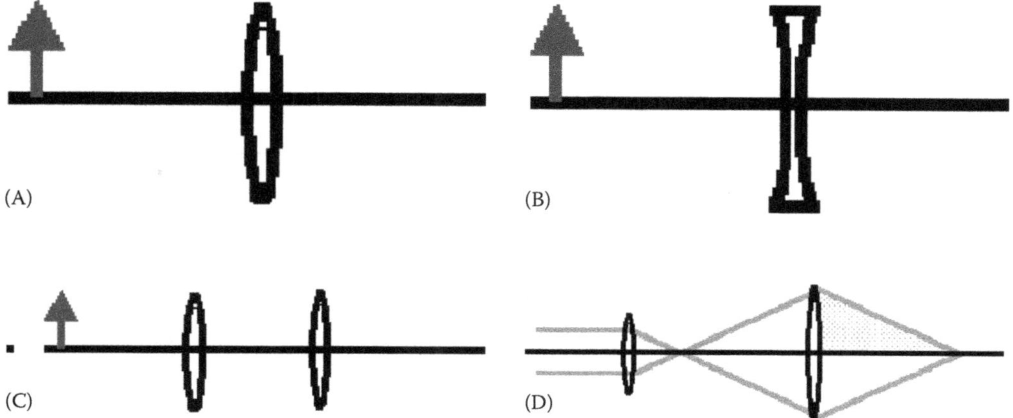

FIGURE P2.4
Sample lenses for Problem 2.2.

When I replace the beamsplitter, the light follows a "dogleg" path, and I must move the detector in a direction transverse to the beam. How far?

2.1a Derive the general equation.

2.1b Check your result for $n' = n$ and for $n'/n \to \infty$. Calculate and plot vs. angle, for $t = 5\,\text{mm}$, $n = 1$, with $n' = 1.5$, ...

2.1c and again with $n' = 4$.

2.2 Some Lenses

2.2a Consider the thin lens in Figure P2.4A. Let the focal length be $f = 10\,\text{cm}$. Plot the image distance as a function of object distance from 50 cm down to 5 cm.

2.2b Plot the magnification.

2.2c Consider the thin lens in Figure P2.4B. Let the focal length be $f = -10\,\text{cm}$. Plot the image distance as a function of object distance from 50 cm down to 5 cm.

2.2d Plot the magnification.

2.2e In Figure P2.4C, the lenses are both thin and have focal lengths of $f_1 = 20\,\text{cm}$ and $f_2 = 10\,\text{cm}$, and the separation is $d = 5\,\text{cm}$. For an object 40 cm in front of the first lens, where is the final image?

2.2f What is the magnification?

2.3 Snow Globe (NS)

Let's consider a snow globe, which is a transparent plastic container full of water with a 3-d scene inside. To make it simple suppose the container is a sphere of radius 3 cm. Pick one axis through the sphere as the axis, and let s vary continuously along the axis inside the sphere. Plot s', the image location, m, the magnification, and m_z as functions of s.

2.4 Oil-Immersion Lens (NS)

Consider an oil-immersion microscope lens where the object is immersed in oil of index of refraction 1.5. The back focal length is $f' = 2\,\text{mm}$, and the magnification is $m = 100$. What is the front focal length, f? Where are the object and image, s and s'. If s is changed by $5\,\mu\text{m}$, what is the change in s'?

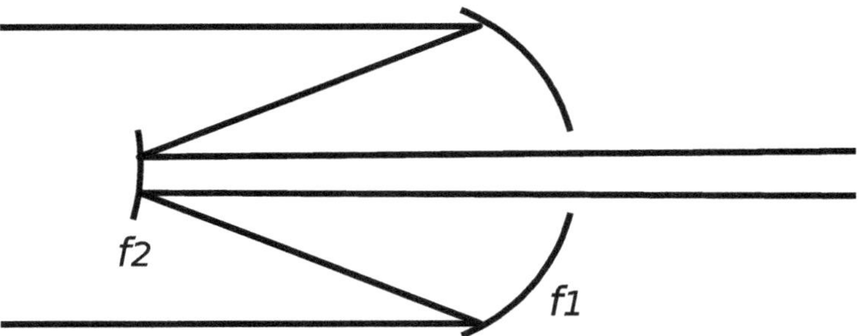

FIGURE P2.5
Figure for Problem 2.5.

2.5 Telescope (NS)

Consider the telescope shown in Figure P2.5, with $f_1 = 100\,\text{cm}$ and $f_2 = -5\,\text{cm}$ and the separation between the mirrors $z = 95\,\text{cm}$. The primary mirror, f_1 has a diameter of 20 cm and the secondary, f_2 has a diameter of 4 cm. The hole in the primary is 5 cm in diameter.

Draw the straight-line layout with the mirrors replaced by lenses. Show the apertures and obscurations that limit rays that go through the telescope. Reminder: The secondary obscures the incoming light. If the object is located at infinity (*e.g.* a star), where is the image? Where is the intermediate image? Show it on your drawing.

2.6 Thin Lenses (NS)

Remember that for short distances between surfaces, refracting powers add. Consider a plano-convex lens with focal length, $f = 10\,\text{cm}$ in air. If $r_1 \to \infty$, what is r_2. If we put two of these lenses close together, what is the focal length of the combination?

3 Matrix Optics

Chapter 2 introduced the tools to solve all geometric optics problems involving both reflection and refraction at plane and curved surfaces. We also learned the principles of systems with multiple surfaces, and we found some simple equations that could be used for thin lenses, consisting of two closely spaced refracting surfaces. However, many optical systems are more complicated, having more surfaces and surfaces which are not closely spaced. How can we apply the simple concepts of Chapter 2 to a system like that in Figure 3.1, without getting lost in the bookkeeping? In this section, we will develop a matrix formulation for geometric optics that will handle the bookkeeping details when we attempt to deal with complicated problems like this, and will also provide a way of describing arbitrarily complicated systems using exactly the same equations we developed for thin lenses.

Fig. 2.13:

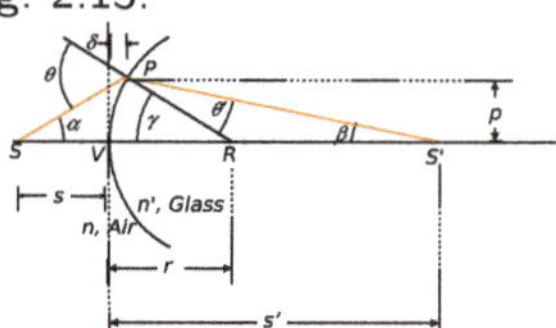

The calculations we will use in this chapter have exactly the same results as those in Chapter 2. Referring back to Figure 2.13, we continue to assume that $\delta = 0$. However, now we want to solve problems of very complicated compound lenses as shown in Figure 3.1A. Letting $\delta = 0$ we draw the vertex planes as shown in Figure 3.1B. Then using the curvature of each surface, the direction of the incident ray, and the indices of refraction, we compute the direction of the refracted ray. In this approach, we assume that the refraction occurs at the vertex plane as shown in Figure 3.1C, instead of at the actual surface as in Figure 3.1A.

Between surfaces in homogeneous media, a ray is a straight line. Because most optical systems have cylindrical symmetry, we usually use a two-dimensional space, x, z for which two parameters define a straight line. Therefore, we can define a ray in terms of a two-element column vector, with one element being the height x at location z along the optical axis, and the other being some measure of the angle as shown in Figure 3.2. We will choose

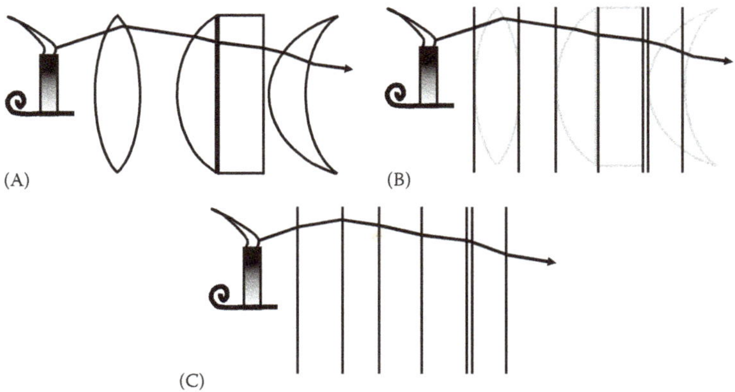

FIGURE 3.1
Compound lens. Matrix optics makes it easier to solve complicated problems with multiple elements. A correct ray trace, would use Snell's law at each interface (A). In our approximation, we find the vertex planes (B) and use the small-angle approximation to Snell's law for the curved surfaces in (A), but applying at the vertex planes (C). We are still assuming $\delta = 0$ in Figure 2.13. The results are reasonably good for paraxial rays.

DOI: 10.1201/9781315157047-3

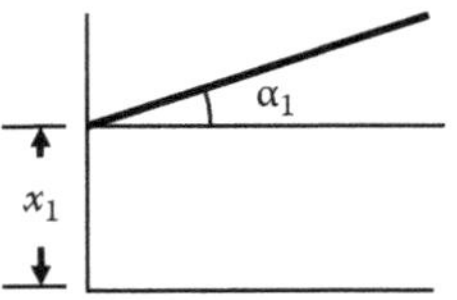

FIGURE 3.2
Ray definition. The ray is defined by its height above the axis and its angle with respect to the axis.

the angle, α, in radians, and write the vector as

$$\mathbf{V} = \begin{pmatrix} x \\ \alpha \end{pmatrix}. \tag{3.1}$$

Ray vector represents a ray: Position and angle.

Snell's law describes the change in direction of a ray when it is incident on the interface between two materials. In the small-angle approximation, the resulting equations are linear and linear equations in two dimensions can always be represented by a two-by-two matrix. This **ray-transfer matrix** is almost always called the **ABCD matrix**, because it consists of four elements:

$$\mathcal{M} = \begin{pmatrix} A & B \\ C & D \end{pmatrix}. \tag{3.2}$$

Ray matrix operates on a ray vector.

Roadmap: Chapter 3: Matrix Optics

Matrix survival guide in Appendix C

We will begin this chapter by exploring the concepts of matrix optics, defining matrices for basic operations, and showing how to apply these techniques to complicated systems in Section 3.1. Then we will discuss the interpretation of the results of the matrices in terms that relate back to the equations of Chapter 2, introducing the new concept of principal planes, and refining our understanding of focal length and the lens equation. We will return to the thick lens that we discussed in Section 2.5, but this time explaining it from a matrix perspective. We will conclude in Section 3.4 with some examples. A brief "survival guide" to matrix mathematics is shown in Appendix C.

3.1 Matrix Optics Concepts

A ray can be defined in two dimensions by two parameters such as the slope and an intercept. As we move through an optical system from source to receiver we would like a description that involves local quantities at each point along the z axis. It is convenient to use the local ray height, x_1, and the angle with respect to the axis, α_1, both shown in Figure 3.2. Some authors choose to use the sine of the angle instead of the angle, but the difference is not important for the purposes of determining the image location and magnification. These two results are obtained using the small-angle approximation. Rays that violate this approximation may lead to aberrations unless corrections are made (Chapter 5).

Having made the decision to use the height and angle in a two-element column vector, the remaining choice is whether to put the position or the angle on top. Again, different authors have adopted different choices. This choice makes no difference in the calculation, provided that we agree on the convention. Of course a different choice will rsult in different matrices. We choose to put the height in the top position, and write the **ray vector**

$$\mathbf{V} = \begin{pmatrix} x \\ \alpha \end{pmatrix}. \tag{3.3}$$

We define matrix operations such that

$$\mathbf{V}_{end} = \mathcal{M}_{start:end}\mathbf{V}_{start}. \tag{3.4}$$

Tracing a ray through an optical system usually involves two alternating tasks, first translating, or moving from one position to another along the optical axis, and second, modifying the ray by interaction with a surface. For this reason, we will use subscript numbers, such as $\mathbf{V}_1$ and $\mathbf{V}_2$, to represent the ray at successive positions, and the "prime" notation, $\mathbf{V}_1'$, to represent the ray after a surface.

Ray tracing: Alternating translation and refraction.

Thus, as we go through a compound lens, we will compute the vectors, $\mathbf{V}_1$, $\mathbf{V}_1'$, $\mathbf{V}_2$ $\mathbf{V}_2'$ ..., and so forth by using matrices to describe refraction, alternating with matrices to describe translation from one vertex plane to the next. We need to develop the matrices for the operations of translation, refraction, and reflection. One way to handle imaging by reflective systems is to unfold the reflective portions into equivalent systems with refractive elements replacing the reflective ones, as discussed in Section 2.7. Therefore, the two fundamental matrices will be translation and refraction. After we describe these two matrices, we can use them to develop complicated optical systems.

Subscript number identifies a surface. Unprimed vector before and primed after.

3.1.1 Basic Matrices

We first want to find a matrix, $\mathcal{T}$, to describe translation such that

$$\mathbf{V}_2 = \mathcal{T}_{12}\mathbf{V}_1, \tag{3.5}$$

where $\mathcal{T}_{12}$ modifies the vector $\mathbf{V}_1$ to become $\mathbf{V}_2$. Looking at the physical operation in Figure 3.3, we see that the ray is really unchanged. After the translation we are looking at the ray at axial location 2 instead of location 1. Thus, the angle α_2 remains the same as α_1,

$$\alpha_2 = \alpha_1 \tag{3.6}$$

and the height increases by αz_{12},

$$x_2 = x_1 + \alpha_1 z_{12}. \tag{3.7}$$

These two equations may be represented in matrix form as

Translation Matrix describes moving through distance z_{12}

$$\mathbf{V}_2 = \mathcal{T}_{12}\mathbf{V}_1 \tag{3.8}$$

$$\begin{pmatrix} x_2 \\ \alpha_2 \end{pmatrix} = \begin{pmatrix} 1 & z_{12} \\ 0 & 1 \end{pmatrix} \begin{pmatrix} x_1 \\ \alpha_1 \end{pmatrix}, \tag{3.9}$$

and the **translation matrix** becomes

$$\mathcal{T}_{12} = \begin{pmatrix} 1 & z_{12} \\ 0 & 1 \end{pmatrix}. \tag{3.10}$$

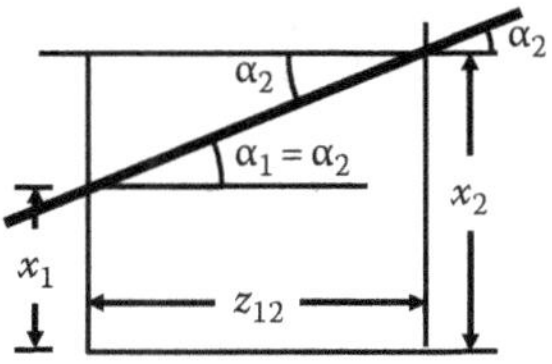

FIGURE 3.3
Translation. In translation through a distance z, the angle remains the same but the height changes.

The next matrix is for refraction at a surface. As before, $\mathbf{V}_1$ describes the ray before interacting with the surface. Then $\mathbf{V}_1'$ describes it afterward, with the equation

$$\mathbf{V}_1' = \mathcal{R}_1 \mathbf{V}_1. \tag{3.11}$$

From Figure 3.4, we can see that the height of the ray is unchanged ($x_1' = x_1$), or

$$x_1' = (1 \times x_1) + (0 \times \alpha_1), \tag{3.12}$$

so we have two elements of the matrix,

$$\begin{pmatrix} x_1' \\ \alpha_1' \end{pmatrix} = \begin{pmatrix} 1 & 0 \\ ? & ? \end{pmatrix} \begin{pmatrix} x_1 \\ \alpha_1 \end{pmatrix}. \tag{3.13}$$

We know from Chapter 2 that the angle changes according to the optical power of the surface. Specifically, looking back at Figure 2.13, the initial angle in the figure is α, exactly as defined here, and the final angle, which we define here as α' is $-\beta$ so Equations 2.22 and 2.23, may be written as

$$\theta = \gamma + \alpha \qquad \theta' = \gamma - \beta = \gamma + \alpha'.$$

In Equation 2.28, we used Snell's law with the small approximation and with angles expressed in terms of their tangents, to obtain the imaging equation. Here we use it with the angles themselves. The height, p, of the ray at the intersection in Figure 2.13 is x in our vector, so remembering that n and n' are the indices of refraction before and after the surface respectively,

$$n\theta = n'\theta' \qquad n(\gamma + \alpha) = n'(\gamma + \alpha'), \tag{3.14}$$

$$n\frac{x}{r} + n\alpha = n'\frac{x}{r} + n'\alpha', \tag{3.15}$$

$$\alpha' = \frac{n - n'}{n'r}x + \frac{n}{n'}\alpha. \tag{3.16}$$

Refraction matrix bends the ray.

Thus, the **refraction matrix** is

$$\mathcal{R} = \begin{pmatrix} 1 & 0 \\ \frac{n-n'}{n'r} & \frac{n}{n'} \end{pmatrix}. \tag{3.17}$$

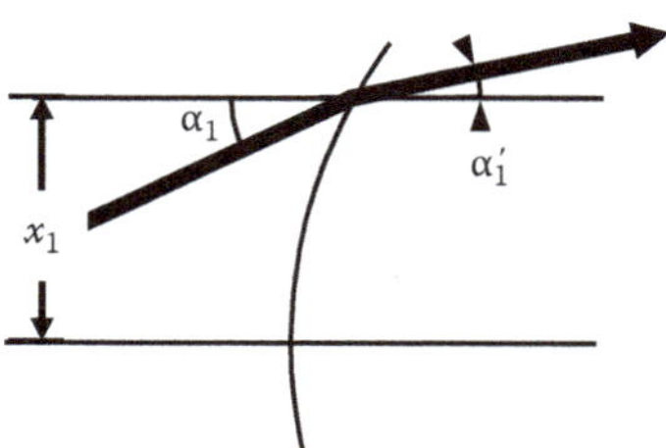

FIGURE 3.4
Refraction. In refraction at a curved interface, the angle changes, but the height remains the same.

We conclude this section on basic matrices by noting that the lower left term involves the refracting power, as defined in Equation 2.33, and by substitution, the matrix becomes

$$\mathcal{R} = \begin{pmatrix} 1 & 0 \\ -\frac{P}{n'} & \frac{n}{n'} \end{pmatrix}. \tag{3.18}$$

3.1.2 Cascading Matrices

The power of the matrix approach lies in the ability to combine as many elements as needed, with the resulting problem being no more complicated than a two-by-two matrix multiplication. Starting with any vector, V_{start}, we multiply by the first matrix to get a new vector. We then multiply the new vector by the next matrix, and so forth. Therefore, as we include new elements from left to right in the system, we introduce their matrices from right to left.* We begin with a simple lens as an example and then generalize to arbitrarily complicated problems.

Each matrix multiplication acts on the result of the previous one. Cascade matrices from right to left.

3.1.2.1 The Simple Lens

A simple lens consists of a refraction at the first surface, a translation from there to the second surface, and a refraction at the second surface, as shown in Figure 3.5. At the first surface,

Refraction at first surface.

$$\mathbf{V}_1' = \mathcal{R}_1 \mathbf{V}_1. \tag{3.19}$$

The resulting vector, $\mathbf{V}_1'$, is then multiplied by the translation matrix to obtain the vector, $\mathbf{V}_2$, just before the second surface:

Translation to next surface.

$$\mathbf{V}_2 = \mathcal{T}_{12} \mathbf{V}_1', \tag{3.20}$$

and finally, the output of the second surface is

Refraction at the second surface.

$$\mathbf{V}_2' = \mathcal{R}_2 \mathbf{V}_2. \tag{3.21}$$

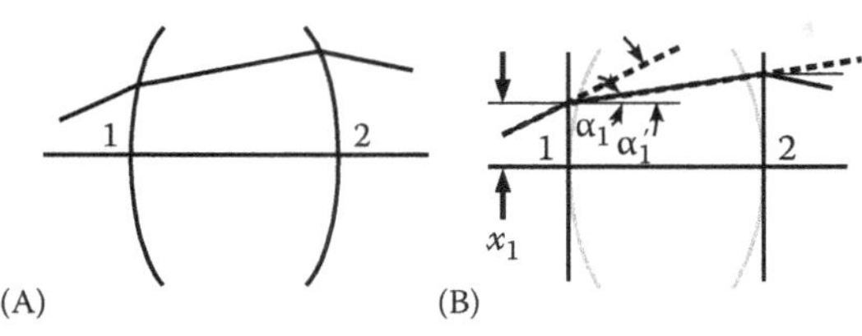

FIGURE 3.5
Simple lens. Light is refracted at the lens surfaces according to Snell's law (A). In our approximation, the refraction is computed using the small-angle approximation and assuming the refraction takes place in the vertex planes (B).

Thus the final output is

Cascaded matrices.

$$\mathbf{V}_2' = \mathcal{R}_2 \mathcal{T}_{12} \mathcal{R}_1 \mathbf{V}_1.$$

We can say

$$\mathbf{V}_2' = \mathcal{L} \mathbf{V}_1,$$

where we have defined the matrix of a simple lens, as

Giving the product a new name.

*The matrix multiplication is shown in Equation C.8 in Appendix C.

$$\mathcal{L} = \mathcal{R}_2 \mathcal{T}_{12} \mathcal{R}_1. \tag{3.22}$$

We need to remember that, because each matrix operates on the output of the preceding one, we cascade the matrices from right to left, as discussed in Equation C.8 in Appendix C.

Writing the component matrices explicitly according to Equations 3.10 and 3.18,

$$\mathcal{L} = \begin{pmatrix} 1 & 0 \\ -\frac{P_2}{n_2'} & \frac{n_1'}{n_2'} \end{pmatrix} \begin{pmatrix} 1 & z_{12} \\ 0 & 1 \end{pmatrix} \begin{pmatrix} 1 & 0 \\ -\frac{P_1}{n_1'} & \frac{n_1}{n_1'} \end{pmatrix}, \tag{3.23}$$

where we have used the fact that $n_2 = n_1'$. The product of these matrices is complicated, but we choose to group the terms in the following way:

$$\mathcal{L} = \begin{pmatrix} 1 & 0 \\ -\frac{P_t}{n_2'} & \frac{n_1}{n_2'} \end{pmatrix} + \frac{z_{12}}{n_1'} \begin{pmatrix} -P_1 & n_1 \\ \frac{P_1 P_2}{n_2'} & -P_2 \frac{n_1}{n_2'} \end{pmatrix},$$

where

$$P_t = P_1 + P_2.$$

Now, we can eliminate some of the subscripts by substituting the initial index, $n_1 = n$, the final index, $n_2' = n'$, and the lens index, $n_1' = n_\ell$, and obtain the matrix for a **simple lens**,

$$\mathcal{L} = \begin{pmatrix} 1 & 0 \\ -\frac{P_t}{n'} & \frac{n}{n'} \end{pmatrix} + \frac{z_{12}}{n_\ell} \begin{pmatrix} -P_1 & n \\ \frac{P_1 P_2}{n'} & -P_2 \frac{n}{n'} \end{pmatrix}. \tag{3.24}$$

Eq. 2.17 $\frac{n}{s} = \frac{n'}{s'}$.

The lens index of refraction, n_ℓ only appears explicitly in one place in this equation, although it is implicit in the optical powers, P_1 and P_2. We recall from Equation 2.17, where we did the fish tank problem, that the apparent thickness of a material is the actual thickness divided by the index of refraction. The expression z_{12}/n_ℓ is the "geometric thickness" of the lens.

We shall return to this equation shortly to develop a particularly simple description of this lens, but for now, one key point to notice is that setting $z_{12} = 0$ leads to the **thin-lens matrix**:

$$\mathcal{L} = \begin{pmatrix} 1 & 0 \\ -\frac{P}{n'} & \frac{n}{n'} \end{pmatrix} \quad \text{(Thin lens)}, \tag{3.25}$$

Powers add in a thin lens.

Eq. 2.48: $f' = \frac{n'}{P_1+P_2}$.

Eq. 2.49: $f = \frac{n}{P_1+P_2}$.

where the power is equal to the sum of the powers of the individual surfaces,

$$P = P_t = P_1 + P_2,$$

in contrast to the power in Equation 3.24, where there was a correction term. Only two indices of refraction appear explicitly in this matrix; n, the index of refraction, of the medium before the lens, and n', that after the lens. Returning to Equations 2.48 and 2.49, the matrix can also be written as

$$\mathcal{L} = \begin{pmatrix} 1 & 0 \\ -\frac{P}{n'} & \frac{n}{n'} \end{pmatrix} = \begin{pmatrix} 1 & 0 \\ -\frac{1}{f'} & \frac{n}{n'} \end{pmatrix} = \begin{pmatrix} 1 & 0 \\ -\frac{n}{n'f} & \frac{n}{n'} \end{pmatrix}. \tag{3.26}$$

If the lens is in air, as is frequently the case, we set $n = n' = 1$, and

Thin lens in air has a simple matrix.

$$\mathcal{L} = \begin{pmatrix} 1 & 0 \\ -P & 1 \end{pmatrix} = \begin{pmatrix} 1 & 0 \\ -\frac{1}{f} & 1 \end{pmatrix} \qquad \text{(Thin lens in air).} \tag{3.27}$$

In this case

$$f = f' = \frac{1}{P}. \tag{3.28}$$

Take-Away

Matrix of Any Lens:

- The matrix of any system can be computed by multiplying alternating refraction and translation matrices.
- The multiplication is done from right to left.
- The matrix of a thin lens is simple; the powers of the two surfaces add.

Example

Consider a germanium lens for use in air with light at a wavelength of $\lambda = 10\,\mu$m and two convex surfaces with radii of curvature of $r_1 = -r_2 = 20$ mm. The index of refraction of germanium at this wavelength is$n = 4$. Write the matrix for this lens assuming it is thin, and assuming a thickness of 6 mm.

From Equation 3.17 we can write the first refraction matrix as

$$\mathcal{R}_1 = \begin{pmatrix} 1 & 0 \\ \frac{1-4}{r_1} & \frac{1}{4} \end{pmatrix} = \begin{pmatrix} 1 & 0 \\ -0.375 & 0.250 \end{pmatrix},$$

and the second as

$$\mathcal{R}_2 = \begin{pmatrix} 1 & 0 \\ \frac{4-1}{-r_1} & \frac{4}{1} \end{pmatrix}. = \begin{pmatrix} 1 & 0 \\ -0.150 & 4 \end{pmatrix}..$$

The Translation matrix is

$$\mathcal{T}_{12} = \begin{pmatrix} 1 & z_{12} \\ 0 & 1 \end{pmatrix}.$$

Performing the matrix multiplication,

$$\mathcal{L} = \mathcal{R}_2\mathcal{T}_{12}\mathcal{R}_1$$

For $z_{12} = 0$

$$\mathcal{L} = \begin{pmatrix} 1 & 0 \\ -0.300 & 1 \end{pmatrix}.$$

and for $z_{12} = 6\,\text{mm}$,

$$\mathcal{L} = \begin{pmatrix} 0.925 & 0.5 \\ -0.289 & 0.925 \end{pmatrix}.$$

3.1.2.2 General Problems

We can extend the cascading of matrices to describe any arbitrary number of surfaces such as those comprising a complicated microscope objective, telescope eyepiece, or modern camera lens. In any case, the general result is

$$\mathbf{V}_{end} = \mathcal{M}_{start:end}\mathbf{V}_{start}$$
$$\begin{pmatrix} x_{end} \\ \alpha_{end} \end{pmatrix} = \begin{pmatrix} m_{11} & m_{12} \\ m_{21} & m_{22} \end{pmatrix} \begin{pmatrix} x_{start} \\ \alpha_{start} \end{pmatrix}. \tag{3.29}$$

It can be shown that the determinant* of any matrix constructed in this way will be

$$\det \mathcal{M} = \frac{n}{n'}, \tag{3.30}$$

where n and n' are the initial and final indices of refraction respectively. The determinant of a two-by-two matrix is the product of the diagonal terms minus the product of the off-diagonal terms,

$$\det \mathcal{M} = m_{11}m_{22} - m_{12}m_{21}.$$

Abbe Sine Invariant

Equation 3.30 results in a very powerful conservation law, sometimes called the **Abbe sine invariant**, the **Helmholtz invariant**, or the **Lagrange invariant**. It states that the angular magnification, m_α, is the inverse of the transverse magnification, m, if the initial and final media are the same, or more generally,

$$n'x'd\alpha' = nxd\alpha. \tag{3.31}$$

This equation tells us that if light from an object of height x is emitted into an angle $\delta\alpha$, then at any other location in an optical system, that light is contained within a height x' and angle $\delta\alpha'$, constrained by this equation. We will make use of this important result in several ways. It will be important when we discuss how optical systems change the light-gathering ability of a system and the field of view in Chapter 4. We will see it again in Chapter 12, where we will see that it simplifies the radiometric analysis of optical systems.

3.1.2.3 Using the Abbe Invariant

We show by example the use of the Abbe sine invariant. This shows that the Abbe sine invariant is often a good starting point for design of a system.

Example

For now, let us consider one simple example of the Abbe sine invariant. Suppose we have an infrared detector with a diameter of $100\,\mu\text{m}$ that collects all the rays of light that are contained within a cone up to 30° from the normal. This detector

*See Section C.5 in Appendix C.

is placed at the receiving end of a telescope having a front lens diameter of 20 cm. What is the maximum field of view of this instrument? One might think initially that we could design the optics to obtain any desired field of view. Equation 3.31 states that the product of detector diameter and the cone angle is 100 μm × 30°. This same product holds throughout the system, and the maximum cone at the telescope lens is contained within the angle

$$FOV_{1/2} = \frac{100 \times 10^{-6}\,\mathrm{m} \times 30^\circ}{20 \times 10^{-2}\,\mathrm{m}} = 0.0150^\circ. \tag{3.32}$$

This result is only approximate, because the matrix optics theory we have developed is only valid for the small-angle approximation. More correctly,

$$\sin\left(FOV_{1/2}\right) = \frac{100 \times 10^{-6}\,\mathrm{m} \times \sin 30^\circ}{20 \times 10^{-2}\,\mathrm{m}},$$
$$FOV_{1/2} = 0.0143^\circ. \tag{3.33}$$

Even if we optimistically assume we could redesign the detector to collect all the light in a hemisphere, the field of view would still only reach 0.0289°.

The important point here is that we obtained this fundamental limit without knowing anything about the intervening optical system. We might try to design an optical system to have a larger field of view. We would then either build the system or analyze its apertures as described in Chapter 4. In either case, we would always find that the design fails to meet our expectations. With more experience, we would use the Abbe invariant as a starting point for an optical design, and judge the quality of the design according to how close it approaches this limit. Any attempt to exceed this field of view will result either in failure to achieve the field of view or in a system which does not use the full aperture of the telescope.

3.1.3 Summary of Matrix Optics Concepts

Here we summarize what we have learned about matrix optics so far. Next, we will begin discussing the interpretation of results to better understand thick and compound lenses.

Take-Away

Generating the Matrix:

- Ray propagation in the small-angle approximation can be calculated using two-element column vectors for the rays and two-by-two matrices to describe basic operations (Equation 3.4).

$$\mathbf{V}_{end} = \mathcal{M}_{start:end}\mathbf{V}_{start}.$$

- A system can be described by alternating pairs of two basic matrices.
- The matrix for translation is given by Equation 3.10.

$$\mathcal{T}_{12} = \begin{pmatrix} 1 & z_{12} \\ 0 & 1 \end{pmatrix}.$$

- The matrix for refraction is given by Equation 3.18.

$$\mathcal{R} = \begin{pmatrix} 1 & 0 \\ -\frac{P}{n'} & \frac{n}{n'} \end{pmatrix}.$$

- These matrices may be cascaded from right to left to make matrices for more complicated systems.
- The matrix for a thin lens is given by Equation 3.25.

$$\mathcal{L} = \begin{pmatrix} 1 & 0 \\ -\frac{P}{n'} & \frac{n}{n'} \end{pmatrix}.$$

- The determinant of any matrix is the product of the transverse and angular magnifications, and is given by n/n'.

Common ABCD Matrices: Tables E.1 and E.2.

Tables E.1 and E.2 in Appendix E show some common ABCD matrices used in matrix optics.

3.2 Interpreting the Results

At this point, we have the basic tools to develop the matrix for any lens system. What remains is to interpret the results. We will make it our goal in this section to find a way to convert the matrix for any lens to the matrix for a thin lens, and in doing so, will define optical power and focal lengths in terms of matrix elements, and introduce the new concept of principal planes.

3.2.1 Principal Planes

We begin with an arbitrary matrix representing some lens system from vertex to vertex,

$$\mathcal{M}_{VV'} = \begin{pmatrix} m_{11} & m_{12} \\ m_{21} & m_{22} \end{pmatrix}. \tag{3.34}$$

E. 3.30: $\det \mathcal{M} = \frac{n}{n'}$.

The only constraint is Equation 3.30, where the determinant must be n/n'. What operations on this matrix would convert it to the form of a thin lens? Do these operations have a physical interpretation? Let us try a simple translation through a distance h before the lens system and another translation, h', after it. We call the plane before the front vertex H, the **front principal plane** and the one after the last vertex H', the **back principal plane** as shown in Figure 3.6. Then

$$\mathcal{M}_{HH'} = \mathcal{T}_{V'H'}\mathcal{M}_{VV'}\mathcal{T}_{HV}. \tag{3.35}$$

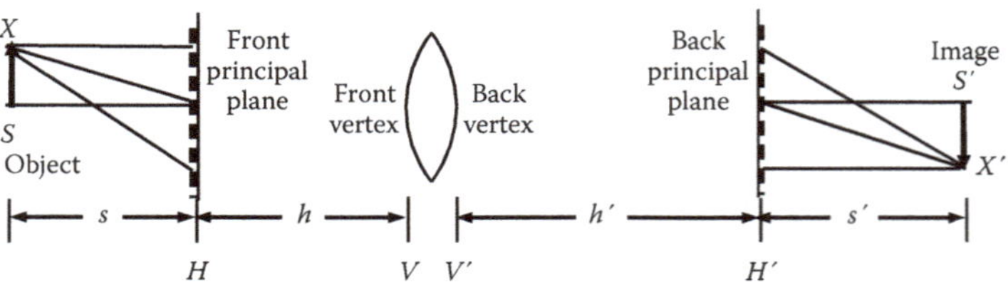

FIGURE 3.6
Principal planes. The front principal plane, H, is in front of the front vertex, V, by a distance h. The distance from the back vertex, V' to the back principal plane, H', is h'. The matrix from H to H' is a thin-lens matrix.

Substituting Equation 3.10 for the two $\mathcal{T}$ matrices, we want to know if there is a solution where $\mathcal{M}_{HH'}$ is the matrix of a thin lens as in Equation 3.25:

$$\begin{pmatrix} 1 & 0 \\ -\frac{P}{n'} & \frac{n}{n'} \end{pmatrix} = \begin{pmatrix} 1 & h' \\ 0 & 1 \end{pmatrix} \begin{pmatrix} m_{11} & m_{12} \\ m_{21} & m_{22} \end{pmatrix} \begin{pmatrix} 1 & h \\ 0 & 1 \end{pmatrix}, \tag{3.36}$$

$$\begin{pmatrix} 1 & 0 \\ -\frac{P}{n'} & \frac{n}{n'} \end{pmatrix} = \begin{pmatrix} m_{11} + m_{21}h' & m_{11}h + m_{12} + m_{21}hh' + m_{22}h' \\ m_{21} & m_{21}h + m_{22} \end{pmatrix}. \tag{3.37}$$

This matrix equation consists of four scalar equations:

$$m_{11} + m_{21}h' = 1, \tag{3.38}$$

$$m_{11}h + m_{12} + m_{21}hh' + m_{22}h' = 0, \tag{3.39}$$

$$m_{21} = -\frac{P}{n'}, \tag{3.40}$$

$$m_{21}h + m_{22} = \frac{n}{n'}, \tag{3.41}$$

with only three unknowns, P, h, and h'. We can solve any three of them, and then check to see that the remaining one is solved, using Equation 3.30. Equation 3.39 looks difficult, so let us solve the other three.

First, from Equation 3.40

$$P = -m_{21}n'. \tag{3.42}$$

Then from Equation 3.41 and Equation 3.38,

$$h = \frac{\frac{n}{n'} - m_{22}}{m_{21}} \qquad h' = \frac{1 - m_{11}}{m_{21}}. \tag{3.43}$$

In many cases, these distances will be negative; the principal planes will be inside the lens. As a result, some authors choose to use negative signs in these equations for h and h', and the reader must be careful to understand the sign conventions. Substituting these results into Equation 3.37 produces Equation 3.25, so we have achieved our goal of transforming the arbitrary matrix to a thin-lens matrix with only two translations.

The concept of principal planes makes our analysis of compound lenses almost as simple as our analysis of the thin lens. The matrix from H to H' is exactly the matrix for a thin lens that has the same power, initial index, and final index as does this thick lens. Thus any result that was true for a thin lens is true for the system represented by $\mathcal{M}_{HH'}$. These planes are therefore of particular importance, because the imaging equation and our simple drawing are correct, provided that we measure the front focal length and object distance from H, and the back focal length and image distance from H'. Using matrix from Equation 3.36,

$$\begin{pmatrix} x_{H'} \\ \alpha_{H'} \end{pmatrix} = \begin{pmatrix} 1 & 0 \\ -\frac{P}{n'} & \frac{n}{n'} \end{pmatrix} \begin{pmatrix} x_H \\ \alpha_H \end{pmatrix}$$

Eq. 3.25: $\mathcal{L} = \begin{pmatrix} 1 & 0 \\ -P & 1 \end{pmatrix} = \begin{pmatrix} 1 & 0 \\ -\frac{1}{f} & 1 \end{pmatrix}.$

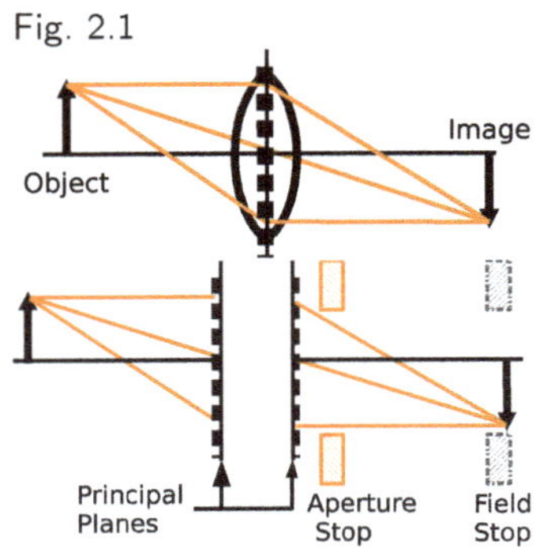

Fig. 2.1

or

$$x_{H'} = x_H,$$

the two principal planes are conjugate (the upper right element of the matrix is zero, so $x_{H'}$ is independent of α_H) and the magnification between them (upper left element) is unity. This result was also true for the thin lens in Equation 3.25. We define the **principal planes** as the pair of conjugate planes having unit magnification. This definition leads to the same result as our analysis of Equation 3.37.

Because the matrix $\mathcal{M}_{HH'}$ is exactly the same as that for a thin lens, we could simply use the thin-lens results we have already developed. As we anticipated in Figure 2.1, we can simply cut the graphical solution of the thin-lens imaging problem through the plane of the lens, and paste the cut edges onto the principal planes we found in Figure 3.6. Nevertheless it is instructive to derive these results using matrix optics, which we will do in the next section. We will then see in Section 3.3 some specific and simple results. For a thick lens, the principal planes can be located quite accurately with a simple rule of thumb. Thus we will be able to treat the thick lens using the same equations as a thin lens with a slight adjustment of the object and image distances. We will follow that in Section 3.4 with examples including compound lenses.

3.2.2 Imaging

Because the matrix from H to H' is a thin-lens matrix, we know that the equations we developed for the thin lens will be valid for any arbitrary lens system provided that we measure distances from the principal planes. Therefore let us consider two more translations: one from the object, S, to the front principal plane, H, and one from the back principal plane, H', to the object, S'. Then, looking back at Figure 3.6, the imaging matrix is

$$\mathcal{M}_{SS'} = \mathcal{M}_{H'S'}\mathcal{M}_{HH'}\mathcal{M}_{SH} = \mathcal{T}_{s'}\mathcal{M}_{HH'}\mathcal{T}_{s}. \tag{3.44}$$

We want to find an equation relating s and s' such that this matrix describes an imaging system, but what are the constraints on an imaging matrix? To answer this question, we return to our definition of the matrix in Equation 3.4, and our understanding of image formation in Section 1.8. Any ray that passes through the object point, X, must pass through the image point, X', regardless of the direction of the ray, α,

$$x' = (? \times x) + (0 \times \alpha)\,.$$

We recall from Chapter 2 that planes related in this way are called conjugate planes. From this equation, we see that a general imaging matrix is constrained by

$$\mathcal{M}_{SS'} = \begin{pmatrix} m_{11} & 0 \\ m_{21} & m_{22} \end{pmatrix}. \tag{3.45}$$

From Equation 3.44 this matrix from S to S' is

$$\mathcal{M}_{SS'} = \begin{pmatrix} 1 & s' \\ 0 & 1 \end{pmatrix}\begin{pmatrix} 1 & 0 \\ -\frac{P}{n'} & \frac{n}{n'} \end{pmatrix}\begin{pmatrix} 1 & s \\ 0 & 1 \end{pmatrix},$$

$$\mathcal{M}_{SS'} = \begin{pmatrix} 1 - \frac{s'P}{n'} & s - \frac{ss'P}{n'} + \frac{s'n}{n'} \\ -\frac{P}{n'} & -\frac{sP}{n'} + \frac{n}{n'} \end{pmatrix}. \tag{3.46}$$

Then Equation 3.45 requires that

$$s - \frac{ss'P}{n'} + \frac{s'n}{n'} = 0,$$

leading to the most general form of the lens equation,

$$\frac{n}{s} + \frac{n'}{s'} = P. \tag{3.47}$$

The form of this equation is exactly the same as Equation 2.47 for a thin lens. However we have not assumed a thin lens here, and this equation is valid for any imaging system, provided that we remember to measure f, f', s, and s' from the principal planes.

As we did in Chapter 2, let us find the magnification. According to the matrix result, the transverse magnification is $m = m_{11}$, or

$$m = 1 - \frac{s'P}{n'} = 1 - \frac{s'}{n'}\left(\frac{n}{s} + \frac{n'}{s'}\right) = -\frac{ns'}{n's}. \tag{3.48}$$

The angular magnification is m_{22} or

$$m_\alpha = -\frac{s}{n'}\left(\frac{n}{s} + \frac{n'}{s'}\right) + \frac{n}{n'} = -\frac{s}{s'}. \tag{3.49}$$

The product of the transverse and angular magnifications is n'/n as required by the determinant condition (Equation 3.30). We can write the imaging matrix as

$$\mathcal{M}_{SS'} = \begin{pmatrix} m & 0 \\ -\frac{P}{n'} & \frac{n'}{n}\frac{1}{m} \end{pmatrix}. \tag{3.50}$$

This important result relates the angular and transverse magnification in an image:

$$m_\alpha = \frac{n'}{nm} \tag{3.51}$$

or, in air

$$m_\alpha = \frac{1}{m} \qquad (n' = n). \tag{3.52}$$

As we did with the thin lens, we can let the object distance approach infinity, and define the resulting image distance as the back focal length,

$$f' = \frac{n'}{P} \tag{3.53}$$

and, alternatively make the image distance infinite to obtain the front focal length,

$$f = \frac{n}{P}. \tag{3.54}$$

The distances s and s' are always measured from their respective principal planes. These definitions of focal length are special cases of object and image distances, so they are also measured from the principal planes.

Instead of treating each surface in the imaging system separately, we can now simply find the matrix of the system from vertex V to vertex V', using Equations 3.42, 3.53, and 3.54

$$P = -m_{21}n',$$

$$f = \frac{n}{P} = -\frac{n}{m_{21}n'} \qquad f' = \frac{n'}{P} = -\frac{1}{m_{21}} \tag{3.55}$$

to find the power and focal lengths, and Equations 3.43,

$$h = \frac{\frac{n}{n'} - m_{22}}{m_{21}} \qquad h' = \frac{1 - m_{11}}{m_{21}},$$

to locate the principal planes. Then, as we concluded in the discussion of principal planes above, we can draw the simple ray-tracing diagram on the left side of Figure 2.1, cut it along the line representing the thin lens, and place the two pieces on their respective principal planes, as shown in the right side of that figure. We must remember that this is just a mathematical construction. The rays of light actually bend at the individual lens surfaces. Furthermore, we must not forget the locations of the actual optical components; in some cases for example, a positive object distance measured from the front principal plane could require placing the object in a location that is physically impossible with the particular lens system. Because h and h' are often negative, this situation can result quite easily.

3.2.3 Summary of Principal Planes and Interpretation

In summary, we can use all the equations from our thin-lens theory with any thick or compound lens, provided that we make all our measurements from the appropriate principal planes. Specifically, we emphasize the following points.

Take-Away

Using the Matrix:

- The front and back principal planes are the planes that are conjugate to each other with unit magnification. This definition is unique; there is only one such pair of planes for a system.
- The focal lengths of the lens can be found by using Equation 3.55,

$$f = -\frac{n}{m_{21}n'} \qquad f' = -\frac{1}{m_{21}}.$$

- The locations of the principal planes can be found from the system's matrix, using Equation 3.43.

$$h = \frac{\frac{n}{n'} - m_{22}}{m_{21}} \qquad h' = \frac{1 - m_{11}}{m_{21}}..$$

- All the imaging results, specifically the lens equation (Equation 3.47),

$$\frac{n}{s} + \frac{n'}{s'} = P = \frac{n}{f} = \frac{n'}{f'},$$

and the magnification equations, (3.48 and 3.51),

$$m = -\frac{ns'}{n's} \qquad m_\alpha = \frac{n'}{nm},$$

hold provided we measure f and s from the front principal plane and f', and s' from the back principal plane.

3.3 The Thick Lens Again

The simple thick lens was discussed in Section 2.5, in terms of imaging, and again in Section 3.1.2 in terms of matrices, ending with Equation 3.24. The simple lens is frequently used by itself and more complicated **compound lenses** are built from multiple simple lenses. There are also some particularly simple approximations that can be developed for this lens. For these reasons, it is worthwhile to spend some time examining the thick lens in some detail. We return to Equation 3.24,

$$\mathcal{L} = \begin{pmatrix} 1 & 0 \\ -\frac{P_t}{n'} & \frac{n}{n'} \end{pmatrix} + \frac{z_{12}}{n_\ell} \begin{pmatrix} -P_1 & n \\ \frac{P_1 P_2}{n'} & -P_2 \frac{n}{n'} \end{pmatrix},$$

which was the result just before we made the thin-lens approximation. From that equation, we use Equation 3.42,

$$P = -m_{21} n',$$

to compute the power and focal lengths,

$$P = P_1 + P_2 - \frac{z_{12}}{n_\ell} P_1 P_2, \tag{3.56}$$

$$f = \frac{n}{P} \qquad f' = \frac{n'}{P},$$

and Equations 3.43 and 3.50 to locate the principal planes,

$$h = -\frac{n'}{n_\ell} \frac{P_2}{P} z_{12} \qquad h' = -\frac{n}{n_\ell} \frac{P_1}{P} z_{12}. \tag{3.57}$$

It is often useful to design a lens of power P such that both P_1 and P_2 have the same sign as P, although there are many exceptions. The front principal plane distance, h, varies linearly with the geometric thickness z_{12}/n_ℓ, and with the power of the back surface, P_2, while the back principal plane distance, h', varies with z_{12}/n_ℓ and the power of the first surface. Thus, if one surface has zero power, then the opposite principal plane is located at the vertex:

For a plano-convex or plano-concave lens one of the principal planes is at the curved surface.

$$h = 0 \quad \text{if} \quad P_2 = 0,$$

and

$$h' = 0 \quad \text{if} \quad P_1 = 0. \tag{3.58}$$

For the special case of the thick lens in air,

$$f = f' = \frac{1}{P}$$

and

$$h = -\frac{1}{n_\ell}\frac{P_2}{P}z_{12} \qquad h' = -\frac{1}{n_\ell}\frac{P_1}{P}z_{12}, \tag{3.59}$$

Fig. 3.6

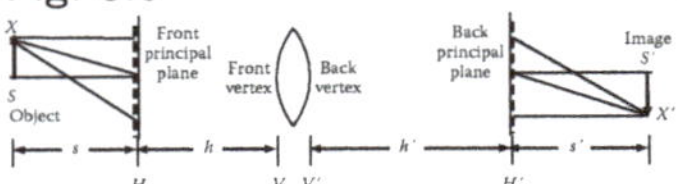

so the spacing of the principal planes (see Figure 3.6) is

$$z_{HH'} = z_{12} + h + h' = z_{12}\left(1 - \frac{P_2 + P_1}{n_\ell P}\right). \tag{3.60}$$

Unless the lens is very thick, we will see that to a good approximation,

$$P \approx P_1 + P_2$$

and

$$z_{HH'} = z_{12} + h + h' \approx z_{12}\left(1 - \frac{1}{n_\ell}\right). \tag{3.61}$$

For a glass lens, $n_\ell \approx 1.5$,

$$z_{HH'} = \frac{z_{12}}{3} \qquad \text{(Simple glass lens in air).} \tag{3.62}$$

We refer to this equation as the "thirds rule." The spacing of the principal planes is about one-third of the spacing of the vertices, provided that the lens is made of glass, and used in air. Other fractions can be determined for other materials. For germanium lenses in air the fraction is 3/4. We will see that this rule is useful not only for biconvex or biconcave lenses, but for most simple lenses that we will encounter.

In Practice

When we study aberrations in Chapter 5, we will see that optimal choices of radii of curvature may be complicated. If we choose to use the optimal radii for every lens in our system, our work will be slow and expensive. The most common simple lenses are biconvex and plano-convex (or biconcave and plano-concave for negative focal lengths), and these are available "off the shelf" in many diameters for many focal lengths, so we frequently try to design with these. For glass lenses in air, the spacing of the principal planes is about one third of the lens thickness. If one surface is plane, one of the principal planes is at the curved surface. If the lens is symmetric (same radii of curvature on both sides), then the principal planes are also symmetric about the lens center.

Example

We want to magnify an object about 10 times with a biconvex glass lens having a focal length of 10 mm onto the chip of a CCD camera. Where do we put the object and chip?

First, we find the object and image distances, s and $s' = 10s$. We note that the image will be inverted.

$$\frac{1}{s} + \frac{1}{10s} = \frac{1}{f} \qquad \frac{1.1}{s} = \frac{1}{f},$$

so

$$s = 1.1f = 11.0\,\text{mm} \qquad s' = 110\,\text{mm}$$

In a catalog, we find a biconvex lens with this focal length and a thickness of 2.4 mm.

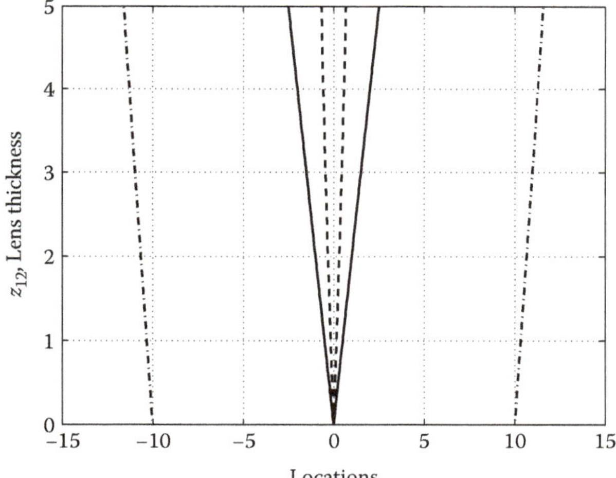

FIGURE 3.7
Principal planes. The principal planes (dashed lines) of a biconvex lens are separated by about one-third of the vertex (solid lines) spacing. The focal length (dash-dotted lines) varies only slightly as the vertex spacing varies from zero to half the focal length, $f = 10\,\text{cm}$.

In Chapter 5, we will learn that we want to put the convex side toward the object because $s << s'$. The front principal plane is at the curved vertex so we place the object,

$$w = s = 11.0\,\text{mm}$$

in front of that vertex. Note that w is the "working distance" from the first vertex whereas s is the object distance from the principal plane, but in this case they are the same.

The thirds rule tells us that the back principal plane is $2.4\,\text{mm}/3$ behind that vertex, so we subtract $2.4\,\text{mm} \times 2/3$ from s' and place the camera chip at

$$w' = 110\,\text{mm} - 2.4\,\text{mm} \times 2/3 = 108.4\,\text{mm}$$

from the flat surface. The working distance to the chip, w' is slightly shorter than the image distance, s'.

This isn't exactly correct; the thirds rule is only an approximation, and there will be tolerances on our measurement ability and on the focal length itself. However, it will be close, and we will want to have some focal adjustment to account for errors. If we build the mechanical setup for these distances with a small amount of adjustment, we can focus to the exact distance by viewing the image for best focus.

Figure 3.7 illustrates some results for a biconvex glass lens in air such that $P_1 + P_2 = 10$ diopters, corresponding to a focal length of $f = 10\,\text{cm}$. Note that we do not distinguish front and back focal length because the lens is in air and the two focal lengths are equal. For the biconvex lens, each surface has the same power, given by Equation 2.33. The actual power of the lens, given by Equation 3.56, varies by less than 10% as the thickness of the lens varies from zero to half the nominal focal length, or 5 cm. The thickness of the lens is plotted on the vertical and the locations of the vertices are plotted as solid lines on the horizontal axis, assuming that zero corresponds to the center of the lens. The dashed lines show the locations of the two principal planes, which appear to follow the thirds rule of Equation 3.62 very well. The locations of the front and back focal planes are shown by the dash-dotted lines. Note that if we were to adjust the optical powers of the two surfaces

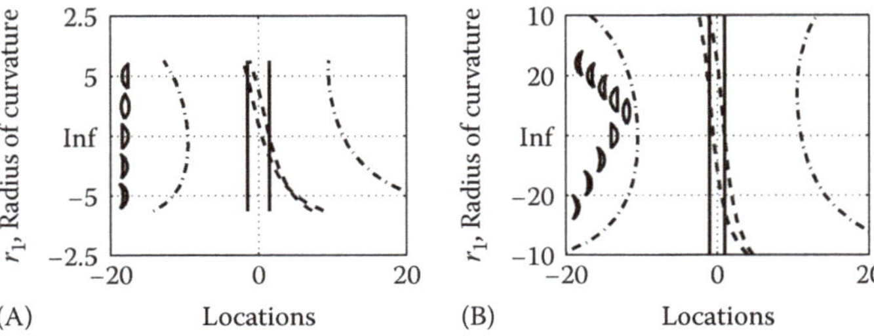

FIGURE 3.8
Bending the lens. The locations of the principal planes move out of the lens with extreme bending. Such bending is not often used in glass (A), but is quite common in germanium (B) in the infrared.

to keep the focal length constant at 10 cm, these lines would both be a distance f from their respective principal planes.

Next, we demonstrate "bending the lens" by varying the radii of curvature in Figure 3.8, while maintaining a constant value of $P_1 + P_2 = 10$ diopters. We vary the radius of curvature of the first surface, and compute P_1 using Equation 2.33. We then compute P_2 to maintain the desired sum. The focal length of the lens will vary because keeping the sum constant is not quite the same as keeping the focal length constant. The vertical axis is the radius of curvature of the first surface. Using Equation 2.33, a radius of curvature of 2.5 cm results in a power of 5 diopters. To obtain $P_1 + P_2 = 10$ diopters, the second surface has the same power, and the lens is biconvex. The lens shape is shown along the vertical axis.

3.3.1 Summary of the Thick Lens

The thick lens, consisting of a single element with two surfaces, is important for its simplicity and low cost. Often we can design a system with thin lenses, select real lenses from a catalog, make small corrections for the thicknesses of the lenses using the concept of principal planes, and find that our design will work well with only simple minor adjustments. The following points are important to remember.

Take-Away

Thick Lens:

- For a thick lens the optical power is approximately the sum of the powers of the individual surfaces, provided the lens is not extremely thick .
- The spacing of the principal planes of a glass lens is approximately one-third of the thickness of the lens.
- For a symmetric biconvex or biconcave lens, the principal planes are symmetrically placed around the center of the lens.
- For a plano-convex or plano-concave lens, one principal plane is at the curved surface.
- More exact results can always be obtained using the calculations in Section 3.1 to generate the matrix and the interpretation in Section 3.2. These should be used if the lens is very thick.

3.4 Compound Lenses

Next, we extend our discussion of matrix optics to compound lenses, consisting of more than two surfaces. Our matrix calculations are very general and allow for such configurations as a cemented doublet, triplet with two or three glass elements, possibly cemented together. Such lenses offer correction for aberrations as we will discuss in Chapter 5. These calculations also allow for the complicated combinations of elements found in a microscope objective, or for combinations such as the primary and secondary of a telescope, or the objective, tube lens, and eyepiece of a modern microscope.

3.4.1 Two Thin Lenses

We begin our examples with a compound lens consisting of two thin lenses in air. After solving the matrix problem, we will then extend the analysis to the case where the two lenses are not thin. The matrix from the front vertex of the first lens to the back vertex of the second is

$$\mathcal{M}_{V_1,V_2'} = \mathcal{L}_{V_2,V_2'}\mathcal{T}_{V_1',V_2}\mathcal{L}_{V_1,V_1'} \qquad \text{(Thin lenses)} \tag{3.63}$$

$$\mathcal{M}_{V_1,V_2'} = \begin{pmatrix} 1 & 0 \\ -\frac{1}{f_2} & 1 \end{pmatrix}\begin{pmatrix} 1 & z_{12} \\ 0 & 1 \end{pmatrix}\begin{pmatrix} 1 & 0 \\ -\frac{1}{f_1} & 1 \end{pmatrix}, \tag{3.64}$$

where the two lens matrices are from Equation 3.26 and the translation is from Equation 3.10. At this point, we choose to keep the equations general, rather than evaluate with specific numbers. Performing the matrix multiplication,

$$\mathcal{M}_{V_1,V_2'} = \begin{pmatrix} 1 & 0 \\ -\frac{1}{f_2} & 1 \end{pmatrix}\begin{pmatrix} 1-\frac{z_{12}}{f_1} & z_{12} \\ -\frac{1}{f_1} & 1 \end{pmatrix},$$

$$\mathcal{M}_{V_1,V_2'} = \begin{pmatrix} 1-\frac{z_{12}}{f_1} & z_{12} \\ -\frac{1}{f_2}+\frac{z_{12}}{f_1 f_2}-\frac{1}{f_1} & 1-\frac{z_{12}}{f_2} \end{pmatrix}. \tag{3.65}$$

We first determine the focal length of the lens, $f = 1/P = f'$, using Equation 3.27:

$$\frac{1}{f} = \frac{1}{f_2} + \frac{1}{f_1} - \frac{z_{12}}{f_1 f_2}. \tag{3.66}$$

Notice the similarity of this equation to the focal length of a thick lens in Equation 3.56. For small separations, $P = P_1 + P_2$ or $1/f = 1/f_1 + 1/f_2$. For example, two lenses with focal lengths of 50 cm can be placed close together to approximate a compound lens with a focal length of 25 cm.

Eq. 3.56:
$P = P_1 + P_2 - \frac{z_{12}}{n_\ell} P_1 P_2$,
$f = \frac{n}{P} \qquad f' = \frac{n'}{P}$.

In Practice

We can substitute a pair of lenses of focal length f_1 and f_2 closely spaced, for a single lens of focal length,

$$\frac{1}{f} = \frac{1}{f_1} + \frac{1}{f_2}. \tag{3.67}$$

We might choose to do this if we don't have the right lens available. One disadvantage is the potential loss of light through an extra lens, and another is the extra challenge of aligning the two. We will see in Chapter 5 that the aberrations may become better or worse, depending on the lenses chosen.

However, if the spacing is increased, this result no longer holds. In particular, if

$$z_{12} = f_1 + f_2 \tag{3.68}$$

then

$$\frac{1}{f} = \frac{f_1}{f_1 f_2} + \frac{f_2}{f_1 f_2} - \frac{z_{12}}{f_1 f_2} = 0. \tag{3.69}$$

An optical system with

$$m_{21} = 0, \qquad \frac{1}{f} = 0, \qquad \text{or} \qquad f \to \infty \qquad \text{(Afocal)} \tag{3.70}$$

Afocal: Note that $h \to \infty$ and $h' \to \infty$. We will discuss this later.

is said to be **afocal**. We will discuss this special but important case shortly.

Next, we turn to the principal planes, using Equation 3.43. With the compound lens being in air in Equation 3.65,

Eq. 3.43:
$h = \frac{\frac{n}{n'} - m_{22}}{m_{21}}$
$h' = \frac{1 - m_{11}}{m_{21}}$.

Eq. 3.65:
$\mathcal{M}_{V_1,V_2'} = \begin{pmatrix} 1 - \frac{z_{12}}{f_1} & z_{12} \\ -\frac{1}{f_2} + \frac{z_{12}}{f_1 f_2} - \frac{1}{f_1} & 1 - \frac{z_{12}}{f_2} \end{pmatrix}$.

$$h = \frac{\frac{z_{12}}{f_2}}{-\frac{1}{f_2} + \frac{z_{12}}{f_1 f_2} - \frac{1}{f_1}} = \frac{z_{12} f_1}{z_{12} - f_1 - f_2},$$

$$h' = \frac{\frac{z_{12}}{f_1}}{-\frac{1}{f_2} + \frac{z_{12}}{f_1 f_2} - \frac{1}{f_1}} = \frac{z_{12} f_2}{z_{12} - f_1 - f_2}. \tag{3.71}$$

If the spacing z_{12} approaches zero, then h and h' also approach zero. If both focal lengths are positive and the spacing is smaller than their sum, then h and h' are negative, and the principal planes are inside the compound lens. If the spacing exceeds the sum of the focal lengths, then h and h' are positive and the principal planes are outside the lens.

3.4.2 2× Magnifier with Compound Lens

As a specific example, consider the goal of building a lens to magnify an object by a factor of two, creating a real inverted image, with an object distance of $s = 100$ mm and an image distance of $s' = 200$ mm. One could simply use a lens of focal length f such that $1/f = 1/s + 1/s'$, or $f = 66.67$ mm. However, as we will see in Chapter 5, a simple lens may suffer from aberrations that prevent the formation of a good image. We could hire a lens designer, and then work with manufacturers to produce the best possible lens. That course might be justified if we are going to need large quantities, but the process is expensive and slow for a few units. On the other hand, lenses are readily available "off-the-shelf" for use with an object or image distance of infinity. Suppose we choose to use two lenses (maybe camera lenses), with focal lengths of 100 and 200 mm, designed for objects at infinity. We reverse the first lens, so that the object is at the front focal point, and we place the lenses close together. For this illustration, we will we choose lenses from a catalog (Thorlabs parts LA1509 and LA1708[225]) using BK7 glass with an index of refraction of 1.515 at 633 nm, appropriate for the red object. Specifications for the lenses from the catalog are shown in Table 3.1. We start with a thin-lens approximation as in Figure 3.9. With this starting point, we will analyze the magnifier using thick lenses and the thirds rule.

The layout of the compound lens is shown in Figure 3.10. From Equation 3.66, we find

$$\frac{1}{f} = \frac{1}{100\,\text{mm}} + \frac{1}{200\,\text{mm}} - \frac{20\,\text{mm}}{100\,\text{mm} \times 200\,\text{mm}} \qquad f = 71.43\,\text{mm}. \tag{3.72}$$

From Equation 3.71,

$$h = \frac{20\,\text{mm} \times 100\,\text{mm}}{20\,\text{mm} - 100\,\text{mm} - 200\,\text{mm}} = -7.14\,\text{mm},$$

$$h' = \frac{20\,\text{mm} \times 200\,\text{mm}}{20\,\text{mm} - 100\,\text{mm} - 200\,\text{mm}} = -14.28\,\text{mm}. \tag{3.73}$$

TABLE 3.1
Two Lenses for 2× Magnifier

Parameter	Label	Value
First lens focal length	f_1	100 mm
First lens front radius (LA1509 reversed)	r_1	Infinite
First lens thickness	$z_{v1,v1'}$	3.6 mm
First lens back radius	r_1'	51.5 mm
First lens "back" focal length	$f_1 + h_1$	97.6 mm
Lens spacing	$z_{v1',v2}$	20 mm
Second lens focal length	f_2	200 mm
Second lens front radius (LA1708)	r_2	103.0 mm
Second lens thickness	$z_{v2,v2'}$	2.8 mm
Second lens back radius	r_2'	Infinite
Second lens back focal length	$f_2' + h_2'$	198.2 mm

The numbers pertaining to the individual lenses are from the vendor's catalog. The spacing has been chosen somewhat arbitrarily to allow sufficient space for lens mounts. The "back" focal length of the first lens from the catalog will be the front focal length for our design, because we are reversing the lens. The back focal length is different here than in our definition. Here it is the distance from the back vertex to the back focal plane, while in our definition it is the distance from the back principal plane to the back focal plane.

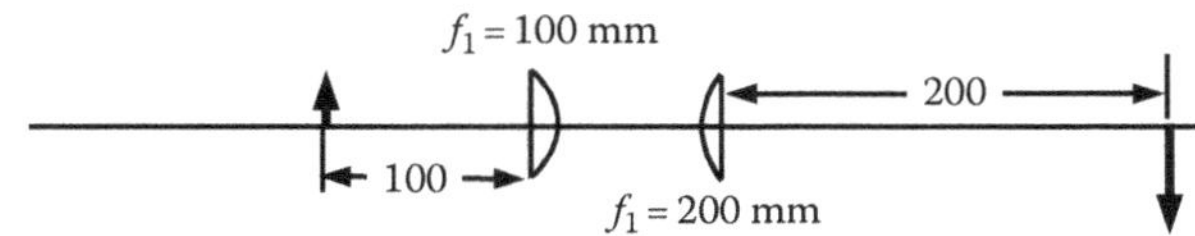

FIGURE 3.9
Thin-lens beam expander. We will use this as a starting point. A better analysis will use thick lenses with the thirds rule. Distances are in millimeters.

Thus the principal planes are both "inside" the compound lens.

To make our magnifier, we want to solve

$$m = -\frac{s'}{s} = -2 \qquad s' = 2s,$$

and

$$\frac{1}{f} = \frac{1}{s} + \frac{1}{s'} = \frac{1}{s} + \frac{1}{2s},$$

with the result

$$s = \frac{3f}{2} = 107.1\,\text{mm} \qquad s' = 3f = 214.3\,\text{mm}. \tag{3.74}$$

To this point, we have still considered the lenses to be thin lenses. Let us now return to Table 3.1, and consider their thicknesses. In constructing the compound lens, distances must be measured from the appropriate principal planes of the individual lenses. In Equation 3.63, we now understand that the matrices are related to principal planes, so the equation reads,

$$\mathcal{M}_{H_1,H_2'} = \mathcal{L}_{H_2,H_2'}\mathcal{T}_{H_1',H_2}\mathcal{L}_{H_1,H_1'}, \tag{3.75}$$

$$\mathcal{M}_{H_1,H_2'} = \begin{pmatrix} 1 & 0 \\ -\frac{1}{f_2} & 1 \end{pmatrix} \begin{pmatrix} 1 & z_{12} \\ 0 & 1 \end{pmatrix} \begin{pmatrix} 1 & 0 \\ -\frac{1}{f_1} & 1 \end{pmatrix}. \tag{3.76}$$

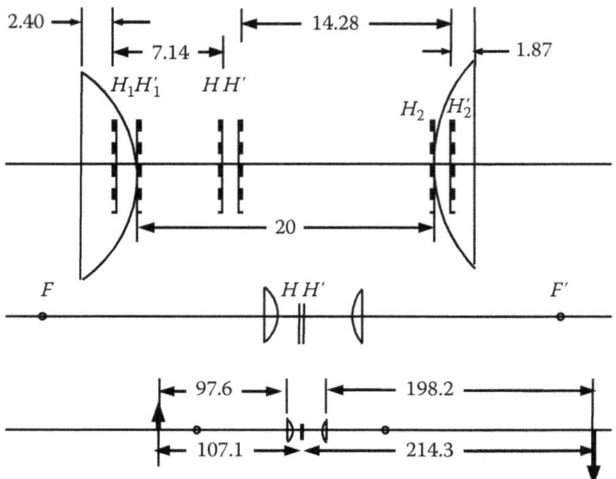

FIGURE 3.10
2× magnifier. The top part shows the principal planes of the individual simple lenses and of the compound lens. The center shows the front and back focal planes, and the lower part shows the object and image distances. Lens diameters are not to scale. All distances are in mm.

and z_{12} is measured between the back principal plane of the first lens and the front principal plane of the second.

Because one of the principal planes of a plano-convex lens is at the convex side, the spacing of the lenses is unchanged:

$$z_{h_1'h_2} = z_{v_1'v_2} = 20\,\text{mm}.$$

However, the distance h to the front principal plane of the compound lens must be measured from the front principal plane of the first lens. Using the thirds rule, the principal planes are spaced by one-third the thickness, so this plane is two-thirds of the thickness inward from the first vertex:

$$h_1 = -\frac{2z_{v_1v_1'}}{3} = -2.4\,\text{mm}. \tag{3.77}$$

Comparing to the table, the front focal length is

$$w_1 = 100\,\text{mm} - 2.4\,\text{mm} = 97.6\,\text{mm},$$

as expected. Likewise, the back principal plane is located relative to the back vertex of the second lens by

$$h_2' = -\frac{2z_{v_2v_2'}}{3} = -1.87\,\text{mm}, \tag{3.78}$$

and the back focal length is 198.1 mm compared to the more exact 198.2 mm in the catalog. The small error results from the assumptions that $n = 1.5$ within the derivation of the thirds rule, and that the optical power of the lens is the sum of the power of its two surfaces.

Now the distance h to the front principal plane of the compound lens must be measured from the front principal plane of the first lens. The distance to this plane from the vertex is (using Equation 3.71)

$$z_{H,V1} = z_{H1,V1} + z_{H,H1} = h_1 + h =$$
$$-2.4\,\text{mm} - 7.14\,\text{mm} = -9.54\,\text{mm}, \tag{3.79}$$

and likewise

$$z_{V2',H'} = z_{V2',H2'} + z_{H2',H'} = h_2' + h' = \\ -1.87\,\text{mm} - 14.28\,\text{mm} = -16.15\,\text{mm}. \tag{3.80}$$

Finally, the object is placed using Equation 3.74

$$w = s + h + h_1 = 107.1\,\text{mm} - 9.54\,\text{mm} = 97.6\,\text{mm} \tag{3.81}$$

in front of the first vertex, and the image is

$$w' = s' + h' + h_2' = 214.3\,\text{mm} - 16.15\,\text{mm} = 198.2\,\text{mm} \tag{3.82}$$

after the back vertex of the second lens.

For this particular case, with the intermediate image at infinity, the object is at the front focal point of the first lens, and the image is at the back focal point of the second lens. However, our approach to the analysis is perfectly general.

Take-Away

Compound Lenses:

- The principal planes of a compound lens are not easily estimated, and are best determined by the matrix calculations in Section 3.1 to generate the matrix and the interpretation in Section 3.2.
- It is possible to estimate the image location by stepping through each lens of the system using the image from the first lens as the object for the second, and so forth until the final image. Comparing this to the result of the full matrix calculation is a good way to minimize the possibility of computational errors.
- In the case of a purchased compound lens, such as a microscope objective or projector lens, the manufacturer may keep the details of the design proprietary. In such cases, the focal planes can be found by autocollimation, and one more imaging experiment is needed to determine the principal planes.

3.4.2.1 Principal Planes by Experiment

Often we purchase expensive and complicated lenses when image quality requirements are very high. Frequently the design of these lenses is proprietary and we cannot learn the properties of the various components to put into our matrix equations. Yet, when we want to incorporate such lenses in systems we design, we may need to know the locations of the principal planes to determine the location of other components. Looking at the system in Figure 3.11, we have unknown some compound lens in the center of the figure. We can define two vertices, which might be the front and back edges of the housing holding the lens.

Assuming the lens is used in air the front and back focal lengths are equal, $f = f'$. Thus we have three unknowns, f, h, and h', so we need to perform three experiments. Let's find the front and back focal planes, F and F' using autocollimation, as shown in Figure 2.20. We now have the distances, u from F to V and u' from V' to F'. Next, we do one imaging experiment with a finite object distance. Let's measure the distances, w from S to V and w' from V; to S. With the four distances, u, u', w, and w, we have four equations but we've added two unknowns, s and s' to f, h, and h'. Our fifth equation is the lens equation for lenses in air. After some mathematical manipulation, we can write this equation with the variables in the figure as 2.51.

Fig. 2.20

Eq. 2.51:
$\frac{1}{s} + \frac{1}{s'} = \frac{1}{f}$.

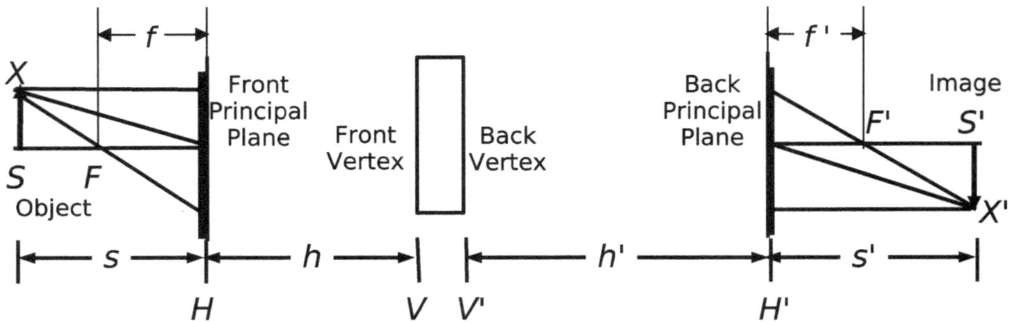

FIGURE 3.11
Distances from principal planes. The principal planes can be located by finding the focal planes and doing one imaging experiment.

$$(w-u)(w'-u') = f^2. \tag{3.83}$$

This is often called the Newtonian form of the lens equation, which was derived by Isaac Newton in the early 1700s[165]. Some find this equation easier than the one derived by Gauss over a century later[74], although now most authors prefer the latter in most cases. Newton's equation provides us an easy solution.

$$f = \sqrt{(w-u)(w'-u')},$$

and knowing the locations of the focal planes, we can immediately calculate the principal plane locations;

$$h = u - f \qquad h' = u' - f' = u' - f.$$

The problem is somewhat more complicated if the focal planes are inside the housing of the lens. In this case we need an auxiliary lens to image the focal plane.

3.4.3 Global Coordinates

The reader will probably notice that even this relatively simple problem, with only two lenses, requires us to pay attention to a large number of distances. We have kept track of these by careful use of a notation involving subscripts and primes. It is often helpful to define a **global coordinate system** so that we can make our drawings more easily. To do so, we simply choose one point as $z = 0$, and then determine the positions of all relevant planes from there. For example, if the first vertex of the first lens is chosen for $z = 0$, then the location of the front principal plane of the first lens is $-h_1$. A good notation is to use the letter z, followed by a capital letter for the plane:

$$zH1 = -h_1. \tag{3.84}$$

To consider one more example, the location of the front principal plane of the system in this global coordinate system is

$$zH = zH1 - h = h_1 - h. \tag{3.85}$$

In the example in Section 3.4.2, if zero is chosen so that the origin of coordinates is at the first vertex, $zV1 = 0$, then the object is at

$$zS = -97.6\,\text{mm}, \tag{3.86}$$

and the front vertex of the second lens should be placed at

$$zV2 = 3.6\,\text{mm} + 20\,\text{mm} = 23.6\,\text{mm}. \qquad (3.87)$$

Likewise, all the other points can be found in the global coordinates by adding appropriate distances.

3.4.4 Telescopes

In the example of the compound lens in Section 3.4.1, we noted one special case which is of great practical interest. In Equation 3.66,

$$\frac{1}{f} = \frac{1}{f_2} + \frac{1}{f_1} - \frac{z_{12}}{f_1 f_2},$$

if the separation of the two lenses (measured from the back principal plane of the first to the front principal plane of the second) is equal to the sum of their focal lengths, then the optical power is zero, and the compound lens is said to be afocal. A telescope is one example of an afocal system. The equations we have developed involving the lens equation and principal planes fail to be useful at this point, because $m_{21} = 0$ in Equation 3.57. We can analyze an afocal system by letting the spacing asymptotically approach the sum of the focal lengths. However, instead, we make use of the matrix methods in a different way. Returning to Equation 3.65, and setting $z_{12} = f_1 + f_2$,

Eq. 3.66: $\frac{1}{f} = \frac{1}{f_2} + \frac{1}{f_1} - \frac{z_{12}}{f_1 f_2}$.

Eq. 3.57: $h = -\frac{n'}{n_\ell}\frac{P_2}{P}z_{12}$
$h' = -\frac{n}{n_\ell}\frac{P_1}{P}z_{12}$.

$$\mathcal{M}_{V_1,V_2'} = \begin{pmatrix} 1 - \frac{f_1+f_2}{f_1} & f_1 + f_2 \\ -\frac{1}{f_2} + \frac{f_1+f_2}{f_1 f_2} - \frac{1}{f_1} & 1 - \frac{f_1+f_2}{f_2} \end{pmatrix} = \begin{pmatrix} -\frac{f_2}{f_1} & f_1 + f_2 \\ 0 & -\frac{f_1}{f_2} \end{pmatrix}. \qquad (3.88)$$

Eq. 3.45: $\mathcal{M}_{SS'} = \begin{pmatrix} m & 0 \\ -\frac{P}{n'} & \frac{n'}{n}\frac{1}{m} \end{pmatrix}$.

Now, following the approach we used in Equation 3.45, the imaging matrix is

$$\mathcal{M}_{SS'} = \begin{pmatrix} 1 & s' \\ 0 & 1 \end{pmatrix} \begin{pmatrix} -\frac{f_2}{f_1} & f_1 + f_2 \\ 0 & -\frac{f_1}{f_2} \end{pmatrix} \begin{pmatrix} 1 & s \\ 0 & 1 \end{pmatrix} = \begin{pmatrix} ? & \mathbf{0} \\ ? & ? \end{pmatrix}. \qquad (3.89)$$

Once again, the requirement that the upper right matrix element be zero (Equation 3.45) indicates that this matrix describes the optical path between conjugate planes. Completing the multiplication,

$$\mathcal{M}_{SS'} = \begin{pmatrix} -\frac{f_2}{f_1} & -s\frac{f_2}{f_1} + f_1 + f_2 - s'\frac{f_1}{f_2} \\ 0 & -\frac{f_1}{f_2} \end{pmatrix} = \begin{pmatrix} ? & 0 \\ ? & ? \end{pmatrix}, \qquad (3.90)$$

with the result that

$$-s\frac{f_2}{f_1} + f_1 + f_2 - s'\frac{f_1}{f_2} = 0.$$

Simplifying by noting that the upper left element of the matrix is the magnification,

Afocal lens pair: $m = -f_2/f_1$ independent of s.

$$m = -f_2/f_1, \qquad \text{(Afocal)} \qquad (3.91)$$

we have

$$ms + f_1 + f_2 + s'/m = 0$$

and the imaging equation is

$$s' = -m^2 s - f_1 m\,(1 + m)\,, \qquad (3.92)$$

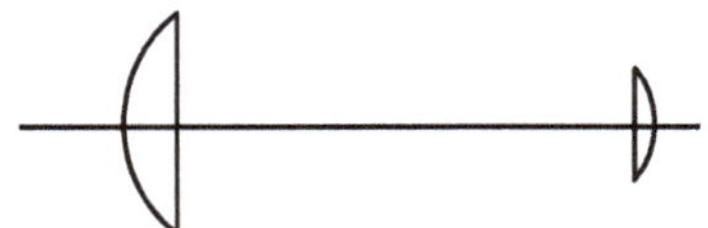

FIGURE 3.12
Telescope. Astronomical telescopes are usually used for large angular magnification. Because $m_\alpha = 1/m$, the magnification, $m = -f_2/f_1$, is small.

The matrix is

$$\mathcal{M}_{SS'} = \begin{pmatrix} m & 0 \\ 0 & \frac{1}{m} \end{pmatrix}. \tag{3.93}$$

For large s,

$$s' \approx -m^2 s. \tag{3.94}$$

For telecentric systems the magnification is independent of the object location.

We note that for this afocal lens pair, the magnification is independent of the object location. We call this a **telecentric** system. We will find such systems very useful, and will encounter them several more times in the remaining chapters.

For an astronomical telescope, generally the **primary**, or first lens, has a large focal length, and the **secondary** or eyepiece has a much shorter one, as shown in Figure 3.12.

In astronomical telescopes $|m| << 1$. The image is smaller than the object but much closer and thus appears larger.

In large telescopes, the primary is generally a mirror rather than a lens, as it is easier to fabricate mirrors than lenses in large sizes. In any case, the magnitude of the magnification is small. The telescope in fact actually results in an image which is smaller than the object, by a factor m. However, because the longitudinal magnification is m^2, the image is much closer, and the angular magnification, as determined by the lower-right matrix element, is $1/m$. Thus distant objects appear bigger through the telescope because they subtend a larger angle. For example, if the telescope magnification is 0.1, then an image is 10 times smaller than the object, but 100 times closer. Because the image has 10 times the angular extent of the object, it covers a larger portion of the retina, and appears larger. Because s' is positive to the right, the image is on the same side of the telescope as the object. Because $m = -f_2/f_1$ is negative and both focal lengths are positive, the image is inverted. The image distance is the object distance divided by the square of the magnification, plus a small offset, $f_1 m(1+m)$.

Telescopes also find application as beam expanders, for example, to increase the size of a laser beam. In this case, we want a telescope with a magnification greater than one, and we simply choose lenses f_1 and f_2 that provide $|m| = f_2/f_1$. A beam 1 mm in diameter can be expanded to 6 mm by a telescope with $f_2/f_1 = 6$. Telescopes are also used to relay an image from one location to another. The periscope and the endoscope can both be implemented through the use of multiple telescopes.

Because the length of a telescope is $f_1 + f_2$, it is tempting to use a negative element. With f_2 negative, the length of the telescope becomes shorter by twice the magnitude of f_2 and the image is upright rather than inverted. Such telescopes may be useful on occasion, but there are certain disadvantages which will be discussed in Section 4.4.1.

Take-Away

Afocal Telescope:

- In the afocal telescope, the two lenses are separated by the sum of their focal lengths. The magnification is $-f_2/f_1$ independent of the object distance.

- The longitudinal magnification is m^2.
- For an astronomical telescope $|m| << 1$, and the image is smaller than the object by m, but closer by m^2, so it subtends a larger angle.
- An afocal telescope is an example of a telecentric imaging system.

Problems

3.1 Focusing a Laser Beam

The configuration in Figure P2.4D can be used to focus a laser beam to couple it into an optical fiber. We want the sine of the half-angle subtended by the last lens (shaded region) to equal the numerical aperture of the fiber, NA = 0.4. Let the second lens have a focal length of 16 mm and assume that the initial laser beam diameter is $D_1 = 1.4$ mm. We want the magnification of the second lens to be $-1/4$.

3.1a What is the focal length of the first lens?

3.1b What is the spacing between the two lenses, assuming thin lenses?

3.1c Now, assume that the lenses are both 5 mm thick, the first is convex-plano and the second is biconvex. What is the spacing between the vertices of the lenses now? What is the distance to the image point from the second lens vertex? Where should the fiber be placed relative to the second lens? You may use an approximation we discussed in class, and the fact that the lenses are glass.

3.1d Use matrices and find the focal length of this pair of lenses and the location of the principal planes. Be sure to specify distances and directions from lens vertices.

3.2 Camera Lens (NS)

Consider the camera lens shown in Figure P3.2. We want to find the focal length and principal planes. For this, of course we need 3 experiments to obtain 3 unknowns, f, h, and h'.

Using autocollimation or just a distant light source, we locate the focal planes and measure their locations relative to two reference surfaces that we call the vertices. If we

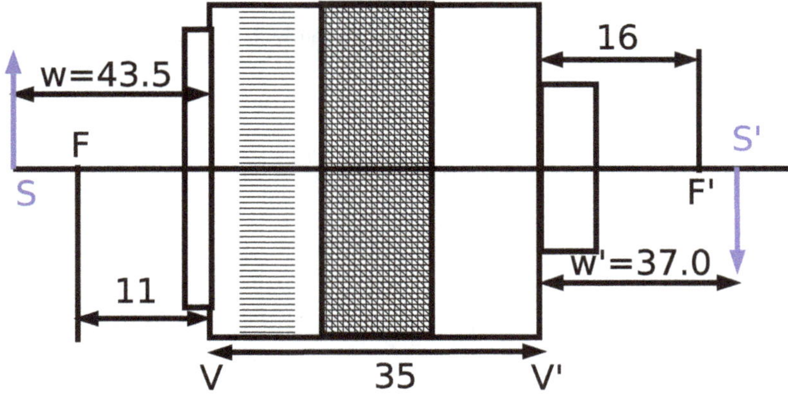

FIGURE P3.2
Figure for Problem 3.2.

FIGURE P3.3
Figure for Problem 3.3.

knew f we would be finished, but we don't, so we do one more experiment where we image an object through the lens. We place the object to the left at S, and move a paper screen on the right until the image is in focus at the position S'. Of course, we can't measure s and s' because we don't know yet where the principal planes are. We instead measured the distances w and w' from the respective vertices.

What are f, h, and h'?

Hint: I found it useful to write equations like

$$Z_{SV} = Z_{SH} + Z_{HV} \qquad \text{or} \qquad w = s + Z_{HV}$$

where, for example, Z_{HV} is the distance from H to V. This helped me to avoid sign errors. Then I used the lens equation with substitutions and eliminated h and h', to solve for f. You may find an easier way to do it.

3.3 Meniscus Lens (NS)

Consider the germanium meniscus lens in Figure P3.3. The index of refraction is 4 and the lens is used in air. The focal length is $f = -30$ mm. We choose to make

$$\frac{r_1 + r_2}{r_1 - r_2} = 5$$

for reasons that we will see in Chapter 5. Use the thin lens assumption to determine the radii of curvature. The thickness is 6 mm.

3.3a. What are the radii of curvature?

3.3b. What is the actual focal length?

3.3c. Where are the principal planes?

3.3d. Where are the focal planes?

3.4 Very Thick Lens (NS)

Consider the very thick glass (or more likely plastic) lens in Figure P3.4. Use $n = 1.5$. The optical power of each surface is 5 diopters and the length of the lens is 50 mm.

3.4a What are the radii of curvature?

FIGURE P3.4
Figure for Problem 3.4.

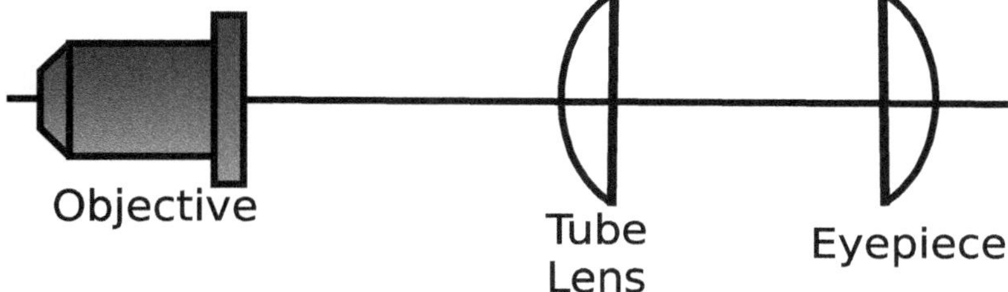

FIGURE P3.5
Figure for Problem 3.5.

3.4b Where are the principal planes?

3.4c Where are the focal planes?

3.5 Compound Microscope (NS)

Figure P3.5 shows a compound microscope with an infinity-corrected objective. That means that the object is placed at the front focal plane of the objective and the intermediate image after the objective is at infinity. That image becomes the object for the tube lens which produces an image at its back focal plane. Then we place an eyepiece with $f_{eyepiece} = 10\,\text{mm}$ after the tube lens. The eyepiece is used as a simple magnifier like you might use to read the telephone book (if anyone knows what a telephone book is; I haven't seen one in years). The object is placed to have an object distance, s slightly less than f and the image is virtual with $s'_{eyepiece} \approx -20\,\text{cm}$.

The focal length of the tube lens is $f_{tube} = 200\,\text{mm}$ and the objective is claimed to have a magnification of 30.

We have found the principal planes of the objective to be located at $h = -1\,\text{mm}$ and $h' = -8\,\text{mm}$ from the vertices of the objective, and the tube lens and eyepiece are simple lenses with thicknesses of 6 and 3 mm respectively.

3.5a What is the focal length of the objective?

3.5b What is the spacing between the objective and tube lens?

3.5c What is the spacing between the tube lens and the eyepiece?

3.5d What is the total magnification as given by the height of the final virtual image divided by the height of the original object?

4 Stops, Pupils, and Windows

Thus far, as we have considered light propagating through an optical system, we have assumed all surfaces to be infinite in the dimensions transverse to the axis. Thus there has never been a question of whether a ray would or would not go through a particular surface. This assumption proved useful in graphical solutions where we worked with three rays in Figure 2.1. As we used the ray parallel to the axis, for example, we did not trouble ourselves by asking whether or not this ray passed through the lens, or was beyond its outer radius. We knew that all the rays from the object, regardless of their angle, would go through the image, so we could determine image location and magnification by using the rays for which we had simple solutions, without concern for whether these rays actually did go through the system. If there was an intermediate image, we simply started tracing new rays from that image through the rest of the system, rather than continuing the rays we used from the original object. While these assumptions allow us to determine image location and magnification, they cannot give us any information about how much light from the object will go into forming the image, nor about which parts of the object can be seen in the image. In this chapter, we address these two issues.

Figure 4.1 shows an optical system with aperture and field stops. This optical system consists of two 4-f 1:1 relay pairs, so-called because the object and image distances are $2f$, for a total length of $4f$. In Figure 4.1-A, a stop is placed near the second lens. This stop could be simply a piece of light-blocking material with a hole in it, or it could be the edge of the lens or the mount holding it. Let's assume the hole is round and centered on the axis, although other shapes are possible. Rays are traced from a point where the object intersects the axis. Rays that make an angle of less than some limit pass through the system, while rays making a greater angle with the axis are blocked. In the figure, after they are blocked, they are shown as continuing with dashed lines to show where they would have gone in the absence of the aperture stop. The wide colored line shows the limiting angle. All rays within this cone pass through the system and all rays beyond it do not. As the stop is made smaller,

Figure 4.1-B shows the same imaging system with a different stop at the intermediate image, which limits the part of the object from which rays can converge toward the image. Comparing the two figures we see that this stop will not block any rays from the point at the base of the object. However it will block rays from parts of the object beyond a certain distance from the axis, and thus limit the field of view. Again, the thick dashed line shows the boundary. In the image, rays that are blocked by the field stop are shown dashed to indicate where they would have gone in the absence of the stop. The image is truncated at a height determined by the field stop. Our goal in this chapter is to develop an understanding of the locations and functions of stops in optical systems.

Not all rays from the object pass through the optical system. Stops and finite apertures of lenses limit the rays.

One approach to understanding finite stops would be to trace rays from every point in the object in every possible direction, stopping when a ray is outside the boundary of a lens or stop. Ray tracing is very useful in designing an optical system, but it is tedious, and does not provide insight in the early stages of design. We will instead discuss the problem in terms of the **aperture stop** and the **field stop**. The aperture stop of an optical system determines the portion of the cone of rays from a point in the object that actually arrives at the image, after passing through all the optical elements. The field stop determines the points in the object from which light passes through the optical elements to the image. The functions of the stops of a system may depend on the location of the

DOI: 10.1201/9781315157047-4

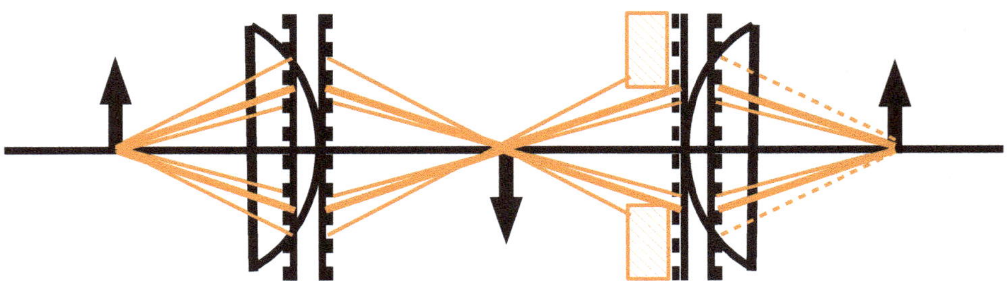

A. Aperture Stop. The wide colored line bounds the cone of rays that passes through the system from the part of the object that is on the axis. The thinner colored lines are other rays. The ones that are blocked are continued beyond the aperture stop with dashes to show that they do not reach the object.

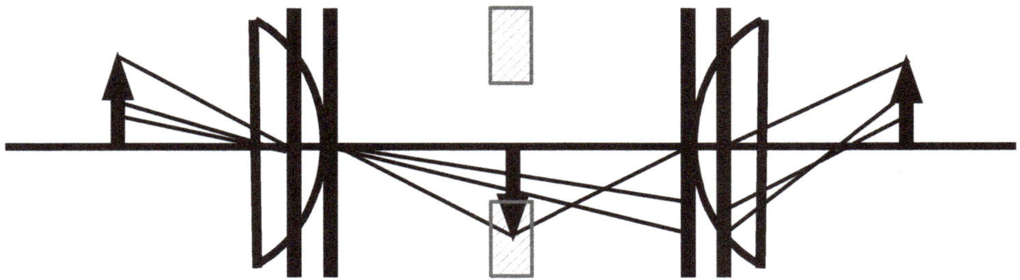

B. Field Stop. The wide dashed line bounds the portions of the object that are visible. The thinner lines are other rays. The ones that are blocked are continued beyond the field stop with dashes to show that they do not reach the object.

FIGURE 4.1
Stops in an optical system. The aperture stop determines which rays from an object on the axis can pass through the system (A), while the field stop limits the portion of the object that can be seen in the image (B).

object and image, and if an optical system is used in a way other than that for which it was designed, then the intended aperture stop and field stop may not serve their functions, and may in fact exchange roles.

The major task addressed in this chapter is the determination of which optical elements form the stops of the system. We begin by discussing definitions and functions of apertures using a simple lens. Next, we determine how to locate them, given an arbitrary optical design. Because readers new to optics often find the study of apertures and stops difficult, we will begin with very simple systems, where the functions of the apertures are easier to understand. Then we will approach more complicated problems, eventually developing a general theory using the notions of "object space" and "image space." We can determine the effect of the apertures in either space, and will make use of whichever is more appropriate to the problem. We will define the terms "pupil" and "window" in these spaces. At the end of this discussion, we will have sufficient background to understand in some detail the operation of telescopes, scanning systems, and microscopes, and we will use these as examples.

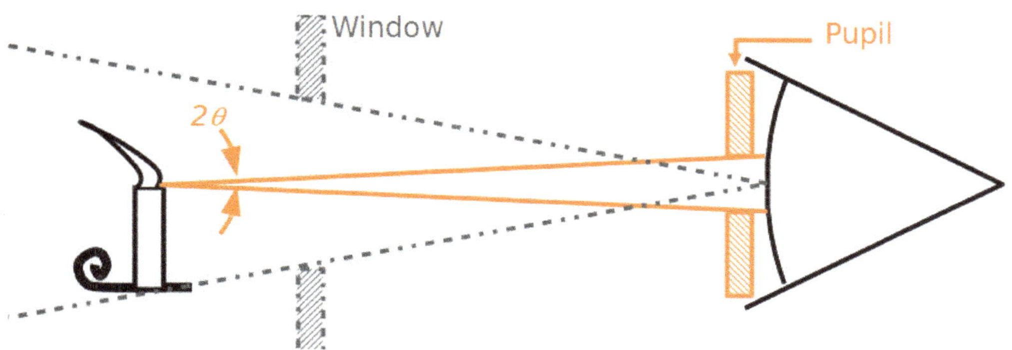

FIGURE 4.2
Stops. In this book a field stop is shown with 45° hatching from upper right to lower left, and an aperture stop is shown with hatching from upper left to lower right in color. The cone of rays passing through the aperture stop is shown by a line with dashes and dots in color. The field of view is shown with dashed lines.

4.1 Aperture Stop

The functions of stops can be understood intuitively from Figure 4.2. The pupil of the eye contains an iris which opens and closes to provide us with vision under a wide dynamic range of lighting conditions. In bright light, the aperture of the iris is approximately 2 mm in diameter, restricting the amount of light that can reach the retina. In dim light, the iris opens to about 8 mm, increasing the area, and thus the amount of light collected, by a factor of 16. One can observe the changing iris diameter by looking into a mirror and changing the lighting conditions. The iris opens or closes on a timescale of seconds, and the change is readily observed. The iris is the aperture stop, and limits the amount of light transferred from the object to the image on the retina. Further adaptations to changing light levels involve chemical processes which happen on a much longer timescale.

The pupil limits the amount of light.

The iris of the eye varies the pupil diameter between about 2 and 8 mm to adjust to light variations.

In this figure, the viewer is looking through a window at objects beyond it. The window determines the **field of view**, which can be expressed as an angle, viewed from the center of the aperture stop, or as a linear dimension across the object at a specific distance.

The window limits the field of view.

Before examining more complicated systems, it is useful to discuss the effects of these two apertures in the context of Figure 4.2. The reader who is unfamiliar with the concept of solid angle may wish to review Appendix B.

Solid Angle: See Appendix B.

4.1.1 Solid Angle and Numerical Aperture

Let us divide the object into small patches of area, each emitting some amount of light. If we choose the patches small enough, they cannot be resolved, and can be considered as points. If the power from a single patch is dP_{source}, uniformly distributed in angle, the power per unit solid angle, or intensity, will be $dI = dP_{source}/4\pi$. We will define intensity in more detail in Section 12.1. The intensity will probably vary with angle, and this equation provides only an average. The power passing through the aperture will be the integral of the intensity over solid angle, Ω, or for a small increment of solid angle,

$$dP_{aperture} = dI \times \Omega. \tag{4.1}$$

If the aperture is circular, we define the **numerical aperture** as the product of the index of refraction and the sine of the angle subtended by the radius of the circle as seen from the base of the object.

Numerical aperture is set by the pupil. It determines how much light goes through the system.

$$NA_{object} = n \sin\theta = n\frac{D/2}{\sqrt{s^2 + (D/2)^2}}. \tag{4.2}$$

Exact Equation for solid angle.

As shown in Appendix B, the solid angle is

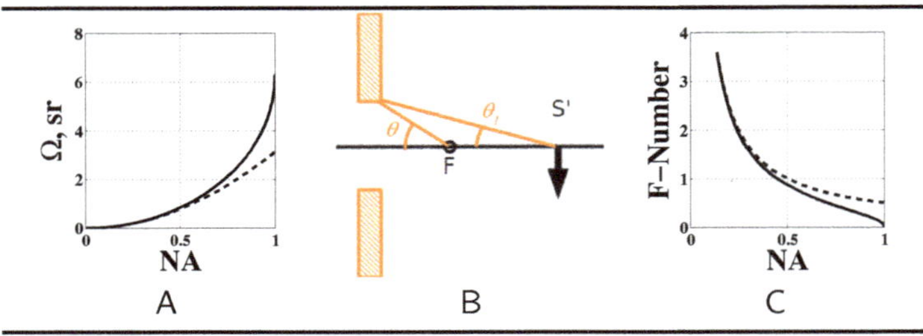

FIGURE 4.3
Numerical aperture, f-number, and solid angle. The solid angle is plotted as a function of numerical aperture in (A). The solid line is Equation 4.3 and the dashed line is the approximation in Equation 4.4 The definitions of numerical aperture and f-number are shown in (B), with $(f/\#) = 1/(2\tan\theta_t)$, and $NA = n\sin\theta$. Their relationships, given by Equation 4.9 or the approximation, Equation 4.10, in (C).

$$\Omega = 2\pi\left(1 - \sqrt{1 - \left(\frac{NA}{n}\right)^2}\right), \tag{4.3}$$

and for small numerical apertures, we can use the approximation

$$\Omega = \frac{\pi}{4}\left(\frac{D}{s}\right)^2 \qquad \text{(Small NA)}, \tag{4.4}$$

Approximation for solid angle with small NA.

where

- D is the aperture diameter
- s is the object distance

The solid angle is plotted in Figure 4.3A.

Eq. 2.2: $n\sin\theta = n'\sin\theta'$.

Why is the numerical aperture defined as $n\sin\theta$ instead of just $\sin\theta$? When light enters a medium at an angle it refracts according to Snell's Law, Equation 2.2. In Figure 4.4, all the rays contained in the cone enclosed by the angle θ in air will be refracted into the cone enclosed by θ' in water. The numerical apertures in air and water are

$$NA_{air} = \sin\theta \qquad NA_{water} = n_{water}\sin\theta' = NA_{air}. \tag{4.5}$$

We note that $NA \le n$.

Example

Let's look back at the thin germanium lens in the example in Section 2.5. Recall that we calculated the focal length to be $f = 5\,\text{cm}$. The object was at $s = 20\,\text{cm}$ and we calculated the image location to be $s' = 6.667\,\text{cm}$. Assume the lens has a 1 cm diameter. What is the numerical aperture as viewed from the object?

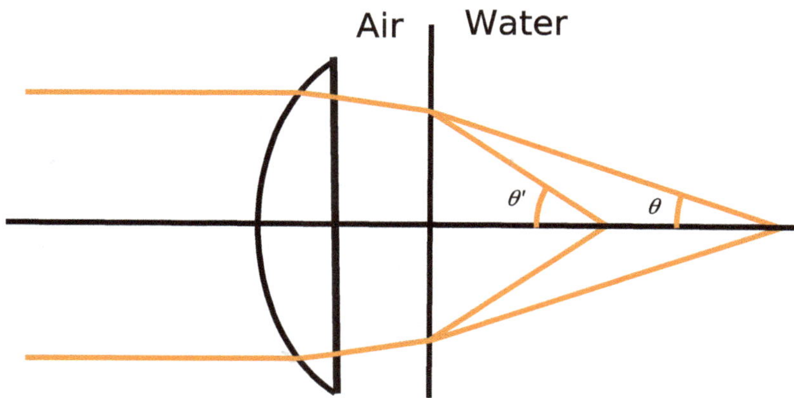

FIGURE 4.4
Numerical aperture in different media. As we go from one medium to another, Snell's Law ensures that the numerical aperture does not change.

From the object, the distance is s and the angle, θ is given by Equation 4.2 as

$$\tan\theta = \frac{1\,\text{cm}/2}{20\,\text{cm}} = 0.025.$$

Then

$$\theta = 1.4\,\text{deg}$$

$$NA = n\sin\theta = 0.025.$$

Note that n is the index of refraction of the medium on the image side, and not that of the lens itself.

4.1.2 f-Number

Alternatively, we may characterize the aperture in terms of the **f-number**,

$$(f/\#) = \frac{f}{D}, \tag{4.6}$$

where f is the focal length of the lens. The relationship between the f-number and numerical aperture is complicated by the fact that the f-number is defined in terms of the focal length while the numerical aperture is defined in terms of the object (or image) distance. The numerical aperture in Equation 4.2 is the object-space numerical aperture. It is also useful to think of the image-space numerical aperture,

$$NA_{image} = n'\sin\theta' = n'\frac{D/2}{\sqrt{(s')^2 + (D/2)^2}}. \tag{4.7}$$

The f-number also determines the size of the aperture stop. We often use term like "f/4" when discussing telescopes or cameras. We more commonly use NA when discussing microscope objectives, but both terms are used in all cases.

We refer to a lens with a low f-number as a **fast** lens, because its ability to gather light means that an image can be obtained on film or on an electronic camera with a short exposure time. The concept of speed is also used with film; a "fast" film is one that requires a small energy density to produce an image. We can also think of a high-NA system as "fast" because the rays converge rapidly as a function of axial distance.

We can define NA in object or image space. The answers are different.

Fast lens: Low f-number and high NA.

Example

What is the image-space numerical aperture from the previous example? What is the f-number?

Using $s' = 6.667\,\text{cm}$, the NA in image space is

$$\tan\theta = \frac{1\,\text{cm}/2}{6.667\,\text{cm}} = 0.0750$$

$$\theta = 4.29\,\text{deg}$$

$$NA = n'\sin\theta = 0.0748.$$

Using Equation 4.6, we find that

$$(f/\#) = \frac{5\,\text{cm}}{1\,\text{cm}} = 5,$$

and we say this is an $f/5$ lens.

Note that this result does not depend on how the lens is used (it is independent of s), in contrast to the numerical aperture.

Figure 4.3B shows the geometry for definitions of the f-number and image-space numerical aperture. The differences between the definitions are (1) the angle, θ for f-number is defined from the focus and θ_1, for the NA, is defined from the object or image, (2) the f-number is defined in terms of the tangent of the angle θ, and the NA is defined in terms of the sine of θ_1, (3) the f-number is inverted, so that large numbers indicate a small aperture, and (4) f-number is expressed in terms of the full diameter, while NA is expressed in terms of the radius of the pupil.

We will now derive an equation relating the two, in terms of the magnification. For this purpose, we return to the lens equation, recalling from Equation 2.54 that $m = -s'/s$, and write

$$\frac{1}{f} = \frac{1}{s} + \frac{1}{s'} \qquad \frac{1}{f} = \frac{-m}{s'} + \frac{1}{s'} \qquad s' = (1-m)f. \tag{4.8}$$

Thus, the numerical aperture in terms of f is

$$NA_{image} = n'\frac{D/2}{\sqrt{|(m-1)f|^2 + (D/2)^2}},$$

or in terms of f-number, $(f/\#)$,

$$NA_{image} = n'\frac{1}{\sqrt{[|m-1| \times 2\,(f/\#)]^2 + 1}}, \tag{4.9}$$

Equation relating image-space NA to f-number.

In the case of small NA, a good approximation is

$$NA_{image} = n'\frac{1}{|m-1| \times 2\,(f/\#)} \qquad \text{(Small } NA\text{, large } (f/\#)\text{)}, \tag{4.10}$$

These equations are shown in Figure 4.3C, for the case $m = 0$. The case of $m = 0$

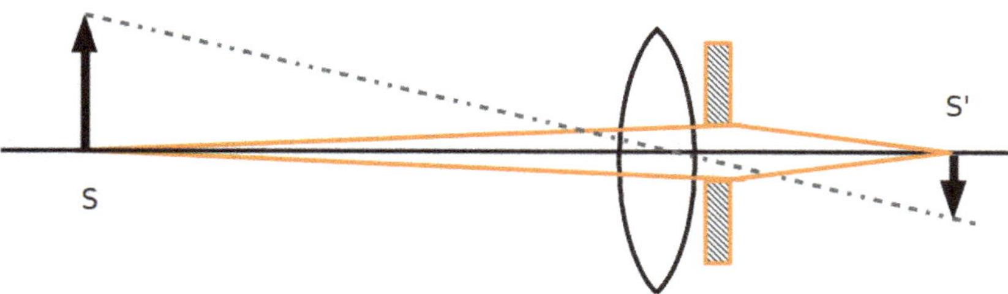

FIGURE 4.5
Camera. The amount of light on a pixel is determined by the object radiance and the numerical aperture.

represents a lens imaging an object at infinity, and is often an appropriate approximation for a camera, where the object distance is usually much larger than the focal length.

The object-space numerical aperture is found similarly with the results,

$$NA_{object} = n\frac{1}{\sqrt{\left[\left|\frac{1}{m} - 1\right| \times 2\,(f/\#)\right]^2 + 1}}, \qquad (4.11)$$

Equation relating object-space NA to f-number.

$$NA_{object} = n\frac{1}{\left|\frac{1}{m} - 1\right| \times 2\,(f/\#)} \qquad \text{(Small } NA\text{, large } (f/\#)\text{)}, \qquad (4.12)$$

The plot in Figure 4.3C is appropriate for the object-space numerical aperture for a microscope objective. The object is placed at or near the front focal plane, and the image is at or near infinity, so $m \to \infty$.

In other cases, the plot in Figure 4.3C can be multiplied by the appropriate function of magnification. One important example is the 1:1 relay lens, for which $s = s' = 2f$. This configuration is often called a "four-f" imager, because the total length is four times the focal length. The image is inverted, so $m = -1$ and

$$NA = \frac{1}{\sqrt{4\,(f/\#)^2 + 1}} \qquad \text{(1:1 relay)} \qquad (4.13)$$

Eq. 4.1 $dP_{aperture} = dI \times \Omega$.

Example

Returning to Equation 4.1, let us suppose that the intensity is that associated with a patch of area, $dA = dx \times dy$. In Section 12.1, we will define the intensity per unit area as the radiance. In Figure 4.5, we can use similar triangles to show that the width of the patch dA is magnified by m to form the patch $dA' = dx' \times dy'$ in the image, and that the numerical aperture is magnified by $1/m$.

Let us consider a specific example. Suppose we view a scene 1 km away ($s = 1000$ m) with an electronic camera having pixels $x_{pixel} = 7.4\,\mu$m on a side. Suppose that we use a lens with a focal length of $f = 9$ mm. The image distance s' is close to f, and the magnification is $f/s \approx 0.009\,\text{m}/1000\,\text{m} = 9 \times 10^{-6}$. If we use the lens at $f/2$, the object-space NA is

$$NA_{Object} \approx \frac{D}{2s} = \frac{f}{2\,(f/\#)\,s} = \frac{0.009\,\text{m}}{4 \times 1000\,\text{m}} = 2.25 \times 10^{-6}, \qquad (4.14)$$

and the solid angle is

$$\Omega \approx \pi NA^2 = 1.6 \times 10^{-11}\,\text{sr}. \tag{4.15}$$

The camera pixel imaged back onto the object has a width of

$$\frac{x_{pixel}}{m} = \frac{7.4 \times 10^{-6}\,\text{m}}{9 \times 10^{-6}} = 0.82\ \text{m},$$

or an area of about $0.68\,\text{m}^2$. Scattered visible sunlight from a natural scene might have an intensity of 50 W/sr from a patch this big (sunlight at earth consists of about $1000/unW/m^2$, with about 1/3 of it in the visible spectrum, and we assume it scatters into π steradians from the given area; $1000/unW/m^2 \times 0.82\,\text{m}^2/3/(\pi\,\text{steradians} \approx 50\,\text{W/steradian})$. Then Equation 4.1 becomes

$$dP_{aperture} = dI\Omega = 50\,\text{W/sr} \times 1.6 \times 10^{-11}\,\text{sr} = 7.9 \times 10^{-10}\,\text{W}. \tag{4.16}$$

Camera specifications often describe the number of electrons to produce a certain amount of signal. If we use the photon energy, $h\nu = hc/\lambda$ of green light, at $\lambda = 500$ nm, the number of electrons is

$$\frac{dP_{aperture}}{h\nu}\eta t, \tag{4.17}$$

where

- η, the quantum efficiency, is the probability of a received photon producing an electron in the detector (for a detailed discussion of detectors, see Chapter 13)
- t is the frame time of the camera

For $\eta = 0.4$ and $t = (1/30)$ s, we expect about 2.7 million electrons.

We note that we could reduce the number of electrons by using a higher f-number. Going from $f/2$ to $f/2.8$ would reduce the number by a factor of two.

Eq. 3.52: $m_\alpha = \frac{1}{m}$.

We have analyzed the system using object-space values, namely NA_{object} and the area of the pixel on the object. We have shown in Equation 3.52 that a magnification from object to image of m requires angular magnification of the NA by $1/m$.

Object $A\Omega$ is equal to image $A'\Omega$. We will see that this is quite general.

Thus, we would find the same answer if we used NA_{image} and the pixel area $(7.4\ \mu\text{m})^2$. The intensity associated with this smaller area is $50\,\text{W/sr} \times m^2$ and the number of electrons is the same. Thus we can compute the camera signal knowing only the radiance (intensity per unit area) of the object, the camera's detector parameters, and the numerical aperture in image space. No knowledge of the focal length of the lens, the object size, or object distance is required.

Camera f-numbers usually correspond to powers of 2 in light-collecting ability.

On a typical camera, the f-number can be changed using a knob with settings in powers of $\sqrt{2}$, some of which are shown in Table 4.1. Because for small NA, the solid angle varies as the square of the NA, each successive f-number corresponds to a change of a factor of two in the amount of light.

In Practice

The f-numbers on a typical camera correspond to factors of 2 in light-collecting ability. Therefore, it is easy to trade aperture against exposure time. If a photo is

TABLE 4.1
f-Numbers

$(f/\#)$, Indicated f-number	1.4	2	2.8	4	5.6	8	11
Actual f-number	$\sqrt{2}^1$	$\sqrt{2}^2$	$\sqrt{2}^3$	$\sqrt{2}^4$	$\sqrt{2}^5$	$\sqrt{2}^6$	$\sqrt{2}^7$
NA	0.3536	0.2500	0.1768	0.1250	0.0884	0.0625	0.0442
Ω, sr	0.3927	0.1963	0.0982	0.0491	0.0245	0.0123	0.0061

Typical f-numbers and the associated numerical apertures in image space are shown for $|m| \ll 1$.

properly exposed at f/16 with an exposure time of 100 msec, then we can reduce the exposure time by 1/4 to 25 μsec if we open the aperture by four stops to f/4.

In Chapter 8, we will see that larger apertures provide better resolution, but in Chapter 5, we will see that faster lenses require more attention to correction of aberrations. Summarizing the current section, we make the following statements:

Take-Away

Aperture Stop:

- The aperture stop determines which rays pass through the optical system from the object to the image.
- The aperture stop can be characterized by the f-number or the numerical aperture.
- Larger apertures allow more light to pass through the system.
- the f-number, $(f/\#)$ and numerical aperture, NA, in image space is different from those in object space.
- We refer to a lens with a low f-number or high NA as a "fast" lens.
- "One stop" on a camera corresponds to a change in aperture diameter of $\sqrt{2}$ or a change in area of a factor of 2.

4.2 Field Stop

Returning to Figure 4.2, we see that the window limits the parts of the object that can be seen. For this reason, we refer to the window in this particular system as the field stop, and the parts of the image that can be seen comprise what we call the field of view.

The field of view can be expressed as an angle, and in that case, is measured from the center of the aperture stop. For example, a 0.5 m window 5 m away subtends an angle of about 0.1 rad or 6°. The angular field of view can be increased by moving the field stop and aperture stop closer together. In defining the field of view, as is so often the case in optics, one must be careful to notice whether a given definition is in terms of the full angle (diameter) or half angle (radius). Both definitions are in common use.

Field of view in terms of angle.

The field of view can also be expressed as a linear dimension, and can, of course, be expressed either in object space or image space. For the simple examples considered here, the angular field of view is the same in both spaces, but this simple result will not be true in general.

Field of view in terms of linear dimension.

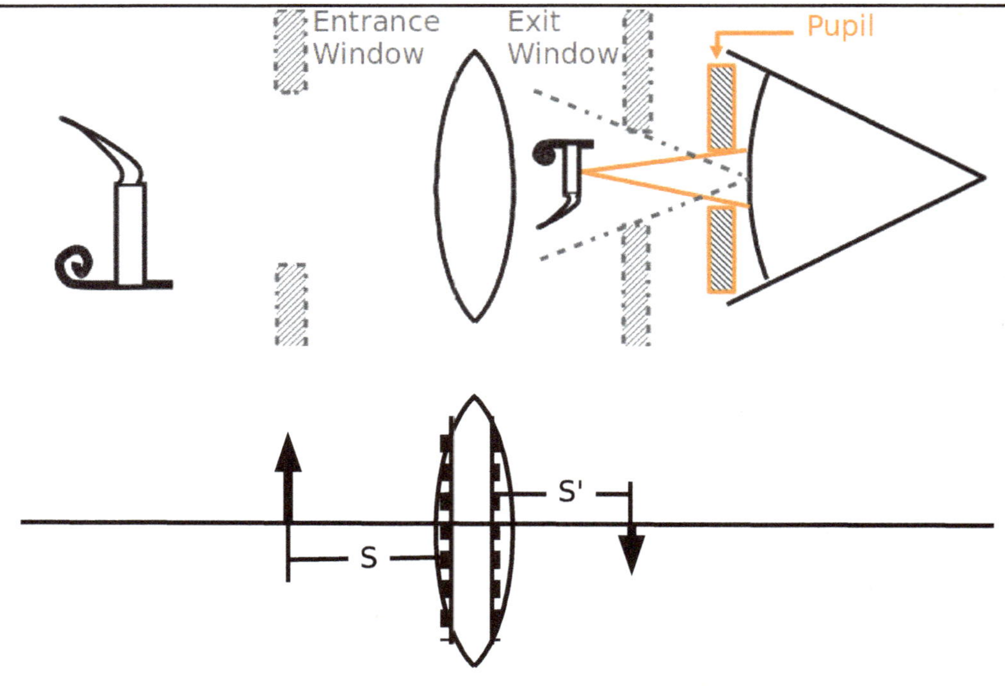

FIGURE 4.6
Exit window. The image of the field stop is the exit window. It determines what parts of the image can be seen. We find the image of the field stop by treating it as an object and applying the lens equation.

Example

The camera in the example above had a pixel size of 7.4 μm on the camera, and 82 cm on the object. These numbers define the field of view of a single pixel. The field of view of the entire camera in either space is the single-pixel number multiplied by the number of pixels. For example, a camera with 1000 of these pixels in each direction will have a square imaging chip 7.4 mm on a side, and with this lens, will have a square field of view in the object plane, 1 km away, that is 820 m on a side.

4.2.1 Exit Window

Fig. 4.2

We return now to Figure 4.2, and add a lens as shown in the top panel of Figure 4.6. From Chapter 2 or 3, we know how to find the image of the object. As shown in the bottom panel of Figure 4.6, we can also treat the window as an object and find its image. We refer to this image as the exit window. The original window, which we called the field stop, is also called the entrance window. Later, we will offer precise definitions of all these terms. Now, as we look into this system, we will see the image of the window, and through it, we will see the image of the object. Just as the physical window and field stop in Figure 4.2 limited the field of view for the object, the exit window limits the field of view for the image. Clearly the same parts of the object (image) must be visible in both cases.

4.2.2 Example: Camera

Later, we will develop a general theory for describing these windows, but first, we consider an example which illustrates the relationship among entrance window, exit window, and field stop. In this case, the field stop is the boundary of the film plane, and its image in object space is the entrance window.

This example is another one involving a camera, now showing the different fields of view that can be obtained with different lenses. The camera is a 1/2.3 in. compact digital camera. The diagonal dimension of the imaging chip is 11 mm. A normal lens for a compact digital camera has a focal length of about 10 mm, and thus a diagonal field of view of

$$FOV = 2\arctan\frac{11\,\text{mm}/2}{10\,\text{mm}} = 58^\circ. \tag{4.18}$$

An image with this lens is shown in the middle of Figure 4.7. Generally a "normal" camera lens will subtend a field of view of something between 40° and 60°, although there is no generally accepted number.

Typical normal camera lens: 40 to 60 deg Field of view.

The top row of the figure shows the same scene viewed through a 5 mm lens. A lens of such a short focal length with a compact digital camera is considered a wide-angle lens. Wide-angle lenses can have angles up to nearly 180°. The field of view using Equation 4.18 is 95°. In this case, the photographer moved forward to half the original distance, so that the linear dimensions of the field of view on the building in the background remained the same. As indicated by the vertical arrows in the drawings, the field of view for objects in the foreground is smaller with the wide-angle lens, because the photographer is closer to them.

Typical Wide-angle lens; Field of view $>$ 60 deg (Short focal length).

The telephoto lens with $f = 20$ mm has a field of view of only 31°. Much more foreground information is included in this picture, and the lens appears to "bring objects closer" to the camera.

Typical telephoto lens; Field of view $<$ 40 deg (Long focal length).

The field of view in image space is the same in each of these images, and as nearly as possible, the field of view expressed in linear dimensions at the object (the building in the background) is the same. However, the different angular field of view means that the linear field of view is different for other object planes in the foreground.

Take-Away

Field Stop:

- The field stop limits the field of view.
- The field of view may be measured in linear units or in angular units, in the space of the object or of the image.
- A "normal" camera lens has an angular field of view of about 50°. The focal length depends on the size of the image plane.
- A "telephoto" lens has a longer focal length and thus a smaller field of view.
- A "wide-angle" lens has a shorter focal length and thus a wider field of view.

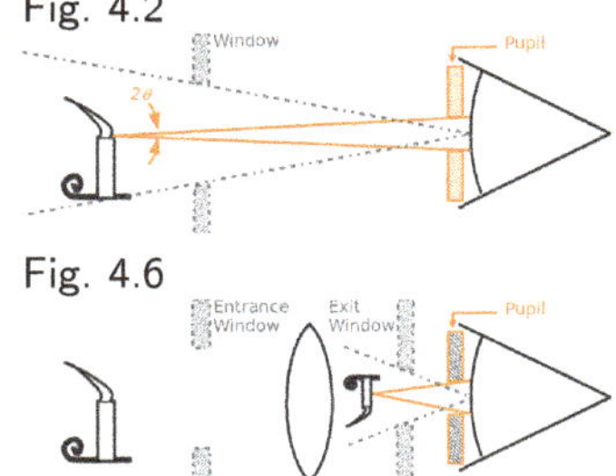

Example

Let's take a look at an example of a 4f relay in image space. In the discussion above using Figures 4.2 and 4.6, we have worked with simple examples. The aperture stop is unambiguously located at the lens, and both the object and the image (on the retina in Figure 4.2 and after the lens in Figure 4.6) are real and have positive

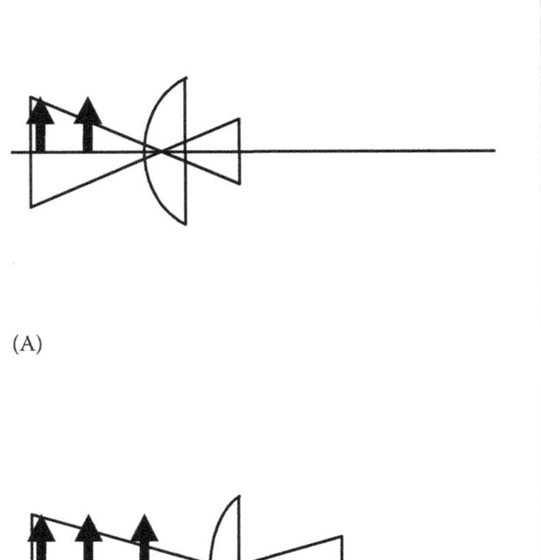

(A)

(B)

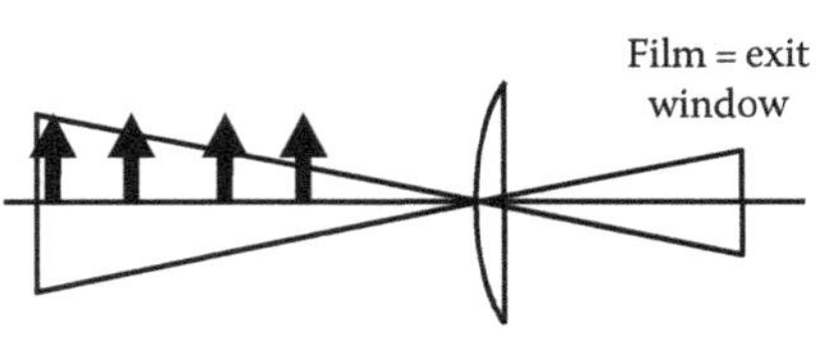

(C)

FIGURE 4.7
Changing focal length. Three lenses were used on a digital camera, and the photographer moved back with increasing focal length, so that the actual field of view remained nearly the same. (A) Wide-angle lens, $f = 5\,\text{mm}$. (B) Normal lens, $f = 10\,\text{mm}$. (C) Telephoto lens, $f = 20\,\text{mm}$.

distances. We have discussed numerical aperture in object space and image space, with an intuitive understanding of the distinction. The equations still work with a more complicated example. In Figure 4.8, an object is imaged in a "four-f" imaging relay with $f = 100\,\text{cm}$, $s = 200\,\text{cm}$, and $s' = 200\,\text{cm}$. The window is placed 80 cm in front of the lens, inside the focal point, resulting in a virtual exit window at $s'_{fieldstop} = -400\,\text{cm}$. This situation may seem peculiar in that the observer sees the real image, but the virtual exit window is 600 cm before it. Does it still limit the field of view? Before we answer this question, let us add one more new concept, the "entrance pupil." We can think of the lens as imaging "in reverse"; it forms an image of the observer's eye in image space. If the observer is far away this image will be near the front focus, as shown in the figure. We must remember that the light in the actual system travels sequentially from the source through the entrance window, the lens, and eventually to the eye. Therefore, if we want to think of the whole process in image space, the light travels from the object forward to the entrance pupil, and then backward to the eye. This analysis is simply an abstract construction that allows us to discuss the system in object space. Likewise, in image space, light travels from the image back to the exit window, and then forward to the eye. Therefore, the virtual exit window does indeed limit the field of view. Any ray that is blocked by the virtual exit window will never reach the eye. Here we assume that the lens is large enough that rays that pass through both stops will always pass within its diameter.

Why would we ever use such a complicated abstraction, rather than straightforward left-to-right propagation? The answer will become more apparent as we study compound lenses. Whatever complexity is involved in describing the system in object space or image space, once we have done so, the refractive behavior of the lenses has been eliminated and in this abstraction light travels in straight lines, which will make our analysis much simpler in the case of complicated compound lenses. For the present case, as shown in Figure 4.9, we know that the iris of the eye forms the aperture stop, because it limits the cone of rays from a point on the image. The angle subtended by the iris is smaller than that subtended by the exit pupil. We refer to the iris of the eye in this case as the exit pupil. The exit pupil is the aperture, or image of an aperture in image space, that limits the cone of rays from a point on the image.

Likewise, the exit window is the aperture or image of an aperture in image space that limits the cone of rays from the center of the exit pupil, as shown in Figure 4.10. We see now that we had defined our exit window correctly.

In order to determine the light-gathering ability and field of view of even this simple system without this level of abstraction, we would have needed to consider multiple rays from different parts of the object in different directions, trace each ray through the lens, and determine which ones were obstructed along the way. Furthermore, if the result is not to our liking, we have little intuition about how to modify the system. Even for a system consisting of only a lens, an aperture, and an observer's eye, the effort to develop an image-space description is simpler than the effort to solve the problem in what would appear at first glance to be a more straightforward way.

However, this approach shows its true strength when we consider a complicated compound lens system. In the next section, we develop a general methodology which can be extended to any system.

4.3 Locating and Identifying Pupils and Windows

Now that we have discussed some aperture examples, let us proceed to the formal definitions of the terms for pupils and windows. Our goal is to determine, for an arbitrary imaging system, which apertures limit the light-gathering ability and the field of view. The first step is to locate all the apertures and lenses in either object space or image space. Implicit in our discussion of compound lenses so far is the idea of sequential propagation, in which every ray propagates from the object to the first surface, then to the second, and so forth, to the image. Nonsequential systems also exist, and their analysis is more complicated. One example of a nonsequential system is a pair of large lenses with a smaller lens between them. Light can propagate through all three lenses in sequence, forming an image at one location, or can propagate through the outer two lenses, missing the inner one, and thus image at a different distance. Such systems are outside the scope of this discussion.

In sequential optical systems, all rays go from one element to another in the same defined order.

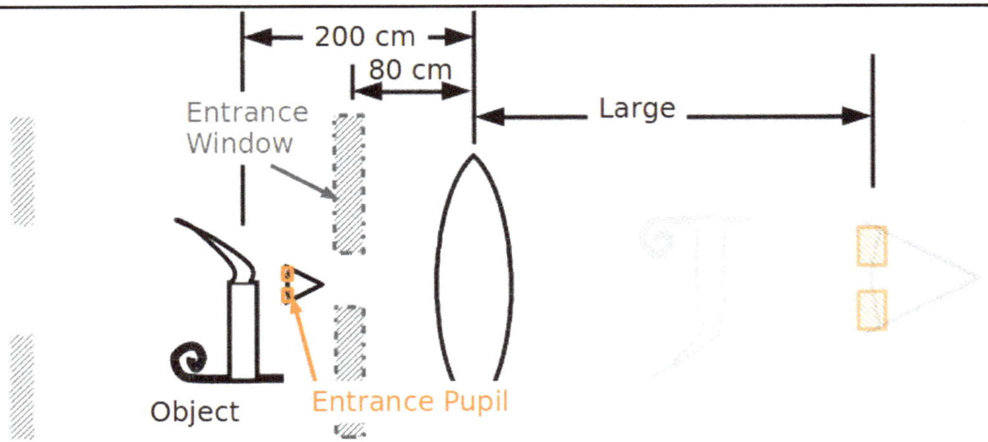

A. Object space: $f = 100\,\text{cm}$, $s = 200\,\text{cm}$, $s_{fieldstop} = 80\,\text{cm}$.

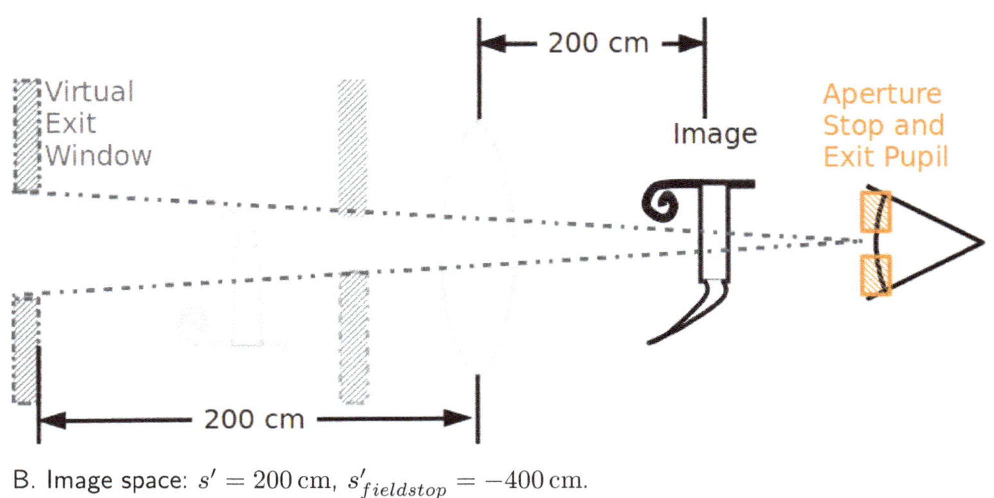

B. Image space: $s' = 200\,\text{cm}$, $s'_{fieldstop} = -400\,\text{cm}$.

FIGURE 4.8
Virtual exit window. The object is imaged through a real entrance window, through a 1:1 imaging system, and viewed by an observer far away (A, $f = 100\,\text{cm}$, $s = 200\,\text{cm}$, $s_{fieldstop} = 80\,\text{cm}$). The entrance pupil is an image of the eye's pupil. In image space (B, $s' = 200\,\text{cm}$, $s'_{fieldstop} = -400\,\text{cm}$), the exit pupil is the eye. The image of the field stop is the exit window, but in this system it is virtual, despite the image of the object being real. Furthermore, the exit window is at a greater distance from the observer than the image. However, it still functions as a window.

Object and image spaces defined for sequential optical systems.

Using the equation for refraction at a surface, the lens equation, or the matrix of any surface or lens, we have solved problems by finding the image location and size as functions of the object location and size. Thus, going through the first element (surface or lens) we produce a mapping from any point, (x_0, z_0) (and y_0 if we do not assume cylindrical symmetry), where z_0 is the absolute axial position of the object and x_0 is its height. Having completed that step, we use the image as an object and solve the problem for the next element, continuing until we reach the image, (x_i, z_i). The range of the variables (x_0, z_0) defines **object space**, and the range of the variables (x_i, z_i) defines **image space**.

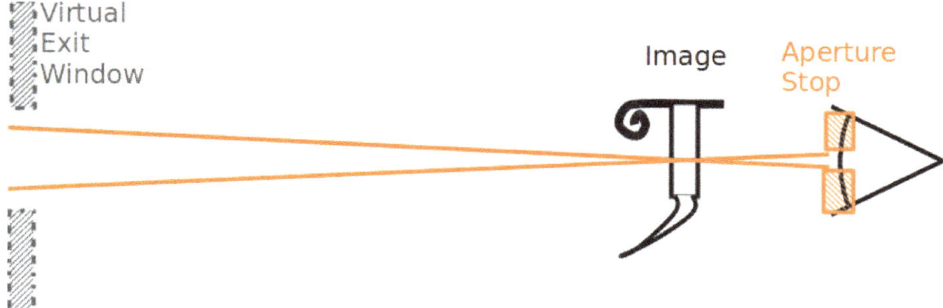

FIGURE 4.9
Exit pupil. The exit pupil limits the cone of rays from a point on the object.

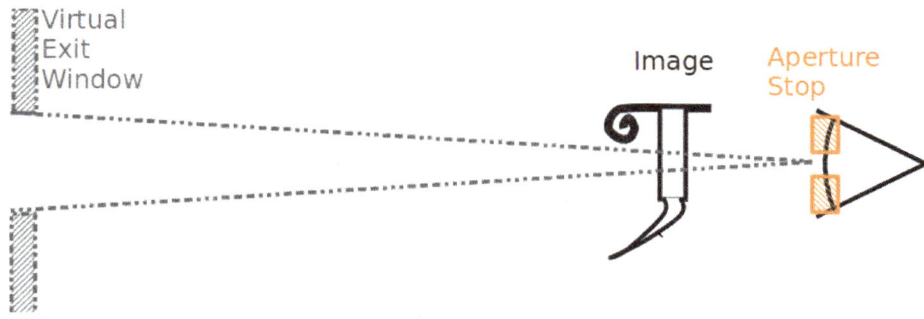

FIGURE 4.10
Exit window. The exit window limits the field of view as seen from a point in the exit pupil.

It is important to understand that image space includes the whole range of (x_i, z_i), whether before or after the last element, and whether the image is real or virtual. Likewise, object space includes the whole range of (x_0, z_0), whether before or after the first element, and whether the object is real or virtual.

When we solve the problem of the last surface or lens, we perform a mapping from (x_{i-1}, z_{i-1}) to (x_i, z_i). An object placed at (x_{i-1}, z_{i-1}) will have its image at (x_i, z_i). We have applied this idea when the current "object" is the image of the original object through all the previous elements. What happens if we use the immediately preceding element as the current object? In that case, we will find the image of that element in image space. We have already seen one example where we transformed a lens location to image space in Figure 4.8.

Fig. 4.8

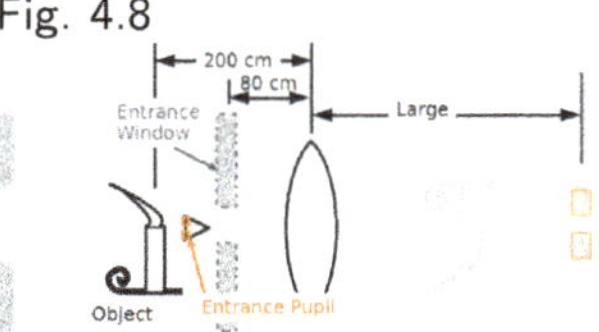

To illustrate the approach in a more complex system, we will consider the example in Figure 4.11. Four simple lenses are used together to form a compound lens. This example is chosen to illustrate the process, and is not intended to be a practical lens design for any application. An object at infinity is imaged in an inverted virtual image to the left of the last lens. The image in this case happens to lie within the system. The only apertures in this system are the boundaries of the lenses themselves or their holders. We will see how to find all the apertures in object space and in image space. Then we will identify the pupils, windows, and stops.

4.3.1 Object-Space Description

The object-space description of the system requires us to locate images of all the apertures in object space. Recall that light travels sequentially from the object through lenses L_1, L_2, L_3, and L_4 to the image. Thus there is no refraction between the object and L_1 and

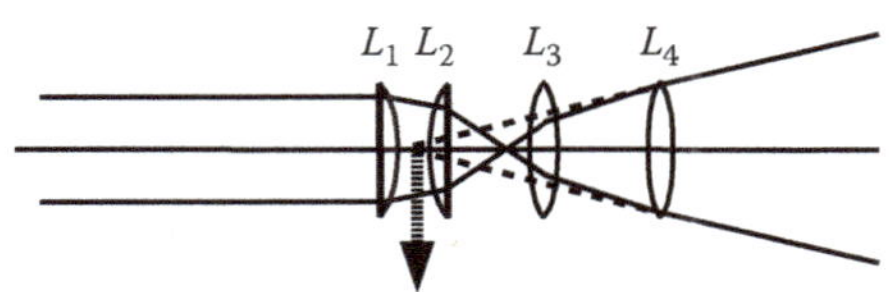

FIGURE 4.11
Example system for apertures. Four simple lenses comprise a particular compound lens which forms a virtual image of an object at infinity. To find the entrance pupil, all lenses are imaged into object space.

we can consider the first aperture to be the outer diameter of this lens. In practice, the diameter could be smaller because of the deliberate introduction of a smaller aperture, or because of the diameter of the hardware used to hold the lens in place. Next, we consider lens L_2. If this lens is close to L_1, it will form a virtual image as shown in Figure 4.12. In any case, it can be located by using the lens equation, Equation 2.51, where s is the distance from L_1 to L_2 (positive for L_2 to the right of L_1, f is the focal length of L_1, and s' is the distance by which the image, called L_2' is to the left of L_1. We defy our usual sign convention of light traveling from left to right, but this approach is consistent in that we are asking for the image of L_2 as seen through L_1. Many experienced lens designers will turn the paper upside-down to solve this problem! If L_1 is a thick lens, we must be sure to compute distances relative to the principal planes. The diameter of L_2' is $|m|$ times the diameter of L_2 where $m = -s/s'$.

Eq. 2.51 $\frac{n}{s} + \frac{n'}{s'}$.

Next, we find the image of L_3 as seen through L_2 and L_1. We could find it through two steps. First we would use L_3 as the object, and L_2 as the lens, and repeat the equations from the previous paragraph. Then we would use the resulting image as an object and L_1 as the lens. Alternatively, we could use the matrix formulation for the combination of L_1 and L_2, find the focal length and principal planes, and then find the image L_3' through the combination. Finally, we find L_4', the image of L_4 through lenses L_3, L_2, and L_1. Figure 4.13 shows all the apertures in object space.

In the previous paragraph all the steps are simple. However, the number of small steps creates an amount of complexity that requires tedious repetition, attention to detail, and careful bookkeeping. Doing these calculations by hand makes sense for a combination of two simple lenses, or in the case where the problem can be approximated well by two simple lenses. Any more complicated situations are best solved with the aid of a computer. The best approach to bookkeeping is to establish a global coordinate system to describe the locations of all lenses, principal planes, apertures and their images, and to compute the local coordinates (h, h', s, s') as needed, following the ideas outlined in Section 3.4.1.

For global coordinates see Section 3.4.3. For example, we used the notation $zV1$ to denote the global position of the first vertex, V_1.

Fortunately, many problems can be simplified greatly. If two lenses are close together, they can often be treated as a single lens, using whichever aperture is smaller. Often a compound lens will have entrance and exit pupils already determined by the above analysis, and these elements can be treated more simply as we shall see in the case of the microscope in Section 4.4.3.

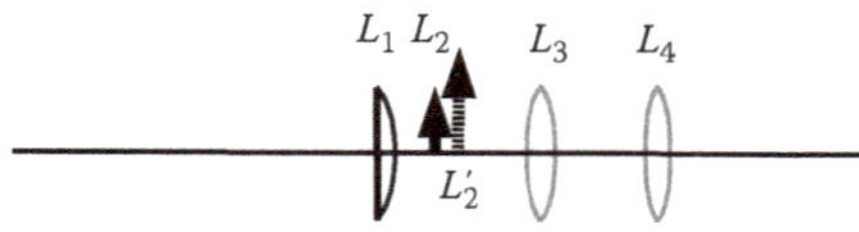

FIGURE 4.12
Lens 2 in object space, L_2' is found by imaging L_2 through L_1. To find the entrance pupil, all lenses are imaged into object space.

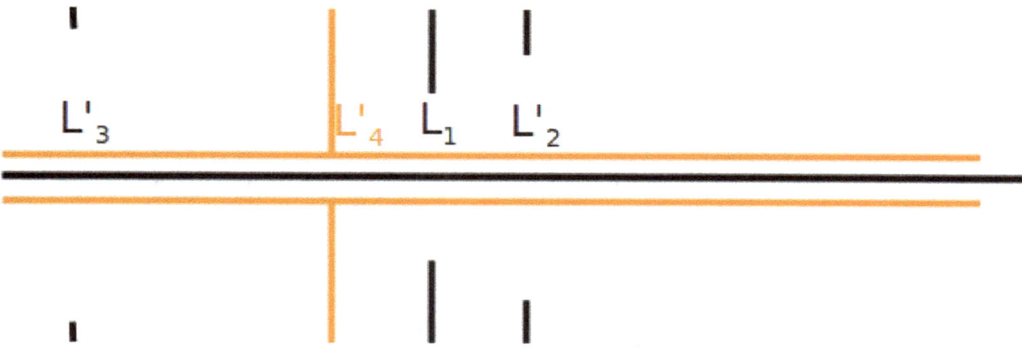

FIGURE 4.13
Example system in object space. L_4' is the entrance pupil and Lens 4 is the aperture stop for an object at infinity.

Now that we have all the apertures in object space, we can find the entrance pupil and entrance window. However before doing so, we will consider the image-space description briefly.

4.3.2 Image-Space Description

The process used in the previous section can be turned around and used to find the apertures in image space. The lens L_4, being the last component, is part of image space. The lens L_3 seen through L_4 is called L_3''. Likewise, we find L_2'' and L_1''. Once we have all locations and diameters, we have a complete image-space description of the system, as shown in Figure 4.14. Like the object-space description, the image-space description is an abstraction that simplifies identification of the stops. The same results will be obtained in either space. We can choose whichever space requires less computation or provides better insight into the system, or we can solve the problem in both spaces to provide a "second opinion" about our conclusions.

It is worthwhile to note that it may sometimes be useful to use spaces other than object and image. For example, we will see that in a microscope with an objective and eyepiece, it is often convenient to perform our calculations in the space between the two.

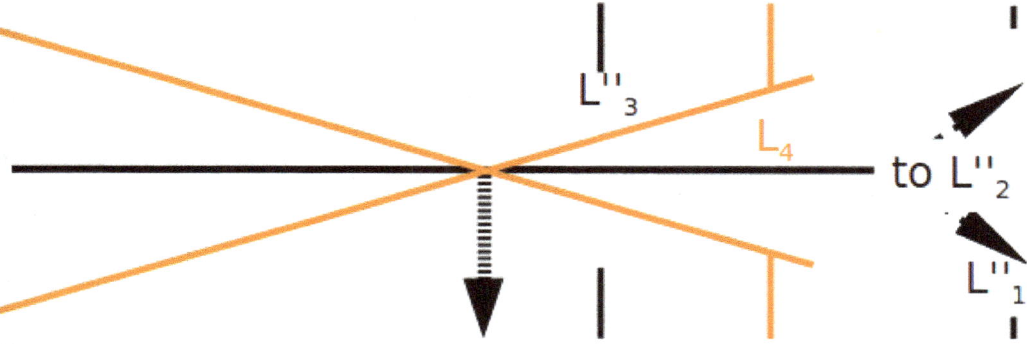

FIGURE 4.14
Example system in image space. The images of the first three lenses are shown. Note that L_2'' is far away to the right, as indicated by the arrows. The exit pupil is L_4, and the dotted lines show the boundary determined by the image-space numerical aperture.

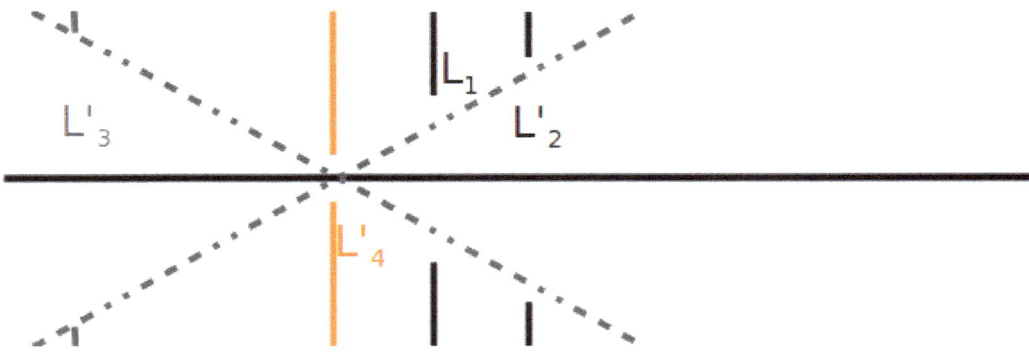

FIGURE 4.15
Finding the entrance pupil. For a different object distance, shown by the arrow, L_3' is the pupil and lens L_3 is the aperture stop. For the object at infinity in Figure 4.13, L_4' was the entrance pupil.

TABLE 4.2
Stops, Pupils, and Windows

Object Space	Physical Component	Image Space
Entrance pupil: Image of aperture stop in object space. Limits cone of rays from object	*Aperture stop:* Limits cone of rays from object which can pass through the system	*Exit pupil:* Image of aperture stop in image space. Limits cone of rays from image
Entrance window: Image of field stop in object space. Limits cone of rays from entrance pupil	*Field stop:* Limits locations of points in object which can pass through system	*Exit window:* Image of field stop in image space. Limits cone of rays from exit pupil

The stops are physical components of the optical system. The pupils and windows are defined in object and image space.

4.3.3 Finding the Pupil and Aperture Stop

Returning to Figure 4.13, we are ready to determine the entrance pupil. The object is at infinity so the "cone" of rays from the object consists of parallel lines. Because these rays are parallel, the limiting aperture is the one which has the smallest diameter. In this example, in the figure, the smallest aperture is L_4'. In Figure 4.2 and other examples with simple lenses earlier in the chapter, the aperture stop was the entrance (and exit) pupil. Even in a system as simple as Figure 4.8, we had to image the iris of the eye through the first lens to obtain the entrance pupil. We now formally state the definition; the **entrance pupil** is the image of the aperture stop in object space. Thus, having found the entrance pupil to be L_4', we know that L_4 is the aperture stop.

We see in Figure 4.15 that the rules for finding the entrance pupil produce different results for different object locations. If the object had been located at the arrow in this figure, then the entrance pupil would have been L_3', and L_3 would have been the aperture stop. The change in the aperture stop could lead to results unexpected by the designer of the system. It is a good principle of optical design to be sure that the pupil is well defined over the entire range of expected object distances.

Exit pupil defined.

Returning to Figure 4.14, the **exit pupil** is defined as the aperture in image space that subtends the smallest angle as seen from the image.

Table 4.2 shows the relationships among the pupils and aperture stop.

A. Typical camera lens.

B. Finding the pupil.

FIGURE 4.16
Finding the exit pupil of a camera lens. A typical camera lens is shown in the photo at the top. Notice the markings for focus adjustment and f/number. The expmeriment in the drawing shows how to find the exit pupil. The auxiliary lens relays an image of the pupil to the screen. We can then calculate the exit pupil location using the lens equation, or by finding the image of a hole in the screen to the left of the auxiliary lens (removing the camera lens).

In Practice

Sometimes we are given a lens and want to find the pupils. Consider the camera lens in Figure 4.16. We can illuminate the lens with a flashlight from a distance (Figure 4.16A), and we should be able to see an image of the aperture stop at the output. However, it is likely to be inside the lens so we cannot located it exactly. By using an auxiliary lens as in Figure 4.16B, we can image the pupil onto a screen. When the image of the pupil is in sharp focus, we can use the lens equation, Equation 2.51, $\frac{n}{s}+\frac{n'}{s'}$ to determine its location, a distance s in front of the lens. Measuring everything carefully we can find the distance from the pupil to the flange, z_{pf}. We can turn the lens around and repeat the experiment to find the entrance pupil.

4.3.4 Finding the Windows

Once we have found the pupil in either space, it is straightforward to find the window, as seen in Figure 4.17. The **entrance window** is the image of the field stop in object space. It is the aperture in object space that subtends the smallest angle from the entrance pupil. For an object at infinity, the portion of the object that is visible is determined by the cone of rays limited by L_3'. Thus, L_3' is the entrance window, and L_3 is the field stop. Likewise, in the bottom part of the figure, L_3'' limits the part of the image that can be seen. The **exit window** is the image of the field stop in image space. It is the aperture in image space that subtends the smallest angle from the exit pupil.

Entrance window defined.

Exit window defined.

As previously noted, the results are the same in both object space and image space. The remaining line in Table 4.2 shows the relationships among the windows and field stop.

The rays passing through the system are limited by these two apertures, and the remaining apertures have little effect. In fact, in this example, lenses L_1 and L_2 are larger than necessary. From Figure 4.17 we see that they are only slightly larger. If they had been considerably larger, they could be reduced in size without compromising the numerical aperture or field of view, and with a potential saving in cost, weight, and size of the system.

Take-Away

Identifying Stops:
The material in this section is summarized compactly in Table 4.2.

- The aperture stop and field stop are physical pieces of hardware that limit the rays that can propagate through the system. The aperture stop limits the light-gathering ability, and the field stop limits the field of view.
- We can describe the apertures in an optical system in object space or image space.
- Light travels in a straight line in either of these abstract spaces, but it may travel along the straight line in either direction.
- Once all the apertures are found in either space, we can identify the pupil and window. The pupil is an image of the aperture stop and the window is an image of the field stop.
- In object space, we call these the entrance pupil and entrance window. In image space, we call these the exit pupil and exit window.

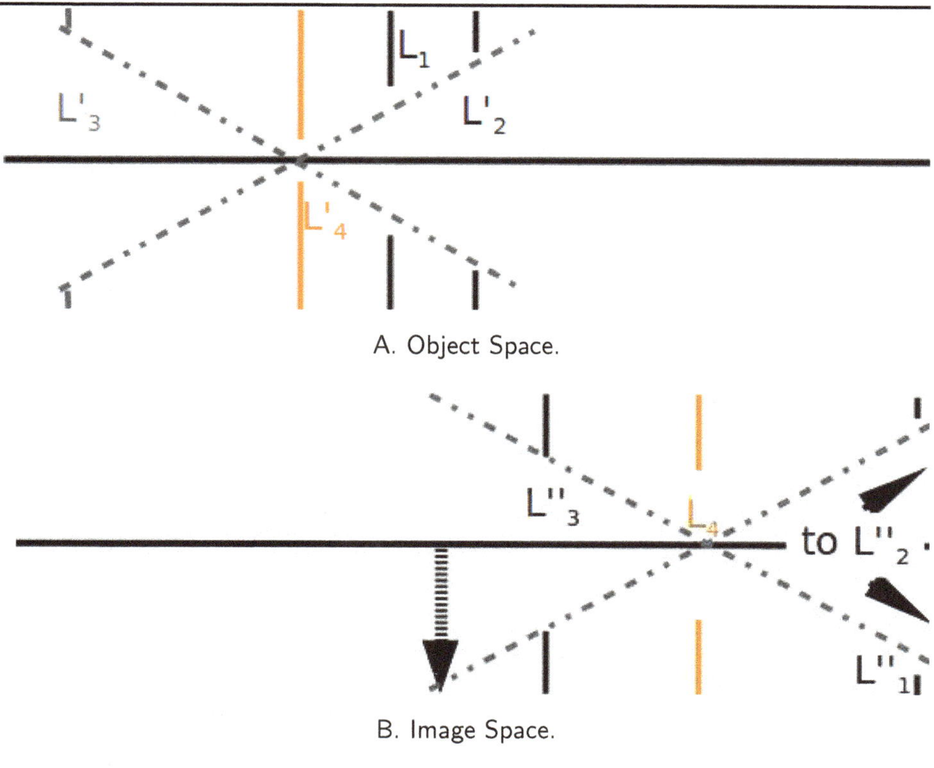

FIGURE 4.17
Entrance and exit windows. In object space (A), the image of Lens L_3 is the entrance window. The object is at infinity. In image space (B), the image of Lens L_3 is the exit window. The window subtends the smallest angle as viewed from the center of the pupil. The field stop is Lens 3, whether determined from the entrance window or the exit window.

- The pupil is the aperture that subtends the smallest angle from a point at the base of the object or image.
- The window is the aperture that subtends the smallest angle from a point in the center of the pupil.
- All other apertures have no effect on rays traveling through the system.
- The functions of the apertures might change if the object is moved. Such a change is usually undesirable.
- In object space or image space, the pupil and window may be before or after the object or image, but light travels sequentially through the system from one real element to the next, so the apertures have an effect regardless of their location.

4.4 Typical Optical Instruments

Two of the most common optical instruments are the telescope and the microscope. In Section 3.4.4, we saw that the telescope normally has a very small magnification, $|m| \ll 1$ which provides a large angular magnification. In contrast, the goal of the microscope is to produce a large magnification. Both instruments were originally designed to include

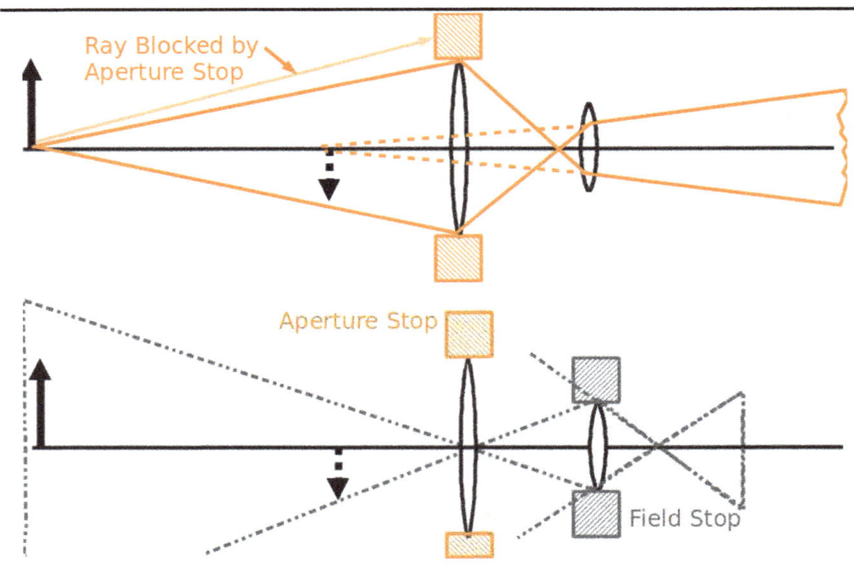

FIGURE 4.18
Telescope. The top panel shows that the primary is the aperture stop, because it limits the cone of rays that passes through the telescope from the object. The bottom panel shows that the eyepiece or secondary is the field stop because it limits the field of view.

an eyepiece for visual observation, and even now the eyepiece is almost always included, although increasingly the image is delivered to an electronic camera for recording, display, and processing.

4.4.1 Telescope

At the end of Section 3.4.4, we discussed the possibility of a telescope with a negative secondary and indicated that there were reasons why such a telescope might not be as good as it appeared. Here we will discuss those reasons in terms of pupils and windows. A telescope is shown in Figure 4.18. Often large telescopes use reflective optics instead of lenses, but it is convenient to show lenses in these figures, and the analysis of stops is the same in either case. The object is shown as a vertical arrow, and the image is shown as a dotted vertical arrow indicating a virtual image. As before, the image is inverted. The top panel shows a cone from the base of the object to show which rays will pass through the telescope. It is readily apparent that the primary is the aperture which limits these rays, and is thus the aperture stop. The lower panel shows the cone from the center of the primary, traced through the system to show the field of view. The image subtends a larger angle than the object by a factor $1/m$, and the field of view is correspondingly larger in image space than in object space ($m \ll 1$). For example, if $m = 1/6$, a field of view of 5° in object space is increased to 30° in image space.

Eq. 3.91
$m = -f_2/f_1$.

In Figure 4.19, the telescope is analyzed in object space, where it consists of the primary and an image of the secondary. From Equation 3.91, very small magnification, $|m| = |f_2/f_1| \ll 1$ requires that the focal length of the primary be much larger than that of the secondary. Thus, the secondary, treated as an object, is close to the focal point of the primary ($f_1 + f_2 \approx f_1$), and its image is thus close to infinity and thus close to the object. The primary is the entrance pupil, and the image of the secondary is the entrance window. Clearly, a large diameter pupil is an advantage in gathering light from faint stars and other celestial objects.

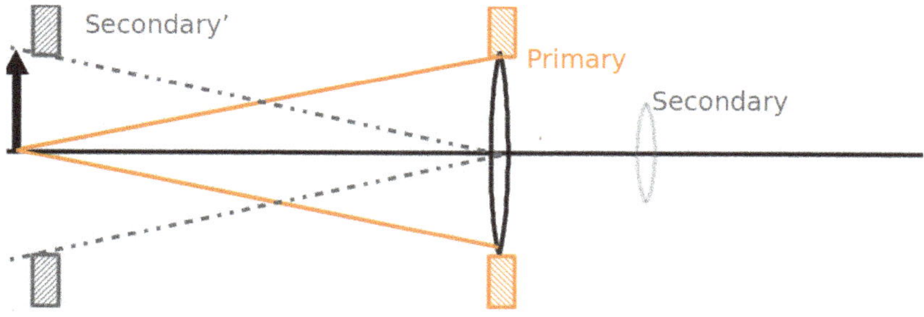

FIGURE 4.19
Telescope in object space. The primary is the entrance pupil and the image of the secondary is the entrance window.

Figure 4.20 shows the telescope in image space, where it is described by the secondary and the image of the primary. Again, the focal length of the primary is large and the distance between lenses is $f_1 + f_2 \approx f_1$, so the image of the primary is near the back focal point of the secondary. The image of the primary is very small, being magnified by the ratio $m_p \approx f_2/(f_1+f_2)$, and it is normally the exit pupil. The secondary becomes the exit window, and these results are all consistent with those in image space, and those found by tracing through the system in Figure 4.18.

If the telescope is to be used by an observer, it is important to put the entrance pupil of the eye close to the exit pupil of the telescope. Indeed, this result is very general; to combine two optical instruments, it is important to match the exit pupil of the first to the entrance pupil of the second. This condition is satisfied in the middle panel of Figure 4.22. The distance from the last vertex of the secondary to the exit pupil is called the **eye relief**. The numerical aperture is limited in image space either by the image of the primary or by the pupil of the eye, whichever is smaller. In any event, they are at the same location. The field of view is limited by the secondary.

In Practice

When two optical systems are combined, it is important to keep the following in mind.

- It is important to match the exit pupil of the first to the entrance pupil of the second. See Figure 4.21 for an example where the first system is a telescope or microscope and the second is the observer's eye.

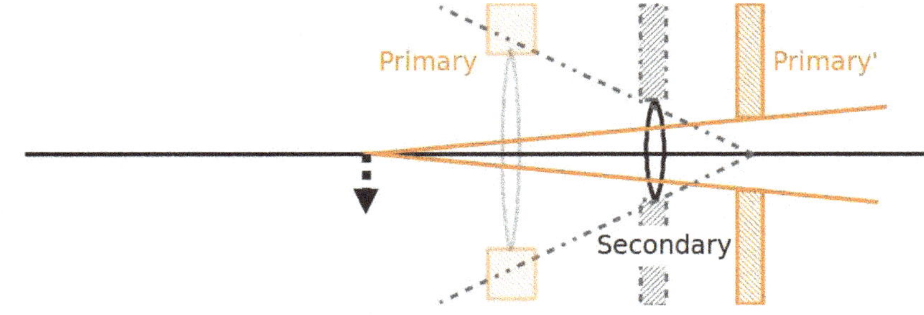

FIGURE 4.20
Telescope in image space. The image of the primary is the exit pupil and the secondary is the exit window.

FIGURE 4.21
Matching the pupils. If the exit pupil of the system is at the same axial location as the entrance pupil of the eye, then the system has a single pupil plane. The NA is determined by the overlap of the two pupils.

- The pupil area of the combination is the overlap of the two pupils, normally the smaller of the two.
- The larger of the two pupils determines how much tolerance can be allowed on the transverse motion of the smaller one. For example, in a head-up display for a pilot, the exit pupil of the system must be large enough to allow for the pilot to move around in the seat (and possibly for pilots of different heights if adjustment is not provided).

If the observer is too far back (A), the pupil of the combined telescope and eye is the pupil of the eye. However, the image of the primary now limits the field of view, and has become the exit window of the telescope. This result is undesirable as it usually results in a very small field of view. If the eye is placed too close to the secondary (C), a similar result is obtained. The intended pupil of the telescope has become the window, and limits the field of view. However, because it is a virtual object to the eye, having a negative object distance, the eye cannot focus on it, and the edge of it appears blurred. New users of telescopes and microscopes frequently place their pupils in this position. The best configuration (B) is to match the pupil locations. In general, when combining two optical systems, the entrance pupil of the second must be at the exit pupil of the first in order to avoid unpredictable results.

A telescope with a negative secondary has negative eye relief. The eye pupil cannot be located at the exit pupil of the telescope. See the end of Section 4.4.2 for details.

Now we can understand the comment at the end of Section 3.4.4. If the secondary has a negative focal length, the image of the primary will be virtual, to the left of the secondary, and it will be impossible to match its location to the entrance pupil of the eye. Such a telescope could still be useful for applications such as beam expanding or beam shrinking, although even for those applications, it has undesirable effects as we shall see in the next section.

Take-Away

Telescope Apertures:

- A telescope used for astronomical purposes normally has a magnification $m \ll 1$.
- The aperture stop is normally the primary of the telescope, and functions as the entrance pupil. The exit pupil is near the back focal point of the secondary.
- The secondary is the field stop and exit window. The entrance window is near infinity in the object space.
- The eye relief is the distance from the back vertex of the secondary to the exit pupil, and is the best position to place the entrance pupil of the eye or other optical instrument.

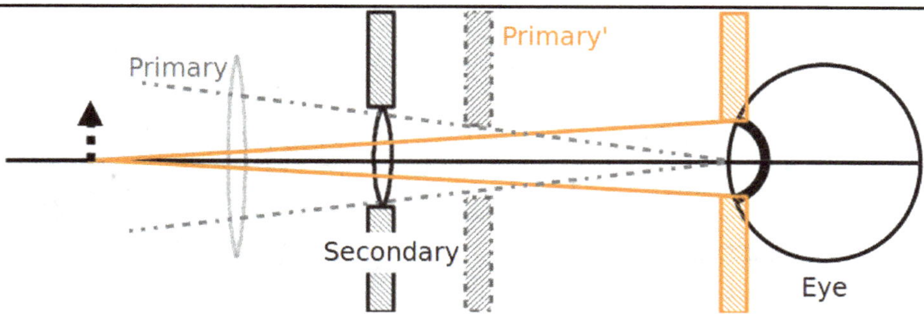

A. Eye Too Far Away.

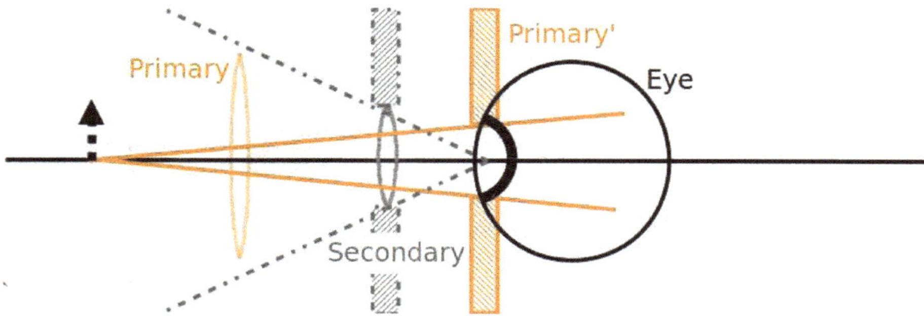

B. Pupils Matched.

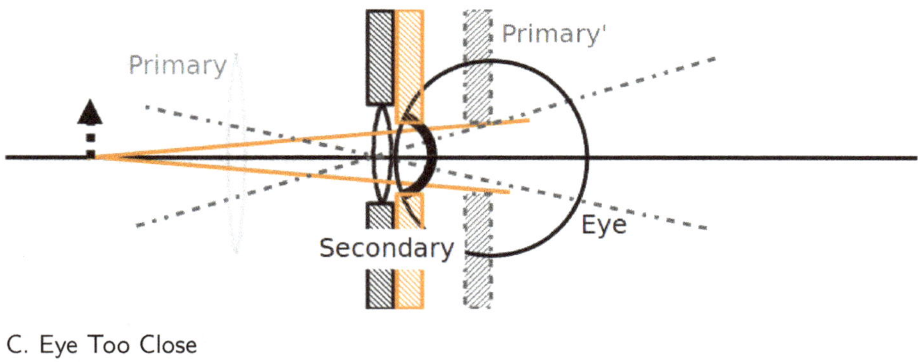

C. Eye Too Close

FIGURE 4.22
Matching pupils. If the eye pupil is too far away from the telescope pupil (A), the eye pupil becomes the pupil of the combined system and the telescope primary becomes the field stop. At the other extreme the eye is too close to the telescope (C), and becomes the pupil. The primary again becomes the field stop. Matching pupils (B) produces the desired result. The telescope pupil is invisible to the observer.

- Telescopes can also be used as beam expanders or reducers in remote sensing, microscopy, and other applications.

4.4.2 Scanning

An understanding of windows and pupils is essential to the design of scanning systems which are common in such diverse applications as remote sensing of the environment and confocal microscopy. Suppose for example that we wish to design a laser radar for

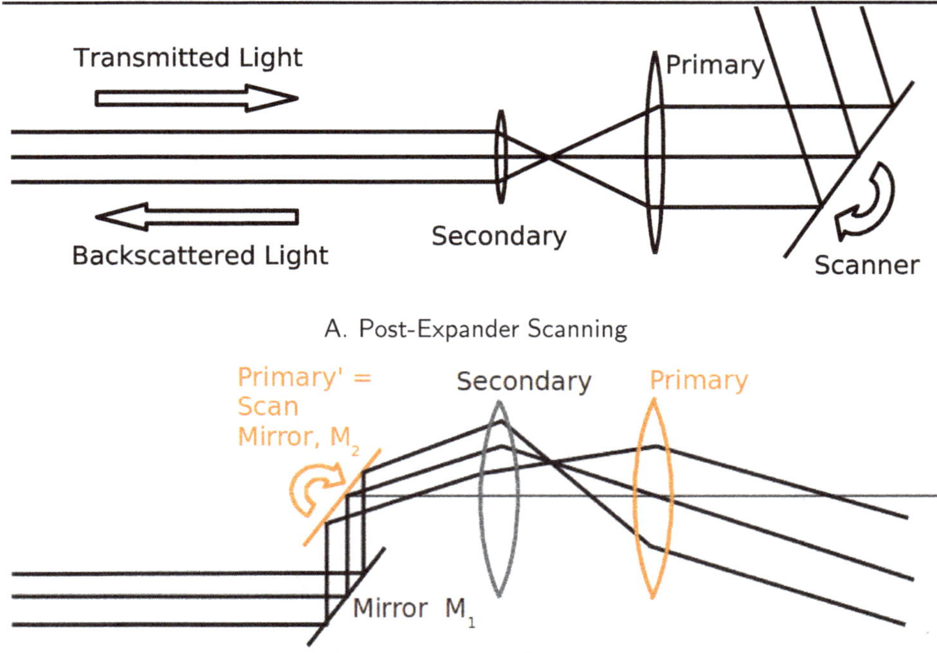

FIGURE 4.23
Telescopes with scanners. In post-expander scanning (A), a large mirror is used after the beam-expanding telescope to direct the laser beam to the target. Alternatively (B), the scanning mirror is located at the entrance pupil of the beam-expanding telescope. Because the entrance pupil of the telescope is an image of the primary, rays that pass through the center of the scan mirror pass through the center of the primary. Thus, the beam is scanned in the primary.

some remote sensing application. We will consider a monostatic laser radar in which the transmitter and receiver share a single telescope. To obtain sufficient backscattered light, the laser beam is typically collimated or even focused on the target with a large beam expanding telescope. The use of a large primary provides the opportunity to gather more light, but also offers the potential to make the transmitted beam smaller at the target as we shall see in Chapter 8. In many cases, atmospheric turbulence limits the effectiveness of beams having a diameter greater than about 30 cm so that diameter is commonly chosen. The telescope, being used in the direction opposite that used for astronomical applications, will have a large magnification, $m \gg 1$. We still refer to the lens at the small end of the telescope as the secondary and the one at the large end as the primary, despite the "unconventional" use.

One approach to scanning is to place a tilting mirror after the primary of the telescope as shown in Figure 4.23A. As the mirror moves through an angle of $\alpha/2$, the beam moves through an angle of α. The obvious disadvantage of this configuration is the need for a large scanning mirror. If the nominal position of the mirror is at 45°, then the longer dimension of the mirror must be at least as large as the projection of the beam diameter, $30\,\text{cm} \times \sqrt{2}$. A mirror with a large area must also have a large thickness to maintain flatness. The weight of the mirror will therefore increase with the cube of the beam size, and moving a heavy mirror requires a powerful motor. A further disadvantage exists if two-dimensional scanning is required. A second mirror placed after the first, to scan in the orthogonal direction, must have dimensions large enough to accommodate the beam diameter and the scanning of the first mirror. Even if the mirrors are placed as close as

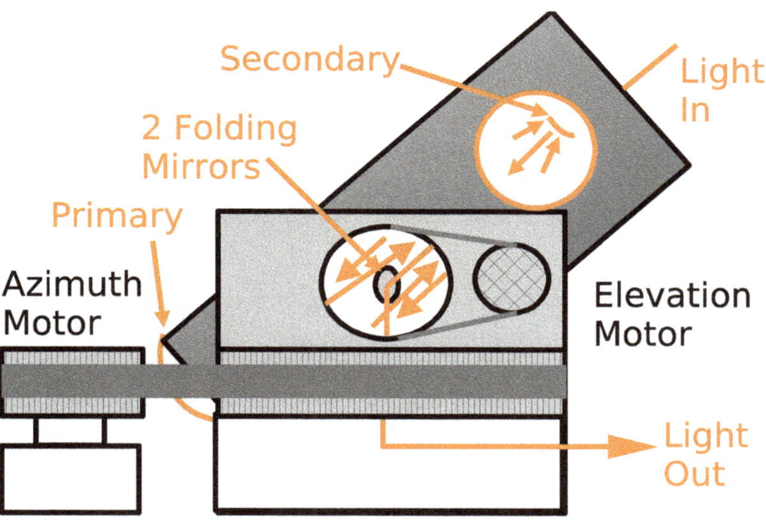

FIGURE 4.24
Coudé telescope. Light reflects from the primary to the secondary and then to a folding mirror which is mounted to the telescope housing. The housing is moved in elevation by a motor. This complete assembly, in turn, is moved in azimuth by another motor with a second folding mirror.

possible to each other, the size of the second mirror must be quite large. Often, alternative configurations such as single gimbaled mirrors or siderostat are used. Alternatively, the entire telescope can be rotated in two dimensions on a gimbaled mount, with the beam coupled through the gimbals in a so-called Coudé mount as shown in Figure 4.24. "Coudé" is the French word for "elbow."

Chief ray defined

A completely different approach is to use pre-expander scanning, as shown in Figure 4.23B. It is helpful to think of scanning in terms of chief rays. The **chief ray** from a point is the one which passes through the center of the aperture stop. Thus, a scanning mirror pivoting about the center of a pupil plane will cause the chief ray to pivot about the center of the aperture stop and of every pupil. In Figure 4.23B, the transmitted laser light reflects from a fixed mirror, M_1, to a scanning mirror, M_2. We might expect that scanning here would cause the beam to move across the pupil of the telescope, requiring an increase in its diameter. However, if we place the scanner at the pupil of the telescope, then the chief ray that passes through the center of the scanner will pass through the center of the primary; the pupil and the primary are conjugate so, although the beam scans across the secondary, it simply pivots about the center of the primary. Now the scan mirror can be smaller than in Figure 4.23A by $1/m$. If the laser beam is 1 cm in diameter, then $m = 30$ and the scan mirror can be reduced in diameter by a factor of 30, or in mass by 30^3. This extreme advantage is offset by some disadvantages. First, the fact that $m \gg 1$ means that $m_\alpha \ll 1$, so the mirror must move through an angle $m\alpha/2$ (instead of $\alpha/2$ for post-expander scanning) to achieve a scan angle of α. Second, the size of the secondary must be increased to accommodate this large scan angle. Third, scanning in two dimensions becomes problematic. A gimbaled mirror is still an option, but is complicated. We could use two scanning mirrors. For example, instead of a fixed mirror at M_1, we could use a mirror scanning in the direction orthogonal to that of M_2. Such an approach is problematic because at best only one of the two mirrors can actually be in the pupil plane. It may be possible to place the two mirrors close enough together, one on each side of the pupil plane, so that their images are both close enough to the location of the primary and to make the primary slightly larger than the expanded beam, so that this configuration can still be used. Otherwise a small 1:1 "relay" telescope can be used between the two scanning mirrors, producing the desired scan at the expense of additional size and weight.

In the previous section, we discussed the disadvantages of a telescope with a negative secondary, often called a Galillean telescope but noted that such a telescope could find use in a beam expander. However, as we have seen, the lack of a real pupil will also prevent the effective use of pre-expander scanning. These telescopes offer an advantage in compactness for situations where scanning is not required, but initial alignment may be difficult. With a positive secondary (such a telescope is called Keplerian), we have a real pupil, and can design a system with an alignment mirror at or near the pupil (like M_2 in Figure 4.23), and one closer to the light source (like M_1). Then M_1 is aligned to center the beam on the primary, and M_2 is adjusted to make the beam parallel to the telescope axis. Even if M_2 is not positioned exactly at the pupil, alignment can be achieved using a simple iterative procedure, but this process is often more difficult with a virtual pupil that lies deep inside the telescope.

In Practice

- Scanning optics are ideally placed in real pupils.
- If two scanning mirrors are needed, it may be necessary to use relay optics to generate an additional pupil.
- It may also be possible to place two scanning mirrors "close" to the pupil.
- An adjustable alignment mirror need not be placed near a pupil, but doing so may make alignment easier.
- One may be tempted to place an adjustable mirror near an image plane, but it is important to remember that dust and imperfections on the mirror will be imaged

4.4.3 Magnifiers and Microscopes

One of the most common tasks in optics is magnification. A single lens can provide sufficient magnification to observe objects that are too small to be observed by the human eye, and modern microscopes are capable of resolving objects smaller than a wavelength of light and observing biological processes inside cells. The history of the microscope is filled with uncertainty and controversy. However, it is widely believed that Johannes (or Hans) and his son Zacharias Janssen (or Iansen or Jansen) invented the first compound microscope (using two lenses), in 1590, as reported in a letter by Borel[27]. Its importance advanced with the physiological observations of Robert Hooke[98] in the next century. Further observations and exceptional lens-making skills, leading to a magnification of almost 300 with a simple magnifier, have tied the name of Antonin (or Anton or Antonie) Leeuwenhoek[140] to the microscope. Some interesting historical information can be found in a number of sources[212;41;70]. Despite the four-century history, advances in microscopy have grown recently with the ever-increasing interest in biomedical imaging, new contrast agents, the laser as a light source for novel imaging modes, and the computer as a tool for image analysis. For further reading on the use of microscopes in biomedical imaging see the recent book by Ober, Ward, and Choi[169]. As evidence of the continued interest in microscopy, the Nobel Prize in Chemistry was awarded to Eric Betzig, Stefan W. Hell, and William E. Moerner, "for the development of super-resolved fluorescence microscopy."

FURTHER READING: Quantitative Bioimaging[169].

The basic microscope has progressed through three generations, the simple magnifier, consisting of a single lens, the compound microscope, consisting of an "objective" and

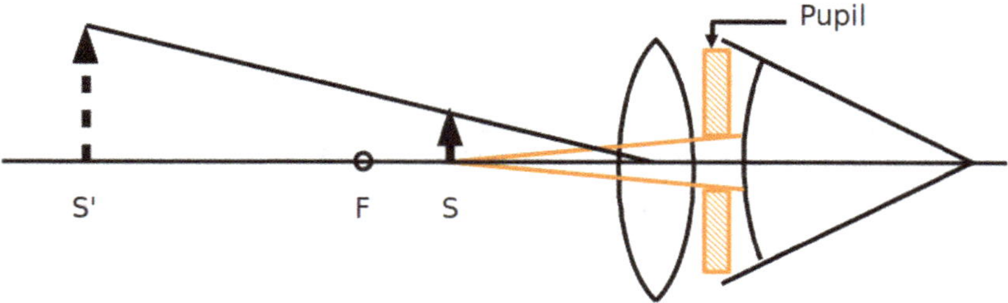

FIGURE 4.25
Simple magnifier. The object is placed just inside the front focal point.

"eyepiece," and finally the infinity-corrected microscope, consisting of the "objective," the "tube lens," and the "eyepiece." We will discuss each of these in the following sections.

4.4.3.1 Simple Magnifier

Let us begin with an example of reading a newspaper from across a room. We can read the headlines and see the larger pictures, but the majority of the text is too small to resolve. Our first step is to move closer to the paper. A 10-point letter (10/72 in. or about 3.53 mm) subtends an angle of about 0.02° at a distance of 10 m, and increases to 0.4° at a distance of 0.5 m. The height of its image on the retina is increased by a factor of 20. However, this improvement cannot continue indefinitely. Short distances require the use of muscles to change the focusing of the eye, and at some point, we reach the minimum distance at which the eye can focus. A person with normal vision can focus on an object at a distance as close as about 20 cm, and can read 10-point type comfortably. If we wish to look closer, to examine the "halftone" pattern in a photograph for example, the focal length of the eye will not reduce enough to produce a sharp image on the retina.

Figure 4.25 shows the use of a simple magnifier, consisting of a positive lens, with the object placed slightly inside the front focal point. The image can be found using the lens equation, Equation 2.51.

Eq. 2.51 $\frac{n}{s} + \frac{n'}{s'}$.

$$\frac{1}{s'} = \frac{1}{f} - \frac{1}{s} \tag{4.19}$$

$$s' = -\frac{fs}{f-s} \approx -\frac{f^2}{f-s}. \tag{4.20}$$

With $f - s$ small but positive, s' is large and negative, resulting a virtual image as shown in the figure. The magnification, from Equation 2.55 is large and positive;

$$m = \frac{-s'}{s} > 0. \tag{4.21}$$

It would appear that we could continue to make m larger by moving the object closer to the focal point, but we do so at the expense of increasing $|s'|$. The angular extent of an object of height x will be $|mx/s'|$. It may surprise us to complete the calculation and realize that the angle is x/s; the image always subtends the same angular extent, and thus the same dimension on the retina. What good is this magnifier? The advantage is that the image can be placed a distance $-s' > 20$ cm in front of our eyes, so we can now focus on it.

Object slightly inside front focus $\rightarrow m$ is positive and large.

How can we define the **magnification** of this lens? We could use the magnification equation, Equation 2.54, but it does not really help, because we can always move the object a little closer to the focus and compute a larger number, but it offers no advantage

Eq. 2.54: $m = \frac{-s'}{s}$.

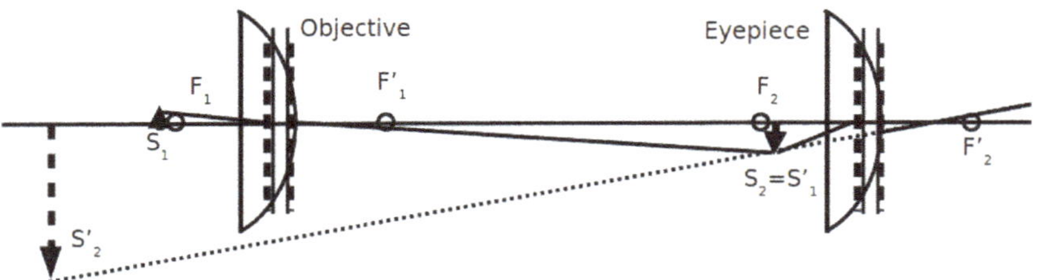

FIGURE 4.26
Compound microscope. The objective produces a real inverted image. The eyepiece is used like a simple magnifier to provide additional magnification.

to the user. It is common practice to define the magnification of a simple magnifier as

$$M = \frac{20\,\text{cm}}{f}, \tag{4.22}$$

Magnification of a simple magnifier.

where we have used the capital letter to distinguish from the true magnification. The distance of 20 cm is chosen somewhat arbitrarily as the shortest distance at which the normal eye can focus.

We can now increase the magnification by decreasing the focal length. However, we reach a practical limit. The f-number of the lens is f/D where D is the diameter. We will see in Chapter 5 that lenses with low f-number are expensive or prone to aberrations. Normally, the pupil of this system is the pupil of the eye. If we reduce D along with f to maintain the f-number, we eventually reach a point where the lens becomes the pupil, and the amount of light transmitted through the system decreases. A practical lower limit for f is therefore about 1 cm, so it is difficult to achieve a magnification much greater than about 20 with a simple magnifier. It is difficult, but not impossible; Leeuwenhoek achieved much larger magnification with a simple lens. Most certainly he did this with a very small lens diameter, and achieved a dim image even with a bright source. These days, for magnification beyond this limit, we turn to the compound microscope.

Take-Away

Simple Magnifier:
The simple magnifier provides an upright image at a sufficient distance for viewing by eye, but has limited magnification.

4.4.3.2 Compound Microscope

The original microscopes consisted of two simple lenses, the **objective** and the eyepiece. We now refer to such a microscope as a **finite-conjugate microscope**; The image formed by the objective is at a finite distance. Until recent years, the same basic layout was used, even as the objectives and eyepieces were changed to compound lenses. For simplicity, the microscope in Figure 4.26 is shown with simple lenses, but distances are measured between principal planes, so one can easily adapt the analysis to any compound objective and eyepiece.

160 mm tube length was once the standard of the Royal Microscope Society. At the time the "RMS" standard included not only tube length, but such details as specifications for the screw threads of the objective.

The object is placed at S_1, slightly outside the front focal point of the objective lens, and a real inverted intermediate image is produced at S_1', a long distance, s_1' away. In a typical microscope, the tube length is 160 mm, and the magnification is approximately $m = 160\,\text{mm}/f$ where f is the focal length of the objective. Thus, a 10×, 20×, or 40× objective would have a focal length of about 16, 8, or 4 mm, respectively. There are some microscopes with different tube lengths such as 150 or 170 mm.

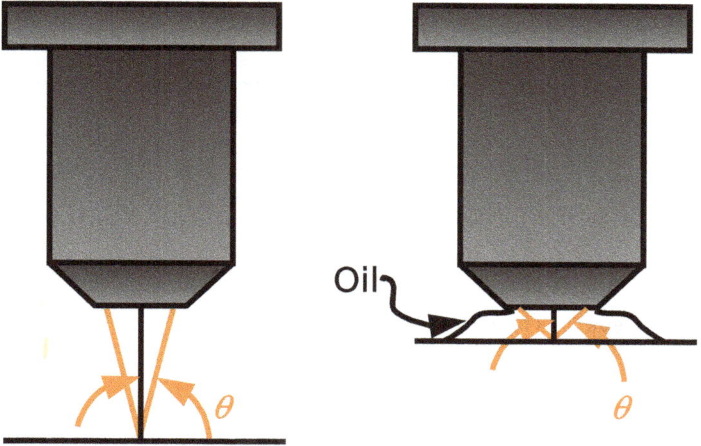

FIGURE 4.27
Microscope objectives. Low magnification objectives typically have low NA, long focal lengths, and long working distances (Left). High magnification objectives have shorter focal lengths, NA, and working distances. They are often oil- or water-immersion objectives. Most objectives are compound lenses with many elements and the aperture stop is inside the objective.

The eyepiece is used as a simple magnifier, using the image from the objective at S_1' as an object, S_2, and producing a larger virtual image at S_2'. Because the eyepiece in this configuration does not invert ($m_2 > 0$), and the objective does ($m_1 < 0$), the final image is inverted ($m = m_1 m_2 < 0$).

Microscope in object space.

Figure 4.28 shows the apertures in object space. Because the focal length of the objective is short, the eyepiece may be considered to be at almost infinite distance, and it will have an image near the front focal point of the objective. Because this location is near the object, this aperture will not limit the cone of rays from the object, and the objective is the aperture stop. Thus, the objective is also the entrance pupil, and determines the object-space numerical aperture of the system. Normally the objective is a rather complicated compound lens, and deliberately contains an aperture between some of the elements to limit the numerical aperture. Microscope objectives, unlike most camera lenses and telescope lenses, are designed to have a high numerical aperture in object space. Recalling that the numerical aperture is $n \sin\theta$, the upper limit is n, the index of refraction of the medium in which the object is immersed, and modern microscope objectives can have numerical apertures approaching n. Of course, a large numerical aperture requires a large diameter relative to the focal length. For this reason, high numerical apertures are usually associated with short focal lengths and high magnifications. For example, a 10× objective might have a numerical aperture of 0.45, while a 20× might have NA = 0.75, and a 100× oil-immersion objective could have NA = 1.45, with the oil having an index of refraction slightly above 1.5. Figure 4.27 illustrates some examples.

Objective determines aperture stop.

The image of the eyepiece, near the front focal point of the objective, thus becomes the entrance window, limiting the field of view. If the eyepiece has a diameter of $D_e = 10\,\text{mm}$, then the field of view in linear units will be

The eyepiece can function as the field stop. Often a separate aperture is deliberately introduced as the field stop. When using a camera, the camera normally is the field stop.

$$FOV = \frac{D_e}{m} = \frac{10\,\text{mm}}{10} = 1\,\text{mm} \qquad (10\times \text{ objective}) \tag{4.23}$$

$$\frac{10\,\text{mm}}{60} = 0.17\,\text{mm} \qquad (60\times \text{ objective}).$$

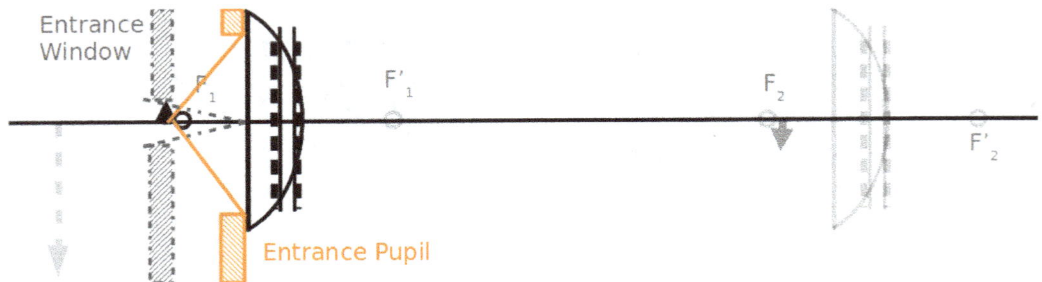

FIGURE 4.28
Microscope in object space. The entrance window is near the object. The objective is both the entrance pupil and aperture stop.

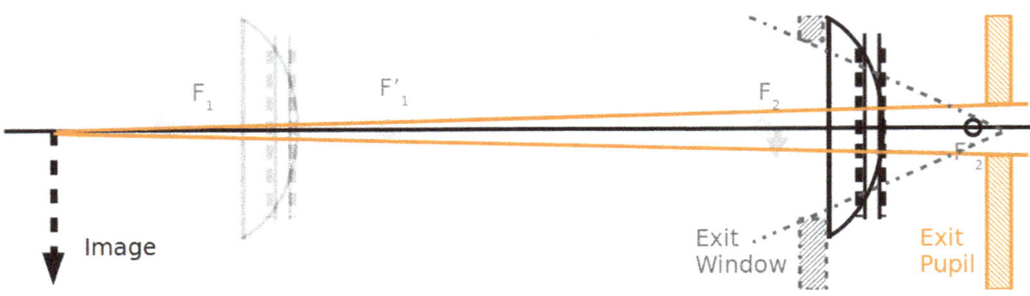

FIGURE 4.29
Microscope in image space. The exit pupil is near the back focus of the eyepiece, providing appropriate eye relief. The eyepiece forms the field stop and exit window.

Microscope in image space.

In image space, the eyepiece is the field stop and the exit window. The exit pupil is the image of the objective, as shown in Figure 4.29. Because the focal length of the eyepiece is short compared to the tube length, the objective can be considered to be at infinity, and the exit pupil is near the back focal plane of the eyepiece, providing convenient eye relief, as was the case with the telescope. With a 100× objective and a 10×-20× eyepiece, total magnification of 1000-2000 is possible. A camera can be placed at the intermediate image between the objective and the eyepiece. If an electronic camera that has a thousand pixels, each 5 μm^2, then with a 10× objective, the pixel size on the object will be 0.5 μm, and the field of view in object space will be 500 μm.

In a typical microscope, several objectives can be mounted on a turret so that the user can readily change the magnification. Focusing on the sample is normally accomplished by axial translation of the objective, which moves the image, and requires readjustment of the eyepiece. Many modern applications of microscopy use cameras, multiple light sources, filters, and other devices in the space of the intermediate image. We will see in Chapter 5 that inserting beam-splitting cubes and other optics in this space will introduce aberrations. To minimize the aberrations and the effect of moving the objective, modern microscopes use infinity-corrected designs, which are the subject of the next section.

Take-Away

- In a compound microscope with an objective and eyepiece, the magnification is given by the ratio of the tube length to the focal length of the objective.

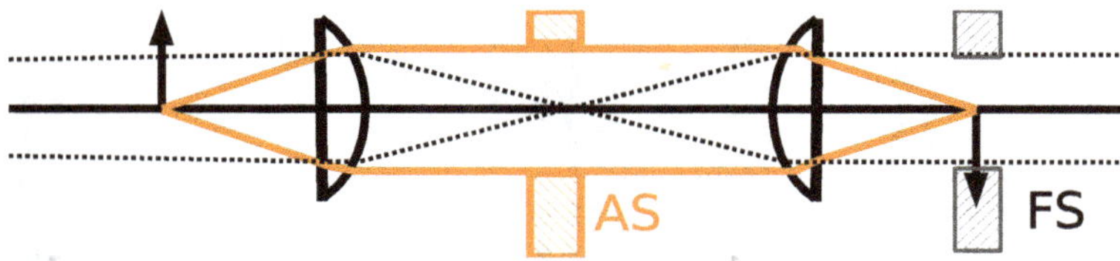

FIGURE 4.30
Telecentric imaging. The lenses are separatec by the sum of their focal lengths. The object is at the front focal plane of the first lens, the image is at the back focal plane of the second, and the pupil is at the common intermediate focal plane.

- The aperture stop is normally located inside the objective and limits the numerical aperture.
- Higher magnification objectives have shorter focal lengths, and typically have shorter working distances and higher numerical aperture.
- The field stop is either the eyepiece, the camera, or a deliberately introduced stop. It limits the field of view.

4.4.3.3 Infinite-Conjugate Compound Microscope

In recent decades, the basic compound microscope design has undergone extraordinary changes. Along with a change to infinity-corrected designs, pupil and window sizes have increased, and the ability to integrate cameras, lasers, scanners, and other accessories onto the microscope has grown. Along the way, the standardization that was part of the old microscope design has been lost, and such standards as the 160 mm tube length, and even the standard screw threads on the objectives have been lost. The name **infinite-conjugate** is used because the image formed by the objective is at infinity. We say the objective is **infinity-corrected**. We will see in Chapter 5 that for high-resolution imaging, a lens must be corrected for aberrations at a specific object and image distance. Formerly the correction was for an image distance of 160 mm. The new standard is to correct the lens for an infinite image distance, and then to use a **tube lens** to generate the real image, as shown in Figure 4.30. We briefly introduced the idea of telecentric systems in Section 3.4.4, when we discussed telescopes, and showed that the magnification is given by the ratio of the focal lengths of the two lenses in Equation 3.91. In the microscope, we will introduce an aperture at the back image plane for a field stop, and another between the lenses at the common intermediate focal plane as an aperture stop. We apply this to the microscope as shown in Figure 4.31. We now define the magnification in conjunction with a particular tube lens as

Infinity-corrected objective defined.

A telecentric lens pair, objective and tube lens, forms the basis for the modern infinity-corrected microscope.

$$M = \frac{f_{tube}}{f_{objective}}, \tag{4.24}$$

and the focal length of the tube lens, f_{tube} may be different for different microscopes, depending on the manufacturer. Typical tube lenses have focal lengths of up to 200 mm or even more. These longer focal lengths for the tube lenses lead to longer focal lengths for objectives at a given magnification, and correspondingly longer working distances. The working distance is important for imaging thick samples, and even for working through a glass cover-slip placed over a sample. It is important to note that if an infinity-corrected microscope objective is used with a tube lens other than the one for which it was designed, a good image will still be obtained, but the actual magnification will be

$$m = M\frac{f_{tube(actual)}}{f_{tube(design)}}. \tag{4.25}$$

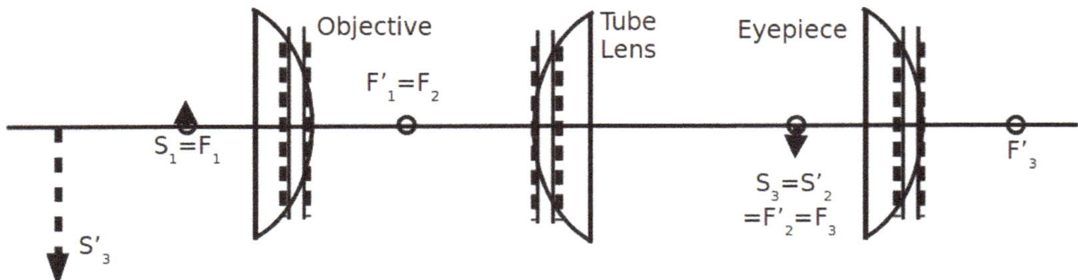

FIGURE 4.31
Microscope with infinity-corrected objective. The objective can be moved axially for focusing without introducing aberrations.

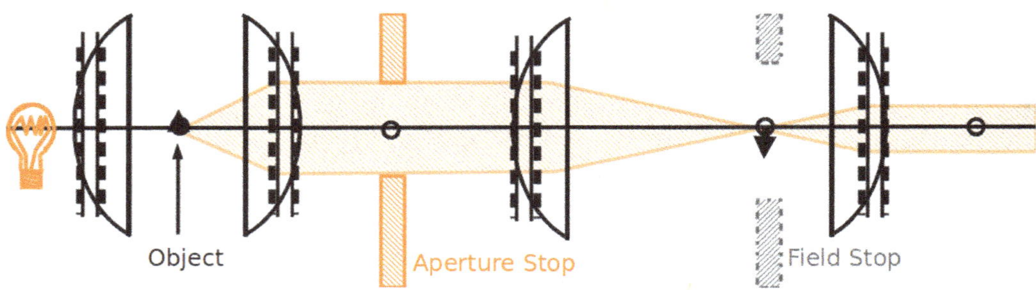

FIGURE 4.32
Rays from object. The object is imaged at the back focus of the tube lens and then at or near infinity after the eyepiece.

After the real image is formed, it can be imaged directly onto a camera, passed through spectroscopic instrumentation to determine the wavelength content of the light, or further magnified with an eyepiece for a human observer.

Infinity space defined.

A cone of light rays from a point in the object is shown in Figure 4.32. Because the rays are parallel in the so-called **infinity space** between the objective and the tube lens, beam-splitting cubes can be placed there to split the light among multiple detectors. In this case, multiple tube lenses may also be used. In this figure we have added a light source and a "condenser lens" to collimate the light onto the object. These will be discussed at the end of this section.

In Practice

Here are a few practical points to be considered when using microscope components in a system.

- One might be tempted to use a finite-conjugate objective in place of an infinite-conjugate one (or *vice versa*) under the assumption that it is only necessary to move the object distance a small amount. However aberrations (see Chapter 5) will degrade the image. The aberrations are more severe for larger NA.
- An infinity-corrected objective can usually be used with a tube lens other than the one intended, but the magnification will not be as expected.
- Some manufacturers use the tube lens to correct for aberrations that are not corrected in the objective. Such objectives can only be used with the appropriate tube lens.

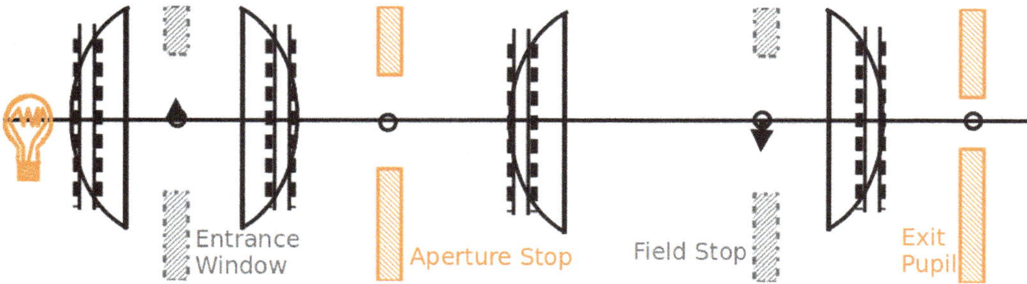

FIGURE 4.33
Microscope apertures. The entrance window is near the object, and the entrance pupil is at infinity. The exit pupil is at the back focus of the eyepiece and the exit window is at infinity, as is the image.

An aperture stop is placed at the back focal plane of the objective to limit the size of this cone of rays. The aperture stop is part of the objective lens, and may have different diameters for different lenses. The tube lens produces an inverted image, and a field stop is placed in the plane of this image, limiting the field of view in this plane to a fixed diameter, 24 mm for example. Because the magnification is M and the field stop is exactly in the intermediate image plane, the entrance window in this case has a diameter of 24 mm/M, limiting the field of view in the object plane, as shown in Figure 4.33.

Aperture stop and field stop for an infinite-conjugate microscope.

The aperture stop is located at the back focal plane of the objective, so the entrance pupil is located at infinity. Thus, the object-space description of the microscope has a window of size 24 mm/M in this example, and a pupil at infinity, subtending a half angle of $\arcsin(NA/n)$, where n is the index of refraction of the immersion medium.

Infinite-conjugate microscope in object space.

In image space, the exit pupil is the image of the aperture stop and has a diameter equal to that of the aperture stop times $f_{eyepiece}/f_{tube} \ll 1$. The eye relief (from the back of the eyepiece to the exit pupil) is still approximately $f_{eyepiece}$ as in the previous microscope example, but now the exit window is located at infinity, because the field stop is at the front focal plane of the eyepiece. Thus, the exit window is in the same plane as the image.

Infinite-conjugate microscope in image space.

If we wish to build a microscope, we may use objectives and an eyepiece from manufacturers that will not want to provide the details of their proprietary designs. Thus we may not know the locations of principal planes or any of the components that make up these complicated compound lenses. However, all we need are the locations of the focal points, and we can assemble an infinite-conjugate microscope.

Let us briefly discuss the illumination problem. Assume we want **transillumination** (through the sample) as opposed to **epi-illumination** (illuminating and imaging from the same side of the sample). If we put a tungsten light source before the sample, and the sample is reasonably clear, we will see the filament of the light source in our image. To avoid this problem, we can put some type of diffuser such as ground glass between the source and the sample. It will scatter light in all directions, leading to more uniform illumination at the expense of lost light. **Ground glass** can be made by sand-blasting a sheet of glass to create a rough surface, and is commonly used in optics for its scattering behavior.

Ground-glass viewing screen.

There are ample opportunities to lose light in an optical system, without incurring a large loss at the diffuser. Can we do better? If we place the light source in a pupil plane, then it will spread more or less uniformly over the object plane. For an observer, the image of the filament will be located at the pupil of the eye, with the result that the user cannot focus on it. Thus, it is not expected to adversely affect the image. Viewed in object space, the light source is at infinity, and thus provides uniform illumination. The approach, called **Köhler illumination**, is shown in Figure 4.34. Light from the filament uniformly illuminates the object plane, and every plane conjugate to it: the intermediate image, the image through the eyepiece, the image on the retina, and any images on cameras in other

Köhler Illumination.

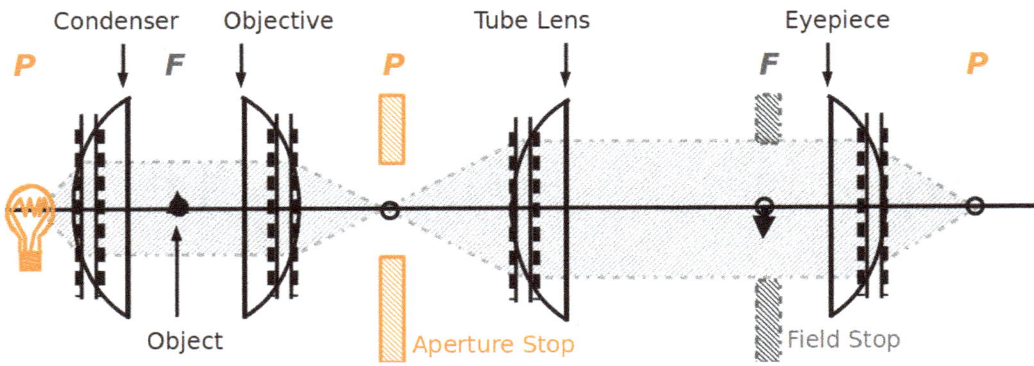

FIGURE 4.34
Köhler illumination. In this telecentric system, field planes (**F**) and pupil planes (**P**) alternate. The light source is placed in a pupil plane. In object space, the light source appears infinitely far to the left and the entrance pupil appears infinitely far to the right. The system is telecentric until the eyepiece. The intermediate image is not quite at the front focal plane of the eyepiece, and the final (virtual) image is not quite at infinity.

image planes. We will address Köhler illumination again in Chapter 11, when we study Fourier optics.

Telecentric defined.

Figure 4.34 is a good example of a **telecentric** chain of lens pairs. The lenses are all positioned so that the back focal plane of one is the front focal plane of the next. In a telecentric system, the chief rays (see Section 4.4.2) from all points in the object are parallel to each other so the entrance pupil must be at infinity. This condition can be achieved by placing the object (and the entrance window) at the front focal plane of a lens and the aperture stop at the back focal plane. One of the advantages of a telecentric system is that the magnification is not changed if the object is moved in the axial direction. This constraint is important in imaging three-dimensional structures without introducing distortion. In Figure 4.34. In the figure, there is one image plane conjugate to the object plane, and there are two planes conjugate to the aperture stop, the light source and the exit pupil. We will call the object plane and its conjugates **field planes**, labeled with F, and the aperture stop and its conjugates **pupil planes**, labeled with P.

Field planes and pupil planes.

Example

Let's determine the pupil size and object field of view in each space for the microscope in Figure 4.34. Let's assume the following focal lengths:

Condenser	$f_{condenser} = 15\,\text{mm}$
Objective	$f_{objective} = f_{tube}/M = 20\,\text{mm}$ NA=0.25
Tube Lens	$f_{tube} = 200\,\text{mm}$
Eyepiece	$f_{eyepiece} = 10\,\text{mm}$ with 2 mmField Stop

The filament of the light source is $x = 10\,\text{mm}$ long. The image of the filament in the pupil plane in the infinity space is $x' = x\frac{f_{objective}}{f_{condenser}} = 13.3\,\text{mm}$. This image is further magnified to $x'' = x'\frac{f_{eyepiece}}{f_{tube}} = 0.65\,\text{mm}$.

The real pupil must be about $f_{objective} * NA = 0.5\,\text{mm}$ in radius, or 10 mm diameter. See Table 4.3. We notice that the filament overfills the pupil.

Now let's put a camera at the image plane after the tube lens. The camera has 1280 (horizontal) by 1024 (vertical) pixels that are 4.8 square. The eyepiece contains a 10 mm diameter field stop. What is the field of view and pixel size on the object? Compare the view through the eyepiece and the camera. The magnification going from the field stop to the object is just $1/M = \frac{f_{objective}}{f_{tube}}$. To compare the view through the eyepeice we really should think about the angle, assuming the image from the eyepiece is at infinity. The angle is approximately the height of the object at the field stop divided by the focal length of the eyepiece. See Table 4.4

TABLE 4.3
Pupil Data for Example

	Filament Length	Pupil Diameter
At light source	$\ell_{filament} = 10\,\text{mm}$ given	$d_{pupil}\frac{f_{condenser}}{f_{objective}} = 3.75\,\text{mm}$
At pupil	$\ell_{filament}\frac{f_{objective}}{f_{condenser}} = 13.3\,\text{mm}$	$d_{pupil} = 10\,\text{mm}$ given
At eye	$\ell_{filament}\frac{f_{objective}}{f_{condenser}}\frac{f_{eyepiece}}{f_{tube}} = 0.67\,\text{mm}$	$d_{pupil} = 10\,\text{mm}\frac{f_{eyepiece}}{f_{tube}} = 0.5\,\text{mm}$

Take-Away

We have discussed the basic concepts of magnifiers and microscopes.

- A simple magnifier magnifies an object but the image distance increases with magnification.
- A simple magnifier can achieve a magnification of about 20.
- Two lenses can be combined to make a microscope, and large magnifications can be achieved.
- Modern microscopes have objectives with very large numerical apertures.
- Most modern microscopes use infinity-corrected objectives and a tube lens. The magnification is the ratio of the focal lengths of the tube lens and objective.
- The aperture stop is located at the back focal plane of the objective, the entrance pupil is at infinity, and the exit pupil is at the back focal plane of the secondary, providing eye relief.
- The field stop is located in the back focal plane of the tube lens, the entrance window is at the sample, and the exit window is at infinity.
- Köhler illumination provides uniform illumination of the object.

Problems

4.1 Coaxial Lidar: Pupil and Image Conjugates (G)

Consider the coaxial lidar (laser radar) system shown in Figure P4.1. The laser beam is 1 in. in diameter, and is reflected from a 45° mirror having an elliptical shape so that its projection along the axis is a 2 in. circle. (This mirror, of course, causes an obscuration of light returning from the target to the detector. We will make good use of this feature.) The

TABLE 4.4
Field-Plane Data for Example

	Camera Pixel	Camera Field of View	Window
At Field Stop	$4.8\,\mu$m given	H $1280 \times 4.8\,\mu m = 614\,\mu m$ V $1024 \times 4.8\,\mu m = 492\,\mu m$	10 mm
At object	$4.8\,\mu m \frac{f_{objective}}{f_{tube}} = 0.48\,\mu m$	H $1280 \times 4.8\,\mu m \frac{f_{objective}}{f_{tube}} = 61.4\,\mu m$ V $1024 \times 4.8\,\mu m \frac{f_{objective}}{f_{tube}} = 49.2\,\mu m$	$10\,mm \frac{f_{objective}}{f_{tube}}$ $= 1$ mm
At eye	$\frac{4.8\,\mu m}{f_{eyepiece}} = 480\,\mu$radians	H $1280 \times 4.8\,\mu m / f_{eyepiece} = 614$ mrad V $1024 \times 4.8\,\mu m = / f_{eyepiece} = 492$ mrad	$20\,mm / f_{eyepiece}$ $= 1$ rad

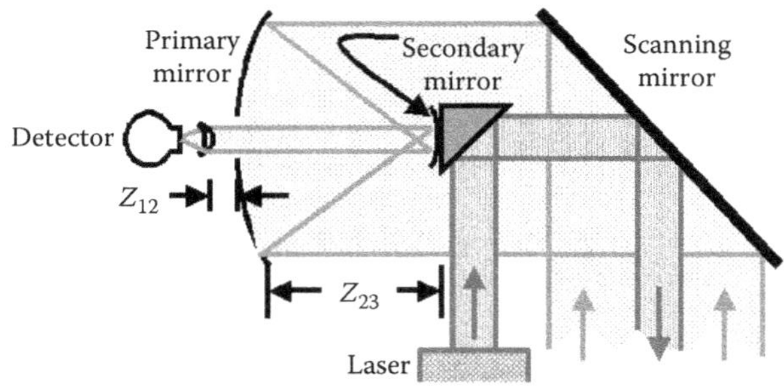

FIGURE P4.1
Coaxial lidar configuration.

light is then reflected by a scanning mirror to a distant target. Some of the backscattered light is returned via the scanning mirror, reflected from the primary mirror to a secondary mirror, and then to a lens and detector. The focal lengths are 1 m for the primary, 12.5 cm for the secondary, and 5 mm for the detector lens. The diameters are, respectively, 8, 2, and 1 in. The detector, with a diameter of 100 μm, is at the focus of the detector lens, which is separated from the secondary by a distance of 150 cm. The primary and secondary are separated by the sum of their focal lengths.

4.1a Locate all the optical elements in object space, i.e., the space containing the target, and determine their size; the detector, detector lens, secondary lens, primary, obscuration caused by the transmitter folding mirror.

4.1b Identify the aperture stop and the field stop for a target at a long distance from the lidar.

4.1c As mentioned earlier, the folding mirror for the transmitter causes an obscuration of the received light. It turns out that this is a very useful feature. The scanning mirror surface will always contain some imperfections, scratches, and dirt, which will scatter light back to the receiver. Because it is very close, the effect will be strong and will prevent detection of weaker signals of interest. Now, if we place the scanning mirror close enough, the obscuration and the finite diameter of one or more lenses will block light scattered from the relatively small transmitter spot at the center of the mirror. Find the "blind distance," z_{blind}, inside which the scan mirror can be placed.

4.2 Microscope (NS)

A microscope objective is labeled 20X 0.5NA, meaning magnification of 20 and numerical aperture of 0.5. The manufacturer uses a 200 mm tube lens and a 20 mm field stop.

4.2a What is the focal length of the objective?

4.2b Where is the aperture stop located relative to the back principal plane of the objective?

4.2c What is the diameter of the aperture stop?

4.2d What is the field of view of the image?

4.2e What is the numerical aperture in image space?

4.3 Scanning Laser Radar (NS)

We want to design a telescope for a laser radar so that we can transmit a 1 cm diameter laser beam out through the primary, from a 1 mm laser beam we put into the secondary. The primary has a 200 mm focal length and a 12.5 mm (half inch) diameter. Assume thin lenses.

We wish to place a scanner at the entrance pupil of this beam-expanding telescope. Notice that we would call this the exit pupil if we were using the telescope in the normal way to view a star.

4.3a What is the focal length of the secondary?

4.3b Where is the entrance pupil? In other words, where should we place the scanner mirror?

4.3c Suppose we incorrectly place the scanner 2 cm before this location. Sketch the aperture of the lens and the laser beam as two circles for various scan angles from zero to an angle such that no laser light passes through the telescope. Hint: It may be easier to calculate the locations in object space using the entrance pupil and then apply the magnification equation. Just find the center of the laser beam and make a circle around it. That isn't quite correct as the off-axis spot will be elliptical, but it's close enough.

5 Aberrations

Eq. 3.47, $\frac{n}{s} + \frac{n'}{s'} = P$.

Eq. 3.48, $m = -\frac{ns'}{n's}$.

In developing the paraxial imaging equations in Chapters 2 and 3 with curved mirrors, surfaces, and lenses, we assumed that for any reflection or refraction, all angles, such as those in Figures 2.10, 2.13, and 3.4, are small, so that for any angle, represented here by θ,

$$\sin\theta = \theta = \tan\theta \qquad \text{and} \qquad \cos\theta = 1. \tag{5.1}$$

As a result, we obtained the lens equation, Equation 2.51 in air or, more generally, Equation 3.47, and the equation for magnification, Equation 2.54 or 3.48. Taken together, these equations define the location of "the image" of any point in an object. Any ray from the point (x, s) in the object passes exactly through this image point (x', s'). Indeed, this is in keeping with our definition of an image based on Fermat's principle in Figure 1.17. Specifically, the position of a ray at the image was independent of the angle at which it left the object. Another interpretation of the last sentence is that the location of the ray in the object was independent of the position at which it passed through the pupil in the geometry shown in Figure 5.1. The small-angle assumptions lead to the conclusion that the object and image distances are related by the lens equation, and that the image height is given by $x' = mx$ with m defined by Equation 3.48. From the wavefront perspective, the assumption requires that the optical path length (OPL) along any ray from X to X' be independent of the location of the point X_1 in the pupil.

What happens if we remove these assumptions? To begin to understand the answer to this question, we will trace rays through a spherical surface, using Snell's law. This approach gives us the opportunity to discuss the concepts behind computational **ray tracing** programs, and gives us some insight into the behavior of real optical systems. After that example, we will consider a reflective system and discuss the concepts of aberrations in general. Then we will develop a detailed description of the aberrations in terms of changes in optical path length, followed by a more detailed discussion of one particular type of aberration, and a brief description of the optical design process.

We will see that, in general, the image of a point in an object will have a slightly blurred appearance and possibly a displacement from the expected location. These effects are called **aberrations**. We will see that these aberrations are fundamental properties of imaging systems. While there may be aberrations from imperfections in the manufacturing

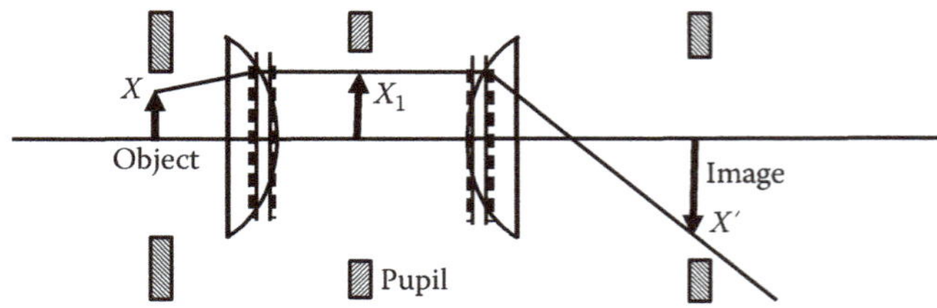

FIGURE 5.1
Object, pupil, and image. One ray is shown leaving the object at a point X in the front focal plane of the first lens. This ray passes through the pupil at a location, P, a distance p away from the axis. It then passes through an image point X' at the back focal plane of the second lens.

DOI: 10.1201/9781315157047-5

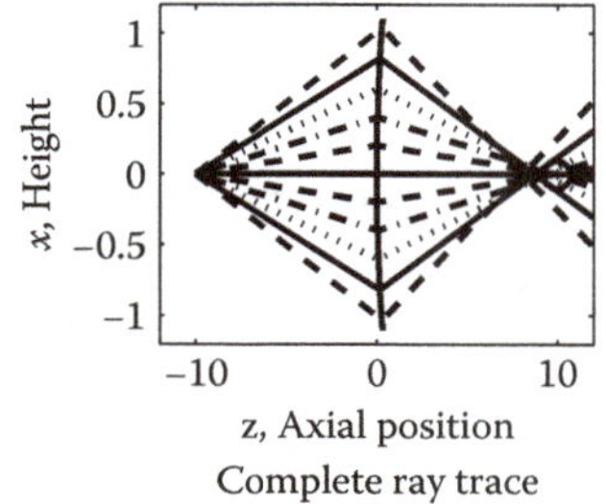

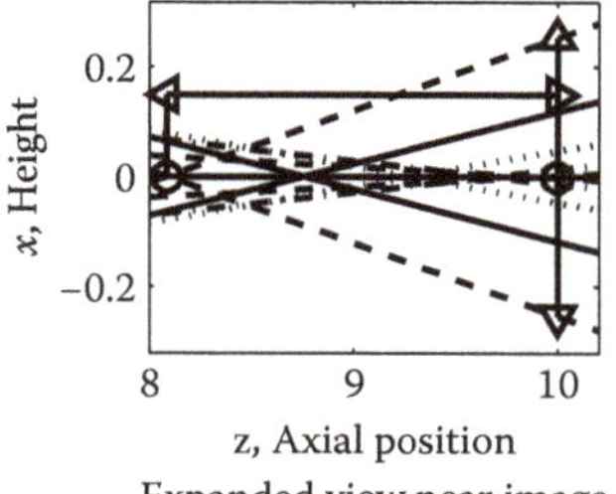

FIGURE 5.2
Rays through an air-glass interface. The rays are traced exactly using Snell's Law at the interface. The rays intersect at a distance shorter than the paraxial image distance. In the expanded view, the horizontal arrow represents Δz, and the vertical one represents Δx. The object and image distances are $s = s' = 10$, the index of refraction is $n = 1.5$, and the radius of curvature is determined from Equation 2.29, $r = 2$. Note that the scales of x and z are not the same, so the surface does not appear spherical.

process, they are not directly the subject of this chapter. We will consider aberrations from less-than-optimal design, for example, using spherical surfaces instead of optimized ones. However, we note that even the best imaging systems have aberrations. An imaging system with a nonzero field of view and a nonzero numerical aperture will always have aberrations in some parts of the image. The best we can hope is to minimize the aberrations so that they remain below some acceptable level. When we study diffraction in Chapter 8, we will find that for a given aperture, there is a fundamental physical limit on the size of a focused spot, which we will call the "diffraction limit." The actual spot size will be approximately equal to the diffraction limit or the size predicted by geometric optics, including aberrations, whichever is larger. Once the aberrated spot size is smaller than the diffraction limit, there is little benefit to further reduction of the aberrations.

Take-Away

Aberrations in General:
The image of a point will always be blurred because of diffraction as discussed in Chapter 8. Any imaging system with a nonzero field of view and nonzero numerical aperture will have aberrations that can be predicted by geometric optics. If these aberrations are much smaller than the diffraction limit, the imaging system cannot be made better.

5.1 Exact Ray Tracing

Let us perform an exact ray trace through a single convex spherical interface between air and glass, as shown in Figure 5.2. We will learn the mechanics of ray tracing and begin to study some simple aberrations.

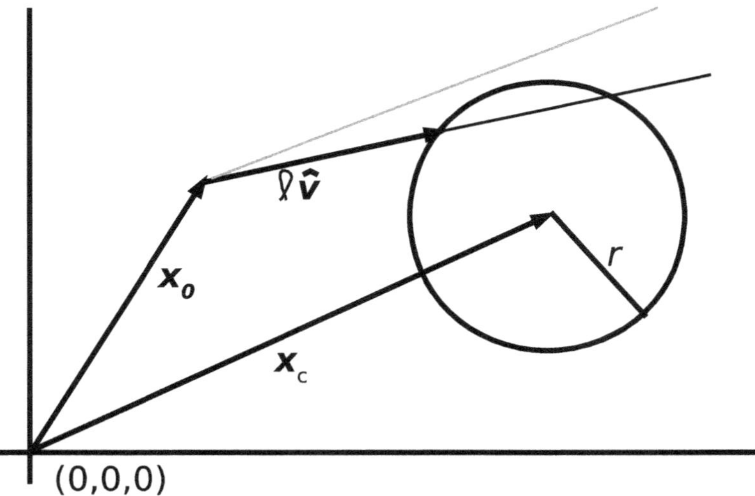

FIGURE 5.3
Intersection of a ray and a sphere. The quadratic equation for ℓ, Equation 5.4, has at most two solutions. If two solutions exist, only one is the correct one.

5.1.1 Ray Tracing Computation

Define a ray

We start by considering one ray from a point in the object. The **ray** will be defined by a three-dimensional vector, $\mathbf{x}_o$, from the origin of coordinates to the point, and a three-dimensional unit vector, $\hat{\mathbf{v}}$. We call the point at $\mathbf{x}_o$ the ray origin and the unit vector $\hat{\mathbf{v}}$ the **ray direction**. A point on the ray is given parametrically as

$$\mathbf{x} = \mathbf{x}_0 + \ell\hat{\mathbf{v}}$$
$$\begin{pmatrix} x \\ y \\ z \end{pmatrix} = \begin{pmatrix} x_0 \\ y_0 \\ z_0 \end{pmatrix} + \ell \begin{pmatrix} u \\ v \\ w \end{pmatrix}, \tag{5.2}$$

where ℓ is a distance along the ray. Normally, the ray will be one of a **fan of rays**, defined in terms of angles, points in the pupil, or some other parameters.

Define a surface

We describe the **spherical interface** in terms of its center $\mathbf{x}_c = (0, 0, z_c)$ and **radius of curvature**, r. The equation of the sphere is thus

$$(\mathbf{x} - \mathbf{x}_c) \cdot (\mathbf{x} - \mathbf{x}_c) = r^2$$
$$x^2 + y^2 + (z - z_c)^2 = r^2. \tag{5.3}$$

Compute intersection of ray and surface

Combining these two equations, we obtain a quadratic equation in ℓ for the intersection of the ray with the surface as in Figure 5.3:

$$(\mathbf{x}_0 + \ell\hat{\mathbf{v}} - \mathbf{x}_c) \cdot (\mathbf{x}_0 + \ell\hat{\mathbf{v}} - \mathbf{x}_c) = r^2$$
$$(x_0 + \ell u)^2 + (y_0 + \ell v)^2 + (z_0 + \ell w - z_c)^2 = r^2. \tag{5.4}$$

Expanding this equation,

$$\hat{\mathbf{v}} \cdot \hat{\mathbf{v}}\ell^2 + 2(\mathbf{x}_0 - \mathbf{x}_c) \cdot \hat{\mathbf{v}}\ell + (\mathbf{x}_0 - \mathbf{x}_c) \cdot (\mathbf{x}_0 - \mathbf{x}_c) - r^2 = 0$$

$$u^2\ell^2 + v^2\ell^2 + w^2\ell^2 + 2x_0 u\ell + 2y_0 v\ell + 2(z_0 - z_c) w\ell \tag{5.5}$$

$$+x_0^2 + y_0^2 + (z_0 - z_c)^2 - r^2 = 0. \tag{5.6}$$

Using the usual notation for the quadratic equation,

$$a_q\ell^2 + b_q\ell + c_q = 0, \tag{5.7}$$

$$a_q = \hat{\mathbf{v}} \cdot \hat{\mathbf{v}} = 1, \qquad b_q = 2(\mathbf{x}_0 - \mathbf{x}_c) \cdot \hat{\mathbf{v}},$$

$$c_q = (\mathbf{x}_0 - \mathbf{x}_c) \cdot (\mathbf{x}_0 - \mathbf{x}_c) - r^2, \tag{5.8}$$

with up to two real solutions,

$$\ell = \frac{-b_q \pm \sqrt{b_q^2 - 4a_q c_q}}{2a_q}. \tag{5.9}$$

We need to take some care in selecting the right solution. For the present case, the object point is to the left of the sphere. We have also stated that we are considering a convex surface, and $\hat{\mathbf{v}}$ is defined with its z component positive, so a quick sketch will show that both solutions, if they exist, will be positive, and that the smaller solution is the correct one. If the angle is chosen too large, then there will be no real solutions because the ray will not intersect the sphere. The intersection is at

$$\mathbf{x}_A = \mathbf{x}_0 + \ell\hat{\mathbf{v}}. \tag{5.10}$$

New ray origin

Next, we construct the mathematical description of the refracted ray. To do so, we need the **surface normal**:

$$\hat{\mathbf{n}} = \frac{\mathbf{x} - \mathbf{x}_c}{\sqrt{(\mathbf{x} - \mathbf{x}_c) \cdot (\mathbf{x} - \mathbf{x}_c)}}. \tag{5.11}$$

For a point on the ray, we choose the point of intersection, $\mathbf{x}_A$. The ray direction is given by

New ray direction

$$\hat{\mathbf{v}}' = \frac{n}{n'}\hat{\mathbf{v}} + \left[\sqrt{1 - \left(\frac{n}{n'}\right)^2 \left[1 - (\hat{\mathbf{v}} \cdot \hat{\mathbf{n}})^2\right]} - \frac{n}{n'}\hat{\mathbf{v}} \cdot \hat{\mathbf{n}}\right] \hat{\mathbf{n}}, \tag{5.12}$$

which can be proven from Snell's law [16;251]. This **vector form of Snell's law** can be very useful in situations such as this; the added complexity of the vector form is often offset by removing the need for careful attention to coordinate transforms.

Now, if we had multiple optical elements, we would proceed to handle the next element in the same way using Equations 5.2 through 5.11. A ray tracing program moves from one surface to the next until the ray is closed and then repeats the process for each ray in the fan.

"Close" the ray for a useful result

The final step for this ray is to **close** the ray by finding some terminal point. There are many ways in which to close a ray, depending on our goals. We might simply compute the intersection of the ray with the paraxial image plane, determined by s' in Equation 2.29. We might also compute the location where the ray approaches nearest the axis. For display purposes in Figure 5.2, we choose to compute the intersection with a plane slightly beyond the image. The intersection of a ray with a plane can be calculated by following the same procedure outlined above for the intersection of a ray and a sphere. For the present example, the plane is defined by

$$\mathbf{x}_B \cdot \hat{z} = z_{close}, \tag{5.13}$$

and

$$(\mathbf{x}_A + \ell\hat{\mathbf{v}}_1) \cdot \hat{z} = z_{close}. \tag{5.14}$$

5.1.2 Aberrations in Refraction

In Figure 5.2, the paraxial image is at $s' = 10$. We will characterize rays by their position in the image, $\mathbf{x}_0$, and in the pupil, $\mathbf{x}_1$. We see that **paraxial rays**, those close to the axis ($x_1 \approx 0$), intersect very close to the paraxial image, s', which we will now call $s'(0)$. In contrast, the **edge rays** , those near the edge of the aperture (large x_1), intersect at locations, $s'(x_1)$, far inside the paraxial image distance. Although the small-angle approximation showed all rays from the object passing through a single image point, the correct ray traces show that rays passing through different parts of the spherical interface behave as though they had different focal points with three key results:

- There is no axial location at which the size of this imaged point is zero.
- The size of the spot increases with increasing aperture at the interface.
- The smallest spot is not located at the paraxial image.

Take-Away

Ray Tracing Approach:

- Ray tracing is a computational process for determining the exact path of a ray without the simplification that results from the small-angle approximation.
- In ray tracing, each of a fan of rays is traced through the optical system from one surface to the next, until some closing condition is reached.
- The ray, after each surface, is defined by a point, a direction, and an unknown distance parameter.
- The intersection of the ray with the next surface is located by solving the equation for that surface, using the ray definition, for the unknown distance parameter.
- The new direction is determined using Snell's law or Equation 5.12.
- In general, the rays will not pass through the paraxial image point but will suffer aberrations.

5.2 Ellipsoidal Mirror

As another example of aberration, let us consider a curved mirror. We start with a point at a position S, which we wish to image at S'. From Fermat's principle (Section 1.8.2), we know that if our goal is to have every ray from S pass through S', then every path from S to S' must have minimal optical path length. The only way this equality is possible if all paths have the same length. At this point the reader may remember a procedure for drawing an ellipse. The two ends of a length of string are tacked to a sheet of paper

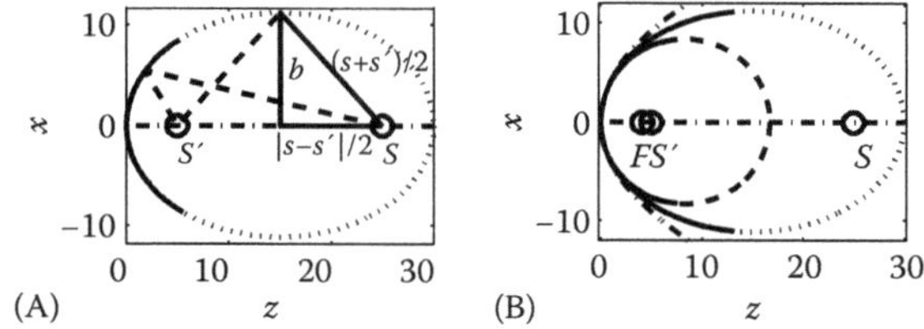

FIGURE 5.4
Ellipsoidal mirror. The foci are at $s = 25$ and $s' = 5$. All rays from the object to the image have the same length. The rays that intersect the ellipse at the end of the minor axis can be used to determine b using the Pythagorean theorem on the triangle shown with solid lines in (A). The best-fit sphere and parabola are shown in (B).

at different places. A pencil holding the string tight can then be used to draw an ellipse. Revolving this ellipse around the z axis (which contains S and S'), we produce an ellipsoid. In this case, the two minor axes x and y have the same length.

Figure 5.4A shows an ellipsoidal mirror imaging a point S to another point S'. The actual mirror, shown by the solid line, is a segment of the complete ellipsoid shown by the dashed line. We draw the complete ellipsoid because doing so will simplify the development of the equations we need. The mathematical detail is included here for the interested reader. The important conclusions are that a specific ellipsoidal mirror is exactly right to ensure that all the rays from one object point pass through its image point and that for any other pair of object and image points, a different ellipsoid is the right one.

Because all of the paths reflecting from the ellipsoid have the same length, we can develop our equations with any of them, whether they reflect from the part that is actually used or not. Three example paths from S to S', by way of the ellipsoid, are shown. We choose one path from S to the origin with a distance s, and then to S' with an additional distance s'. We recall from Figure 2.5 that both s and s' are positive on the source side of the reflective surface in contrast to the sign convention used with a refractive surface. The total distance is thus $s + s'$. The second path is an arbitrary one to a point x, y, z, for which the total distance must be the same as the first:

$$\sqrt{(z-s)^2 + x^2 + y^2} + \sqrt{(z-s')^2 + x^2 + y^2} = s + s' \tag{5.15}$$

The third point is on the minor axis.

We may recall that the equation for an ellipsoid passing through the origin with foci on the z axis is

$$\left(\frac{z-a}{a}\right)^2 + \left(\frac{x}{b}\right)^2 + \left(\frac{y}{b}\right)^2 = 1, \tag{5.16}$$

where the major axis has length $2a$ and the other two axes have length $2b$.

The ellipsoid as drawn passes through the origin and is symmetric about the point $z = a$. The half-length of the major axis is thus the average of s and s':

$$a = \frac{s + s'}{2}. \tag{5.17}$$

In the triangle shown by solid lines in Figure 5.4A, the height is the half-length of the minor axis, b, the base is half the distance between S and S', and the hypotenuse is half the total optical path, so

$$b^2 = \left(\frac{s+s'}{2}\right)^2 - \left(\frac{s-s'}{2}\right)^2 = ss'. \tag{5.18}$$

This mirror is the perfect choice to image the point S to the point S'. For these two conjugate points, there are no aberrations. We note for later use that we can define the focal length from the lens equation (Equation 2.51 in air) by

$$\frac{1}{f} = \frac{1}{s} + \frac{1}{s'} = \frac{s'+s}{s's} = \frac{2a}{b^2}$$

$$f = \frac{b^2}{2a}. \tag{5.19}$$

Summarizing the equations for the ellipsoidal mirror,

$$\left(\frac{z-a}{a}\right)^2 + \left(\frac{x}{b}\right)^2 + \left(\frac{y}{b}\right)^2 = 1 \qquad a = \frac{s+s'}{2} \qquad b = \sqrt{ss'}. \tag{5.20}$$

For $s = s'$, the ellipsoid becomes a sphere with radius $a = b = s$:

$$z^2 + x^2 + y^2 = r^2 \qquad r = s = 2f. \tag{5.21}$$

For $s \to \infty$, the equation becomes a paraboloid:

$$(z-a)^2 + \left(\frac{xa}{b}\right)^2 + \left(\frac{ya}{b}\right)^2 = a^2,$$

$$\left(z - \frac{s+s'}{2}\right)^2 + x^2\frac{(s+s')^2}{4ss'} + y^2\frac{(s+s')^2}{4ss'} = \left(\frac{s+s'}{2}\right)^2,$$

$$z^2 - (s+s')\,z + x^2\frac{(s+s')^2}{4ss'} + y^2\frac{(s+s')^2}{4ss'} = 0,$$

$$\frac{z^2}{s+s'} - z + x^2\frac{(s+s')}{4ss'} + y^2\frac{(s+s')}{4ss'} = 0,$$

$$z = \frac{x^2}{4s'} + \frac{y^2}{4s'}, \tag{5.22}$$

and of course, for $s' \to \infty$,

$$z = \frac{x^2}{4s} + \frac{y^2}{4s}, \tag{5.23}$$

so the paraboloid is defined as

$$z = \frac{x^2}{4f} + \frac{y^2}{4f}. \tag{5.24}$$

Take-Away

Focusing with Ellipsiodal Mirrors:
For a point object and image at distances s and s' an ellipsiodal mirror (The surface equation is given by Equation 5.20) provides perfect imaging according to geometric optics, meaning that all the rays from the object point that reflect from the mirror pass through the image point.

If the object and image distances are equal, the ellipsoid becomes a sphere (Equation 5.21).

The ellipsoid becomes a paraboloid, if the object distance $s \to \infty$ (Equation 5.22, or if the image distance $s' \to \infty$ (Equation 5.23).

5.2.1 Aberrations and Field of View

Aberrations increase with field of view.

For the particular pair of points, S at $(0, 0, s)$ and S' at $(0, 0, s')$, we have, in Equation 5.20, a design which produces an aberration-free image. However, an optical system that images only a single point is of limited use. Let us consider another point, near S, for example, at $(x, 0, s)$. Now, using our usual imaging equations, we expect the images of these two points to be at $(x, 0, s')$ and $(mx, 0, s')$ where $m = -s'/s$. However, the ellipsoid in Equation 5.15 is not perfect for the new points. We need another ellipsoid to image this point without aberration. It is not possible to construct a surface which will produce aberration-free images of all points in an object of finite size. However, if x is small enough, then the error will also be small. In general, as the field of view increases, the amount of aberration also increases. The statements above also hold for refraction. We could have modified our refractive surface in Section 5.1 to image the given point without aberration, but again the perfect result would only be obtained for one object point.

5.2.2 Design Aberrations

For the refractive surface in Section 5.1, we saw that a spherical surface produced aberrations even for an on-axis point. We commented that we could eliminate these aberrations by a careful design of the surface shape. However, spherical surfaces are the simplest to manufacture. Two surfaces will only fit together in all orientations if they are both spheres with the same radii of curvature. As a result, if two pieces of glass are rubbed together in all possible directions with an abrasive material between them, their surfaces will approach a common spherical shape, one concave and the other convex. Because of the resulting low cost of manufacturing, spherical surfaces are often used. In recent decades, with improved optical properties of plastics, the injection molding of plastic aspheric lenses has become less expensive, and these lenses can be found in disposable cameras and many other devices. Modern machining makes it possible to turn large reflective elements on a lathe, and machined parabolic surfaces are frequently used for large telescopes. Nevertheless, spherical surfaces remain common for their low cost, and because a sphere is often a good compromise, not eliminating aberration at any one point, but distributing aberrations throughout the field of view.

A spherical surface is often a good compromise.

5.2.3 Aberrations and Aperture

Aberrations increase with numerical aperture.

With these facts in mind, can we use a spherical mirror as a substitute for an elliptical mirror? We want to match the sphere to the ellipsoid at the origin as well as possible.

Having done so, we will then be able to calculate the aberrations as a function of distance away from the origin (*i.e.* from the axis).

We can first ensure that both surfaces pass through the origin. Next, we examine the first derivatives of z with respect to x (or y):

$$2\left(\frac{z-a}{a}\right)\frac{dz}{a}+2\left(\frac{x}{b}\right)\frac{dx}{b}=0.$$

$$\frac{dz}{dx}=-\left(\frac{a}{b}\right)^2\frac{x}{z-a}.$$

The first derivative is zero at the origin, $x=z=0$.

The second derivative is

$$\frac{d^2z}{dx^2}=-\left(\frac{a}{b}\right)^2\frac{(z-a)-x\frac{dz}{dx}}{(z-a)^2},$$

and evaluating it at the origin,

$$\left.\frac{d^2z}{dx^2}\right|_{x=0}=\frac{a}{b^2}=\frac{s+s'}{2ss'}. \tag{5.25}$$

The radius of curvature of the sphere with the same second derivative as the ellipsoid is r obtained by differentiating Equation 5.21 :

$$\frac{1}{r}=\left.\frac{d^2z}{dx^2}\right|_{x=0},$$

where

$$\frac{2}{r}=\frac{1}{s}+\frac{1}{s'}, \tag{5.26}$$

so the focal length is given by

$$f=r/2, \tag{5.27}$$

in agreement with our earlier calculation in Equation 2.12.

Not surprisingly, this equation is the same as Equation 5.21, where we discussed the sphere as a special case of the ellipsoid for $s=s'$. It is a good approximation for $s\approx s'$, but reflective optics are often used when $s\gg s'$, as in an astronomical telescope or $s\ll s'$, in a reflector to collimate a light source. Then a paraboloid in Equation 5.24 is a better approximation.

We have chosen to match the shape of the sphere or paraboloid up to the second order around the origin. If we keep the aperture small so that all the rays pass very close to the origin, the approximation is good, and either the sphere or the paraboloid will perform almost as well as the ellipsoid.

Shortly, we will want to study the aberrations quantitatively, and we will need the optical-path-length (OPL) error, Δ, between the spherical mirror and the ideal ellipsoidal one, or between the paraboloid and the ellipsoid.

We have discussed the fact that aberrations are minimized when all rays from an object point pass through a single image point. Along with Fermat's principle, this means that all rays from object to image are of equal length. Therefore, using the wave theory of light,

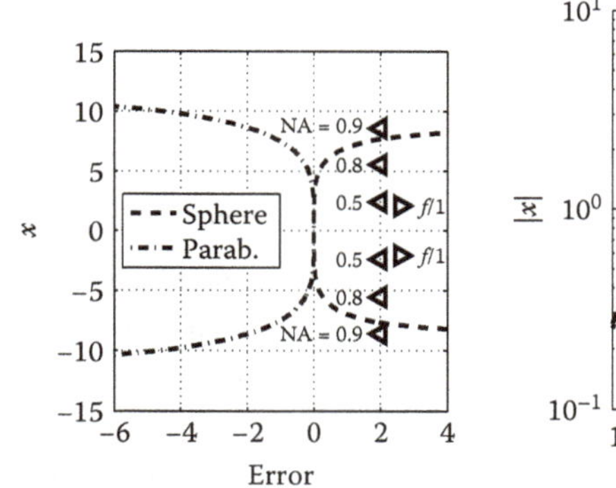

FIGURE 5.5
Optical path length errors. The plot on the left shows the errors in optical path length, Δ, for the sphere and paraboloid using the object and image distances in Figure 5.4. The plot on the right shows the absolute values of these errors on a logarithmic scale. The errors vary according to the fourth power of the radial distance from the axis. Because the errors are in z, it is convenient to plot them on the abscissa with x on the ordinate. Units are arbitrary but are the same on all axes. The focal length is 4.167 units.

all wave contributions add in phase at the image, resulting in constructive interference (see Chapter 8). The OPL error, Δ, measures the extent to which there are differences in phase between wave contributions arriving with different rays. If these phase differences are large, then the brightness of the image is reduced and partial constructive interference occurs in a region about the expected image. As a result the image is blurred.

We write the equations for the perfect ellipsoid using x, y, and z so that $z_{sphere} = z + \Delta_{sphere}/2$ and $z_{para} = z + \Delta_{para}/2$:

$$
\begin{aligned}
(z-a)^2 &= a^2 - \left(\frac{a}{b}\right)^2 (x^2+y^2) && \text{(Ellipsoid)}\\
\left(z + \frac{\Delta_{sphere}}{2} - r\right)^2 &= r^2 - x^2 - y^2 && \text{(Sphere)}\\
z + \frac{\Delta_{para}}{2} &= \frac{x}{4f} && \text{(Paraboloid)},
\end{aligned}
\tag{5.28}
$$

where in all three equations, z describes the surface of the ideal ellipsoid. Equations for Δ can be computed in closed form by comparing the sphere or paraboloid equation here to Equation 5.21 or 5.24 respectively. However, the calculations are messy, so instead we choose to do the computation numerically and plot the results of a specific example, using the parameters of Figure 5.4. The OPL errors, Δ, are plotted in Figure 5.5. The errors vary according to the fourth power of $|x|$, the radial distance from the axis. For this example, if the units are centimeters, then the errors are less than 0.1 μm for $x < 0.5$ cm, corresponding to an $f/4$ aperture. Because the errors increase so drastically with radial distance and because the usual circular symmetry dictates that the increment of power in an annulus of width dr at a distance r from the axis is the irradiance times $2\pi r dr$, the regions of worst error contribute significantly to the formation of an image. To achieve diffraction-limited operation, we want the OPL error to be a fraction of the wavelength. For many demanding applications, we would like an error of $\lambda/20$, $\lambda/40$, or better.

Eq. 5.21
$z^2 + x^2 + y^2 = r^2$
$r = s = 2f$.

Eq. 5.24 $z = \frac{x^2}{4f} + \frac{y^2}{4f}$.

OPL errors, Δ, vary with the fourth power of the distance from the axis.

Take-Away

Mirror Aberrations:

- An ellipsoidal mirror is ideal to image one point to another.
- For any other pair of points, aberrations will exist.
- A paraboloidal mirror is ideal to image a point at its focus to infinity.
- Spherical and paraboloidal mirrors may be good approximations to ellipsoidal ones for certain situations, and the aberrations can be computed as surface errors.
- Aberrations are fundamental to optical systems. Although it is possible to correct them perfectly for one object point, in an image with nonzero pupil and field of view, there will always be some aberration.
- Aberrations generally increase with increasing field of view and increasing numerical aperture.

In Practice

To evaluate a spherical mirror for a specific application, compute the OPL error, Δ, at the edge of the aperture (maximum x). If the result is much less than a wavelength, then the spherical mirror will suffice. If not, there will be aberrations.

The same can be done to evaluate a parabolic mirror for finite object and image distances, by measure the OPL error between the two surfaces.

5.3 Seidel Aberrations and OPL

We have noted that aberrations are a result of the failure of the approximation

$$\sin\theta \approx \theta. \tag{5.29}$$

We could develop our discussion of aberrations in terms of the "third-order theory" based on

$$\sin\theta \approx \theta - \frac{\theta^3}{3!}, \tag{5.30}$$

and we would then define the so-called Seidel aberrations.

Instead, we will consider aberrations in terms of wave-front errors in the pupil of an optical system. From Section 1.8, we recall that wavefronts are perpendicular to rays, so the two approaches must be equivalent. For example, we can define a tilt as an error in angle, $\delta\theta$, or as a tilt of a wave front. If the OPL error is Δ, then

$$\frac{d\Delta_{tilt}}{dx_1} = a_1 \approx \delta\theta. \tag{5.31}$$

The idea is illustrated in Figure 5.6. We consider an imaging system consisting of two lenses separated by their focal lengths. The object is placed at the front focal point of the first and the image at the back focal point of the second. We consider a pupil plane placed at the back focal plane of the first lens (front focal plane of the second). In the pupil plane of Figure 5.6A, the rays are parallel to the axis and the wavefronts are perpendicular to the axis, and the image is on the axis, just like the object. In Figure 5.6B, we imagine a prism placed in the pupil to tilt the wavefronts (or rays). In this case, the image is moved

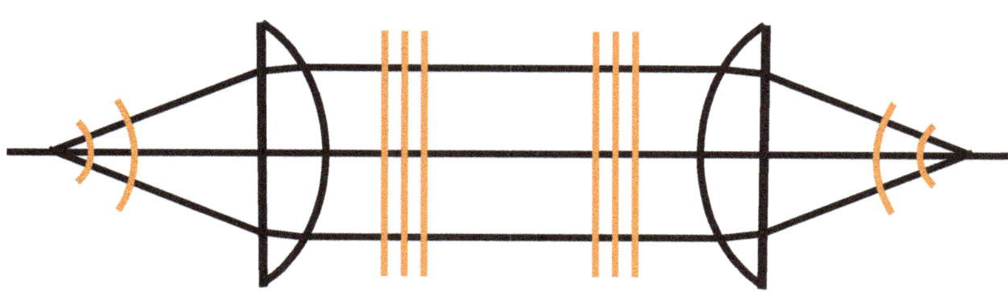

A. Two–lens imaging system.

B. Tilting the wavefront in the pupil plane.

FIGURE 5.6
Tilt. If the wavefront is delayed linearly with transverse distance, the result is a transverse displacement of the image. Here we assume that the lenses are aberration-free. This figure illustrates that changes in the wavefront in the pupil are related to changes in the image.

off the axis. A delay of the wavefronts that is linear in transverse direction and increasing toward the top of the figure, results in upward displacement of the image.

We will take the approach of describing aberrations in terms of the OPL error. Returning to Figure 5.1, let us describe our optical system starting with an ideal imaging system with no aberrations, and then introducing an OPL error, Δ, for each ray. In Figure 5.6 we showed that a linear variation in phase with x_1 in the pupil plane displaced the image in the x direction. Now we will let the OPL error be a function of both the location of the ray in the field plane, x, y and of the position in the pupil plane, x_1, y_1. That is, we can imagine the prism in Figure 5.6B being replaced by a glass plate having an optical thickness that varies with x_1 and y_1, with the understanding that there is a different plate for each image point location, x, y. We will then explore the types of aberrations that result from different functional dependence of Δ on x, y, x_1, and y_1. Using ray–tracing computer programs or techniques such as we developed in Section 5.2.3, we can compute the functional dependence and predict the aberrations for a specific system. Because the functions can be arbitrary, we choose to write an expression for the OPL error as a series expansion in x, x_1, and y_1. We choose for simplicity to consider the object point to be on the x axis, eliminating the need to deal with y. We can employ circular symmetry to interpret the results in two dimensions. We define a normalized radius in the pupil, ρ, and an associated angle so that $x_1 = \rho d_p/2 \cos\phi$ and $y_1 = \rho d_p/2 \sin\phi$, where d_p is the pupil diameter. Then we consider the zeroth-, second-, and fourth-order terms in the expansion, neglecting the odd orders which correspond to tilt:

Fig. 5.1

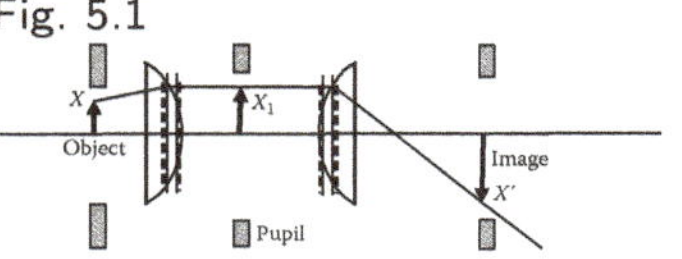

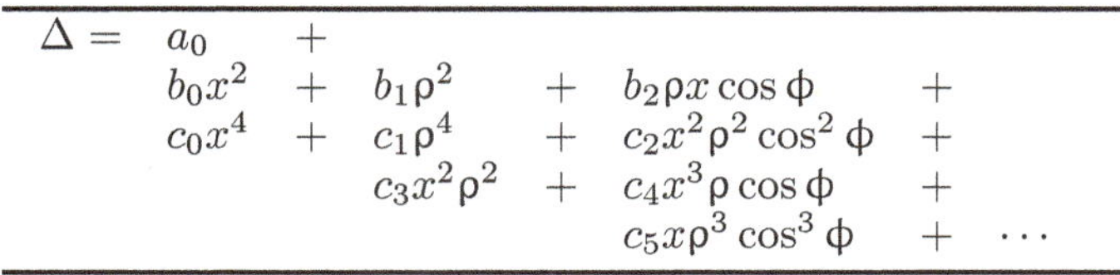

$$\begin{array}{lllllllll}\Delta = & a_0 & + & & & & & & \\ & b_0 x^2 & + & b_1 \rho^2 & + & b_2 \rho x \cos\phi & + & & \\ & c_0 x^4 & + & c_1 \rho^4 & + & c_2 x^2 \rho^2 \cos^2\phi & + & & \\ & & & c_3 x^2 \rho^2 & + & c_4 x^3 \rho \cos\phi & + & & \\ & & & & & c_5 x \rho^3 \cos^3\phi & + & \cdots & \end{array} \tag{5.32}$$

This expansion is perfectly general for a circularly symmetric system, and will allow us to describe any system if we consider enough terms. If we keep the aperture and field of view sufficiently small, then only a few terms are required. For a given system, the task is to determine the coefficients a_n, b_n, and c_n that define the amount of aberration. For now, we want to study the effects of these aberrations.

a_0 describes phase change (no aberration)

b_1 characterizes change in focus (no aberration)

The first term a_0 characterizes overall OPL change, which does not affect the image. This term can be represented by placing a sheet of glass in the place of the prism in Figure 5.6B, which does not affect the image. Likewise, second-order terms do not affect the quality of the image. For example, $b_1\rho^2$ corresponds to a change in the focal length. In our model in Figure 5.6B, we replace the prism by a glass sheet with thickness varying quadratically with ρ. The plane wavefronts approaching the pupil will be spherical after the pupil. The wavefronts converging to the image will be spherical but with different radii of curvature. While this change is significant, it can be corrected by appropriate use of the lens equation. The image will form in a slightly different plane than expected, but the quality will not be affected.

5.3.1 Spherical Aberration

c_1 characterizes spherical aberration. It does not depend on location in the field plane.

The first real aberrations are in the fourth-order terms in position coordinates. The angle of a ray is proportional to the tilt in wavefront, or the derivative of OPL, so these fourth-order errors in OPL correspond to third-order errors in angles of rays. The first aberration is the term, $c_1\rho^4$. Recalling that ρ^2 represents a change in focus, we write this term as $\left(c_1\rho^2\right)\rho^2$. Thus the focus error increases in proportion to ρ^2. This is exactly the error that we encountered in Figure 5.2. The larger the distance, ρ, the more the focus departed from the paraxial value. This term is called **spherical aberration**:

$$\Delta_{sa} = c_1\rho^4 \qquad \text{(Spherical aberration).} \tag{5.33}$$

Eq. 5.24: $z = \frac{x^2}{4f} + \frac{y^2}{4f}$.

In Figure 5.6B, this aberration can be represented by replacing the prism by a glass plate with a thickness that varies according to ρ^4. The waves converging toward the image will not be spherical.

Example

In Section 5.2, we looked at the error between a spherical mirror and an ellipsoidal one, as well as a parabolic mirror and an ellipsoidal one. Let's compare a spherical mirror to a parabolic one. We know the parabolic mirror is the right shape for an object (or image) at infinity. Equation 5.24 describes the parabolic mirror, and a sphere through the origin of coordinates is

$$z = 2f - \sqrt{2f^2 - x^2 - y^2}.$$

Let's characterize the error by the optical path error at the value of x at the edge. Because of the spherical symmetry we can assume $y = 0$. The error is

$$\Delta = 2\left[\frac{x^2}{4f} - 2 * 2f + \sqrt{2f^2 - x^2 - y^2}\right]. \tag{5.34}$$

We'll consider a mirror with 1 m focal length reflecting infrared light at 10 μm. Let's consider the error in proportion to the wavelength. If the error at the edge is more than a fraction of a wavelength, we can be assured that the resulting image will have aberrations.

FIGURE 5.7
Distortion. The left panel shows an object consisting of straight lines. The center panel shows barrel distortion, and the right panel shows pincushion distortion.

The error is about one tenth of a wavelength at a radius of about $x = 7.5\,\text{cm}$ or a diameter of 15 cm Spherical mirrors smaller than this will meet the $\lambda/10$ specification for an object or image at infinity. We note that this specification is not usually considered very severe.

5.3.2 Distortion

Next, we consider the terms

$$\Delta_d = c_4 x^3 \rho \cos\phi \qquad \text{(Distortion)}. \tag{5.35}$$

The wavefront tilt, $\rho\cos\phi$, is multiplied by the cube of the image height. In our model, this can be represented by a prism in Figure 5.6 that varies in apex angle with the distance x of the image from the axis. Wavefront tilt in the pupil corresponds to a shift in location in the image plane. The excess tilt, Δ_d, results in distortion. Objects at increasing distance, x, from the axis suffer a greater shift in their position, proportional to the cube of x. The result is a **distortion** of the image; the magnification is not uniform across the field of view. In this case, the apparent magnification varies with radial distance from the center. Straight lines in the object are not imaged as straight lines. Figure 5.7 shows an example of two kinds of distortion. In barrel distortion, the tilt toward the center increases with radial distance ($c > 0$), giving the appearance of the staves of a barrel. On the other hand, if the tilt away from the center increases with radial distance ($c < 0$), the result is pincushion distortion.

c_4 characterizes distortion.

Distortion is more of a problem in lenses with large fields of view because of the dependence on x^3. Wide-angle lenses, with their short focal lengths, are particularly susceptible.

Distortion is to be expected in wide-angle lenses.

5.3.3 Coma

The term

$$\Delta_c = c_5 x \rho^3 \cos^3\phi \qquad \text{(Coma)}, \tag{5.36}$$

describes an aberration called **coma** because of the comet-like appearance of an image (Figure 5.8). We rewrite the equation to emphasize the tilt of the wavefront, $\rho\cos\phi$,

$$\Delta_c = \left(c_5 x \rho^2 \cos^2\phi\right) \rho\cos\phi, \tag{5.37}$$

so the tilt varies linearly in object height, x and quadratically in $\rho\cos\phi$. Because the cosine is even, the same tilt is obtained from the top of the pupil and the bottom. In our model, the prism is replaced with a plate of thickness that varies according to x_1^3 with the proportionality constant increasing with transverse image distance x.

c_5 characterizes coma.

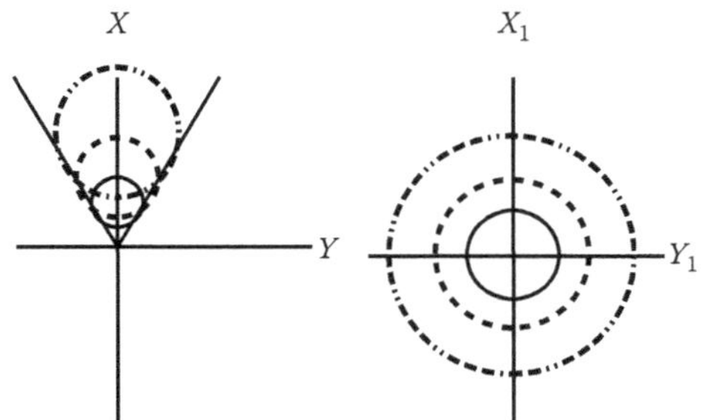

FIGURE 5.8
Coma. The image is tilted by an amount proportional to $h\rho^2 \cos^2 \phi$. The paraxial image is at the vertex of the V.

5.3.4 Field Curvature and Astigmatism

We take the remaining two terms together,

$$\Delta_{fca} = c_2 x^2 \rho^2 \cos^2 \phi + c_3 x^2 \rho^2$$
$$\text{(Field curvature and astigmatism).} \tag{5.38}$$

recognizing again that ρ^2 represents an error in focus. Grouping terms, we obtain

$$\left(c_2 x^2 \cos^2 \phi + c_3 x^2\right) \rho^2$$

c_2 and c_3 characterize field curvature and astigmatism

Now the error in focal length changes quadratically with x, which means that the z location of the image varies quadratically with x. The image thus lies on the surface of an ellipsoid, instead of on a plane. This term is called **field curvature**. Because of the presence of the $\cos^2 \phi$, the focus error also depends on the angular position in the pupil. If we trace rays in the plane that contains the object, image, and axis, $\cos^2 \phi = 1$, and the two terms sum to $(c_2 + c_3) x^2 \times \rho^2$. This plane is called the **tangential plane**. If we consider the plane orthogonal to this one, that also contains the object and image, called the **sagittal plane**, then $\cos \phi = 0$, and the result is $c_3 x^2 \times \rho^2$. Thus, there is a focal error which grows quadratically with x, but this error is different between the two planes. This effect is called **astigmatism**. The result is that the focused spot from a point object has a minimum height at one z distance and a minimum width at another, as shown in Figure 5.9.

Figure 5.10 shows a ray-trace which demonstrates astigmatism. At the tangential and sagittal image locations, the widths in x and y, respectively, are minima. The image is considerably more complicated than the ideal shown in Figure 5.9 because the ray tracing, includes all the other aberrations that are present. There is distortion, which would make the plots hard to visualize. Therefore the center of each plot is adjusted to be at the center of the image. However, for comparison, the same scale factor is used on all plots. There is also coma as evidenced by the asymmetric pattern.

Astigmatism is inherent in off-axis imaging but can be introduced into a system deliberately for on-axis objects using an ellipsoidal or cylindrical lens in place of a spherical one. While a spherical lens focuses a collimated light beam to a point, a cylindrical lens focuses it to a line. A combination of a spherical and cylindrical lens as in Figure 5.11 can focus light to a vertical line in one plane and a horizontal line in another.

For example, consider a spherical thin lens of focal length, f_s, followed immediately by a cylindrical lens of focal length, f_c, so that the axis of the cylinder is horizontal. The combination has a focal length of $f_y = f_s$ for rays in the horizontal plane, but

$$\frac{1}{f_x} = \frac{1}{f_s} + \frac{1}{f_c} \tag{5.39}$$

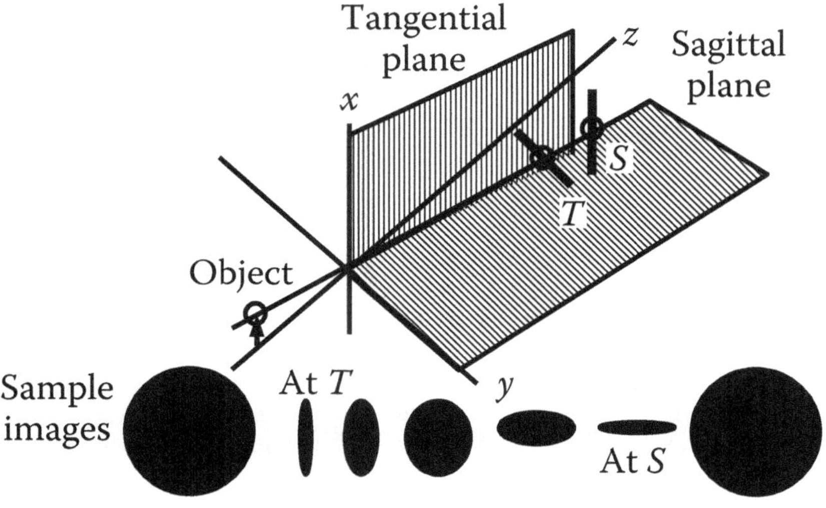

FIGURE 5.9
Astigmatism. The object is located at a height x above the z axis. Viewed in the tangential, x, z, plane, the lens appears to have one focal length. Viewed in the sagittal plane, orthogonal to the tangential one, it appears to have another focal length.

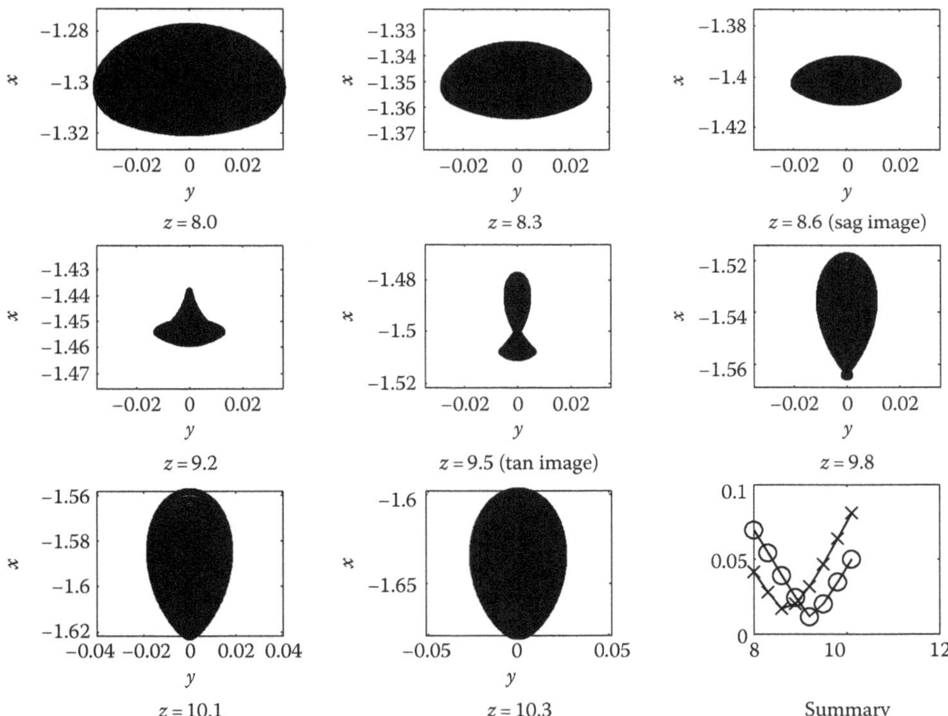

FIGURE 5.10
Astigmatism. The size scales of the image plots are identical. The portion of the field of view shown, however, differs from one plot to the next. The tangential and sagittal images are evident in the individual plots for different z values and in the summary, where it is seen that the minimum values of x and y occur at different values of z. Other aberrations are also in evidence.

for rays in the vertical plane. Thus, collimated light will be focused to a vertical line at a distance f_y and to a horizontal line at f_x.

In Practice

Field curvature and astigmatism can often be reduced by using telecentric designs.

An ellipsoidal lens with different radii of curvature in the x and y planes could be designed to achieve the same effect. In fact, the lens of the eye is often shaped in this way, leading to a vision problem called **astigmatism**. Just as the focal length can be corrected with eyeglasses, astigmatism can be corrected with an astigmatic lens. Frequently, a spherical surface is used on one side of the lens to correct the overall focal length and a cylindrical surface is used on the other side to correct astigmatism. Astigmatism is described by the difference in focusing power in diopters and the orientation of the x or y axis.

Example

Let's look at the lens configuration in Figure 5.11. Suppose that the focal length of the spherical lens, on the right, is $f_1 = 5\,\text{mm}$ and the focal length of the cylindrical lens, on the left, is 30 mm. The beam diameter coming out of the lens is $D(0) = 2\,\text{mm}$ How do the height in the plane of the page and the width perpendicular to the page vary?

By similar triangles, the height of the beam as a function of distance, z, from the lens is

$$x = D(0)\left|\frac{z-f_1}{f_1}\right| = 2\,\text{mm}\left|\frac{z-5\,\text{mm}}{5\,\text{mm}}\right|.$$

In the lower part of the figure, with the two lenses close together the inverse focal lengths add (Equation 3.67), so $f_2 = 6\,\text{mm}$. Then the width is

$$y = D(0)\left|\frac{z-f_2}{f_2}\right| = 2\,\text{mm}\left|\frac{z-6\,\text{mm}}{6\,\text{mm}}\right|.$$

Let's find the value of z where $x = y$ the Solving simultaneously for z is somewhat complicated because of the absolute values. We need to try both the positive and negative fractions. The result is $z = 5.4545$ and $x = y = 181.8\,\mu\text{m}$. From this point, if we change z, we find the following:

Add 100 μm to z	$x = 222\,\mu\text{m}$	$y = 148\,\mu\text{m}$
Subtract 100 μm	$x = 142\,\mu\text{m}$	$y = 216\,\mu\text{m}$

Collecting the light on a quadrant detector as shown in Figure 5.12 will provide a sensitive measure of distance. The lenses could be mounted on a voltage controlled stage controlled by the signal $a+c-b-d$ from this detector to keep the light focused. One example application is the device that maintains focus in a CD player.

Eq: 3.67:
$\frac{1}{f} = \frac{1}{f_1} + \frac{1}{f_2}$.

Take-Away

Seidel Aberrations:
The Seidel aberration terms in Equation 5.32 are summarized in Table 5.1.

- Aberrations can be characterized by phase errors in the pupil.

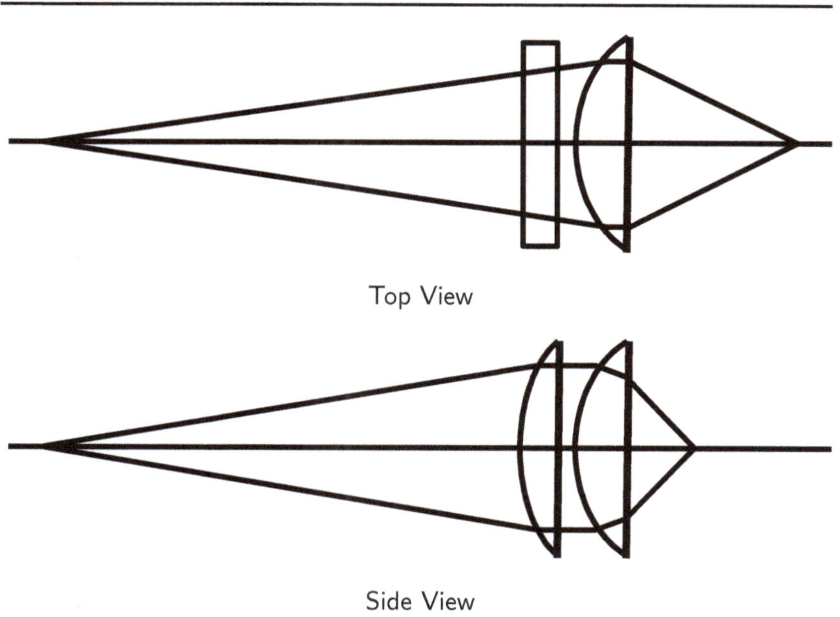

FIGURE 5.11
Deliberate astigmatism. The cylindrical lens introduces astigmatism to the system, even for an on-axis object.

TABLE 5.1
Expressions for Aberrations

	x^0	x^1	x^2	x^3
ρ^1	Tilt			Distortion
ρ^2	Focus		F. C. and astigmatism	
ρ^3		Coma		
ρ^4	Spherical			

Aberrations are characterized according to their dependence on x and ρ.

- These phase errors are functions of the position in the object (or image) and the pupil.
- They can be expanded in a Taylor's series of parameters describing these positions.
- Each term in the series has a specific effect on the image.
- The aberrations increase with increasing aperture and field of view.
- Spherical aberration is the only one which occurs for a single-point object on axis ($x = 0$).

5.4 Spherical Aberration for a Thin Lens

For a thin lens, it is possible to find closed-form expressions for the aberrations. Here we will develop equations for computing and minimizing spherical aberration. The result of this analysis will be a set of useful design rules that can be applied to a number of real-world problems. In Chapter 2, we computed the focal length of a thin lens from the

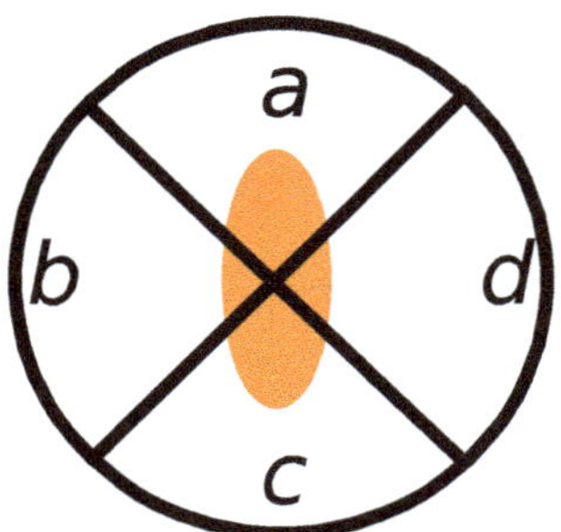

FIGURE 5.12
Quadrant detector. If the lens in Figure 5.11 is focused too close, the pattern will look like the colored shape. The signal $a + c - b - d$ will be positive. If it is focused too far away the signal will be negative. This signal can be used as feedback to a motor or voice coil to position the lens to maintain focus.

Eq. 2.46 $\frac{n}{s} + \frac{n'}{s'} = \frac{n' - n_\ell}{r_2} + \frac{n_\ell - n}{r_1}$.

Eq. 3.24 $\mathcal{L} = \begin{pmatrix} 1 & 0 \\ -\frac{P_t}{n'} & \frac{n}{n'} \end{pmatrix} + \frac{z_{12}}{n_\ell} \begin{pmatrix} -P_1 & n \\ \frac{P_1 P_2}{n'} & -P_2 \frac{n}{n'} \end{pmatrix}$.

radii of curvature and index of refraction using Equation 2.46. As a tool for designing a lens, this equation was incomplete because it provided only one equation for the two variables r_1 and r_2. In Chapter 3, we found a similar situation in Equation 3.24 for a thick lens. We can imagine tracing rays from object to image, and we might expect that some combinations of r_1 and r_2 will result in smaller focused spots than others. Let us consider specifically the case of a point object on axis so that the only aberration is spherical.

5.4.1 Coddington Factors

Before we consider the equations, let us define the useful parameters. Given a lens of focal length, f, there are an infinite number of ways in which it can be used. For each object location, s, there is an image location, s', that is determined by the lens equation (Equation 2.51). We need a parameter that will define some relationship between object and image distance for a given f. We would like this parameter to be independent of f, and we would like it to be well behaved when either $s \to \infty$ or $s' \to \infty$. We define the **Coddington position factor** as

Coddington position factor describes how the lens is used

$$p = \frac{s' - s}{s' + s}. \tag{5.40}$$

Thus, for positive lenses, $f > 0$,

$p > 1$ for	$0 < s < f,$	$s' < 0$	(Virtual image)
$p = 1$	$s = f,$	$s' \to \infty$	
$p = 0$	$s' = s = 2f$		(1:1, four-f relay)
$p = -1$	$s \to \infty,$	$s' = f$	
$p < -1$	$s < 0,$	$0 < s' < f$	(Virtual object)

For negative lenses, $f < 0$,

$p > 1$	for $f < s < 0,$	$s' > 0$	(Real image)
$p = 1$	$s = f,$	$s' \to \infty$	
$p = 0$	$s' = s = 2f$		(1:1, virtual object and image)
$p = -1$	$s \to \infty,$	$s' = f$	
$p < -1$	$s > 0,$	$f < s' < 0$	(Real object)

Just as the Coddington position factor defines how the lens is used, the **Coddington shape factor**,

Coddington shape factor describes how the lens is made.

$$q = \frac{r_2 - r_1}{r_2 + r_1}, \tag{5.41}$$

defines the shape. Lenses of different shapes were shown in Figure 3.8, although in that chapter, we had no basis for choosing one over another. If we can determine an optimal q for a given p, then we have a complete specification for the lens; the focal length, f, is determined by the lens equation (Equation 2.51), and the radii of curvature are obtained from f and q by simultaneously solving Equation 5.41 and the thin-lens version of the lensmaker's equation, Equation 2.52.

Fig. 3.8

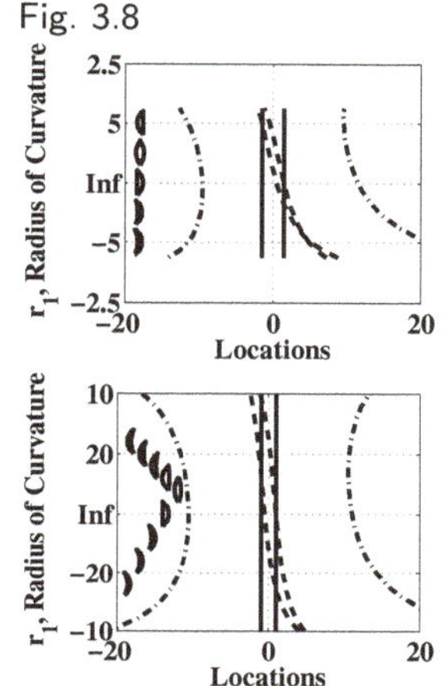

$$\frac{1}{f} = (n-1)\left(\frac{1}{r_1} - \frac{1}{r_2}\right)$$

$$r_1 = 2f\frac{q}{q+1}(n-1) \qquad r_2 = -2f\frac{q}{q-1}(n-1). \tag{5.42}$$

Eq. 2.52 $\frac{1}{f} = \frac{1}{f'} = P_1 + P_2 = (n_\ell - 1)\left(\frac{1}{r_1} - \frac{1}{r_2}\right).$

5.4.2 Analysis

Can we find an equation relating q to p so that we can search for an optimal value? Yes; closed form equations have been developed for this case[90]. The longitudinal spherical aberration is defined as

$$L_s = \frac{1}{s'(x_1)} - \frac{1}{s'(0)}, \tag{5.43}$$

where

- $s'(0)$ is the paraxial image point given by the lens equation
- $s'(x_1)$ is the image point for a ray through the edge of a lens of radius x_1

$$L_s = \frac{x_1^2}{8f^3}\frac{1}{n(n-1)} \times \left(\frac{n+2}{n-1}q^2 + 4(n+1)pq + (3n+2)(n-1)p^2 + \frac{n^3}{n-1}\right). \tag{5.44}$$

L_s describes a change of focal length with distance of ray from the axis at the pupil.

The definition of longitudinal spherical aberration in Equation 5.43 may seem strange at first, but it has the advantage of being useful even when $s' \to \infty$. The result, expressed in diopters, describes the change in focusing power over the aperture of the lens. In cases where s' is finite, it may be useful to consider the displacement of the image formed by the edge rays from the paraxial image:

$$\frac{1}{s'(x_1)} - \frac{1}{s'(0)} = \frac{s'(0) - s'(x_1)}{s'(x_1)s'(0)} \approx \frac{s'(0) - s'(x_1)}{s'(0)^2}$$

$$\Delta s'(x_1) \approx \left[s'(0)\right]^2 L_s(x_1). \tag{5.45}$$

Transverse spherical aberration is useful if s' is not infinite.

Using similar triangles in Figure 5.2, it is possible to define a transverse spherical aberration as the distance, Δx, of the ray from the axis at the paraxial image:

$$\Delta x(x_1) = x_1 \frac{\Delta s'(x_1)}{s'(0)}. \tag{5.46}$$

Find shape factor, q for minimum spherical aberration. Equate derivative to zero.

Now that we can compute the spherical aberration, let us see if we can minimize it. We take the derivative of Equation 5.44 with respect to q and set it equal to zero:

$$\frac{dL_s}{dq} = 0.$$

Optimal q.

The derivative is easy because the equation is quadratic in q, and the result is

$$q_{opt} = -\frac{2\left(n^2 - 1\right)p}{n+2}, \tag{5.47}$$

Minimum transverse spherical aberration.

leading to

$$\Delta x(x_1) = \frac{x_1^3 s'(0)}{8f^3}\left[\frac{-np^2}{n+2} + \left(\frac{n}{n-1}\right)\right]. \tag{5.48}$$

Now we can see in Equation 5.44 that lenses of different shapes will have different aberrations, and with the optimal q from Equation 5.47, we can use Equations 5.42 as the second constraint on our designs.

The optimal lens shape given by Equation 5.47 does not depend on x_1, so the best choice of surface curvatures is independent of the aperture. However, the amount of aberration depends strongly on the aperture. Writing Equation 5.48, in terms of the f-number, $F = f/D$, where the maximum value of x_1 is $D/2$,

$$\Delta x\left(\frac{D}{2}\right) = \frac{s'(0)}{64F^3}\left[\frac{-np^2}{n+2} + \left(\frac{n}{n-1}\right)\right],$$

or substituting $s' = 2f/(p-1)$,

$$\Delta x\left(\frac{D}{2}\right) = \frac{2f}{64F^3(p-1)}\left[\frac{-np^2}{n+2} + \left(\frac{n}{n-1}\right)\right]. \tag{5.49}$$

Spherical aberration varies with the inverse cube of f/number.

The inverse cubic dependence on F means that spherical aberration will become a serious problem for even moderately large apertures.

Figure 5.13 shows the transverse spherical aberration as a function of q for various lenses. In all cases, the object and image distances are the same, as is the height of the edge ray, but three different indices of refraction are used. We will see in Equation 8.85 that there exists a fundamental physical limit, called the diffraction limit, on the size of the focused spot from a lens of a given size, and that this limit is inversely proportional to the size of the lens:

Compare transverse aberration to the diffraction limit.

$$\Delta x_{DL} = 1.22\frac{\lambda}{D}s' = 1.22\lambda F, \tag{5.50}$$

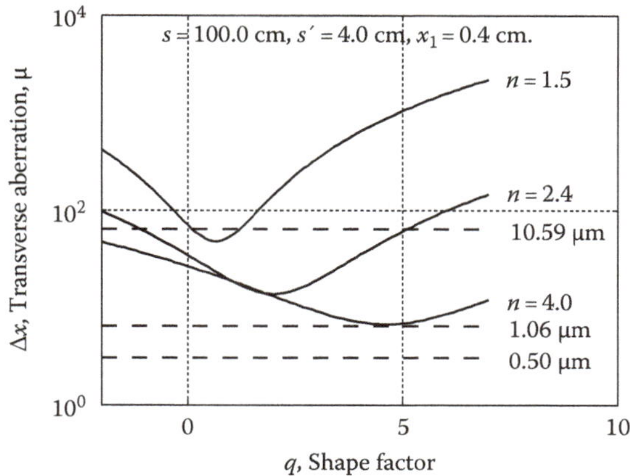

FIGURE 5.13
Transverse spherical aberration. Three curves show the transverse spherical aberration as a function of q for three different indices of refraction. The three horizontal lines show the diffraction-limited spot sizes at three different wavelengths.

for a lens of diameter $D = 2x_1$. The actual spot size will be approximately the larger of $\Delta x\,(D/2)$ predicted by geometric optics, including aberration, and Δx_{DL} predicted by diffraction theory. If $\Delta x\,(D/2)$ is smaller, we say that the system is diffraction-limited; there is no need to reduce aberrations beyond this limit as we cannot make the spot any smaller. In the figure, the diffraction limits are plotted for three common wavelengths, 500 nm in the middle of the visible spectrum, 1.064 μm (Nd:YAG laser) and 10.59 μm (CO_2 laser). Plotting Equation 5.46, we find a minimum for each index of refraction. We can use these curves along with the diffraction limits to evaluate the performance of this lens in this application.

In Practice

A good way to get an idea about aberrations is to use Equation 5.49 to determine the minimum spherical aberration, or Equation 5.46 with 5.45 to determine the actual spherical aberration. In either case, the aberration can be compared either with some desired performance goal, or with the diffraction limit (Equation 5.50). If it is good enough, then it is worthwhile to evaluate the other aberrations, perhaps with a ray–tracing program. If not, then a simple lens with spherical surfaces will not work.

Example

Focusing a Carbon-Dioxide Laser Beam
For example, if we are trying to focus light from a CO_2 laser, at $\lambda = 10.59$ μm, diffraction-limited performance can be achieved with both the zinc-selenide lens, with $n = 2.4$ (see the $n = 2.4$ aberration curve and the 10.59 μm diffraction-limit line in Figure 5.13. Note that the aberration curve goes below the diffraction limit for a wide range of shape factors.), and the germanium lens, with $n = 4$, although the latter is better than the diffraction limit over the whole range of shapes shown in the figure. The same zinc-selenide lens is considerably worse than the diffraction limit at the

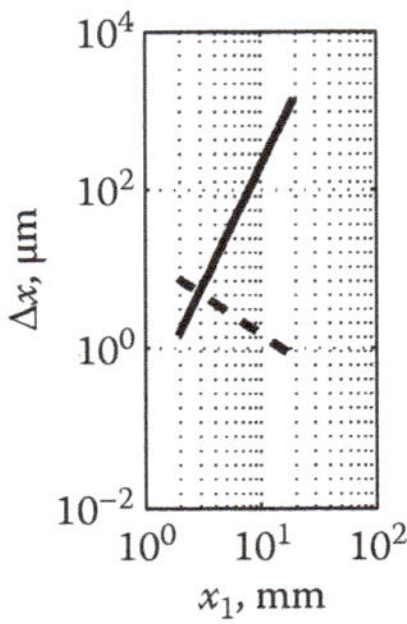

FIGURE 5.14
Transverse spherical aberration with height solid line. The transverse aberration is proportional to h^3. The diffraction limit varies as $1/x_1$ dashed line, so the ratio varies as x_1^4.

two shorter wavelengths (compare the $n = 2.4$ curve to both of the remaining two diffraction limits), even for the optimal shape. We note that we can also meet the diffraction limit with $n = 1.5$, but glass is not transparent beyond about 2.5 μm, so this is not a practical approach, and designers of infrared optics must use the more costly materials. Although there are some exceptions, it is at least partly true that, "Far infrared light is transmitted only by materials that are expensive."

Example

Focusing at Shorter Wavelengths
For the Nd:YAG laser at 1.06 μm, an optimized zinc-selenide lens is still not quite diffraction-limited; the $n = 2.4$ curve is above the 1.06 μm diffraction limit, even at its lowest point. For visible light, the glass lens produces aberrations leading to a spot well over an order of magnitude larger than the diffraction limit, and a good lens this fast ($F = f/D = 3.8\,\text{cm}/(2 \times 0.8\,\text{cm})$) cannot be made with a single glass element with spherical surfaces. Although the curve for $n = 4$ comes closer to the diffraction limit, we must remember that germanium does not transmit either of the shorter two wavelengths.

From Equations 5.44 and 5.46, the transverse spherical aberration varies with the f-number, $F = f/D$ (where D is the maximum value of $2x_1$), in proportion to F^{-3}, or in terms of numerical aperture, $NA \approx D/(2f)$, according to NA^3. Figure 5.14 shows an example of the transverse aberration and its dependence on the height, x_1. The transverse aberration is proportional to x_1^3, butt in contrast, the diffraction limit varies with $1/D$, so the ratio varies with D^4 or NA^4.

The ratio of spot size caused by spherical aberration to diffraction limit varies like the fourth power of D or NA.

Figure 5.15 shows the optimal Coddington shape factor, q_{opt}, given by Equation 5.47 as a function of the Coddington position factor, p. Each plot is linear and passes through the origin, and the slope increases with increasing index of refraction. For a "four-f" imager ($s = s' = 2f$) with unit magnification, $p = 0$ and $q_{opt} = 0$, represented by the cross at the origin. A good rule is to "share the refraction," as equally as possible between the two surfaces. This rule makes sense because the aberrations result from using first-order approximation for trigonometric functions. Because the error increases rapidly with angle (third-order), two small errors are better than one large one.

Share the refraction among available refracting surfaces to reduce aberrations.

A plano-convex lens is useful to collimate light from a point. A convex-plano lens is useful to focus collimated light to a point.

For glass optics, it is common to use a plano-convex lens with the flat surface first ($q = -1$) to collimate light from a point ($s' \to \infty$, or $p = 1$) and to use the same lens with the flat surface last ($q = 1$) to focus light from a distant source ($p = -1$). These

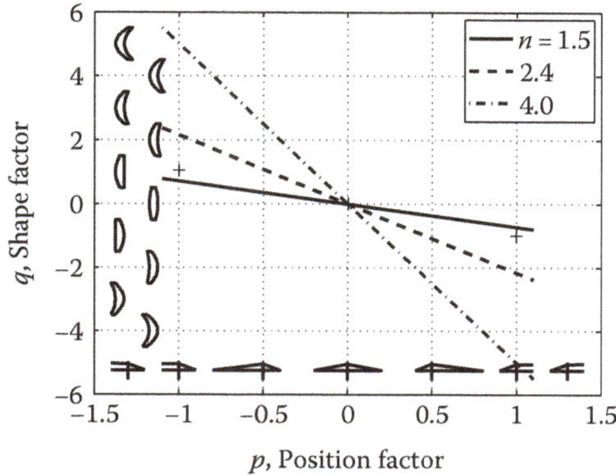

FIGURE 5.15
Optimal Coddington shape factor. The optimal q is a linear function of p, with different slopes for different indices of refraction.

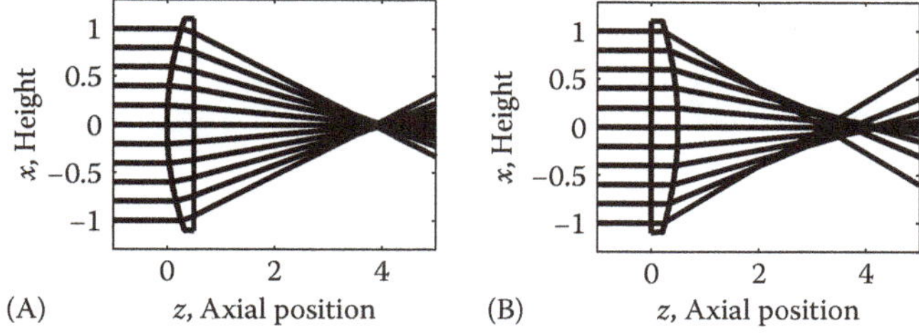

FIGURE 5.16
Plano-convex lens for focusing. If the refraction is shared (A), then collimated rays focus to a small point. If the planar side of the lens is toward the source (B), all the refraction occurs at the curved surface, and spherical aberration is severe. If we wanted to use the lens to collimate light from a small object, then we would reverse the lens.

points are shown by crosses as well, and although they are not exactly optimal, they are sufficiently close that it is seldom worth the expense to manufacture a lens with the exactly optimized shape. Figure 5.16 traces collimated input rays exactly through a plano-convex lens oriented in both directions. In Figure 5.16A, there is some refraction at each surface, and the focused spot is quite small. Figure 5.16B shows the lens reversed so that there is no refraction at the first surface. The larger angles of refraction required at the second surface result in greater aberrations, and the increased size of the focused spot is evident. If we wanted to use this lens to collimate light from a small point, we can imagine reversing Figure 5.16A. For this application, the planar surface would be toward the source. These results are consistent with the rule to "share the refraction."

Take-Away

Spherical Aberration in a Thin Lens:

- Spherical aberration exists even for an object on the axis of an optical imaging system.
- The ratio of the spot size caused by spherical aberration to the diffraction limit varies with the fourth power of the aperture size.

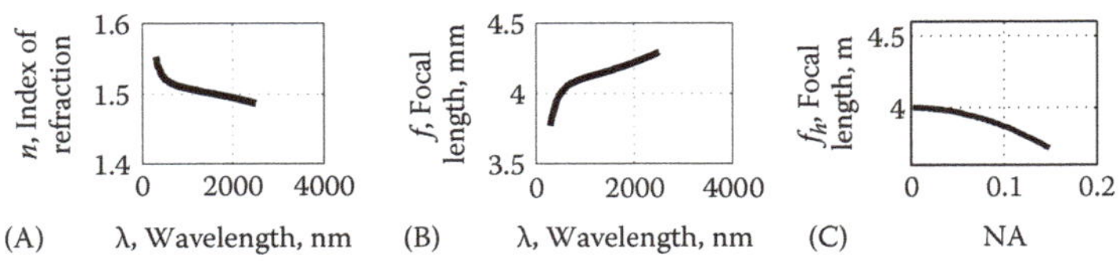

FIGURE 5.17
Chromatic aberration the dispersion of BK7 (Figure 5.8) is significant in the visible spectrum (A). The resulting chromatic aberration of a lens optimized for spherical aberration is close to $\pm 100\ \mu\text{m}$ of focal shift over the visible spectrum (B). However, spherical aberration is still larger than chromatic aberration for numerical apertures larger than 0.1 (C).

- The choice of lens shape depends on the object and image location through the Coddington position and shape factors, p and q.
- Even a lens with an optimized shape may not be diffraction limited.

In Practice

Minimum aberrations are achieved when the refraction is shared among all the available surfaces.

- A plano-convex lens is typically used if there is a big disparity in object and image distance.
- If the object distance is much shorter than the image distance, the flat side is placed toward the object, as when a point source is to be collimated.
- If the object distance is much longer than the image distance, the flat side is placed toward the image. In both cases the refraction is shared between the two surfaces.
- For object and image distances nearly matched, a biconvex lens is a good choice.
- For negative lenses, the same rules hold for plano-concave and bi-concave lenses.
- For large apertures, these choices may still result in aberrations that exceed the diffraction limit and a more complicated system may be needed.

Fig. 1.8

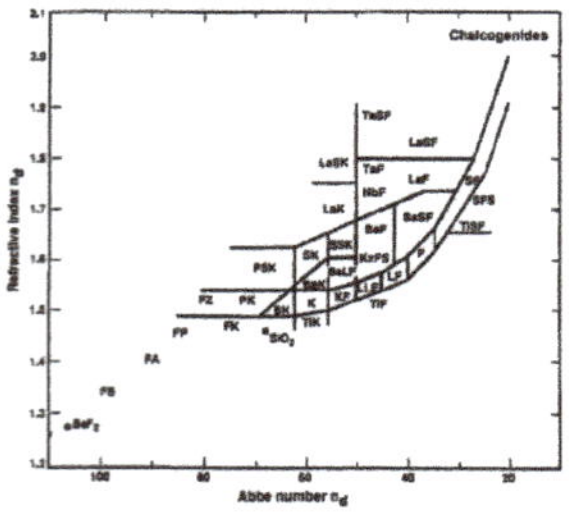

Eq. 2.52 $\frac{1}{f} = \frac{1}{f'} = P_1 + P_2 = (n_\ell - 1)\left(\frac{1}{r_1} - \frac{1}{r_2}\right)$.

Eq. 3.17 $\mathcal{R} = \begin{pmatrix} 1 & 0 \\ \frac{n-n'}{n'r} & \frac{n}{n'} \end{pmatrix}$.

5.5 Chromatic Aberration

The final aberration we will consider is **chromatic aberration**, which is the result of dispersion in the glass elements of a lens. As discussed in Figure 1.8, all optical materials are at least somewhat dispersive; the index of refraction varies with wavelength. By way of illustration, the focal length of a thin lens is given by Equation 2.52. The analysis can also be extended to the matrix optics we developed in Chapter 3, provided that we treat each refractive surface separately, using Equation 3.17. If we are using a light source that covers a broad range of wavelengths, chromatic aberration may become significant.

Example

Chromatic Aberration

We consider one numerical example to illustrate the problem. In this example, the focal length is $f = 4\,\text{mm}$, which would be appropriate for a $40\times$ microscope objective using a tube length of 160 mm. We will try a lens made from a common glass, BK7. The index of refraction of this glass, shown in Figure 5.17A, varies strongly in the visible spectrum. Let us make the object distance infinite and locate the image as if we were focusing laser light in a confocal microscope. On the detection side of the microscope, this lens will collimate light from the object, and the results we are about to obtain will tell us how far we would move the object to bring it into focus at different wavelengths.

We will choose the "nominal" wavelength as 500 nm, for which $n = 1.5214$. This wavelength is a good choice for a visible-light system because it is near the sensitivity peak of the eye, as we shall see in Figure 12.9, and is in the middle of the visible spectrum. We then calculate the radii of curvature using Equations 5.42. Next, we compute the focal length (and thus the image distance in this example) at each wavelength, using Equation 2.52. The result, in Figure 5.17B, shows that the focal plane varies by almost 100 μm or 0.1 mm over the visible spectrum. Despite this severe chromatic aberration, spherical aberration may still be more significant. Figure 5.17C shows the focal shift caused by spherical aberration as a function of numerical aperture for the same lens. The transverse displacement of the edge ray in the focal plane is given by Equation 5.46:

$$\Delta x\left(x_1\right) = x_1 \frac{0.1\,\text{mm}}{4\,\text{mm}}. \tag{5.51}$$

For an aperture of radius $x_1 = 1\,\text{mm}$, $(NA \approx 0.25)$ the transverse aberration is $\Delta x\left(x_1\right) = 25\,\mu\text{m}$, while the diffraction limit in Equation 5.50 is about $2.5\lambda = 1.25\,\mu\text{m}$. Even for this modest numerical aperture, the spot is 50 times the diffraction limit in the middle of the spectrum.

The transverse distance, $\Delta x\left(x_1\right)$, would determine the resolution, and the longitudinal distances in Figure 5.17B would determine the focal shifts with wavelength. These distances are unacceptably large, even though the numerical aperture is much smaller than is commonly used in modern microscopes. Clearly, the potential for making a good microscope objective with spherical refracting surfaces on a single glass element is small.

5.6 Design Issues

In view of the example in the previous section, how can we make a useful microscope objective? In fact, how can we even make any useful imaging system faster than $f/3$ or so, without introducing aberrations? Figure 5.18 shows some possible approaches to reducing aberrations. First, if one lens is not enough, try two. The rule about sharing the refraction applies as well to more than two surfaces. The more small refracting steps, the smaller the total error can become. Second, use two multielement commercial lenses. For example, to achieve an object distance of 25 mm and an image distance of 75 mm, we could use two camera lenses. Normally, camera lenses are carefully designed to minimize aberrations when the object is at infinity. Using a 25 mm lens backward, with the object where the film would normally be, will then result in a good image at infinity. A 75 mm lens used in the normal way will then produce a good image at its back focal plane. The hard work of design is all done by the designer of the camera lenses. Of course, we are probably using more glass elements than we would have in a custom design and, therefore, are losing more light to reflections at surfaces and to absorption. However, this disadvantage may be small compared to the saving of design costs if we require only a few units. Finally, aspheric surfaces may be used on the lenses. The aspheric lens may be costly to design, but aspheres are available off the shelf for a number of object and image distances.

Compound lenses can reduce aberrations beyond what is possible with simple lenses.

Combining commercial lenses is one way to reduce aberrations with minimal design effort.

Aspheres can reduce aberrations.

In discussing Figure 5.17, we saw that for this particular lens, as for many simple lenses, the chromatic aberrations were small compared to spherical aberrations, even for

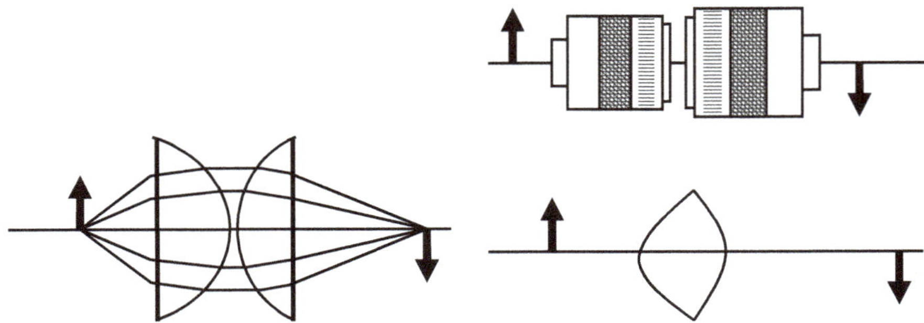

FIGURE 5.18
Reducing aberrations. If one simple lens does not provide the desired performance, we can try adding an extra lens (left), using multiple commercial lenses (upper right), or an aspheric lens (lower right).

very modest numerical apertures. However they were still large compared to the diffraction limit, and if we make efforts to correct the Seidel aberrations with multiple elements and possibly the use of aspheres, chromatic aberration may become the dominant remaining aberration. A laser has an extremely narrow range of wavelengths, and chromatic aberration will not be a major problem with such a source. Even if the laser is not at the design wavelength of the lens, the most significant result will be a simple shift in focus, which can usually be corrected by a simple adjustment of some optical component. A light emitting diode typically spans 20–40 nm of wavelength and, in most cases, we would not expect much chromatic aberration with such a source. In contrast, a tungsten lamp has an extremely wide range of wavelengths. The wavelengths of a system using visible light may be limited more by the detector or by the transmission spectra of the optical components, but in any case, we must then pay attention to chromatic aberration in our designs.

Chromatic aberration is usually not important with light from narrow–band sources such as lasers and LEDs (Femtosecond pulsed laser beams are an exception).

Current trends in microscopy are toward ever-increasing ranges of wavelengths with some techniques spanning a factor of two to three in wavelength. In general, a microscope objective will be well corrected for Seidel aberrations across wavelengths. Dispersion has a small effect on these corrections. However, a microscope corrected for chromatic aberrations in the visible may not be well corrected in the infrared or ultraviolet. For laser illumination at a single wavelength not too far from the band over which the lens was designed, the major effect will be a shift in focus. Comparison of sequential images taken at different wavelengths may require an adjustment of the position of the objective to correct for this shift, and an image obtained with a broadband infrared source, or with multiple laser wavelengths simultaneously, may be blurred by chromatic aberrations.

Another situation in which chromatic aberration may be important is in systems with femtosecond pulses. Mode-locked laser sources can produce these short pulses, which are useful for many applications such as two-photon-excited fluorescence microscopy[50]. We will discuss these sources briefly in our discussion of coherence around Figure 10.1. The bandwidth of a short pulse is approximately the inverse of the pulse width, so a pulse consisting of a few optical cycles can have a very large bandwidth and, thus, can be affected by chromatic aberration.

Femtosecond pulses from a laser may be lengthened in time by chromatic aberration.

Take-Away

Chromatic Aberration:

Chromatic aberrations may affect broadband imaging systems, but are less important with narrow–band illumination. One exception is that chromatic aberration can lengthen a very short (femtoseconds) laser pulse.

The use of reflective optics provides an advantage in that the angle of reflection is independent of the index of refraction. Thus, large reflective telescopes and similar instruments are free of chromatic aberrations and can be used across the entire wavelength band over which the surface is reflective. Gold (in the IR and at the red end of the visible spectrum), silver, and aluminum mirrors are commonly used. With the increasing interest in broadband and multiband microscopy, reflective microscope objectives have become available. Although their numerical apertures are not as large as the most expensive refractive objectives, the lack of chromatic aberration is an advantage in many applications.

In Practice

Reflective optics may be used over an extremely wide range of wavelengths, and is commonly used in large telescopes. Reflective microscope objectives exist but not yet with the high magnification and numerical aperture of refractive ones.

5.7 Lens Design

Modern optical design is a complicated process and is the subject of many excellent books[215;204;214;75;130]. Camera lenses, microscope objectives, and other lenses requiring large apertures, large fields of view, and low aberrations are designed by experts with a combination of experience and advanced computational ray tracing programs. The intent of this section is to provide some useful information for the system designer who will seek the assistance of these experts on challenging designs, or will use the available tools directly in the design of simpler optics.

The first step in any optical design is usually a simple sketch in which the locations of objects, images, and pupils are sketched, along with mirrors, lenses, and other optical elements. At this point, all the lenses are treated as perfect thin lenses. As this design is developed, we consider the size of the field of view at each field plane (object or image plane), the size of each pupil, and lens focal lengths interactively, until we arrive at a basic design.

Next, we determine initial lens designs, selecting radii of curvature and thicknesses for the lenses. We may try to choose standard lens shapes, biconvex, plano-convex, etc., or we may try to optimize shapes using Equation 5.47. At this point, if the system requirements are not too demanding, we may look for stock items in catalogs that have the chosen specifications. Then using the thicknesses and radii of curvature, we determine the principal planes of the lenses, and make corrections to our design. For a very simple system with a small aperture and field of view, we may stop at this point.

Eq. 5.47:
$$q_{opt} = -\frac{2(n^2-1)p}{n+2}.$$

For a more complicated system, we then proceed to enter this design into a ray tracing program and determine the effects of aberrations. Now we can also consider chromatic aberrations, incorporating the wavelengths we expect to use in the system. The more advanced ray tracing programs contain lists of stock lenses from various manufacturers, and we may be able to select lenses that will improve the design from such catalogs. These programs also allow customized optimization. We can decide on some figure of merit, such as the size of the focused spot, in a spot diagram like those shown in Figure 5.10, and declare certain system parameters, such as the spacing of some elements, to be "variable." The computer will then adjust the variable parameters to produce an optimal figure of merit. These optimizations are subject to finding local minima, so it is important to resist

Fig. 5.10

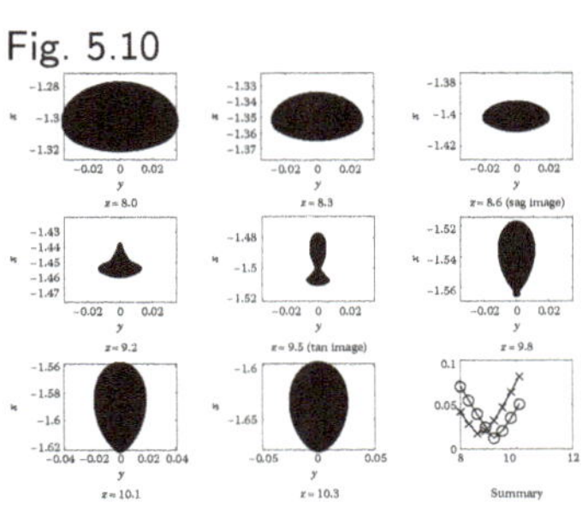

the temptation to let the computer do all the work. We usually must give the computer a reasonably good starting point. If the optimization still fails, we may solve the problem by using two lenses in place of one. Here is where the lens designer's skill becomes critical. An experienced designer can often look at a failed design, find the offending element, and suggest a solution based on years of working with a wide range of optical systems. The less experienced user may still be able to locate the problem by isolating each component, tracing rays from its object to its image, and comparing to expectations such as the diffraction limit.

If these steps fail, the use of more complicated commercial lenses such as camera lenses or microscope objectives may help, although potentially at the expense of increased complexity and reduced light transmission. Another disadvantage is that vendors will usually not divulge the designs of these lenses so they cannot be incorporated into the ray tracing programs. The advantage of using these components is that they are the result of more significant investment in optical design, and thus probably well corrected for their intended use.

In addition to optimization of spacing of elements, the programs will optimize the radii of curvature. This sort of optimization is used in the design of those high-performance commercial lenses. The programs usually have catalogs of different glasses and of standard radii of curvatures for which different manufacturers have tools already available. They can use this information to optimize a design based on a knowledge of which vendors we wish to use. The programs contain a variety of other options, including aspheres, holographic elements, and other components which can optimize a design.

Many of the programs have the ability to determine tolerances for manufacturing and alignment, the ability to analyze effects of polarization, temperature, multiple reflections between surfaces, and even to interface with mechanical design programs to produce drawings of the final system.

Problems

5.1 Spherical Aberrations

Here we consider how the focus moves as we use larger parts of an aperture. The lens has a focal length of 4 cm, and the object is located 100 cm in front of it. We may assume a thin lens with a diameter of 2 cm.

5.1a What is the f-number of this lens?

5.1b Plot the location of the focus of rays passing through the lens at a height, h, from zero to $d/2$, assuming the lens is biconvex.

5.1c Repeat assuming it is plano-convex.

5.1d Repeat assuming it is convex–plano.

5.1e Repeat for the optimal shape for minimizing spherical aberrations.

5.2 Low-Cost Microscope (NS)

What would happen if we made a microscope with simple lenses? We will use plano-convex lenses mounted in the "right" direction. Make the tube lens 200 mm. We want to achieve a magnification of 40X at $NA = 0.5$.

5.2a What is the focal length of the objective?

5.2b What is the transverse spherical aberration? Hint: L_S being an inverse focal length, adds for the two lenses.

5.2c Now assume we have a perfectly corrected objective ($L_s = 0$). What is the transverse aberration now?

5.2d Compare to the diffraction limit for 500 nm wavelength green light. Comment on the importance of correcting aberrations and your impression of the difficulty of doing so for the two lenses.

Note that this analysis is pessimistic on one hand because only a few rays will be at the edge of the aperture, but optimistic on the other hand as it applies to objects in the center of the field of view. Other aberrations may become more important at the edges.

5.3 Spherical Mirror (NS)

We want to use a spherical mirror to view an object at infinity, because it is cheaper than a parabolic one which would be correct for such a situation. What is the error, Δ in optical path length as a function of the radius of the lens? Plot the results. What is the maximum diameter to keep this error less than $\lambda/10$ for green light and for light from a carbon-dioxide laser ($10.59\,\mu$m)?

6 Polarized Light

In the 1600s, Huygens[102] recognized that there are "two kinds of light." Snell's law of refraction, which had recently been discovered empirically, related the bending of light across a boundary to a quantity called the index of refraction. In some materials Huygens observed light refracting in two different directions. He assumed, correctly, that the two different kinds of light had different indices of refraction. Almost 200 years later, Young[258] provided an understanding in terms of transverse waves that shortly preceded Maxwell's unification of the equations of electromagnetics.

Today, the principles of polarization are used in applications ranging from high-bandwidth modulation for communication devices and optimization of beamsplitters for optical sensing to everyday applications such as liquid crystal displays and sunglasses. In this chapter, we will begin with the fundamentals of polarized light (Section 6.1) and then show how devices can manipulate the polarization (Section 6.2). Next in preparation for a discussion of how polarizing devices are made we discuss the physics of light interacting with materials (6.3), and the boundary conditions leading to polarization effects at interfaces (6.4). We then address the physics of specific polarizing devices in common use (6.5). Systems with multiple polarizing devices are best handled with one of two matrix methods. Section 6.6 discusses the Jones calculus [115], based on two-dimensional vectors representing the field components in some orthogonal coordinate system, with two-by-two matrices representing polarizing components. In Section 6.7, partial polarization is discussed with coherency matrices and Mueller calculus. The Mueller calculus [163], also developed in the 1940s, uses four-by-four matrices to manipulate the Stokes vectors[218]. The Stokes vectors provide a more intuitive treatment of partial polarization than do the Jones vectors. Finally the Poincaré sphere provides a visualization of the Stokes vectors. For a more detailed discussion of polarization, several textbooks are available [209;17;80].

FURTHER READING:
More on polarization [209;17;80]

6.1 Fundamentals of Polarized Light

By now, we understand light as a transverse wave, in which only two orthogonal components are needed to specify the vector field, and we will consider alternative basis sets for these two.

Eq. 1.1 $\nabla \times \mathbf{E} = -\frac{\partial \mathbf{B}}{\partial t}$.

6.1.1 Light as a Transverse Wave

Maxwell's equations, specifically the Faraday Equation (Eq. 1.1) and the definition of the Poynting vector, (Eq. 1.45) imply angle relationships that lead to two triplets of mutually orthogonal vectors for plane waves with $\boldsymbol{\nabla} \times \mathbf{E} = j\,\mathbf{k} \times \mathbf{E}$ and $\frac{\partial}{\partial t} = -j\omega$

Eq. 1.45 $\mathbf{S} = \mathbf{E} \times \mathbf{H}$.

$$\mathbf{k} \times \mathbf{E} = -\omega \mathbf{B} \quad \rightarrow \quad (1)\ \mathbf{B} \perp \mathbf{k} \quad (2)\ \mathbf{B} \perp \mathbf{E} \tag{6.1}$$

$$\mathbf{k} \times \mathbf{H} = \omega \mathbf{D} \quad \rightarrow \quad (1)\ \mathbf{D} \perp \mathbf{k} \quad (1)\ \mathbf{D} \perp \mathbf{H} \tag{6.2}$$

$$\mathbf{E} \times \mathbf{H} = \mathbf{S} \quad \rightarrow \quad (2)\ \mathbf{S} \perp \mathbf{E} \quad (2)\ \mathbf{S} \perp \mathbf{H}. \tag{6.3}$$

$B \parallel H$

At optical wavelengths, μ is a scalar equal to μ_0 (Section 1.4.1), and thus $\mathbf{H}$ is parallel to $\mathbf{B}$. From the angle relationships noted by (1) in the above three equations, $\mathbf{H}$, $\mathbf{D}$, and

DOI: 10.1201/9781315157047-6

k, are all mutually perpendicular, and from those labeled (2), **E**, **B** (and thus **H**), and **S** are mutually perpendicular. Thus, for a given direction of **k** we can represent any of the vectors **E**, **D**, **B**, or **H** with only two dimensions, and if we specify one of the vectors, we can determine the others. It is customary to treat **E** as fundamental. For example, the expression "vertical polarization," is understood to mean polarization such that the electric field is in the vertical plane that contains the **k** vector.

Polarization can be described by a 2–dimensional vector.

Take-Away

Polarization Definition:

- Polarization of light is generally defined by the direction of the electric field vector.
- Given a direction of propagation, the polarization can be defined by a 2–dimensional vector (*i.e.* two complex numbers.

6.1.2 Linear Polarization

For most people, the most intuitive coordinate system for polarized light uses two linear polarizations in orthogonal directions. Depending on the context, these may be called "vertical" and "horizontal" or some designation related to the orientation of polarization with respect to some surface. Thus the complex electric field can be described by

$$\mathbf{E} = \left[E_v\hat{\mathbf{v}} + E_h\hat{\mathbf{h}}\right] e^{j(\omega t - kz)}, \tag{6.4}$$

where $\hat{\mathbf{v}}$ and $\hat{\mathbf{h}}$ are unit vectors in the vertical and horizontal directions respectively, or

Linear–polarization basis directions; Vertical/Horizontal, x/y, *etc.*

$$\mathbf{E} = \left[E_x\hat{\mathbf{x}} + E_y\hat{\mathbf{y}}\right] e^{j(\omega t - kz)}, \tag{6.5}$$

in a Cartesian coordinate system in which light is propagating in the $\hat{\mathbf{z}}$ direction. In this case, using Equation 1.43,

$$H_x = \frac{-E_y}{Z} \qquad H_y = \frac{E_x}{Z}$$

$$\mathbf{H} = \left[-\frac{E_y}{Z}\hat{\mathbf{x}} + \frac{E_x}{Z}\hat{\mathbf{y}}\right] e^{j(\omega t - kz)}, \tag{6.6}$$

where Z is the impedance of the medium.

The reader who is familiar with electromagnetic theory may be comfortable with the terms "transverse magnetic (TM)" and "transverse electric (TE)." These terms define the polarization relative to the **plane of incidence**, which contains the normal to the surface and the incident ray, and thus must also contain the reflected and refracted rays. This plane is the one in which the figure is normally drawn. For example, for light reflecting from the ocean surface, the incident ray is in a direction from the sun to the point of incidence, and the normal to the surface is vertical. The TM wave's magnetic field is perpendicular to the plane of incidence and thus horizontal, as is the TE wave's electric field. In optics, the preferred terms are **P-polarized** for waves in which the electric field vector is *parallel* to the plane of incidence and **S-polarized** for cases where it is *perpendicular (or senkrecht)* to the plane of incidence. Thus "P" is the same as "TM" and "S" is "TE." Figure 6.1 illustrates the two cases. By analogy to Equation 6.4, the complex field can be represented by

Plane of incidence defined.

P and S polarization states are useful bases when describing effects of a surface.

$$\mathbf{E} = \left[E_s\hat{\mathbf{s}} + E_p\hat{\mathbf{p}}\right] e^{j(\omega t - kz)}. \tag{6.7}$$

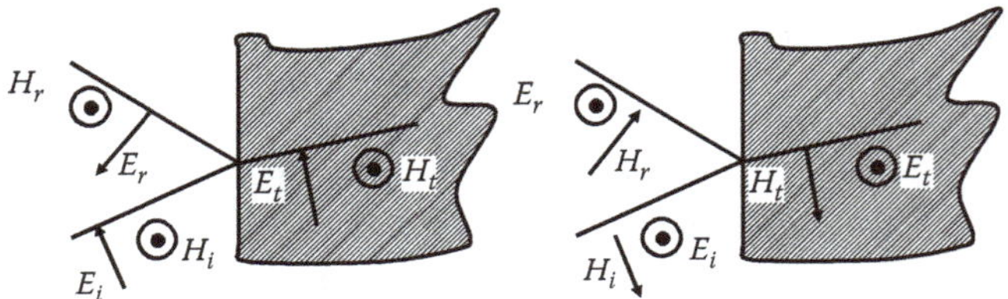

FIGURE 6.1
Polarization labels. Light that has the electric field parallel to the plane of incidence is P, or TM polarized (left) and that which has the electric field perpendicular to the plane of incidence is S, or TE polarized. The plane of incidence contains the normal to the surface and the incident ray. Here it is the plane of the paper.

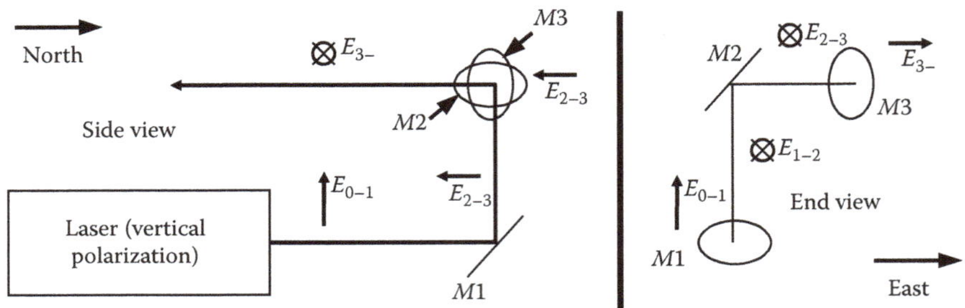

FIGURE 6.2
Polarization with mirrors. In this example, the polarization labels change as light propagates along the path, illustrating the complexity of linear polarization labels. Table 6.1 lists the labels at each point along the path. To help us understand the directions in three dimensions, we will use compass points such as "north" and "east" to define the horizontal directions. The polarization states are summarized in Table 6.1.

Both of these representations present a lack of consistency. For example, when light propagating horizontally is reflected by a mirror to propagate vertically, both polarizations become horizontal. The situation in Figure 6.2 shows that light which is, for example, S polarized at mirror $M1$ is P polarized at mirror $M2$. Nevertheless, because they are so intuitive, linear-polarization basis sets are most frequently used.

Take-Away

Basis Vectors:
Orthogonal linear polarization states are often used as basis vectors. Different definitions can be used depending on the configuration. At an interface, the use of S and P definitions is common.

In Practice

When buying a beamsplitter, it is most common to specify the desired parameters of reflectance and transmittance (to be discussed later) in terms of S and P polarization.

TABLE 6.1
Polarization Labels

Location	Label	Propagation	Polarization	Label
Before $M1$	E_{0-1}	North	Up	Vertical
At $M1$				P
After $M1$	E_{1-2}	Up	South	Horizontal
At $M2$				S
After $M2$	E_{2-3}	West	South	Horizontal
At $M3$				P
After $M3$	E_{3-}	South	East	Horizontal

The linear polarization labels, P and S change from one location to another throughout the path in Figure 6.2. Note that after $M1$ the labels of "vertical" and "horizontal" are insufficient, because all polarization is horizontal beyond this point.

6.1.3 Circular Polarization

Circular polarization provides a more consistent, but less intuitive, basis set. In Equation 6.4, the field components, E_v and E_h can be complex numbers. If, for example, $E_x = E_0/\sqrt{2}$ is real, and $E_y = jE_0/\sqrt{2}$ is purely imaginary, then the complex field is

$$\mathbf{E}_r = \frac{E_0}{\sqrt{2}} \left[\hat{\mathbf{x}} + j\hat{\mathbf{y}}\right] e^{j(\omega t - kz)}. \tag{6.8}$$

At $z = 0$, the time-domain field is

$$\mathbf{E}_r = \frac{E_0}{\sqrt{2}} \Re\left[\hat{\mathbf{x}}\left(e^{j(\omega t)} + e^{-j(\omega t)}\right) + j\hat{\mathbf{y}}\left(e^{j(\omega t)} + e^{-j(\omega t)}\right)\right]$$

$$= \hat{\mathbf{x}}\ E_0\sqrt{2}\cos\omega t + \hat{\mathbf{y}}\ E_0\sqrt{2}\sin\omega t. \tag{6.9}$$

This vector rotates sequentially through the $\hat{\mathbf{x}}$, $\hat{\mathbf{y}}$, $-\hat{\mathbf{x}}$, and $-\hat{\mathbf{y}}$ directions, which is the direction of a right-hand screw, or clockwise as viewed from the source. For this reason, we call this right-hand-circular (RHC) polarization, and have given the field the subscript r. Likewise, left-hand-circular (LHC) polarization is represented by

RHC and *LHC* basis vectors.

$$\mathbf{E}_\ell = \left[E_0\hat{\mathbf{x}} - jE_0\hat{\mathbf{y}}\right] e^{j(\omega t - kz)}. \tag{6.10}$$

Any field can be represented as a superposition of RHC and LHC fields. For example,

$$\frac{1}{\sqrt{2}}\mathbf{E}_r + \frac{1}{\sqrt{2}}\mathbf{E}_\ell = E_x\hat{\mathbf{x}} \tag{6.11}$$

is linearly polarized in the $\hat{\mathbf{x}}$ direction as can be seen by substitution of Equations 6.8 and 6.10.

The two states of circular polarization can be used as a basis set, and one good reason for using them is that they do not require establishment of an arbitrary coordinate system. The only Cartesian direction that matters is completely specified by $\mathbf{k}$. As we shall see later, circularly polarized light is also useful in a variety of practical applications.

Take-Away

Circular Polarization:
Left- and right-hand circular polarization states offer an alternative basis set.

6.1.4 Note about Random Polarization

Natural light is unpolarized or randomly polarized, at least in part.

Beginning with Maxwell's equations, it is easy to talk about a vector sum of two states of polarization. However, most natural light sources are unpolarized. Describing this type of light is more difficult. In fact, what we call **unpolarized light** would better be called randomly polarized, and can only be described statistically. In the x–y coordinate system, randomly polarized light is described by

$$\langle E_x \rangle = \langle E_y \rangle = 0 \qquad \langle E_x E_x^* \rangle = \langle E_y E_y^* \rangle = \frac{I}{2} Z$$
$$\langle E_x E_y^* \rangle = 0, \tag{6.12}$$

where

- I is the irradiance*
- $\langle \cdot \rangle$ denotes the ensemble average

In any situation with random variables, the **ensemble average** is obtained by performing an experiment or calculation a number of times, each time with the random variables assuming different values, and averaging the results. Later, we will formulate a precise mathematical description of partial polarization that accommodates all possibilities.

Take-Away

Unpolarized Light:
Natural light is at least partially randomly polarized. We often refer to this light as being unpolarized.

In Practice

Calculations with unpolarized light are in general more complicated and will be discussed later. However, a system illuminated with unpolarized light often begins with a linear polarizer. In such a case, we can often start our calculations after that first polarizer with half the incident light. For example, if we have $P_{in} = 1\,\mathrm{W}$ of sunlight falling on a pair of perfect polarizers with an angle ζ between them, then the output of the first polarizer is $P/2 = 0.5\,\mathrm{W}$, and the output of the second is

$$P_{out} = P_{in} T = \frac{P_{in}}{2} \cos^2 \zeta,$$

or half the result predicted by Malus' Law in Equation 6.19.

6.2 Behavior of Polarizing Devices

Roadmap: Chapter 6

To complete our discussion of polarization fundamentals, we discuss the functions of basic devices that manipulate the state of polarization. We will discuss each of the major devices in this section, and then after some background discussion of materials (Section 6.3) and Fresnel reflection (Section 6.4), return in Section 6.5, to how the devices are made and mathematical methods to describe them in Sections 6.6 and 6.7.

*In this chapter, we continue to use the variable name, I, for irradiance, rather than the radiometric E, as discussed in Section 1.4.4.

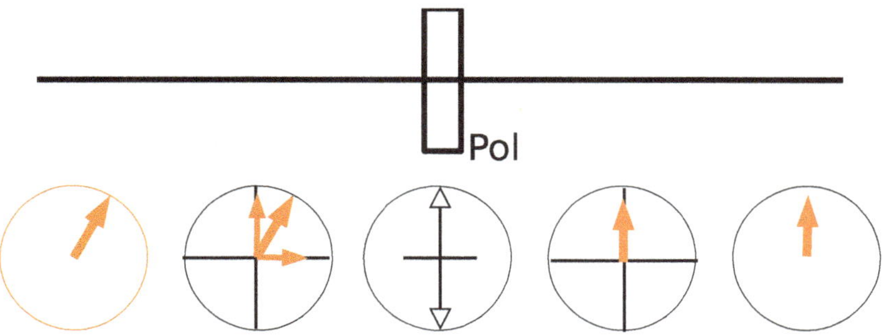

FIGURE 6.3
Linear Polarizer. The top part of the figure shows the layout. Below, states of polarization are shown in end-view. Note: The lines with open triangles represent the pass axis of the polarizer.

6.2.1 Linear Polarizer

The first, and most basic, element is the linear polarizer, which ideally blocks light with one linear polarization, and passes the other. If the device is chosen to pass x-polarized light and block y-polarized light, then with an input polarization at an angle ζ to the x axis,

$$\mathbf{E}_{in} = E_x\hat{\mathbf{x}} + E_y\hat{\mathbf{y}} = E_o\left[\cos\left(\zeta\right)\hat{\mathbf{x}} + \sin\left(\zeta\right)\hat{\mathbf{y}}\right], \tag{6.13}$$

the output will be

$$\mathbf{E}_{out} = 1 \times E_x\hat{\mathbf{x}} + 0 \times E_y\hat{\mathbf{y}} = E_o\cos\left(\zeta\right)\hat{\mathbf{x}}. \tag{6.14}$$

Later when we develop a matrix formulation in Section 6.6, we will refer to the $\hat{\mathbf{x}}$ and $\hat{\mathbf{y}}$ as the eigenvectors of the polarizer and the 1 and 0 as the eigenvalues. If the input polarization is one of these eigenvectors, then the output is in the same state of polarization, multiplied by the scalar eigenvalue. For an input vector which is not one of the eigenvectors, the output will generally be a different state of polarization.

The irradiance (Equation 1.46) of the input is

Eq. 1.46;
$I = |\mathbf{S}| = \frac{|\mathbf{E}|^2}{Z}$.

$$|\mathbf{E}_{in}|^2 = E_o^2, \tag{6.15}$$

and that of the output is

$$|\mathbf{E}_{out}|^2 = E_o^2\cos^2\zeta. \tag{6.16}$$

Figure 6.3 shows an interpretation of this result, and illustrates an approach to visualizing polarization problems.

The layout is shown in the top of the figure, and an "end view" of the polarization state is shown below. In this case, we begin with linear polarization at some angle with respect to the vertical. We know that the polarizer "passes" vertical polarization and "blocks" horizontal, so we resolve the input vector into these two components. When we develop vector formulations for polarization in Section 6.6, we will call these the eigenvectors of the polarizer. After the polarizer, the vertical component is the same as before, but the horizontal component is set to zero. The output field is vertically polarized with an amplitude equal to the input multiplied by the cosine of the angle as we have seen in the equations. It is convenient for us to discuss the **transmission**

$$T = \frac{|\mathbf{E}_{out}|^2}{|\mathbf{E}_{in}|^2}, \tag{6.17}$$

or in the present case

$$T = \cos^2 \zeta. \tag{6.18}$$

Malus' Law describes transmission through a polarizer.

We have assumed in this equation that the input and output are in the same material. This equation, called Malus' law[148],* allows us to compute the amount of irradiance or power through a polarizer:

$$P_{out} = P_{in}T = P_{in}\cos^2 \zeta. \tag{6.19}$$

For a more realistic polarizer, we could replace Equation 6.14 with

$$\mathbf{E}_{out} = \tau_{pass} \times E_x\hat{\mathbf{x}} + \tau_{block} \times E_y\hat{\mathbf{y}}, \tag{6.20}$$

where

- τ_{pass} is less than, but close to, 1
- τ_{block} is more than, but close to, 0

Insertion loss and extinction (or extinction ratio) measure the performance of a polarizer.

We can call $1 - |\tau_{pass}|^2$ the **insertion loss** and $|\tau_{block}|^2$ the extinction. Alternatively, we can define the **extinction ratio** as $|\tau_{pass}|^2 / |\tau_{block}|^2$. A good insertion loss might be a few percent or even less, and a good extinction might be 10^{-5}. Specific devices will be discussed later.

Take-Away

Linear Polarizers:
An ideal polarizer passes one state of linear polarization and blocks the orthogonal one. The performance of a polarizer is measured by its ability to perform these functions, as measured by insertion loss and extinction.

In Practice

Some people characterize polarizers by using two identical polarizers and rotating them for maximum and minimum transmission of unpolarized light. The maximum transmission is the square of the insertion loss and the minimum is the product of the insertion loss and extinction. The ratio of minimum to maximum is the extinction ratio.

6.2.2 Waveplate

Suppose that in Equation 6.20 we change the values of τ to complex numbers such that both have unit magnitude but their phases are different. This will cause one component of the polarization to be retarded relative to the other. Later, we will see that birefringent materials can be used to achieve this condition. For now, let us consider two examples. First, consider

$$\tau_x = 1 \qquad \tau_y = -1 \tag{6.21}$$

*Translated in [193].

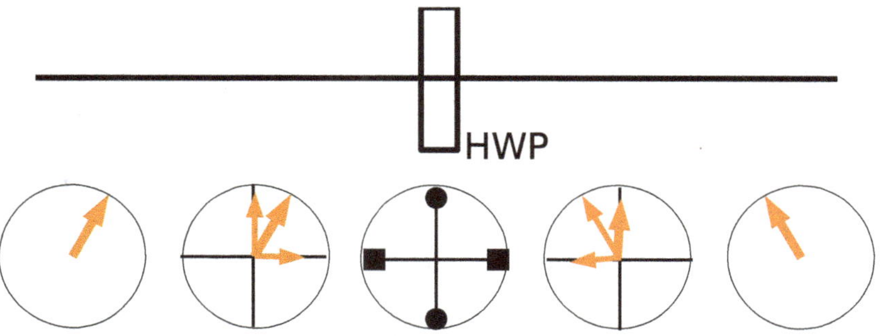

FIGURE 6.4
Half-wave plate. The half-wave plate reflects the state of polarization across one axis. For linearly polarized light, rotating the half-wave plate rotates the linear polarization by twice the angle. Note: The line with circles represents the polarization state of the HWP with eigenvalue 1 and the one with squares represents the axis with eigenvalue −1.

which corresponds to retarding the y component by a half cycle, and is thus called a **half-wave plate (HWP)**. Now with the input polarization described by Equation 6.13,

$$\mathbf{E}_{in} = E_x\hat{\mathbf{x}} + E_y\hat{\mathbf{y}} = E_o\left[\cos(\zeta)\,\hat{\mathbf{x}} + \sin(\zeta)\,\hat{\mathbf{y}}\right],$$

the output will be

A half-wave plate changes the angle of linear polarization.

$$\mathbf{E}_{out} = E_o\left[\cos(\zeta)\,\hat{\mathbf{x}} - \sin(\zeta)\,\hat{\mathbf{y}}\right] \qquad \angle\mathbf{E}_{out} = -\zeta. \tag{6.22}$$

In other words, the half-wave plate reflects the direction of polarization about the x axis. Thus, as the waveplate is rotated through an angle ζ, the direction of polarization rotates through 2ζ, as shown in Figure 6.4. The waveplate reverses the direction of the y component while leaving the x component unchanged.

For example, a HWP oriented at 45° is frequently used to “rotate” vertical polarization 90° to horizontal. Strictly the effect should not be called rotation, as it depends on the input polarization. If the incident light were polarized at 20° from the vertical, then the

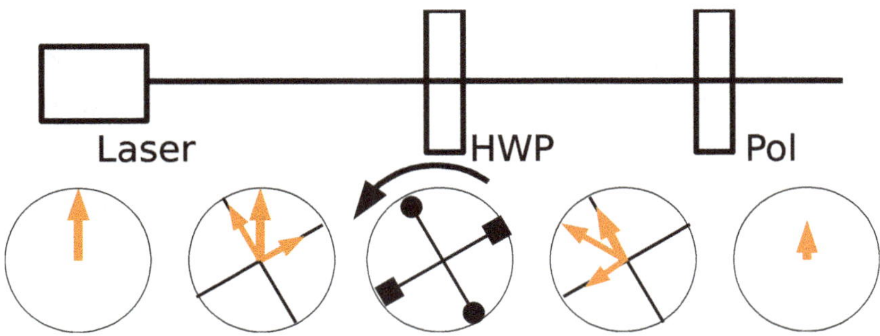

FIGURE 6.5
Adjustable attenuator. Rotating the half-wave plate (HWP) changes the amount of light out of the polarizer (POL). As the angle of the HWP rotates from zero to 45 degrees, the output varies from the laser output down to zero. Note that the variation is non-linear. Note: The lines with open triangles represent the “pass” axis of the polarizer. The line with circles represents the polarization state of the HWP with eigenvalue 1 and the one with squares represents the axis with eigenvalue −1.

output would be "reflected" to 70°, rather than "rotated" to 110. In Section 6.2.3, we will discuss a true rotator.

Example

Power Control Suppose that we have a laser with a vertically polarized output beam and a fixed power. We would like to be able to adjust the power from its maximum value down to zero. We can accomplish this task with a rotating half-wave plate and a vertically oriented polarizer as shown in Figure 6.5. In the figure, the half-wave plate is rotated about 30 degrees counterclockwise, and thus the polarization is rotated about 60 degrees. The output polarizer remains fixed in the vertical position, so the laser's field amplitude is reduced by a factor of $(\cos 60^\circ) = 0.5$ or the transmission of the waveplate and polarizer is $(\cos 60^\circ)^2 = 0.25$. If the insertion loss is significant, we need to include it in the calculation. Also, if the extinction is poor, the direction of the output polarization could change. These issues are easier to handle using the computational techniques of Section 6.6.

A **quarter-wave plate (QWP)** has

$$\tau_x = 1 \qquad \tau_y = j, \tag{6.23}$$

and for the same input polarization described by Equation 6.13, the output will be

$$\mathbf{E}_{out} = E_o \left[\cos(\zeta)\,\hat{\mathbf{x}} + j\sin(\zeta)\,\hat{\mathbf{y}}\right], \tag{6.24}$$

which is circular polarization.

A quarter-wave plate converts linear polarization to circular polarization.

Other waveplates are also useful, but half-wave and quarter-wave plates are the most common.

Finally we note that waveplates may have some insertion loss, in which case, the magnitudes of the complex field transmissions can be less than one. Often the loss is independent of direction of the polarization, and can be included as a scalar.

Take-Away

Waveplates:

- Waveplates change the relative phase of two states of linear polarization.
- The two most common waveplates are half-wave to change the direction of linear polarization, and quarter-wave to convert linear polarized light to circularly polarized light.
- Ideal waveplates are lossless; the output power is equal to the input power. Actual waveplates will have some insertion loss.

In Practice

Often higher-order waveplates are used. For example, a quarter-wave plate might actually have a phase change of 9/4 of a wave. This difference may or may not matter, and is discussed in more detail in Section 6.5.2.1.

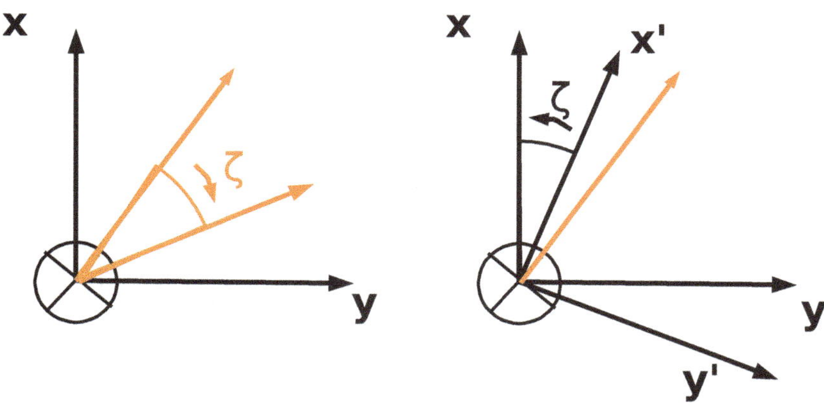

FIGURE 6.6
Rotation. The left panel shows rotation of linear polarization. Later we will see that rotation of polarization in one direction is mathematically equivalent to rotation of the coordinate system in the other (right panel).

6.2.3 Rotator

Some devices can rotate the direction of polarization. These are not so easily described by the coefficients τ, although later we will see that they can be, with a suitable change of basis vectors. The output of a rotator is best defined by a matrix equation. For example, the equation*

$$\begin{pmatrix} E_{x:out} \\ E_{y:out} \end{pmatrix} = \begin{pmatrix} \cos\zeta_r & -\sin\zeta_r \\ \sin\zeta_r & \cos\zeta_r \end{pmatrix} \begin{pmatrix} E_{x:in} \\ E_{y:in} \end{pmatrix}, \tag{6.25}$$

describes a rotation through an angle ζ_r, as shown in Figure 6.6. We will later see that this matrix equation is an example of the Jones calculus, which is widely applied in the analysis of polarized light.

A rotator rotates the direction of linearly polarized light.

A rotator may also have some loss, which can be expressed by multiplying the rotation matrix by a scalar less than one.

6.2.3.1 Summary of Polarizing Devices

This section has discussed three key devices.

- Linear polarizers ideally pass one linear polarization and block the orthogonal one. Real polarizers have some insertion loss and a finite extinction ratio.
- Waveplates retard one state of linear polarization with respect to the orthogonal one.
- Rotators rotate linear polarization through some angle around the axis of propagation.
- Real waveplates and rotators may also have some insertion loss, usually incorporated as multiplication by a scalar.

6.3 Interaction with Materials

Although Maxwell's equations allow for the interaction of light with matter through the constitutive relations, they provide no insight into the mechanism of the interaction. We

*See Appendix C for a brief overview of matrix multiplication and other operations.

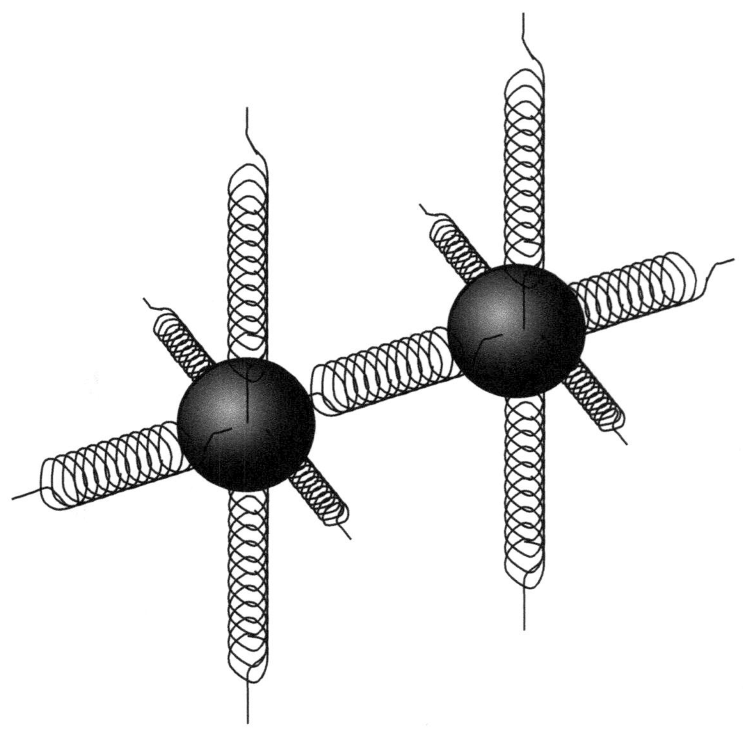

FIGURE 6.7
Lorentz oscillator. The basic interactions between light and materials can be explained with a model that considers electrons as charged masses attached to springs, and light as the driving force.

now know that the interaction of light with materials occurs primarily with electrons. Hendrik Antoon Lorentz[143;144] proposed a simple model of an electron as a mass on a set of springs, as shown in Figure 6.7. For this work, Lorentz was awarded the Nobel Prize in physics in 1902, along with Pieter Zeeman, whose most famous discovery was the magnetic splitting of spectral lines.

Lorentz model: Electron on springs.

In the **Lorentz model**, the springs provide a restoring force that returns the electron to its equilibrium position. Under an electric field $\mathbf{E}_r = E_0\hat{\mathbf{x}}\left(e^{j\omega t} + e^{-j\omega t}\right)$, the electron experiences a force $-e\mathbf{E}_r$, where $-e$ is the charge on the electron. This force results in an acceleration,

$$\frac{d^2x}{dt^2} = -\mathbf{E}_r e/m - \kappa_x x/m, \tag{6.26}$$

where

- m is the mass of the electron
- κ_x is the "spring constant" of the springs in the $\hat{\mathbf{x}}$ direction
- x is the position with respect to equilibrium

This differential equation,

$$\frac{m}{e}\frac{d^2x}{dt^2} + \frac{m}{\kappa_x e/m}x = -\mathbf{E}_r \cdot \hat{\mathbf{x}}, \tag{6.27}$$

can be solved for x. The moving electrons produce a varying polarization, $\mathbf{P}_r(t) = -N_v e x(t)\hat{\mathbf{x}}$, where N_v is the number of electrons per unit volume. The polarization can be expressed as

$$\mathbf{P}_r = (\epsilon_0 \chi \mathbf{E}_r \cdot \hat{\mathbf{x}})\,\hat{\mathbf{x}}, \tag{6.28}$$

leading to a displacement vector in the $\hat{\mathbf{x}}$ direction,

$$\mathbf{D}_r = \epsilon_0 \mathbf{E}_r + \mathbf{P}_r = \epsilon \mathbf{E}_r = \epsilon_0\,(1+\chi)\,\mathbf{E}_r. \tag{6.29}$$

The polarization, $\mathbf{P}_r = \chi\epsilon_0\mathbf{E}_r$, depends on the amount of motion of the electrons, which is determined by the "spring constant" of the springs. In Equations 1.5 through 1.8, this polarization affects ϵ, and the index of refraction, $n = \sqrt{\epsilon/\epsilon_0}$. Thus, the n is related to the "spring constant" in the Lorentz model. We have developed this analysis for a field in the $\hat{\mathbf{x}}$ direction. If the material is isotropic, all "spring constants" are equal, and the analysis applies equally to any direction.

If the material is anisotropic, then the "spring constants" in different directions have different values. In this case, Equation 6.29 becomes

$$\mathbf{D}_r = \epsilon_0 \mathbf{E}_r + \mathbf{P}_r = \boldsymbol{\mathcal{E}}\ \mathbf{E}_r = \epsilon_0\,(1+\chi)\,\mathbf{E}_r, \tag{6.30}$$

where the dielectric constant (now called the dielectric tensor) is a three-by-three matrix and χ is called the susceptibility tensor.

$$\boldsymbol{\mathcal{E}} = \begin{pmatrix} \epsilon_{xx} & \epsilon_{xy} & \epsilon_{xz} \\ \epsilon_{yx} & \epsilon_{yy} & \epsilon_{yz} \\ \epsilon_{zx} & \epsilon_{zy} & \epsilon_{zz} \end{pmatrix}. \tag{6.31}$$

In the example shown in Figure 6.7, with springs only along the directions defining the coordinate system,

$$\boldsymbol{\mathcal{E}} = \begin{pmatrix} \epsilon_{xx} & 0 & 0 \\ 0 & \epsilon_{yy} & 0 \\ 0 & 0 & \epsilon_{zz} \end{pmatrix}. \tag{6.32}$$

In fact, in any material, the matrix in Equation 6.31 can always be made diagonal by a suitable transformation of coordinates. Along each of the three axes of such a coordinate system, the displacement, $\mathbf{D}_r$, will be parallel to $\mathbf{E}_r$, and each polarization in Equation 6.29 can be treated independently, albeit with a different index of refraction equal to the square root of the appropriate component of $\boldsymbol{\mathcal{E}}/\epsilon_0$. Fields in these directions are said to be eigenvectors of the dielectric tensor. If light has polarization parallel to one of these axes, then no coupling to the other polarization will occur. The directions correspond to eigenvectors of the dielectric tensor.*

However, if the applied field has, for example, components in the $\hat{\mathbf{x}}$ and $\hat{\mathbf{y}}$ directions, each of these components will be met with a different restoring force. The result is that the displacement will be in a direction other than that of the force, and $\mathbf{D}_r$ will now be in a different direction from $\mathbf{E}_r$. We will see that this anisotropy forms the basis for waveplates and other polarization-changing devices.

*Eigenvectors are discussed in Section C.8 of Appendix C.

Take-Away

Birefringence:

- In isotropic media, $\mathbf{D}_r$ is parallel to $\mathbf{E}_r$, and no coupling occurs between orthogonal states of polarization.
- In anisotropic media, if light has a polarization parallel to one of the principal axes, no coupling to the other polarization will occur.
- In general, coupling will occur. We can treat these problems by resolving the input into two components along the principal axes, solving each of these problems independently, and summing the two components at the end.

6.4 Fresnel Reflection and Transmission

Eq. 1.38: $\nabla^2\mathbf{E} = -\omega^2\frac{n^2}{c^2}\mathbf{E}$.

Reflection and refraction at an interface can be understood through the application of boundary conditions. On each side of the interface, the wave is a solution to Maxwell's Equations, and thus also to the vector wave equation, Equation 1.38. We can thus write expressions for plane waves on each side of the boundary, and match the boundary conditions. We expect reflected and refracted waves, so we need two waves on one side of the boundary (the incident and reflected) and one on the other (the refracted or transmitted).

We consider an interface in the x–y plane, so that $\hat{\mathbf{z}}$ is normal to the interface. For S polarization, the three waves are the incident

$$\mathbf{E}_i = E_i\hat{\mathbf{x}}e^{-jkn_1(\sin\theta_i y+\cos\theta_i z)}, \tag{6.33}$$

reflected

$$\mathbf{E}_r = E_r\hat{\mathbf{x}}e^{-jkn_1(\sin\theta_r y-\cos\theta_i z)}, \tag{6.34}$$

and transmitted

$$\mathbf{E}_t = E_t\hat{\mathbf{x}}e^{-jkn_2(\sin\theta_t y+\cos\theta_t z)}. \tag{6.35}$$

The corresponding magnetic field equations are given by (see Equation 1.41)

$$\mathbf{H}_i = \frac{E_i}{Z_0/n_1}\left(\sin\theta_i\hat{\mathbf{z}} - \cos\theta_i\hat{\mathbf{y}}\right)e^{-jkn_1(\sin\theta_i y+\cos\theta_i z)}, \tag{6.36}$$

$$\mathbf{H}_r = \frac{E_r}{Z_0/n_1}\left(\sin\theta_r\hat{\mathbf{z}} + \cos\theta_r\hat{\mathbf{y}}\right)e^{-jkn_1(\sin\theta_r y-\cos\theta_i z)}, \tag{6.37}$$

$$\mathbf{H}_i = \frac{E_t}{Z_0/n_2}\left(\sin\theta_t\hat{\mathbf{z}} - \cos\theta_t\hat{\mathbf{y}}\right)e^{-jkn_2(\sin\theta_t y+\cos\theta_t z)}, \tag{6.38}$$

The boundary conditions are

$$\nabla\cdot\mathbf{D} = \rho = 0 \qquad \rightarrow \qquad \Delta D_{normal} = 0, \tag{6.39}$$

$$\nabla\times\mathbf{E} = -\frac{\partial\mathbf{B}}{\partial t} \qquad \rightarrow \qquad \Delta E_{tangential} = 0, \tag{6.40}$$

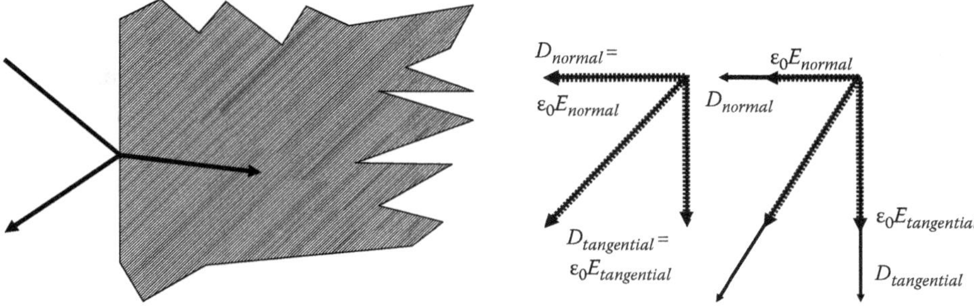

FIGURE 6.8
Fresnel boundary conditions. The normal component of $\mathbf{D}$ is constant across a boundary, as is the tangential component of $\mathbf{E}$. Because these two are related by $\epsilon = \epsilon_0 n^2$, the direction of the vectors changes across the boundary.

$$\nabla \cdot \mathbf{B} = 0 \quad \rightarrow \quad \Delta B_{normal} = 0, \tag{6.41}$$

$$\nabla \times \mathbf{H} = \mathbf{J} + \frac{\partial \mathbf{D}}{\partial t} = \frac{\partial D}{\partial t} \quad \rightarrow \quad \Delta H_{tangential} = 0. \tag{6.42}$$

Figure 6.8 shows the relationships among the vectors, $\mathbf{D}$ and $\mathbf{E}$.

6.4.1 Snell's Law

When we apply the boundary conditions, we recognize that they must apply along the entire boundary, which means they are independent of y. Therefore the transverse components of $\mathbf{k}n$ should be constant across the boundary,

$$kn_1 \sin\theta_i = kn_1 \sin\theta_r = kn_2 \sin\theta_t, \tag{6.43}$$

with the results that

$$\theta_i = \theta_r \qquad n_1 \sin\theta_i = n_2 \sin\theta_t. \tag{6.44}$$

Thus, Snell's law can be derived directly from the boundary conditions, and specifically from the preservation of $nk \sin\theta$ across the boundary.

6.4.2 Reflection and Transmission

Equation 6.40 requires that there be no change in the transverse component of $\mathbf{E}$. For S polarization, $\mathbf{E}$ is completely transverse, and we can immediately write

$$E_i + E_r = E_t. \tag{6.45}$$

The magnetic-field boundary requires that there be no change in the transverse component of $\mathbf{H}$. The transverse field here is only the $\hat{\mathbf{y}}$ component so

$$\frac{E_r}{Z_0/n_1}\cos\theta_i - \frac{E_i}{Z_0/n_1}\cos\theta_i = \frac{E_t}{Z_0/n_2}\cos\theta_t. \tag{6.46}$$

There are several ways to manipulate these equations to arrive at the Fresnel reflection and transmission coefficients in different forms. We choose to use a form which eliminates θ_t from the equations, employing Snell's law in the form

$$\cos\theta_t = \sqrt{1 - \sin^2\theta_i \left(\frac{n_1}{n_2}\right)^2}, \tag{6.47}$$

to obtain

$$E_i - E_r = E_t \frac{\sqrt{\left(\frac{n_2}{n_1}\right)^2 - \sin^2\theta_i}}{\cos\theta_i}. \tag{6.48}$$

Dividing the difference between Equations 6.45 and 6.48 by their sum, we obtain the **Fresnel reflection coefficient**, $\rho = E_r/E_i$. We give it the subscript s as it is valid only for S-polarized light:

Fresnel reflection coefficient for field amplitude for S polarization.

$$\rho_s = \frac{\cos\theta_i - \sqrt{\left(\frac{n_2}{n_1}\right)^2 - \sin^2\theta_i}}{\cos\theta_i + \sqrt{\left(\frac{n_2}{n_1}\right)^2 - \sin^2\theta_i}}, \tag{6.49}$$

and we can immediately find the **Fresnel transmission coefficient** using Equation 6.45 as

Fresnel transmission coefficient for field amplitude for S polarization.

$$\tau_s = 1 + \rho_s. \tag{6.50}$$

The reader may be troubled by the sum in this equation when thinking of conservation, but we must remember that fields are not conserved quantities. Shortly we will see that a conservation law does indeed apply to irradiance or power.

The coefficients for P polarization can be derived in a similar way with the results

Fresnel reflection coefficient for field amplitude for P polarization.

Fresnel transmission coefficient for field amplitude for P polarization.

$$\rho_p = \frac{\sqrt{\left(\frac{n_2}{n_1}\right)^2 - \sin^2\theta_i} - \left(\frac{n_2}{n_1}\right)^2 \cos\theta_i}{\sqrt{\left(\frac{n_2}{n_1}\right)^2 - \sin^2\theta_i} + \left(\frac{n_2}{n_1}\right)^2 \cos\theta_i} \tag{6.51}$$

and

$$\tau_p = (1 + \rho_p)\frac{n_1}{n_2}. \tag{6.52}$$

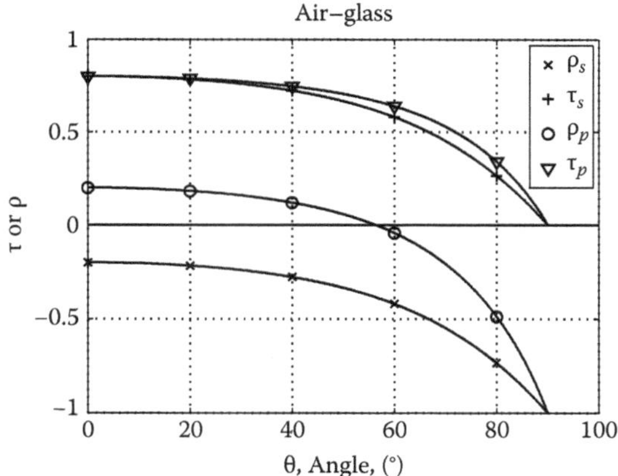

FIGURE 6.9
Fresnel field coefficients. The field transmission coefficients for an air–glass interface are real and include a sign change at a zero for P polarization.

Figure 6.9 shows an example for an air-to-glass interface. One of the most interesting features of this plot is that ρ_p passes through zero and changes sign, while the other coefficients do not. Examining Equation 6.51 we find that this zero occurs at the angle θ_B given by

Brewster's Angle: The reflection coefficient goes to zero for P polarization at this angle.

$$\tan\theta_B = \frac{n_2}{n_1}, \tag{6.53}$$

which we call **Brewster's angle**, plotted in Figure 6.10. At this angle, all P-polarized light is transmitted into the medium. This property can be extremely useful, for example in making a window to contain the gas in a laser tube. Laser operation is very sensitive to internal loss, and a window at normal incidence could cause a laser not to function. Placing the windows on the tube at Brewster's angle as shown in Figure 6.11 allows efficient transmission of light, the only losses being absorption, and also determines the state of polarization of the laser.

In the examples we have seen so far, the coefficients have all been real. Equations 6.49 and 6.51 require them to be real, provided that n is real and the argument of the square root is positive, which means the angle of incidence is less than the **critical angle**, θ_C, given by

Critical Angle: 100% reflection occurs beyond this angle. The angle is only defined for light going from a high index of refraction to a lower one.

$$\sin\theta_C = \frac{n_2}{n_1}. \tag{6.54}$$

If $n_2 > n_1$, then there is no critical angle. We will discuss the critical angle in greater detail in Section 6.4.4. Figure 6.10 shows, along with Brewster's angle, a plot of Equation 6.54. Brewster's angle is shown as a function of the index of the medium for light passing from air to the medium and from the medium to air. The critical angle is shown, of course, for light passing from the medium to air.*

*More correctly, we would say "vacuum" instead of "air," but the index of refraction of air is 1.0002–1.0003[201], and so the behavior of the wave at the interface will be very similar.

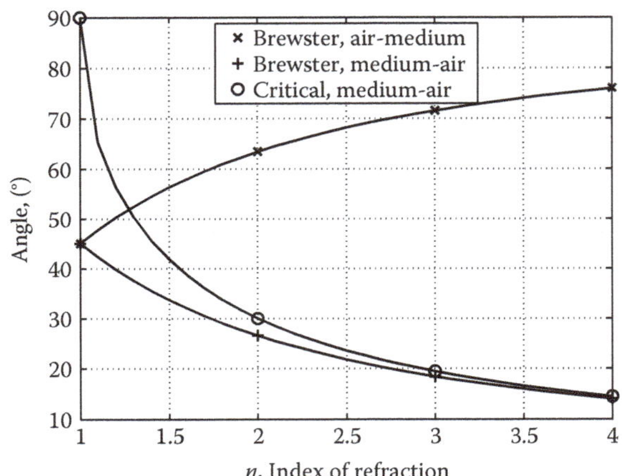

FIGURE 6.10
Special angles. Brewster's angle is plotted as a function of the index of a medium for both the air-medium interface and the medium-air interface. The critical angle is also shown.

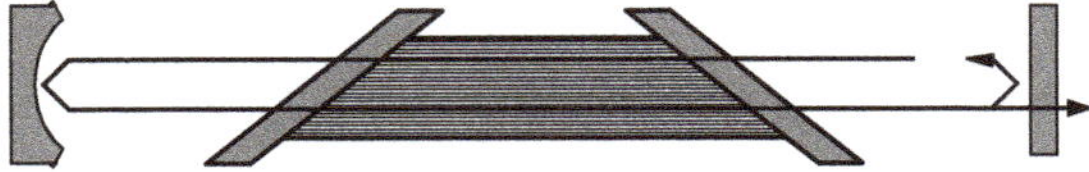

FIGURE 6.11
Gas laser. Two Brewster windows are used to seal the gas tube. This provides lower loss than windows normal to the beam. Low loss is critical in many laser cavities.

6.4.3 Power

The Fresnel coefficients are seldom presented as shown in Figure 6.9, because we cannot measure the fields directly. Of greater interest are the coefficients for reflection and transmission of irradiance or power.

Eq. 1.46:
$I = |\mathbf{S}| = \frac{|\mathbf{E}|^2}{Z}$.

The irradiance is given by the magnitude of the Poynting vector $\mathbf{S} = \mathbf{E} \times \mathbf{H}$, (See Equation 1.46)

$$I = \frac{|\mathbf{E}|^2}{Z}, \tag{6.55}$$

and is the power per unit area A' measured transversely to the beam direction. The area, A, on the surface is larger than the area, A' transverse to the beam by $1/\cos\theta$, where θ is the angle between the incident beam direction and the normal to the surface. Therefore,

$$I = \frac{dP}{dA'} = \frac{dP}{\cos\theta\, dA}. \tag{6.56}$$

Thus, A' is the same for incident at reflected light and the fraction of the incident irradiance that is reflected is

The power or irradiance reflection coefficient is the squared magnitude of the amplitude reflection coefficient. Use ρ_s or ρ_p in this equation.

$$\frac{I_r}{I_i} = R = \rho\rho^*. \tag{6.57}$$

The case of transmission is a bit more complicated for two reasons. First, the irradiance is related to the impedance and the square of the field, so the transmission coefficient for irradiance must include not only τ but also the impedance ratio. Second, the power that passes through an area A on the surface is contained in an area $A' = A\cos\theta$ transversely to the direction of propagation. For the reflection case the impedance was Z_1 for both incident and reflected waves, and the angles were equal. For transmission, we have the slightly more complicated equation,

$$\frac{I_t}{I_i} = T = \tau\tau^* \frac{Z_1 \cos\theta_t}{Z_2 \cos\theta_i} = \tau\tau^* \frac{n_2}{n_1} \frac{\sqrt{\left(\frac{n_2}{n_1}\right)^2 - \sin^2\theta_i}}{\cos\theta_i}. \tag{6.58}$$

With a little effort, it can be shown that

$$T + R = 1, \tag{6.59}$$

an expected result since it is required by conservation of energy.

Example

How does the fraction of sunlight transmitted through the ocean surface change as the solar zenith angle changes from 0 to 30 to 60 degrees?

When the sun is directly overhead, the Fresnel reflection coefficient using Equation 6.61, is about 2%, and the transmission is thus about 98%.

At non-normal angles the answer depends on polarization. Sunlight is nearly unpolarized, so half the light is in each polarization and our approach is to average the reflection coefficients over the two states of polarization. We use Equation 6.57 first with $\rho = \rho_s$ from Equation 6.49 and then with with $\rho = \rho_p$ from Equation 6.51 and average the results.

$$R = \frac{R_s + R_p}{2} = \frac{\rho_s^2 + \rho_p^2}{2}$$

For 30° the results are $(0.0305 + 0.0117)/2 = 0.021$ and for 60° the results are $(0.1139 + 0.0044)/2 = 0.0591$. At 30 degrees, the transmission ($T = 1 - R$) is 97.9%, almost unchanged from the zenith value. At 60 degrees it is reduced to 94% or about 96% of the zenith value. We conclude that most of the incident sunlight enters the water and that there is very little change with angle. Of course, at 60 degrees the reflected light is nearly triple the zenith value.

The reflection and transmission coefficients for irradiance or power are shown for an air-to-water interface in Figure 6.12. At normal incidence,

$$R = \left|\frac{(n_2/n_1) - 1}{(n_2/n_1) + 1}\right|^2, \tag{6.60}$$

or, if the first medium is air,

$$R = \left|\frac{n-1}{n+1}\right|^2, \tag{6.61}$$

The equation for normal incidence is simple and the same for S and P polarizations. It is about 4% going from air to glass or glass to air. Substituting water for the glass, it is about 2%.

which is about 2% for water. The reflection grows from the value in Equation 6.61 monotonically for S polarization, following Equation 6.57 with ρ given by Equation 6.49, reaching 100% at 90°. For P polarization, starting from Equation 6.61, the reflectivity decreases, reaching zero at Brewster's angle of 53°, and then rises rapidly to 100% at 90°. We note that the S reflectivity is always higher than the P reflectivity. When sunlight is incident on the surface of water, more light will be reflected in S polarization than P. The plane of

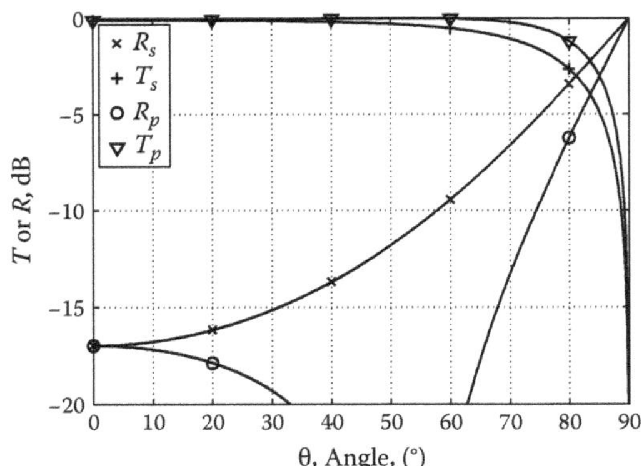

FIGURE 6.12
Air–water interface. Fresnel power coefficients in dB. Reflectivity of water at normal incidence is about 2%. The reflectivity of S-polarized light is always greater than that of P.

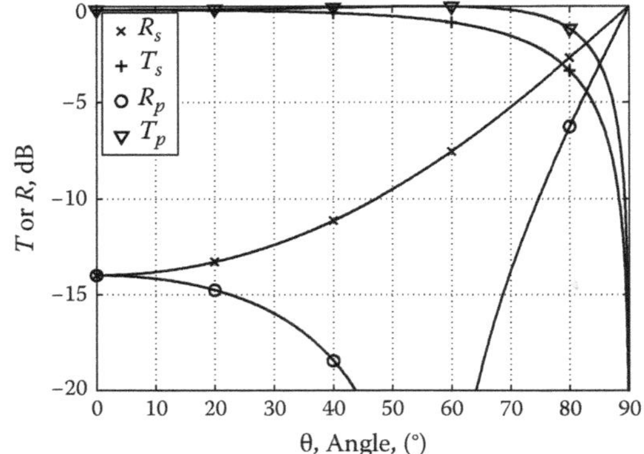

FIGURE 6.13
Air–glass interface. Fresnel coefficients for glass interface are similar to those for water (T and R in dB).

incidence contains the normal (vertical) and the incident ray (from the sun), and thus the reflected ray (and the observer). Thus S polarization is horizontal. The strong reflection can be reduced by wearing polarizing sunglasses consisting of vertical polarizers.

Figure 6.13 shows the same T and R curves for typical glass with $n = 1.5$. The normal reflectivity is about 4%, and Brewster's angle is 56°. The general shapes of the curves are similar to those for water.

Figure 6.14 shows a reflection from a polished tile floor, which is mostly S polarized. As with the water surface described earlier, the reflected light is strongly S, or horizontally, polarized. We can determine if a pair of sunglasses is polarized by first holding it in the usual orientation, and second rotating it 90°, while looking at a specular reflection from a shiny floor or table top. If the specular reflection is stronger in the second case than the first, then the glasses are polarized.

Germanium is frequently used as an optical material in the infrared. Although it looks like a mirror in the visible, it is quite transmissive for wavelengths from 2 to beyond 10 μm. With its high index of refraction ($n = 4$), the normal reflectivity is about 36%, and Brewster's angle is almost 76° (Figure 6.15). Another material used frequently in the infrared is zinc selenide with an index of $n = 2.4$, which is relatively transparent between about 600 nm and almost 20 μm. It is particularly useful because it passes some visible light along with infrared, making alignment of infrared imagers somewhat easier than it

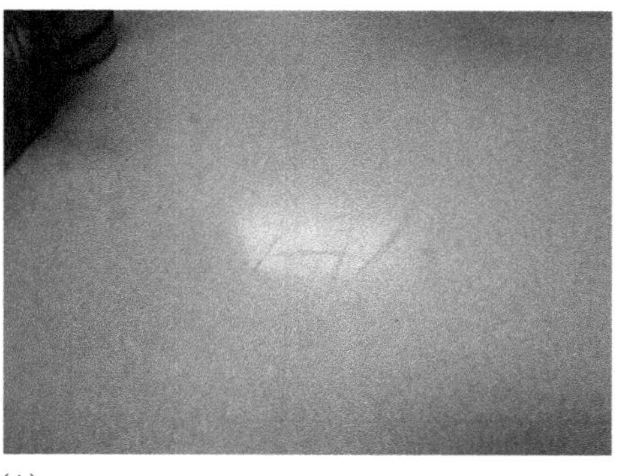
(A)

(B)

(C)

FIGURE 6.14
Fresnel reflection. The reflection from a polished tile floor (A) is polarized. A polarizer blocks more than half the diffusely scattered light (B,C). If the polarizer oriented to pass horizontal, or S, polarization with respect to the floor, then the specular reflection, is bright (B). If the polarizer is rotated to pass vertical, or P polarization, the specular reflection is greatly decreased (C).

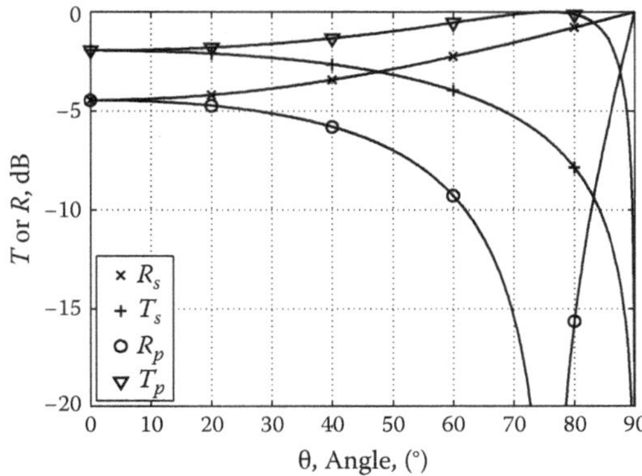

FIGURE 6.15
Air–germanium interface. Germanium is often used as an optical material in the infrared around 10 μm, where it has an index of refraction of about 4. The normal reflectivities are much higher, and the Brewster angle is large.

is with germanium. Zinc selenide has a yellow appearance because it absorbs the shorter visible wavelengths. Its Fresnel coefficients are between those of glass and germanium.

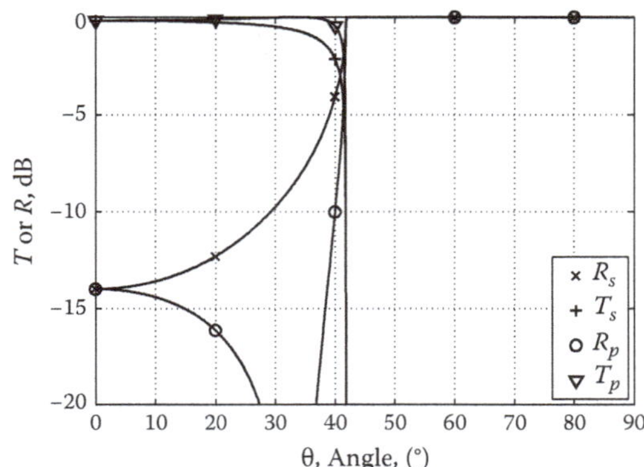

FIGURE 6.16
Glass–air interface. Fresnel coefficients show a shape similar to those for air-to-glass, but compressed so that they reach 100% at the critical angle. Beyond the critical angle, there is no transmission.

In Practice

Because the transmission coefficients are mathematically more complicated than those for reflection, it is often easier to use $T = 1 - R$ and compute transmission coefficients from reflection coefficients.

Example

Windows in buildings are typically uncoated, as are eyeglasses and many other glass optical elements. Why is this not true for materials like zinc selenide and germanium in the infrared?

Using Equation 6.61, reflectivities, R, for glass ($n = 1.5$), zinc selenide ($n = 2.4$), and germanium ($n = 4.0$) are 4%, 17% and 36% respectively. The transmission of a window or lens with two surfaces is $(1 - R)^2$ are 92%, 69% and 41%. Because the higher losses are significant, the last two materials are normally coated with an anti-reflection layer, as discussed in Section 7.7.2.

6.4.4 Total Internal Reflection

Returning to the Fresnel equations, beyond the critical angle we find that the argument of the square root in Equations 6.49 and 6.51 becomes negative. In this case, the square root becomes purely imaginary, and the reflectivity is a fraction in which the numerator and denominator are a complex conjugate pair. Thus the magnitude of the fraction is one, and the phase is twice the phase of the numerator. Specifically,

$$\tan \phi_s = -2\frac{\sqrt{\sin^2 \theta_i - \left(\frac{n_2}{n_1}\right)^2}}{\cos \theta_i} \tag{6.62}$$

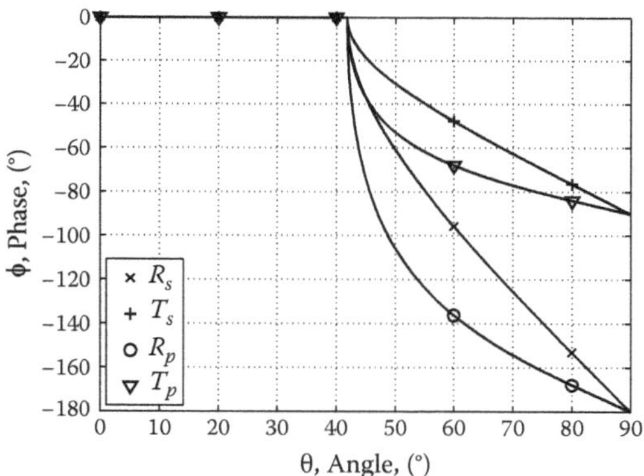

FIGURE 6.17
Glass–air interface. Total-internal reflection. When the magnitude of the reflection is 100%, the phase varies.

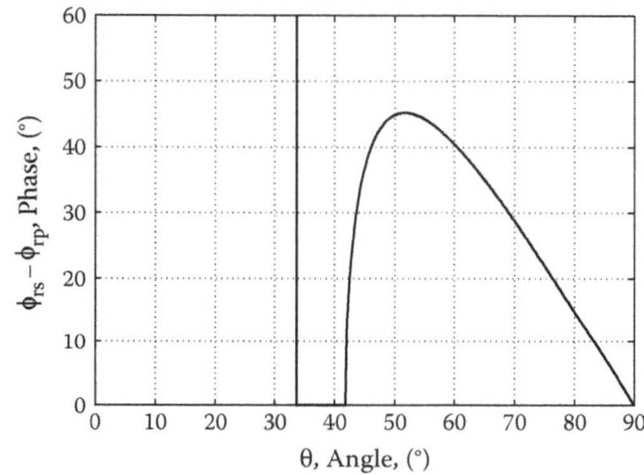

FIGURE 6.18
Glass–air interface. Phase difference. The difference between phases of S and P components has a peak at above 45°. Using two reflections at appropriate angles could produce a 90° phase shift.

and

$$\tan \phi_p = 2\frac{\left(\frac{n_2}{n_1}\right)^2 \cos \theta_i}{\sqrt{\sin^2 \theta_i - \left(\frac{n_2}{n_1}\right)^2}}. \qquad (6.63)$$

Figure 6.16 shows the reflection and transmission coefficients for irradiance for light passing from glass to air. As expected, the reflectivities are 100% beyond the critical angle. Figure 6.17 shows the phases. It will be noted that the Fresnel coefficients for the field are real below the critical angle (phase is 0° or 180°), but they become complex, with unit amplitude and a difference between the phases for larger angles. This unit reflection is the basis of total internal reflection such as is used in the retroreflectors shown in Figure 2.9C.

Figure 6.18 shows the difference between the phases. The difference has a maximum slightly above 45°. Later, we will see how to make a "Fresnel rhomb" which uses two total internal reflections to produce a phase shift of 90°, equivalent to that of a quarter-wave plate.

Take-Away

Total Internal Reflection:
Total internal reflection occurs at angles greater than the critical angle. It only happens for light going from a higher to lower index of refraction. The reflections have different phases and this difference can be exploited in devices such as the Fresnel rhomb.

6.4.5 Transmission through a Beamsplitter or Window

It can be shown that for an air-to-glass interface at any angle, θ_{air}, in Figure 6.13, R and T are the same for the glass-to-air interface in Figure 6.16 at the value of θ_{glass} that satisfies Snell's law, $n_{glass}\sin\theta_{glass} = \sin\theta_{air}$.

$$T_{n_1,n_2}(\theta_1) = T_{n_2,n_1}(\theta_2) \qquad R_{n_1,n_2}(\theta_1) = R_{n_2,n_1}(\theta_2)\,. \tag{6.64}$$

Thus, if light passes through a sheet of glass with parallel faces, the transmission will be T^2. However, the field transmission coefficients for the two surfaces, τ, are not the same (see Equation 6.58). If we use Equations 6.49 through 6.52 and apply Snell's law to determine the angles, we can be assured of calculating the correct result.

Eq. 6.59 $T + R = 1$.

However, we can often take a short-cut. Once we have calculated, or been given, the reflectivities, R_s and R_p, we can use Equation 6.59 to obtain the power transmission. For example, at Brewster's angle, the reflectivity of an air-to-glass interface for S polarization is about $R_{S1} = 0.15$, and the transmission is thus about $T_{s1} = 0.85$, and the transmission through the two surfaces of a slab is $T_s = 0.85^2 \approx 0.72$. Figure 6.19 illustrates the transmission of a window, along with two reflections. In fact, there are an infinite number of progressively weaker reflected and transmitted rays.

There is one important difference between the behaviors at the two surfaces. Inspection of Equations 6.49 through 6.52 shows that τ always has the same sign, whether light is transmitted from a high index to a low index or vice versa. However, the signs of ρ for the two cases are opposite. We will make use of this fact in balanced mixing in Section 7.1.5.

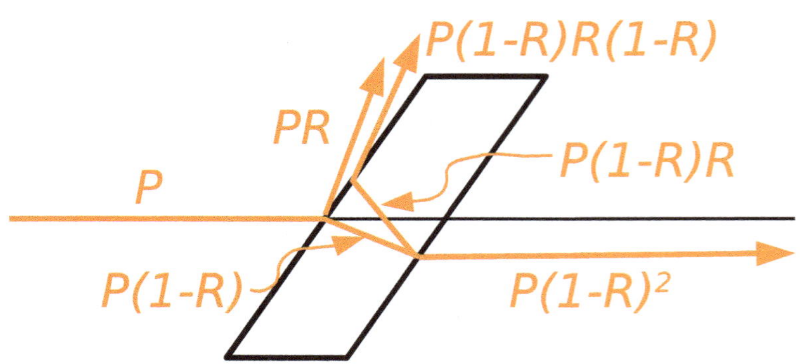

FIGURE 6.19
Transmission and reflection at a window. There are multiple transmissions and reflections. The irradiance or power of each can be calculated using the Fresnel reflection coefficient, R.

6.4.6 Complex Index of Refraction

Thus far, we have dealt with materials having a real index of refraction. A complex index of refraction describes a material that we call **Lossy**. In such a material, the amplitude of the field decreases with distance. A plane wave is characterized by an electric field

$$Ee^{j(\omega t-n\mathbf{k}\cdot\mathbf{r})}. \tag{6.65}$$

For a complex index of refraction, $n = n_r - jn_i$,

$$Ee^{j(\omega t-n_r\mathbf{k}\cdot\mathbf{r}+jn_i\mathbf{k}\cdot\mathbf{r})} = Ee^{j(\omega t-n_r\mathbf{k}\cdot\mathbf{r})}e^{-n_i\mathbf{k}\cdot\mathbf{r}}. \tag{6.66}$$

This equation represents a sinusoidal wave with an exponential decay in the $\mathbf{k}$ direction.

Now, reconsidering the interface, we note that Equation 6.43,

$$kn_1 \sin\theta_i = kn_1 \sin\theta_r = kn_2 \sin\theta_t,$$

requires both the real and imaginary parts of $n\mathbf{k}$ to be preserved. If the first medium has a real index of refraction, then the imaginary part of $n\mathbf{k}$ is zero there, and the transverse component of the imaginary part of $n\mathbf{k}$ must be zero in the second medium as well. Thus, the wave in the medium is

$$Ee^{j(\omega t-n_r\mathbf{k}\cdot\mathbf{r})}e^{-(n_ik)\hat{\mathbf{z}}\cdot\mathbf{r}}, \tag{6.67}$$

where $\mathbf{k}$ is in the direction determined from Snell's law. This makes intuitive sense, because the incident wave is sinusoidal and has no decay as a function of y. Therefore, the transmitted wave at $z = 0$ must also have no decay in that direction.

A metal can also be characterized by a complex index of refraction. It has a high conductivity, so that electrons are free to move under the force of an electric field. We can incorporate this high conductivity into a high imaginary part of the dielectric constant. In this model, the dielectric constant approaches a large, almost purely imaginary, number, and the index of refraction, being the square root of the dielectric constant, has a phase angle that approaches 45°. The Fresnel coefficients of a metal with $n = 4 + 3j$ are shown in Figure 6.20. We see a high normal reflectivity of 53%. As with dielectric interfaces, R_p is always less than R_s. There is a **pseudo-Brewster's angle**, where R_p reaches a minimum, but it never goes to zero. Again, the reflectivities at grazing angles approach 100%. Because of the complex index, the reflectivities are in general complex, with phases varying with angle, as shown in Figure 6.21.

Take-Away

Fresnel Coefficients:
Inspection of the Fresnel equations shows several general rules:

- Reflections of both polarizations are equal at normal incidence, as must be the case, because there is no distinction between S and P at normal incidence.
- There is always a Brewster's angle at which the reflectivity is zero for P polarization, if the index of refraction is real.
- If the index is complex, there is a pseudo-Brewster's angle, at which R_p reaches a nonzero minimum.
- The reflectivity for P polarization is always less than that for S polarization.
- All materials are highly reflective at grazing angles, approaching 100% as the angle of incidence approaches 90°.
- The sum $T + R = 1$ for each polarization independently.

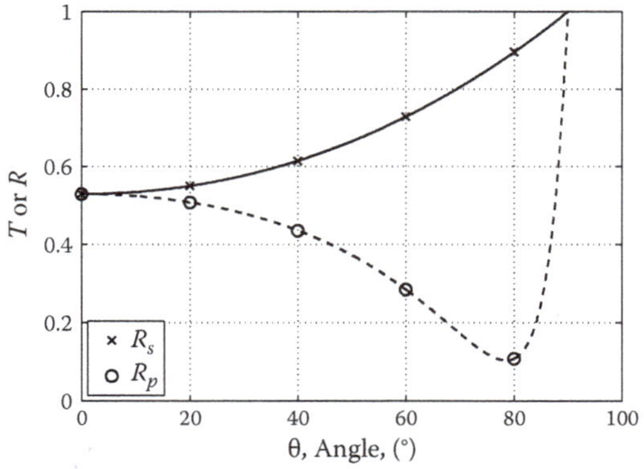

FIGURE 6.20
Air–metal interface. Reflection from a metal. Although the basic shape of the reflectivity curves does not change, the normal reflectivity is large, and the reflectivity for P polarization has a nonzero minimum at what we now call the pseudo-Brewster angle.

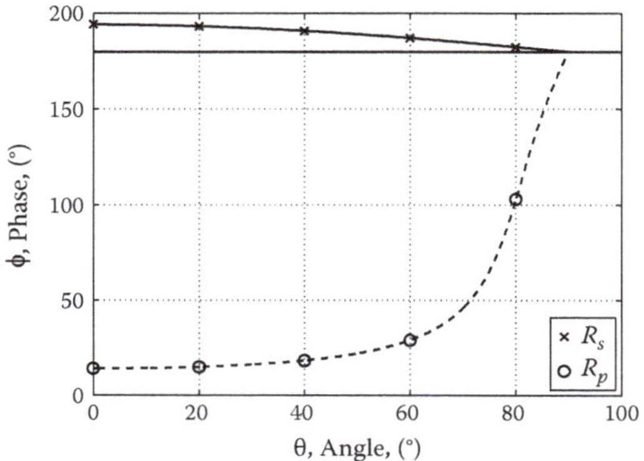

FIGURE 6.21
Air–metal interface. Phase on reflection from a metal. The reflectivities are in general always complex numbers, and the phase varies continuously in contrast to dielectrics, such as water. In Figure 6.9, the numbers were real everywhere, so the phase was either 0° or 180°.

- For a slab of material with parallel sides, the values of T are the same at each interface, as are the values of R.
- If the second index is lower, light incident beyond the critical angle is reflected 100% for both polarizations. There is an angle-dependent phase difference between the polarizations.

6.5 Physics of Polarizing Devices

Although the concepts of polarizers, waveplates, and rotators discussed in Section 6.2 are simple to understand, there are many issues to be considered in their selection. There are a large number of different devices, each with its own physical principles, applications, and limitations. Here we discuss some of the more common ones.

6.5.1 Polarizers

Eq. 6.14: $\mathbf{E}_{out} = 1 \times E_x\hat{\mathbf{x}} + 0 \times E_y\hat{\mathbf{y}} = E_o \cos(\zeta)\,\hat{\mathbf{x}}$.

An ideal polarizer will pass one direction of linear polarization and block the other. Thus, it has eigenvalues of one and zero as shown by Equation 6.14. Real devices will have some "insertion loss" and imperfect extinction. Further issues include wavelengths at which the devices are usable, size, surface quality, light scattering, and power-handling ability. In the following paragraphs we will discuss various types of polarizers.

6.5.1.1 Brewster Plate, Tents, and Stacks

Eq. 6.49: $\rho_s = \dfrac{\cos\theta_i - \sqrt{\left(\frac{n_2}{n_1}\right)^2 - \sin^2\theta_i}}{\cos\theta_i + \sqrt{\left(\frac{n_2}{n_1}\right)^2 - \sin^2\theta_i}}$.

Eq. 6.51: $\rho_p = \dfrac{\sqrt{\left(\frac{n_2}{n_1}\right)^2 - \sin^2\theta_i} - \left(\frac{n_2}{n_1}\right)^2\cos\theta_i}{\sqrt{\left(\frac{n_2}{n_1}\right)^2 - \sin^2\theta_i} + \left(\frac{n_2}{n_1}\right)^2\cos\theta_i}$.

A plane dielectric interface at Brewster's angle provides some polarization control. The P component will be 100% transmitted through the surface. We can compute the S component by examining Equations 6.49 and 6.51. At Brewster's angle, the numerator of the latter must be zero, so the square root must be equal to $(n_2/n_1)^2 \cos\theta_i$. Substituting this expression for the square root in Equation 6.49,

$$\rho_s = \frac{\cos\theta_i - \left(\frac{n_2}{n_1}\right)^2\cos\theta_i}{\cos\theta_i + \left(\frac{n_2}{n_1}\right)^2\cos\theta_i} = \frac{1 - \left(\frac{n_2}{n_1}\right)^2}{1 + \left(\frac{n_2}{n_1}\right)^2}, \tag{6.68}$$

so

$$R_s = \rho_s\rho_s^* = \left(\frac{1 - \left(\frac{n_2}{n_1}\right)^2}{1 + \left(\frac{n_2}{n_1}\right)^2}\right)^2. \tag{6.69}$$

A slab with two parallel sides oriented at Brewster's angle, such as one of the plates in Figure 6.22 will transmit all the P component except for absorption in the medium, and the S component transmitted will be

$$T_{sbp}^2 = (1 - \rho_s\rho_s^*)^2 = \left[1 - \left(\frac{1 - \left(\frac{n_2}{n_1}\right)^2}{1 + \left(\frac{n_2}{n_1}\right)^2}\right)^2\right]^2 = \frac{16}{\left[\frac{n_1}{n_2} + \frac{n_2}{n_1}\right]^4}, \tag{6.70}$$

where the subscript on T_{sbp} reminds us that this transmission is specific to S-polarized light for a Brewster plate. This transmission becomes very small if there is a large contrast between the indices of refraction in either direction. The actual transmission will be further reduced by absorption, which will be the same as for the P component. However, the absorption will cancel in the "extinction ratio"

Brewster plates perform best with high index contrast.

$$\frac{T_{pbp}}{T_{sbp}} = \frac{\left[\frac{n_1}{n_2} + \frac{n_2}{n_1}\right]^4}{16}. \tag{6.71}$$

For glass ($n = 1.5$), a single reflection, R_s, is about 0.15, and the two surfaces provide a 1.38 extinction ratio. In contrast, for germanium in the infrared, $n = 4$, the reflection is $R_s = 0.78$, and the extinction ratio for the plate is 20.4.

For a real Brewster plate, the actual extinction is limited by the extent to which the surfaces are parallel. If they are not perfectly parallel, at least one surface will not be at Brewster's angle. The best compromise will be to allow a small angular error at each surface. In addition, performance will be limited by the user's ability to align both Brewster's angle, and the desired angle of polarization. Suitable holders must be used to hold the plate at the correct angle, and to allow precise angular motion for alignment.

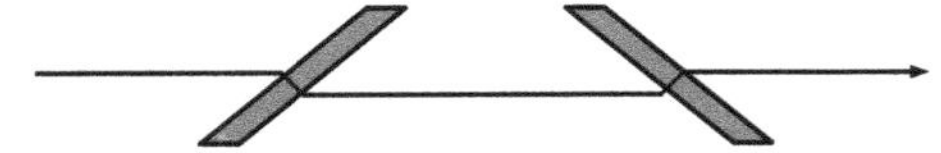

FIGURE 6.22
Tent polarizer. Two Brewster plates are mounted with opposite orientations. One advantage is the elimination of the "dogleg" that occurs in a stack of plates. A major disadvantage is the total length.

Brewster plates are quite useful in the longer IR wavelengths. The are less useful in the visible and Near IR because of the lower index of refraction of most materials there.
Brewster plates can handle high power laser beams.

Tents and stacks can improve the extinction.

The extinction can be increased by using a stack of Brewster plates, with small air gaps between them. The extinction ratio for m plates is simply

$$(T_{pbp}/T_{sbp})^m . \tag{6.72}$$

Thus 10 glass plates provide an extinction ratio of 24.5, while only two germanium plates provide over 400. The disadvantages of a stack of plates include the obvious ones of size, and the **dogleg** the stack introduces into a beam. (Although the beam exits at the same angle, it is displaced laterally, much like the leg of a dog.) A further disadvantage is the requirement for precise, relative alignment of both the angle of incidence and the orientation of the plane of incidence across all components.

For infrared, a germanium tent polarizer is attractive. Two Brewster plates are arranged in a tent configuration as shown in Figure 6.22. This configuration has the advantage that the dogleg introduced by one plate is compensated by the other, minimizing alignment issues when the tent is used in a system. Unfortunately the tent has the added disadvantage of requiring a large amount of space in the axial direction, particularly for germanium. With an index of 4, it is a good polarizer, but Brewster's angle is nearly 76°, so the minimum length is about four times the required beam diameter. Double tents with two plates on each side can, in principle, provide an extinction ratio of $400^2 = 160,000$. Beyond this, practical limitations of alignment may limit further enhancement.

Brewster plates and their combinations are particularly attractive in the infrared, because the availability of high-index materials leads to high extinction ratios, and because the extremely low absorption means that they are capable of handling large amounts of power without damage.

In Practice

Brewster plates, including stacks, and tents, are very useful polarizers for high power in the infrared. They are quite commonly used with carbon dioxide lasers around $10\,\mu$m. The unused S-polarized light is reflected rather than absorbed, so it is important to be careful to use a beam block to capture that unwanted power.

6.5.1.2 Wire Grid

Wire-grid polarizers are useful, especially in the infrared, for low powers.

An array of parallel conducting wires can serve as a polarizer. Intuitively, the concept can be explained by the fact that a perfect conductor reflects light because electrons are free to move within it, and thus cannot support an electric field. A thin wire allows motion of the electrons along its axis, but not across the wire. A more detailed explanation requires analysis of light scattering from cylinders. In any case, the diameter of the wires and the pitch of the repetition pattern must be much less than the wavelength to provide good extinction and to reduce diffraction effects, which will be discussed in Section 8.7. Thus wire grids are particularly useful in the infrared, although they are also available for visible light. They are fabricated on an appropriate substrate for the desired wavelength range. Diffraction of light from the wires is one issue to be considered, particularly at shorter wavelengths. Another limitation is that the conductivity is not infinite, so some heating does occur, and thus power-handling ability is limited. Smaller wires provide better beam quality, but have even less power-handling ability. Extinction ratios can be as large as 300, and two wire grid polarizers used together can provide extinction ratios up to 40,000.

A common misperception is to think of the wire grid as a sieve, which blocks transverse electric fields and passes ones parallel to the wires. As described earlier, the actual performance is exactly the opposite. Even the analysis here is overly simplified. Quantitative analysis requires computational modeling.

6.5.1.3 Polaroid H-Sheets

Polaroid H-sheets, invented by Edwin Land prior to the camera for which he is most famous, are in many ways similar to wire grids, but are used primarily in the visible wavelengths. Briefly, a polyvinyl alcohol polymer is impregnated with iodine, and stretched to align the molecules in the sheet[134;133]. H-sheets are characterized by a number indicating the measured transmission of unpolarized light. A perfect polarizer would transmit half the light, and would be designated HN-50, with the "N" indicating "neutral" color. However, the plastic is not anti-reflection coated, so there is an additional transmission factor of 0.96^2 and the maximum transmission is 46%. Typical sheets are HN-38, HN-32, and HN-22. Land mentions the "extinction color." Because they have an extinction ratio that decreases at the ends of the visible spectrum, a white light seen through two crossed HN-sheets will appear to have either a blue or red color. It should be noted that outside the specified passband, their performance is extremely poor. However, HR-sheets are available for work in the red and near infrared.

Polaroid H-sheets are low-cost polarizers of reasonable extinction, but the optical quality and power-handling are poor.

H-sheets can be made in very large sizes, and a square 2 ft on a side can be purchased for a reasonable price. They can be cut into odd shapes as needed. Figure 6.14 shows a polaroid H-sheet cut from such a square. They are frequently used for sunglasses and similar applications. The extinction can be quite high, but the optical quality of the plastic material may cause aberrations of a laser beam or image. Power-handling ability is low because of absorption.

6.5.1.4 Glan–Thompson Polarizers

Glan–Thompson polarizers are made from two calcite prisms. Calcite is birefringent. Birefringence was introduced in Section 6.2.2 and will be discussed further in Section 6.5.2. By carefully controlling the angle at which the prisms are cemented together, it is possible to obtain total internal reflection for one polarization and good transmission for the other, and extinction ratios of up to 10^5 are possible. To achieve the appropriate angle, the prisms have a length that is typically three times their transverse dimensions, which makes them occupy more space than a wire grid or H-sheet, but typically less than a tent polarizer. They are useful through the visible and near infrared. Because of the wide range of optical adhesives used, power handling ability may be as low as a Watt per square centimeter, or considerably higher.

Glan–Thompson polarizers provide good extinction and maybe high damage thresholds depending on how they are manufactured.

6.5.1.5 Polarization-Splitting Cubes

Polarization-splitting cubes are fabricated from two 45° prisms with a multilayer dielectric coating on the hypotenuse of one of the prisms. The second prism is cemented to the first, to make a cube, with the advantage that the reflecting surface is protected. The disadvantage is that there are multiple flat faces on the cube itself, normal to the propagation direction, which can cause unwanted reflections, even if anti-reflection coated. Beam-splitting cubes are useful where two outputs are desired, one in each polarization. If the incident light is randomly polarized, linearly polarized at 45°, or circularly polarized, the two outputs will be approximately equal.

These cubes can be used with moderate laser power. The multilayer coating uses interference (Section 7.7), and is thus fabricated for a particular wavelength. By using a different combination of layers, it is possible to produce broad-band beam-splitting polarizers, but the extinction ratios are not as high as with those that are wavelength specific. Often they can be used in combination with linear polarizers to achieve better extinction.

In Practice

Polarizing cubes can split a light wave into reflected S and transmitted P components. Generally the reflection for S polarization can exceed 90% (not including reflection losses at the cube faces), and P transmission can approach or exceed 95%.

Take-Away

There are a variety of different polarizers for different applications in different wavelength bands. Generally the important parameters are insertion loss, extinction, damage threshold, optical quality, and cost. It is generally a good idea to determine the required parameters and compare them to catalog data to meet a particular need.

Fig. 6.7

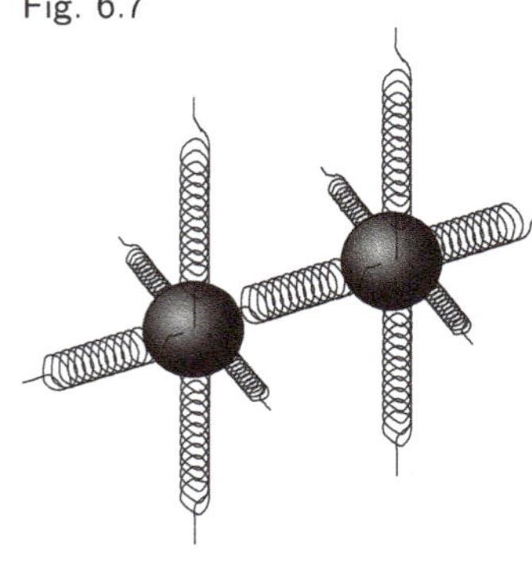

6.5.2 Birefringence

The function of a waveplate is to retard one component of linear polarization more than its orthogonal component. If the springs in Figure 6.7 have different spring constants along different axes, then the index of refraction will be different for light polarized in these directions. Huygens recognized from observation of images through certain materials in the 1600s that there are two different kinds of light, which we know today to be the two states of polarization. For example, Figure 6.23 shows a double image through a quartz crystal, caused by its **birefringence**.

Eq., 6.31:

$$\boldsymbol{\varepsilon} = \begin{pmatrix} \epsilon_{xx} & \epsilon_{xy} & \epsilon_{xz} \\ \epsilon_{yx} & \epsilon_{yy} & \epsilon_{yz} \\ \epsilon_{zx} & \epsilon_{zy} & \epsilon_{zz} \end{pmatrix}.$$

Eq., 6.32:

$$\boldsymbol{\varepsilon} = \begin{pmatrix} \epsilon_{xx} & 0 & 0 \\ 0 & \epsilon_{yy} & 0 \\ 0 & 0 & \epsilon_{zz} \end{pmatrix}.$$

No matter how complicated the crystal structure, because **D** and **E** are three-dimensional vectors, the dielectric tensor can always be defined as a three-by-three matrix as in Equation 6.31, and can be transformed by rotation to a suitable coordinate system into a diagonal matrix as in Equation 6.32, which is the simplest form for a **biaxial crystal**. In many materials, called **uniaxial crystals**, two of the indices are the same

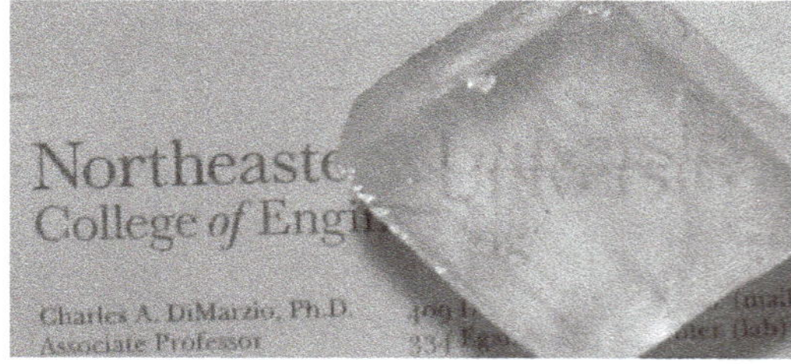

FIGURE 6.23
Birefringence. A "double image" is seen through a quartz crystal. Quartz is birefringent, having two indices of refraction for different polarizations. Thus the "dogleg" passing through this crystal is different for the two polarizations.

and

$$\underline{\epsilon} = \begin{pmatrix} \epsilon_{xx} & 0 & 0 \\ 0 & \epsilon_{yy} & 0 \\ 0 & 0 & \epsilon_{yy} \end{pmatrix}. \tag{6.73}$$

In this case, we say the **ordinary ray** is the ray of light polarized so that its index is $n_{yy} = \sqrt{\epsilon_{yy}\epsilon_0}$, and the **extraordinary ray** is the one polarized so that its index is $n_{xx} = \sqrt{\epsilon_{xx}\epsilon_0}$.

In either case, for propagation in the $\hat{\mathbf{z}}$ direction, the input field to a crystal of length ℓ starting at $z = 0$ is

$$\mathbf{E}_{in} = (E_{xi}\hat{\mathbf{x}} + E_{yi}\hat{\mathbf{y}})\, e^{j(\omega t - kz)} \qquad (z < 0). \tag{6.74}$$

Inside the crystal, assuming normal incidence, the field becomes

$$\mathbf{E} = \tau_1 \left(E_{xi}\hat{\mathbf{x}} e^{j(\omega t - kn_{xx}z)} + E_{yi}\hat{\mathbf{y}} e^{j(\omega t - kn_{yy}z)} \right) \qquad (0 < z < \ell), \tag{6.75}$$

and specifically at the end,

$$\mathbf{E} = \tau_1 \left(E_{xi}\hat{\mathbf{x}} e^{j(\omega t - kn_{xx}\ell)} + E_{yi}\hat{\mathbf{y}} e^{j(\omega t - kn_{yy}\ell)} \right).$$

After the crystal, the field becomes

$$\mathbf{E} = \tau_1\tau_2 \left(E_{xi}\hat{\mathbf{x}} e^{j[\omega t - kn_{xx}\ell - k(z-\ell)]} + E_{yi}\hat{\mathbf{y}} e^{j[\omega t - kn_{yy}\ell - k(z-\ell)]} \right)$$

$$(\ell < z).$$

Rewriting,

$$\mathbf{E} = \tau_1\tau_2 \left(E_{xi} e^{-jk(n_{xx}-1)\ell}\hat{\mathbf{x}} + E_{yi} e^{-jk(n_{yy}-1)\ell}\hat{\mathbf{y}} \right) e^{j(\omega t - kz)}$$

$$(\ell < z),$$

or more simply,

$$\mathbf{E}_{out} = (E_{xo}\hat{\mathbf{x}} + E_{yo}\hat{\mathbf{y}})\, e^{j(\omega t - kz)} \qquad (\ell < z), \tag{6.76}$$

with

$$E_{xo} = \tau_1\tau_2 E_{xi} e^{-jk(n_{xx}-1)\ell} \qquad E_{yo} = \tau_1\tau_2 E_{yi} e^{-jk(n_{yy}-1)\ell}. \tag{6.77}$$

In these equations, τ_1 is the Fresnel transmission coefficient for light going from air into the material at normal incidence, and τ_2 is for light leaving the material. From Equation 6.64, $|\tau_1\tau_2| = T$. The irradiance in the output wave is thus T^2 times that of the input, as would be the case if the crystal were not birefringent, and the phase shift occurs without additional loss. In practice, the surfaces may be anti-reflection coated so that $T \to 1$.

In most cases, we do not measure z with sufficient precision to care about the exact phase,

$$\phi_{x0} = \angle E_{x0},$$

and what is important is the difference between the phases.

The difference is

$$\delta\phi = k\ell\,(n_{yy} - n_{xx}). \tag{6.78}$$

We call $\delta\phi$ the phase **retardation**.

Eq. 1.29:
$k = \frac{2\pi}{\lambda}$.

Eq. 1.77:
$\ell = OPL = \int_A^B n d\ell_p$.

Eq. 6.24:
$\mathbf{E}_{out} = E_o \left[\cos(\zeta)\,\hat{\mathbf{x}} + j\sin(\zeta)\,\hat{\mathbf{y}}\right]$.

6.5.2.1 Half- and Quarter-Wave Plates

Two cases are of particular importance as discussed in Section 6.2.2. The half-wave plate reflects the direction of linear polarization, with

$$\delta\phi_{hwp} = \pi = k\ell\left(n_{yy} - n_{xx}\right), \tag{6.79}$$

or in terms of the optical path lengths (Equation 1.77)

$$\delta\phi_{hwp} = n_{yy}\ell - n_{xx}\ell = \frac{\lambda}{2}, \tag{6.80}$$

resulting in a reflection of the input polarization as shown in Equation 6.22. Likewise a quarter-wave plate to convert linear polarization to circular according to Equation 6.24 requires

$$\delta\phi_{qwp} = \frac{\pi}{2} = k\ell\left(n_{yy} - n_{xx}\right) \qquad n_{yy}\ell - n_{xx}\ell = \frac{\lambda}{4}. \tag{6.81}$$

For example, in quartz at a wavelength of 589.3 nm,

$$n_{yy} - n_{xx} = 1.5534 - 1.5443, \tag{6.82}$$

Higher-order waveplates are useful for some applications where the environment is fairly benign and the optical bandwidth is not too wide.

and a quarter-wave plate must have a thickness of 16.24 μm. Normally, this birefringent material will be deposited onto a substrate. It is easier to manufacture a thicker waveplate, so frequently **higher-order waveplates** are used. For example, a five-quarter-wave plate, 81 μm thick in this example, is quite common, and considerably less expensive than a **zero-order** quarter-wave plate in which the phase difference is actually one quarter wave. According to the equations, the resulting field should be the same for both. However, the thicker plate is five times more susceptible to variations in retardation with temperature (T),

$$\frac{d\delta\phi_{qwp}}{dT} = k\ell\frac{d\left(n_{yy} - n_{xx}\right)}{dT} \qquad \frac{d\delta\phi_{5qwp}}{dT} = 5k\ell\frac{d\left(n_{yy} - n_{xx}\right)}{dT}, \tag{6.83}$$

or angle, for which the equation is more complicated. Often, one chooses to avoid perfect normal incidence to prevent unwanted reflections, so the angular tolerance may be important.

Two limitations on bandwidth exist in Equation 6.78. First, the wavelength appears explicitly in k, so dispersion of the phase retardation must be considered.

$$\frac{d\delta\phi_{qwp}}{d\lambda} = \frac{2\pi}{\lambda}\ell\left(n_{yy} - n_{xx}\right) \qquad \frac{d\delta\phi_{5qwp}}{d\lambda} = 5\frac{2\pi}{\lambda}\ell\left(n_{yy} - n_{xx}\right). \tag{6.84}$$

Example

For a source having a bandwidth of 100 nm at a center wavelength of 800 nm, the phase in a zero-order quarter-wave plate will be 6°, and for a five-quarter-wave plate it will be 30°.

Second, there is **dispersion of birefringence**; the index difference varies with wavelength,

$$\delta\phi\left(\lambda\right) = \frac{2\pi}{\lambda}\ell\left(n_{yy}\left(\lambda\right) - n_{xx}\left(\lambda\right)\right). \tag{6.85}$$

Recently high-bandwidth waveplates have become commercially available.

This dispersion has been turned into an advantage by manufacturers who have used combinations of different birefringent materials with the material dispersion properties chosen to offset the phase dispersion inherent in the equation. It is now possible to purchase a broadband quarter- (or half-) wave plate.

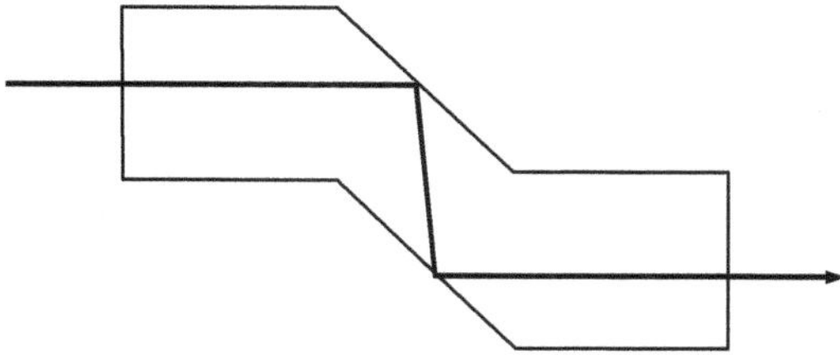

FIGURE 6.24
Fresnel rhomb. This device, made of glass, produces a 90° phase difference between the two states of polarization. The only wavelength dependence is in the index of refraction.

6.5.2.2 Electrically Induced Birefringence

We have considered crystals in which the birefringence is a consequence of the crystal structure to produce waveplates. However, some crystals can be made birefringent through the application of a DC electric field[254]. The birefringence is proportional to the applied voltage. The effect is called the **Pockels effect**, or **linear electro-optical effect** and the device is called an electro-optical modulator. Also of interest here is the **Kerr effect**, with similar applications but in which the birefringence is proportional to the square of the applied voltage. Either effect provides the opportunity to modulate light in several ways. If the crystal is placed so that one axis is parallel to the input polarization (the input polarization is an eigenvector* of the device), then the output will have the same polarization but will undergo a phase shift proportional to the applied voltage. We will see some applications for this **phase modulator** when we cover interference. Alternatively, if we align the crystal with axes at 45° to the incident polarization, it changes the state of polarization. Following the crystal with a polarizer orthogonal to the input polarization produces an amplitude modulation, with maximum transmission when the applied voltage produces a half-wave shift. The crystal may be characterized by the voltage required to produce this half-wave shift, V_π, and then the phase shift will be

Application of an external voltage to some materials can control birefringence. Electro-optical modulators use this effect in a variable waveplate to control phase or polarization.

$$\delta\phi = \pi\frac{V}{V_\pi}. \tag{6.86}$$

Applications will be discussed later, after the introduction of Jones matrices.

6.5.2.3 Fresnel Rhomb

We noted in Equation 6.84 that a quarter-wave plate is only useful over a narrow band of wavelengths, depending on the tolerance for error. The problem is that the waveplate retards the field in time rather than phase, so the retardance is only a quarter wave at the right wavelength. However, in Figure 6.17, there is a true phase difference between the two states of polarization in total internal reflection. Selecting the angle carefully, we can choose the phase difference. With two reflections, we can achieve a 90° retardation as shown in Figure 6.24. The **Fresnel rhomb** is a true quarter-wave retarder over a broad range of wavelengths. Because the phase retardation is not dependent on birefringence, the important parameter is the index itself (about 1.5), rather than the difference in indices, which is small (see Equation 6.82). The small dispersion in the index has a small effect on phase difference. The advantages of the Fresnel rhomb are this inherent wide bandwidth and high power-handling capability. The disadvantage is the severe "dogleg" or translation of the beam path, which makes alignment of a system more difficult, particularly if one wishes to insert and remove the retarder.

Fig. 6.17

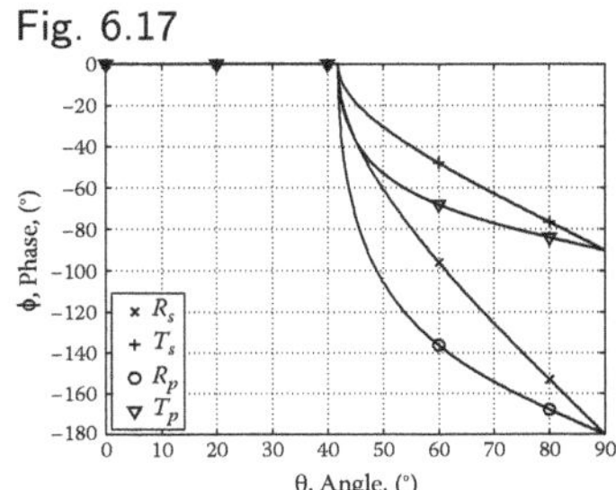

*See Section C.8 in Appendix C.

Take-Away

Waveplates:

The key points we have learned about waveplates are

- Birefringence causes retardation of one state of linear polarization with respect to the orthogonal one.
- Zero-order waveplates have the least dispersion and lowest temperature and angle dependence.
- Higher-order waveplates are often less expensive, and can be used in narrow-band light situations.
- Wideband waveplates are available, in which material dispersion compensates for the phase dispersion.
- A Fresnel rhomb is a very wideband quarter-wave plate, but it may be difficult to install in a system.

6.5.3 Polarization Rotator

Rotators are devices that rotate the direction of linear polarization. We will see in Section 6.6 that rotation of linear polarization is equivalent to introduction of different retardations for the two states of circular polarization. There are two common underlying mechanisms for rotation with different applications.

6.5.3.1 Reciprocal Rotator

Following the Lorentz model of the electron as a mass on a spring in considering linear polarization, the first mechanism to consider for a rotator is a molecular structure such as that of a benzene ring, which provides circular paths for electrons to travel, but with additional elements that cause a difference in the electrons' freedom to move between clockwise and counterclockwise paths. Such structures are common in sugars, and a mixture of sugar and water is an example of a rotator that is a common sight at almost every high-school science fair. The amount of rotation, $\delta\zeta$, is given by

$$\delta\zeta = \kappa C\ell, \tag{6.87}$$

where

- κ is the **specific rotary power** of the material
- C is the concentration
- ℓ is the thickness of the material

Sugar-water solutions are reciprocal rotators.

The specific rotary power may have either a positive or negative sign, and refers to the direction of rotation as viewed from the source. Thus, the sugar, dextrose (Latin *dexter* = right) rotates the polarization to the right or clockwise as viewed from the source. In contrast, levulose (Latin *laevus* = left) rotates the polarization to the left or counterclockwise. If light goes through a solution of dextrose in water in the $\hat{\mathbf{z}}$ direction, $\hat{\mathbf{x}}$ polarization is rotated toward the right, or $\hat{\mathbf{y}}$ direction, by an amount determined by Equation 6.87. Upon reflection and returning through the solution, the polarization is again rotated to the right, but this time, as viewed from the mirror, so the rotation is exactly back to the original $\hat{\mathbf{x}}$ direction. Thus, these materials are called **Reciprocal** rotators.

6.5.3.2 Faraday Rotator

The second mechanism that creates a circular path for an electron is the **Faraday effect**, which uses a magnetic field along the direction of light propagation. This results in acceleration, **a** of an electron of charge $-e$ and mass m, moving with a velocity **v** by the Lorentz force,

$$\mathbf{a} = -\frac{e}{m}\mathbf{v} \times \mathbf{B}. \tag{6.88}$$

Thus, if the optical **E** field causes the electron to move in the $\hat{\mathbf{x}}$ direction, and the **B** field is in the $\hat{\mathbf{z}}$ direction, the electron will experience acceleration in the $-\hat{\mathbf{x}} \times \hat{\mathbf{z}} = \hat{\mathbf{y}}$ direction, leading to a rotation to the right. The overall effect is characterized by the **Verdet constant**[233], v, in degrees (or radians) per unit magnetic field per unit length.

Faraday rotators are non-reciprocal devices based on a static magnetic field.

$$\delta\zeta = v\mathbf{B} \cdot z\ell. \tag{6.89}$$

Now, upon reflection, the polarization will be rotated in the same direction, because Equation 6.89 is independent of the direction of light propagation. Instead of canceling the initial rotation, this device doubles it.

Faraday rotators are useful for **Faraday isolators**, sometimes called optical isolators. A Faraday isolator consists of a polarizer followed by a Faraday rotator with a rotation angle of 45°. Light reflected back through the isolator from optical components later in the system is rotated an additional 45°, and blocked from returning by the polarizer. The Faraday rotator is used to protect the source from reflections. As we will note in Section 7.4.1, some lasers are quite sensitive to reflections of light back into them, so isolators are often used with such lasers. Faraday isolators can be built for any wavelength, but for longer infrared wavelengths the materials are difficult to handle in the manufacturing process, are easily damaged by excessive power, and require high magnetic fields, leading to large, heavy devices. Considerable effort is required to design satisfactory isolators for high powers at these wavelengths[121].

A Faraday rotator can be used in an optical isolator, which prevents light from being reflected back to a source.

Take-Away

Rotators:

- Natural rotators such as sugars may rotate the polarization of light to the left or right.
- These rotators are reciprocal devices. Light passing through the device, reflecting, and passing through again, returns to its original state of polarization.
- Faraday rotators are based on the interaction of light with an axial magnetic field.
- They are nonreciprocal devices; the rotation is doubled upon reflection.
- Faraday rotators can be used to make optical isolators, which are used to protect laser cavities from excessive reflection.

6.6 Jones Vectors and Matrices

By now we have developed an understanding of polarized light and the physical devices that can be used to manipulate it. Many optical systems contain multiple polarizing devices, and we need a method of bookkeeping to analyze them. For a given direction of propagation, we saw in Section 6.1 that we can use two-element vectors, **E**, to describe polarization. Thus, we can use two-by-two matrices, $\mathcal{J}$, to describe the effect of polarizing devices. The formalization of these concepts is attributed to Jones[115], who published the work in 1941.

A two-element Jones vector describes a state of polarization.

A 2×2 Jones matrix multiplies a vector to describe the effect of a device on the polarization.

In this section we will develop a rigorous approach for potentially complicated polarization problems using Jones matrices. Let us look ahead at the approach we will use.

(1) Cascading matrices: The output of a single device is

$$\mathbf{E}_1 = \mathcal{J}\mathbf{E}_0. \tag{6.90}$$

for an input $\mathbf{E}_0$. Then the effect of multiple devices can be determined by simple matrix multiplication. The output vector of the first device becomes the input to the next, so that

$$\mathbf{E}_m = \mathcal{J}_m \mathcal{J}_{m-1} \dots \mathcal{J}_2 \mathcal{J}_1 \mathbf{E}_0. \tag{6.91}$$

Matrices are multiplied from right to left.

Matrices are multiplied from right to left.* We need to find matrices for each of our basic components: polarizers, waveplates, and rotators.

(2) Coordinate Transforms: We have only defined these components in Section 6.2 in terms of a specific $\hat{\mathbf{x}}$–$\hat{\mathbf{y}}$ coordinate system in which they are easy to understand. What is the matrix of a polarizer oriented at 17° to the $\hat{\mathbf{x}}$ axis? To answer a question like this we need to find matrices for coordinate transforms.

(3) Quantitative Analysis: Usually the last step in a polarization problem is to calculate the irradiance, I, or power, P. The irradiance is $|\mathbf{E}|^2/Z$, so in Jones-matrix notation

$$P = IA = \frac{\mathbf{E}^\dagger \mathbf{E} A}{Z}, \tag{6.92}$$

where

- A is the area
- the † represents the **Hermitian Adjoint** or the transpose of the complex conjugate†

Usually we normalize Jones vectors to input power or irradiance

In this way, we can define the input vector in terms of actual electric-field values in volts per meter, but this is awkward and very seldom done. Instead, it makes sense to define the input polarization to have one unit of irradiance or power, and then calculate the transmission of the system and multiply by the actual initial irradiance or power. In this case then, we choose the input vector so that

$$\mathbf{E}_0^\dagger \mathbf{E}_0 = 1, \tag{6.93}$$

and compute the transmission through the system from the output vector

$$T = \mathbf{E}_{out}^\dagger \mathbf{E}_{out}, \tag{6.94}$$

from which

$$I_{out} = T I_{in} \qquad P_{out} = T P_{in}. \tag{6.95}$$

Basically, we are expressing the electric field in some unusual units. The price we pay for this is the inability to calculate the actual electric field directly from the Jones vector, and the inability to relate the electric field in a dielectric to that in air, because of the different impedances. In most polarization problems we are not interested in the field, and if we are, we can compute it correctly from the irradiance.

θ for angle of incidence.
ζ for rotation angle around the beam axis.
ϕ for phase.

(4) Notation: In the analysis of polarizing devices, where we are going to consider rotation of polarization and rotation of components about the propagation axis, and where we may be using Fresnel reflection at different angles of incidence, the notation can become confusing. We will continue to use the Greek letter theta, θ, to represent the angle of incidence, as in the Fresnel equations. We will use zeta, ζ, to represent angles of rotation about the beam axis, whether they are of polarization vectors or devices. Finally, we will use phi, ϕ, to represent the phase of an electric field.

*See Appendix C for a brief overview of matrix multiplication and other operations.

†See Section C.3 in Appendix C for a discussion of the Hermitian adjoint matrix.

6.6.1 Basic Polarizing Devices

With the above-mentioned background, the matrix of each basic device is quite easy to understand. We will begin by assuming that the device is rotated in such a way that the $\hat{\mathbf{x}}$ and $\hat{\mathbf{y}}$ directions are eigenvectors. Then we know the behavior of the device for each of these polarizations, and our matrices will be diagonal.[‡] The input polarization vector is described by

$$\mathbf{E}_0 = \begin{pmatrix} E_{x0} \\ E_{y0} \end{pmatrix}. \tag{6.96}$$

Then the output will be

$$\mathbf{E}_1 = \begin{pmatrix} j_{11} & 0 \\ 0 & j_{22} \end{pmatrix} \begin{pmatrix} E_{x0} \\ E_{y0} \end{pmatrix}. \tag{6.97}$$

6.6.1.1 Polarizer

A perfect $\hat{\mathbf{x}}$ polarizer is

Perfect $\hat{\mathbf{x}}$ polarizer.

$$\mathcal{P}_x = \begin{pmatrix} 1 & 0 \\ 0 & 0 \end{pmatrix}. \tag{6.98}$$

If we wish, we can define a $\hat{\mathbf{y}}$ polarizer as

$$\mathcal{P}_y = \begin{pmatrix} 0 & 0 \\ 0 & 1 \end{pmatrix}, \tag{6.99}$$

which we will find convenient although not necessary.

We can describe an $\hat{\mathbf{x}}$ polarizer more realistically by

Actual $\hat{\mathbf{x}}$ polarizer.

$$\mathcal{P}_x = \begin{pmatrix} \tau_{pass} & 0 \\ 0 & \tau_{block} \end{pmatrix}, \tag{6.100}$$

where τ_{pass} (slightly less than unity) accounts for the insertion loss, and τ_{block} (slightly larger than zero) for imperfect extinction. The transmissions may or may not be complex, to describe changes in phase between the two states of polarization. This equation is the matrix form of Equation 6.20. We could develop an optical system design using an ideal polarizer, and then determine design tolerances using a more realistic one.

Eq. 6.20: $\mathbf{E}_{out} = \tau_{pass} \times E_x\hat{\mathbf{x}} + \tau_{block} \times E_y\hat{\mathbf{y}}$.

Example

As an example, let us consider 5 mW of light with linear polarization at an angle ζ with respect to the x axis, passing through a polarizer with an insertion loss of 8% and an extinction ratio of 10,000. Then $\tau_x = \sqrt{1-0.08}$ and $\tau_y = \tau_x/\sqrt{10{,}000}$. One must remember the square roots, because the specifications are for irradiance or power, but the τ values are for fields. Light is incident with polarization at an angle ζ. The incident field is

$$\mathbf{E}_0 = \begin{pmatrix} \cos\zeta \\ \sin\zeta \end{pmatrix}. \tag{6.101}$$

[‡]See Section C.8 in Appendix C.

The matrix for the polarizer is given in Equation 6.100, and the output is

$$\mathbf{E}_1 = \mathcal{P}_x \mathbf{E}_0 = \begin{pmatrix} \tau_x & 0 \\ 0 & \tau_y \end{pmatrix} \begin{pmatrix} \cos\zeta \\ \sin\zeta \end{pmatrix} = \begin{pmatrix} \tau_x \cos\zeta \\ \tau_y \sin\zeta \end{pmatrix}. \tag{6.102}$$

The transmission of the system in the sense of Equation 6.94 is

$$T = \mathbf{E}_1^\dagger \mathbf{E}_1 \tag{6.103}$$

Recall that the Hermitian adjoint of a product is

$$(\mathcal{AB})^\dagger = \mathcal{B}^\dagger \mathcal{A}^\dagger. \tag{6.104}$$

Thus

$$\mathbf{E}_1^\dagger \mathbf{E}_1 = \mathbf{E}_1^\dagger \mathcal{P}_x^\dagger \mathcal{P}_x \mathbf{E}_1, \tag{6.105}$$

and

$$T = \begin{pmatrix} \cos\zeta & \sin\zeta \end{pmatrix} \begin{pmatrix} \tau_x^* & 0 \\ 0 & \tau_y^* \end{pmatrix} \begin{pmatrix} \tau_x & 0 \\ 0 & \tau_y \end{pmatrix} \begin{pmatrix} \cos\zeta \\ \sin\zeta \end{pmatrix} =$$

$$T_x \cos^2\zeta + T_y \sin^2\zeta. \tag{6.106}$$

Figure 6.25 shows the power with a comparison to Malus' law. The power in our realistic model is lower than predicted by Malus' law, for an input angle of zero because of the insertion loss, and higher at 90° (y-polarization) because of the imperfect extinction. For the angle of polarization, we return to Equation 6.102 and compute

$$\tan\zeta_{out} = \frac{E_{yout}}{E_{xout}} = \frac{\tau_x \cos\zeta}{\tau_y \sin\zeta}. \tag{6.107}$$

The angle is also plotted in Figure 6.25. For most input angles the polarization remains in the $\hat{\mathbf{x}}$ direction as expected for a good polarizer. When the input is nearly in the $\hat{\mathbf{y}}$ direction, then the small amount of leakage is also polarized in this direction.

6.6.1.2 Waveplate

Arbitrary waveplate.

A waveplate with phase retardation, $\delta\phi$ may be defined by

$$\mathcal{W} = \begin{pmatrix} e^{j\delta\phi/2} & 0 \\ 0 & e^{-j\delta\phi/2} \end{pmatrix}. \tag{6.108}$$

Many matrices for the same device can be used. They differ from each other by a constant phase.

Many variations are possible, because we usually do not know the exact phase. One will frequently see

$$\mathcal{W} = \begin{pmatrix} e^{j\delta\phi} & 0 \\ 0 & 1 \end{pmatrix}, \tag{6.109}$$

Quarter-wave plate.

which differs from the above by a phase shift. Specifically, a quarter-wave plate is

$$\mathcal{Q} = \begin{pmatrix} e^{-j\pi/4} & 0 \\ 0 & e^{j\pi/4} \end{pmatrix}, \tag{6.110}$$

Quarter-wave plate.

or more commonly

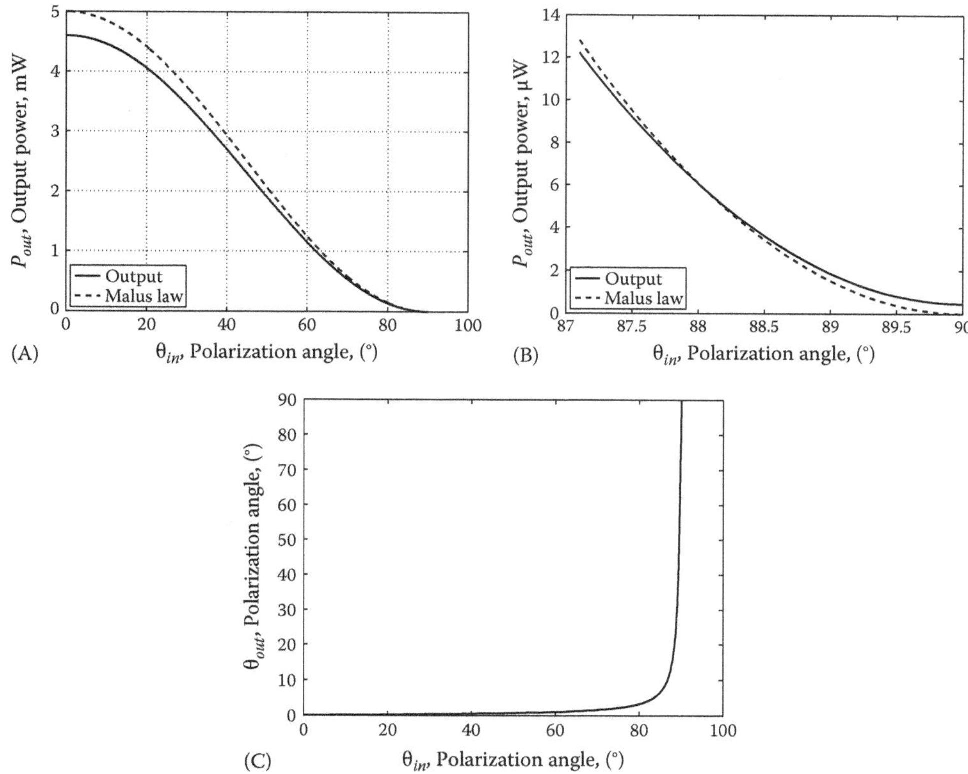

FIGURE 6.25
Imperfect polarizer. The output of this device behaves close to that of a perfect polarizer, except for the small insertion loss and a slight leakage of $\hat{\mathbf{y}}$ polarization. Transmitted power with a source of 5 mW is shown for all angles (A). The last few degrees are shown on an expanded scale (μW instead of mW) (B). The angle of the output polarization as a function of the angle of the input changes abruptly near 90° (C).

$$\mathcal{Q} = \begin{pmatrix} 1 & 0 \\ 0 & j \end{pmatrix}. \tag{6.111}$$

A half-wave plate is usually defined by

Half-wave plate.

$$\mathcal{H} = \begin{pmatrix} 1 & 0 \\ 0 & -1 \end{pmatrix}. \tag{6.112}$$

6.6.1.3 Rotator

Finally, a rotator with a rotation angle ζ is defined by

Polarization Rotator.

$$\mathcal{R}(\zeta) = \begin{pmatrix} \cos\zeta & -\sin\zeta \\ \sin\zeta & \cos\zeta \end{pmatrix}, \tag{6.113}$$

where ζ is positive to the right, or from $\hat{\mathbf{x}}$ to $\hat{\mathbf{y}}$.

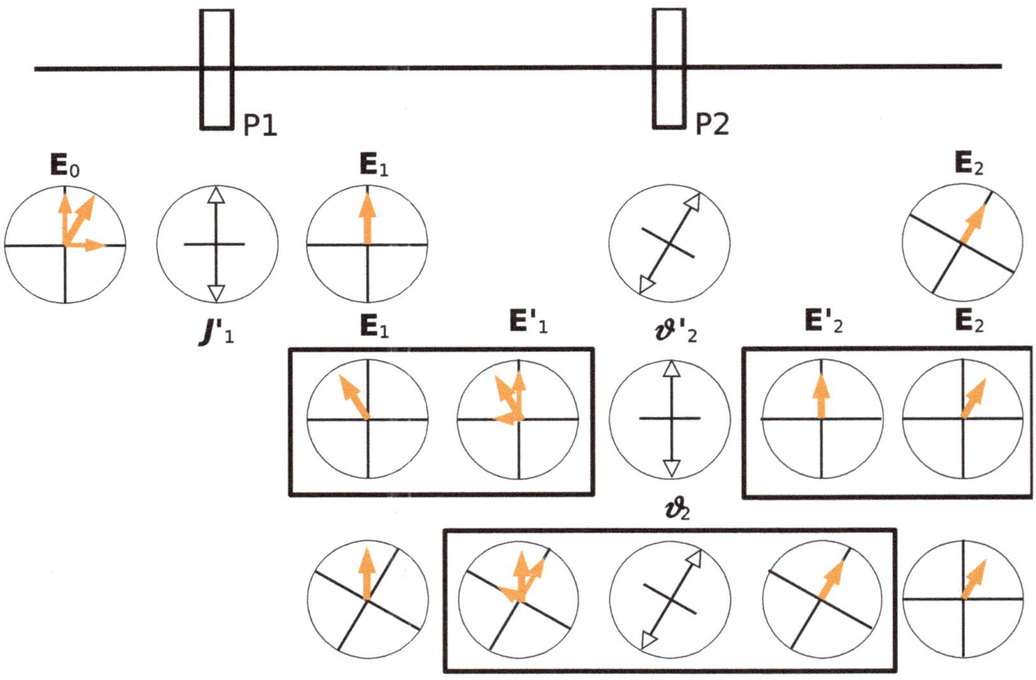

FIGURE 6.26
Rotation of coordinates. The layout at the top shows a system consisting of two polarizers. The first row of polarization figures shows the orientations. The second row shows the mathematical description of changing coordinates. The third row shows the construction of a single matrix for the second polarizer. Note: The lines with open triangles represent the "pass" axis of the polarizer.

6.6.2 Coordinate Transforms

We now have the basic devices, but we still cannot approach general polarization problems, as we have only defined devices where the axes match our coordinate system. With a single device, we can always choose the coordinate system to match it, but if the second device is rotated to a different angle, then we need to convert the coordinates. The first device is easy; we can work in the coordinates in which the matrix is diagonal, and write

$$\mathbf{E}_1 = \mathcal{J}_1 \mathbf{E}_0.$$

6.6.2.1 Rotation

If the second device is defined in terms of coordinates $\hat{\mathbf{x}}'$ and $\hat{\mathbf{y}}'$, rotated from the first by an angle ζ_r, and the matrix in this coordinate system is $\mathcal{J}_2'$, (the prime denotes that the matrix is in the x',y' coordinate system) then we could approach the problem in the following way, as shown in the second row of polarization states in Figure 6.26. After solving for the output of the first device, we could rotate the output vector through an angle $-\zeta_r$, so that it would be correct in the new coordinate system. Thus the input to the second device would be the output of the first, multiplied by a rotation matrix,

$$\mathcal{R}(-\zeta_r)\,\mathbf{E}_1 = \mathcal{R}^\dagger(\zeta_r)\,\mathbf{E}_1. \tag{6.114}$$

Then, we can apply the device matrix, $\mathcal{J}_2'$ to obtain

$$\mathcal{J}_2'\mathcal{R}^\dagger(\zeta_r)\,\mathbf{E}_1. \tag{6.115}$$

The result is still in the coordinate system of the device. We want to put the result back into the original coordinate system, so we must undo the rotation we used at the beginning. Thus,

$$\mathbf{E}_2 = \mathcal{R}(\zeta_r)\,\mathcal{J}_2'\mathcal{R}^\dagger(\zeta_r)\,\mathbf{E}_1. \tag{6.116}$$

Thus, we can simplify the problem to

$$\mathbf{E}_2 = \mathcal{J}_2\mathbf{E}_1, \tag{6.117}$$

where

Rotating a polarizing device.

$$\mathcal{J}_2 = \mathcal{R}(\zeta_r)\,\mathcal{J}_2'\mathcal{R}^\dagger(\zeta_r). \tag{6.118}$$

This is the procedure for **rotation of a polarizing device** in the Jones calculus, represented in the third row of polarization states in Figure 6.26.

Let us consider as an example a wave with arbitrary input polarization,

$$\mathbf{E}_{in} = \begin{pmatrix} E_x \\ E_y \end{pmatrix}, \tag{6.119}$$

incident on a polarizer with its $\hat{\mathbf{x}}$ axis rotated through an angle ζ_r,

$$\mathcal{P}_{\zeta_r} = \mathcal{R}(\zeta_r)\,\mathcal{P}_x\mathcal{R}^\dagger(\zeta_r). \tag{6.120}$$

Then

$$\mathbf{E}_{out} = \mathcal{P}_{\zeta_r}\mathbf{E}_{in} = \mathcal{R}(\zeta_r)\,\mathcal{P}_x\mathcal{R}^\dagger(\zeta_r)\begin{pmatrix} E_x \\ E_y \end{pmatrix}. \tag{6.121}$$

We can think of this in two ways. Multiplying from right to left in the equation, first rotate the polarization through $-\zeta_r$ to the new coordinate system. Then apply $\mathcal{P}_x$. Then rotate through ζ_r to get to the original coordinates. Alternatively we can do the multiplication of all matrices in Equation 6.120 first to obtain

$$\mathcal{P}_{\zeta_r} = \begin{pmatrix} \cos\zeta_r & -\sin\zeta_r \\ \sin\zeta_r & \cos\zeta_r \end{pmatrix}\begin{pmatrix} 1 & 0 \\ 0 & 0 \end{pmatrix}\begin{pmatrix} \cos\zeta_r & \sin\zeta_r \\ -\sin\zeta_r & \cos\zeta_r \end{pmatrix} = \begin{pmatrix} \cos^2\zeta_r & \cos\zeta_r\sin\zeta_r \\ \cos\zeta_r\sin\zeta_r & \sin^2\zeta_r \end{pmatrix}, \tag{6.122}$$

and refer to this as the matrix for our polarizer in the original coordinate system. Then, the output is simply

$$\begin{pmatrix} \cos^2\zeta_r & \cos\zeta_r\sin\zeta_r \\ \cos\zeta_r\sin\zeta_r & \sin^2\zeta_r \end{pmatrix}\begin{pmatrix} E_x \\ E_y \end{pmatrix} = \begin{pmatrix} E_x\cos^2\zeta_r + E_y\cos\zeta_r\sin\zeta_r \\ E_x\cos\zeta_r\sin\zeta_r + E_y\sin^2\zeta_r \end{pmatrix}. \tag{6.123}$$

For x-polarized input, the result is Malus's law; the power is

$$\mathbf{E}_{out}^\dagger\mathbf{E}_{out} = \begin{pmatrix} E_x\cos^2\zeta_r & E_x\cos\zeta_r\sin\zeta_r \end{pmatrix}\begin{pmatrix} E_x\cos^2\zeta_r \\ E_x\cos\zeta_r\sin\zeta_r \end{pmatrix}$$

$$= E_x\cos^2\zeta_r\left(\cos^2\zeta_r + \sin^2\zeta_r\right) = E_x\cos^2\zeta_r. \tag{6.124}$$

For y-polarized input, the transmission is $\cos^2(\zeta_r - 90°)$. In fact, for any input direction, Malus's law holds, with the reference taken as the direction of the input polarization. The tangent of the angle of the output polarization is given by the ratio of the y component to the x component:

$$\tan\zeta_{out} = \frac{E_x \cos\zeta_r \sin\zeta_r + E_y \sin^2\zeta_r}{E_x \cos^2\zeta_r + E_y \cos\zeta_r \sin\zeta_r} =$$

$$\frac{\sin\zeta_r (E_x \cos\zeta_r + E_y \sin\zeta_r)}{\cos\zeta_r (E_x \cos\zeta_r + E_y \sin\zeta_r)}$$

$$= \frac{\sin\zeta_r}{\cos\zeta_r}.$$

$$\tan\zeta_{out} = \tan\zeta_r \qquad \rightarrow \qquad \zeta_{out} = \zeta_r$$

Eq. 6.100: $\mathcal{P}_x = \left(\tau_{pass} \; 0 0 \; \tau_{block}\right)$.

For this perfect polarizer, the output is always polarized in the ζ_r direction, regardless of the input, as we know must be the case. The reader may wish to complete this analysis for the realistic polarizer in Equation 6.100.

Now, consider the experiment shown in Figure 6.27. Vertically–polarized light is passed through a very fast lens between two crossed polarizers. The angles in this experiment can be confusing, so the drawing in Figure 6.28 may be useful. The origin of coordinates is at the center of the lens, and we will choose the $\hat{\mathbf{x}}$ direction to be toward the top of the page, $\hat{\mathbf{y}}$ to the right, and $\hat{\mathbf{z}}$, the direction of propagation, into the page in the top view. The lens surface is normal to the incident light at the origin. At any other location on the lens (x, y), the plane of incidence is vertical and oriented so that it contains the origin and the point (x, y). It is thus at an angle ζ in this coordinate system (top view). Thus at this point, the Jones matrix is that of a Fresnel transmission,

Transmission in the P,S coordinates is given by τ_p and τ_s for an angle of incidence θ.

$$\mathcal{F}'(\theta, 0) = \begin{pmatrix} \tau_p(\theta) & 0 \\ 0 & \tau_s(\theta) \end{pmatrix}, \tag{6.125}$$

with P oriented at an angle ζ, or

$$\mathcal{F}(\theta, \zeta) = \mathcal{R}(\zeta)\,\mathcal{F}'(\theta, 0)\,\mathcal{R}^\dagger(\zeta). \tag{6.126}$$

Here θ is the angle of incidence as seen in the side view. We need to rotate the coordinates through an angle ζ, from this P,S system to the x, y one about the $\hat{\mathbf{z}}$ axis (top view). If the incident light is unpolarized, after it passes through the first polarizer, it will be reduced by 50% and will be x polarized as we discussed in Section 6.1.4. Then,

The P,S coordinate system is related to the x,y one by a rotation about $\hat{\mathbf{z}}$ through an angle ζ.

$$\mathbf{E}_{out} = P_y \mathcal{R}(\zeta)\,\mathcal{F}'(\theta, 0)\,\mathcal{R}^\dagger(\zeta)\begin{pmatrix} 1 \\ 0 \end{pmatrix}. \tag{6.127}$$

We can immediately compute the transmission, but we gain some physical insight by solving this equation first.

If the $\hat{\mathbf{x}}$ input is aligned with P or S, the transmission through the lens does not change the state of polarization; the image is dark.

$$\mathbf{E}_{out} = \begin{pmatrix} 0 & 0 \\ 0 & 1 \end{pmatrix}\begin{pmatrix} \tau_p(\theta)\cos^2\zeta - \tau_s(\theta)\sin^2\zeta \\ \tau_p(\theta)\cos\zeta\sin\zeta - \tau_s(\theta)\cos\zeta\sin\zeta \end{pmatrix}$$

$$= \begin{pmatrix} 0 \\ \tau_p(\theta)\cos\zeta\sin\zeta - \tau_s(\theta)\cos\zeta\sin\zeta \end{pmatrix}. \tag{6.128}$$

Transmission depends on $\tau_p - \tau_s$ and $\cos\zeta\sin\zeta$. The transmission difference increases with θ.

τ_p and τ_s differ most for steep angles of incidence, θ away from the $\hat{\mathbf{z}}$ axis.

If the matrix $\mathcal{R}(\zeta)\,\mathcal{F}\mathcal{R}^\dagger(\zeta)$ is diagonal, then the incident polarization is retained in passing through the lens, and is blocked by the second polarizer. There will be four such angles, where ζ is an integer multiple of 90° and thus $\cos\zeta\sin\zeta$ is zero. At other angles, the matrix is not diagonal, and conversion from x to y polarization occurs. The coupling reaches a

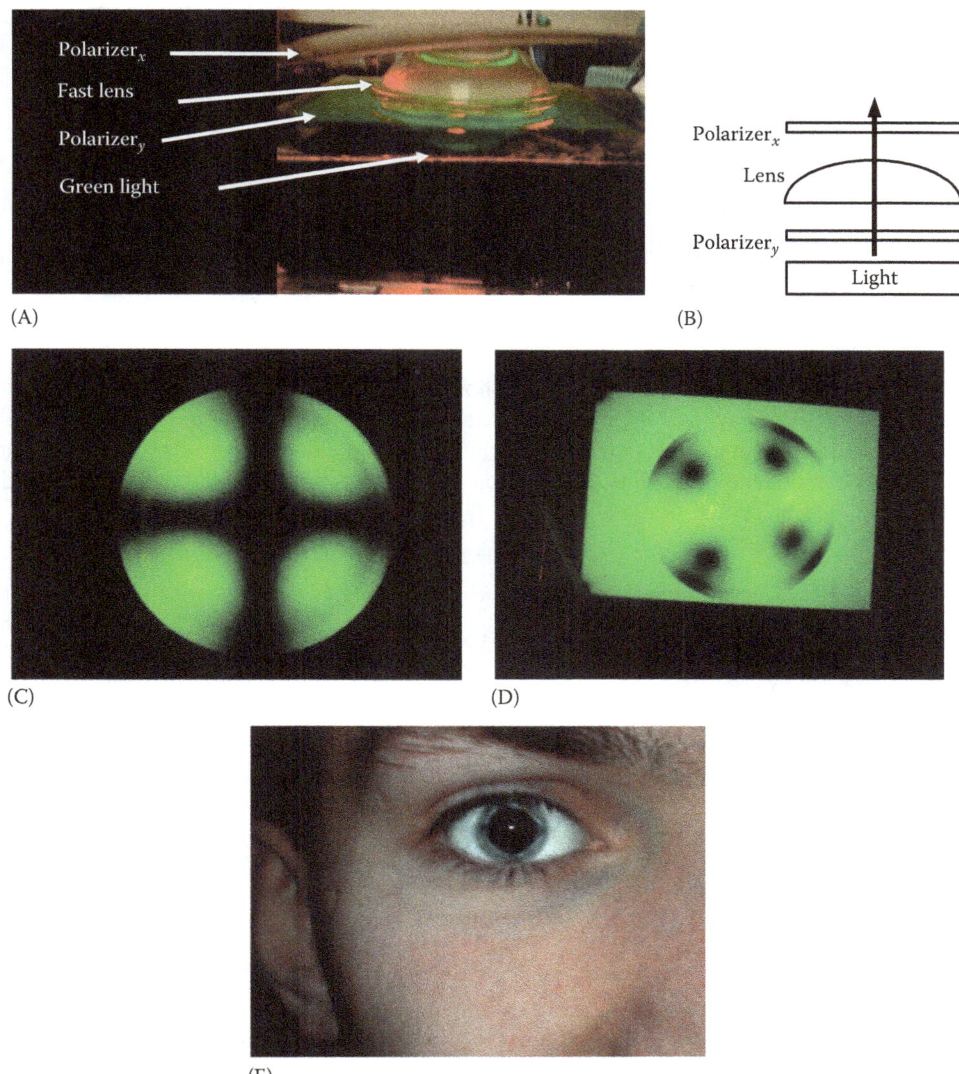

FIGURE 6.27
Maltese cross. The experiment is shown in a photograph (A) and sketch (B). Results are shown with the polarizers crossed (C) and parallel (D). A person's eye is imaged with crossed polarizers over the light source and the camera (E). The maltese cross is seen in the blue iris of the eye.

maximum in magnitude at four angles, at odd-integer multiples of 45°, and light is allowed to pass. Thus we observe the so-called Maltese cross pattern.

If $|\tau_s| = |\tau_p|$, as is always the case at normal incidence (at the center of the lens), then no light passes. For slow lenses the angle of incidence, θ, is always small and the Maltese cross is difficult to observe with slow lenses.

The opposite effect can be observed with parallel polarizers, as shown in Figure 6.27D.

The Maltese cross can be observed in a photograph of a human eye, taken with crossed polarizers over the light source and camera, respectively. This effect is often attributed erroneously to birefringence in the eye. If it were birefringence, one would expect the locations of the dark lines not to change with changing orientation of the polarizers. In fact, it does change, always maintaining alignment with the polarizers. There is birefringence

The Maltese cross can be observed in the human eye with cross-polarized illumination. It is cause by curvature and not by birefringence.

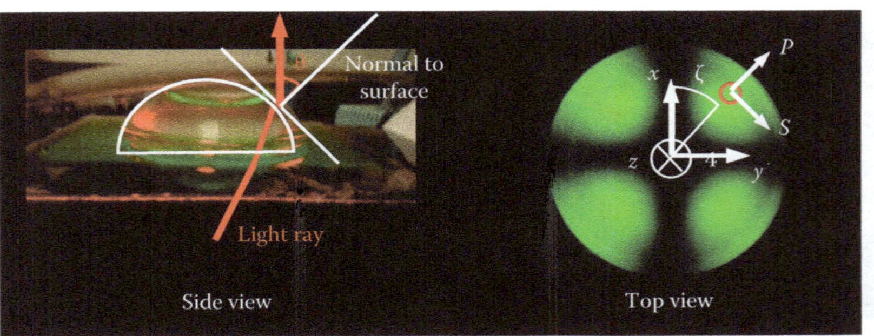

FIGURE 6.28
Angles in the Maltese cross. The left panel shows a side view with the lens and the normal to the surface for one ray, with light exiting at an angle of θ with respect to the normal. The right panel shows the angle of orientation of the plane of incidence, $\zeta = \arctan(y/x)$. The S and P directions depend on ζ. The polarization of the incident light is a mixture of P and S, except for $\zeta = 0°$ or $90°$.

in the collagen that comprises the cornea of the eye, but it is not the cause of the Maltese cross.

In Practice

Polarization effects that lead to the Maltese cross occur in any fast optical system. Although the pattern may change in going from the surface that caused it to an image plane, it is important to recognize that such effects are to be expected. For example, a high-NA objective may cause errors in polarization measurements in microscopy. Some objectives are designed to minimize such effects.

6.6.2.2 Circular/Linear

We can also transform to basis sets other than linear polarization. In fact, any orthogonal basis set is suitable. We consider, for example, transforming from our linear basis set to one with left- and right-circular polarization as the basis vectors. The transformation proceeds in exactly the same way as rotation. A quarter-wave plate at 45° to the linear coordinates converts linear polarization to circular:

$$\mathcal{Q}_{45} = \mathcal{R}_{45}\mathcal{Q}\mathcal{R}_{45}^{\dagger}. \tag{6.129}$$

A quarter-wave plate at 45 degrees is often used to convert linear polarization to circular.

The matrix $\mathcal{Q}_{45}$ occurs so frequently that it is useful to derive an expression for it:

$$\mathcal{Q}_{45} = \mathcal{R}_{45}\mathcal{Q}\mathcal{R}_{45}^{\dagger} = \frac{1}{\sqrt{2}}\begin{pmatrix}1 & j\\ j & 1\end{pmatrix}. \tag{6.130}$$

For any Jones matrix, $\mathcal{J}$ in the linear system, conversion to a circular basis set is accomplished for a vector $\mathbf{E}'$ and for a device matrix, $\mathcal{J}'$ by

$$\mathbf{E} = \mathcal{Q}_{45}^{\dagger}\mathbf{E}' \qquad \mathcal{J} = \mathcal{Q}_{45}\mathcal{J}'\mathcal{Q}_{45}^{\dagger}, \tag{6.131}$$

where $\mathcal{Q}$ is defined by Equation 6.24.

As an example, if x polarization is represented in the linear basis set by

$$\begin{pmatrix}1\\ 0\end{pmatrix}, \tag{6.132}$$

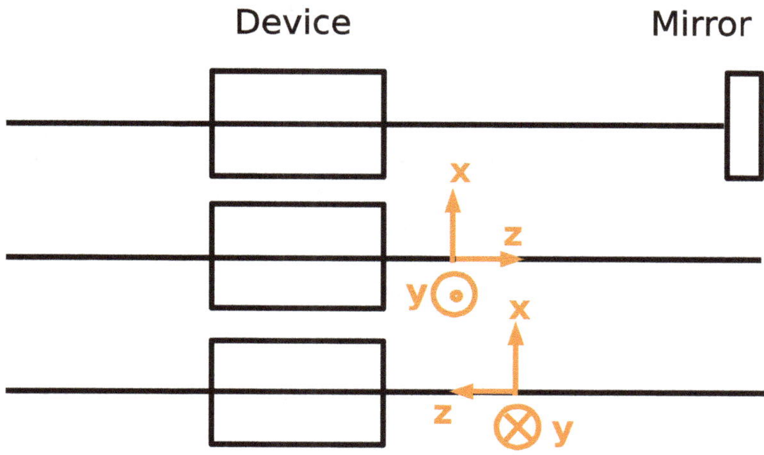

FIGURE 6.29
Reflection. Systems such as laser radars and epi-illumination microscopes present a particular problem with regard to coordinates.

then in circular coordinates it is represented by

$$\frac{1}{\sqrt{2}}\begin{pmatrix}1 & j\\ j & 1\end{pmatrix}\begin{pmatrix}1\\0\end{pmatrix} = \frac{1}{\sqrt{2}}\begin{pmatrix}1\\j\end{pmatrix}, \tag{6.133}$$

where in the new coordinate system, the top element of the matrix represents the amount of right-hand circular polarization, and the bottom represents left:

$$\frac{1}{\sqrt{2}}\begin{pmatrix}1\\j\end{pmatrix} = \frac{1}{\sqrt{2}}\hat{\mathbf{E}}_r + \frac{j}{\sqrt{2}}\hat{\mathbf{E}}_\ell. \tag{6.134}$$

To confirm this, we substitute Equations 6.8 and 6.10 into this equation and verify that the result is linear polarization.

Eq. 6.8:
$\mathbf{E}_r = \frac{E_0}{\sqrt{2}}\left[\hat{\mathbf{x}} + j\hat{\mathbf{y}}\right] e^{j(\omega t - kz)}$.

Eq. 6.10:
$\mathbf{E}_\ell = \left[E_0\hat{\mathbf{x}} - jE_0\hat{\mathbf{y}}\right] e^{j(\omega t - kz)}$.

In Practice

A circular basis set may be attractive because it is independent of the orientation of mirrors and beamsplitters, depending only on the direction of propagation. Of course, it then becomes necessary to convert the matrices of the polarizing devices to circular polarization, and it is easy to make a mistake. In many cases circular basis vectors seem less intuitive than linear ones. Circular basis vectors are not suitable, for example, to specify a beamsplitter.

6.6.3 Mirrors and Reflection

Mirrors or other materials that reflect the light back along the path of incidence as shown in Figure 6.29 present a particular problem in establishing sign conventions in the study of polarization. For example, the direction of linear polarization of light passing through a solution of dextrose sugar will be rotated clockwise as viewed from the source. Specifically the polarization will rotate from the $+x$ direction toward the $+y$ direction. If the light is reflected from a mirror, and passed back through the solution, we know that the rotation will be clockwise, but because it is traveling in the opposite direction, the rotation will be back toward the $+x$ direction. We faced a similar problem in the equations of Section 6.4, where we had to make decisions about the signs of **H** and **E** in reflection. These reflection problems are common in laser radars and in epi-illumination microscopes to cite two examples.

We are confronted here with two questions. First, what is the Jones matrix for a mirror? Second, if we know the Jones matrix of a device for light propagating in the forward direction, what is the matrix for propagation in the reverse direction? We choose to redefine our coordinate system after rotation so that (1) propagation is always in the $+z$ direction and (2) the coordinate system is still right-handed. This choice requires a change of either the x or y direction.

We choose to change the y direction, and write the Jones matrix for a perfect mirror as[234]

$$\mathcal{M} = \begin{pmatrix} 1 & 0 \\ 0 & -1 \end{pmatrix}. \tag{6.135}$$

Of course, we can multiply this matrix by a scalar to account for the finite reflectivity.

Now to answer our second question, given the matrix of a device in the forward direction, we construct the matrix for the same device in the reverse direction by taking the transpose of the matrix and negating the off-diagonal elements:

Matrices for a device in forward and reverse directions using the conventions of this book.

$$\mathcal{J}_{forward} = \begin{pmatrix} j_{11} & j_{12} \\ j_{21} & j_{22} \end{pmatrix} \quad \rightarrow \quad \mathcal{J}_{reverse} = \begin{pmatrix} j_{11} & -j_{21} \\ -j_{12} & j_{22} \end{pmatrix}. \tag{6.136}$$

In Practice

Systems with reflections such as described in this section can be confusing. Different authors use different conventions with regard to coordinate systems. When reading papers, we must be careful to make sure we are aware of the authors' choice. Such situations are common in laser radar, epi-illumination microscopy, and other applications.

6.6.4 Matrix Properties

Transformation matrices simply change the coordinate system, and thus must not change the power. Therefore,*

$$\mathbf{E}_{out}^{\dagger}\mathbf{E}_{out} = \mathbf{E}_{in}^{\dagger}\mathcal{J}^{\dagger}\mathcal{J}\mathbf{E}_{in} = \mathbf{E}_{in}^{\dagger}\mathbf{E}_{in}, \tag{6.137}$$

for all $\mathbf{E}_{in}$. This condition can only be satisfied if

Coordinate transforms are always represented by unitary matrices.

$$\mathcal{J}^{\dagger}\mathcal{J} = \mathcal{I} = \begin{pmatrix} 1 & 0 \\ 0 & 1 \end{pmatrix}. \tag{6.138}$$

A matrix with this property is said to be unitary. This condition must also be satisfied by the matrix of any device that is lossless, such as a perfect waveplate or rotator. The unitary condition can often be used to simplify matrix equations to obtain the transmission.

The matrix for a real waveplate can often be represented as the product of a unitary matrix and a scalar. For example, a waveplate without anti-reflection coatings can be

*See Appendix C for a discussion of matrix mathematics.

described by the unitary matrix of the perfect waveplate multiplied by a scalar to account for the Fresnel transmission coefficients at normal incidence.

Often, the eigenvectors and eigenvalues* of a matrix are useful. In a diagonal matrix, the two diagonal elements are the eigenvalues, and the eigenvectors are $\begin{pmatrix}1\\0\end{pmatrix}$ and $\begin{pmatrix}0\\1\end{pmatrix}$. The eigenvectors of a polarizing device are those inputs for which the state of polarization is not changed by the device.

In general, if the eigenvectors are orthogonal, there is a transform that converts the basis set to one in which the matrix is diagonal.

Example

Rotator

It can be shown that the eigenvectors of a rotator are right- and left-circular polarization. The eigenvectors are the two states of circular polarization. What are the eigenvalues? We can find them through

$$\tau_{rhc}\begin{pmatrix}1\\j\end{pmatrix} = \begin{pmatrix}\cos\zeta & -\sin\zeta\\ \sin\zeta & \cos\zeta\end{pmatrix}\begin{pmatrix}1\\j\end{pmatrix}$$

$$\tau_{rhc}\begin{pmatrix}1\\j\end{pmatrix} = \begin{pmatrix}\cos\zeta - j\sin\zeta\\ \sin\zeta + j\cos\zeta\end{pmatrix} = \begin{pmatrix}e^{-j\zeta}\\ je^{-j\zeta}\end{pmatrix} = e^{j\zeta}\begin{pmatrix}1\\j\end{pmatrix}. \tag{6.139}$$

Repeating the analysis for LHC polarization, we find

$$\tau_{rhc} = e^{j\zeta} \qquad \tau_{lhc} = e^{-j\zeta}. \tag{6.140}$$

Thus the effect of a rotator is to change the relative retardation of the two states of circular polarization. The reader may wish to verify that a sum of these two states, with equal amplitudes, but different retardations results in linear polarization with the appropriate angle.

Example

Circular Polarizer

Here we consider a device consisting of two quarter-wave plates with axes at 90° to each other, and a polarizer between them at 45°. The Jones matrix is

$$\mathcal{J} = \mathcal{Q}_{90}\mathcal{P}_{45}\mathcal{Q} =$$

$$\begin{pmatrix}j & 0\\ 0 & 1\end{pmatrix}\frac{1}{2}\begin{pmatrix}1 & -1\\ -1 & 1\end{pmatrix}\begin{pmatrix}1 & 0\\ 0 & j\end{pmatrix} = \frac{1}{2}\begin{pmatrix}j & 1\\ -1 & j\end{pmatrix}, \tag{6.141}$$

where we have simply exchanged x and y coordinates for the $\mathcal{Q}_{90}$ matrix and used Equation 6.122 for the polarizer. The eigenvectors and eigenvalues are

$$\mathbf{E}_1 = \frac{1}{\sqrt{2}}\begin{pmatrix}1\\ i\end{pmatrix} = \mathbf{E}_{RHC} \qquad \mathbf{E}_2 = \frac{1}{\sqrt{2}}\begin{pmatrix}1\\ -i\end{pmatrix} = \mathbf{E}_{LHC}$$

$$v_1 = 1 \qquad v_2 = 0. \tag{6.142}$$

This device is a perfect circular polarizer that passes RHC, and blocks LHC.

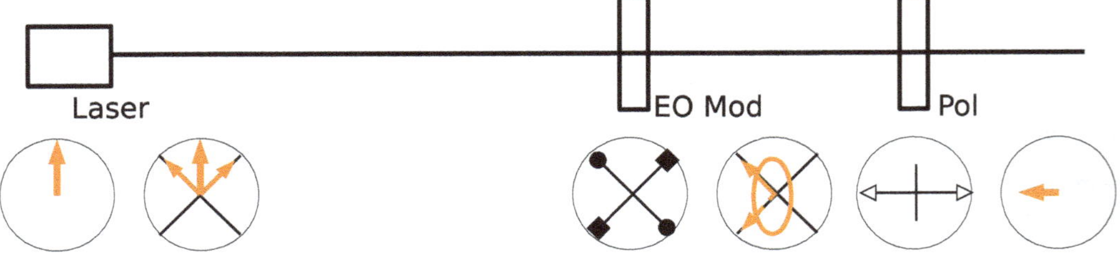

FIGURE 6.30
Amplitude modulator. An electro-optical modulator and a polarizer are combined to modulate the transmission of vertically polarized laser light as a voltage is applied to the crystal. Note: The lines with open triangles represent the "pass" axis of the polarizer. The line with circles represents the polarization state of the modulator with a positive phase in the eigenvalue and the one with squares represents the axis with a negative phase. The amount of phase is proportional to the voltage applied to the crystal.

Example

Two Polarizers

Consider two perfect polarizers, one oriented to pass $\hat{\mathbf{x}}$ polarization, and one rotated from that position by $\zeta = 10°$. The product of the two matrices is

$$\mathcal{M} = \mathcal{R}_{10}\mathcal{P}\mathcal{R}_{10}^{\dagger}\mathcal{P} = \begin{pmatrix} \cos^2\zeta & 0 \\ \cos\zeta\sin\zeta & 0 \end{pmatrix}. \tag{6.143}$$

The eigenvectors are at $\zeta_1 = 10°$ with a $\cos\zeta_1$ eigenvalue, and at $\zeta_2 = 90°$ with a zero eigenvalue. This is as expected. For light entering at 10°, the first polarizer passes the vertical component, $\cos\zeta$, and the second polarizer passes the 10° component of that with $T = \cos^2(10°)$. For horizontal polarization, no light passes the first polarizer. As is easily verified, the output is always polarized at 10°. One can use these two non-orthogonal basis vectors in describing light, although the computation of components is not a straightforward projection as it is with an orthogonal basis set. Although problems of this sort arise frequently, in practice, these basis sets are seldom used.

6.6.5 Applications

We consider two examples of the use of Jones matrices in applications to optical systems.

6.6.5.1 Amplitude Modulator

As mentioned in Section 6.5.2.2, an electro-optical modulator can be used to modulate the amplitude of an optical beam. Suppose that we begin with x-polarized laser light, and pass it through a modulator crystal rotated at 45°. We then pass the light through a horizontal polarizer as shown in Figure 6.30. The vector output is

$$\mathbf{E}_{out} = \mathcal{P}_y\mathcal{R}_{45}\mathcal{M}(V)\mathcal{R}_{45}^{\dagger}\begin{pmatrix} 1 \\ 0 \end{pmatrix}, \tag{6.144}$$

Eq. 6.86: $\delta\phi = \pi\frac{V}{V_\pi}$.

where $\mathcal{M}(V)$ is a waveplate with phase given by Equation 6.86;

$$\mathcal{M} = \begin{pmatrix} e^{j\pi V/(2V_\pi)} & 0 \\ 0 & e^{-j\pi V/(2V_\pi)} \end{pmatrix}, \tag{6.145}$$

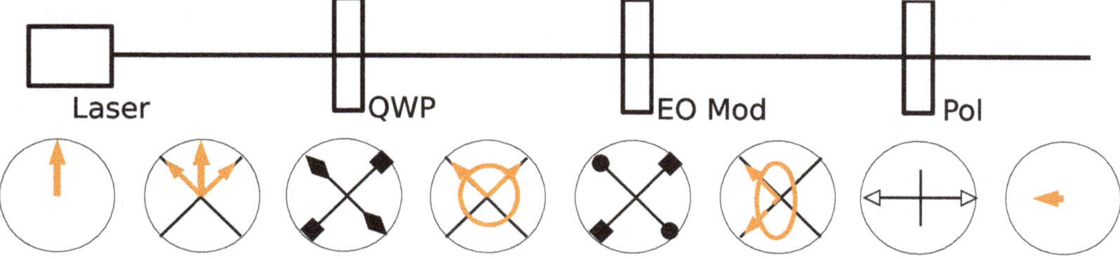

FIGURE 6.31
Modified amplitude modulator. With a quarter-wave plate, the DC voltage bias is no longer needed. Note: The diamond tips on the QWP axis represent the direction that is shifted by a quarter wave.

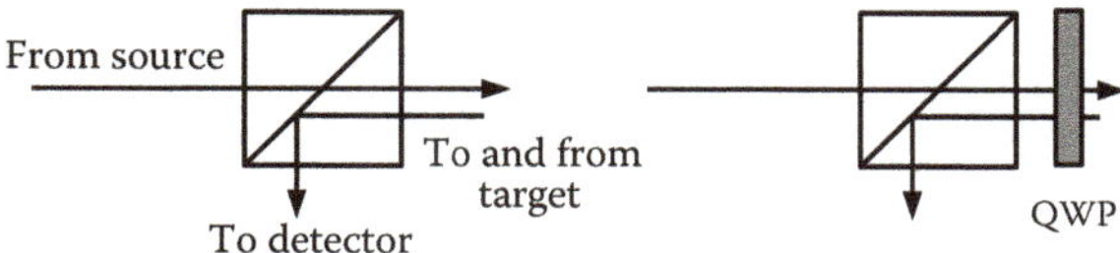

FIGURE 6.32
The optical T/R switch. We want to transmit as much light as possible through the beam-splitter and detect as much as possible on reflection.

where

- V is the applied voltage
- V_π is the voltage required to attain a phase difference of π radians

When $V = 0$, the matrix, $\mathcal{M}$ is the identity matrix, and

$$\mathbf{E}_{out} = \mathcal{P}_y \mathcal{R}_{45} \mathcal{R}_{45}^\dagger \begin{pmatrix} 1 \\ 0 \end{pmatrix} = \mathcal{P}_y \begin{pmatrix} 1 \\ 0 \end{pmatrix} = \begin{pmatrix} 0 \\ 0 \end{pmatrix}, \tag{6.146}$$

and the transmission is $T = 0$. When the voltage is $V = V_\pi$ the device becomes a half-wave plate and $T = 1$. Choosing a nominal voltage, $V_{DC} = V_\pi/2$, the modulator will act as a quarter-wave plate, and the nominal transmission will be $T = 1/2$. One can then add a small signal, V_{AC}, which will modulate the light nearly linearly. The problem is the need to maintain this quarter-wave DC voltage. Instead, we can modify the system to include a real quarter-wave plate in addition to the modulator as shown in Figure 6.31. The new matrix equation is

$$\mathbf{E}_{out} = \mathcal{P}_y \mathcal{R}_{45} \mathcal{Q} \mathcal{M}(V) \mathcal{R}_{45}^\dagger \begin{pmatrix} 1 \\ 0 \end{pmatrix}. \tag{6.147}$$

Looking at the inner two matrices

$$\mathcal{Q}\mathcal{M}(V) = \begin{pmatrix} e^{j\pi V/(2V_\pi)+\pi/4} & 0 \\ 0 & e^{-j\pi V/(2V_\pi)-\pi/4} \end{pmatrix}, \tag{6.148}$$

and now, at $V = 0$ this matrix is that of the quarter-wave plate and $T = 1/2$. As V becomes negative, the transmission drops and as V becomes positive, it increases. Now modulation is possible with only the AC signal voltage.

Use a QWP to bias the modulator to $T = 1/2$ at $v = 0$, avoiding the need for high DC voltage.

6.6.5.2 Transmit/Receive Switch (Optical Circulator)

Next, we consider a component for a coaxial laser radar or reflectance confocal microscope, as shown in the left panel of Figure 6.32. Light from a source travels through a beam-splitting cube to a target. Some of the light is reflected, and we want to deliver as much of it as possible to the detector. The first approach is to recognize that $T + R = 1$ from Equation 6.59. Then the light delivered to the detector will be

$$(1 - R)\, F_{target} R, \tag{6.149}$$

where F_{target} is the fraction of the light incident on the target that is returned to the system. The goal is to maximize $(1 - R)\, R$. Solving $d\left[(1 - R)\, R\right]/dR = 0$, we obtain a maximum at $R = 1/2$ where $(1 - R)\, R = 1/4$, which is a loss of 6 dB. However, the conservation law only holds for a given polarization. We have already seen polarization-splitting cubes, where T_p and R_s both approach unity. All that is needed is a quarter-wave plate in the target arm, as shown in the right panel. Now, we need to write the above equation for fields in Jones matrices as

$$\mathcal{J}_{tr} = \mathcal{R}_{pbs}\mathcal{Q}_{45}\mathcal{F}_{target}\mathcal{Q}_{45}\mathcal{T}_{pbs}. \tag{6.150}$$

If the target maintains polarization, $\mathcal{F}_{target}$ is a scalar, $\mathcal{F}_{target} = \sqrt{F}$, F is the fraction of the incident power returned by the target, and if the input is $\hat{\mathbf{p}}$ polarized, the output is

$$\mathcal{J}_{tr}\hat{\mathbf{x}} = \sqrt{F}\mathcal{R}_{pbs}\mathcal{Q}_{45}\mathcal{Q}_{45}\mathcal{T}_{pbs}\hat{\mathbf{p}} = \sqrt{F}\mathcal{R}_{pbs}\mathcal{H}_{45}\mathcal{T}_{pbs}\hat{\mathbf{p}}. \tag{6.151}$$

Assuming 2% insertion loss and a leakage of the wrong polarization of 5%,

$$\mathcal{J}_{tr}\hat{\mathbf{x}} =$$

$$\sqrt{F}\begin{pmatrix}\sqrt{0.05} & \\ 0 & \sqrt{0.98}\end{pmatrix}\begin{pmatrix}0 & -1\\ -1 & 0\end{pmatrix}\begin{pmatrix}\sqrt{0.98} & \\ 0 & \sqrt{0.05}\end{pmatrix}\begin{pmatrix}1\\ 0\end{pmatrix} =$$

$$\sqrt{F}\begin{pmatrix}0\\ 0.98\end{pmatrix}. \tag{6.152}$$

Thus, the output is S-polarized, and the loss is only about $1 - 0.98^2$ or 4%. The Jones formulation of the problem allows us to consider alignment, extinction, and phase errors of the components in analyzing the loss budget. In carbon dioxide laser radars, the beam-splitting cube is frequently replaced by a Brewster plate.

In Practice

A transmit/receive switch is an important part of any "monostatic" optical system, where transmission and reception occur along the same lines. When the received light is distinguishable from the transmitted by its wavelength (such as in fluorescence microscopy), only a dichroic mirror is needed. If the reflecting target is a long distance away, it may be possible to switch from transmitting to receiving mode by moving parts or electro-optical devices. In other cases, the pupil of the system may be split into transmitting and receiving areas. However, if none of these is possible, then the transmit/receive switch shown here is usually the best solution.

Often practical constraints dictate that the quarter-wave plate be placed between the beamsplitter and some optics. For example, in a laser radar, the telescope aperture is often many centimeters in diameter, and such a large quarter-wave plate would be impractical. In a microscope there is likely to be insufficient space between the objective and object for a quarter-wave plate. In such cases, one must be very careful about reflections from optics beyond the quarter–wave plate.

Finally, it must be noted that if the target depolarizes the light, as a highly scattering medium might, this configuration is not effective.

Take-Away

Jones Calculus:

The Jones calculus provides a good bookkeeping method for keeping track of polarization in complicated systems.

- Two-dimensional vectors specify the state of polarization.
- Usually we define the two elements of the vector to represent orthogonal states of linear polarization.
- Sometimes we prefer the two states of circular polarization, or some other orthogonal basis set.
- Each polarizing device can be represented as a simple two-by-two matrix.
- Matrices can be transformed from one coordinate system to another using a unitary transform.
- Matrices of multiple elements can be cascaded to describe complicated polarizing systems.
- The eigenvectors of the matrix provide a convenient coordinate system for understanding the behavior of a polarizing device or system.
- The eigenvectors need not be orthogonal.

6.7 Partial Polarization

Many light sources, including most natural ones, are in fact not perfectly polarized. Thus, we cannot write a Jones vector, or solve problems involving such light sources using Jones matrices alone. In cases where the first device in a path is assumed to be a perfect linear polarizer, we have started with the polarized light from that element, which is half the incident light but polarized, and have been quite successful in solving interesting problems. For example, we used this approach to the first polarizer in the Maltese Cross problem in Figure 6.27. However, in many cases we want a mathematical formulation that permits us to consider partial polarization explicitly. For example, the incident light might be partially polarized or the first element might not be a perfect polarizer. Although the Stokes vectors, which address partial polarization, preceded the Jones vectors in history, it is convenient to begin our analysis by an extension of the Jones matrices, called coherency matrices, and then derive the Stokes vectors.

6.7.1 Coherency Matrices

We have used the inner product of the field with itself, to obtain the scalar irradiance,

$$\mathbf{E}^\dagger\mathbf{E}.$$

We now consider the outer product,*

$$\mathbf{E}\mathbf{E}^\dagger, \tag{6.153}$$

which is a two-by-two matrix,

$$\begin{pmatrix} E_x \\ E_y \end{pmatrix} \begin{pmatrix} E_x^* & E_y^* \end{pmatrix} = \begin{pmatrix} E_xE_x^* & E_xE_y^* \\ E_yE_x^* & E_yE_y^* \end{pmatrix}. \tag{6.154}$$

*See Section C.4 in Appendix C for a definition of the outer product.

In cases where the fields are well-defined, this product is not useful, but in cases where the polarization is random, the expectation value of this matrix, called the **coherency matrix**, holds information about statistical properties of the polarization. Although the coherency matrices here are derived from Jones matrices, the idea had been around some years earlier [250]. The terms of

Coherency matrix.

$$\mathcal{C} = \left\langle \mathbf{E}\mathbf{E}^\dagger \right\rangle = \begin{pmatrix} \left\langle E_x E_x^* \right\rangle & \left\langle E_x E_y^* \right\rangle \\ \left\langle E_y E_x^* \right\rangle & \left\langle E_y E_y^* \right\rangle \end{pmatrix} \tag{6.155}$$

provide four statistical parameters. These include two real numbers: the amount of irradiance or power having x polarization,

a indicates the amount of power having $\hat{\mathbf{x}}$ polarization.

$$a = \left\langle E_x E_x^* \right\rangle, \tag{6.156}$$

and the amount having y polarization,

c indicates the amount of power having $\hat{\mathbf{y}}$ polarization.

$$c = \left\langle E_y E_y^* \right\rangle, \tag{6.157}$$

b indicates correlation of $\hat{\mathbf{x}}$ and $\hat{\mathbf{y}}$ polarizations.

and a complex number indicating the cross-correlation between the two,

$$b = \left\langle E_x E_y^* \right\rangle. \tag{6.158}$$

A positive real part of b indicates that the $\hat{\mathbf{x}}$ and $\hat{\mathbf{y}}$ components are in phase with each other leading to 45° polarization. Negative real b -45° polarization. The imaginary part of b indicates circular polarization with a positive value for RHC.

For example, for the polarization state

$$\begin{pmatrix} 1 \\ 0 \end{pmatrix}, \tag{6.159}$$

$a = 1$, $b = c = 0$, and we see that the light is completely x-polarized. For

$$\frac{1}{\sqrt{2}} \begin{pmatrix} 1 \\ 1 \end{pmatrix}, \tag{6.160}$$

$a = b = c = 1$, and the light is polarized at 45°.

In another example, right-circular polarization is given by

$$\frac{1}{\sqrt{2}} \begin{pmatrix} 1 \\ j \end{pmatrix}, \tag{6.161}$$

so $a = c = 1$ and $b = j$.

Because the equation for a device described by the Jones vector $\mathcal{J}$ acting on an input field $\mathbf{E}_{in}$:

$$\mathbf{E}_{out} = \mathcal{J}\mathbf{E}_{in}, \tag{6.162}$$

and correspondingly*

$$\mathbf{E}_{out}^\dagger = \mathbf{E}_{in}^\dagger \mathcal{J}^\dagger, \tag{6.163}$$

the equation for the coherency matrix of the output is obtained from

$$\left\langle \mathbf{E}_{out}\mathbf{E}_{out}^\dagger \right\rangle = \left\langle \mathcal{J}\mathbf{E}_{in}\mathbf{E}_{in}^\dagger \mathcal{J}^\dagger \right\rangle. \tag{6.164}$$

*See Appendix C for a discussion of matrix mathematics.

This equation allows for the device Jones matrix $\mathcal{J}$ to be a random variable, as it would be in the case of scattered light from moving particles. It also allows $\mathcal{J}$ to be a function of wavelength with a broad-band light source. However, in the most common cases, $\mathcal{J}$ is a constant, and then the expectation value just applies to the field matrices and

The effect of a device with matrix $\mathcal{J}$ on the coherency matrix.

$$\mathcal{C}_{out} = \mathcal{J}\mathcal{C}\mathcal{J}^{\dagger}. \tag{6.165}$$

We will consider two examples. In the first, the polarizing device is an air–water interface, which is constant given by the Fresnel coefficients and the incident light is unpolarized. We can use Equation 6.165. In the second, the device is a waveplate with a Jones matrix that depends on wavelength, but the incident light has a broad range of wavelengths, even though it may be perfectly polarized. In this case, because the Jones matrix of the device is variable, we must include it in the expectation value and use Equation 6.164.

Example

What is the coherency matrix of sunlight reflected from a smooth water surface?

Direct sunlight is nearly unpolarized, so $a = c = 1$ and $b = 0$. Reflecting from the surface of water,

$$\mathcal{C}_{out} = \begin{pmatrix} \rho_p & 0 \\ 0 & \rho_s \end{pmatrix} \begin{pmatrix} 1 & 0 \\ 0 & 1 \end{pmatrix} \begin{pmatrix} \rho_p^* & 0 \\ 0 & \rho_s^* \end{pmatrix} = \begin{pmatrix} R_p & 0 \\ 0 & R_s \end{pmatrix}. \tag{6.166}$$

Because $R_s > R_p$, this light is more strongly S-polarized. Because the water surface is horizontal, this S-polarized light is horizontal. The sunlight will still be at least partially randomly polarized, but with a preference for horizontal.

Example

We consider light incident on a half-wave plate at 45 deg. We know from Section 6.5.2.1 that we expect the output polarization to be reflected to −45 deg. However, our incident light is uniformly distributed across a wide spectrum. Specifically, the center wavelength is at the argon ion laser green wavelength of $\lambda_0 = 514$ nm, from $\lambda_0 - 50$ nm to $\lambda_0 + 50$ nm. Furthermore, we may use a $\frac{1}{2}\lambda$ or a $\frac{5}{2}\lambda$, and from Section 6.5.2.1 we know these issues will affect the phase. How much?

First, let's look at the coherency matrix of the incident light.

$$\mathcal{C}_{in} = \mathbf{E}_{in}\mathbf{E}_{in}^{\dagger} = \begin{pmatrix} 0.5 & 0.5 \\ 0.5 & 0.5 \end{pmatrix}$$

The Jones matrix of the waveplate is

$$\mathcal{H} = \begin{pmatrix} e^{j\ d\phi/2} & 0 \\ 0 & e^{j\ d\phi/2} \end{pmatrix} \qquad d\phi = \pi.$$

Let's quickly check the output for narrow-band light at the design wavelength, λ_0.

$$\mathcal{C}_{out} = \begin{pmatrix} 0.5 & -0.5 \\ -0.5 & 0.5 \end{pmatrix},$$

as expected. Only the sign of b has changed, indicating that the light is polarized at -45 deg. We'll record the results in the table below.

Next, let's choose a wavelength, $\lambda_!$, other than the design one. The optical path difference for the waveplate is

$$OPD = \pi \times \lambda_0 2\pi = \frac{\lambda_0}{2}$$

and the phase difference at λ_1 is

$$d\phi = 2\pi \frac{OPD}{\lambda_1} = \pi \frac{\lambda_0}{\lambda_1}. \tag{6.167}$$

For our shortest wavelength, $\lambda_1 = \lambda_0 - 50\,\text{nm} = 464\,\text{nm}$ we compute

$$C_{out} = \mathcal{H} C_{in} \mathcal{H}^\dagger,$$

and show the results in the third row of our table. Although b is complex, indicating some polarization other than linear, $|b|$ is unchanged. The light is perfectly polarized, but not at the desired polarization.

Now, let's consider the broad-band input. We use Equation 6.167 for each wavelength in turn. Then according to Equation 6.164, we compute the output coherency matrix for each wavelength and average them. If the spectrum were not uniform we'd do a weighted average. The results are in the fourth line of the table. There is a slight imaginary component to b, and there is about a 2% loss in $|b|$ indicating some depolarization, but the result is "pretty good."

Now, let's try the 5/2 waveplate. The only change is that Equation 6.167 becomes

$$d\phi = 2\pi \frac{OPD}{\lambda_1} = 5\pi \frac{\lambda_0}{\lambda_1}.$$

The final line in the table shows that this result is not so good. The light is strongly depolarized.

	a	c	b	$\lvert b \rvert$
Input	0.5	0.5	0.5	0.5
Narrow–Band Output	0.5	0.5	-0.5	0.5
464 nm Output	0.5	0.5	-0.47-0.27j	0.5
Broad-Band Output	0.5	0.5	-0.49-0.005j	0.49
Broad-Band 5/2	0.5	0.5	-0.32	0.32

It may be helpful to look at the averaging process graphically. Figure 6.33 shows the upper left element of $\mathcal{H}$ in the complex plane. The larger the band of wavelengths, the lower the average value of $|b|$. We note that because the band is symmetric about the design wavelength, the average phase shift is still correct.

Take-Away

Coherency Matrix:

The elements of the coherency matrix provide a statistical description of randomly polarized light. Such an approach is needed when dealing with unpolarized or partially polarized light. In the next section we will consider linear combinations of these parameters that provide an alternative but equivalent description. The effect of a device on a coherency matrix is given by Equation 6.164, or 6.165.

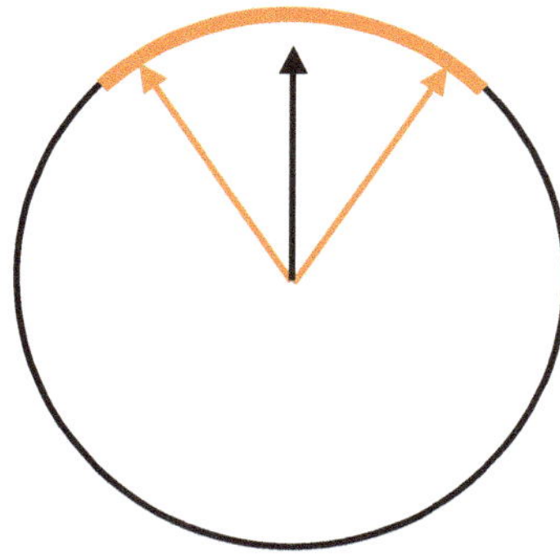

FIGURE 6.33
Matrix element of $\mathcal{H}$. As the wavelength changes, the complex number sweeps along the unit circle. The average (black line) is smaller than 1.

6.7.2 Stokes Vectors and Mueller Matrices

The Stokes parameters make the analysis of the components of the coherency matrix more explicitly linked to an intuitive interpretation of the state of polarization. Various notations are used for these parameters. We will define the **Stokes vector** in terms of the first vector in

Stokes Vector Defined.

$$\begin{pmatrix} I \\ M \\ C \\ S \end{pmatrix} = \begin{pmatrix} S_0 \\ S_1 \\ S_2 \\ S_3 \end{pmatrix} = \begin{pmatrix} I \\ Q \\ U \\ V \end{pmatrix} = \begin{pmatrix} a+c \\ a-c \\ b+b^* \\ (b-b^*)/j \end{pmatrix}. \tag{6.168}$$

The second and third vectors in this equation are used by many other authors. The fourth defines the Stokes parameters in terms of the elements of the coherency matrix. Obviously I is the total power. The remaining three terms determine the type and degree of polarization. All vary from $-I$ to $+I$. Specifically M determines the preference for x vs. y, C the preference for $+45°$ vs. $-45°$, and S the preference for RHC vs. LHC. Thus, in the notation of Stokes vectors,

Sample Stokes Vectors

$$\mathbf{E}_x = \begin{pmatrix} 1 \\ 1 \\ 0 \\ 0 \end{pmatrix} \quad \mathbf{E}_y = \begin{pmatrix} 1 \\ -1 \\ 0 \\ 0 \end{pmatrix} \quad \mathbf{E}_{45} = \begin{pmatrix} 1 \\ 0 \\ 1 \\ 0 \end{pmatrix}$$

$$\mathbf{E}_{-45} = \begin{pmatrix} 1 \\ 0 \\ -1 \\ 0 \end{pmatrix} \quad \mathbf{E}_{RHC} = \begin{pmatrix} 1 \\ 0 \\ 0 \\ 1 \end{pmatrix} \quad \mathbf{E}_{LHC} = \begin{pmatrix} 1 \\ 0 \\ 0 \\ -1 \end{pmatrix}.$$

$$\mathbf{E}_{unpolarized} = \begin{pmatrix} 1 \\ 0 \\ 0 \\ 0 \end{pmatrix} \quad \mathbf{E}_{sun-reflected-from-water} = \begin{pmatrix} R_s + R_p \\ R_s - R_p \\ 0 \\ 0 \end{pmatrix}, \tag{6.169}$$

The **degree of polarization** (DOP) is given by

Degree of polarization (DOP) defined.

$$V = \frac{\sqrt{M^2 + C^2 + S^2}}{I} \le 1, \tag{6.170}$$

where $V = 1$ represents complete polarization, and a situation in which the Jones calculus is completely appropriate, $V = 0$ represents random polarization, and intermediate values indicate partial polarization. A mixture of a large amount of polarized laser light with a small amount of light from a tungsten source would produce a large value of V.

Example

Let's go back to our example on the coherency matrices and compute the Stokes parameters using Equations 6.169, and 6.170.

	I	M	C	S	V
Input	1	0	1	0	1
Narrow-Band Output	1	0	−1	0	1
464 nm Output	1	0	−0.94	0.33	1
Broad-Band Output	1	0	0.98	0.01	0.98
Broad-Band 5/2	1	0	−0.64	0.02	0.65

We notice that $I = 1$ in all cases, so there is no loss. This is expected for a waveplate; it only affects the phase, and not the amplitude. The degree of polarization, $V = 1$, for the narrow–band cases, even if the wavelength is not λ_0. The polarization is wrong, because $C \neq -1$ indicating imperfect −45 deg polarization and $S = 0.33$ indicating some preference for RHC polarization. The 1/2 waveplate is quite effective, only losing about 2% of the desired polarization, while the 5/2 waveplate is only 65% polarized.

Take-Away

Stokes Vectors and Coherency Matrix:
The Stokes vectors provide the same information as the elements of the coherency matrix, but in a form that is perhaps more intuitive. The Degree of Polarization (DOP) is a commonly used measure of polarized light.

6.7.3 Mueller Matrices

Sample Mueller Matrices.

Stokes vectors, developed by Stokes[218] in 1852, describe states of partial polarization. Mueller matrices, attributed to Hans Mueller[163], describe devices. The Mueller matrices are quite complicated, and only a few will be considered. Polarizers are

$$\mathcal{P}_x = \frac{1}{2}\begin{pmatrix} 1 & 1 & 0 & 0 \\ 1 & 1 & 0 & 0 \\ 0 & 0 & 0 & 0 \\ 0 & 0 & 0 & 0 \end{pmatrix} \qquad \mathcal{P}_y = \frac{1}{2}\begin{pmatrix} 1 & -1 & 0 & 0 \\ -1 & 1 & 0 & 0 \\ 0 & 0 & 0 & 0 \\ 0 & 0 & 0 & 0 \end{pmatrix}. \tag{6.171}$$

Obviously not all parameters in the Mueller matrices are independent, and it is mathematically possible to construct matrices that have no physical significance. Mueller matrices can be cascaded in the same way as Jones matrices. The matrix

$$\mathcal{P}_r = \frac{1}{2}\begin{pmatrix} 1 & 0 & 0 & 0 \\ 0 & 0 & 0 & 0 \\ 0 & 0 & 0 & 0 \\ 0 & 0 & 0 & 0 \end{pmatrix} \tag{6.172}$$

describes a device that randomizes polarization. While this seems quite implausible at first, it is quite possible with broadband light and rather ordinary polarizing components, as shown by the example above.

Take-Away

Stokes Vectors and Mueller Matrix:

The Stokes parameters are intuitively useful to describe partial polarization, but the Mueller matrices are often considered unintuitive, in comparison to the Jones matrices, in which the physical behavior of the device is easy to understand from the mathematics. Many people find it useful to use the formulation of the coherency matrices to compute the quantities a, b, and c, and then to compute the Stokes vectors, rather than use the Mueller matrices. The results will always be identical.

In the case of highly polarized light and well-characterized devices, it normally makes most sense to use Jones matrices in computation. For partially polarized light, either of the two approaches in the above paragraph can be used. Mueller matrices can be measured experimentally by measuring the state of the output polarization for various inputs, and the Mueller matrices can be used to represent depolarization more easily than Jones matrices. In the end, the choice is flexible and may be guided in part by the user's experience and intuition.

6.7.4 Poincaré Sphere

The Poincaré sphere [181] can be used to represent the Stokes parameters. If we normalize the Stokes vector, we only need three components

$$\frac{1}{I}\begin{pmatrix} M \\ C \\ S \end{pmatrix}. \tag{6.173}$$

Now, plotting this three-dimensional vector in Cartesian coordinates, we locate the polarization within the volume of a unit sphere. If the light is completely polarized ($V=1$) in Equation 6.170, the point lies on the surface. If the polarization is completely random ($V=0$), the point lies at the center. For partial polarization, it lies inside the sphere, the radial distance telling the degree of polarization, V, and the nearest surface point telling the type of polarization. The Poincaré sphere is shown in Figure 6.34.

Eq. 6.170: $V = \frac{\sqrt{M^2+C^2+S^2}}{I} \leq 1.$

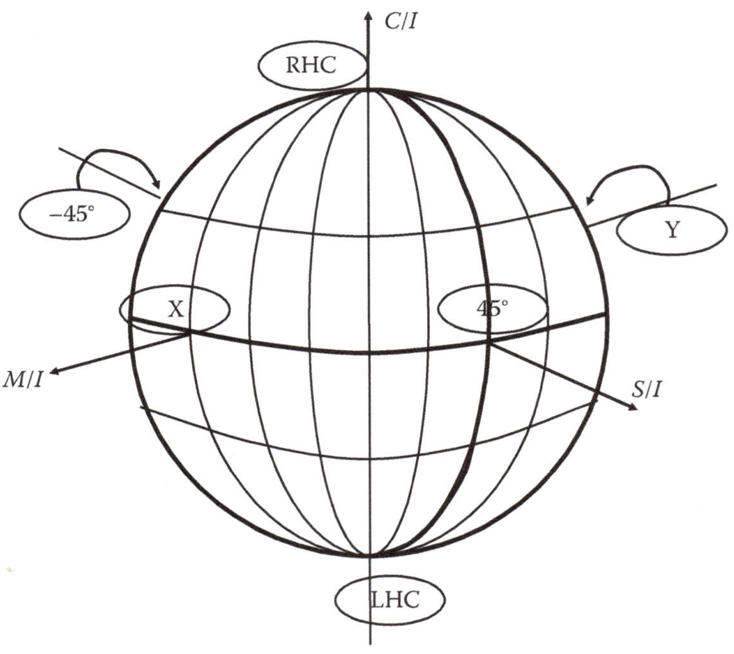

FIGURE 6.34
Poincaré sphere. The polarization state is completely described by a point on the sphere. A surface point denotes complete polarization, and in general the radius is the degree of polarization.

Take-Away

Summary of Partial Polarization:

- Most light observed in nature is partially polarized.
- Partially polarized light can only be characterized statistically.
- Coherency matrices can be used to define partially polarized light and are consistent with the Jones calculus.
- The Stokes vector contains the same information as the coherency matrix.
- Mueller matrices represent devices and operate on Stokes vectors, but are somewhat unintuitive.
- Many partial-polarization problems can best be solved by using the Jones matrices with coherency matrices and interpreting the results in terms of Stokes vectors.
- The Poincaré sphere is a useful visualization of the Stokes vectors.

Problems

6.1 Polarizer in an Attenuator

Polarizers can be useful for attenuating the level of light in an optical system. For example, consider two polarizers, with the second one aligned to pass vertically polarized light. The first can be rotated to any desired angle, with the angle defined to be zero when the two polarizers are parallel. The first parts of this problem can be solved quite simply without the use of Jones matrices.

6.1a What is the output polarization?

6.1b Plot the power transmission of the system as a function of angle, from 0° to 90°, assuming the incident light is unpolarized.

6.1c Plot the power transmission assuming vertically polarized incident light.

6.1d Plot the power transmission assuming horizontally polarized incident light.

6.1e Now, replace the first polarizer with a quarter-wave plate. I recommend you use Jones matrices for this part. In both cases, the incident light is vertically polarized. Plot the power transmission as a function of angle from 0° to 90°.

6.1f Repeat using a half-wave plate.

6.2 Electro-Optical Modulator

Suppose that an electro-optical modulator crystal is placed between crossed polarizers so that its principal axes are at 45° with respect to the axes of the polarizers. The crystal produces a phase change between fields polarized in the two principal directions of $\delta\phi = \pi V/V_\pi$, where V_π is some constant voltage and V is the applied voltage.

6.2a Write the Jones matrix for the crystal in the coordinate system of the polarizers.

6.2b Calculate and plot the power transmission, T, through the whole system, polarizer–crystal–polarizer, as a function of applied voltage.

6.2c Next, add a quarter-wave plate with axes parallel to those of the crystal. Repeat the calculations of Part (b).

6.2d What is the range of voltages over which the output power is linear to 1%?

6.3 Laser Cavity

Consider a laser cavity with a gain medium having a power gain of 2% for a single pass. By this we mean that the output power will be equal to the input power multiplied by 1.02. The gain medium is a gas, kept in place by windows on the ends of the laser tube, at Brewster's angle. This is an infrared laser, and these windows are ZnSe, with an index of refraction of 2.4. To make the problem easier, assume the mirrors at the ends of the cavity are perfectly reflecting.

6.3a What is the Jones matrix describing a round trip through this cavity? What is the power gain for each polarization? Note that if the round-trip multiplier is greater than one, lasing can occur, and if not, it cannot.

6.3b What is the tolerance on the angles of the Brewster plates to ensure that the laser will operate for light which is P-polarized with respect to the Brewster plates?

6.3c Now assume that because of sloppy manufacturing, the Brewster plates are rotated about the tube axis by 10° with respect to each other. What states of polarization are eigenvectors of the cavity? Will it lase?

6.4 Depolarization in a Fiber (G)

Consider light passing through an optical fiber. We will use as a basis set the eigenvectors of the fiber. The fiber is birefringent, because of imperfections, bending, etc., but there are no polarization-dependent losses. In fact, to make things easier, assume there are no losses at all.

6.4a Write the Jones matrix for a fiber of length ℓ with birefringence Δn in a coordinate system in which the basis vectors are the eigenvectors of the fiber. Note that your result will be a function of wavelength. The reader is advised to keep in mind that the answer here is a simple one.

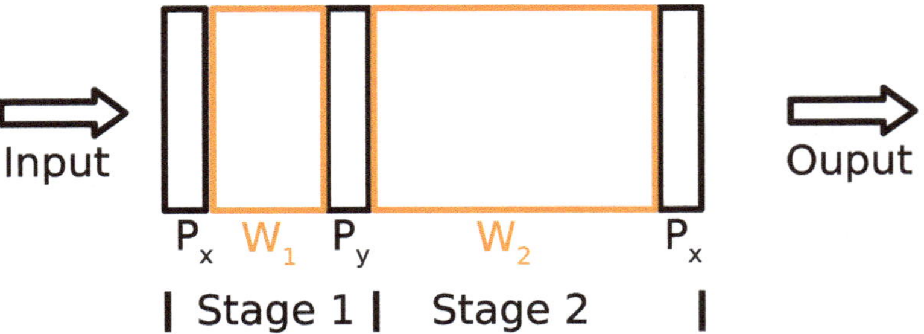

FIGURE P6.5
Figure for Problem 6.5.

6.4b Suppose that the coherency matrix of the input light in this basis set is

$$\begin{pmatrix} a & b \\ b^* & c \end{pmatrix}. \tag{6.174}$$

What is the coherency matrix of the output as a function of wavelength?

6.4c Assume now that the input light is composed of equal amounts of the two eigenvectors with equal phases, but has a linewidth $\Delta\lambda$. That is, there are an infinite number of waves present at all wavelengths from $\lambda - \Delta\lambda/2$ to $\lambda + \Delta\lambda/2$. Provided that $\Delta\lambda$ is much smaller than λ, write the coherency matrix in a way that expresses the phase as related to the center wavelength, λ, with a perturbation of $\Delta\lambda$.

6.4d Now let us try to average the coherency matrix over wavelength. This would be a hard problem, so let us approach it in the following way: Show the value of b from the coherency matrix as a point in the complex plane for the center wavelength. Show what happens when the wavelength changes, and see what effect this will have on the average b. Next, write an inequality relating the length of the fiber, the line width, and the birefringence to ensure that depolarization is "small."

6.5 Lyot Filter (NS)

Here we consider a Lyot (Pronounced like "Leo") filter, which can be used to select a wavelength band from an image. The filter consists of multiple stages as shown in Figure P6.5. The first stage consists of crossed polarizers separated by a liquid crystal layer, that can be tuned by adjusting a voltge to have a particular optical path difference (OPD) between x and y polarizations, that is rotated by 45 degrees to the coordinate system of the polarizers. The second stage is similar but with a different optical path length and with the polarizations reversed from the first stage. For some wavelength, the OPD in the liquid crystal will be an odd multiple of half that wavelength, so that light will be completely transmitted through the (assumed perfect) crossed polarizers. For other wavelengths, the transmission will be less. Because the two liquid-crystal layers have different thicknesses the wavelength behavior is different.

6.5a Write the Jones matrices for P_x, P_y, and the waveplates in terms of OPD and wavelength. Then generate the matrix for each stage. Use the following optical path differences for the two stages: $OPD_1 = 600\,\text{nm} \times 1.5$ and $OPD_2 = 600\,\text{nm} \times 3.5$.

6.5b Calculate the transmission, T_1 of the first stage for x–polarized input and T_2 for the second stage with y–polarized input. Plot these for wavelengths from 400 to 800 nm.

6.5c Calculate the transmission of the whole system for x–polarized input, and include on the same plot.

6.5d How would your results in part c be different if the input were randomly polarized.

6.6 T/R Switch

Consider the T/R switch discussed in class. Assume that the matrix for the target is the identity matrix. Assume that the polarizing beamsplitter transmits 95% of P-polarized light and 2% of S-polarized light (Irradiance fractions).

6.6a Write the Jones matrices for the components and for the complete T/R switch from source to detector.

6.6b If the input is perfectly polarized in the x direction, find the output. What is the efficiency of the switch?

6.6c Do the same calculation using coherency matrices.

6.6d Assume the input is 80% polarized in the x direction and 20% in y. What is the output coherency matrix and what is the efficiency.

6.6e Repeat for unpolarized input.

6.7 Waveplate Orientation (NS)

We have made the claim that we often don't know or care which axis of a waveplate is which. Prove this by repeating the solution with the QWP in the opposite orientation. That is, exchange the eigenvectors of the QWP matrix.

7 Interference

This chapter and the next both address the **coherent addition** of light waves. Maxwell's equations by themselves are linear in the field parameters, **E**, **D**, **B**, and **H**. Provided that we are working with linear materials, in which the relationships between **D** and **E**, and between **B** and **H**, are linear, $\mathbf{D} = n^2\epsilon_0\mathbf{E}$, and $\mathbf{B} = \mu_0\mathbf{H}$, the resulting wave equation (1.15),

$$\nabla^2\mathbf{E} = \mu\epsilon\frac{\partial^2\mathbf{E}}{\partial t^2}.$$

is also linear. This linearity means that the principle of **linear superposition** holds; the product of a scalar and any solution to the differential equations is also a solution, and the sum of any solutions is a solution. In Chapter 14, we will explore some situations where **D** is not linearly related to **E**.

We begin with a simple example of linear superposition. In Figure 7.1A, one wave is incident from the left on a dielectric interface at 45°. From Section 6.4, we know that some of the wave will be reflected and propagate downward, while some will be transmitted and continue to the right, as E_1. Another wave is introduced propagating downward from above. Some of it is reflected to the right as E_2, and some is transmitted downward. The refraction of the waves at the boundary is not shown in the figure. At the right output, the two waves are added, and the total field is

$$E = E_1 + E_2. \tag{7.1}$$

A similar result occurs as the downward waves combine.

Combining two waves from different sources is difficult. Consider, for example, two helium-neon lasers, at $\lambda = 632.8\,\text{nm}$. The frequency is $f = c/\lambda = 473.8\,\text{THz}$. To have a constant phase between two lasers for even 1 s, would require keeping their bandwidths within 1 Hz. In contrast, the bandwidth of the helium-neon laser gain line is about 1.5 GHz. As a result, most optical systems that use coherent addition derive both beams from the same source, as shown in Figure 7.1B. The **beamsplitters** $BS1$ and $BS2$, may be dielectric interfaces or other components that separate the beam into two parts. Often they will be slabs of optical substrate with two coatings, on one side to produce a desired reflection, and on the other side to inhibit unwanted reflection. More details on beamsplitters will be discussed in Sections 7.6 and 7.7. Regardless of the type of beamsplitters used, this configuration is the **Mach–Zehnder interferometer**[259;145], with which we begin our discussion of interferometry in Section 7.1, where we will also discuss a number of basic techniques for analyzing interferometers. We will also discuss the practical and very important issue of alignment, and the technique of balanced mixing, which can improve the performance of an interferometer. We will discuss applications to laser radar in Section 7.2. Ambiguities that arise in laser radar and other applications of interferometry require special treatment, and are discussed in Section 7.3. We will then turn our attention to the Michelson and the Fabry-Perot interferometers in Sections 7.4 and 7.5. Interference occurs when it is wanted, but also when it is unwanted. We will discuss the interference effects in beamsplitters, and some rules for minimizing them in Section 7.6. We will conclude with a discussion of thin-film dielectric coatings, which could be used to control the reflectance of beamsplitter faces in Section 7.7. These coatings are also used in high-reflectivity mirrors, anti-reflection coatings for lenses, and in optical filters.

Roadmap: Chapter 7

DOI: 10.1201/9781315157047-7

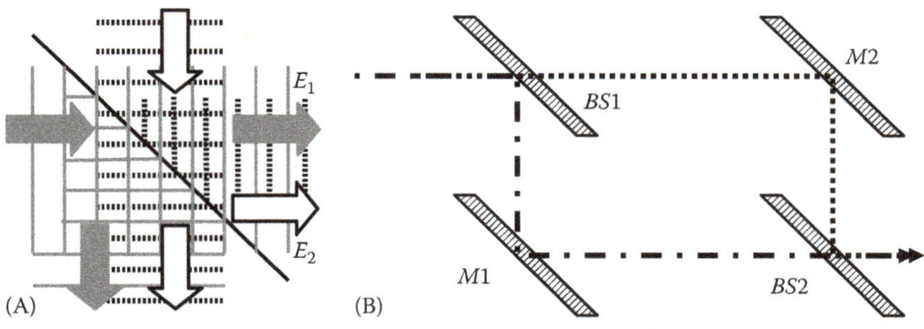

FIGURE 7.1
Coherent addition. Two waves can be recombined at a dielectric interface (A). Normally the two waves are derived from a single source, using an interferometer (B).

The distinction between the topics of this chapter, interference, and the next, diffraction, is that in the present case, we split, modify, and recombine the beam "temporally" using beamsplitters as shown in Figure 7.1B, whereas in diffraction we do so spatially, using apertures and lenses as will be seen in Figure 8.1A in Chapter 8.

Interference *vs.* Diffraction.
Fig. 8.1A

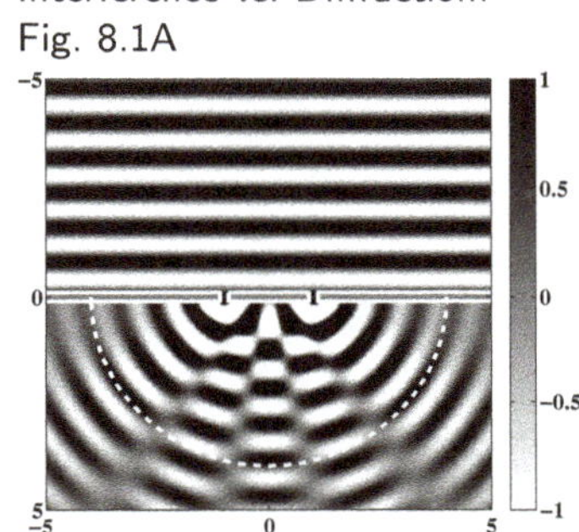

7.1 Mach–Zehnder Interferometer

The Mach–Zehnder interferometer[259;145] presents us with a good starting point, because of its conceptual simplicity. As shown in Figure 7.1B, light from a single source is first split into two separate paths. The initial beamsplitter, $BS1$, could be a dielectric interface, or other device. The mirrors can be adjusted in angle and position so that the two waves recombine at $BS2$, which is often called the **recombining beamsplitter**. We can place transparent, or partially transparent, materials in one path and measure their effect on the propagation of light. Materials with a higher index of refraction will cause this light wave to travel more slowly, thereby changing the relative phase at which it recombines with the other wave.

The Mach–Zehnder interferometer is thus useful for measuring the refractive index of materials, or the length of materials of known index. After some background, we will discuss, as an example, the optical quadrature microscope in Section 7.3.3. Later, in Section 7.2, we will discuss a modification of the Mach–Zehnder to image reflections from objects, and show how this modification can be applied to a Doppler laser radar.

7.1.1 Basic Principles

In some waves we can measure the field directly. For example, an oceanographer can measure the height of ocean waves. An acoustical engineer can measure the pressure associated with a wave. Even in electromagnetics, we can measure the electric field at radio frequencies. However, at optical frequencies it is not possible to measure the field directly. The fundamental unit we can measure is energy, Q, so with a detector of known area, A measuring for a known time, t, we can find $I = Q/(At)$, the magnitude of the Poynting vector, from Equation 1.45 which is the irradiance (Section 1.4.4)*:

Eq. 1.45: $\mathbf{S} = \mathbf{E} \times \mathbf{H}$.

$$I = \frac{|E|^2}{Z} = \frac{EE^*}{Z}. \tag{7.2}$$

For the two fields summed in Equation 7.1,

Eq. 7.1:
$E = E_1 + E_2$.

*In this chapter, we continue to use the variable name, I, for irradiance, rather than the radiometric E, as discussed in Section 1.4.4.

$$I = \frac{(E_1^* + E_2^*)(E_1 + E_2)}{Z}.$$

Expanding this equation, we obtain the most useful equation for irradiance from these two beams.

$$I = \frac{E_1^* E_1 + E_2^* E_2 + E_1^* E_2 + E_1 E_2^*}{Z}. \tag{7.3}$$

The first two terms are often called the **DC terms**, a designation that is most meaningful when the two fields are at different frequencies. In any case, these two terms vary only if the incident light field amplitude varies on one or both inputs. The last two terms have amplitudes dependent on the amplitudes of both fields and on their relative phase. We call these last two terms the **mixing terms**. If the two fields, E_1 and E_2 are at different frequencies, these terms will be AC terms at the difference frequency. In an optical detector (see Chapter 13), a current is generated in proportion to the power falling on the detector, and thus to the irradiance. In Chapter 13, we will discuss the constant of proportionality. For now, we consider the irradiances,

$$I_{mix} = \frac{E_1 E_2^*}{Z} \qquad \text{and} \qquad I_{mix}^* = \frac{E_1^* E_2}{Z}. \tag{7.4}$$

The reader who is familiar with radar will recognize these as the outputs of an RF mixer when the two inputs have different frequencies. In fact, some in the laser-radar community refer to an optical detector as an "optical mixer." The magnitude of either mixing term is

$$|I_{mix}| = |I_{mix}^*| = \sqrt{I_1 I_2}. \tag{7.5}$$

The mixing term, I_{mix} is complex, but it always appears in equations added to I_{mix}^*; their imaginary parts cancel in the sum and their real parts add, so

$$I = I_1 + I_2 + 2\sqrt{I_1 I_2}\cos(\phi_2 - \phi_1), \tag{7.6}$$

where $E_1 = \sqrt{I_1 Z} e^{j\phi_1}$, $E_1 = \sqrt{I_2 Z} e^{j\phi_2}$ and each phase is related to the optical path length, ℓ by $\phi = k\ell$.

Let us first consider adding gas under pressure to the cell of length ℓ_c, in Figure 7.2. As we do, the index of refraction will increase, and we will observe a change in the axial irradiance at the detector. What is the sensitivity of this measurement to the index of refraction? The change in optical path length will be

$$\Delta = \delta(n\ell_c), \tag{7.7}$$

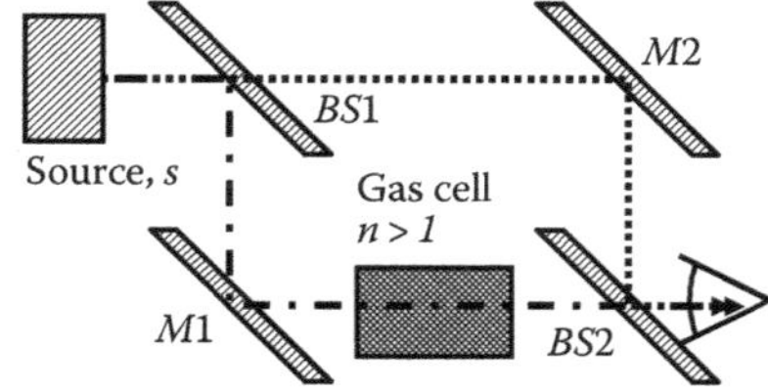

FIGURE 7.2
Mach–Zehnder interferometer. The index of refraction of the gas in the cell causes the phase to change. The index can be determined by counting the changes in the detected signal as the gas is introduced, or by examining the fringes in the interference pattern.

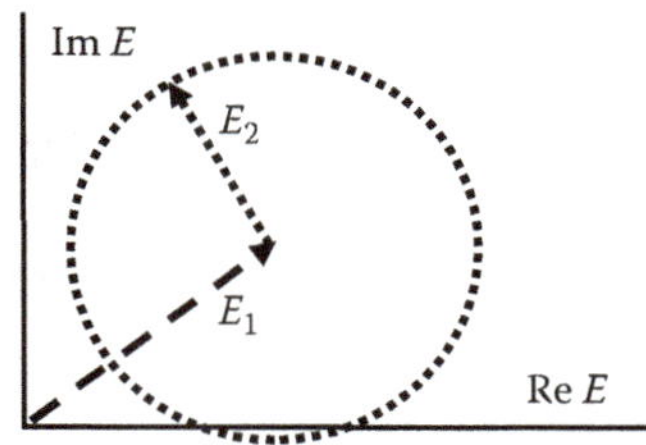

FIGURE 7.3
Complex sum. The sum of two complex numbers varies in amplitude, even when the amplitudes of the two numbers themselves are constant. The maximum contrast occurs when the two numbers are equal.

If the cell's physical length, ℓ_c, is constant, the phase change will be

$$\delta\phi_1 = k\Delta = 2\pi\frac{\Delta}{\lambda} = 2\pi\frac{\ell_c}{\lambda}\delta n. \tag{7.8}$$

The phase will change by 2π when the optical path changes by one wavelength. As the phase changes, the irradiance will change according to Equation 7.3. The coherent addition is illustrated in Figure 7.3, where the sum of two complex numbers is shown. One complex number, E_1, is shown as having a constant value. The other, E_2, has a constant amplitude but varying phase. When the two are in phase, the amplitude of the sum is a maximum, and when they are exactly out of phase, it is a minimum. The contrast will be maximized when the two irradiances are equal. Then the circle passes through the origin. If the phase change is large, we can simply count cycles of increase and decrease in the detected signal.

If the phase change is smaller, we must then make quantitative measurements of the irradiance and relate them to the phase. Beginning at $\phi_2 - \phi_1 = 0$ is problematic, because then

$$\frac{dI}{d(n\ell_c)} = 2\sqrt{I_1 I_2}\left[-\sin(\phi_2 - \phi_1)\right]\frac{2\pi}{\lambda} = 0 \qquad \phi_2 - \phi_1 = 0 \tag{7.9}$$

If we could keep the interferometer nominally at the so-called **quadrature point**,

$$\phi_2 - \phi_1 = \pi/2 \tag{7.10}$$

or 90°, then the sensitivity would be maximized.

The maximum sensitivity of an interferometer occurs at the quadrature point.

Example

For example, if $\lambda = 500\,\text{nm}$ and $\ell_c = 1\,\text{cm}$, then from Equation 7.8, $d\phi/dn = 1.25 \times 10^5$, a change in index of 10^{-6} produces a 125 mrad phase change. If $I_1 = I_2 = I/2$ then

$$\frac{dI}{d(n\ell_c)} = \frac{2\pi}{\lambda}I$$

$$\delta(n\ell_c) = \frac{\lambda}{2\pi}\frac{dI}{I}. \tag{7.11}$$

In our example,

$$\frac{dI}{dn} = I \times 2\pi\frac{\ell_c}{\lambda} = 1.25 \times 10^5 \times I, \tag{7.12}$$

and, with the index change of 10^{-6},

$$\delta I = 1.25 \times 10^5 \times I \times \delta n = 0.125 \times I; \tag{7.13}$$

the irradiance changes by about 12% of the DC value. This change is easily detected.

We can use the interferometer to detect small changes in index or length, using Equation 7.11, or

$$\delta\left(n\ell_c\right) = \ell_c \delta n + n\delta\ell_c = \frac{\lambda}{2\pi}\frac{dI}{I} \tag{7.14}$$

The practicality of this method is limited first by the need to maintain the $\phi_2 - \phi_1 = \pi/2$ condition, and second by the need to calibrate the measurement. We will discuss some alternatives in Sections 7.3.2 and 7.3.3.

Example

In the Mach–Zehnder interferometer, temperature fluctuations can cause variations in the index of refraction of air in the two arms, leading to changes in the fringe pattern. How much average temperature change between the two arms 10 cm long will cause a phase change of π radians for Helium–Neon laser light ($\lambda = 633$ nm)?

From the literature[219] (and the website,

http://emtoolbox.nist.gov/Wavelength/Ciddor.asp,

at atmospheric pressure of 101.325 kPa, and 50% relative humidity, the index of refraction at $T = 20\,°\text{C}$ is 1.000271374 and at $T = 21\,°\text{C}$ is 1.000270422. Thus,

$$\frac{dn}{dT} = \frac{1.000270422 - 1.000271374}{(21-20)\ °\text{C}} = -9.52 \times 10^{-7}\,(°\text{C})^{-1}.$$

Then the change in index is $\Delta n =$ and the change in optical path length is

$$\Delta OPL = 0.1\,\text{m} \times \Delta n = 0.1\,\text{m} \times dn/dT\Delta T.$$

For a phase change of π, $\Delta OPL = \lambda/2$. Thus,

$$\Delta T = \frac{\lambda/2}{0.1\,\text{m} \times -9.52 \times 10^{-7}\,(°\text{C})^{-1}} = -3.32\,°\text{C}.$$

The phase change of π radians associated with this change in temperature would cause the fringe pattern to changed from bright to dark. Even a fraction of a degree Celsius will cause a noticable change.

In this discussion, we have considered that the incident light consists of plane waves, or at least we have only considered the interference on the optical axis. We can often make more effective use of the interferometer by examining the interference patterns that are formed, either as we view an extended source, or as we collect light from a point source on a screen. To understand these techniques, it is useful to consider a straight-line layout for the interferometer.

Fig. 7.2:

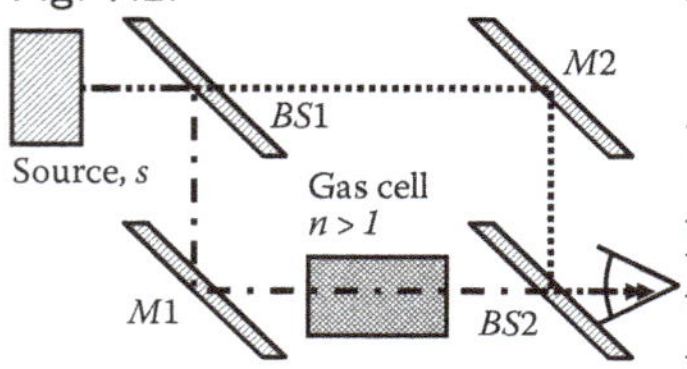

7.1.2 Straight-Line Layout

If we look into the interferometer in Figure 7.2, from one of the outputs, blocking the path that includes $M1$ and the sample, we will see $BS2$, $M2$ reflected from $BS2$, $BS1$ reflected from $M2$ and $BS2$, and finally the source, through $BS1$. The optical path will look straight,

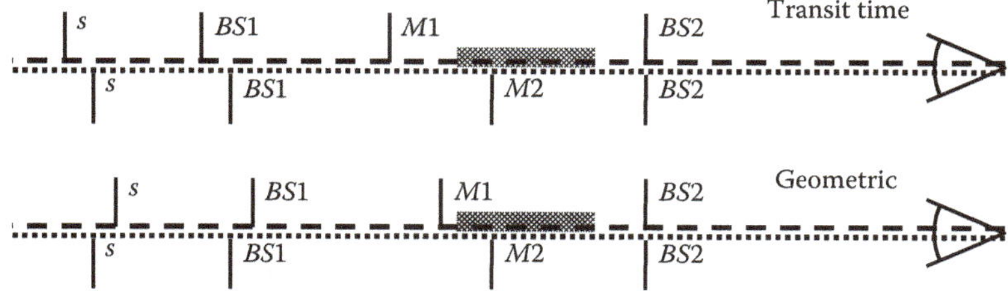

FIGURE 7.4
Straight-line layout. Looking back through the interferometer, we see the source in two places, one through each path. Although the source appears closer viewed through the path with increased index of refraction (bottom), the transit time of the light is increased (top).

and we can determine the distance to the source by adding the component distances. If we now block the other path near $M2$, we will see a similar view of the source, but this time *via* transmission through $BS2$, reflection from $M1$, and reflection from $BS1$. The apparent distance will be different than in the first case, depending on the actual distance and on the index of refraction of the sample. There are in fact two different measures of distance. One is the geometric-optics distance, determined by imaging conditions as in Equation 2.17, and the other is the OPL from Equation 1.77. Figure 7.4 shows the two paths. The part of the OPL contributed by the cell is given by

There are two important measures of distance in an interferometer; the apparent distance by geometric optics and the OPL.

Eq. 2.17:
$\frac{n}{s} = \frac{n'}{s'}$.
Eq. 1.77:
$\ell = OPL = \int_A^B n d\ell_p$.

$$\ell = \ell_c n, \tag{7.15}$$

and the contribution of the cell to the geometric-optics distance is

$$\ell_{geom} = \frac{\ell_c}{n}. \tag{7.16}$$

Eq. 7.1:
$E = E_1 + E_2$.
Eq. 7.3:
$I = \frac{E_1^* E_1 + E_2^* E_2 + E_1^* E_2 + E_1 E_2^*}{Z}$.

With light arriving via both paths, we expect interference to occur according to Equations 7.1 and 7.3. Our next problem then is to determine how changes, Δ, in OPL, vary with x and y for a given ℓ. These changes will result in a fringe pattern, the details of which will be related to ℓ, and thus to the index of refraction.

7.1.3 Viewing an Extended Source

We first consider a source that is extended in the transverse directions, that we view as consisting of individual point sources, each with a phase that is random with respect to the others, but with the same wavelength. Although the phases for different points are random, the difference in phase between the two paths, for a single point, is not random, but depends on the optical path difference for that point.

7.1.3.1 Fringe Numbers and Locations

From the viewer's point of view, the distance to a point in an extended source is

$$r = \sqrt{x^2 + y^2 + z^2}. \tag{7.17}$$

A small change, ℓ in z results in a change, Δ to the optical path r given by

$$\Delta = \frac{\ell}{\cos\theta} = \ell\sqrt{\frac{x^2 + y^2 + z^2}{z}}, \tag{7.18}$$

where θ is the angle shown in Figure 7.5.

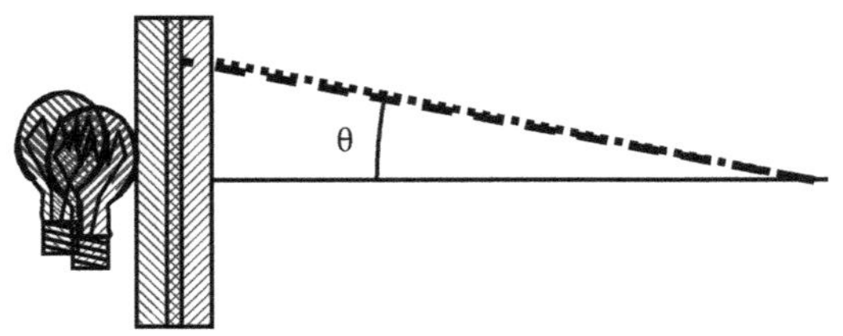

FIGURE 7.5
Viewing an extended source. The same source is seen at two different distances. The phase shift between them is determined by the differences in optical pathlength.

Paraxial approximation.

A significant simplification is possible provided that x and y are much smaller than z. We will find numerous occasions to use this **paraxial approximation** in interferometry and in diffraction. We expand the square root in Taylor's series to first order in $(x^2+y^2)/z^2$, with the result,

$$\sqrt{\frac{x^2+y^2+z^2}{z}} \approx 1 + \frac{1}{2}\frac{x^2+y^2}{z^2}, \tag{7.19}$$

and obtain

$$r \approx z + \frac{x^2+y^2}{2z} \tag{7.20}$$

or

$$\phi_1 = kr = \frac{2\pi}{\lambda}r \approx \frac{2\pi z}{\lambda} + 2\pi\frac{x^2+y^2}{2\lambda z} \qquad \frac{x^2+y^2}{2\lambda z} \ll 1 \tag{7.21}$$

With this approximation,

$$\delta\phi = \phi_2 - \phi_1,$$
$$\delta\phi = k\Delta = \frac{2\pi\Delta}{\lambda} = \frac{2\pi\ell}{\lambda}\left(1 + \frac{1}{2}\frac{x^2+y^2}{z^2}\right) \tag{7.22}$$

This quadratic phase shift means that an observer looking at the source will see alternating bright and dark circular fringes with ever-decreasing spacing as the distance from the axis increases (Figure 7.6).

We can compute the irradiance pattern, knowing the beamsplitters, the source wavelength, and the optical paths. If the irradiance of the source is I, then along each path separately, the observed irradiance will be

$$I_1 = I_0 R_1 T_2 \qquad I_2 = I_0 T_1 R_2 \tag{7.23}$$

where $R_1 = |\rho_1|^2$, R_2, T_1 and T_2 are the reflectivities and transmissions of the two beamsplitters. The fields are

$$E_1 = E_0 \rho_1 \tau_2 e^{jknz_1} \qquad E_2 = E_0 \tau_1 \rho_2 e^{jknz_2}$$

and the interference pattern is

$$I = I_0\left(R_1T_2 + T_1R_2 + 2\sqrt{R_1T_2T_1R_2}\cos\delta\phi\right) \tag{7.24}$$

The mixing term adds constructively to the DC terms when $\cos\delta\phi = 1$, or $\delta\phi = 2\pi N$,

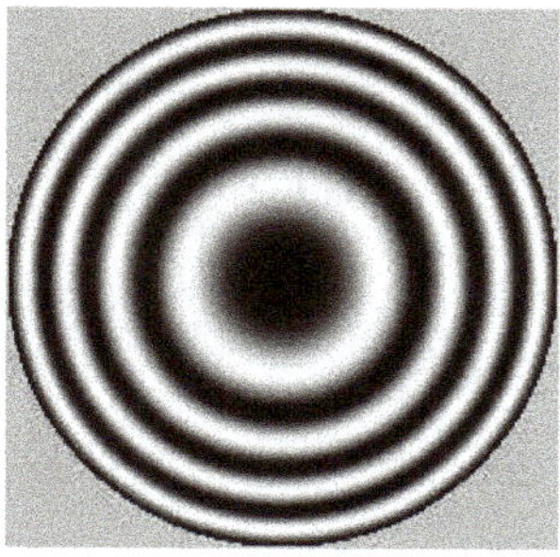

FIGURE 7.6
Fringe pattern. Interference of differently curved wavefronts results in a circular pattern.

where N is an integer that counts the number of bright fringes. From Equations 7.22 and 7.24, we see that the number of fringes to a radial distance ρ in the source is

$$N = \left(\frac{\rho}{z}\right)^2 \frac{\delta\ell}{\lambda}, \tag{7.25}$$

recalling that $\delta\ell$ represents either a change in the actual path length, ℓ_c, or in the index of refraction, n, or in a combination of the two as given by Equation 7.14. For example, if we count the number of fringes, we can compute

Eq. 7.14: $\delta(n\ell_c) = \ell_c\delta n + n\delta\ell_c = \frac{\lambda}{2\pi}\frac{dI}{I}$.

$$\ell = \lambda\left(\frac{z}{\rho}\right)^2 N, \tag{7.26}$$

with a sensitivity

$$\frac{d\ell}{\ell} = \frac{dN}{N}. \tag{7.27}$$

Deliberately introducing a large path difference, we can then detect small fractional changes. While this approach can measure the index of refraction absolutely, it is still not as sensitive to changes in the relative index as is the change in axial irradiance in Equation 7.14.

Take-Away

Fringe Patterns:
Interference of two spherical waves with different distances produces a pattern of circular fringes which appear on the image of the extended source. Their spacing decreases as the distance from the center increases. Note that if the transverse alignment is poor, we will see only fringes far from the center, which could appear as straight lines.

7.1.3.2 Fringe Amplitude and Contrast

In analyzing the amplitudes, two parameters are of importance: the overall strength of the signal and the fringe contrast. The overall strength of the mixing term,

Fringe contrast.

$$I_m = I_{max} - I_{min}, \tag{7.28}$$

may be characterized by the term multiplying the cosine in Equation 7.24,

$$\sqrt{R_1 T_1 R_2 T_2} I_0. \tag{7.29}$$

For a lossless beamsplitter, $T = 1 - R$, and this function has a maximum value at $R_1 = R_2 = 0.5$, with a value of $0.25 I_0$. This calculation is a more complicated version of the one we saw in Section 6.6.5.

The fringe contrast may be defined as

$$V = \frac{I_{max} - I_{min}}{I_{max} + I_{min}} \qquad (0 \le V \le 1). \tag{7.30}$$

As the DC terms are constant, the maximum and minimum are obtained when the cosine is 1 and −1, respectively, and the contrast is

$$V = 2\frac{\sqrt{R_1 T_2 T_1 R_2}}{R_1 T_2 + T_1 R_2}. \tag{7.31}$$

If the beamsplitters are identical ($R_1 = R_2 = R$) and ($T_1 = T_2 = T$), then

$$I = I_0 2RT\left(1 + \cos \delta\phi\right) \tag{7.32}$$

and the irradiance assumes a value of $I_{min} = 0$ when the cosine becomes −1, regardless of the value of R. If $I_{min} = 0$, then the visibility is $V = 1$ according to Equation 7.30. Achieving maximum visibility requires only matching the beamsplitters to each other, rather than requiring specific values.

Take-Away

Fringe Contrast:
The fringe contrast is a maximum when the intensities are equal. The peaks will be bright and the valleys dark. This will normally be the case in this type of interferometer.

In Practice

One can use uncoated glass beamsplitters, and obtain maximum contrast, but only on one output. Furthermore, with uncoated glass, the reflectivity is low and light will be reflected equally from both surfaces of the beamsplitter, which may cause unwanted fringes as discussed in Section 7.6.

7.1.4 Viewing a Point Source: Alignment Issues

The equations for fringes from a point source on a viewing screen are the same as for the fringes seen in an extended source.

The Mach–Zehnder interferometer can also be used with a point source. The point source produces spherical waves at a distant screen. Because the source appears at two different distances, the spherical waves have different centers of curvature. If they are in phase on the axis, they will be out of phase at some radial distance, and then back in phase again at a larger distance. Figure 7.7 shows the configuration. The phase difference between the two waves depends on the optical path difference along the radial lines shown. Although

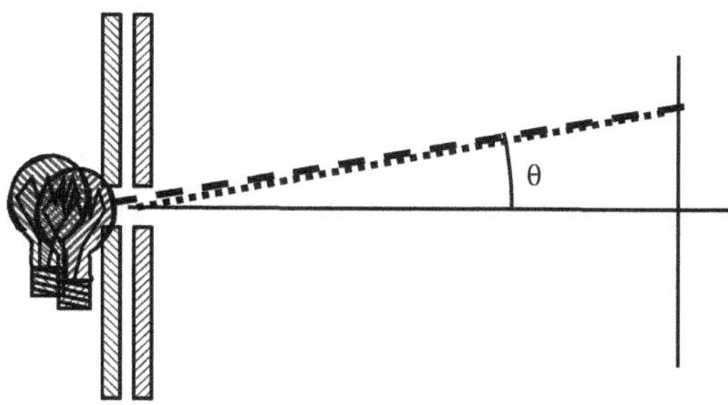

FIGURE 7.7
Viewing on a screen. The point source created by the small aperture appears at two different distances, so circular fringes are seen on the screen.

the interpretation is different, the analysis is the same as for Figure 7.5, and the equations of Section 7.1.3 are valid.

This discussion of the Mach–Zehnder interferometer presents an ideal opportunity to discuss alignment issues. Interferometry presents perhaps the greatest alignment challenge in the field of optics. From the equations in Section 7.1.3, we can see that variations in the apparent axial location of the source of a fraction of a wavelength will have serious consequences for the interference fringes. For example, with mirrors and beamsplitters at 45° as shown in Figure 7.2, a mirror motion of 50 nm in the direction normal to the mirror will produce a variation in path length of $(5\,\mathrm{nm} \times \sqrt{2}) \approx 7.1\,\mathrm{nm}$. At a wavelength of 500 nm, this will result in a phase change of 0.09 rad. Interferometry has the potential to measure picometer changes in optical path, so even such seemingly small motions will limit performance.

In terms of angle, a transverse separation of the apparent sources will result in an angular misalignment of the wavefronts at the screen. As shown in Figure 7.8, careful adjustment of the mirror angles is required to achieve alignment. In this figure, we've used a straight-line layout similar to that discussed in Figure 7.4, but drawn so that the alignment points can be visualized easily. The interferometer is the one shown in Figure 7.2. We look back into the interferometer, and on one path we see $BS2$, $M1$, and $BS1$, while on the other we see $BS2$, $M2$, and $BS1$. Because of the misalignment, the source points appear to be in different places. Our goal in alignment is to adjust the angles of beamsplitters and mirrors until the paths are coaxial.

Fig. 7.2:

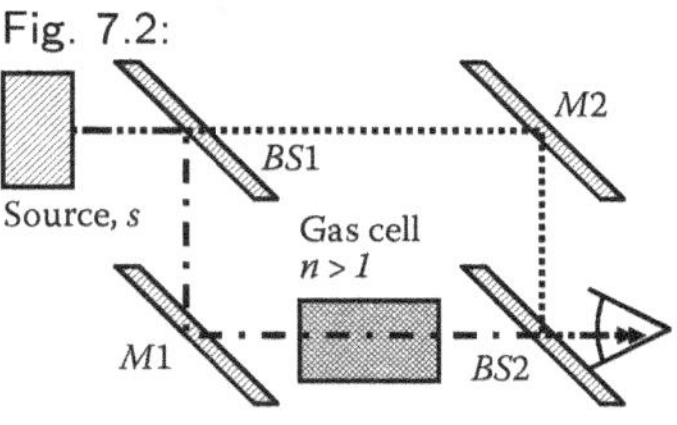

If the axial distance from source to screen is large, then a transverse misalignment by an angle $\delta\theta$ will result in straight-line fringes with a separation,

$$d_{fringe} = \frac{\lambda}{\delta\theta} \tag{7.33}$$

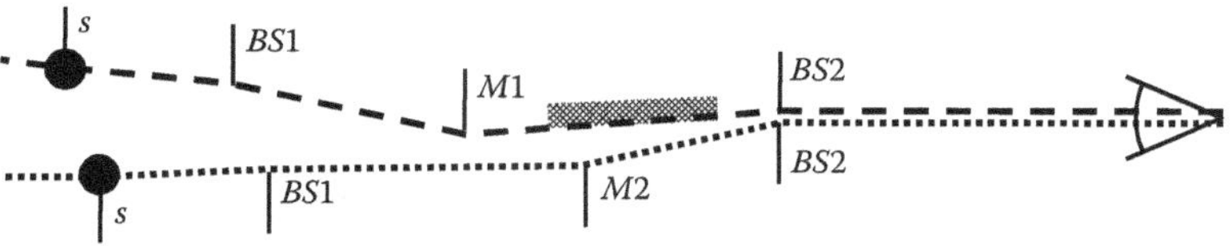

FIGURE 7.8
Aligning the interferometer. The goal is to bring the two images of the point source together. We can bend the path at each mirror or beamsplitter by adjusting its angle.

Example

For example, if the pattern on the screen has a diameter of $d = 1\,\text{cm}$, then with green light at $\lambda = 500\,\text{nm}$, an error of one fringe is introduced by an angular misalignment of

$$\delta\theta = \frac{\lambda}{d} = \frac{500 \times 10^{-9}\,\text{m}}{0.01\,\text{m}} = 5 \times 10^{-5}\,\text{rad}. \tag{7.34}$$

One fringe is a serious error, so this constraint requires a tolerance on angles of much better than 50 μrad. In Figure 7.8, if the distance from screen to source is 30 cm then the apparent source positions must be within

$$\delta\theta \times 0.30\,\text{m} = 3 \times 10^{-6}\,\text{m}. \tag{7.35}$$

A small angular misalignment can result in fringes too closely spaced to observe.

A misalignment of 500 μrad will result in 10 fringes across the pattern, and we can observe the pattern as we adjust the mirrors to reduce the number of fringes to 1 and thus achieve alignment. However, at a misalignment of 1°, the fringe spacing is less than 30 μm, and we will not be able to resolve the fringes visually. Therefore, to begin our alignment, we need to develop a different procedure. In this example, we might place a lens after the interferometer to focus the source images onto a screen, and adjust the interferometer until the two images overlap. Hopefully then some fringes will be visible, and we can complete the alignment by optimizing the fringe pattern. In order to ensure that both angle and position are matched, we normally align these spots at two distances.

Eq. 7.22: $\delta\phi = \frac{2\pi\ell}{\lambda}\left(1 + \frac{1}{2}\frac{x^2+y^2}{z^2}\right).$

Considering sources at finite distance, we examine Equation 7.22 to find the density of fringes at different radial distances from the center of the pattern. We can set $y = 0$, and consider x as the radial coordinate. Differentiating,

$$\frac{d(\delta\phi)}{dx} = \frac{2\pi\ell}{\lambda}\frac{x}{z^2}. \tag{7.36}$$

The interference pattern will go through one complete cycle, from bright to dark to bright, for $d(\delta\phi) = 2\pi$. Thus the fringe spacing is

$$dx = \frac{\lambda z^2}{x\ell}. \tag{7.37}$$

For an axial path-length difference of $\ell = 5\,\text{mm}$, and a screen $z = 30\,\text{cm}$ from the sources, and green argon-ion laser light with $\lambda = 514\,\text{nm}$, the fringe spacing is 1.9 mm at the edge of the 1 cm diameter fringe pattern ($x = 0.005\,\text{m}$). From this point, we can adjust the axial position to bring the interferometer into the desired alignment. The center of the fringe pattern is along the line connecting the two source images in Figure 7.8, at an angle, θ_c given by

$$\tan\theta_c = \frac{\delta x_{source}}{\ell}. \tag{7.38}$$

Before we begin our alignment, let us suppose that both δx_{source} and ℓ are 2 cm. Then the center of the circular pattern in Figure 7.6 is displaced by $z \tan 45° = 30\,\text{cm} \times \sqrt{2}$, and we see a small off-axis part of it where the fringe spacing is less than 8 μm. Because only a small portion of the fringe pattern is visible, it will appear to consist of parallel lines with more or less equal spacing (assuming that we can even resolve such closely spaced lines). Again, we must make use of some other alignment technique before we can begin to "fine-tune" the interference pattern.

Section 15.3.5 discusses interferometry experiments.

Because of the severe alignment constraints, interferometry requires very stable platforms, precisely adjustable optical mounts, and careful attention to alignment procedures in the design stage. In the research laboratory, interferometry is usually performed on tables with air suspensions for vibration isolation, often with attention to the effects of temperature on the expansion of materials, and even being careful to minimize the effects of air currents with the attendant changes in index of refraction. For other applications, careful mechanical design is required to maintain alignment. It is important to note that the alignment constraints apply to the relative alignment of the two paths, and not to the alignment of the source or observer with respect to the interferometer. After carefully designing and building our interferometer on a sufficiently stable platform, we can, for example, carelessly balance the light source on a pile of books and we will still observe stable fringes.

Take-Away

Alignment:

- An interferometer must be carefully aligned so that the beams are coaxial. This means that they must be aligned in position at two different points, or equivalently, that they must be aligned at one point in both position and angle.
- Stable platforms are usually required for interferometry.
- The straight-line layout can help us design a system that will be easier to align, and to develop procedures for the alignment.
- The goal of coarse alignment is to get the beams roughly aligned and is usually done by overlapping spots.
- Fine alignment is done by minimizing the number of fringes.

In Practice

It may be useful to add a temporary lens to aid in alignment. For example, in Figure 7.8, we might align mirrors $M1$ and M to make the beams coaxial. While aligning $M1$, we can image $M2$ onto a card with a lens at the output of the interferometer. We then adjust $M1$ so that the beams overlap on the image of $M2$. Then we can adjust the lens to image $M1$ and adjust $M2$. The procedure changes with every instrument, but this idea can often be used.

7.1.5 Balanced Mixing

Thus far we have only discussed one of the two outputs of the interferometer. In Figure 7.1B, we looked at the output that exits $BS2$ to the right. However, as shown in Figure 7.1A, the beamsplitter has two outputs, with the other being downward. Is there anything to be gained by using this output? Let us begin by computing the fringe pattern.

As we began this chapter, we considered that the beamsplitters could be implemented with Fresnel reflection and transmission at a dielectric interface. In Section 6.4.2, we developed equations for these coefficients in Equations 6.49 through 6.52. For real indices of refraction, regardless of polarization, τ is always real and positive. However, ρ is positive if the second index of refraction is higher than the first, and negative in the opposite case. Thus, if Equation 7.1,

$$E = E_1 + E_2,$$

describes the output to the right, and

$$E_{alt} = E_{1alt} + E_{2alt}, \tag{7.39}$$

Eq. 6.49:
$$\rho_s = \frac{\cos\theta_i - \sqrt{\left(\frac{n_2}{n_1}\right)^2 - \sin^2\theta_i}}{\cos\theta_i + \sqrt{\left(\frac{n_2}{n_1}\right)^2 - \sin^2\theta_i}}.$$

Eq. 6.50:
$\tau_s = 1 + \rho_s$..

Eq. 6.51:
$\rho_p = \frac{\sqrt{\left(\frac{n_2}{n_1}\right)^2 - \sin^2\theta_i} - \left(\frac{n_2}{n_1}\right)^2 \cos\theta_i}{\sqrt{\left(\frac{n_2}{n_1}\right)^2 - \sin^2\theta_i} + \left(\frac{n_2}{n_1}\right)^2 \cos\theta_i}.$

Eq. 6.52:
$\tau_p = (1+\rho_p)\frac{n_1}{n_2}.$

Eq. 7.24:
$I = I_0\left(R_1T_2 + T_1R_2 + 2\sqrt{R_1T_2T_1R_2}\cos\delta\phi\right).$

describes the alternate, downward, signal, then E_{2alt} and E_2 must be 180° out of phase. Then, by analogy to Equation 7.24,

$$E_{1alt} = E_0\rho_1\rho_2 e^{jknz_1} \qquad E_{2alt} = -E_0\tau_1\tau_2 e^{jknz_2}$$
$$I_{alt} = I_0\left(T_1T_2 + R_1R_2 - 2\sqrt{T_1T_2R_1R_2}\cos\delta\phi\right). \tag{7.40}$$

The DC terms may be different in amplitude from those in Equation 7.24, but the mixing terms have equal amplitudes and opposite signs. Therefore, by electrically subtracting the outputs of the two detectors, we can enhance the mixing signal. With a good choice of beamsplitters, we can cancel the DC terms and any noise associated with them.

7.1.5.1 Analysis

First, we consider the sum of the two outputs. Assuming lossless beamsplitters, so that $T + R = 1$,

$$\begin{aligned} I_{total} &= I + I_{alt} = I_0\left(R_1T_2 + T_1R_2 + T_1T_2 + R_1R_2\right) \\ &= I_0\left[R_1(1-R_2) + (1-R_1)R_2 + \right. \\ &\quad \left.(1-R_1)(1-R_2) + R_1R_2\right] = I_0. \end{aligned} \tag{7.41}$$

We expect this result, because of conservation of energy. If energy is to be conserved, then in steady state, the power out of a system must be equal to the power into it. Because the beamsplitters do not change the spatial distribution of the power, we also expect the irradiance to be conserved. This mathematical exercise has a very interesting and practical point. Although the relationship between the fields in Equations 7.24 and 7.40 were derived from Fresnel reflection coefficients at a dielectric interface, they hold regardless of the mechanism used for the beamsplitter. Thus the balanced mixing concept we are about to develop will work whether the beamsplitter uses dielectric interfaces, coated optics, or even fiber-optical beamsplitters.

In Practice

Any low–loss beamsplitter mechanism that splits the beam into exactly two components can be used for balanced mixing. Because the balanced mixing is predicted by conservation of energy, it will occur at a dielectric-interface, a multi-layer coating (Section 7.7), or a fiber beamsplitter. We can trust the beamsplitter for balanced mixing even if we don't understand its principles.

Note that if there are more than two outputs, such as would be the case if we used a diffraction grating (Section 8.8)for a beamsplitter we cannot simply use two of the outputs for balanced mixing, because the conservation of energy applies to the total output.

If we subtract the two outputs,

$$I_{dif} = I - I_{alt} =$$
$$I_0\left[R_1(T_2 - R_2) + T_1(R_2 - T_2) + 4\sqrt{T_1T_2R_1R_2}\cos\delta\phi\right]$$
$$I_{dif} = I_0 \times \ldots$$
$$\left[(1-2R_1)(1-2R_2) - 4\sqrt{(1-R_1)(1-R_2)R_1R_2}\cos\delta\phi\right]. \tag{7.42}$$

Now, we have enhanced the mixing signal that carries the phase information. Furthermore, if we use at least one beamsplitter with $R_1 = 50\%$ or $R_2 = 50\%$, then the residual DC term vanishes, and we have just the mixing terms. If both beamsplitters are 50%, then

$$I_{dif} = I_0 \cos \delta\phi \qquad \left(\text{for } R_1 = R_2 = \frac{1}{2}\right). \tag{7.43}$$

This irradiance can be either positive or negative. It is not a physical irradiance, but a mathematical result of subtracting the two irradiances at the detectors, proportional to the current obtained by electrically subtracting their two outputs. The resulting signal is proportional to the pure mixing signal, in a technique familiar to radar engineers, called **balanced mixing**. Our success is dependent upon one of the beamsplitters being exactly 50% reflecting, the detectors being perfectly balanced, and the electronic implementation of the subtraction being perfect. In practice, it is common to have an adjustable gain on one of the detector circuits, which can be adjusted to null the DC terms. If we multiply I_{alt} by this compensating gain term, we will have a scale factor on I_{dif}, but this error is a small price to pay for removing the DC terms. We will usually choose to make $R_2 = 50\%$, in case there is unequal loss along the paths that we have not considered in our equations.

In practice, we need a variable gain on one channel to correct for imperfections in the splitting ratios, detectors and amplifiers.

Example

As an illustration of the importance of balanced mixing, we consider an interferometer with a light source having random fluctuations, I_n with a mean value equal to 1% of the nominal irradiance, $\bar{I}_0$:

$$I_0 = \bar{I}_0 + I_n. \tag{7.44}$$

Then the output of a single detector, using Equation 7.24 with 50% beamsplitters is

$$I = \frac{1}{2}\left(\bar{I}_0 + I_n\right)(1 + \cos \delta\phi)$$

$$I = \frac{1}{2}\bar{I}_0 + \frac{1}{2}\bar{I}_0 \cos \delta\phi + \frac{1}{2}I_n + \frac{1}{2}I_n \cos \delta\phi. \tag{7.45}$$

The first term is the DC term, the second term is the desired signal, and the last two terms are noise terms. Suppose the interferometer is working near the optimal point where $\cos \delta\phi \approx 0$. Then the variation in the signal term is given by

$$\frac{\partial I}{\partial (\delta\phi)} = -\frac{1}{2}I_0 \delta\phi, \tag{7.46}$$

and for a 0.1° change in phase, the signal amplitude in the second term of Equation 7.45 is

$$\text{Signal} = \frac{1}{2}I_0 \times 0.1\frac{\pi}{180} \approx \frac{\bar{I}_0}{1200} \qquad (1 \text{ detector}, \delta\phi = 0.1°). \tag{7.47}$$

If I_n has a RMS value of 1% of $\bar{I}_0$, then the third term in Equation 7.45 is

$$\text{Noise} = \frac{\bar{I}_0}{200} \qquad (\text{Single detector}), \tag{7.48}$$

or six times larger than the signal. The fourth term in Equation 7.45 is small compared to either of these, and can be ignored.

In contrast, for balanced mixing, in Equation 7.43,

$$I_{dif} = \bar{I}_0 \cos \delta\phi + I_n \cos \delta\phi$$

and for the same parameters,

$$\text{Signal} = I_0 \times 0.1\frac{\pi}{180} \approx \frac{\bar{I}_0}{600} \qquad (\text{Balanced: } \delta\phi = 0.1^\circ), \tag{7.49}$$

and the noise is

$$\text{Noise} = I_0 \times \frac{1}{100} \times 0.1\frac{\pi}{180} \approx \frac{\bar{I}_0}{60,000} \qquad (\text{Balanced}). \tag{7.50}$$

The significant results are an improvement in the signal of a factor of 2, and a complete cancellation of the third term in Equation 7.45, leading to a reduction of a factor of 300 in the noise. Balanced mixing is also known as "common-mode rejection," because it cancels the part of the noise term that is common to both signal and reference. In principle, one could digitize the detector outputs and implement the subtraction in software. However, often the DC terms are large compared to the variations we wish to detect in the mixing terms, and thus a large number of bits would be required in the digitization.

Take-Away

Mach–Zehnder Interferometer:

- The Mach–Zehnder interferometer splits a light wave into two separate waves on different paths and recombines them to recover information about the optical path length of one path, using the other as a reference.
- The quantity of interest is usually a distance or an index of refraction.
- The interferometer can detect very small phase shifts, corresponding to changes in optical path length of picometers.
- Measurement of shifts in brightness on the optical axis can be used to observe temporal changes.
- Measurement of the interference pattern in the transverse direction can be used for stationary phenomena.
- Balanced mixing uses both outputs of the interferometer, subtracting the detected signals.
- Balanced mixing doubles the signal.
- Balanced mixing can be used to cancel the noise that is common to signal and reference beams.
- Practical implementation usually requires a slight gain adjustment on one of the two channels.
- Conservation of energy dictates that the two mixing outputs have opposite signs.

7.2 Doppler Laser Radar

Fig. 7.2:

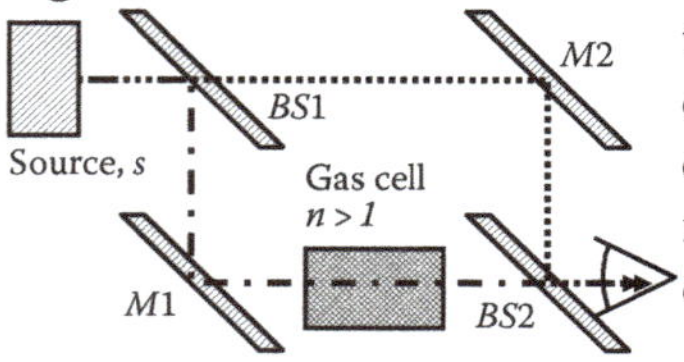

The Mach–Zehnder interferometer is used in situations where the light is transmitted through the object, to measure its thickness or index of refraction. A slight modification of Figure 7.2, shown in Figure 7.9 can be used to observe light reflected from an object, for example, to measure its distance. We will discuss this configuration in the context of laser radar, although it is useful for other applications as well. A similar configuration will be discussed in the context of biomedical imaging in Section 10.3.4. The reader interested in

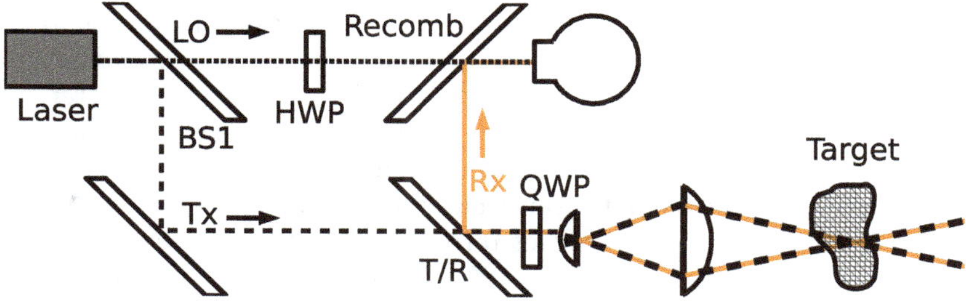

FIGURE 7.9
CW Doppler laser radar. This interferometer measures reflection to determine distance and velocity. The waveplates enhance the efficiency as discussed in Section 6.6.5. The telescope expands the transmitted beam and increases light collection, as discussed in Chapter 8.

further study of laser radar may wish to explore a text on the subject[112]. The figure shows some new components that will be discussed shortly, but for now the essential difference from Figure 7.2 is the change in orientation of the upper-right beamsplitter.

7.2.1 Basics

Now, the signal path consists of the transmitter (Tx) path from the laser to the target via the lower part of the interferometer, and the receiver path (Rx) back to the detector. The reference beam, also called the **local oscillator** or LO, by analogy with Doppler radar, uses the upper path of the interferometer. The LO and Rx waves are combined at the recombining beamsplitter (Recomb.), and mixed on the detector. The equations can be developed by following the reasoning that led to Equation 7.24. However, it is easier in the analysis of laser radar to think in terms of power rather than irradiance, especially to compare with microwave radar. The LO power is

Eq. 7.24: $I = I_0\left(R_1T_2 + T_1R_2 + 2\sqrt{R_1T_2T_1R_2}\cos\delta\phi\right)$.

$$P_{LO} = P_0 T_{BS1} T_{Recomb}. \tag{7.51}$$

The power delivered to the target is

$$P_t = P_0 R_{BS1} T_{T/R}, \tag{7.52}$$

and the received signal power at the detector is

$$P_{sig} = P_t F R_{T/R} R_{Recomb}, \tag{7.53}$$

where F includes the reflection from the target, the geometry of the collection optics, and often some factors addressing coherence, atmospheric extinction, and other effects. Because F is usually very small, we must make every effort to maximize the mixing signal strength and contrast in Equations 7.28 and 7.30. For a carbon dioxide laser radar, the laser power is typically a few Watts, and the detector requires only milliwatts, so we can direct most of the laser power toward the target with $R_{BS1} \to 1$ and collect most of the returned signal with $R_{Recomb} \to 1$. Typically these beamsplitters are between 90% and 99% reflecting.

Eq. 7.28: $I_m = I_{max} - I_{min}$.

Eq. 7.30: $V = \frac{I_{max} - I_{min}}{I_{max} + I_{min}}$ $(0 \le V \le 1)$.

The amount of transmitted light backscattered from the target and then collected is

$$F = \frac{\pi D^2}{4z^2} \times \rho(\pi) \times 10^{-2\alpha z/10}, \tag{7.54}$$

where the first term represents the fraction of the energy collected in an aperture of diameter D at a distance z, the second term, $\rho(\pi)$, is called the diffuse reflectance of the target, and the third term accounts for atmospheric extinction on the round trip to and from the target. For a distance of tens of kilometers, and a telescope diameter in tens of centimeters, the first term is about 10^{-10}. For atmospheric backscatter, the second term is related to the thickness of the target, but a typical value may be 10^{-6}. Atmospheric extinction is given by the extinction coefficient, α, which depends on precipitation, humidity, and other highly variable parameters. For light at 10 μm, $\alpha = 1\,\mathrm{dB/km}$ is a reasonable choice on a typical dry day, so this term may be 10^{-2} or worse. Therefore, a good estimate is $F = 10^{-18}$, so for a few kilowatts of transmitted power, the signal received may be expected to be a few femtowatts. The coherent detection process amplifies the signal through multiplication with the local oscillator, as shown in Table 7.1. Although the primary reason for using coherent detection is to measure velocity as discussed in the following, the coherent advantage in signal is significant. The increased signal is easier to detect above the electronic noise associated with the detector.

In practice, the laser radar usually includes a telescope to expand the transmitter beam before it is delivered to the target, and to collect as much of the returned light as possible. The effect of this telescope is captured in F in the equations. In addition, a quarter-wave plate is included, and the T/R beamsplitter is a polarizing device, as discussed in Section 6.6.5, to avoid the constraint $R_{T/R} + T_{T/R} \le 1$.

The phase of the signal relative to the LO is

$$\phi = \frac{2\pi}{\lambda}(2z - z_0), \tag{7.55}$$

where

- z is the distance from the laser radar to the target
- z_0 accounts for distance differences within the interferometer

The factor of 2 multiplying z accounts for the round trip to and from the target.

Take-Away

Doppler Lidar:

- Doppler lidar mixes the signal from a target with a reference beam.
- Pulsed Doppler lidar provides both Doppler velocity and distance.

TABLE 7.1
Laser Radar Power Levels

Signal (P_1)	10^{-15} W
Reference (P_2)	10^{-3} W
Detected	10^{-9} W

The typical signal power from a laser radar may be difficult to detect. Interference works to our advantage here. The coherent detection process "amplifies" the signal by mixing it with the reference. The resulting detected power is the geometric mean of the signal and reference.

In Practice

Incoherent lidar is considerably more common that coherent, and provides distance information, but the Doppler velocity requires coherent lidar. Incoherent lidar is now available in compact packages at prices affordable for many consumers.

7.2.2 Mixing Efficiency

The mixing power detected is

$$P_{mix} = \int I_{mix} dA = \int \frac{E_{Sig}(x_d, y_d) E_{LO}^*(x_d, y_d)}{Z} dA, \tag{7.56}$$

where x_d and y_d are coordinates in the detector. If the signal field $E_{sig}(x_d, y_d)$ is proportional to the local oscillator $E_{LO}(x_d, y_d)$, then we can show that the integral becomes

$$P_{mix} = \sqrt{P_{sig} P_{LO}} \qquad \text{Maximum mixing power.} \tag{7.57}$$

If the proportionality relationship between the fields differs in any way, the result will be smaller. For example, if the two waves are plane waves from different directions, then they add constructively at some points in the detector and destructively at others. The mixing term then has many different phases, and the resulting integral is small. We can examine this situation mathematically by noting that

$$P = \int \frac{E(x_d, y_d) E^*(x_d, y_d)}{Z} dA$$

for both waves, and writing

$$P_{mix} = \eta_{mix} \sqrt{P_{sig} P_{LO}} \tag{7.58}$$

where

$$\eta_{mix} = \int \frac{E_{sig}(x_d, y_d) E_{LO}^*(x_d, y_d)}{\sqrt{P_{sig} P_{LO}}} dA$$

where

$$P_{sig} = \int E_{sig}(x_d, y_d) E_{sig}^*(x_d, y_d) dA$$

and

$$P_{LO} = \int E_{LO}(x_d, y_d) E_{LO}^*(x_d, y_d) dA \tag{7.59}$$

We define η_{mix} as the **mixing efficiency**. The mixing efficiency is low if the local oscillator wavefront is badly matched to the signal in terms of amplitude, wavefront, or both. If the signal arises from a rough surface, for example, it will be a random field, and we will find it difficult to construct a matched local oscillator.

Any difference in wavefront shape, location, or tilt between signal and LO reduces the mixing ratio.

Take-Away

Mixing Efficiency:
The mixing efficiency depends on the relative shape of the signal and reference wavefronts. Perfect matching produces $\eta_{mix} = 1$, and any mismatch reduces this value.

7.2.3 Doppler Frequency

The most common use of this configuration is to measure the velocity of a moving object. Dust particles, fog droplets, rain, snow, smoke, or other aerosols in the atmosphere provide sufficient scattering that carbon dioxide laser radars can be used for measuring wind velocity in applications such as meteorology and air safety[54]. Such applications require high power, sensitive detectors, and large precisely–aligned optics. They are generally expensive and, to date, not produced in significant quantities. For more highly reflecting objects such as aircraft, the ground, vehicles, and other objects, incoherent laser radars that can measure distance to centimeters are available for a few hundred dollars.

The frequency of light transmitted from one platform to another is shifted if there is relative motion between the platforms by the Doppler frequency, f_d[56],

$$2\pi f_d = \mathbf{k} \cdot \mathbf{v} \tag{7.60}$$

where

- $\mathbf{k}$ is the original propagation vector with magnitude $2\pi/\lambda$ and direction from source to detector
- $\mathbf{v}$ is the velocity vector between the source and detector

Doppler shift for relative motion between transmitter and receiver.

If the distance is increasing, the Doppler frequency is negative. Thus,

$$f_d = \frac{v_{parallel}}{\lambda} \qquad \text{(Moving source or detector)} \tag{7.61}$$

where $v_{parallel}$ is the component of the velocity vector of the source toward the detector, parallel to the line connecting the two. For the laser radar, the light is Doppler shifted in transit from the source to the target, and then shifted by the same amount on return to the detector. Therefore, in radar or laser radar, the Doppler frequency is

Doppler shift for relative motion between monostatic transmitter/receiver and reflecting target.

$$f_{DR} = 2\frac{v_{parallel}}{\lambda} \qquad \text{(Moving reflector).} \tag{7.62}$$

Example

As two examples of the Doppler shift for a carbon dioxide laser radar with $\lambda = 10.59\,\mu\text{m}$, we consider two different target speeds.

For a slow target consider a wind velocity component $v_{parallel} = 0.54\,\text{m/s}$, the Doppler frequency is

$$f_{DR} = 2\frac{v_{parallel}}{\lambda} = 100\,\text{kHz}.$$

As an example of a faster target, consider a satellite in low–earth orbit, with $v_{parallel} = 7.5\,\mathrm{km/s}$. Then

$$f_{DR} \approx 1.5\,\mathrm{GHz}.$$

These targets and others with velocities between these extremes all produce Doppler frequencies accessible to modern electronics.

Take-Away

Doppler Shift:
The Doppler shift is proportional to the velocity component of the moving target along the line of sight. It is often in the radio frequency range, and is usually high enough that several cycles of the Doppler shift occur during a pulse.

7.2.4 Range Resolution

Laser radar, like radar, can measure the distance, or **range**, to a "hard target" such as an aircraft, a vehicle, or the ground. Also like a radar, it can be used to restrict the distances over which the instrument is sensitive to atmospheric aerosols, so that we can map the velocities at various distances along the beam. We can then, if we like, sweep the beam transversely, and produce two- or three-dimensional maps of velocity.

One approach to the ranging task is focusing. Instead of collimating the transmitted beam, the telescope can be adjusted to focus it at a prescribed distance by moving the telescope secondary slightly. Scattering particles near the transmitter focus will produce images at infinity, so their waves will be plane and will mix efficiently with the LO wave. The matched wavefronts will produce the same mixing term in Equation 7.4

$$I_{mix} = \frac{E_{sig}E_{LO}^*}{Z}$$

across the entire LO, and the detector will integrate this result over the width of the beam. In contrast, a particle out of focus will produce an image at some finite distance, s' relative to the detector, and the signal wavefront will be curved:

$$E_{sig} = |E_{sig}|\, e^{jk\sqrt{x^2+y^2+s'^2}}. \tag{7.63}$$

This curved wavefront, mixing with the plane-wave E_{LO}, will result in I_{mix} having different phases at different x, y locations on the detector. The resulting integral will be small, so these scattering particles will not contribute significantly to the signal. The transmitted beam is larger away from the focus so there are many more particle scattering light, but this increase is not sufficient to overcome the decrease caused by the wavefront mismatch, and the laser radar will be most sensitive to targets near its focus. The **range resolution** or depth of the sensitive region will be discussed in Section 8.6.5.

We will see in Chapter 8 that strong focusing with a system having an aperture diameter D is only possible at distances much less than D^2/λ. Large telescopes are expensive, and in any case atmospheric turbulence often limits the useful aperture for focusing to about 30 cm, suggesting an upper limit of a few kilometers for a laser radar at $\lambda = 10\,\mu\mathrm{m}$. In practice, the range resolution degrades rapidly starting around $z = 1000\,\mathrm{m}$ or less.

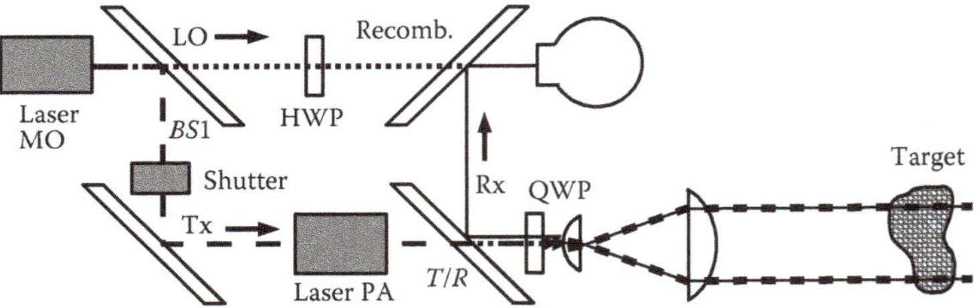

FIGURE 7.10
Pulsed Doppler laser radar. Starting with Figure 7.9, a shutter is added to produce pulses, and a laser power amplifier (PA) is added to increase the output power of the pulses. The original laser is called the master oscillator (MO), and the configuration is therefore called MOPA. Other pulsed laser radar approaches are also possible.

Take-Away

Focusing:
Focusing the beam of a coherent Doppler lidar decreases the mixing efficiency rapidly with distance from focus, producing a sectioning effect such the the signal is mostly from the focal region. This effect is similar to confocal microscopy, except that the process of mixing with the reference beam acts as the pinhole.

In Practice

Focusing a CW Doppler laser radar is the preferred method of achieving range resolution at short distances. Time of flight is measured for resolving range at longer distances. Typically the change occurs somewhere around 1 km.

At longer ranges, a pulsed laser radar can be used. One approach is shown in Figure 7.10. The major conceptual addition is a shutter that can be opened or closed to generate the pulses. Of course, the pulses must be short, because the range resolution is

$$\delta r = \frac{c\tau}{2}, \tag{7.64}$$

where τ is the pulse length. A 1 μs pulse provides a range resolution of 150 m. For these short pulses, mechanical shutters are not practical and an electro-optical modulator (Section 6.6.5) is often used for this purpose. If we expect signals from as far as $r_{max} = 10$ km, we must wait at least $2r_{max}/c = 67$ μs before the next pulse to avoid introducing **range ambiguities**, in which the return from one pulse arrives after the next pulse, and is thus assumed to be from a much shorter distance than it is. Therefore, the **pulse repetition frequency**, PRF, must be

$$PRF < \frac{c}{2r_{max}}. \tag{7.65}$$

The laser is turned off most of the time, so the average power,

$$P_{avg} = P_{laser} \times \tau \times PRF. \tag{7.66}$$

A 5 W laser will thus provide an average power of less than 5 W × 1/67. Normally, a laser power amplifier (PA) of some sort will be used to increase the laser power to a more useful level. In this case, the source laser is called the master oscillator (MO). This configuration is called a MOPA (master-oscillator, power-amplifier) laser radar.

High power can be achieved with a MOPA configuration.

With a pulse length of 1 µs, we can measure frequencies with a resolution of about 1 MHz (the Fourier transform of a pulse has a bandwidth of at least the inverse of the pulse length), so our **velocity resolution** is a few meters per second. Velocity and range resolution issues at laser wavelengths are quite different from those of microwave Doppler radar. In the microwave case, the Doppler frequencies are extremely low, and a train of pulses is used to produce a useful Doppler signal. A **Doppler ambiguity** arises because of this "sampling" of the Doppler signal. In laser radar, the Doppler ambiguity is limited by the sampling of the Doppler signal within each pulse, rather than by the pulses as is the case with most radars. The sampling can be chosen to be faster than twice the bandwidth of an anti-aliasing filter in the electronics after the detector. The Doppler ambiguity interval and the range ambiguity interval are not related in laser radar as they are in radar.

There is no inherent Doppler ambiguity in laser radar.

Example

A Doppler lidar uses a carbon-dioxide laser at a wavelength of $\lambda = 10.59\,\mu$m, with a pulse width of $\tau = 8\,\mu$s and pulse repetition frequency of $PRF = 1000\,$Hz. What are the range resolution, the velocity resolution, and the range ambiguity?

The range resolution from Equation 7.64 is

$$\delta r = \frac{c\tau}{2} = 1.2\,\text{km}.$$

The velocity resolution is limited by the pulse width. The Doppler frequency is given by Equation 7.62.

$$f_{DR} = 2\frac{v_{parallel}}{\lambda}.$$

The resolution of frequency is about $\frac{1}{2\pi\tau}$ so the resolution of velocity is approximately

$$\delta v_{parallel} = \frac{\lambda}{4\pi\tau} = 0.11\,\text{m/s}.$$

We can invert Equation 7.65 to obtain the maximum unambiguous range,

$$r_{max} = \frac{c}{2PRF} = 300\,\text{km}.$$

In Practice

Part of the reason is that accurate measurement of the Doppler frequency (and thus velocity) requires long times and a pulsed laser radar receiver cannot be used until after the transmitted pulse has ended.

Simple range-only (incoherent) laser radars can use nano-second pulses and high-speed electronics for measurements at short distances.

The designs for Doppler laser radar in Figures 7.9 and 7.10 have one glaring conceptual deficiency; although they can measure the magnitude of the parallel velocity component with exquisite precision and accuracy, they cannot tell if it is positive or negative. This sign ambiguity will be addressed along with other ambiguities in the next section.

Eq. 7.6:
$I = I_1 + I_2 + 2\sqrt{I_1 I_2}\cos(\phi_2 - \phi_1)$.

Eq. 7.10:
$\phi_2 - \phi_1 = \pi/2$.

7.3 Resolving Ambiguities

Because the phase we want to measure appears inside the cosine term in Equation 7.6, an interferometer will achieve its potential sensitivity only at the quadrature point defined by Equation 7.10. However, the interferometer is so sensitive to changes in distances and indices of refraction that it is usually difficult to find and maintain this operating condition. In general, even keeping I_2 constant, we still face several ambiguities:

1. It is difficult to determine whether measured irradiance variations arise from variations in I_1 or in the phase.
2. It is difficult to find and maintain the quadrature point for maximal sensitivity.
3. The cosine is an even function, so even if we know both I_1 and I_2, there are two phases with each cycle that produce the same irradiance. In laser radar, this ambiguity manifests itself as a sign ambiguity on the velocity.
4. The cosine is a periodic function so our measurement of $n\ell_c$ is "wrapped" into increments of λ.

We will address items 1 thorough 3 using two different techniques in Sections 7.3.1 through Section 7.3.4. Item 4 will be addressed in Section 7.3.6.

7.3.1 Phase-Shifting Interferometry

To find the quadrature point, we can deliberately and systematically adjust some distance such as the path length of the reference, producing a variable bias phase so that the phase shift is Equation 7.6, becomes

$$\delta\phi = \phi_2 - \phi_1 + \phi_{bias} \tag{7.67}$$

Eq. 7.4:
$I_{mix} = \frac{E_1 E_2^*}{Z}$.

If we find the maximum and minimum of the resulting irradiance, then we know the quadrature point, $\phi_{bias} = \pi/2$, lies exactly halfway between them. We also then know the magnitude of the complex number, I_{mix} in Equation 7.4. If the individual irradiances do not change, then any variations in measured irradiance at the quadrature point can be attributed to the desired phase variations.

Eq. 7.6:
$I = I_1 + I_2 + 2\sqrt{I_1 I_2}\cos(\phi_2 - \phi_1)$.

A more common approach is **phase-shifting interferometry**,[86;91]. in which we collect data with different bias phases and solve simultaneous equations for the unknowns in Equation 7.6. Let us consider the unknowns to be $I_1 + I_2$, $\sqrt{I_1 I_2}$, and $\cos(\phi_2 - \phi_1)$. We make measurements, C_n, with different phase biases, $\phi_{bias:n}$;

$$C_n = I_1 + I_2 + 2\sqrt{I_1 I_2}\cos(\phi_2 - \phi_1 + \phi_{bias:n}), \tag{7.68}$$

where $n = 0, 1, 2 \cdots N-1$. We need at least 3 measurements to recover the three unknowns. Let us consider one special case where $N = 3$ and the bias phases are $\phi_{bias} = 0, \pi * 2/3, \pi * 4/3$. we can write the cosine using the usual formula for the cosine of a sum, but it is much easier to work in exponential notation. Therefore we write

$$C_n = I_1 + I_2 +$$

$$\sqrt{I_1 I_2} e^j(\phi_2 - \phi_1 + \phi_{bias:n}), + \sqrt{I_1 I_2} e^- j(\phi_2 - \phi_1 + \phi_{bias:n}), \tag{7.69}$$

where $M = \sqrt{I_1 I_2} e^j(\phi_2 - \phi_1)$, or

$$C_0 = I_1 + I_2 + M + M^*, \tag{7.70}$$

$$C_1 = I_1 + I_2 + Me^{j2\pi/3}, + M^* e^{-j2\pi/3}, \tag{7.71}$$

and

$$C_2 = I_1 + I_2 + Me^{j4\pi/3}, + M^* e^{-j4\pi/3}, \tag{7.72}$$

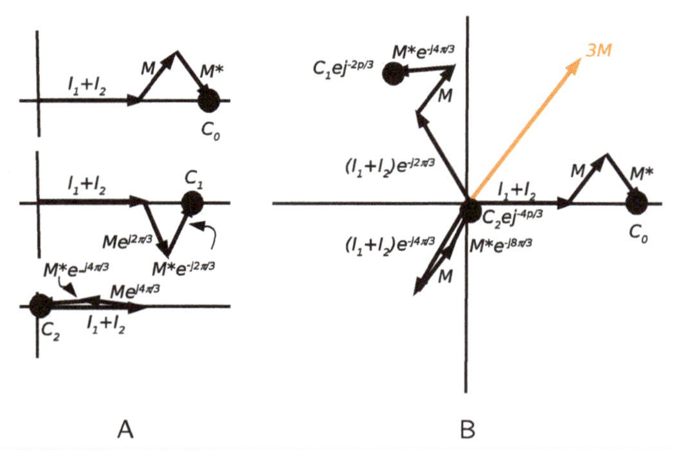

FIGURE 7.11

Processing phase-shifting interferometry signals. In A, each measurement, C is a real number, determined by Equation 7.69, with the appropriate phase shifts. Multiplying each equation by an appropriate phase-shifting term and adding, we obtain the desired mixing term (B), provided that we choose the phase shifts appropriately. Any unique phase shifts can be used, but the equations become more complicated.

If we multiply each equation by $e^{-j\phi_{bias:n}}$, then because we have chosen our bias phases well, the $I_1 + I_2$ terms cancel, as do the M^* terms (the conjugate mixing terms), while the mixing terms, M, all have the same phase, and thus add, as shown in Figure 7.11. If we do not chose these three phase shifts, symmetrically arranged between 0 and 2π, the cancellation will not occur. We need to do more work in the solution of the simultaneous equations, but there is still a solution. Another advantage of the symmetrical arrangement of phase shifts is that the errors in the answers are the same in both directions in the complex plane.

Take-Away

Phase-Shifting Interferometry:

Phase-shifting interferometry makes use of measurements with different phase shifts in the reference beam to solve for the DC components and the complex mixing term.

As mentioned above, a minimum of three phases is required to remove the ambiguity. Of course it is important that neither the sample nor the interferometer change during the time taken by the measurements.

7.3.2 Offset Reference Frequency

We can produce a continuously and linearly changing bias phase by shifting the frequency of the reference wave. If

$$E_1 = E_{10}e^{j2\pi\nu t} \qquad \text{and} \qquad E_2 = E_{20}e^{j(2\pi\nu - 2\pi f_o)t}, \tag{7.73}$$

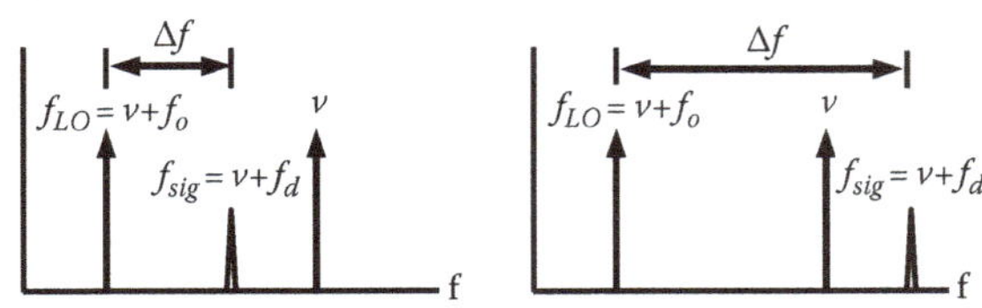

FIGURE 7.12
Offset local oscillator. Mixing the signal with a local oscillator offset in frequency by f_o resolves the ambiguity in the Doppler velocity. In this example, $f_o < 0$ so that frequencies above f_o represent positive Doppler shift and those below f_o represent negative Doppler shift.

where ν is the laser frequency, or "optical carrier" frequency, c/λ, and f_o is an offset frequency, ranging from Hz to GHz, then the mixing terms will be

$$I_{mix} = \frac{1}{Z} E_{10} E_{20}^* e^{j2\pi f_o t} \tag{7.74}$$

and

$$I_{mix}^* = \frac{1}{Z} E_{10}^* E_{20} e^{-j2\pi f_o t}, \tag{7.75}$$

so that

$$I_{mix} + I_{mix}^* =$$
$$\frac{1}{Z}\left(E_{10}E_{20}^* + E_{10}^* E_{20}\right)\cos\left(2\pi f_o t\right) + j\left(E_{10}E_{20}^* - E_{10}^* E_{20}\right)\sin\left(2\pi f_o t\right)$$
$$I_{mix} + I_{mix}^* = \frac{2}{Z}\Re\left(E_{10}E_{20}^* + E_{10}^* E_{20}\right)\cos\left(2\pi f_o t\right) +$$
$$\Im\left(E_{10}E_{20}^* - E_{10}^* E_{20}\right)\sin\left(2\pi f_o t\right) \tag{7.76}$$

If we choose f_o in the radio frequency band, we can easily measure this sinusoidal wave and extract its cosine and sine components to obtain the real and imaginary parts of $E_{10}E_{20}^*$. With some suitable reference measurement, we can find the phase of E_{10}. As with most interferometers, it is usually not possible to know all the distances exactly to obtain absolute phase measurements, but it is quite easy to make very precise relative phase measurements.

In the case of the laser radar, the result is particularly simple, as shown in Figure 7.12. If the transmitter frequency is f, then the Doppler-shifted frequency will be $f + f_d$. Now, the signal from a stationary target at f will produce a signal in the detector at exactly f_o, and any target will produce a signal at $|\,f_o + f_d|$. If we choose the offset frequency f_o larger than any expected value of $|\,f_d|$, then we can always determine the Doppler velocity unambiguously.

Take-Away

Offset Local Oscillator:
An offset local oscillator frequency removes the sign ambiguity on the Doppler shift, provided that the frequency offset is larger than the maximum Doppler frequency expected. This technique is useful for laser radar, vibrometers, and other types of interferometry.

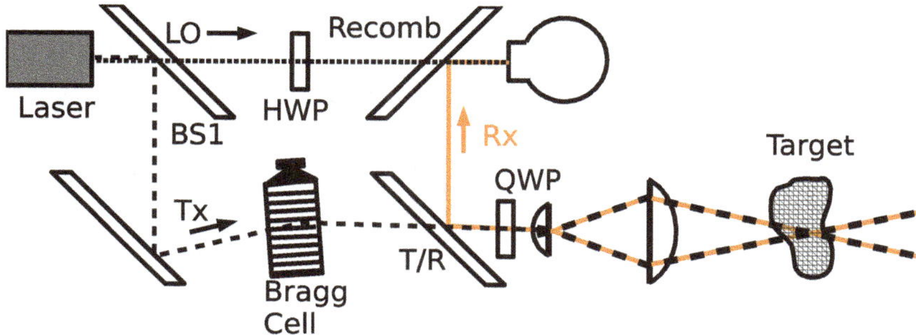

FIGURE 7.13
Offset local oscillator. A Bragg cell can be used to shift the transmitter beam in frequency. The Bragg cell could be placed in the LO arm, but "leakage" of unshifted light would mix with the shifted light, producing a mixing term that could be larger than that of the signal with the desired offset LO.

In Practice

An offset local oscillator can be implemented by using a Bragg cell in either the LO or transmitter arm of the interferometer. If the Bragg cell is placed in the LO arm then a small fraction of unshifted LO light may mask the very weak signal. For this reason it is best to put the Bragg cell in the transmitter arm as in Figure 7.13. Doing so requires a Bragg cell of high efficiency to avoid loss of light.

An offset local oscillator can also be implemented with a second laser offset from the first, but maintaining a constant frequency offset between the two lasers may be challenging.

See Section 8.8.8 for discussion of the Bragg Cell.

7.3.3 Optical Quadrature

Knowing that the ambiguities in interferometry arise because of the cosine in Equation 7.24, we may ask whether it is possible to obtain a similar measurement with a sine. Having both the cosine and the sine terms would be equivalent to having the complex value of I_{mix}. **Optical quadrature detection** implements the optical equivalent of an electronic quadrature mixer to detect these two channels simultaneously. It has been demonstrated in a microscope[103] and has been proposed for laser radar[97]. In electronics, the quadrature mixer is implemented with two mixers[36]. The reference is delivered to one of them directly, and to the other through a 90° phase shifter. The problem in optics is that the short wavelength makes it difficult to maintain the phase difference. The approach taken here makes use of polarized light, and implements phase-shifting interferometry with four phases detected simultaneously.

Eq. 7.24: $I = I_0\left(R_1T_2 + T_1R_2 + 2\sqrt{R_1T_2T_1R_2}\cos\delta\phi\right)$.

Analysis of optical quadrature detection provides an excellent opportunity to examine the equations of interference for polarized light using Jones vectors and matrices. The layout is shown in Figure 7.14. Let us briefly discuss the principles of operation before we go through the mathematics. Light from the source is linearly polarized at 45°. The interferometer is a Mach–Zehnder with two additions. First, in the reference path is a quarter-wave plate with one axis in the plane of the paper to convert the reference to circular polarization. It is assumed that the sample does not change the 45° polarization of the incident light. If it does, a polarizer placed after the sample could restore the polarization, which is critical to optical quadrature. The recombining beamsplitter, $BS2$, is unpolarized, and reflects 50% of the incident light. The second addition consists of a polarizing beamsplitter and additional detector. The transmitted light, detected

Optical quadrature detection is phase-shifting interferometry with four phases detected simultaneously.

The signal is linearly polarized at 45 degrees.

by the I detector, consists of the horizontal component of the signal, and the horizontal component of the circularly polarized reference, which we will define as the cosine part.* The Q detector measures the vertical components. The two components of the signal are the same, but the vertical component of the reference is the sine, rather than the cosine. In complex notation, the vertical component of the reference has been multiplied by $e^{j\pi/2} = j$.

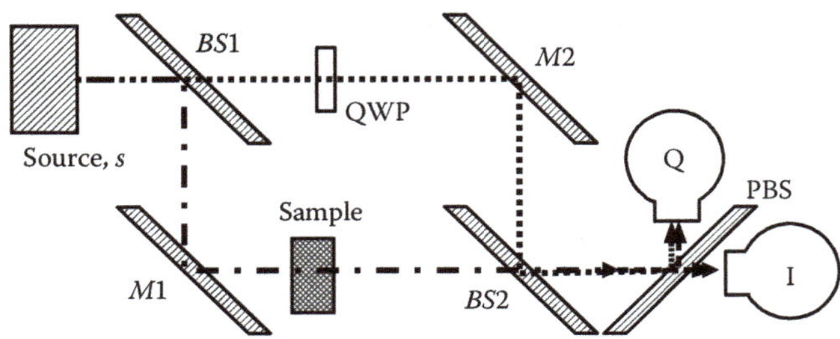

FIGURE 7.14
Optical quadrature detection. The interferometer is a Mach–Zehnder with two additions. A quarter-wave plate converts the reference wave to circular polarization, and two detectors are used to read out the two states of linear polarization, corresponding to the so-called I and Q signals.

Eq. 7.1:
$E = E_1 + E_2$.

Equation 7.1 is valid in vector form,

$$\mathbf{E} = \mathbf{E}_{sig} + \mathbf{E}_{LO}. \tag{7.77}$$

The normalized input field from the laser, following the approach of Section 6.6 is

$$\mathbf{E}_0 = \begin{pmatrix} 1 \\ 1 \end{pmatrix} \times \frac{1}{\sqrt{2}}. \tag{7.78}$$

The local oscillator at the output of $BS2$ is (assuming the mirrors are perfect)

$$E_{LO} = \mathcal{R}_2 \mathcal{Q} \mathcal{T}_1 \mathbf{E}_0 \tag{7.79}$$

and the signal field is

$$E_{sig} = \mathcal{T}_2 \mathcal{J}_{sample} \mathcal{R}_1 \mathbf{E}_0, \tag{7.80}$$

where we use the form

$$\mathcal{Q} = \begin{pmatrix} 1 & 0 \\ 0 & j \end{pmatrix}, \tag{7.81}$$

for the quarter-wave plate. Here, $\mathcal{R}$ and $\mathcal{T}$ are matrices for beamsplitters, and $\mathcal{J}_{sample}$ is the Jones matrix for the sample.

The fields at the two detectors are

$$\mathbf{E}_I = \mathcal{P}_x (\mathbf{E}_{sig} + \mathbf{E}_{LO}) \quad \text{and} \quad \mathbf{E}_Q = \mathcal{P}_y (\mathbf{E}_{sig} + \mathbf{E}_{LO}) \tag{7.82}$$

where

$$\mathcal{P}_X = \begin{pmatrix} 1 & 0 \\ 0 & 0 \end{pmatrix} \quad \text{and} \quad \mathcal{P}_Y = \begin{pmatrix} 0 & 0 \\ 0 & 1 \end{pmatrix}. \tag{7.83}$$

*Because we are almost never sure of the absolute phase, we can choose the zero of phase to make this component the cosine.

From Equation 6.94, the two signals are given by

$$P_I = \left(\mathbf{E}_{sig}^\dagger + \mathbf{E}_{LO}^\dagger\right) \mathcal{P}_x^\dagger \mathcal{P}_x \left(\mathbf{E}_{sig} + \mathbf{E}_{LO}\right) P_0,$$
$$P_Q = \left(\mathbf{E}_{sig}^\dagger + \mathbf{E}_{LO}^\dagger\right) \mathcal{P}_y^\dagger \mathcal{P}_y \left(\mathbf{E}_{sig} + \mathbf{E}_{LO}\right) P_0, \tag{7.84}$$

where the laser power is $P_0 = E_0^\dagger E_0$.

For simplicity, we will assume that the beamsplitter matrices are adequately described as scalars such as $\mathcal{T}_1 = \tau_1$ with $T_1 = \tau_1^* \tau_1$, and that the sample matrix is also a scalar, given by

$$\mathcal{J}_{sample} = Ae^{i\phi}. \tag{7.85}$$

With no significant loss of generality, we can say

$$\tau_1 = \sqrt{T_1} \qquad \tau_2 = \sqrt{T_2}$$
$$\rho_1 = \sqrt{R_1} \qquad \rho_2 = \sqrt{R_2}$$

Then

$$P_I = A^2 T_2 R_1 P_0/2 + R_2 T_1 P_0/2 + \sqrt{T_2 R_1 R_2 T_1} Ae^{j\phi} P_0/2 + \sqrt{T_2 R_1 R_2 T_1} Ae^{-j\phi} P_0/2 \tag{7.86}$$

$$P_Q = A^2 T_2 R_1 P_0/2 + R_2 T_1 P_0/2 - j\sqrt{T_2 R_1 R_2 T_1} Ae^{j\phi} P_0/2 + j\sqrt{T_2 R_1 R_2 T_1} Ae^{-j\phi} P_0/2. \tag{7.87}$$

Both of these results are real and positive. In both cases, the mixing terms are complex conjugates and their imaginary parts cancel. In Equation 7.86, the two mixing terms sum to $\sqrt{T_2 R_1 R_2 T_1} A \cos\phi P_0/2$. In Equation 7.87, they sum to $\sqrt{T_2 R_1 R_2 T_1} A \sin\phi P_0/2$. The mixing terms provide the information we want about the object: the amplitude, A, and the phase, ϕ. Having collected the data, we can compute the sum

$$P_I + jP_Q = (1+j)\left(A^2 T_2 R_1 P_0/2 + R_2 T_1 P_0/2\right) + \sqrt{T_2 R_1 R_2 T_1} Ae^{j\phi} P_0. \tag{7.88}$$

The fourth terms in the two equations have canceled, and the third terms (first mixing terms) have added, providing us with the complex representation of the field. This result is not quite the good news that it appears to be, because our measurements, P_I and P_Q, contain not only the mixing terms, but also the DC terms. We need at least one more measurement to remove the DC terms. One approach is to measure these two DC terms independently, by blocking first the LO and then the signal path. We can subtract these terms from Equations 7.86 and 7.87 before combining them in Equation 7.88. Alternatively, we could add two more detectors on the other output of the interferometer in a balanced mixer configuration. We recall from Section 7.1.5 that this approach removes the DC terms. We can write equations for the other two outputs as

$$P_{-I} = T_2 R_1 P_0/2 + R_2 T_1 P_0/2 - \sqrt{T_2 R_1 R_2 T_1} Ae^{j\phi} P_0/2 - \sqrt{T_2 R_1 R_2 T_1} Ae^{-j\phi} P_0/2 \tag{7.89}$$

$$P_{-Q} = T_2R_1P_0/2 + R_2T_1P_0/2 + j\sqrt{T_2R_1R_2T_1}Ae^{j\phi}P_0/2 - j\sqrt{T_2R_1R_2T_1}Ae^{-j\phi}P_0/2. \quad (7.90)$$

Then we can write the complex signal as

$$P_I + jP_Q + j^2P_{-I} + j^3P_{-Q} = 2\sqrt{T_2R_1R_2T_1}Ae^{j\phi}P_0. \quad (7.91)$$

The DC terms have multipliers 1, j, -1, and $-j$, so that they sum to zero. Imperfect beamsplitters, cameras, or electronics will result in imperfect cancellation and, as was the case with balanced detection, we must be careful to adjust the "zero point" of the signal calculated with this equation. If we allow for realistic polarizing elements and alignment errors, the scalars in these equations all become matrices, and the calculation looks more complicated. Numerical computations may be useful in determining tolerances.

Others have also developed polarization-based techniques for phase imaging with multiple phases simultaneously. One example uses a pixelated array of wire-grid polarizers at the camera.[46].

Take-Away

Optical Quadrature:
Optical quadrature detection and other similar techniques achieve phase-shifting interferometry with four phases simultaneously. The four phases are separated by polarization techniques. Such techniques are often useful in coherent imaging.

7.3.4 Other Imaging Approaches

If we are examining fringe patterns in an image, one approach is to introduce a **deliberately tilted reference wave**. Then if the signal and reference are both plane waves, we expect parallel fringes in the direction orthogonal to the tilt, with a spacing that increases with the tilt angle. A change in the phase of the signal will cause the pattern to shift in one direction or the other. Provided that the phase shifts from pixel to pixel are small enough, we can then recover the complex signal by comparing the fringe pattern to the one predicted for a constant signal phase. Examples will be shown in Section 7.4.3 on optical testing. Examples image of the interference between reflection from a mirror and a tilted reference are shown in Figure 7.15.

Take-Away

Tilted Wavefront:
A tilted wavefront can be used to resolve the ambiguities. Higher pixel density is required to capture the resulting fringes, but the approach is simpler than the use of multiple cameras and associated optics in optical quadrature, and faster than phase-shifting.

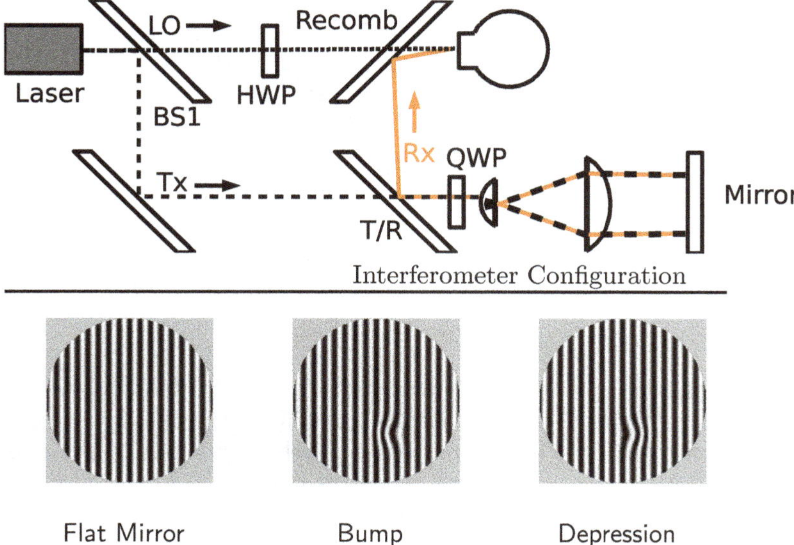

FIGURE 7.15
Tilted reference. The reference wave is tilted with respect to the wavefronts returning from a mirror. If we know the direction of the tilt, we can distinguish between a bump and a depression on the surface.

Another variation is to derive the local oscillator from a portion of the signal beam. One example is **phase-contrast microscopy**. Light that is not scattered by the specimen is used as a reference and the scattered light is used as the signal. Frits Zernike developed one technique in 1934[260], for which he was awarded the Nobel Prize in Physics in 1953. Basically, the source light passes through a small annulus in the pupil plane. On the imaging side, the image of the annulus is phase-shifted by 90 degrees, relative to the remainder of the pupil. One advantage is that the signal and reference pass through a common path eliminating the phase variations across the two paths of a Mach–Zehnder interferometer, providing exceptional stability. Another advantage is that the interfering beams are nominally at the quadrature point, maximizing the signal.

Nomarski[167;168] invented a method in which the reference is the same as the signal except for a slight shear or spatial shift. The technique is commonly called **Differential Interference Microscopy** (DIC), and combines the signal and reference using polarization. The result is an image of the directional derivative of the phase in the direction of the shear. Again the common path leads to exceptional stability at the quadrature point.

Some more recent work with common-path techniques is the imaging of biological materials with full-field quantitative phase imaging[182;104].

Take-Away

Common Path Interferometer:
Common-path interferometers can provide exceptional stability. The reference wave is derived in some way from the signal wave.

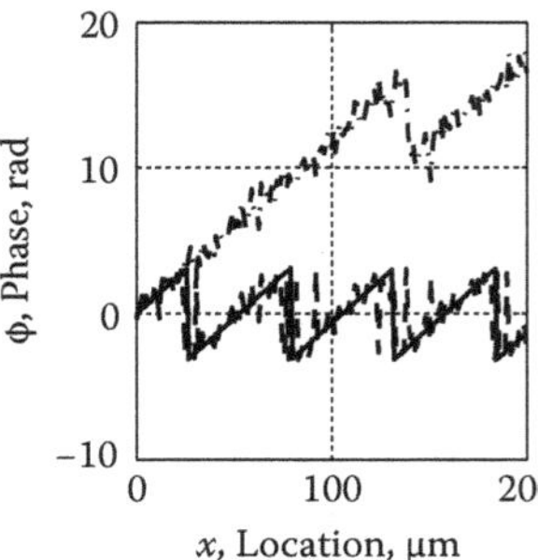

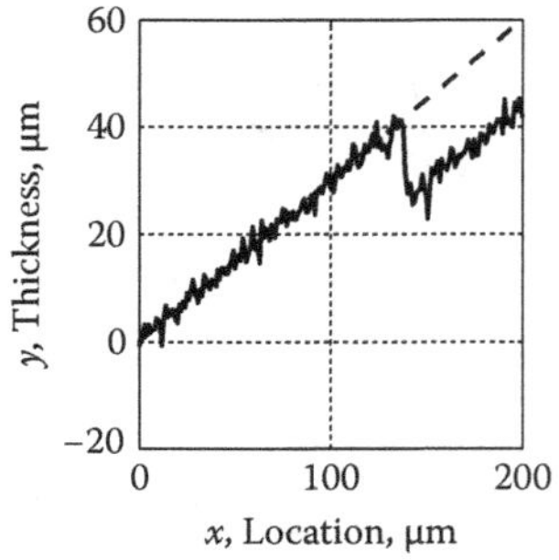

FIGURE 7.16
Phase unwrapping. The object here is a prism with an index of refraction 0.04 above that of the background medium. The left panel shows the phase (solid line), the phase with noise added to the field (dashed line), and the unwrapped phase (dash-dot). There is a phase-unwrapping error around $x = 130\,\mu\text{m}$. The right panel shows the original prism thickness and the reconstructed value obtained from the unwrapped phase.

7.3.5 Comparison

We have considered several different approaches to removing the ambiguities that arise in interferometry. Because interferometry holds the promise of measuring very small changes in distance or index of refraction, and because the ambiguities are so fundamental to the process, many efforts have been made to develop solutions.

In phase-shifting interferometry, it is important that neither the sample nor the interferometer change during the time taken by the measurements. Because the change usually involves mechanical motion, it is slow.

In the offset LO technique, different phase shifts are obtained at different times because the reference frequency is shifted. The signal must not vary too fast, although it can generally vary much faster than in phase-shifting interferometry. Specifically in Figure 7.12, the Doppler spectrum must not contain frequencies higher than the offset frequency or else aliasing will occur. Offset frequencies of tens to hundreds of Megahertz are common. Because of the need for RF signal processing on each receiver channel, most lidar systems work with a single point or at most a small imaging array. Scanning is often used to produce images with more pixels.

Optical quadrature is similar to phase-shifting, except that four phases are obtained at the same time. The price for this simultaneity is the need to divide the light among four detectors or cameras and the increased size, complexity, and cost of the equipment.

In the tilted wavefront technique, different phase shifts are obtained at different locations in the image. We can think of the tilt as introducing an "offset spatial frequency." By analogy with the offset LO technique, the sample must not produce phase variations too fast in comparison to than those introduced by the tilt. The concept of spatial frequency will be discussed more in Section 8.5.1, and in Chapter 11.

7.3.6 Periodicity Issues

We have addressed three of the four ambiguities in our list. The last ambiguity is the result of the optical waves being periodic. A phase shift of ϕ produces an irradiance which is indistinguishable from that produced by a phase shift of $\phi + 2\pi$. If the range of possible phases is known, and subtends less than 2π, then very precise absolute measurements are possible. If the actual value is not known, but the range of values is less than 2π, very precise relative measurements are possible. However, if the phase varies over a larger range, then resolving the ambiguities presents greater challenges.

Phase unwrapping techniques may be of some use. The success of such techniques depends on the phase change from one measurement to the next being less than π. Moving from one sample in time (or one pixel in an image) to the next, we test the change in phase. If it is greater than π, we subtract 2π, and if it is less than $-\pi$, we add 2π. In this way, all phase changes are less than π. Other more sophisticated algorithms have also been developed. An example is shown in Figure 7.16. A prism with an index of refraction 0.04 above that of the background was simulated. The thickness varies from zero at the left to 60 µm at the right. The light source is a helium neon laser with $\lambda = 632.8\,\text{nm}$. The solid line shows the phase of the wave after passing through the prism. The field was computed as $e^{j2\pi ny/\lambda}$. Complex random noise was added, and the phase plotted with the dashed line. The MATLAB® unwrap function was used to produce the dash-dot line. The thickness was estimated from the unwrapped phase,

Phase unwrapping is useful if noise is reasonably low and phase change from one measurement to the next is less than π.

$$\hat{y} = \frac{\lambda\phi}{2\pi n}, \tag{7.92}$$

and plotted as a solid line in the right panel, along with the actual thickness as a dashed line. At one point, the phase noise was too large for correction, and an error occurred. From that point forward, the error propagated. Local relative phases are still correct, but the phase relationship to the rest of the object has been lost.

For two-dimensional data, such random errors will occur at different points across the image. It is possible to use the extra information provided in two dimensions to improve the phase unwrapping, but at the expense of added complexity[77].

2-D phase unwrapping is harder to do but has some error correction.

Another technique for removing phase ambiguities is to use a second wavelength to resolve some of the ambiguities. The ambiguity interval is then determined by the difference wavelength resulting from combining the two:

$$\frac{1}{\lambda_{dif}} = \frac{1}{\lambda_2} - \frac{1}{\lambda_1}, \tag{7.93}$$

and the resolution limit is obtained from the wavelength:

$$\frac{1}{\lambda_{sum}} = \frac{1}{\lambda_2} + \frac{1}{\lambda_1}. \tag{7.94}$$

Multiple wavelengths can be used to resolve further ambiguities. Ultimately, combining enough different wavelengths, we can synthesize a pulse.

Multiple wavelengths can be used to reduce phase ambiguity.

Multiple wavelengths can also be produced by modulation. Various techniques can be adapted from radar. To mention but one example, the frequency of the laser can be "chirped" using methods explained in Section 7.5.4. If the laser transmitter frequency is $f = f_1 + f't$, then a target with a Doppler frequency f_{DR} at a distance z will have a return frequency of $f_{up1} + f_{DR} + 2f'z/c$. If the direction of chirp is changed, then $f_{down1} + f_{DR} - 2f'z/c$. Measuring the two frequency shifts, we can recover both the velocity and distance. In Section 7.5, we will discuss how the frequency modulation can be achieved.

Take-Away

Phase Unwrapping:
After Ambiguities between amplitude and phase as well as ambiguities of the sign of the phase have been removed, the remaining ambiguity arises from the 2π periodicity of the trigonometric functions. To some extent it can be removed by computational unwrapping techniques or multiple phase measurements with different wavelengths.

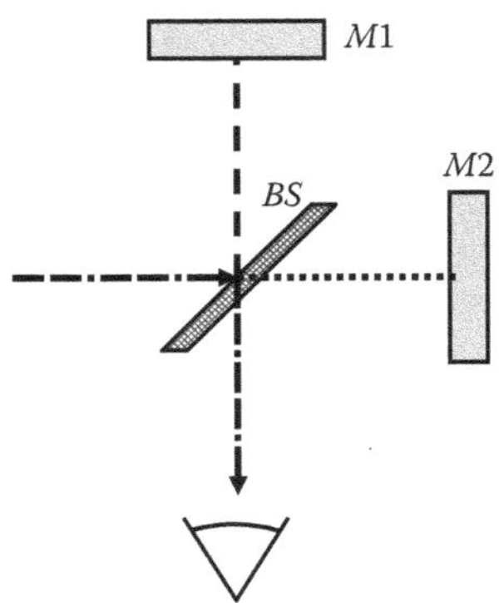

FIGURE 7.17
Michelson interferometer. Light from the source is split into two paths by the beamsplitter. One path is reflected from each mirror, and the two are recombined at the same beamsplitter.

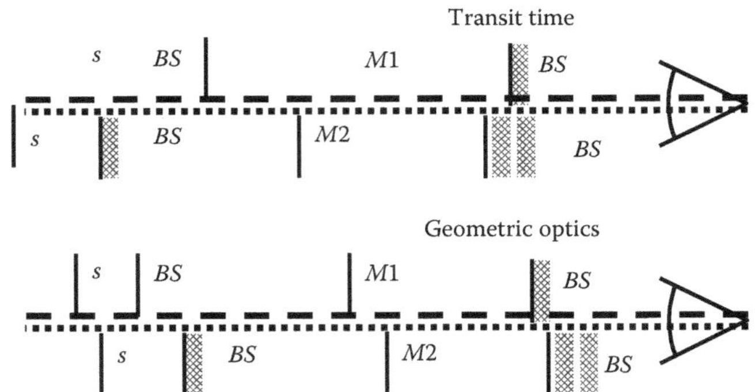

FIGURE 7.18
Straight-line layout. The path lengths can be determined from the distances and indices of refraction. Light on one path passes through the substrate of the beamsplitter once, but light on the other path does so three times. Increased index of refraction lengthens the optical path, but shortens the geometric path.

7.4 Michelson Interferometer

The next interferometer we will consider is the Michelson interferometer, shown in Figure 7.17, made famous in experiments by Albert Michelson and Edward Morley[156] that ultimately refuted the existence of the "luminous aether." For this and other work, Michelson was awarded the Nobel Prize in Physics in 1907. An historical account of this research was published in 1928 as proceedings of a conference attended by Michelson, Lorentz, and others[155].

7.4.1 Basics

The Michelson interferometer is a simple device consisting at a minimum of only three components: two mirrors and a single beamsplitter, which serves the functions of both $BS1$ and $BS2$ in the Mach–Zehnder interferometer. Light from the source is split into two paths by the beamsplitter. One path serves as a reference and the other as the signal. A sample may be placed in the signal arm as was the case with the Mach–Zehnder interferometer. In the Michelson interferometer, the light traverses the sample twice, doubling the phase shift. The Michelson interferometer is particularly useful for measuring linear motion by attaching one mirror to a moving object.

The signal at the detector is

$$I = I_0 \left[R_{BS} R_{M1} T_{BS} + T_{BS} R_{M2} R_{BS} \right. \\ \left. + \sqrt{R_{BS}^2 R_{M1} R_{M2} T_{BS}^2} \cos(\delta\phi) \right]. \tag{7.95}$$

The phase difference can be determined as it was for the Mach–Zehnder, using the straight-line layout, shown in Figure 7.18. This figure can be constructed by analogy with Figure 7.4. It is important to consider the optical path length introduced as the light passes through the substrate material of the beamsplitter. Assuming that the reflection occurs at the front surface, light that follows the path containing mirror $M1$ passes through the beamsplitter once, en-route from $M1$ to the observer. The other path passes through the beamsplitter three times: once from the source to $M2$ and twice on reflection from $M2$ to the observer. Compensation for the resulting problems is discussed in Section 7.4.2.

Fig. 7.4

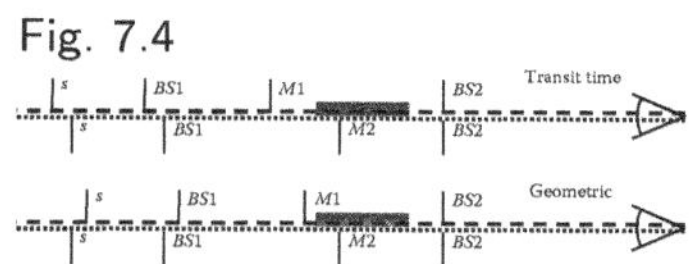

The equations for length follow the same approach used in Figure 7.4. With the Michelson interferometer, the use of curved mirrors is common. Computing the length changes produced by a mirror with a radius of curvature r is easily accomplished by placing a perfect lens of focal length $r/2$ at the location of the mirror in Figure 7.18.

Alignment of the interferometer is somewhat easier than it is for the Mach–Zehnder. Alignment is accomplished by aligning the two angles, azimuth and elevation, of the two mirrors, $M1$ and $M2$. A good approach is to align each mirror so that light from the source is directed back toward the source. In fact, when the interferometer is properly aligned, the light will be directed exactly back to the source. The Mach–Zehnder interferometer has two outputs, and we have seen that the balance between these two outputs is required by conservation of energy. The second output of the Michelson interferometer reflects directly back into the source. This reflection may be problematic, particularly in the case of a laser source. For this reason, the use of the Michelson interferometer in applications such as laser radar is less attractive than would be suggested by its simplicity.

In Practice

Alignment is often done by adjusting the two mirrors until light from both goes directly back to the source.

When a Michelson interferometer is used with a laser source, a Faraday isolator (Section 6.5.3) is often employed to prevent the reflections from reaching the laser.

7.4.2 Compensator Plate

The fact that the light passes through the substrate of the beamsplitter once on one path and three times on the other may be problematic, as it results in changes of transit time and radius of curvature. In the top part of Figure 7.18, the path involving mirror $M2$ has a longer transit time. Cannot we simply move $M1$ slightly further away from the beamsplitter to compensate for this additional path? Yes, we can, but only at the expense of making the difference in geometric paths in the bottom part of the figure worse. In this case the temporal match will be perfect, but the density of fringes will be increased, perhaps so much that the fringes can no longer be observed. If we shorten the path with $M1$ we match wavefronts, but at the expense of increasing the temporal disparity.

The temporal disparity is an obvious concern if we use a pulsed source. If the pulse is shorter than the temporal disparity, then the pulse will arrive at two different times via the two paths, and no interference will occur, as shown in Figure 7.19. In Chapter 10, we will discuss temporal coherence in some detail, and will see that interference will not occur if the path difference exceeds the "coherence length" of the source.

Fig. 7.18:

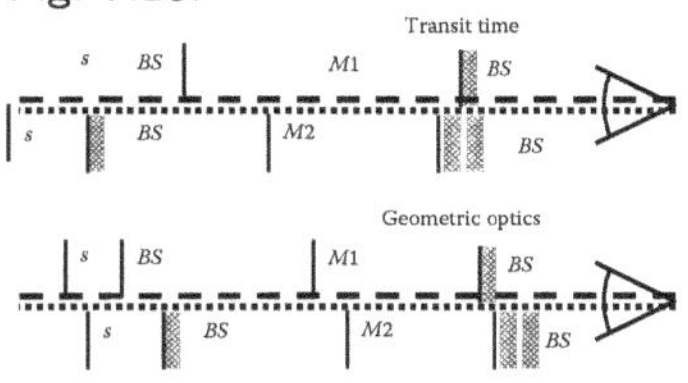

We need to match both wavefronts (geometric-optics path lengths) and optical path lengths to see fringes.

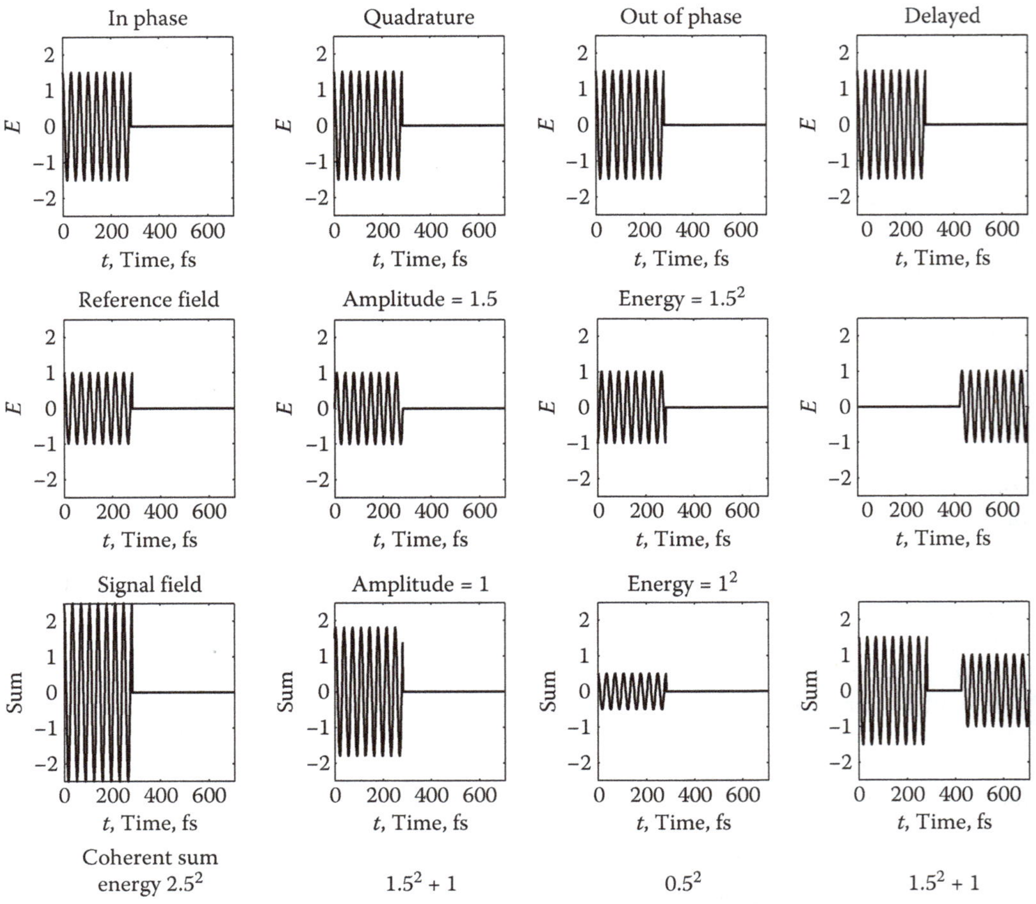

FIGURE 7.19
Interference with a pulsed source. In the left column, the reference (top) and signal (center) are in phase, and constructive interference occurs in the sum (bottom). In the second column, the reference is a cosine and the signal is a sine. The waves add in quadrature, and the total energy is the sum of the individual energies. In the third column, the waves are out of phase and destructive interference occurs. In the final column, the waves do not overlap in time, so no interference occurs.

Example

We will discuss coherence length in some detail in Chapter 10, but let's see if we can make a decent guess about the coherence length of a light-emitting diode (LED). Assume the linewidth of the LED is 20 nm at a center wavelength of 635 nm.

The LED emits light at wavelengths between $\lambda = 625\,\text{nm}$ and $645\,\text{nm}$ or $f = c/\lambda = 480.00\,\text{THz}$ and $465.12\,\text{THz}$. The bandwidth is the difference between these or 14.9 THz. The coherence time is roughly the inverse of the bandwidth or $\tau = 67.1\,\text{fs}$. Light travels $c\tau = 20\,\mu\text{m}$ in that time, so if the OPL difference between the two paths is greater than that, interference will not occur. Even a thin beamsplitter will exceed this limit.

To match both the wavefronts and times, we can insert a **compensator plate** in the path between the beamsplitter and $M1$, as shown in Figure 7.20. As shown in Figure 7.21,

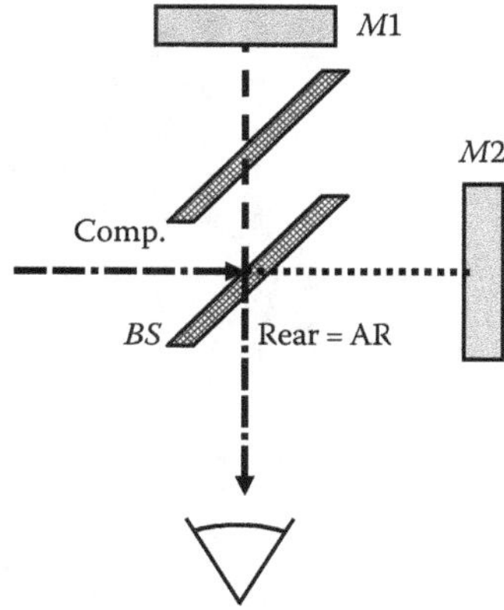

FIGURE 7.20
Compensator plate. The thickness of the compensator plate (Comp) is matched to that of the beamsplitter (BS), to match both the optical paths and the geometric paths of the two arms of the interferometer.

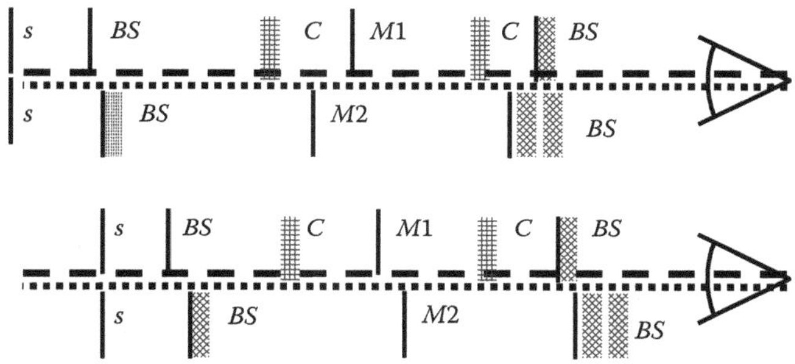

FIGURE 7.21
Compensator plate straight-line layout.

if $M1$ is adjusted so that its distance from the beamsplitter matches that of $M2$, then the matching will be perfect.

In Practice

With low coherence sources, it may be necessary to use a compensator plate to match both the pathlengths and the glass thickness in the two paths. If the coherence length of the source is sufficiently long, we can adjust the interferometer to match the geometric-optics distance, and not use the compensator. The concept of coherence length is discussed quantitatively in Section 10.3.2.

7.4.3 Application: Optical Testing

One common application of interferometers based on the Michelson interferometer is in the field of optical testing. Using a known reference surface as $M2$ and a newly fabricated surface, intended to have the same curvature as $M1$, we expect a constant phase across

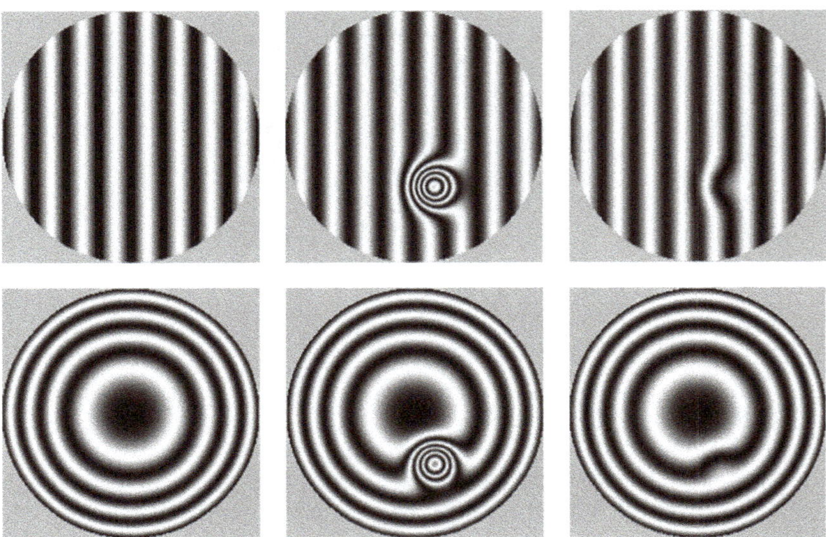

FIGURE 7.22
Synthetic fringe patterns in optical testing. In the top row, the mirror curvatures are matched, but one of the mirrors is deliberately tilted. In the bottom row, the tilt is removed, but one of the mirrors is displaced to generate circular fringes. In the first column, both mirrors are perfect. In the other two columns, a bump is present on the mirror. The phase shift is doubled because of the reflective geometry.

the interference pattern. If one mirror is tilted with respect to the other, we expect a pattern of straight lines, and if one mirror is displaced axially with respect to the other, we expect circular fringes. The left column of Figure 7.22 shows some synthetic fringe patterns for these two cases. Now if there are fabrication errors in the test mirror, there will be irregularities in the resulting fringe pattern. The second column shows fringes that would be expected with a bump on the mirror having a height of two wavelengths. Because the round-trip distance is decreased by twice the height of the bump, the result is a departure by four fringes from the expected pattern.

In an experiment, we could measure this height by drawing a line along the straight portions of the fringes and looking for departures from this line. The right column shows the fringes for a bump with a height of 0.2 wavelengths (0.4 fringes).

In the fabrication of a large mirror, the results of such a measurement would be used to determine further processing required to bring the surface to the desired tolerance. In this case, a matching reference surface is not practical. An interferometer can be designed with multiple lenses so that a large mirror under fabrication can be tested against a smaller standard one[83].

Take-Away

Michelson Interferometer:

- The Michelson interferometer consists of two mirrors and a single beamsplitter used both for splitting and recombining the beams.
- To match both the wavefronts and the time delays, a compensator plate may be required.
- Flat or curved mirrors can be used.

- A Faraday isolator may be needed to avoid reflection back into the source, especially if the source is a laser.
- Variations on Michelson interferometers are useful in optical testing.

7.5 Fabry–Perot Interferometer

The final interferometer we will consider is the Fabry–Perot. Of the three, this is physically the simplest, and conceptually the most complicated. It consists of two mirrors, as shown in Figure 7.23. Phenomena in the Mach–Zehnder and Michelson interferometers could be explained by the sum of two waves. In the Fabry–Perot, the sum is infinite. A light wave enters with an irradiance I_1 from the left, is transmitted through the first beamsplitter. Some of it is transmitted through the second, and some is reflected. Of the reflected portion, some is reflected by the first, and then transmitted by the second, in a manner similar to what is seen by an observer sitting in a barber's chair with a mirror in front and behind. If the mirrors are sufficiently parallel, the observer sees his face, the back of his head, his face again, and so forth without end.

A Fabry–Perot interferometer sums an infinite number of waves. This fact leads to improved performance.

The Fabry–Perot interferometer is used in various ètalons or filters, and in laser cavities. The concept forms the basis for thin-film coatings used for beamsplitters and mirrors. In this section, we will discuss the first two of these, and the last will be the subject of its own section later.

Before we begin our mathematical discussion, let us try to understand the operation from an empirical point of view. If we did not know about interference, we might expect the transmission of the device to be the product, $T = T_1T_2$. However, some of the light incident on the second surface will reflect back to the first, where some of it will be reflected forward again. If the round-trip optical path length between the reflectors is an integer multiple of the wavelength,

$$2\ell = N\lambda$$

Here the optical path is a round-trip, 2ℓ for constructive interference.

then constructive interference will occur. In fact, if the reflectivities are large, most of the light will recirculate between the mirrors. This equation indicates that the resonant frequencies of the interferometer will be linearly spaced:

$$f = Nf_0 \qquad f_0 = \frac{c}{2\ell}, \tag{7.96}$$

where f_0 is the fundamental resonant frequency. The spacing of adjacent resonant frequencies is also f_0 and this frequency is often called the **free spectral range** or **FSR**:

Free Spectral Range (FSR) defined.

$$FSR = \frac{c}{2\ell}. \tag{7.97}$$

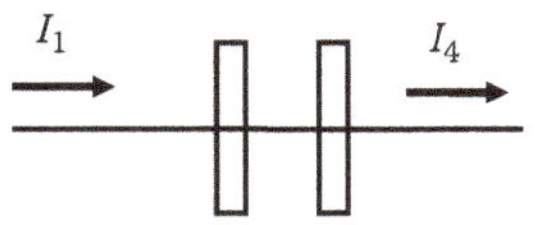

I_1 I_4

FIGURE 7.23
Fabry–Perot interferometer. This interferometer consists of two partial reflectors facing each other. Multiple reflections provide very narrow bandwidth.

When the source light is turned on, if it is at a resonant frequency, the **recirculating power** inside the interferometer will increase on every pass, until steady state is reached. Neglecting losses, the small fraction of this power that is extracted in steady state will be equal to the input by conservation of energy. Therefore, the recirculating power must be

$$P_{recirculating} = \frac{P_{out}}{T_2} = \frac{P_{out}}{1 - R_2} = \frac{P_0}{1 - R_2}. \tag{7.98}$$

Recirculating power can be high.

If the output mirror reflectivity is $R_2 = 0.999$, then the recirculating power is 1000 times the input power. The beamsplitters must be able to handle this level of power.

The multiple reflections lead to a narrower bandwidth relative to the free spectral range.

The outputs of the other interferometers varied sinusoidally with the phase difference between their two paths. Like all interferometers, the FSR is determined by the difference in path lengths, in this case, $c/2\ell$. But because the Fabry–Perot interferometer uses multiple paths, the response is not sinusoidal, and very narrow bandwidths can be obtained. We can think of the probability of a photon remaining in the system after one round trip as R_1R_2. After N round trips, the probability of a photon remaining in the interferometer is

$$P_N = (R_1R_2)^N. \tag{7.99}$$

There is a 50% probability that a photon will still not have left after

$$N = -\log 2 / \log (R_1R_2) \tag{7.100}$$

For good performance, high reflectivity mirrors are important.

trips. For $R_1 = R_2 = 0.999$, $N = 346$. Thus, in some sense we expect some of the behavior of an interferometer of length $N\ell$. The result is that, while the passbands of the Fabry–Perot are separated by FSR, their width can be much smaller.

7.5.1 Basics

Three derivations of the Fabry–Perot equations illustrate different concepts.

There are at least three different derivations of the equations for a Fabry–Perot interferometer. The first, shown in Figure 7.24A, is to compute the infinite sum discussed earlier. Although the rays are shown at an angle for illustrative purposes, the light is usually incident normally on the surface. The second approach (Figure 7.24B) is useful when the reflections are the results of Fresnel reflections at the surfaces of a dielectric. The electric field on the left of the first surface is considered to be the sum of the incident and reflected fields, and the field on the right of the second surface is considered to be the transmitted field. Magnetic fields are treated the same way, and the internal fields are matched using boundary conditions, as was done in Section 6.4. We will use this in Section 7.7, when we study thin films. The third approach (Figure 7.24C) is a network approach, in which we compute the electric field in each direction on each side of each surface. For now we will consider the first and third approaches.

7.5.1.1 Infinite Sum Solution

An observer looking at a point source through a Fabry–Perot interferometer will see the source repeated an infinite number of times as in Figure 7.25, with decreasing amplitude. We recall the discussion of the barber's chair with which we began our discussion of the Fabry–Perot interferometer. In general, the computation of the irradiance from this source is going to be difficult, because for each image, we must consider the loss of power resulting from the reflections and transmissions, and the changing irradiance caused by the increasing distance, and the change in wavefronts.

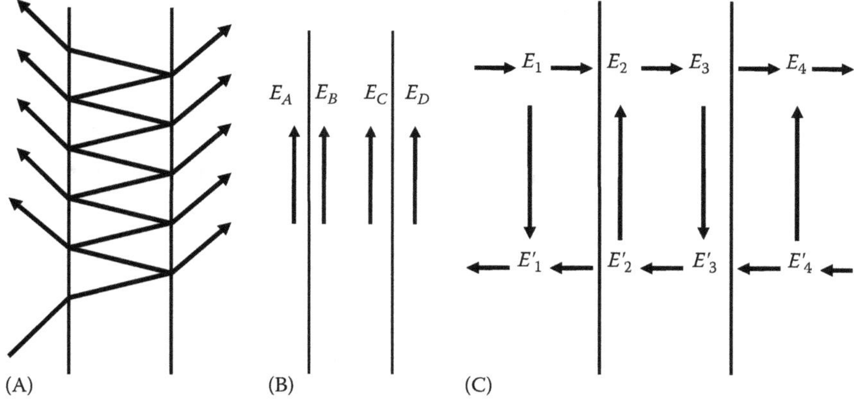

FIGURE 7.24
Analysis of Fabry–Perot interferometer. Three approaches are often used in deriving equations. An infinite sum can be computed either in closed form, or using numerical methods (A). If the reflectors are Fresnel reflections at a dielectric interface, an approach based on boundary conditions works best (B). Finally, a network approach is often useful (C).

If the incident light consists of plane waves, however, the computation is easier, and we quickly arrive at the infinite sum

$$E_t = E_0 \left[\tau_1 e^{jk\ell}\tau_2 + \tau_1 e^{jk\ell}\rho_2 e^{jk\ell}\rho_1' e^{jk\ell}\tau_2 + \cdots\right]$$

$$E_t = E_0\tau_1\tau_2 e^{jk\ell}\sum_{m=0}^{\infty}\left(\rho_1'\rho_2 e^{2jk\ell}\right)^m \tag{7.101}$$

for transmission, and

$$E_r = E_0$$

$$\left[\rho_1 + \tau_1 e^{jk\ell}\rho_2 e^{jk\ell}\tau_1' + \tau_1 e^{jk\ell}\rho_2 e^{jk\ell}\rho_1' e^{jk\ell}\rho_2 e^{jk\ell}\tau_1' + \cdots\right] \tag{7.102}$$

$$E_r = E_0\rho_1 + \tau_1\tau_1' e^{2jk\ell}\sum_{m=0}^{\infty}\left(\rho_1'\rho_2 e^{2jk\ell}\right)^m .$$

In the notation here, primed quantities are used when the incident light is traveling from right to left. The field definitions are shown in the figure. The reflectivity of the first surface is ρ_1 for light incident from the left, and ρ_1' for light incident from the right.

Now, for any real variable, x such that $0 < x < 1$,

$$\sum_{m=0}^{\infty} x^m = \frac{1}{1-x}. \tag{7.103}$$

FIGURE 7.25
Fabry–Perot straight-line layout. The observer sees the source repeated an infinite number of times, with ever decreasing amplitude and increasing distance.

The equations for transmission and reflection become

$$E_t = E_0 \tau_1 \tau_2 e^{jk\ell} \frac{1}{1 - \rho_1' \rho_2 e^{2jk\ell}}, \tag{7.104}$$

and

$$E_r = E_0 \rho_1 + \tau_1 \tau_1' e^{2jk\ell} \frac{1}{1 - \rho_1' \rho_2 e^{2jk\ell}}. \tag{7.105}$$

The transmission,

$$T = \left| \frac{E_t}{E_0} \right|^2,$$

assuming the same index of refraction before and after the interferometer, is

$$T = \tau_1 \tau_2 \tau_1^* \tau_2^* \frac{1}{1 - \rho_1' \rho_2 e^{2jk\ell}} \frac{1}{1 - (\rho')_1^* \rho_2^* e^{-2jk\ell}}$$

$$T = T_1 T_2 \frac{1}{1 - 2\sqrt{R_1 R_2} \cos(2k\ell) + R_1 R_2}.$$

Loss along the path can be included as a "lumped" quantity $e^{-2\Im nk\ell_p}$

We have made two assumptions in this last step that are usually not troublesome. First we have assumed no loss in the material ($\Im n = 0$, as discussed in Section 6.4.6). Such loss can easily be included by first substituting

$$\ell = n\ell_p$$

with a complex n to account for the loss

$$\eta_{rt} = R_1 R_2 e^{-2\Im nk\ell_p}. \tag{7.106}$$

Note that if $\Im n < 0$ the exponential is greater than one, indicating gain rather than loss. If $\eta rt > 1$ then power can increase with successive round trips. This is the principle behind laser operation. The gain per unit length is $\ln nk$.

The physical path length of the Fabry–Perot interferometer is ℓ_p. The other assumption is that ρ_1' and ρ_2 are real. If the reflections introduce a phase shift, it can be included along with $2jk\ell$ inside the cosine. Usually we would adjust ℓ experimentally, and this additional phase term is unlikely to be noticed. Turning our attention to the cosine, we substitute $\cos 2x = 1 - 2\sin^2 x$, to obtain

$$T = T_1 T_2 \frac{1}{1 - 2\sqrt{\eta_{rt}} + 4\sqrt{\eta_{rt}} \sin^2(k\ell) + \eta_{rt}}$$

$$T = \frac{T_1 T_2}{R_1 R_2} \frac{\eta_{rt}}{\eta_{rt} - 1} \frac{1}{1 + F \sin^2(k\ell)}, \qquad (7.107)$$

where we define the **Finesse** as

Finesse is a measure of interferometer performance. High finesse means that the bandwidth is a small fraction of the free-spectral range.

$$F = \frac{4\sqrt{\eta_{rt}}}{1 - \eta_{rt}}. \qquad (7.108)$$

The infinite-sum solution can incorporate effects such as mirror misalignment by summing tilted wavefronts. Other realistic imperfections can also be modeled. Most such calculations are best done by the use of computers, rather than analytical solutions. We may be able to make some rough estimates of the tolerances on the interferometer by understanding the infinite sum. We noted in Equation 7.100 that the probability of a photon remaining in the interferometer is 1/2 after N round trips for

$$N = -\log 2/\log(R_1 R_2).$$

The sum to N, or even better, to a few times N, is thus a good first approximation to the infinite sum. Recalling the analogy with the mirrors in front and in back of the barber's chair in our introduction to the Fabry–Perot interferometer, if one of the mirrors is not exactly parallel to the other, each image will be displaced by a small amount from the previous. After a number of reflections the image would be far away from the original. In order for interference to occur in the interferometer, the beam must not be displaced by more than a small fraction of its width in the N round trips required to add all the wave contributions. Thus the tolerance on mirror tilt, $\delta\theta$ is the width of the beam divided by the total distance in N round trips,

$$\delta\theta \ll 2\frac{D}{N\ell_p}, \qquad (7.109)$$

where

- D is the diameter of the beam
- ℓ_p is the physical length of the interferometer

Eq. 7.97: $FSR = \frac{c}{2\ell}$.

Eq. 7.108: $F = \frac{4\sqrt{\eta_{rt}}}{1-\eta_{rt}}$.

Eq. 7.106: $\eta_{rt} = R_1 R_2 e^{-2\Im n k \ell_p}$.

Example

Let's consider a Fabry–Perot 1 mm thick, with two mirrors having 99% reflectivity. What are the FSR and finesse? The FSR is given by Equation 7.97. Thus for $\ell =$ 1 mm,

$$FSR = 150\,\text{GHz}.$$

The finesse is given by Equation 7.108. with Equation 7.106. Assuming no bulk loss ($\Im n = 0$),

$$F = \frac{4\sqrt{R_1 R_2}}{1 - R_1 R_2} = \frac{4\sqrt{0.99^2}}{1 - 0.99^2} = 49.$$

Later, in Section 7.5.1.3, we will relate the finesse to the spectral resolution.

Take-Away

Infinite-Sum for Fabry–Perot:

- The infinite-sum analysis of the Fabry–Perot interferometer tells us that the bandwith can be very narrow because of the multiple passes.
- The performance of the interferometer is characterized by the finesse.

In Practice

Alignment tolerances become tighter as finesse becomes higher, because the light is reflected more times.

In Practice

As in the case of electrical filters, the response time is inversely related to the bandwidth. The transmission of a narrow-band Fabry–Perot interferometer will respond slowly to change. When the first light of a pulse is transmitted, the transmission will be determined by the first term in the infinite series. After one more round-trip time, the second term is added, and so forth asymptotically approaching steady state. Likewise, when the source is turned off, the output will decay. This effect can be important with short pulses.

Fig. 7.24:

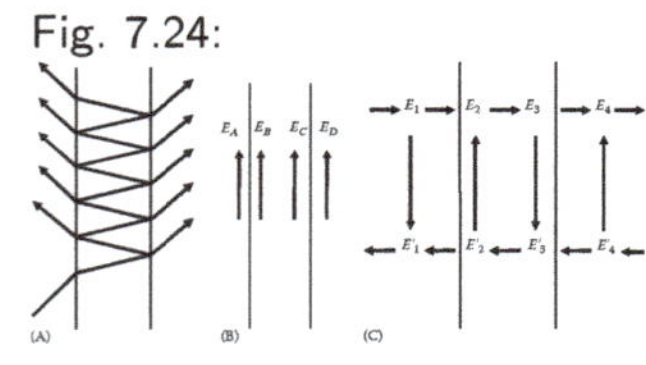

7.5.1.2 Network Solution

Next, we consider the third derivation suggested in Figure 7.24. We relate the fields using field reflection and transmission coefficients, ρ and τ, regardless of how these coefficients are obtained. As in the infinite-sum derivation, primed quantities are used when the incident light is traveling from right to left. The field definitions are shown in the figure. We have eight fields to compute. We assume that the incident fields, E_1 and E_4', are known, and so we are left with six unknowns, and six equations:

$$E_1' = E_1\rho_1 + E_2'\tau_1' \tag{7.110}$$

$$E_2 = E_1\tau_1 + E_2'\rho_1' \tag{7.111}$$

$$E_3 = E_2 e^{jk\ell} \tag{7.112}$$

$$E_2' = E_3' e^{jk\ell} \tag{7.113}$$

$$E_3' = E_4' + E_3\rho_2 \tag{7.114}$$

$$E_4 = E_3\tau_2 + E_4'\rho_2'. \tag{7.115}$$

We can assume that $E_4' = 0$. If there are fields incident from both sides, linear superposition allows us to solve the problem with $E_4' = 0$, reverse the diagram, solve with the input from the right, and add the two solutions. Using Equations 7.113 and 7.114, Equation 7.110 becomes

$$E_1' = E_1\rho_1 + E_2 e^{2jk\ell}\tau_1', \tag{7.116}$$

Equation 7.111 becomes

$$E_2\left(1 - \rho_1'\rho_2 e^{2jk\ell}\right) = E_1\tau_1, \tag{7.117}$$

and Equation 7.115 becomes

$$E_4 = E_2 e^{jk\ell}\tau_2. \tag{7.118}$$

The results are

$$E_1' = E_1 \left(\rho_1 + \frac{\tau_1\tau_1'}{1-\sqrt{\eta_{rt}}e^{2jk\ell}}\right)$$

$$E_4 = E_1 \frac{\tau_1\tau_2 e^{jk\ell}}{1-\sqrt{\eta_{rt}}e^{2jk\ell}}$$

Finally, we can define a reflectivity for the interferometer, $\rho = E_1'/E_1$, and a transmission, $\tau = E_4/E_1$, so that

$$\rho = \rho_1 + \frac{\tau_1\tau_1'}{1-\sqrt{\eta_{rt}}e^{2jk\ell}} \tag{7.119}$$

$$\tau = \frac{\tau_1\tau_2 e^{jk\ell}}{1-\sqrt{\eta_{rt}}e^{2jk\ell}}. \tag{7.120}$$

in agreement with Equations 7.104 and 7.105.

7.5.1.3 Spectral Resolution

The equations we have derived are for the fields. For the irradiances or powers,

$$T = \tau^*\tau \qquad R = \rho^*\rho, \tag{7.121}$$

assuming the same medium on both sides.

Neglecting multiple reflections, we would expect the transmission of the interferometer to be $T = T_1 T_2 e^{-\Im nk\ell}$.

The peak output from Equation 7.107 is

$$T_{max} = \frac{T_1 T_2}{R_1 R_2}\frac{\eta_{rt}}{\eta_{rt}-1}. \tag{7.122}$$

The value at half maximum is obtained when

$$F \sin^2(nk\ell) = 1, \tag{7.123}$$

so that

$$k\ell = \frac{2\pi}{\lambda}\ell = \frac{2\pi f}{c}\ell = \arcsin\frac{1}{\sqrt{F}}. \tag{7.124}$$

The full width at half maximum (FWHM) is

$$\delta f = 2\frac{c}{2\pi\ell}\arcsin\frac{1}{\sqrt{F}}. \quad (7.125)$$

More commonly this result is written in terms of the free spectral range

$$\frac{\delta f}{FSR} = \frac{2}{\pi}\arcsin\frac{1}{\sqrt{F}}. \quad (7.126)$$

If we were to measure a spectrum that consisted of two waves closely spaced in frequency, we would be able to determine that two frequencies were present if they were sufficiently far apart relative to δf, and we would see them as a single wavelength if not. The wavelength difference at which it is just possible to distinguish the two would depend on signal strength, noise, and sample time. We will pursue such a rigorous definition for spatial resolution in Section 8.7, but here we will simply define δf as the **spectral resolution**.

The fractional spectral linewidth approximation is easy to remember and produces good results unless the bandwidths are very large compared to the central frequency.

We can also approximate this width in wavelength units using the **fractional-linewidth approximation** as

$$\left|\frac{\delta\lambda}{\lambda}\right| = \left|\frac{\delta f}{f}\right|, \quad (7.127)$$

or

$$\frac{\delta\lambda}{\lambda} = \frac{\lambda}{\ell}\frac{\arcsin\frac{1}{\sqrt{F}}}{\pi}. \quad (7.128)$$

This dimensionless equation states that the linewidth in ratio to the central wavelength is proportional to the inverse of the interferometer length in wavelengths, and that the constant of proportionality is $\frac{\arcsin\frac{1}{\sqrt{F}}}{\pi}$. A Fabry–Perot interferometer many wavelengths long, if it also has a high finesse, can have a very small **fractional linewidth**, $\delta\lambda/\lambda$.

In Practice

Equation 7.127 is a very useful approximation. We will see this again when discussing linewidth in Chapters 10 and 12, and elsewhere. It is of particular importance because we usually think of light sources in terms of wavelength, while equations are often easier to understand in terms of frequency.

We will discuss two applications of the Fabry–Perot interferometer, first as a filter, and second as a laser cavity. Then we will conclude the discussion of this interferometer by considering a laser cavity with an intra-cavity filter.

7.5.2 Fabry–Perot Ètalon

Eq. 7.107: $T = \frac{T_1T_2}{R_1R_2}\frac{\eta_{rt}}{\eta_{rt}-1}\frac{1}{1+F\sin^2(k\ell)}$.

By definition, any Fabry–Perot interferometer can be called an **ètalon**, but the word is most often used to describe a narrow band-pass filter, often, but not always, tunable by changing length. We will consider an ètalon with a spacing of $\ell = 10\ \mu\text{m}$ as an example.

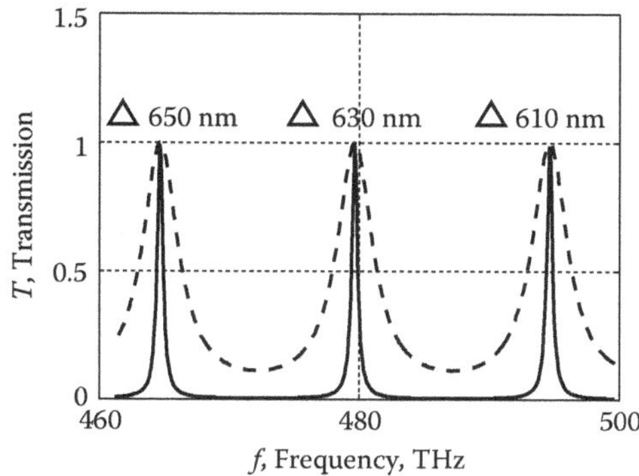

FIGURE 7.26
Ètalon transmission. An ètalon with a spacing of $10\,\mu\text{m}$ has a free spectral range of $c/2\ell = 15\,\text{THz}$. The width of the transmission peaks depends on the mirror reflectivities. The solid line shows the curve for a pair of 90% reflecting mirrors and the dashed line for a pair with reflectivity of 50%.

The free spectral range is $c/2\ell = 15\,\text{THz}$. We first consider the transmission of this ètalon as a filter, using Equation 7.107. Assuming no loss in the medium between the reflectors,

$$T = \frac{(1 - R_1)(1 - R_2)}{R_1 R_2 - 1} \frac{1}{1 + F \sin^2(k\ell)}, \tag{7.129}$$

with

$$F = \frac{4\sqrt{R_1 R_2}}{R_1 R_2 - 1}. \tag{7.130}$$

Figure 7.26 shows that the peaks are spaced by the expected 15 THz. The spacing in wavelength is given by

$$\lambda f = c \qquad \frac{d\lambda}{\lambda} \approx -\frac{df}{f}$$

$$\lambda_{FSR} \approx \lambda \frac{FSR}{f} = \frac{\lambda^2}{c} FSR \tag{7.131}$$

The spacing of modes in wavelengths is about $20\,\mu\text{m}$, but it is not exactly constant.

Two different sets of reflectivities are used. With 50% reflectors, the finesse is 2.67, the peaks are broad, and the valleys are not particularly deep. For 90% reflectors, the performance is substantially improved, with a finesse of 19. The FWHM from Equation 7.126 is

$$\frac{\delta f}{FSR} = \frac{2}{\pi} \arcsin \frac{1}{\sqrt{F}}.$$

These values have been chosen to illustrate the behavior in Figure 7.26. Typical ètalons have much better performance. With mirrors of $R = 0.999$, the finesse is 2000, and the FWHM is 1.4% of the FSR.

In Practice

An étalon can be used as a variable filter. By varying the thickness, the filter can be tuned. We must be careful about the all the passbands at multiples of the free-spectral range.

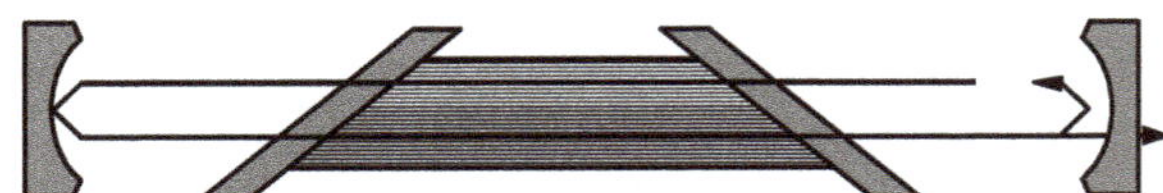

FIGURE 7.27
Laser cavity. The gain medium and the resonant cavity modes of the Fabry–Perot interferometer determine the frequencies at which the laser can work.

7.5.3 Laser Cavity as a Fabry–Perot Interferometer

A typical laser cavity, as shown in Figure 7.27, is also a Fabry–Perot interferometer. The cavity consists of a pair of partially reflecting mirrors with a gain medium between them. The mirrors are curved for reasons that will be discussed in Section 9.6. For now, the curvature is unimportant. The light is generated inside the cavity, but the resonance condition is the same as given by Equation 7.96.

$$f = N f_0 \qquad \text{where} \qquad f_0 = \frac{c}{2\ell}, \tag{7.132}$$

where

- N is the **laser mode number**
- ℓ is the optical path length of the cavity

Eq. 7.106: $\eta_{rt} = R_1 R_2 e^{-2\Im nk\ell_p}$.

Because of the gain medium, $\Im n$ in Equation 7.106 is negative, $\Im nk = -g$. Usually the gain medium provides spontaneous emission, which is then amplified by g through the process of stimulated emission. The gain only applies over the length of the gain medium, which we call l_g. If the round-trip efficiency, $\eta_{RT} = R_1 R_2 e^{2g\ell_g}$, is greater than one, then the spontaneous emission will be amplified on each pass. We define the **threshold gain**, g_{th}, by

Threshold gain defined.

$$g_{threshold} = \frac{1}{2\ell_g} \log \frac{1}{R_1 R_2}. \tag{7.133}$$

If the gain is below threshold, then any light produced will decay rapidly. If the gain is above threshold and the phase is also matched, so that the resonance condition is satisfied, then the cavity will function as a laser. Eventually the gain will saturate and the laser will achieve steady-state operation.

Figure 7.28 explains the frequencies at which a laser will operate. The range of possible operating frequencies of a laser is determined by the gain medium. For example, a helium-neon laser has a gain line centered at 632.8 nm with about 1.5 GHz linewidth and a carbon dioxide laser has a number of gain lines around 10 μm with line widths that depend upon pressure. The argon ion laser has a variety of lines in the violet through green portion of the visible spectrum, the strongest two being green and blue at 514.5 and 488.0 nm. The broad dashed line shows an approximate gain curve for an argon-ion laser at the green wavelength of 514.5 nm. The specific frequencies of operation are determined by the longitudinal cavity modes, which we address in the next section.

7.5.3.1 Longitudinal Modes

For the example shown in Figure 7.28, the length of the cavity is 30 cm, and so the free spectral range is

$$FSR = \frac{c}{2\ell} = 500\,\text{MHz}. \tag{7.134}$$

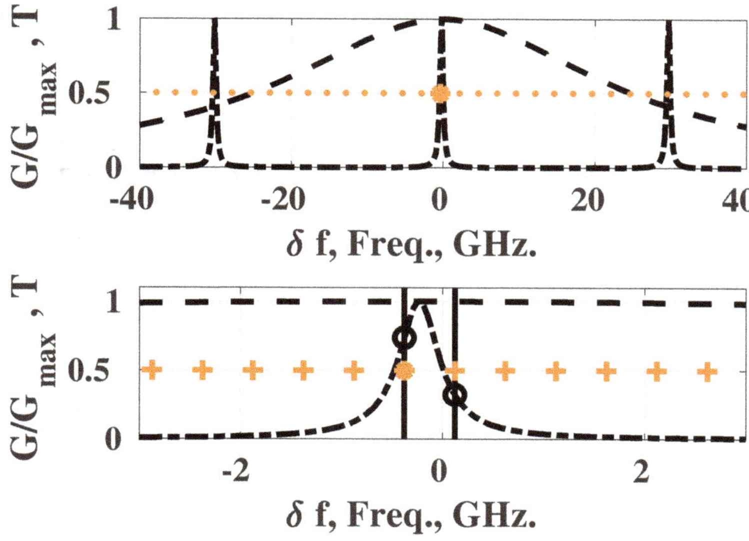

FIGURE 7.28
Laser cavity modes. The upper plot shows the gain line of an argon ion laser around 514.5 nm, with the dashed line. The colored dots show every fifth cavity mode (Equation 7.127 $N = 5, 10, 15, \ldots$), because the modes are too dense to show every cavity on these plots. The bottom plot shows an expanded view, in which all consecutive cavity modes are shown. The ètalon (Section 7.5.5) picks out one cavity mode, but fine-tuning is essential, to match the ètalon to one of the cavity modes. The dash-dot line shows the transmission of an ètalon with a spacing of 5.003 mm.

The colored dots across the middle of the plot show the cavity modes of the Fabry–Perot cavity. For clarity, only every fifth mode is shown, because if all modes were shown, the line would appear solid.

As we can see in this figure, there are a large number of possible frequencies at which the gain exceeds the threshold gain. Which of these lines will actually result in laser light? There are two possibilities. In the first the laser line is inhomogeneously saturated. By this we mean that the gain line can saturate at each frequency independently of the others. In this case, all cavity modes for which $\eta_{RT} > 1$, for which the gain is above threshold (Equation 7.133), will operate. We say that this laser is operating on multiple longitudinal modes.

If the laser gain line is inhomogeneously saturated, each cavity mode will saturate independently. The laser will run on multiple logitudinal modes.

In the second case, saturation at one frequency lowers the gain line at all frequencies. We say that this line is homogeneously saturated. The laser will operate on the cavity mode nearest the peak, unless this mode is below threshold, in which case the laser will not operate. We can find the mode number by computing

If the laser gain line is homogeneously saturated, the entire line is saturated as one, so the laser will run on only one logitudinal mode.

$$N = \text{Round}\left(\frac{f_g}{FSR}\right) \qquad f_{op} = N \times FSR, \tag{7.135}$$

where

- $f_g = c/\lambda_g$ is the **gain-line center frequency** or the frequency at the peak of the gain line
- λ_g is the **gain-line center wavelength**
- $\text{Round}(x)$ is the value of x rounded to the nearest integer

We say that this laser operates on a **single longitudinal mode** or **SLM**. In fact, there are other reasons why a laser will operate on a single longitudinal mode. For example, if the free spectral range is greater than the width of the gain line, the laser will operate on one mode if that mode happens to be under the gain line ($g > g_{threshold}$), and will not operate at all otherwise. To ensure operation of such a laser we may need to actively tune the cavity length with a stabilization loop as discussed later in this section.

The frequency of an SLM laser can be tuned over a free spectral range or the width of the gain line, whichever is smaller. Varying the cavity length, for example, using a piezoelectric transducer,

$$f_{op} = N \times FSR = N \times \frac{c}{2\ell} \qquad df_{op} = -N\frac{c}{2\ell^2}d\ell$$

$$df_{op} = -FSR\frac{d\ell}{\lambda_g/2}. \tag{7.136}$$

Example

Returning to the example shown in Figure 7.28, the length of the cavity is 30 cm, and so the free spectral range is $FSR = \frac{c}{2\ell} = 500\,\text{MHz}$, from Equation 7.134.

Varying ℓ by $d\ell = \lambda_g/2$ would change the frequency, f_{op} by one FSR, and varying ℓ by $\lambda/20$ would change the frequency by a 10th of the FSR, or 50 MHz. However, if f_{op} is adjusted more than half the FSR to one side of the gain peak, then the next cavity mode will become operational, and the mode number, N will change by one. The laser can be tuned from $FSR/2$ below the gain peak to $FSR/2$ above it. For greater distances, the frequency will repeat as in Figure 7.29.

7.5.3.2 Frequency Stabilization

The change in ℓ to tune this laser through 50 MHz is only 51.4 nm, and the laser length will probably vary by much more as a result of thermal changes in the lengths of materials. A piezoelectric transducer, combined with an optical detector sampling the laser power and a servo system can be used for **frequency stabilization** of a laser. However, if the cavity length changes by more than the range of the piezoelectric transducer, then the stabilization loop must "break lock" and reacquire the peak with a new value of the mode number, N.

Figure 7.30 shows the conceptual design of a stabilization loop. The length of the cavity is varied slightly or "dithered" using an AC signal generator connected through a high-voltage amplifier to a piezoelectric transducer (PZT). Typical frequencies are in the kilohertz. A sample of the laser light is detected and mixed with the oscillator. If the laser is operating below the center frequency of the gain line, then decreasing the cavity length (increasing frequency) increases the laser power, the signal is in phase with the oscillator, and the output of the mixer is positive. The output is low-pass filtered and amplified to drive the PZT to further decrease the cavity length. If the frequency is too high, the opposite effect occurs. When the laser is operating at the peak, positive or negative. The output of the mixer is zero, and no correction occurs.

7.5.4 Frequency Modulation

In Section 7.3.6, we discussed the ability to make distance and velocity measurements using a chirped or frequency-modulated (FM) laser. Now that we understand how the

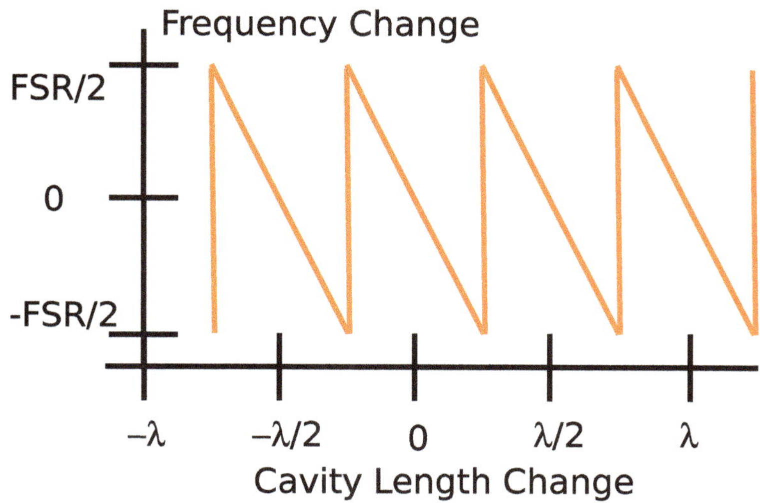

FIGURE 7.29
Tuning laser frequency. Changing the length of the cavity causes a change in the center frequency. A length change of one half wavelength results in a frequency change of one FSR.

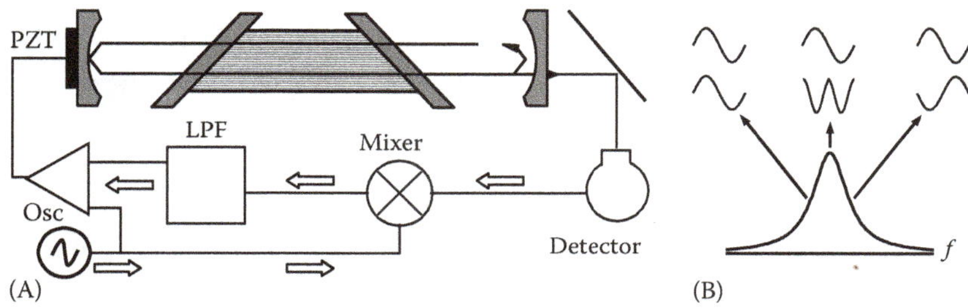

FIGURE 7.30
Stabilization loop. A PZT can be used to control the cavity length and thus the frequency (A). An oscillator (Osc) generates a dither signal that moves the PZT, while a detector samples the output and controls a servo circuit. Sample oscillator and detector signals are shown (B). If the laser is to the left of the center of the gain line, decreasing cavity length increases the signal. On the right, the opposite effect occurs. At line center, the sample signal is frequency-doubled, because the signal drops when the cavity is tuned in either direction.

frequency of a laser is determined, we can discuss the use of an electro-optical modulator to produce the frequency modulation. In Section 6.5.2, we discussed using an electro-optical modulator to modulate phase. Equation 6.86 can be used to determine the amount of phase modulation produced. If the phase modulator is placed inside the laser cavity, it becomes a frequency modulator.

Eq. 6.86: $\delta\phi = \pi\frac{V}{V_\pi}$.

With the phase modulator in place, the round-trip phase change is

$$\phi_{rt} = 2k\ell + 2\pi\frac{V}{V_\pi}. \tag{7.137}$$

The result of the extra phase is equivalent to making a slight change in the cavity length. The modified equation of laser frequency for mode N is

$$\phi_{rt} = N2\pi$$
$$2\frac{2\pi f}{c}\ell + 2\pi\frac{V}{V_\pi} = N2\pi \tag{7.138}$$
$$f = N\frac{c}{2\ell} - \frac{V}{V_\pi}\frac{c}{2\ell}.$$

The nominal laser frequency is still given by $f_N = Nf_0$ with $f_0 = c/(2\ell)$, and the modulation frequency is

$$f - f_N\delta f = -\frac{V}{V_\pi}f_0, \tag{7.139}$$

where f is the actual frequency. In other words, a voltage change sufficient to generate a phase change of π in one direction or 2π for the round trip will produce a frequency shift of one free spectral range. Applying an AC voltage $|V_{AC}| < V_\pi$, will modulate the phase. For example, a voltage of

$$V_{AC} = \frac{V_\pi}{4}\cos 2\pi f_m t$$

modulates the frequency of the laser $\pm FSR/4$ about the nominal $1/f_m$ times per second. The modulation frequency must be slow enough that the laser can achieve equilibrium in a time much less than $1/f_m$.

Take-Away

Phase Modulation:
If a voltage, V_π on the modulator changes the cavity length by a half wavelength (phase of π), then the frequency shift for a voltage, V is $\delta f = \frac{V}{V_{pi}}FSR$, provided the change of V is slow enough to allow many round trips of light in the cavity.

In Practice

An intra-cavity electro-optical modulator can be used to vary the optical length of the cavity and thus the laser frequency. Using a modulator is more expensive and complicated than moving a mirror with a piezoelectric transducer but can be done at much higher speeds. A modulator will introduce some loss into the laser cavity so it may not work with a laser medium that has a low gain.

Fig. 7.28:

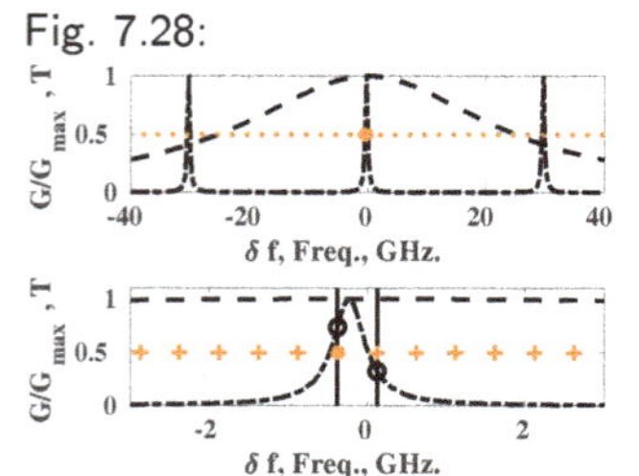

7.5.5 Ètalon for Single-Longitudinal Mode Operation

Multiple longitudinal modes in a laser may be a problem for the user. For example, the argon ion laser in Figure 7.28 will produce beat signals between the different modes, with a beat frequency of 500 MHz. It may be desirable to have the laser operate on a single longitudinal mode. We might be tempted to simply reduce the length of the laser cavity,

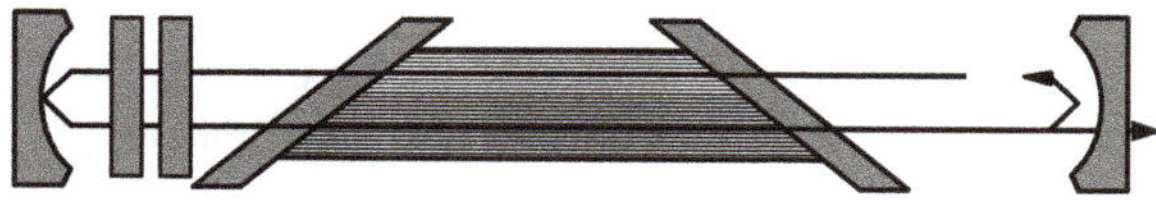

FIGURE 7.31
Intra-cavity ètalon. The ètalon can be used to select only one of the cavity modes.

but doing so would decrease the possible length of the gain medium, and we need $e^{g\ell}$ to be at least above threshold. As a result, it may not be practical to have only one cavity mode under the gain line. We can remove the neighboring lines with an intra-cavity ètalon, as shown in Figure 7.31.

The effect of the ètalon is shown in Figure 7.28 with the dash-dot line. The resonances of the ètalon are wider and more widely spaced than those of the cavity. Tuning the ètalon carefully, we can match one of its peaks with one of the cavity modes. The lower part of the figure shows an expanded view of just a few modes. The ètalon has selected one of the two modes and the other modes are all below threshold. In order to obtain the result shown in the figure, the ètalon spacing was tuned to exactly 5.003 mm. It is evident that precise tuning of the laser cavity, the ètalon, or both, is required.

A laser with very low gain, such as a helium-neon laser, requires special consideration. Mirror reflectivities must be extremely high to maintain the gain above threshold, and high-reflectivity coatings, to be discussed in Section 7.7, are often required. An ètalon will inevitably introduce some loss, and may not be tolerated inside the cavity of such a laser. In order to maintain single-mode operation, the laser cavity length must be kept short in these cases. However, a short cavity means a short gain medium, which results in low power. For this reason, high-power, single-mode helium-neon lasers are uncommon.

Take-Away

Laser Cavity:

- A Fabry–Perot interferometer has a periodic transmission spectrum with a period $c/(2\ell)$.
- The width of each component in the spectrum can be made extremely narrow by using high-reflectivity mirrors.
- A laser cavity consists of a gain medium between two reflectors.
- The range of frequencies over which operation is possible is determined by the properties of the gain medium.
- The laser may operate at one or more discrete frequencies determined by the length of the cavity.
- Depending on the saturation properties of the gain medium, the laser will operate on all possible modes or on the one nearest the gain peak.
- An intra-cavity ètalon can select one of the cavity modes in a multiple-mode laser.

7.6 Beamsplitter

A key component of the interferometers that are discussed in this chapter, and of many other optical systems, is the beamsplitter. In many cases, a beamsplitter will be a flat slab of optical material with two parallel surfaces. We have thus far assumed that one of these

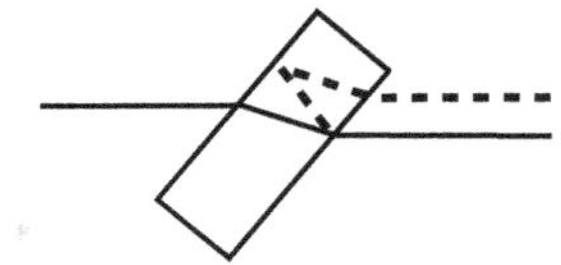

FIGURE 7.32
Beamsplitter. Ideally we would like one reflected and one transmitted beam. However, there is always some reflection from the second surface. If this reflection adds coherently, undesirable fringes may result.

Dielectric coatings will be discussed in Section 7.7.

surfaces may be given any reflectivity, R_1 we desire, and that the other may be made to have $R_2 = 0$. In Section 7.7, we will discuss how to achieve the desired reflectivities, but we will also see that it is difficult to produce a very low reflectivity for the second surface. Now, let us consider the effect of the nonzero reflectivity of the second surface. We will often find that the beamsplitter becomes an "accidental interferometer," and we need to develop some clever solutions to avoid the undesirable fringes.

We consider the beamsplitter shown in Figure 7.32. We will choose one surface to have the desired reflectivity, and make the other as low as possible. First, if our goal is to make a low-reflectivity beamsplitter, then the desired reflection from one surface and the unwanted reflection from the other will be nearly equal, and interference effects will be strong.

Our computation should include an infinite number of reflections within the beamsplitter, but because at least one reflectivity is normally low, we will only use the first two terms in this analysis. If the power in the beam represented by the solid line in Figure 7.32 is

$$P_{T1} = P_0 (1 - R_1)(1 - R_2) \tag{7.140}$$

and that represented by the dashed line is

$$P_{T2} = P_0 (1 - R_1) R_2 R_1 (1 - R_2). \tag{7.141}$$

Then the total power is

$$P_T = \left| \sqrt{P_{T1}} + \sqrt{P_{T2}} e^{j\delta\phi_T} \right|^2$$

$$P_T = P_{T1} + P_{T2} + 2\sqrt{P_{T1} P_{T2}} \cos(\delta\phi_T). \tag{7.142}$$

The power reaches its maximum for $\cos(\delta\phi) = 1$ and its minimum for $\cos(\delta\phi) = -1$. The fringe visibility is

$$\frac{\Delta P_T}{P_T} = \frac{2\sqrt{P_{T1} P_{T2}}}{P_{T1} + P_{T2}} = \frac{2\sqrt{R_1 R_2}}{1 + R_1 R_2}. \tag{7.143}$$

The direct reflected beam is

$$P_{R1} = P_0 R_1 \tag{7.144}$$

and the beam reflected with one bounce is

$$P_{R2} = P_0 (1 - R_1) R_2 (1 - R_1) = P_0 (1 - R_1)^2 R_2 \tag{7.145}$$

with the result

$$P_R = P_{R1} + P_{R2} + 2\sqrt{P_{R1} P_{R2}} \cos(\delta\phi_R). \tag{7.146}$$

TABLE 7.2
Beamsplitter Fringe Visibility

R_1	R_2	V_T	V_R
1.0%	10.0%	6.3%	58.0%
10.0	1.0	6.3	52.7
1.0	90.0	18.8	21.1
90.0	1.0	18.8	2.1

The goal is to keep the fringe visibility low. High reflectivity (90%) beamsplitters are better than low (10%), and beamsplitters with the reflective surface first are better than those with the anti-reflective (1%) side first.

We consider two beamsplitters, one with $R_1 = 90\%$ and the other with $R_1 = 10\%$. In both cases, the other face, intended to be anti-reflective, has a reflectivity of 1%. The results are in Table 7.2. Ideally, we want the fringe visibility to be $V = 0$.

The 10% beamsplitter is in the first two rows. With the reflecting side away from the source (first row), the fringe visibility for the reflected light is $V_R = 58\%$. Reversing the beamsplitter only reduces this value slightly. Furthermore, 99% of the source power passes through the substrate if the anti-reflective side is first and the value is only reduced to 90% if the reflecting side is first. If the source has a high power, and the substrate is slightly absorbing, the center of the beamsplitter will become hot, expand, and become a lens. The 90% beamsplitter is in the last two rows. The worst-case fringe visibilities are much lower than for the 10% beamsplitter. Assuming we have a choice, when we wish to split a beam into two very unequal parts, it is to our advantage to design the system so that the more powerful beam is the reflected one.

Low-reflectivity beamsplitters are more prone to multiple reflection problems.

For the high-reflectivity beamsplitter, the visibility of the transmitted fringes is almost 19%. If the beamsplitter is oriented with the reflecting side toward the source, the visibility of fringes in the reflection is only 2%, but if the reflecting side is away from the source, the visibility is 21.1%. Furthermore with the reflecting side away from the source, the power in the substrate is 99% of the incident power, and the heating effects mentioned in the previous paragraph may occur. If the reflective side is toward the source, the power in the substrate is only 10% of the source power. Clearly, placing the reflecting side toward the source is best.

Keep the reflective side of the beamsplitter toward the source.

If the coherence length of the source is small compared to the thickness of the beamsplitter, these effects will be less serious. This issue will be discussed more extensively in Section 10.3.3. Nevertheless, in an imaging system, the second-surface reflection may still be observed as a **ghost image**. In this case, incoherent addition applies, and the effect is weaker than in the coherent case, but it may still be objectionable. An alternative is the **beamsplitter cube**, which consists of a pair of 45° prisms, glued together, with a coating on the hypotenuse of one. There will be no ghost image but the reflections from the cube face may be problematic.

See Section 10.3.3 for more about coherent effects in beamsplitters.

Finally, we note that we must consider aberrations in the use of beamsplitters. A collimated beam of light passing through a flat substrate will not suffer aberrations, but a curved wavefront will. Therefore, if a curved wavefront is to be split into two parts, if one of the parts requires diffraction-limited imaging, then it is best to design the system with that part being the reflected one. If aberrations will be a problem for both beams, then it is best to design the system with additional optics so that the wavefront is planar at the beamsplitter.

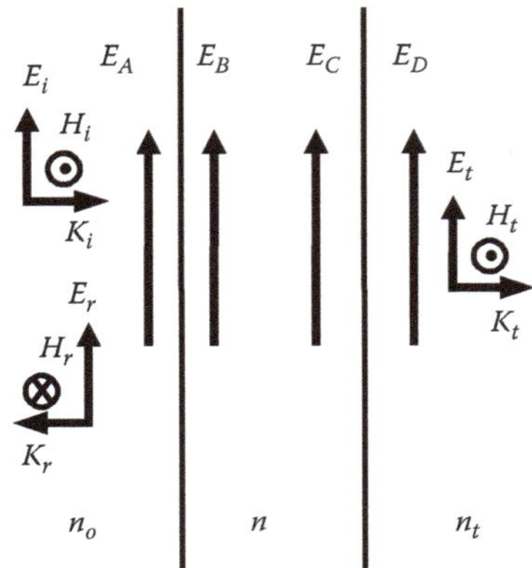

FIGURE 7.33
Thin film. The boundary conditions for the electric and magnetic field lead to a simple equation for this film.

> **Take-Away**
>
> **Beamsplitters:**
>
> - The reflection from the antireflection-coated side of a beamsplitter may produce unwanted effects such as ghost images and interference patterns.
> - Highly coherent sources exacerbate the problem because coherent addition enhances the unwanted contrast.
> - It is good practice to design systems with beamsplitters so that the stronger beam is the reflected one if possible.
> - It is good practice to place the reflective side toward the source.
> - It may be helpful to tilt the second surface with respect to the first.
> - The reflected wave will have weaker fringes than the transmitted one.
> - A beamsplitter cube is an alternative to eliminate the ghost images but the flat surfaces may introduce undesirable reflections.
> - Aberrations may be an issue in the transmitted direction if the wavefront is not planar at the beamsplitter. They will always be an issue for a beamsplitter cube.

7.7 Thin Films

Fig. 7.24

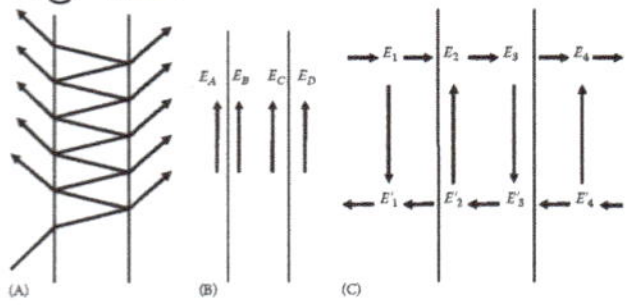

Throughout our discussion of interferometry, we have discussed beamsplitters with arbitrary reflectivities. In this section, we explore the basics of designing beamsplitters to meet these specifications. We will also explore the design of filters to select specific wavelength bands. Although we will derive the equations for normal incidence, they are readily extensible to arbitrary angles, and in these extensions, polarization must be considered.

Here we use the derivation suggested in the center portion of Figure 7.24, reproduced in Figure 7.33 with vector components shown. We know that the transverse components of the electric field, **E**, and the magnetic field, **H**, are preserved across a boundary. Because we limit ourselves to normal incidence, these fields are the total fields. Thus in Figure

7.33,

$$E_B = E_A \qquad E_C = E_D$$
$$H_B = H_A \qquad H_C = H_D. \tag{7.147}$$

If the incident wave is from the left, then

$$E_A = E_i + E_r \qquad E_D = E_t, \tag{7.148}$$

where

- E_i is the incident field
- E_r is the reflected field
- E_t is the transmitted field

Likewise, for the magnetic field,

$$H_A = H_i - H_r \qquad H_D = H_t, \tag{7.149}$$

where the negative sign on the reflected component is to maintain a right-handed coordinate system upon reflection.

The boundary conditions on H can be expressed in terms of E as

$$\frac{E_B}{nZ_0} = \frac{E_A}{n_0 Z_0} \qquad \frac{E_c}{nZ_0} = \frac{E_D}{n_t Z_0}$$

using Equation 1.41, or, multiplying by Z_0,

$$\frac{E_B}{n} = \frac{E_A}{n_0} \qquad \frac{E_C}{n} = \frac{E_D}{n_t}. \tag{7.150}$$

The field inside the material consists of a wave propagating to the right,

$$E_{right} e^{jnkz} \qquad \text{and} \qquad H_{right} e^{jnkz} = \frac{1}{nZ_0} E_{right} e^{jnkz} \tag{7.151}$$

and one propagating to the left

$$E_{left} e^{-jnkz} \qquad \text{and} \qquad H_{left} e^{-jnkz} = -\frac{1}{nZ_0} E_{left} e^{-jnkz}. \tag{7.152}$$

At the boundaries,

$$E_B = E_{left} + E_{right} \qquad E_c = E_{left} e^{-jnk\ell} + E_{right} e^{jnk\ell}. \tag{7.153}$$

Combining all these equations from (7.147) to (7.153), we can show that

$$E_i + E_r = E_t \cos(nk\ell) - E_t \frac{n_t}{n} \sin(nk\ell)$$
$$n_0 E_i - n_0 E_r = -jnE_t \sin(nk\ell) + n_t E_t \cos(nk\ell). \tag{7.154}$$

Writing these two equations as a matrix equation,

$$\begin{pmatrix} 1 \\ n_0 \end{pmatrix} E_i + \begin{pmatrix} 1 \\ -n_0 \end{pmatrix} E_r = \begin{pmatrix} \cos(nk\ell) & -\frac{j}{n}\sin(nk\ell) \\ -jn\sin(nk\ell) & \cos(nk\ell) \end{pmatrix} \begin{pmatrix} 1 \\ n_t \end{pmatrix} E_t, \quad (7.155)$$

we find that the two-by-two matrix depends on the index of refraction, n, of the layer, and its thickness, ℓ, that the terms in the column vector on the right contain only the final layer's index, n_t, and the transmitted field, E_t, and the terms on the left contain only parameters in front of the layer.

Characteristic matrix defined. Single layer.

We define this matrix as the **characteristic matrix**,

$$\mathcal{M} = \begin{pmatrix} \cos(nk\ell) & -\frac{j}{n}\sin(nk\ell) \\ -jn\sin(nk\ell) & \cos(nk\ell) \end{pmatrix}, \quad (7.156)$$

and write the matrix equation as

$$\begin{pmatrix} 1 \\ n_0 \end{pmatrix} + \begin{pmatrix} 1 \\ -n_0 \end{pmatrix} \rho = \mathcal{M} \begin{pmatrix} 1 \\ n_t \end{pmatrix} \tau. \quad (7.157)$$

Multiple layers.

Now for any multilayer stack as shown in Figure 7.34, we can simply combine matrices:

$$\mathcal{M} = \mathcal{M}_1\mathcal{M}_2\mathcal{M}_3\ldots \quad (7.158)$$

and with the resulting matrix, we can still use Equation 7.157. Because the matrix acts on the transmitted field to compute the incident and reflected ones, matrix multiplication proceeds from left to right instead of the right-to-left multiplication we have encountered in earlier uses of matrices. Solving this equation for ρ and τ, if

$$\mathcal{M} = \begin{pmatrix} A & B \\ C & D \end{pmatrix} \quad (7.159)$$

Reflection.

then

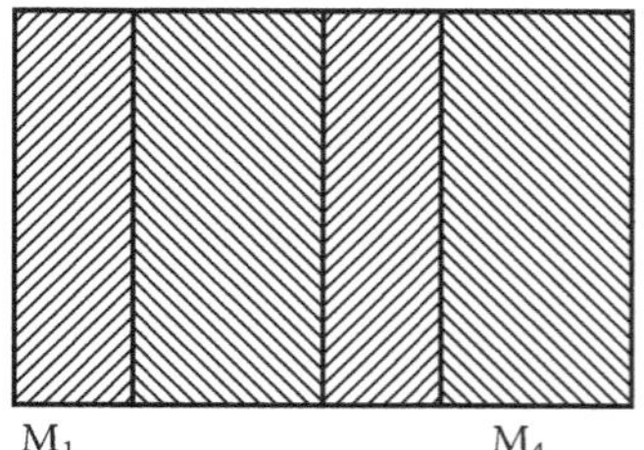

FIGURE 7.34
Multilayer stack. Layers of different materials can be combined to produce desired reflectivities. The stack can be analyzed using a matrix to describe each layer, and then multiplying matrices.

$$\rho = \frac{An_0 + Bn_t n_0 - C - Dn_t}{An_0 + Bn_t n_0 + C + Dn_t} \quad (7.160)$$

and

Transmission.

$$\tau = \frac{2n_0}{An_0 + Bn_t n_0 + C + Dn_t}. \quad (7.161)$$

We can use these equations to determine the reflectivity and transmission of any stack of dielectric layers at any wavelength. We can also use them to explore design options. Among the components that can be designed in this way are band-pass filters that provide a narrower bandwidth than is possible with the alternative of colored-glass filters, long-pass filters that transmit long wavelengths and reflect short ones, and short-pass filters that do the opposite. A multilayer coating that reflects infrared and transmits visible is sometimes called a **hot mirror**, and one that does the opposite is called a **cold mirror**. A dentist's lamp reflects the tungsten light source with a cold mirror. The visible light which is useful is reflected toward the patient, but the infrared light which would create heat without being useful is transmitted out the back of the mirror. In the next sections, we will consider two of the most common goals, high reflectivity for applications such as laser mirrors, and antireflection coatings for lenses and the back sides of beamsplitters.

Take-Away

Dielectric Coatings:

- Interference among reflections from layer boundaries in a dielectric stack controls reflection and transmission.
- Matrix equations can be used to predict performance.
- It is a little more complicated to write equations for non-normal incidence. Polarization must be considered then.
- Because the behavior of a dielectric stack is based on interference, it varies with wavelength.

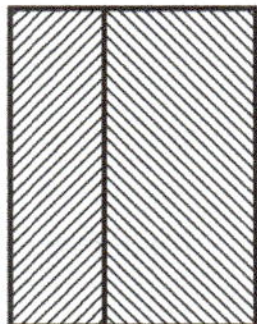

FIGURE 7.35
Layer pair. The left layer has a higher index of refraction and smaller thickness. The optical paths are the same.

7.7.1 High-Reflectance Stack

Intuitively, reflection is a function of contrast in the index of refraction. Let us start with a pair of contrasting materials as shown in Figure 7.35. In this figure, both layers have the same optical path length. The first one has a higher index and thus a shorter physical length. Our goal is to stack several such pairs together as shown in Figure 7.36. We recall that ρ is negative when light is incident from a high index onto an interface with a lower index. Thus, the alternating reflections change sign, or have an additional phase shift of π, as shown in Figure 7.37. If the optical thickness of each layer is $\lambda/4$, then round-trip transit through a layer introduces a phase change of π, and all the reflections add coherently for this particular wavelength.

A quarter wave thickness results in a phase change of 90 deg.

For a layer one quarter wavelength thick, with index n_i,

$$\mathcal{M}_i = \begin{pmatrix} 0 & -j/n_i \\ jn_i & 0 \end{pmatrix}. \quad (7.162)$$

For a pair of high- and low-index layers, n_h and n_ℓ, respectively,

$$\mathcal{M}_p = \begin{pmatrix} 0 & -j/n_h \\ jn_h & 0 \end{pmatrix} \begin{pmatrix} 0 & -j/n_\ell \\ jn_\ell & 0 \end{pmatrix} = \begin{pmatrix} -n_\ell/n_h & 0 \\ 0 & -n_h/n_\ell \end{pmatrix} \quad (7.163)$$

Because this matrix is diagonal, it is particularly easy to write the matrix for N such pairs. Each term in the matrix is raised to the power N:

$$\mathcal{M}_N = \begin{pmatrix} (-n_\ell/n_h)^N & 0 \\ 0 & (-n_h/n_\ell)^N \end{pmatrix}. \quad (7.164)$$

Reflectivity approaches 100% for a large number of layers at the right wavelength.

The reflectivity is

$$R = \left(\frac{\left(\frac{n_\ell}{n_h}\right)^{2N} - \frac{n_t}{n_0}}{\left(\frac{n_\ell}{n_h}\right)^{2N} + \frac{n_t}{n_0}} \right)^2 \quad (7.165)$$

As the number of layers increases, $R \rightarrow 1$, and neither the substrate index, n_t nor the index of the ambient medium, n_0 matters. If the low and high layers are reversed, the result is the same.

As an example, we use zinc sulfide and magnesium fluoride with

$$n_h = 2.3 \qquad n_\ell = 1.35$$

respectively. Some reflectivities are

$$N = 4 \rightarrow R = 0.97 \text{ (8 layers)}$$

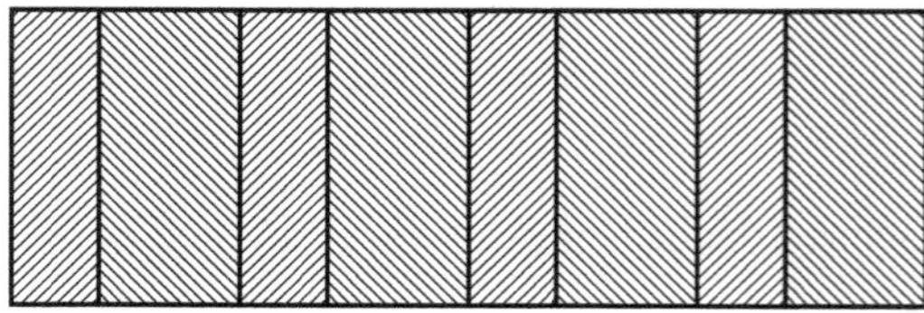

FIGURE 7.36
High-reflectance stack. Repeating the layer of Figure 7.35 multiple times produces a stack that has a high reflectivity. The more layer pairs used, the higher the reflectivity.

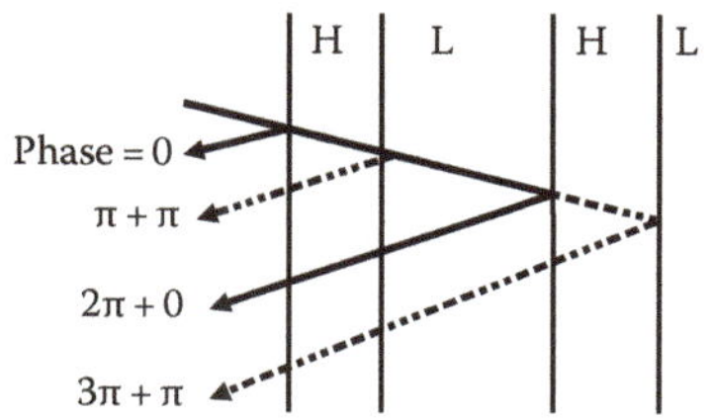

FIGURE 7.37
High-reflectance principle. Each layer is $\lambda/4$ thick, resulting in a round-trip phase change of π. Alternate interfaces have phase shifts of π because the interface is from a higher to lower index of refraction.

and

$$N = 15 \rightarrow R = 0.999\,(30\text{ layers}).$$

A dielectric stack can have a higher reflectivity than a metal mirror does.

Metal mirrors often have reflectivities of about 96%, and the light that is not reflected is absorbed, leading to potential temperature increases, expansion, and thus increasing curvature in applications involving high powers. Dielectric mirrors using stacks of alternating layers make better mirrors for applications such as the end mirrors of lasers. Not only is the reflectivity higher, but the light that is not reflected is transmitted. With the right choice of materials, absorption is very low. Some disadvantages of dielectric coatings are that for non-normal incidence they are sensitive to angle of incidence and polarization, that coatings may be damaged, and of course that they are wavelength sensitive. In fact, some lasers such as helium-neon are capable of operating on different wavelengths, but mirrors tuned to the wavelength of interest are required. The titanium-sapphire laser operates over a wide range of wavelengths from near 700 nm to near 1150 nm, but two different sets of mirrors are required to cover the entire band. For the argon ion laser a number of lines in the blue and green parts of the spectrum are accessible with a single set of mirrors.

Take-Away

High Reflectance Dielectric Coatings:

- Multi-layer dielectric stacks can have higher reflectivity than metal over a range of wavelengths.
- Dielectric stacks are less susceptible to heating than metals are because absorption is low.
- They are often chosen for laser mirrors.

In Practice

To predict the reflection spectrum of a dielectric stack, use the design wavelength to determine the layer thicknesses (quarter wave OPL), and then keep the thicknesses fixed and vary the wavelength.

7.7.2 Antireflection Coatings

On several occasions we have noted that a beamsplitter has two sides, and we would like to consider reflection from one of them, while ignoring the other. Intuitively we can

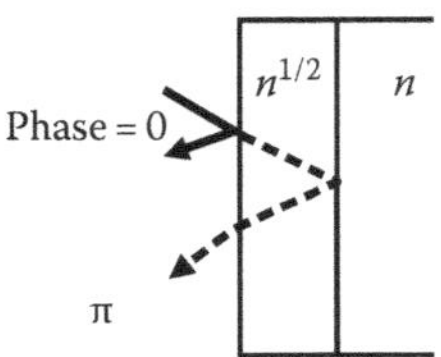

FIGURE 7.38
Principle of anti-reflectance coating. The two reflections are the same, but they add with opposite signs, and so the two waves cancel each other.

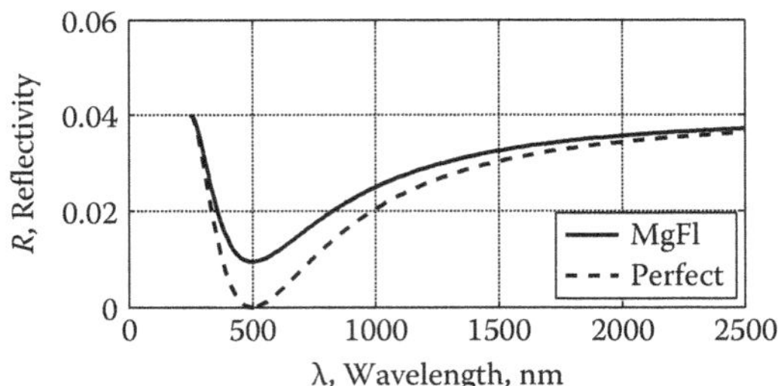

FIGURE 7.39
Anti-reflectance coating. A magnesium fluoride coating is applied to glass with $n_t = 1.5$. The coating optical path length is a quarter of 500 nm. A perfect coating having $n = \sqrt{n_t}$ is shown for comparison. Although the improvement at the design wavelength is perfect, for white-light applications, even just across the visible spectrum, the improvement is marginal.

imagine using the coating shown in Figure 7.38. The two reflections will be equal, and if the thickness adds a quarter wave each way, they will be out of phase, and thus cancel the reflection. This analysis is overly simplistic, as it does not account for the multiple reflections. Let us analyze this layer numerically using the matrix equations.

With only one layer, this problem is particularly simple. Let us assume that the substrate is glass with an index of refraction $n = 1.5$. We will design a coating for $\lambda = 500$ nm. For the ideal **antireflection (AR) coating** the index of refraction is $\sqrt{1.5}$ and the physical thickness is $\lambda/4/\sqrt{1.5}$. The result is shown in Figure 7.39, with the dashed line. We see that the coating is perfect at its design wavelength, and varies up to about the 4% Fresnel reflection that would be expected for uncoated glass at the extrema of 300 nm and 2.5 μm, which are at the ends of the typical wavelength range for glass transmission (see Table 1.2).

The challenge in designing a coating is to find a material with the appropriate index of refraction, reasonable cost, and mechanical properties that make it suitable as a coating, and good properties for easy and safe manufacturing. The solid line in Figure 7.39 shows the results for the same substrate with a quarter-wave coating of magnesium fluoride ($n = 1.35$), with the same design wavelength. The index match is not perfect, but the material is low in cost, durable, and easy to handle. The reflectivity is about 1% at the design wavelength. Over a wider bandwidth, even just the visible spectrum, the difference between the perfect coating and the magnesium fluoride one is small, although for narrow-band light at the design wavelength, the 1% reflection can create some interference artifacts as discussed in Section 7.6.

The performance of an AR coating can be improved with multiple layers. Figure 7.40 shows a sketch of some possible results.

An optical component that is AR coated for visible light can be identified easily by viewing the reflection of a white light source. Many such coatings will cause the source image to appear pink or purple in the reflection, because of the relative absence of green light, as in the case of the one-layer and two-layer examples in Figure 7.40. On the other

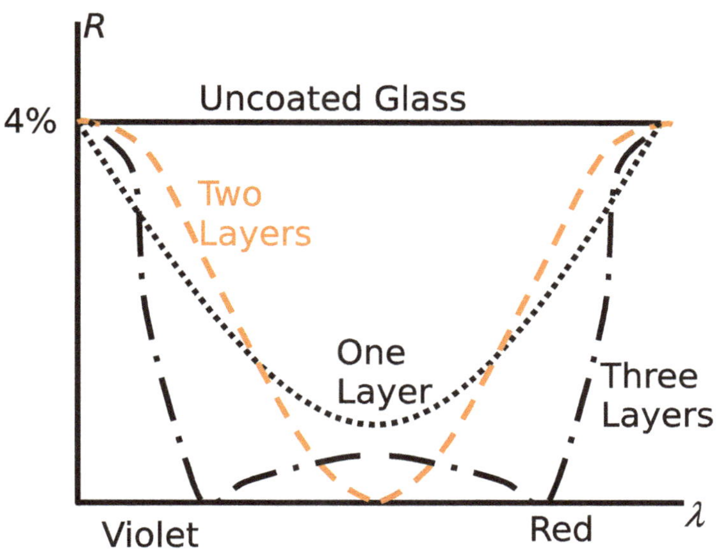

FIGURE 7.40
Anti-reflectance examples. Multiple layers can improve performance of an AR coating, either by reducing the minimum reflection, or by widening the bandwidth of performance.

hand, some AR coatings will appear green because, in an effort to minimize the reflectivity across the visible spectrum, the green reflectivity is slightly increased as shown in the three-layer example. Coatings for wavelengths outside the visible region often have bright colors, but because the layers are designed for non-visible wavelengths, these colors are not useful clues to the functions of the coatings.

Take-Away

Anti-Reflection Coatings:
Dielectric coatings may be used to reduce reflectivity.

- The performance is wavelength dependent.
- A compromise between index of refraction and mechanical properties is needed in choosing the coating material.
- Magnesium oxide is often used on glass with a reflectivity of about one percent.
- Multiple layers can achieve lower reflectivity and/or wider bandwidths of performance.

Take-Away

Summary of Dielectric Coatings:

- Thin dielectric films can be used to achieve specific reflectivities.
- Thin-film reflectors are sensitive to wavelength, angle of incidence, and polarization, so all these parameters must be specified in the design.

- Alternating quarter-wave layers of high and low index of refraction produce a high reflectivity, which can easily exceed that of metals.
- A single quarter-wave coating of magnesium fluoride results in an AR surface for glass with a reflectivity of about 1% at the design wavelength.
- Multiple layers can improve AR coating performance for any application.
- White light reflected from AR-coated optics designed for visible light may have a pink, purple, or green appearance depending on the type of coating used.

Problems

7.1 Laser Tuning

Suppose that I place the back mirror of a laser on a piezoelectric transducer that moves the mirror axially through a distance of 10 μm with an applied voltage of 1 kV, and that the relationship between position and voltage is linear. The laser has a nominal wavelength of 633 nm. The cavity length is 20 cm.

7.1a What voltage is required to sweep the cavity through one free spectral range?

7.1b What is the free spectral range in frequency units?

7.1c What voltage is required to tune the laser frequency by 50 MHz?

7.2 Interference Filters

Design a pair of layers for a high-reflection stack for 633 nm light, using $n = 1.35$ and $n = 2.3$. Assume light coming from air to glass. This problem is very similar to the one solved in the text. Try to solve it without looking at that solution.

7.2a Write the matrix for the layer.

7.2b Plot the reflection for a stack with five pairs of layers. Plot from 400 to 1600 nm.

7.2c Repeat for nine pairs.

7.3 Quadrature Interference (G)

Consider a coherent laser radar in which the signal beam returning from the target is linearly polarized. We place a quarter-wave plate in the local oscillator arm so that this beam becomes circularly polarized. Now, after the recombining beamsplitter, we have a polarizing beamsplitter that separates the beam into two components at ±45° with respect to the signal polarization. A separate detector is used to detect each polarization.

7.3a Show that we can use the information from these two detectors to determine the magnitude and phase of the signal.

7.3b What happens if the sensitivities of the two detectors are unequal. Explain in the frequency domain, as quantitatively as possible.

7.3c Suppose now that both beams are linearly polarized, but at 45° with respect to each other. I pass the combination through a quarter-wave plate with one axis parallel to the reference polarization. I then pass the result to a polarizing beamsplitter with its axes at 45° to those of the waveplate and detect the two outputs of this beamsplitter with separate detectors. Write the mixing terms in the detected signals. Is it possible to determine the phase and amplitude?

7.4 Coherent Lidar (NS)

Here we consider a FMCW lidar. In this case, the frequency of the laser is modulated with

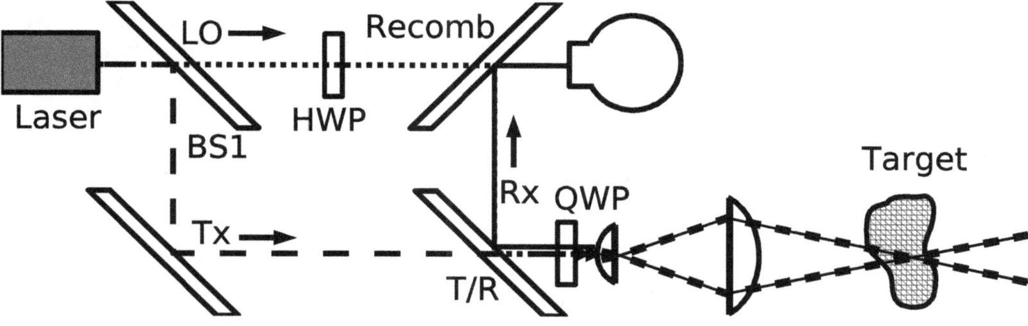

FIGURE P7.5
Figure for Problem 7.5.

a triangle wave that goes linearly from the center frequency of the laser minus f_{max} to the center frequency of the laser plus f_{max}, and then back down. The laser wavelength is 2.1 μm and the frequency modulation is $f_{max} = 20\,\text{MHz}$ at a modulation period of 40 μs.

The light is reflected from a target at $R = 5000\,\text{m}$ moving with a velocity of $v_{parallel} = 10\,\text{m/s}$. The returning light combines with the reference beam or local oscillator.

7.4a What is the Doppler frequency?

7.4b Plot the frequency of the local oscillator, the returning signal, and the measured difference frequency, as functions of time for 100 microseconds. Remember that the detector is only sensitive to irradiance and cannot detect the phase.
Hint: If you are using Matlab, type “help sawtooth.”

7.4c Discuss how you can use this data to determine the sign of the velocity.

7.5 Balanced Mixing (NS)

In the lidar of Figure P7.5, the laser power is 1 W, and the beamspltters are $R_1 = 0.99$, $R_{recomb} = 0.5$, the transmit/receive beamsplitter and waveplate are perfect, and $F = 10^{-9}$ of the transmitted light is returned to the lidar from the target, including reflection, absorption, and geomtric loss of the scattered light.

7.5a What are the signal and local oscillator powers to the detector? What is the mixing power?

7.5b Now, add a detector above the recombining beamsplitter so we can do balanced mixing. What is the mixing power at this detector?

7.5c What is the difference signal between the two detectors?

7.6 Laser Cavity (NS)

A helium–neon laser has a gain line with a center wavelength of 632.8 nm and a linewidth of about 1.5 GHz. We design such a laser with a cavity length of 35 cm.

7.6a What is the free-spectral range of this laser? How many modes are under the gain line?

7.6b Let’s assume incorrectly that the laser will only operate on the cavity mode closest to the peak of the gain line? On what cavity mode is it operating? In other words how may wavelengths are contained in the round trip? What is the frequency difference between the operating frequency and the gain line center frequency?

7.6c In fact the laser will run on all modes above the gain line. In order to ensure single-longitudinal-mode operation, we need to make sure that only one mode is under the gain line. What is the maximum length of the laser cavity to ensure this?

7.6d What problem does this present if we want to make a high-power single-logitudinal-mode laser?

7.7 Anti-Reflection Coating (NS)

In Section 7.7.2, we saw that a perfect anti-reflection coating for glass in air can be made with a single layer of material having an index of refraction equal to the square root of that of the glass. We also noted that it is often hard to find such a material with the needed mechanical properties and we often accept something that is close. This reference, http://eceweb1.rutgers.edu/õrfanidi/ewa/, derives a two-layer coating that will provide zero reflection at one wavelength. Specifically it shows that the following combination will work at wavelength λ_0 with glass having $n = 1.5$.

First Layer: $n_1 = 1.38$,	$\ell_1 = 0.3294 * \lambda_0/n_1$
Second Layer: $n_2 = 2.45$,	$\ell_2 = 0.0453 * \lambda_0/n_2$

7.7a What are the layer thicknesses for $\lambda_0 = 555\,\text{nm}$?

7.7b What is R at the designed wavelength of $\lambda_0 = 555\,\text{nm}$? Is this a good AR coating? I chose this wavelength because it is the peak of the visual spectrum.

7.7c Plot R as a function of λ from 200 to 2500 nm which covers most of the transmission spectrum of glass.

8 Diffraction

Let us begin this chapter by considering a laser pointer with a beam 1 mm in diameter, as shown in the left side of Figure 8.1. If we point it at a wall a few centimeters away, the beam will still be about the same size as it was at the exit of the pointer. Even a few meters away, it will only be slightly larger. The increased size will not be easily noticed. However, if we point it toward the moon, we know that the beam will diverge so that the light is spread over a large area. In 2011, a website encouraged owners of laser pointers to aim at the moon at a particular time to see if the total effect of a large number of pointers would be visible from Earth. The question became a homework assignment in at least one graduate optics course, where students correctly predicted it would not work. Nevertheless, it is likely that many stepped outside at the appointed time to be sure.

The beam divergence is the result of diffraction. Through a study of diffraction we will be able to answer these questions: At what angle, α, does the beam diverge? At what distance is the divergence noticeable? Can we control these numbers? Geometric optics offers us no help, as it predicts that a **collimated beam** (one with rays parallel to each other, or equivalently, wavefronts that are planar) will remain collimated at all distances. In this chapter, we answer these and other questions through the study of diffraction. We will define the near field as the region where the divergence of a collimated beam is not noticeable. We will later see that in the near field, a beam can be focused to a spot with a diameter smaller than its initial diameter, as shown on the right in Figure 8.1. In the far field, the **beam divergence** angle is proportional to λ/D where D is some characteristic dimension of the beam at the source.

The boundary between these regions is necessarily ill-defined, as there is an intermediate region where the beam expands but not at a constant angle. In Section 8.4.3, we will develop a mathematical formulation for the distances we call near and far field. For now, the near field includes distances much less than the **Rayleigh range**, D^2/λ,* and the far field includes those distances which are much greater than this value. Sometimes the Rayleigh range is called the **far-field distance**.

For our laser pointer, assuming a red wavelength of 670 nm, the Rayleigh range is about 1.5 m, and the beam divergence in the far field is 670 μrad. Thus, at the moon, some 400,000 km away, the beam would be at least 270 km across. If the laser's 1 mW power (approximately 10^{16} photons per second) were spread uniformly across this area, the irradiance would be about 17 fW/m^2. In the area of a circle with diameter equal to the original beam, 1 mm^2, one photon would be incident every 11 s. This low level of light is well below a few photons per second, and would be impossible to detect. Clearly we must consider diffraction in such problems.

In the study of interference in the previous chapter, we split a light wave into two or more components, modified at least one of them by introducing a temporal delay, recombined them, and measured the temporal behavior of the wave. The study of diffraction follows a similar approach, but here we are more interested in the spatial, as opposed to the temporal, distribution of the light. Perhaps the simplest diffraction problem is that of two point sources. Figure 8.2A provides an example, where the point sources have been

*Here we use the simple equations λ/D for the divergence and D^2/λ for the Rayleigh range. Each of these expressions is multiplied by a constant depending on our definitions of width and on the profiles of the waves. We will see some specific examples later in this chapter and Chapter 9. However, these equations are good approximations.

DOI: 10.1201/9781315157047-8

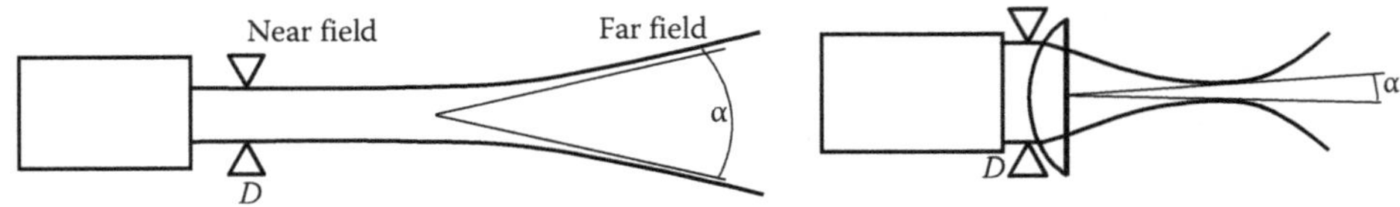

FIGURE 8.1
Near and far fields. A collimated wave remains nearly collimated in the near field, but diverges at an angle α in the far field. A converging wavefront can focus to a small spot in the near field. The angular width of the focused spot, viewed from the source, is also α. Diffraction predicts an inverse relationship between D and α.

simulated by passing a plane wave that travels down the page and then passes through two small apertures at $(\pm d/2, 0, 0)$. The two point sources radiate spherical waves,

$$E_n = \frac{e^{j(\omega t - kr_n)}}{r_n}, \tag{8.1}$$

where

- E_n is the scalar electric field (Section 1.4.3)
- r_n is the distance from the source having the label n,

as shown in Figure 8.3.

Here we have assumed that the apertures are small enough that each one is "unresolvable," or indistinguishable from a point source. Later we will discuss a precise definition of this term.

The sum of 2 spherical waves leads to a regular pattern of bright and dark fringes.

The sum of these two waves, where $r_1^2 = (x + d/2)^2 + y^2 + z^2$ and $r_2^2 = (x - d/2)^2 + y^2 + z^2$, produces the pattern seen in the lower portion of Figure 8.2A. Along the z axis, the waves are in phase with each other, and the field has a high value, with black areas representing positive fields and white areas representing negative fields. At certain angles, the fields from the two source points interfere destructively, and the total field is nearly zero, represented by gray in the image. The dashed curve is a semicircle centered on the origin. On this curve, the phase changes by 180° at each of the nulls.

The situation becomes a bit more complicated when a "large" aperture is involved. Figure 8.2B shows the diffraction pattern for a single large aperture. Following the approach of Huygens (see Chapter 1), we sum the contributions from multiple point sources distributed throughout the aperture. We see that for small positive values of z, in the near field, the plane wave propagates through the aperture almost as we would expect from geometric optics, provided that we ignore the edges. In the far field, we see a fringe pattern of black positive-field peaks and white negative-field valleys separated in angle by gray zero-field spokes. The dashed semicircle is centered at the origin, and at this distance, phase changes as we cross the null lines, just as in the left panel.

Huygens' Principle: Light through an aperture is modeled as a sum of a large number of spherical wavefronts. We will consider this idea with more rigor later.

Huygens' idea of building these diffraction patterns from spherical waves works quite well, but we need to address some issues to obtain quantitative results. One of our first tasks is to determine the amplitudes of the spherical waves. Another is to address a subtle but important directional issue. A true spherical wave would propagate backward as well as forward, contrary to common sense. In Figure 8.2, we have arbitrarily set the backward wave amplitude to zero. We will develop a mathematical theory in which the sum becomes an integral, the amplitudes are computed correctly, and the backward wave does not exist. We will then examine a number of special cases and applications.

In the study of diffraction, one equation stands out above all others. Given any distribution of light of wavelength λ having a "width," D, any measure of the angular width of the diffraction pattern will be

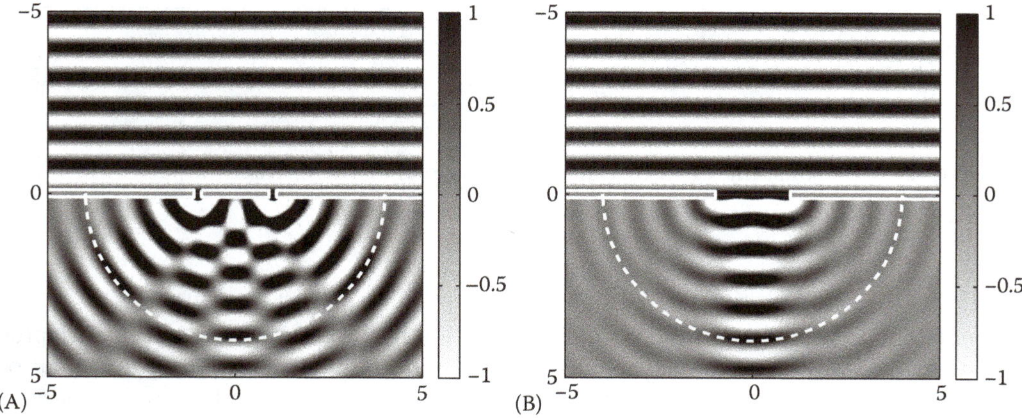

FIGURE 8.2
Diffraction. A wave from the top of the figure is incident on a wall with two small openings (A). A snapshot of the scalar field is shown in arbitrary units with black indicating a large positive field and white a large negative field. The wavelength is $800\,\mathrm{nm}$ and the units on the axes are $\mu\mathrm{m}$. The dashed curve is a semicircle with a radius of five wavelengths in the medium. The field on this semicircle has peaks of alternating phases separated by nulls. Diffraction from an aperture of width 2 µm (B) illustrates the near- and far-field regions. Peaks of alternating sign and nulls are still evident. The divergence of the wave can be seen at the bottom of the figure.

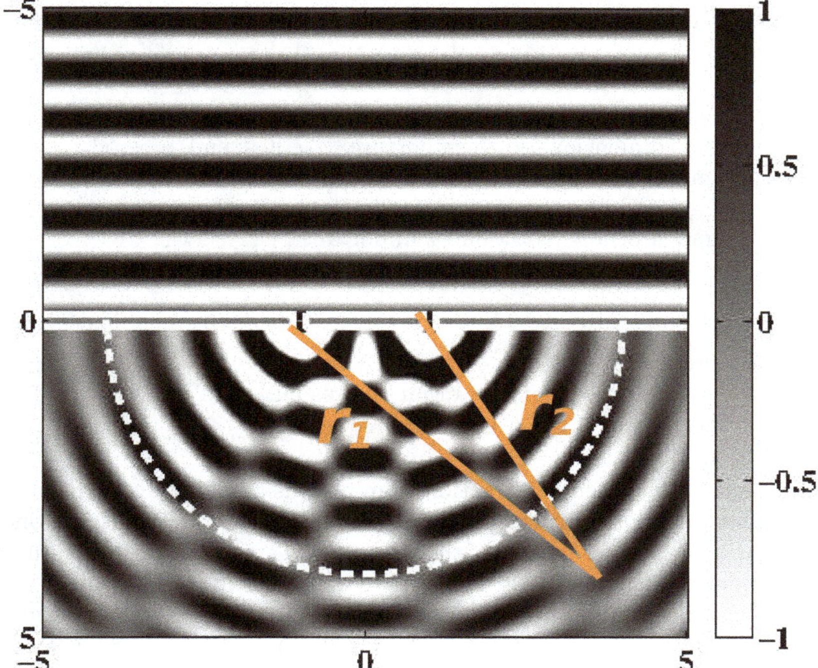

FIGURE 8.3
Summation of waves. At a point, the wave contributions from the slits change phase according to the radial distance, r_n.

$$\alpha = C\frac{\lambda}{D}, \tag{8.2}$$

where C is some constant. At a distance, z, the width of the diffraction pattern will be

$$d = C\frac{\lambda}{D}z.$$

For now, we are being deliberately vague about what we mean by "width" and "angular extent." The topics related to Fraunhofer diffraction in the early sections of this chapter, are all, in one way or another, devoted to precise definitions of D and α, and the evaluation of C. Before we turn to Fraunhofer diffraction, we will investigate the physical basis of diffraction. We can make a good start by noting that the constant C is typically pretty close to unity and so to a first approximation,

$$\alpha \approx \frac{\lambda}{D}, \qquad \text{and} \qquad d \approx \frac{\lambda}{D}z.$$

Take-Away

Diffraction Basics:
Diffraction dictates that light passing through a smaller aperture diverges more rapidly. The angle of divergence is given roughly by

$$\alpha \approx \frac{\lambda}{D},$$

where λ is the wavelength and D is some characteristic dimension of the aperture. We'll spend a good amount of time determining the constant of proportionality for different situations, but this equation stands among others as the most important.

As a result, the minimum spot size at a distance, z, is

$$d \approx \frac{\lambda}{D}z.$$

Reciprocity dictates that the resolution of an imaging system is given by similar equations.

The next few sections discuss diffraction from various points of view. In Section 8.1, we offer two different arguments for the physical basis of diffraction, a classical one based on Maxwell's equations and one based on the uncertainty principal of quantum mechanics. These are provided for the interest of the reader and reading this section is not essential for understanding the remaining sections. In Section 8.2, we lay the foundations of diffraction and Fourier optics in a way this is hopefully easy to understand. Specifically we show that spatial frequency and wavefront tilt in the object or image plane, and position in the pupil plane are all proportional. Understanding these relationships will be useful in quantitative analysis of numerical aperture, resolution, spatial filtering, and other applications of diffraction. In Section 8.3, we examine the mathematics behind the Fresnel–Kirchoff integral, which provides us with a quantitative relationship between the electric fields in any two planes including in particular the object or image plane and the pupil plane. The remainder of the chapter will discuss specific Fraunhofer diffraction patterns, resolution in imaging systems, diffraction gratings, and fresnel diffraction.

8.1 Physics of Diffraction

Our first approach to understanding the physics of diffraction, is based on waves, using Maxwell's equations. We construct a plane wave propagating in the $\hat{z}$ direction, but blocked by an obstacle for positive values of y, as shown in Figure 8.4,

$$\mathbf{E} = A\hat{x}e^{jkz+\phi} \qquad y \le 0$$

$$\mathbf{E} = 0 \qquad y > 0.$$

Samples of the electric field, which is in the $\hat{x}$ direction, are shown by black arrows. We recall Faraday's law, Equation 1.1. A harmonic wave requires a magnetic field $\mathbf{B}$ to have a component normal to a contour such as one of the colored rectangles in the figure, proportional to the integral of $\mathbf{E}$ around the contour. Around the red contour, the integral is nonzero, and produces $\mathbf{B}$ and $\mathbf{H}$ fields in the $\hat{y}$ direction. Around the green contour, the electric field is everywhere perpendicular to the contour, and the integral is zero. Along the solid blue contour, the field has equal and opposite contributions on the left and right, with a net result of zero. Thus with $\mathbf{E}$ in the $\hat{x}$ direction and $\mathbf{B}$ in the $\hat{y}$ direction, the wave propagates in the $\hat{z}$ direction.

Eq. 1.1: $\nabla \times \mathbf{E} = -\frac{\partial \mathbf{B}}{\partial t}$.

Fig. 8.4

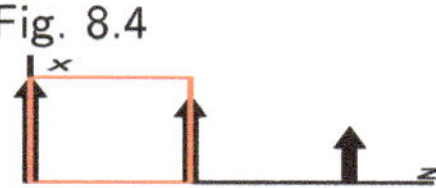

The integral of $\mathbf{E}$ around the contour is the y component of the curl. There is a $\mathbf{B}$ field in the $\hat{y}$ direction.

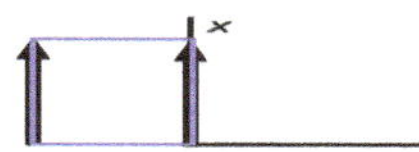

The z component of the curl is zero. $\mathbf{B}$ has no z component.

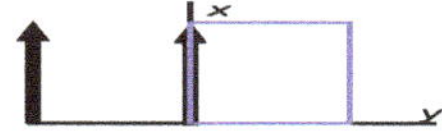

The z component of the curl at the edge is not zero. $\mathbf{B}$ does have a z component.

At the boundary of the obstacle, however, along the dashed blue contour, the upward $\mathbf{E}$ component on the left is not canceled because there is no field on the right. Thus, $\mathbf{B}$ or $\mathbf{H}$ has a $-\hat{z}$ component in addition to the one in the $\hat{y}$ direction, and the direction of propagation, determined by the Poynting vector, $\mathbf{S} = \mathbf{E} \times \mathbf{H}$ (Equation 1.45), now has components in the $\hat{z}$ and $\hat{y}$ directions. As z increases, the wave will diffract.

Quantum mechanics provides another approach to predicting diffraction. The uncertainty principle states that the product of the uncertainty, δx, in a component of position, x, and the uncertainty, δp, in its corresponding component of momentum, p_x, has a minimum value equal to Planck's constant, h. The total momentum of a photon is a constant $\mathbf{p} = h\mathbf{k}/(2\pi)$, and $k_x = |\mathbf{k}| \sin\alpha = k \sin\alpha$. The magnitude, $|\mathbf{k}|$, depends only on λ and doesn't change in passing trough the aperture, so the uncertainty principle,

The uncertainty principle of quantum mechanics also predicts diffraction.

$$\delta x \; \delta p_x \ge h,$$

can be stated in terms of position, x and angle, α, as

$$\delta x \; \delta k_x \ge 2\pi \qquad \delta x \; k \sin\delta\alpha \ge 2\pi$$

$$\sin\delta\alpha \ge \frac{2\pi}{\delta x \; k} \qquad \sin\delta\alpha \ge \frac{\lambda}{\delta x}. \tag{8.3}$$

In Figure 8.5, rays of light are shown, with starting positions at the left in each plot having a Gaussian probability distribution with the specified width, and the angle relative to the $z$axis, having another Gaussian probability distribution according to Equation 8.3. The rays remain mostly inside the initial diameter within the near field, and the divergence, inversely proportional to the diameter, becomes evident in the far field beyond the Rayleigh range.

In quantum mechanics, the probability amplitude of a photon is proportional to the classical field.

Before continuing our analysis, we note that the probability density for momentum is the Fourier transform of that for position. Because the Fourier transform of a Gaussian is a Gaussian, our choice in this example is consistent. It is interesting to note that for this particular case, if we compute the RMS uncertainties,

A Gaussian profile has the minimum uncertainty product.

$$(\delta x)^2 = \langle x^2 \rangle - \langle x \rangle^2 \qquad (\delta p_x)^2 = \langle p_x^2 \rangle - \langle p_x \rangle^2, \tag{8.4}$$

the equality in Equation 8.3 is satisfied;

$$\sin\delta\alpha = \frac{\lambda}{\delta x}. \tag{8.5}$$

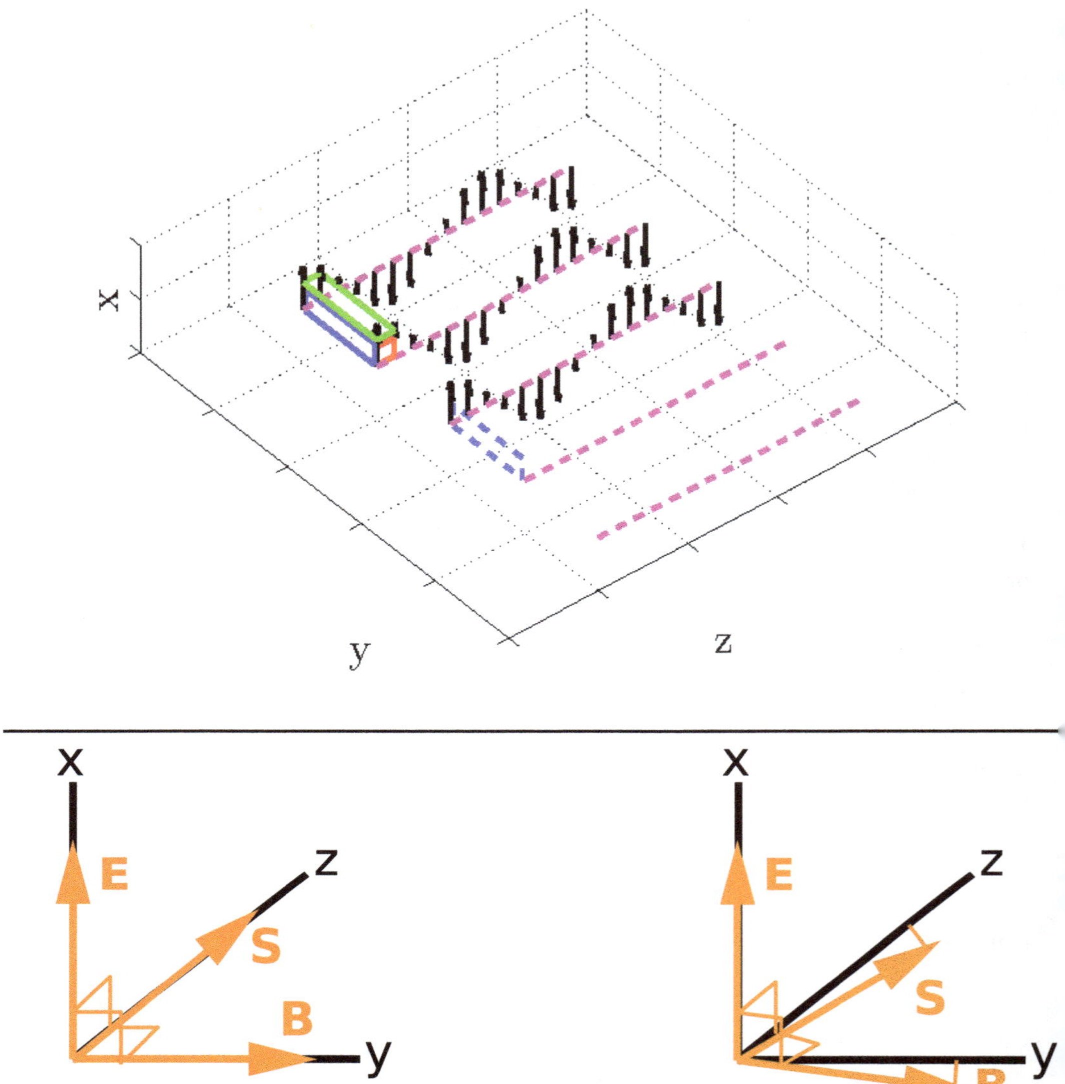

FIGURE 8.4
Diffraction from Maxwell's equations. Maxwell's equations predict diffraction at a boundary. Bla[ck] arrows show samples of the electric field. The integral of **E** around the green or blue solid line[s is] zero, so there is no **B** field perpendicular to these. The **B** field is in the y direction, perpendicu[lar] to the red contour, and **S** = **E** × **B** is in the z direction as expected. At the boundary, the integ[ral] of **E** around the dashed blue line is not zero. Thus, there is a **B** field in the $-z$ direction as w[ell] as in the y direction. Now **S** has components in the z and y directions, and the wave will diffra[ct.] The vectors are shown in the bottom of the figure.

For any other distributions, the inequality,

$$\sin \delta\alpha > \frac{\lambda}{\delta x} \tag{8.6}$$

is satisfied. Thus, the Gaussian distribution that we have chosen is the minimum-uncertainty wave. The limitations of diffraction are least severe for the Gaussian wave,

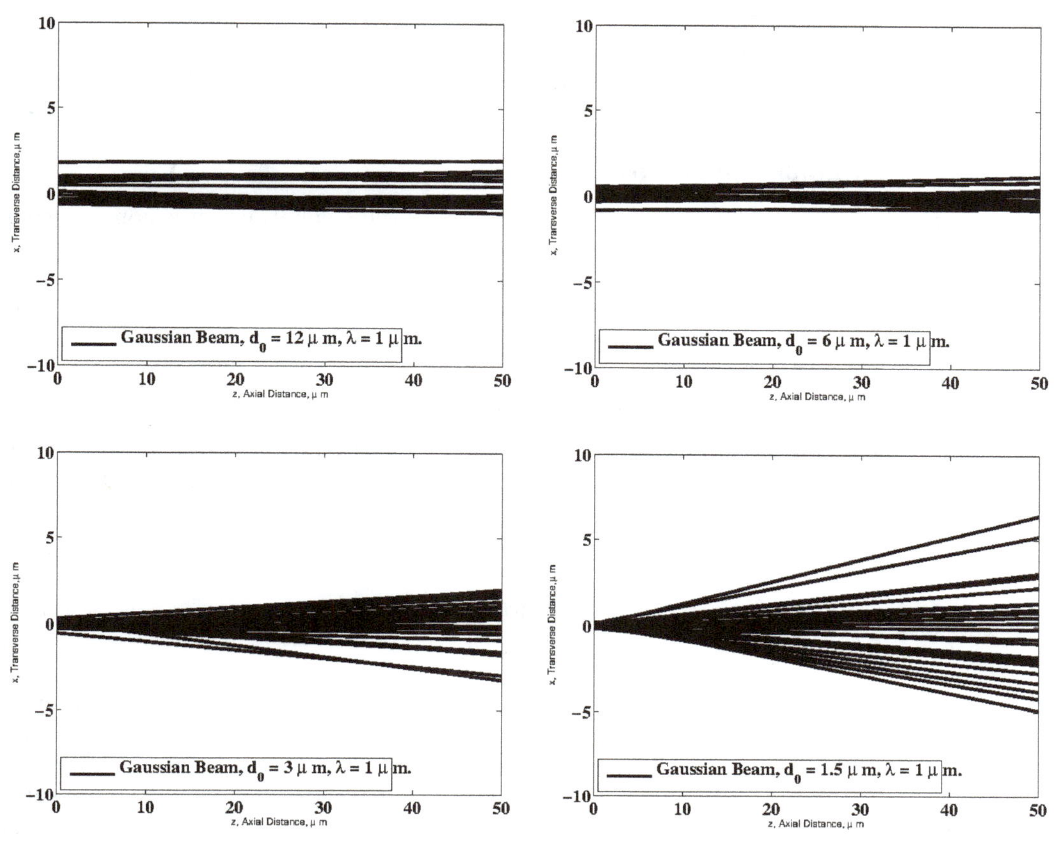

FIGURE 8.5
The uncertainty principle. The more precisely the transverse position of a photon is known (as a result of having passed through a known aperture), the less the transverse momentum, and thus the angle, is known. The rays are generated with position and angle following Gaussian probability distributions.

and it is therefore highly desirable. We will later see that this wave has many interesting and useful properties. Had we chosen a uniform distribution of position over a finite aperture, we would have used a *sinc* function (Fourier transform of a rectangular function) for the angular distribution, and we would have observed concentrations of rays in certain directions in the far field, as we saw in Figure 8.2B.

In summary, diffraction is predicted by quantum mechanics. Specifically a minimum divergence angle is predicted which is inversely proportional to the size of the aperture.

Take-Away

Diffraction Physics:
Diffraction is rooted in fundamental physics, and is thus unavoidable, no matter what care we take in designing and fabricating an optical system. We cannot achieve a smaller spot size or a better resolution than that predicted by diffraction. The most important expression of diffraction is λ/D. The minimum divergence angle is

proportional to λ/D with a constant that depends on the details of the problem. At a distance, z this means the smallest spot size is proportional to $z\lambda/D$.

8.2 The Angular Spectrum

Eq: 1.16: $\mathbf{E} = \hat{\mathbf{x}}E_0 e^{j(\omega t - nkz)}$.

nk is the wavenumber in the medium. k by itself is the vacuum wavenumber in Eq. 1.29.

Eq. 1.29: $k = \frac{2\pi}{\lambda}$ where λ is the vacuum wavelength.

We have seen in Chapter 1 that one solution of Maxwell's equations in a homogeneous isotropic medium is an infinite plane wave (Equation 1.16). We note that k represents the vacuum wavenumber so the wavenumber in the medium is nk. This equation represents a wave traveling in the z direction with an electric field in the $\hat{\mathbf{x}}$ direction and magnetic field in the $\hat{\mathbf{y}}$ direction. For a more general wave, we allow the wave to propagate in an arbitrary direction;

$$\mathbf{E} = \mathbf{E}_0 e^{j(\omega t - nk(ux+vy+wz))}, \tag{8.7}$$

where u, v, and w are direction cosines ($u^2+v^2+w^2 = 1$). Defining θ as the angle between the z axis and the direction of propagation, $u^2 + v^2 = \sin\theta$, and defining ζ as the angle from the x–z plane,

$$\begin{aligned} u &= \sin\theta\cos\zeta \\ v &= \sin\theta\sin\zeta \\ w &= \cos\theta \end{aligned} \tag{8.8}$$

A sum of plane waves is also a solution.

The vector $\mathbf{E}_0$ may be in any direction perpendicular to the propagation direction, $u\hat{\mathbf{x}} + v\hat{\mathbf{y}} + w\hat{\mathbf{z}}$, and the magnetic field, $\mathbf{H}$ must be perpendicular to both $\mathbf{E}$ and the propagation direction. Now the sum of a number of such waves, with different values of $\mathbf{E}_0$, u, v, and w, is also a solution to Maxwell's equations, and we can construct complicated fields as sums of plane waves in a method analogous to Fourier analysis of a signal that is a function of time.

For example, Figure 8.6 shows two plane waves with different directions. Each wave will propagate to the right toward infinity at the given angle. In this example, the angles are given by

$$\theta = 10^\circ \qquad u = \sin\theta = 0.1736 \qquad v = 0$$

and

$$\theta = 20^\circ \qquad u = \sin\theta = 0.3420 \qquad v = 0.$$

Eq. 2.51: $\frac{1}{s} + \frac{1}{s'} = \frac{1}{f}$.

To visualize the field at infinity, we could place a screen a long distance away. However, the plane at infinity, $s \rightarrow \infty$, can be imaged at the back focal plane of a lens. Thus it is often more practical to place a screen at the back focal plane of a lens $s' = f$ as shown in Figure 8.7 using the lens equation (Equation 2.51). The field then looks like that shown in Figure 8.8. The light will focus to a point at a height, x_1 above the z axis, given by

$$x_1 = f\tan\theta \approx fu, \tag{8.9}$$

See Sec. 8.4.4 for discussion of the Fraunhofer lens.

where the approximation is valid for small u. A lens used this way is called a Fraunhofer lens. We will discuss the Fraunhofer lens in detail in Section 8.4.4.

It is interesting to examine Equation 8.7 at $z = 0$ (colored line), which we can use to define a **spatial frequency**,*

$$\mathbf{E}(x, y, 0) = \mathbf{E}_0 e^{j(\omega t - nk(ux+vy))} = \mathbf{E}_0 e^{j[\omega t - 2\pi(f_x x + f_y y)]}, \tag{8.10}$$

where

$$2\pi f_x = nku \qquad 2\pi f_y = nkv. \tag{8.11}$$

*Here we use f_x and f_y for the spatial frequencies. We must remember to avoid confusing f, the focal length, with frequency.

FIGURE 8.6
Plane-wave propagation. Plane waves in the x, z plane at 10 (top) and 20 (bottom) degrees with respect to the z axis (horizontal). The plots on the left show the electric field horizontally as a function of x vertically at the colored line ($z = 0$).

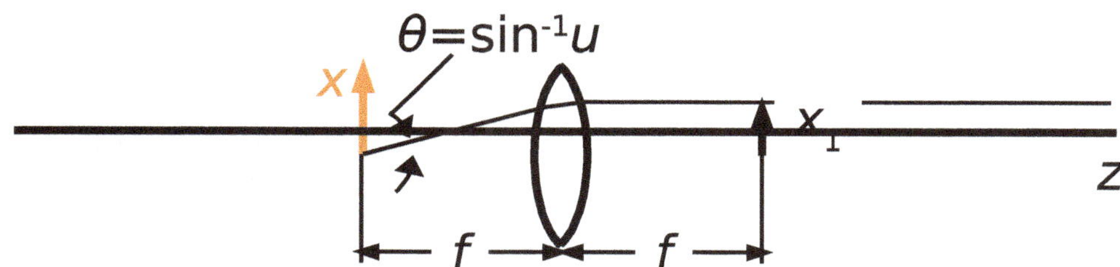

FIGURE 8.7
Fraunhofer lens configuration. The height x_1 in the back focal plane is a measure of the angle of the wavefront in front of the lens.

The fields,

$$\mathbf{E}(x, 0, 0) = \mathbf{E}_0 e^{j2\pi f_x x},$$

are plotted as functions of x at the left in the figures. As we see, the spatial frequency of this function of x is proportional to the tilt (Equations 8.11). The height in the back focal plane of the lens is proportional to the spatial frequency. This relationship is the basis of Fourier optics.

Spatial frequency is proportional to wavefront tilt. Height in the back focal plane is proportional to spatial frequency and tilt in the front.

By superposition, we can combine two waves of the form in Equation 8.10 as shown in Figure 8.9. The field at $z = 0$ consists of a sum,

$$E(x, y, 0) = \mathbf{E}_{01} e^{j(\omega t - nk(u_1 x + v_1 y))} + \mathbf{E}_{02} e^{j(\omega t - nk(u_2 x + v_2 y))} \tag{8.12}$$

which we can now write in terms of spatial frequencies as

$$E(x, y, 0) = \mathbf{E}_{01} e^{j[\omega t - 2\pi(f_{x1} x + f_{y1} y)]} + \mathbf{E}_{02} e^{j[\omega t - 2\pi(f_{x2} x + f_{y2} y)]}. \tag{8.13}$$

Thus the field at $z = 0$ is given by a sum of harmonic functions of x and y with the amplitude at each spatial frequency related to the field at the corresponding point in the back focal plane of the lens. Note that spatial frequency as used here is a two-dimensional variable.

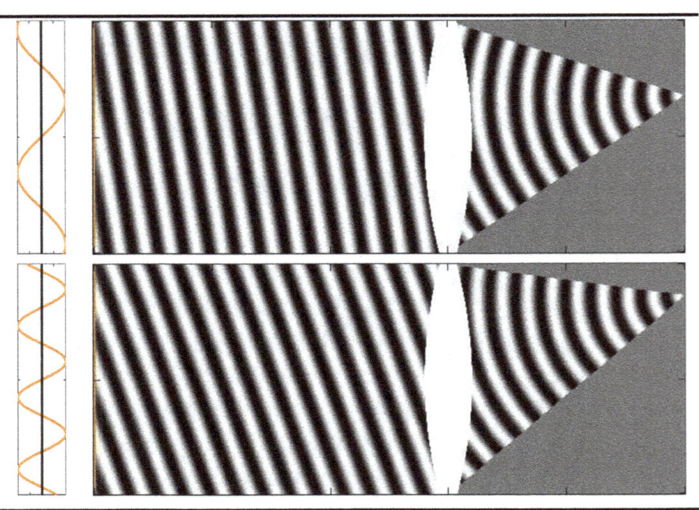

FIGURE 8.8
Using a fraunhofer lens. The waves in Figure 8.6 are focused by a lens to points in the back focal plane of the lens. The x (vertical) location of each point is proportional to its spatial frequency.

As we sum more fields over a tightly spaced range of spatial frequencies (or angles), the solution becomes an integral,

$$\mathbf{E}(x, y, 0) = \int\int \mathbf{E}_F(f_x, f_y)\, e^{j2\pi(f_x x + f_y y)} df_x df_y. \tag{8.14}$$

This equation is similar to the more familiar Fourier transform used by electrical engineers which relates a signal's spectrum to its time history. Here, x and y are analogous to time, f_x and f_y are analogous to frequency, and the Fourier transform is in two dimensions rather than one. Just as a temporal signal can be described as an integral of weighted sinusoids with different frequencies, a light wave in space can be described as a sum of weighted contributions with different spatial frequencies.

If we are given the field, $\mathbf{E}(x, y, 0)$, we can then obtain the angular spectrum by inverse Fourier transformation;

$$\mathbf{E}_F(f_x, f_y) = \int\int \mathbf{E}(x, y, 0)\, e^{-j2\pi(f_x x + f_y y)} dx dy. \tag{8.15}$$

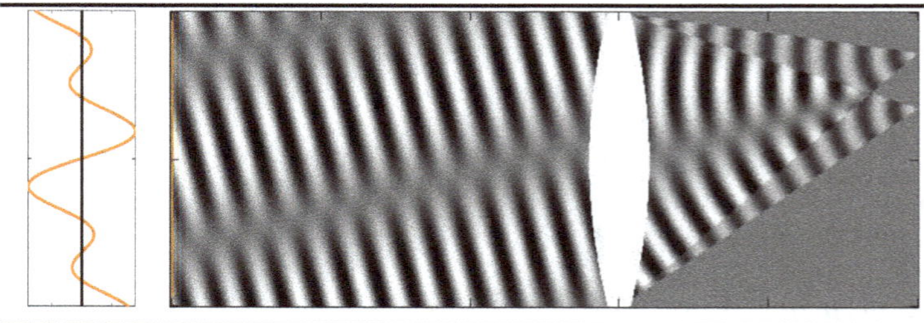

FIGURE 8.9
Superposition of two waves. The two waves in Figure 8.8 superimposed. The field at $z = 0$ is a superposition of waves with two spatial frequencies, and the diffraction pattern consists of two points.

TABLE 8.1
Variables in Fraunhofer Diffraction

	Spatial Frequency	Pupil Location	Direction Cosines
Spatial frequency		$x_1 = \lambda z f_x$	$u = f_x\lambda$
		$y_1 = \lambda z f_y$	$v = f_y\lambda$
Pupil location	$f_x = \frac{x_1}{\lambda z}$		$u = \frac{x_1}{z}$
	$f_y = \frac{y_1}{\lambda z}$		$v = \frac{y_1}{z}$
Direction cosines	$f_x = \frac{u}{\lambda}$	$x_1 = uz$	
	$f_y = \frac{v}{\lambda}$	$y_1 = vz$	
Angle			$u = \sin\theta\cos\zeta$
			$v = \sin\theta\sin\zeta$

The relationships among spatial frequency, location in the pupil, and direction cosines are shown.

Take-Away

Angular Spectrum:
A field can be described by its value as a function of position coordinates or angle coordinates. The two descriptions are Fourier–transform pairs.

The position in the back focal plane is proportional to the angle in the front focal plane, and *vice versa.*

Equivalently we can write these integrals in terms of direction cosines, u and v, using Equation 8.11

8.11
$2\pi f_x = nku \quad 2\pi f_y = nkv.$

$$\mathbf{E}(x, y, 0) = \left(\frac{nk}{2\pi}\right)^2 \int\int \mathbf{E}_U(f_x, f_y)\, e^{jnk(ux+vy)} du\, dv, \tag{8.16}$$

and its dual,

$$\mathbf{E}_U(u, v) = \left(\frac{2\pi}{nk}\right)^2 \int\int \mathbf{E}(x, y, 0)\, e^{-jnk(ux+vy)} dxdy. \tag{8.17}$$

We can also write the equations in terms of position, $x_1, y_1,$ in the pupil plane using Equation 8.9;

8.9:
$x_1 = f\tan\theta \approx fu.$

$$\mathbf{E}(x, y, 0) = \left(\frac{nk}{2\pi}\right)^2 \int\int \mathbf{E}(x_1, y_1, f)\, e^{j\frac{nk}{f}(x_1x+y_1y)} dx_1\, dy_1, \tag{8.18}$$

and

$$\mathbf{E}(x_1, y_1, f) = \left(\frac{2\pi}{nk}\right)^2 \int\int \mathbf{E}(x, y, 0)\, e^{-j\frac{nk}{f}(x_1x+y_1y)} dxdy. \tag{8.19}$$

The distances, x, x_1, *etc.*, were shown in Figure 8.7.

The relationships among the various spatial-frequency variables are shown in Table 8.1.

Take-Away

Fourier Analysis:
The electric field as a function of x and y in the field plane may be written as a Fourier transform using spatial frequency components, f_x and f_y, or as a scaled Fourier transform of the angular distribution of electric field in u and v, or a scaled Fourier transform of the electric field at x_1 and y_1 in the pupil plane. Understanding these concepts will help us understand diffraction theory.

Example

Let's consider a uniform plane wave with field, $\mathbf{E}$, passing through a square aperture of width, D. The spatial spectrum is given by

$$\mathbf{E}_F\left(f_x, f_y\right) = \int_{-D/2}^{D/2}\int_{-D/2}^{D/2} \mathbf{E} e^{-j2\pi(f_x x + f_y y)} dx dy. \tag{8.20}$$

Taking the two-dimensional Fourier transform,

$$\mathbf{E}_F\left(f_x, f_y\right) = D^2 \mathbf{E} \frac{\sin\left(Df_x\right)}{Df_x}\frac{\sin\left(Df_y\right)}{Df_y}. \tag{8.21}$$

In terms of angle then,

$$\mathbf{E}_A\left(u, v\right) = \frac{n^2}{\lambda^2} D^2 \mathbf{E} \frac{\sin\left(nkDu\right)}{nkDu}\frac{\sin\left(nkDv\right)}{nkDv}, \tag{8.22}$$

where we have made use of

$$\mathbf{E}_A\left(u, v\right) dudv = \mathbf{E}_F\left(f_x, f_y\right) df_x df_y$$

and

$$f_x = \frac{u}{\lambda} \qquad f_y = \frac{v}{\lambda}$$

$$\frac{df_x}{du} = \frac{df_y}{dv} = \frac{n}{\lambda}$$

Finally, in the back focal plane of the lens,

$$\mathbf{E}_1\left(x_1, y_1\right) \approx \frac{n^2}{f^2\lambda^2} D^2 \mathbf{E} \frac{\sin\left(\frac{nkD}{f}x_1\right)}{\frac{nkD}{f}x_1}\frac{\sin\left(\frac{nkD}{f}y_1\right)}{\frac{nkD}{f}y_1}, \tag{8.23}$$

where again, the differentials are related by

$$\mathbf{E}_1\left(x_1, y_1\right) dx_1 dy_1 = \mathbf{E}_A\left(u, v\right) dudv$$

$$x_1 \approx fu \qquad y_1 \approx fv$$

$$\frac{dx_1}{du} = \frac{dy_1}{dv} = f$$

For $f = 10\,\text{cm}$ and $u = 0.5$ (30 degree angle), $x_1 = 5\,\text{cm}$.

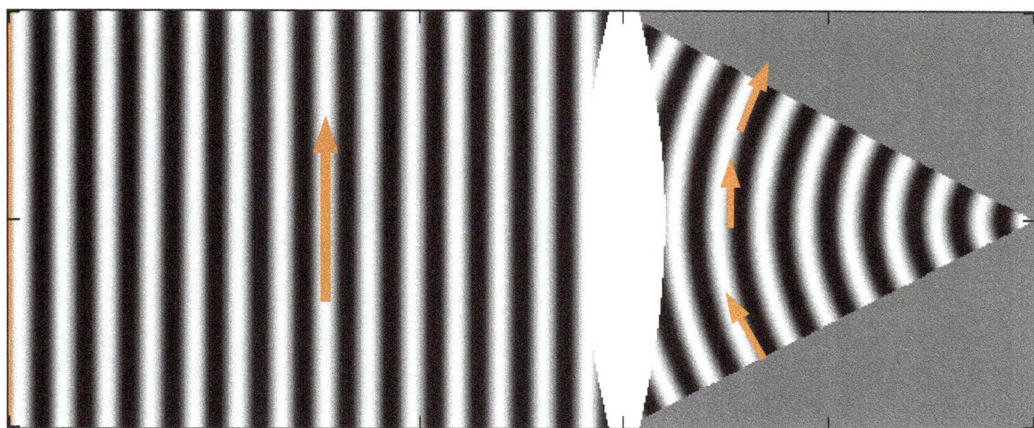

FIGURE 8.10
Polarization in a spherical wave. The figure shows some examples of x polarization. Because the polarization is different at different locations in the spherical wave, scalar wave theory is not valid. Nevertheless, it works quite well except at very large numerical apertures.

We have made some assumptions here which will fail when we consider large angles. There is a relationship among the orientations of the wave vector, $nk\left(u\hat{\mathbf{x}}+v\hat{\mathbf{y}}+w\hat{\mathbf{z}}\right)$, the electric field $\mathbf{E}$ and the magnetic field, $\mathbf{H}$, discussed in Section 6.1.1, which we have ignored here. Over a sufficiently small range of angles, we can pick a direction for $\mathbf{E}$ and ignore this complication. In fact, we can obtain useful quantitative results over quite a large range of angles.

8.3 Fresnel–Kirchhoff Integral

In this section, we will develop an integral equation for solving very general diffraction problems using the approach of Green's functions. This equation will provide us with the ability to solve complicated problems by numerical integration. Furthermore, we will discover simplifying approximations that can be used for most diffraction problems.

We use the Helmholtz equation, which assumes scalar fields. Figure 8.10 shows that the polarization direction is different at different locations on a spherical wave. We expect errors if we assume the field is scalar. However, the errors are usually only significant at very large numerical apertures, so our scalar theory will still be very useful.

We assume harmonic time dependence, and thus use the frequency-domain representation (Equation 1.52).

Eq. 1.52: $\mathbf{E}_r = \mathbf{E}e^{\sqrt{-1}\omega t} + \mathbf{E}^*e^{-\sqrt{-1}\omega t}$.

We begin with the drawing shown on the left side of Figure 8.11. Starting with the field, E, known on the surface A, the goal is to find the field at the point P.

We use a spherical wave, known to be a solution of the Helmholtz equation as an auxiliary field,

$$V\left(x,y,z,t\right)=\frac{e^{j(\omega t+kr)}}{r},\qquad \text{Spherical Wave} \tag{8.24}$$

where r is the distance from the point P. We employ the positive sign on kr, because the wave is assumed to be moving toward the point.

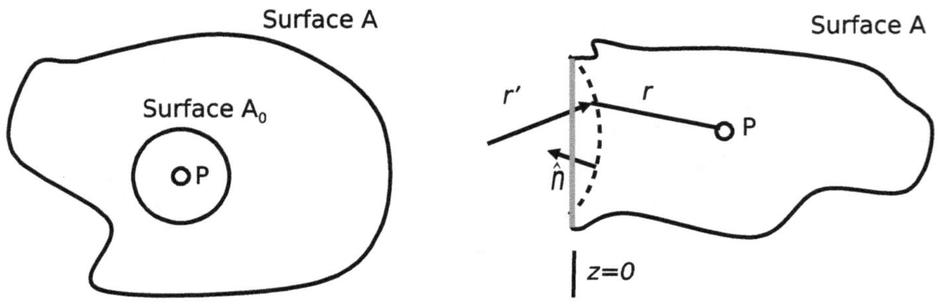

FIGURE 8.11
Geometry for Green's theorem. The surface consists of the combination of two surfaces, A and A_0. The goal is to find the field at the point, P. On the right, the specific case of light propagating through an aperture is considered. The known field is on the plane $z = 0$.

We next write the field E as

$$E\,(x, y, z, t) = E_s\,(x, y, z)\, e^{j\omega t}, \tag{8.25}$$

where E_s represents the spatial part of E. Now, we apply Green's theorem,

$$\iint_{A+A_0} (V\nabla E - E\nabla V) \cdot \hat{n} d^2 A = \\ \iiint_{A+A_0} \nabla \cdot (V\nabla E) - \nabla \cdot (E\nabla v)\, d^3 V, \tag{8.26}$$

where

- d^2A is an increment of area on the surface that includes A and A_0
- $\hat{n}$ is a unit vector normal to the surface
- d^3V is an increment of volume contained by the same surface

Expanding the integrand in the right-hand side of this equation,

$$\nabla \cdot (V\nabla E) - \nabla \cdot (E\nabla V) = \\ (\nabla V)(\nabla E) + V\nabla^2 E - (\nabla E)(\nabla V) - E\nabla^2 V = 0. \tag{8.27}$$

Because both V and E, defined by Equations 8.24 and 8.25, respectively, satisfy the Helmholtz equation,

$$\nabla^2 E = \frac{\omega^2}{c^2} E \qquad \nabla^2 V = \frac{\omega^2}{c^2} V, \tag{8.28}$$

and the volume integral on the right-hand side of Equation 8.26 is zero so the left side must also be zero. We split the integral into two parts, one over A and one over A_0;

$$\iint_A (V\nabla E - E\nabla V) \cdot \hat{n} d^2 A + \iint_{A_0} (V\nabla E - E\nabla V) \cdot \hat{n} d^2 A_0 = 0. \tag{8.29}$$

We then evaluate the second term as the radius of the sphere A_0 goes to zero;

$$\iint_{A_0} (V\nabla E - E\nabla V) \cdot \hat{n} d^2 A = 4\pi E(x, y, z). \tag{8.30}$$

The result is the **Helmholtz–Kirchhoff integral theorem:**

$$E(x, y, z) = -\frac{1}{4\pi} \iint_A \left[E \left(\nabla \frac{e^{jkr}}{r} \right) \cdot \hat{n} - \frac{e^{jkr}}{r} (\nabla E) \cdot \hat{n} \right] d^2 A. \tag{8.31}$$

Examining the right-hand term in the integrand, we are faced with the unpleasant prospect of having a solution which includes both E and its spatial derivative. We would like to eliminate this extra complexity.

Let us turn to the more specific case shown in the right panel of Figure 8.11. We let all but the flat surface (shown in color) of A extend to infinity, so that $E \to 0$ everywhere on the boundary except the flat region, which we define as the plane $z = 0$. Furthermore, the field E arises from a point source a distance, r' to the left of the surface. Then

$$E = E_0 \frac{e^{-jkr'}}{r'}, \tag{8.32}$$

with E_0 a constant. In fact, we do not need to be quite so restrictive. If E_0 varies slowly enough with spatial coordinates, we can neglect ∇E_0, and after some manipulation,

$$(\nabla E) \cdot \hat{n} = \left(jkE - \frac{E}{r'} \right) \hat{n} \cdot \hat{r}'. \tag{8.33}$$

Now, making the further assumptions that $r \gg \lambda$, $r' \gg \lambda$ and that terms in $1/r^2$ and $1/(r')^2$ can be neglected, we obtain the **Fresnel–Kirchhoff integral formula**;

$$E(x_1, y_1, z_1) = \frac{jk}{4\pi} \iint_{Aperture} E(x, y, z) \frac{e^{jkr}}{r} \left(\hat{n} \cdot \hat{r} - \hat{n} \cdot \hat{r'} \right) dA. \tag{8.34}$$

In Practice

The only assumption we have used is that we have a scalar field that solves the Helmholtz equation. This equation can be integrated numerically in very complicated situations for a correct answer. However, it does fail at vary large numerical apertures where the vector nature of the field is important.

We use subscripts on coordinates (x_1, y_1, z_1) for the locations where we want to calculate the field, and no subscripts on $(x, y, 0)$, for locations where we know the field. We will find it useful to think of the plane where we know the field (with coordinates $(x, y, 0)$) as the object plane and the one where we want to calculate the field (with coordinates (x_1, y_1, z_1)) as the pupil plane.

With Equation 8.34, we have a form that is amenable to numerical integration, but we also can make further approximations to develop a few closed-form solutions.

The obliquity factor makes the Huygens approach rigorous by eliminating the backward wave.

The expression $\left(\hat{n} \cdot \hat{r} - \hat{n} \cdot \hat{r'}\right)$ in this integrand is called the **obliquity factor**. In the opening remarks in this chapter, we discussed the approach of Huygens to diffraction, and commented that the approach produces backward as well as forward waves every time it is applied. In the present formulation, we see that the backward wave does not exist because the obliquity factor becomes zero. The obliquity factor contains the cosines, $\hat{n} \cdot \hat{r}$, and $\hat{n} \cdot \hat{r}'$ of two angles, and, for forward propagation, generally along the $\hat{z}$ direction, we can make the assumption that these angles are zero. Then

$$\left(\hat{n} \cdot \hat{r} - \hat{n} \cdot \hat{r}'\right) = 2. \tag{8.35}$$

For very large angles, this assumption will, in principle, produce some errors, but results compare surprisingly well with more rigorous analysis and experiments.

8.3.1 Summary of the Fresnel–Kirchhoff Integral

- Most diffraction problems can be solved by treating the electromagnetic wave as a scalar, despite its inherent vector nature. The exception is in systems with a large numerical aperture where the vector nature cannot be neglected.
- The scalar wave equation is the Helmholtz equation, Equation 1.39.
- The Fresnel–Kirchhoff integral, Equation 8.34, is a solution to the Helmholtz equation.
- The Fresnel–Kirchhoff integral includes an obliquity factor, Equation 8.35, which correctly prevents generation of a backward wave, which was present in Huygens' concept.
- For many cases the obliquity factor is considered constant.

Eq. 8.34:
$E\left(x_1, y_1, z_1\right) = \frac{jk}{4\pi} \iint_{Aperture} E\left(x, y, z\right) \frac{e^{jkr}}{r} \left(\hat{n} \cdot \hat{r} - \hat{n} \cdot \hat{r'}\right) dA.$
Eq. 8.35 $\left(\hat{n} \cdot \hat{r} - \hat{n} \cdot \hat{r}'\right) = 2.$

8.4 Paraxial Approximation

Setting the obliquity factor to 2 as in Equation 8.35, Equation 8.34 becomes

$$E\left(x_1, y_1, z_1\right) = \frac{jk}{2\pi} \iint_{Aperture} E\left(x, y, z\right) \frac{e^{jkr}}{r} dA. \tag{8.36}$$

Now we can further simplify this equation, by making some assumptions about r. We will define three different coordinate systems, and then make the approximations, leading to some particularly useful conclusions.

8.4.1 Coordinate Definitions and Approximations

To handle the diversity of problems we expect to see in our study of diffraction, we would like to consider multiple coordinate systems, as we did in Section 8.2 and in Table 8.1. We summarize these here.

- In some cases, we will want to think in terms of pupil coordinates, x_1 and y_1.
- In other cases, we may want to think of the pupil as very far away and consider either a polar coordinate system with angles, θ, from the axis and ζ around the axis in the transverse plane. We often will prefer to consider direction cosines, $u = \sin\theta\cos\zeta$ and $v = \sin\theta\sin\zeta$. For example, in a microscope, the aperture stop is positioned at the back focus of the objective, and thus the entrance pupil distance is infinite. We note that the maximum value of $\sqrt{u^2 + v^2}$ is the numerical aperture if the systems is in vacuum.

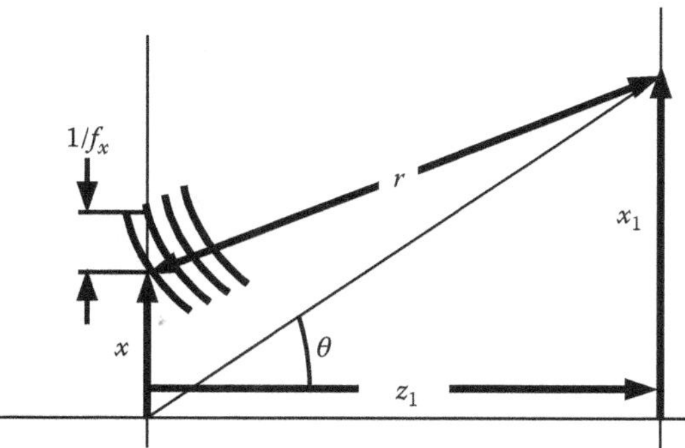

FIGURE 8.12
Coordinates for Fresnel–Kirchhoff integral. The image coordinates are $(x, y, 0)$ and the pupil coordinates are (x_1, y_1, z_1). We define the angle ζ (not shown) as the angle around the z axis from x to y.

- Finally, when we study Fourier optics in Chapter 11, we will want to think in terms of spatial frequency, so that the field of an arbitrary plane wave in the image plane is $E(x, y, 0) = E_0 e^{j2\pi(f_x x + f_y y)}$. Each point in the pupil will correspond to a unique pair of spatial frequencies.

Figure 8.12 shows the geometry. The field plane has coordinates $(x, y, 0)$ and the pupil plane has coordinates (x_1, y_1, z_1). This is similar to Figure 8.6 or 8.8, except that now the point (x_1, y_1, z_1) can be anywhere rather than at $z_1 \to \infty$ or at the back focus of the Fraunhofer lens. The distance from a point in the field (object or image) plane to a point in the pupil plane is

$$r = \sqrt{(x - x_1)^2 + (y - y_1)^2 + z_1^2} \tag{8.37}$$

In Equation 8.36, for the r that appears in the denominator of the integrand, we make a very small error if we set $r = z_1$. Because $z_1 = r\cos\theta$, the error in the field contributions will be small except for very large angles. Having made this approximation, we can bring the $1/z_1$ outside the integral.

We can let $r \to z$ in the denominator of Equation 8.36 $\iint_{Aperture} E(x, y, z) \frac{e^{jkr}}{r} dA$.

If we were to make the same assumption for r in the exponent, we would then find that the integrand is independent of x_1 and y_1, and our goal is to find the field as a function of these coordinates. The issue is that a small fractional error between z_1 and r might still be several wavelengths, and so the integrand could change phase through several cycles. Instead, we will make the first-order approximation using Taylor's series. We write

Taylor Series: $\sqrt{1+q} \approx 1 + \frac{q}{2} + \ldots$.

We need one more term in Taylor's series for r in the exponent.

$$r = z_1\sqrt{\left(\frac{x - x_1}{z_1}\right)^2 + \left(\frac{y - y_1}{z_1}\right)^2 + 1}, \tag{8.38}$$

and using the first-order terms of Taylor's series where $(x - x_1) \ll z_1$ and $(y - y_1) \ll z_1$,

$$r \approx z_1\left[1 + \frac{(x - x_1)^2}{2z_1^2} + \frac{(y - y_1)^2}{2z_1^2}\right],$$

$$r \approx z_1 + \frac{(x - x_1)^2}{2z_1} + \frac{(y - y_1)^2}{2z_1}. \tag{8.39}$$

We followed the same procedure in the paraxial approximation when we considered the interference pattern in the Mach–Zehnder interferometer in deriving Equation 7.21.

We recall that we want to write our equations, in terms of pupil coordinates

$$x_1 \quad \text{and} \quad y_1,$$

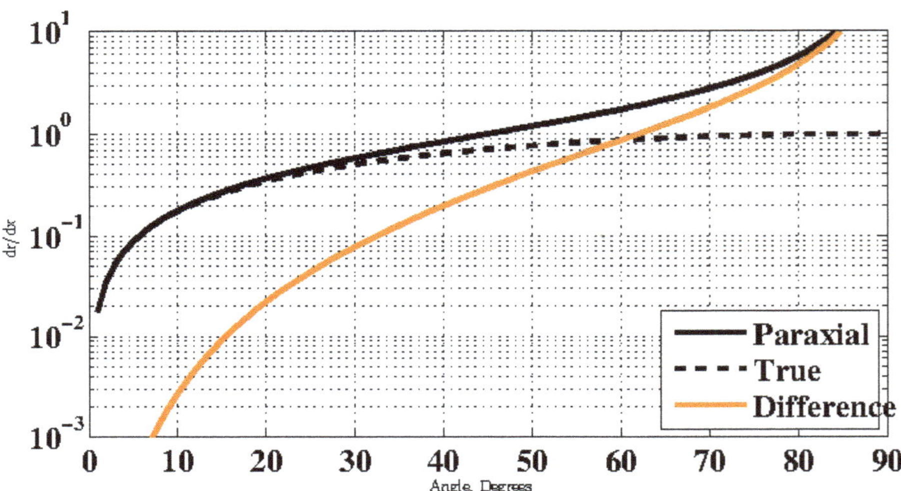

FIGURE 8.13
Paraxial error. The spatial period, $1/f_x$ is the spacing between wavefronts along the x axis. Using the "true" Equation 8.40 and the paraxial approximation, Equation 8.42, the error in dr/dx is less than a 10th of a wavelength up to an angle of about 30°. It only reaches one wavelength at an angle beyond 60°.

in terms of direction cosines (Equation 8.8)

$$u = \sin\theta\cos\zeta \quad \text{and} \quad v = \sin\theta\sin\zeta,$$

or in terms of spatial frequencies

$$f_x \quad \text{and} \quad f_y,$$

Eq. 8.11:
$2\pi f_x = nku$
$2\pi f_y = nkv.$

using Equation 8.11. When we work with spatial frequencies, we will assume that the pupil is at a long distance compared to the dimensions of the object, so we let $x \approx 0$, and

$$f_x \approx \frac{x_1}{r\lambda} = \frac{\sin\theta\cos\zeta}{\lambda} \tag{8.40}$$

and likewise

$$f_y \approx \frac{y_1}{r\lambda} = \frac{\sin\theta\sin\zeta}{\lambda}. \tag{8.41}$$

Now, if we had used the approximation in Equation 8.39 we would have obtained

$$f_x \approx \frac{x_1}{z_1\lambda} = \frac{\tan\theta\cos\zeta}{\lambda}, \tag{8.42}$$

Eq. 8.36:
$E(x_1, y_1, z_1) = \frac{jk}{2\pi}\int\int_{Aperture} E(x, y, z)\frac{e^{jkr}}{r}dA.$.

with a tangent replacing the sine. The error is quite small, except for large angles, as shown in Figure 8.13. The relationships among the variables are shown in Table 8.1.

8.4.2 Computing the Field

Eq. 8.39:
$r \approx z_1 + \frac{(x-x_1)^2}{2z_1} + \frac{(y-y_1)^2}{2z_1}.$
.

Bringing the $1/z_1$ outside the integral in Equation 8.36

$$E(x_1, y_1, z_1) = \frac{jk}{2\pi z_1}\int\int E(x, y, 0)\, e^{jkr} dA \tag{8.43}$$

and making the paraxial approximation in Equation 8.39,

$$E(x_1, y_1, z_1) = \frac{jke^{jkz_1}}{2\pi z_1} \int\int E(x, y, 0)\, e^{jk\frac{(x-x_1)^2}{2z_1} + \frac{(y-y_1)^2}{2z_1}}\, dx\, dy \tag{8.44}$$

The best way to handle the exponent depends on the problem to be solved. For problems where the integrand may have relatively large phase variations, it is best to leave it in this form as we will see in the study of Fresnel diffraction in Section 8.9. For other problems, it is convenient to expand and regroup the transverse-position terms in the exponent.

$$(x - x_1)^2 + (y - y_1)^2 = (x^2 + y^2) + (x_1^2 + y_1^2) - 2(xx_1 + yy_1), \tag{8.45}$$

with the result

$$E(x_1, y_1, z_1) = \frac{jke^{jkz_1}}{2\pi z_1} e^{jk\frac{(x_1^2+y_1^2)}{2z_1}} \iint E(x, y, 0)\, e^{jk\frac{(x^2+y^2)}{2z_1}}\, e^{-jk\frac{(xx_1+yy_1)}{z_1}}\, dx\, dy. \tag{8.46}$$

Eq. 8.19: $\mathbf{E}(x_1, y_1, f) = \left(\frac{2\pi}{nk}\right)^2 \iint \mathbf{E}(x, y, 0)\, e^{-j\frac{nk}{f}(x_1 x + y_1 y)} dx dy$.

This equation looks like Equation 8.19, except for the constants.

Most of our diffraction problems can be solved with either Equation 8.44 or 8.46. In the following, we will see that Equation 8.46 describes the propagation problem as (1) a curvature term at the source plane, (2) a Fourier transform, (3) a curvature term at the final plane, and (4) scaling. We will use this view in our study of Fraunhofer diffraction in Section 8.5, and return to it in Section 11.1, for our study of Fourier optics.

In contrast, Equation 8.44 describes the integrand in terms of transverse distances $x - x_1$ and $y - y_1$ between points in the two planes. The integrand is nearly constant for small values of the transverse distance, and for such cases, the fields in the two planes will be similar except where strong field gradients exist, such as at the edges of shadows.

8.4.3 Fresnel Radius and the Far Field

Terms like $e^{jk\frac{(x^2+y^2)}{2z_1}}$ correspond physically to spherical wavefronts. We call these terms **curvature terms**, and this interpretation may provide some intuition about diffraction. We will explore Equation 8.46 in the study of Fraunhofer diffraction, and then return to Equation 8.44 for the study of Fresnel diffraction. The choice depends on the curvature term in the integrand,

$$e^{jk\frac{(x^2+y^2)}{2z_1}} = e^{j2\pi\left(\frac{r}{\sqrt{2\lambda z_1}}\right)^2} = e^{j2\pi\left(\frac{r}{r_f}\right)^2} \qquad r_f = \sqrt{2\lambda z_1}, \tag{8.47}$$

where we have defined the **Fresnel radius**, r_f, as the radius at which the exponent has gone through one full cycle. Then Equation 8.46 becomes

Fresnel radius defined as the radius at which the integrand goes through a cycle.

$$E(x_1, y_1, z_1) = \frac{jke^{jkz_1}}{2\pi z_1} e^{jk\frac{(x_1^2+y_1^2)}{2z_1}}$$
$$\int\int E(x, y, 0)\, e^{j2\pi\frac{x^2+y^2}{r_f^2}}\, e^{-jk\frac{(xx_1+yy_1)}{z_1}}\, dx\, dy. \tag{8.48}$$

We have plotted the real and imaginary parts of the exponential of Equation 8.47 in Figure 8.14 for $r = \sqrt{x^2 + y^2}$ from zero to $4r_f$, at which point, because the phase varies as

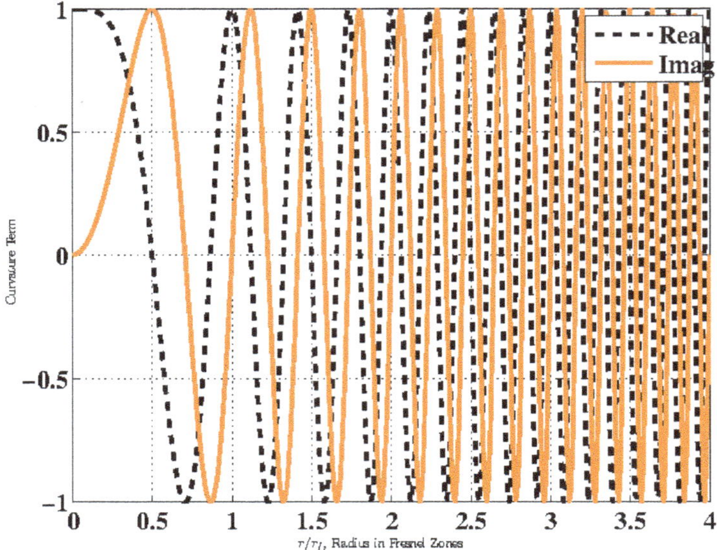

FIGURE 8.14
Curvature term. The real (dashed black line) and imaginary (solid colored line) parts of the curvature term are plotted for the radius normalized to the Fresnel radius.

the square of the distance from the axis, it has gone through 16 cycles, and the oscillation is becoming ever more rapid. If $E(x, y, 0)$ varies slowly in the aperture, we can consider three zones based on the **Fresnel number**

Fresnel Number defined in terms of Fresnel radius.

$$N_f = \left(\frac{r_{max}}{r_f}\right)^2 \tag{8.49}$$

for r_{max} as the largest radius for which the integrand is nonzero. The character of the diffraction pattern and the approach to solving the integral will depend on N_f.

8.4.3.1 Far Field

If z_1 is sufficiently large, so that

$$z_1 \gg D^2/\lambda \qquad \text{(Far-field condition)}, \tag{8.50}$$

then $r_f = \sqrt{2z_1\lambda} \gg \sqrt{2}D$ for all x, y within the aperture, and the Fresnel number is small,

The far field is always in the Fraunhofer zone, where the Fresnel number is small. Fraunhofer diffraction works here.

$$N_f = r/r_f \ll 1 \qquad \text{(Fraunhofer zone)}, \tag{8.51}$$

and

$$e^{j2\pi\frac{\left(x^2+y^2\right)}{r_f}} \approx 1. \tag{8.52}$$

We will see that the integral becomes a two-dimensional Fourier transform of $E(x, y, 0)$, and for many functions, it can be integrated in closed form. We call this region the

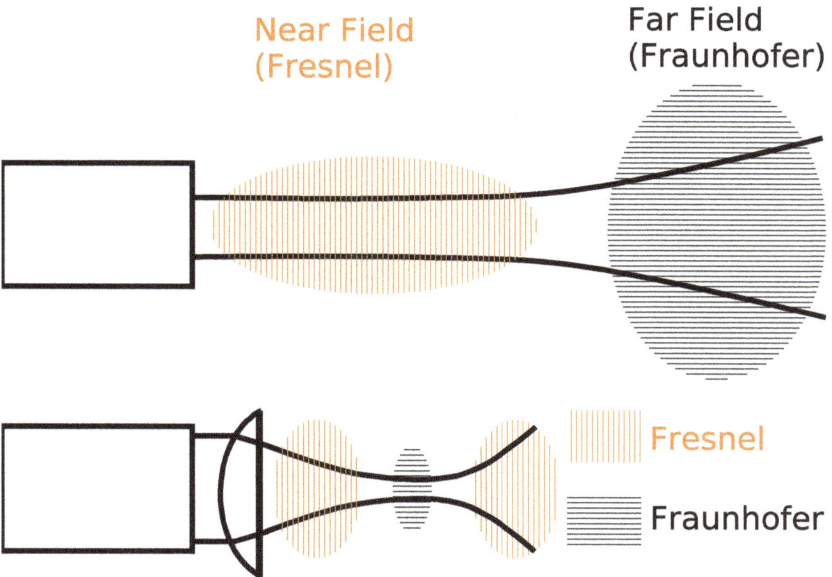

FIGURE 8.15
Fresnel and Fraunhofer zones. For a collimated source, the Fraunhofer zone is the far field, $z_1 \ll D^2/\lambda$, and the Fresnel zone is the near field, $z_1 \gg D^2/\lambda$. For a source focused in the near field, the Fraunhofer zone is near the focus, $|f - z_1| \ll \frac{8z_1^2\lambda}{D^2}$, and the Fresnel zone is far from focus.

Far Field. Because we can use Fraunhofer diffraction in this region, we also call it the **Fraunhofer zone**. We may use Equation 8.46 with $z_1 \to \infty$,

$$E(x_1, y_1, z_1) = \frac{jke^{jkz_1}}{2\pi z_1} e^{jk\frac{\left(x_1^2+y_1^2\right)}{2z_1}} \int\int E(x, y, 0)\, e^{-jk\frac{(xx_1+yy_1)}{z_1}} dx\, dy.$$

Figure 8.15a illustrates these definitions.

8.4.3.2 Near Field

If on the other hand,

$$z_1 \ll D^2/\lambda \qquad \text{(Near-field condition)}, \tag{8.53}$$

then the Fresnel number is large,

The near field of a collimated beam is always in the Fresnel Zone where the Fresnel Number is large. Fresnel diffraction is used here.

$$N_f = r/r_f \gg 1, \tag{8.54}$$

r goes through many Fresnel zones in the aperture, the integration of Equation 8.46 is difficult, and we need other techniques, such as Fresnel diffraction, which we will discuss later. We will call this region the **Near Field**. Because we usually use Fresnel diffraction in this region (Equation 8.44), we call it the **Fresnel zone**.

Take-Away

Collimated Beam:
For a collimated beam of light, we use Fraunhofer diffraction in the far field and Fresnel diffraction in the near field. We can decide which is appropriate using the Fresnel number.

The back focal plane of a Fraunhofer lens is also in the Fraunhofer Zone because the Fresnel number is small. Away from focus the Fresnel number is large.

The situation where the Fresnel Number, $r/r_f \approx 1$ is neither in the near field nor the far field. Neither approximation is valid.

8.4.3.3 Almost Far Field

If the radius of the aperture is comparable to the Fresnel radius, then we are working in neither the near nor far field. In this region, we expect to see characteristics of both the near-field and far-field patterns. We may use Equation 8.46, but the integrand will be complicated enough that we will usually perform the integral numerically.

8.4.4 Fraunhofer Lens

There is one more case where Fraunhofer diffraction may be used. Consider a field given by

$$E(x, y, 0) = E_0(x, y, 0)\, e^{-jk\frac{(x^2+y^2)}{2f}}. \tag{8.55}$$

where f is in the near field of the aperture, $f \ll D^2/\lambda$. If E_0 is a constant, or at least has constant phase, E represents a curved wavefront, converging toward the axis at a distance f. Physically this curvature could be achieved by starting with a plane wave and using a lens of focal length f in the aperture. We refer to the lens as a **Fraunhofer lens**. Now, the integral in Equation 8.46 becomes

$$E(x_1, y_1, z_1) = \frac{jke^{jkz_1}}{2\pi z_1} e^{jk\frac{(x_1^2+y_1^2)}{2z_1}}$$
$$\int\int E(x, y, 0)\, e^{jk\frac{(x^2+y^2)}{2Q}}\, e^{-jk\frac{(xx_1+yy_1)}{z_1}} dx_1\, dy_1, \tag{8.56}$$

where

$$\frac{1}{Q} = \frac{1}{z_1} - \frac{1}{f}, \tag{8.57}$$

and we call Q the **defocus parameter**. This equation is identical to Equation 8.48, provided that we substitute

$$r_f = \sqrt{2\lambda Q}$$
$$\frac{1}{r_f} = \frac{1}{\sqrt{2\lambda z_1}} - \frac{1}{\sqrt{2\lambda f}}. \tag{8.58}$$

Figure 8.15 summarizes the near and far field terms and the Fresnel and Fraunhofer zones. If the source is collimated, the Fraunhofer zone is the far field of the aperture, and the Fresnel zone is the near field. If the source is focused in the near field, then the Fraunhofer zone is near the focus and the Fresnel zone includes regions far in front of and behind the focus.

In the previous section, the fact that $r_f \to \infty$ as $z_1 \to \infty$ allowed us to use Equation 8.53 at large distances. Here the fact that $r_f \to \infty$ as $z \to f$, allows us to use the same equation near the focus of a converging wave.

Fig. 8.15:

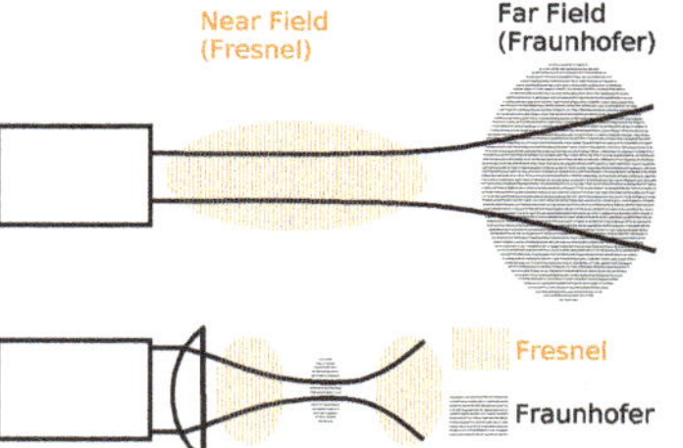

Example

How close must z_1 be to f for this simplification to apply? We need the term in the exponent of Equation 8.56, $k\frac{(x_1^2+y_1^2)}{2Q}$ to be much smaller than 2π for all x and y within the diameter, D. Thus we need $k\frac{(D/2)^2}{2Q} \ll 2\pi$ or

$$Q = \frac{fz_1}{|f - z_1|} \gg \frac{D^2}{8\lambda} \qquad |f - z| \ll \frac{8z_1^2\lambda}{D^2}, \tag{8.59}$$

where we have used the approximation $fz_1 \approx z_1^2$, because we know that $|f - z_1|$ is much smaller than z_1. As an example, consider a beam of diameter, $D = 10\,\text{cm}$, perfectly focused at $f = 10\,\text{m}$. The wavelength is $\lambda = 10\,\mu\text{m}$, typical of a carbon dioxide laser . For this beam, we can use Equation 8.53 if $|f - z_1| \ll 0.8\,\text{m}$. Thus, the beam profile is quite well described by Fraunhofer diffraction, and is close to the diffraction limit for distances z_1 such that $9.2\,\text{m} \ll z_1 \ll 10.8\,\text{m}$. In Section 8.6.5, we will see that this defines the depth of focus.

Take-Away

Focused Beam:
For a focused beam, we use Fraunhofer diffraction near the focus and Fresnel diffraction far from it. The Fresnel number helps us decide which is appropriate.

In Practice

The easy rule is that we use Fraunhofer diffraction when geometric optics predicts a spot size of zero and Fresnel diffraction when geometric optics predicts a large spot size.

8.4.5 Summary of Paraxial Approximation

- The paraxial form assumes that transverse distances in the source plane and the plane of interest are all small compared to z_1, the distance between these planes.
- The paraxial equations are obtained by taking the first terms in a Taylor's series approximation to the distance, r between a point in the source and in the plane of interest (Equation 8.39).
- The resulting equation is the basis for Fresnel diffraction (Equation 8.44).

- Fraunhofer diffraction is an important special case to be discussed in detail in the next section.
- Fraunhofer diffraction is valid in the far field of a collimated wave, or near the focus of a converging wave.

8.5 Fraunhofer Diffraction Equations

Eq. 8.50: $z_1 \gg D^2/\lambda$.

For any finite aperture limiting the extent of x and y, it is always possible to pick a sufficiently large z_1 (Equation 8.50) that all points in the aperture remain well within the first Fresnel zone, $r \ll r_f$, and the curvature can be ignored. Then Equation 8.46 becomes Equation 8.53,

$$E(x_1, y_1, z_1) = \frac{jke^{jkz_1}}{2\pi z_1} e^{jk\frac{\left(x_1^2+y_1^2\right)}{2z_1}} \iint E(x, y, 0)\, e^{-jk\frac{(xx_1+yy_1)}{z_1}}\, dx\, dy. \tag{8.60}$$

8.5.1 Spatial Frequency

Equation 8.60 is simply a scaled two-dimensional Fourier transform:

$$E(f_x, f_y, z_1) = \left(\frac{k}{2\pi z_1}\right)^2 \frac{jke^{jkz_1}}{2\pi z_1} e^{jk\frac{\left(x_1^2+y_1^2\right)}{2z_1}} \iint E(x, y, 0)\, e^{-j2\pi(f_x x+f_y y)}\, dx\, dy$$

$$E(f_x, f_y, z_1) = \frac{j2\pi z_1 e^{jkz_1}}{k} e^{jk\frac{\left(x_1^2+y_1^2\right)}{z_1}} \iint E(x, y, 0)\, e^{-j2\pi(f_x x+f_y y)}\, dx\, dy, \tag{8.61}$$

where we have substituted according to Table 8.1

$$f_x = \frac{k}{2\pi z_1} x_1 = \frac{x_1}{\lambda z_1} \qquad f_y = \frac{k}{2\pi z_1} y_1 = \frac{y_1}{\lambda z_1}, \tag{8.62}$$

and

$$df_x = \frac{k}{2\pi} \frac{dx_1}{z_1} \qquad df_y = \frac{k}{2\pi} \frac{dy_1}{z_1}.$$

8.5.2 Angle and Spatial Frequency

In Equation 8.60, E has units of a field variable. As the distance, z_1, increases, the field decreases proportionally to $1/z_1$. If we then define a new variable, $E_a = Ez_1e^{-jkz_1}$,

$$E_a(u, v) = \frac{jk}{2\pi\cos\theta} \iint E(x, y, 0)\, e^{-jk\frac{xu+yv}{w}}\, dx\, dy. \tag{8.63}$$

We now have three expressions that use the Fourier transform of the field, Equation 8.60, giving the result in terms of the field in the Fraunhofer zone, Equation 8.63 giving it in terms of angular distribution of the field and Equation 8.61 giving the result in terms of spatial frequencies. These three expressions all contain the same information. Thus, a numerical aperture of NA blocks the plane waves that contribute to higher spatial frequencies. The relationships among distances, x_1, y_1, in the aperture, spatial frequencies, f_x, f_y, and direction cosines, u, v, are shown in Table 8.1.

Example

If a camera has discrete pixels with a spacing of 7.4 µm, the sampling frequency is

$$f_{sample} = \frac{1}{7.4 \times 10^{-6}\,\text{m}} = 1.35 \times 10^5\,\text{m}^{-1} \tag{8.64}$$

or 135 cycles per millimeter. Most electrical engineers are familiar with the sampling theorem, first stated in the literature by Shannon[203], but often attributed to Nyquist or Whittaker[247]. In fact, Shannon was well aware of these other works. When he stated that a band-limited signal with Fourier terms at frequencies less than f_{max} is "completely determined by giving its ordinates at a series of points space $T = \pi/\omega_{max}$ seconds apart," or $T = 1/(2f_{max})$, he called $2f_{max}$ the Nyquist frequency[229]. If a digitized signal contains contributions at frequencies higher than f_{max}, half the Nyquist frequency, these higher frequencies will appear at lower frequencies. This result is, of course, true for spatial frequencies. In order to satisfy the sampling theorem, we must ensure that the field incident on the camera does not have spatial frequencies beyond half this value. Thus we need a numerical aperture of

$$NA = \frac{2f_{sample}\lambda}{2} = f_{sample}\lambda \qquad \text{(Coherent imaging)}, \tag{8.65}$$

or, for green light at 500 nm,

$$NA = 0.068. \tag{8.66}$$

The pupil here acts as an anti-aliasing filter. We will discuss this topic more in Section 11.1.4. We note here that this equation is valid only for coherent detection. When we discuss Fourier optics, we will consider both the case of coherent and incoherent detection.

Take-Away

Fraunhofer Diffraction:

- Fraunhofer diffraction is a special case of the paraxial approximation to the Fresnel–Kirchhoff integral, in which the integral becomes a Fourier transform.
- It is useful in the far field $z \gg D^2/(\lambda)$ and at the focus of a Fraunhofer lens.
- The Fraunhofer diffraction pattern can be described as a function of location in the pupil (Equation 8.60), direction cosines, or spatial frequency as shown in Table 8.1.

Eq. 8.60:
$E(x_1, y_1, z_1) = \frac{jke^{jkz_1}}{2\pi z_1} e^{jk\frac{(x_1^2+y_1^2)}{2z_1}} \iint E(x,y,0) e^{-jk\frac{(xx_1+yy_1)}{z_1}} dx\, dy$.

8.6 Some Useful Fraunhofer Patterns

There are a few basic shapes that arise so often that it is useful to examine the results of Equation 8.60 for these special cases: rectangular and circular apertures, and Gaussian beams. These three are the most common shapes, have well-known solutions, and can often be combined or used as a first approximation to solve more complicated problems.

8.6.1 Square or Rectangular Aperture

For a uniformly illuminated rectangular aperture of size D_x by D_y, with power P, the irradiance in the aperture is a constant,

$$E^2(x,y,0) = \frac{P}{D_x D_y}$$

For a rectangular aperture the integrals are separable.

Equation 8.60 becomes

$$E(x_1, y_1, z_1) = \frac{jke^{jkz_1}}{2\pi z_1} e^{jk\frac{(x_1^2+y_1^2)}{2z_1}} \int_{-D_x/2}^{D_x/2} \int_{-D_y/2}^{D_y/2} \sqrt{\frac{P}{D_x D_y}} e^{-jk\frac{(xx_1+yy_1)}{z_1}} dx\, dy \qquad (8.67)$$

This equation is now separable in x and y,

$$E(x_1, y_1, z_1) = \sqrt{\frac{P}{D_x D_y}} \frac{jke^{jkz_1}}{2\pi z_1} e^{jk\frac{(x_1^2+y_1^2)}{2z_1}} \int_{-D_x/2}^{D_x/2} e^{-jk\frac{(xx_1)}{z_1}} dx \int_{-D_y/2}^{D_y/2} e^{-jk\frac{(yy_1)}{z_1}} dy \qquad (8.68)$$

For many readers, this will be familiar as the Fourier transform of a square pulse, commonly called the *sinc* function. The result is

$$E(x_1, y_1, z_1) = \sqrt{P\frac{D_x D_y}{\lambda^2 z_1^2}} \frac{\sin\left(kx_1\frac{D_x}{2z_1}\right)}{kx_1\frac{D_x}{2z_1}} \frac{\sin\left(ky_1\frac{D_y}{2z_1}\right)}{ky_1\frac{D_y}{2z_1}} e^{jk\frac{(x_1^2+y_1^2)}{2z_1}}. \qquad (8.69)$$

Usually, but not always, the field is not important; we want the irradiance.

Normally we are interested in the irradiance:*

$$I(x_1, y_1, z_1) = P\frac{D_x D_y}{\lambda^2 z_1^2} \left(\frac{\sin\left(kx_1\frac{D_x}{2z_1}\right)}{kx_1\frac{D_x}{2z_1}} \frac{\sin\left(ky_1\frac{D_y}{2z_1}\right)}{ky_1\frac{D_y}{2z_1}} \right)^2. \qquad (8.70)$$

We take particular note of

- The on-axis irradiance

$$I_0 = I(0,0,z_1) = P\frac{D_x D_y}{\lambda^2 z_1^2},$$

*In this chapter, we continue to use the variable name, I, for irradiance, rather than the radiometric E, as discussed in Section 1.4.4.

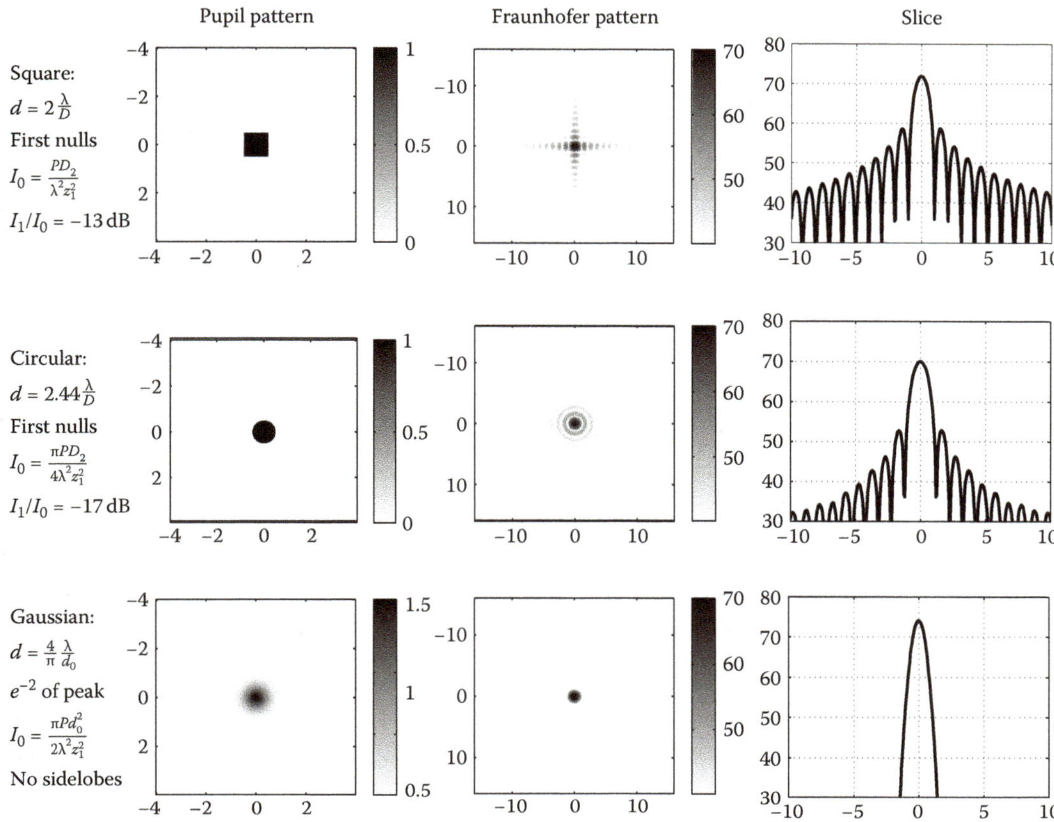

FIGURE 8.16
Common diffraction patterns. Square, circular, and Gaussian patterns are frequently used. The left column shows the burn pattern of the initial beam (linear scale) and the center column shows the diffraction pattern in decibels ($20\log_{10}|E|$). The right column shows a slice through the center of the diffraction pattern.

- The location of the null between the main lobe and the first sidelobe, where the sine function is zero
$$\frac{d_x}{2} = \frac{\lambda}{D_x} z_1$$
- The height of the first sidelobe, $I_1 = 0.0174 I_0$ which is -13 dB.

The special case of a square aperture is shown in the first row of Figure 8.16. The left figure is an image of the irradiance, or a **burn pattern** of the beam at the source. We use the term burn pattern because this is the pattern we would see on a piece of paper or wood illuminated with this beam, assuming sufficient power. The second column shows the burn pattern for the diffracted beam, and the third column shows a plot of the diffraction pattern as a function of x for $y = 0$. The latter two are shown in decibels ($20\log_{10}|E|$).

Example

Let's consider 100 mW of green light ($\lambda = 500$ nm) passing through a 1 cm square aperture at $z = 0$, and focused to a distance of $z_1 = 10$ cm. What is the diameter and on-axis irradiance of the focused spot? The diameter is

$$d_x = \frac{2\lambda}{D_x} z_1 = \frac{2 \times 500 \times 10^{-9}\,\text{m}}{10^{-2}\,\text{m}} \times 10^{-1}\,\text{m} = 10^{-5}\,\text{m}.$$

The distance between first nulls is $10\,\mu$m.

The irradiance on axis is

$$I_0 = I(0,0,z_1) = P\frac{D_x D_y}{\lambda^2 z_1^2} =$$

$$10^{-1}\,\text{W}\frac{\left(10^{-2}\,\text{m}\right)^2}{\left(500 \times 10^{-9}\,\text{m}\right)^2 \left(10^{-1}\,\text{m}\right)^2} = 4 \times 10^9\,\text{W/m}^2,$$

or $4\,\text{GW/m}^2$. For comparison sunlight typically provides about $1\,\text{kW/m}^2$ at the surface of the earth.

Some more complicated problems beyond uniform illumination can be solved using similar techniques, provided that the illumination pattern can be separated into $E(x,y,0) = E_x(x,0)E_y(y,0)$. If the problem cannot be solved in closed form, it can, at worst, still be treated as a product of two Fourier transforms and evaluated numerically.

8.6.2 Circular Aperture

The uniformly–illuminated circular aperture is one of the most common and useful diffraction problems. Many practitioners remember the equation and use it as a first approximation to more complicated problems.

If instead of functions of x and y, the field, $E(x,y,0)$, in Equation 8.60 can be separated in polar coordinates into functions of $r = \sqrt{x^2+y^2}$ and ζ such that $x = r\cos\zeta$ and $y = r\sin\zeta$, another approach can be used. In particular, if the field is independent of ζ, the equation becomes

$$E(r_1,z_1) = \frac{jke^{jkz_1}}{2\pi z_1} e^{jk\frac{\left(x_1^2+y_1^2\right)}{2z_1}} \int_0^{D/2} \int_0^{2\pi} E(r,0)\, e^{-j\frac{k(r\cos\zeta x + r\sin\zeta y)}{z_1}}\, r\, d\zeta\, dr.$$

After performing the inner integral, we are left with a Hankel transform,

$$E(r_1,z_1) = \frac{jke^{jkz_1}}{2\pi z_1} e^{jk\frac{\left(r_1^2\right)}{z_1}} \int_0^{D/2} E(r,0)\, e^{-j\frac{kr^2}{z_1}} J_0\left(\frac{krr_1}{z_1}\right) dr. \tag{8.71}$$

For a uniformly illuminated circular aperture of diameter D, with power, P,

$$E^2(r,0) = \frac{P}{\pi (D/2)^2},$$

and Equation 8.60 becomes

$$E(r_1,z_1) = \frac{jke^{jkz_1}}{2\pi z_1} e^{jk\frac{\left(r_1^2\right)}{z_1}} \int_0^{D/2} \sqrt{\frac{P}{\pi (D/2)^2}} J_0\left(\frac{krr_1}{z_1}\right) dr.$$

Although the integral looks formidable, it can be solved in the form of Bessel functions, with the result

$$E(r_1,z_1) = \sqrt{P\frac{\pi D^2}{4\lambda^2 z_1^2}}\, \frac{J_1\left(kr_1\frac{D}{2z_1}\right)}{kr_1\frac{D}{2z_1}} e^{jk\frac{r_1^2}{2z}}$$

$$I(r_1, z_1) = P\frac{\pi D^2}{4\lambda^2 z_1^2}\left(\frac{J_1\left(kr_1\frac{D}{2z_1}\right)}{kr_1\frac{D}{2z_1}}\right)^2, \tag{8.72}$$

with

- On-axis irradiance $I_0 = P\pi D^2/4\lambda^2 z_1^2$
- Distance between first nulls $2.44 z_1 \lambda D$
- First sidelobe 17 dB down from the peak

These results are also shown in Figure 8.16. This pattern is frequently called the **Airy Pattern**, not to be confused with the Airy function.

Example

Let's consider the same example we used for the square aperture, 100 mW of green light ($\lambda = 500\,\text{nm}$) passing through a 1 cm circular aperture at $z = 0$, and focused to a distance of $z_1 = 10\,\text{cm}$. Thus the circle in the current example is inscribed within the square of the previous one.

Now, the diameter between first nulls is

$$d = 2.44\frac{\lambda}{D}z_1 = 2.44\frac{500 \times 10^{-9}\,\text{m}}{10^{-2}\,\text{m}}10^{-1}\,\text{m} = 1.22 \times 10^{-5}\,\text{m}$$

or 12.2 μm. The on-axis irradiance is

$$I_0 = I(0, 0, z_1) = P\frac{\pi D^2}{4\lambda^2 z_1^2} =$$

$$10^{-1}\,\text{W}\frac{\pi\left(10^{-2}\,\text{m}\right)^2}{4\left(500 \times 10^{-9}\,\text{m}\right)^2\left(10^{-1}\,\text{m}\right)^2} = 3.14 \times 10^9\,\text{W/m}^2,$$

or $3.14\,\text{GW/m}^2$. The diffraction-limited spot is slightly larger and the irradiance slightly smaller than in the case of the square, because the "corners" of the square have been removed.

8.6.3 Gaussian Beam

The Gaussian beam is of particular importance because, as noted earlier, it is the minimum-uncertainty wave, and because it is amenable to closed-form solution (see Chapter 9). The Gaussian beam is also of great practical importance. Many laser beams are closely approximated by a Gaussian beam, and an extension of Gaussian beams can be used to handle more complicated beams. Mathematically, we can take the integrals to infinity, which makes integration easier. In practice, if the aperture is larger than twice the beam diameter, the result will be similar to that obtained with the infinite integral.

Our analysis begins with the field of a plane-wave Gaussian beam. The axial irradiance of a Gaussian beam of diameter d_0 is easy to remember because it is exactly twice that of a uniform circular beam of the same diameter. The field is given by

The peak irradiance of a Gaussian beam is twice the power divided by the area inside a circle of diameter d_0.

$$E(x, y, 0) = \sqrt{\frac{2P}{\pi(w_0)^2}}e^{-r^2/w_0^2}, \tag{8.73}$$

Gaussian beam diameter defined at $1/e^2$ points.

where the beam radius is $w_0 = d_0/2$. We could verify our choice of irradiance by integrating the squared field over x and y. We use the notation, d_0, as opposed to the D we used in the earlier examples because D was the size of an aperture, and here the aperture is mathematically infinite. Note that, at the radius $r = w_0$, the field is $1/e$ of its peak value, and the irradiance is $1/e^2$ of its peak. In the future, we will always use this definition of the **diameter of a Gaussian beam**. Within this diameter, about 86% of the total power is contained. The reader must remember that the remaining 14% is important and any physical aperture must be larger by a factor of 1.5, or preferably more for the approximation to be valid.

The integral can be done in polar,

$$E(r_1, z_1) = \frac{jke^{jkz_1}}{2\pi z_1} e^{jk\frac{\left(r_1^2\right)}{2z_1}}$$

$$\int_0^{\infty}\int_0^{2\pi} \sqrt{\frac{2P}{\pi(w_0)^2}} e^{-r^2/w_0^2} e^{-jkr\frac{(r_1\cos\zeta + r_1\sin\zeta)}{z_1}} r\, d\zeta\, dr, \tag{8.74}$$

or rectangular coordinates,

$$E(r_1, z_1) = \frac{jke^{jkz_1}}{2\pi z_1} e^{jk\frac{\left(r_1^2\right)}{2z_1}}$$

$$\int_{-\infty}^{\infty}\int_{-\infty}^{\infty} \sqrt{\frac{2P}{\pi(w_0)^2}} e^{-r^2/w_0^2} e^{-jkr\frac{(r_1\cos\zeta + r_1\sin\zeta)}{z_1}} dx\, dy \tag{8.75}$$

with the result

$$E(x_1, y_1, z_1) = \sqrt{\frac{2P}{\pi(w)^2}} e^{-r_1^2/w^2}, \tag{8.76}$$

where $w = z_1\lambda/(\pi w_0)$ or in terms of diameters, $d = 2w$,

$$d = \left(\frac{4}{\pi}\right)\left(\frac{\lambda}{d_0}\right) z_1. \tag{8.77}$$

These equations will be discussed in detail in the next chapter. For now, the results are summarized in Figure 8.16.

In Practice

In Chapter 9, we will learn to calculate the size of Gaussian beams as they propagate through an optical system. If we design the system with apertures everywhere at least twice the Gaussian beam diameter, then we can do the integrals to infinity and obtain very accurate results. The calculations are easy, but the performance is not optimal. We may not always want to do this because large-diameter optical components can be expensive. We may instead wish to choose an optimal relationship between aperture and Gaussian beam. This is the subject of the next section.

Example

Again, let's keep the same numbers as in the previous two examples; 100 mW of green light ($\lambda = 500\,\text{nm}$) in a gaussian beam with diameter 1 cm at $z = 0$, and focused to a distance of $z_1 = 10\,\text{cm}$.

$$d = \frac{4}{\pi}\frac{\lambda}{d_0}z_1 = \frac{4}{\pi}\frac{500 \times 10^{-9}\,\text{m}}{10^{-2}\,\text{m}}10^{-1}\,\text{m} = 6.37 \times 10^{-6}\,\text{m}$$

or 6.36 μm. The on-axis irradiance is

$$I_0 = I(0, 0, z_1) = P\frac{ud_0^2}{\pi\lambda^2 z_1^2} =$$

$$10^{-1}\,\text{W}\frac{4\left(10^{-2}\,\text{m}\right)^2}{\pi\left(500 \times 10^{-9}\,\text{m}\right)^2\left(10^{-1}\,\text{m}\right)^2} = 5.09 \times 10^9\,\text{W/m}^2,$$

or $5.09\,\text{GW/m}^2$.

The diameter of the beam is much smaller than the case of the uniformly illuminated circular aperture, but we must remember that the definition of beam diameter is different here. There are no nulls so the diameter is defined in terms of the $1/e^2$ points. The irradiance is higher. The beam diameter at the source is strictly infinite here, because the Gaussian never goes to zero. In practice, if we pass the Gaussian beam through an aperture $D \approx 2d_0$, the results at the focused spot will be close to those we find for the whole Gaussian beam.

8.6.4 Gaussian Beam in an Aperture

We consider one more special case which is of frequent interest. It is quite common in many applications to have a circular aperture and a Gaussian laser beam. The laser beam is to be expanded using a telescope, and then passed through the aperture. A large-diameter telescope is expensive. In designing the telescope, we wish to pick the right diameter Gaussian beam to satisfy some optimization condition. One common optimization is to maximize the axial irradiance. We consider the beam diameter to be hD where D is the aperture diameter, and h is the **fill factor**. Then we need to compute

The "best" Gaussian beam diameter to fit a circular aperture depends on the criterion chosen for optimization.

$$E(x_1, y_1, z_1) = \frac{jke^{jkz_1}}{2\pi z_1}e^{jk\frac{\left(x_1^2+y_1^2\right)}{2z_1}}\int_0^{D/2}\int_0^{2\pi}\sqrt{\frac{2P}{\pi(hD/2)^2}}e^{-\frac{4r^2}{h^2D^2}}$$
$$\cdot\, e^{-jkr\frac{(x_1\cos\zeta + y_1\sin\zeta)}{z_1}}r\,d\zeta\,dr.$$

We can complete the ζ integral in closed form, and the remaining r integral will be in the form of a Hankel transform. However, the results presented here were obtained numerically using the two-dimensional Fourier transform. There are two competing effects. For small h the integral may be taken to infinity, and the Gaussian results are obtained. The small Gaussian beam diameter leads to a large diffraction pattern,

Making the Gaussian beam too small leads to a large diffraction-limited spot with low irradiance.

$$d = \frac{4}{\pi}\frac{\lambda}{d_0}z_1 = \frac{4}{\pi}\frac{\lambda}{hD}z_1 \qquad h \ll 1 \tag{8.78}$$

and thus a low axial irradiance,

$$I_0 = P\left(\frac{\pi hD}{\lambda z_1}\right)^2 \qquad h \ll 1. \tag{8.79}$$

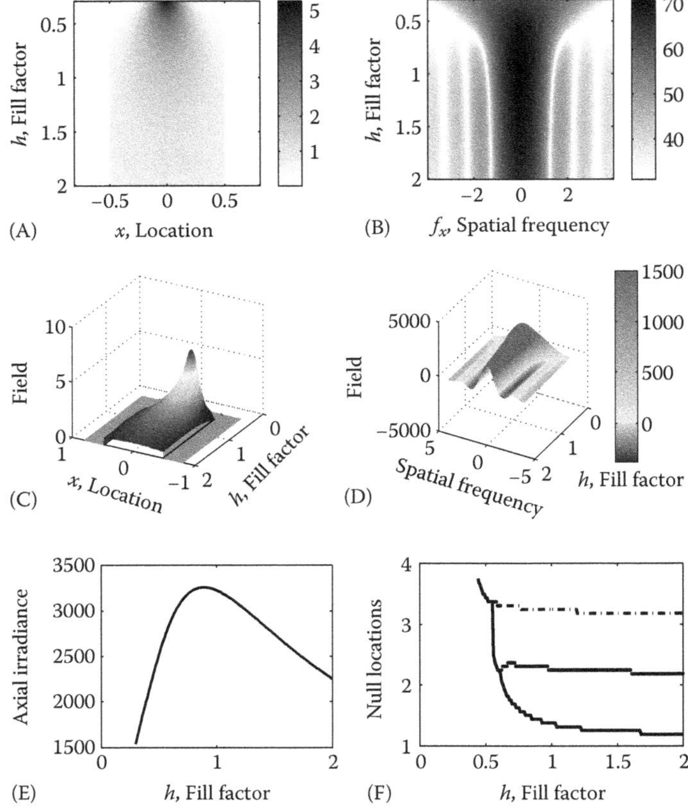

FIGURE 8.17
The Gaussian beam in an aperture. (A, C) A finite aperture of diameter, D is illuminated with a Gaussian beam of diameter hD producing (B, D) a diffraction pattern. The top two panels show irradiance as a function of x on the horizontal and h on the vertical. The next two show the field. (E) The axial irradiance maximizes at about $h = 0.9$ and (F) the locations of nulls vary with h.

Making the Gaussian beam too large results in loss of power beyond the edges of the aperture.

For large h, the aperture is nearly uniformly illuminated, so we can use Equation 8.72, but the power through the aperture is only

$$\frac{2P}{\pi(d_0)^2/4}\pi\frac{D^2}{4} = \frac{2P}{\pi(hD)^2/4}\pi\frac{D^2}{4} = \frac{2P}{h^2} \quad h \gg 1. \tag{8.80}$$

Then using Equation 8.72, we obtain

$$I_0 = \frac{P}{2\pi}\left(\frac{\pi D}{h\lambda z_1}\right)^2 \quad h \gg 1, \tag{8.81}$$

and of course, the distance between first nulls is $2.44 z_1\lambda/D$, as for the circular aperture. In between, the results are more complicated, and are shown in Figure 8.17. The top panels show the irradiance in the aperture and in the Fraunhofer zone as functions of h and position. For a given value of h, the diffraction pattern can be seen along a horizontal line in these two plots. The next two panels show the fields. Note that in both cases, the nulls are visible for large values of h, but not for small ones. The nulls are at the light regions in Panel D and in Panel B, they are the white stripes between gray ones.

Panel E shows the axial irradiance as a function of h. This plot is a slice along the $f_x = 0$ axis in panel B, or the square of the amplitude in Panel D. Finally, Panel F shows the location of the first null and additional nulls out to the third. We note that the axial irradiance reaches its peak at $h = 0.9$. Many other figures of merit are possible. For example, in a coherent laser radar, the maximum signal from a diffuse target is obtained with $h = 0.8$[55].

Example

We wish to make a light source to maximize the axial irradiance in the far field. We are allowed to build a beam expanding telescope with an exit pupil 20 cm in diameter. The light source is a laser with a Gaussian beam diameter of 5 mm. What should be the magnification of the telescope?

We want a Gaussian beam diameter of $0.9D = 0.9 \times 20\,\text{cm} = 18\,\text{cm}$. Thus the magnification required is

$$m = \frac{0.18\,\text{m}}{0.005\,\text{m}} = 36.$$

If we choose to use f/3 lenses, then the telescope primary (output lens) has a focal length of $3 \times 20\,\text{cm} = 60\,\text{cm}$ and the secondary (input lens) has a focal length of $60\,\text{cm}/36 = 1.67\,\text{cm}$.

Take-Away

Fraunhofer Patterns:
Each illumination pattern produces a particular diffraction pattern. In every case, the width of the pattern in angle is close to λ/D, where D is some characteristic dimension of the illumination. Several common patterns are shown in Figure 8.16.

8.6.5 Depth of Field: Almost-Fraunhofer Diffraction

We have considered Equation 8.56 where $z_1 = f$. We now consider cases where z_1 is near, but not exactly f. In other words, we consider cases slightly away from focus, so that the integrand goes through only a few Fresnel zones. Thus, the difficulty of integration, at least numerically, is not severe. The ability to compute the integral in this region is important, as doing so allows us to determine the defocusing, and thus the **depth of field**, also called **depth of focus**. Loosely, the depth of field is that depth over which the beam can be considered focused. Typically we choose some criterion to evaluate the departure from focus such as a drop of a factor of two in the axial irradiance.

Figure 8.18 shows an x–z section of the irradiance at $y = 0$ using Equation 8.56. A circular aperture is used with a numerical aperture of $NA = 0.25$, and a wavelength of 600 nm. Increasing the numerical aperture has two effects. First, it narrows the waist at focus because of diffraction. Second, it results in faster divergence along the axial direction. Therefore, the depth of field varies as

$$DOF \approx \frac{\lambda}{NA^2}. \qquad (8.82)$$

Often this equation appears with a small numerical scale factor, which depends on the exact definition of depth of field and on the beam shape.

Eq. 8.56:
$E(x_1, y_1, z_1) =$
$\frac{jke^{jkz_1}}{2\pi z_1} e^{jk\frac{(x_1^2+y_1^2)}{2z_1}}$
$\int\int E(x, y, 0)\, e^{jk\frac{(x^2+y^2)}{2Q}}$
$e^{-jk\frac{(xx_1+yy_1)}{z_1}}\, dx_1\, dy_1.$

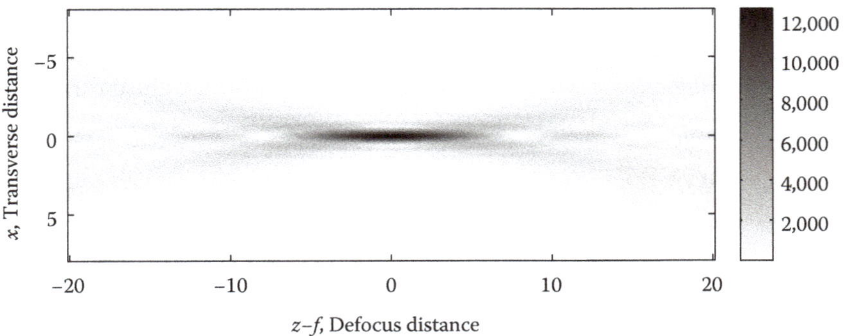

FIGURE 8.18
Near-focus beam. Near the focus of the Fraunhofer lens, Fraunhofer diffraction can still be used, provided that the curvature term is considered in Equation 8.56.

Take-Away

Diffraction Limited Depth of Field:
Recalling that the spot size is λ/NA, the depth of field is λ/NA^2. Note that the equations change in a medium with a higher refractive index. Recalling that $NA = n\sin\theta$, the spot size λ/NA approaches the wavelength in the medium, λ/n for large NA, the equation for depth of field becomes $n\lambda/NA^2$ so in the limit it approaches the same size as the spot, λ/n.

In Practice

To a first approximation, the depth of field in wavelengths is the square of the spot size in wavelengths.

Example

Light from an argon ion laser at 488 nm wavelength is focused through a microscope objective with $NA = 0.25$. What is the size of the focal volume?

The inverse NA is 4, so the spot size is approximately 4 wavelengths or almost 2 μm and the depth of field is 4^2 wavelengths or 8 μm.

8.6.6 Summary of Special Cases

- Closed-form expressions exist for several common aperture shapes.
- The diffraction integral is easily computed numerically in other cases.
- If the field in the aperture is separable into functions of x and y, the solution is at worst the product of two Fourier transforms.

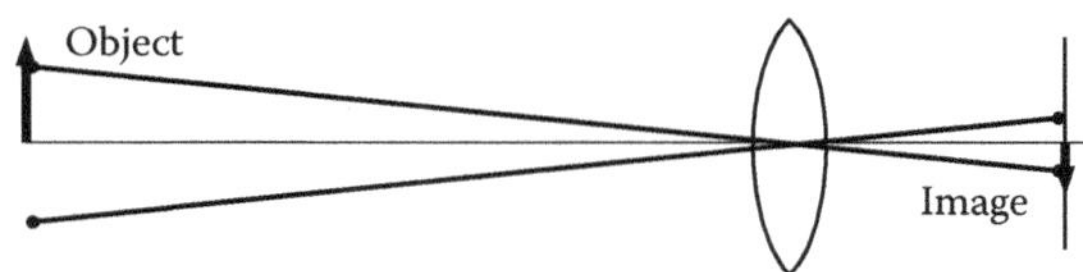

FIGURE 8.19
Imaging two points. This imaging configuration will be used to describe the diffraction limit of an optical system.

8.7 Resolution of an Imaging System

Thus far in this chapter, we've discussed diffraction as it affects our ability to focus light to a small spot. We have seen that the smallest possible spot has a dimension proportional to λ/NA with the proportionality constant depending on the details. We now turn our attention to imaging systems and ask what we can resolve when imaging a scene. We will see that the same equations apply. In an imaging system, as shown in Figure 8.19, light is collected from an object, passed through an optical system, and delivered to an image plane. In Chapters 2 and 3, we used the lens equation, matrix optics, and the concept of principal planes to locate the image and determine its size. In Chapter 4, we learned that the pupil truncates the wavefront, limiting the amount of light that reaches the image and in Chapter 5 we considered the effects of aberrations on the size of the point-spread function. Now, we see that a finite aperture also affects the point-spread function through diffraction. Consider a camera imaging a scene onto a film or electronic imaging device. Each point in the object produces a spherical wave that passes through the entrance pupil, is modified in phase, and directed toward the corresponding point in the image. Assuming that aberrations have been made sufficiently small, the limitation on the resolution is determined by diffraction. We define this lower limit as the **diffraction limit**, and say that a system is **diffraction-limited** if the aberrations alone would produce a smaller spot than the diffraction limit. The diffraction limit is so fundamental that we often use it as a standard; if aberrations produce a point-spread function five times larger than that predicted by diffraction, we say that the system is "five times diffraction limited," or "5XDL."

8.7.1 Definitions

We began this chapter with a discussion of diffraction from two points, each of which is "unresolvable," shown in Figure 8.2a. We now begin a more detailed discussion of the concept of resolution. We begin by turning to the dictionary.

Fig. 8.2a.

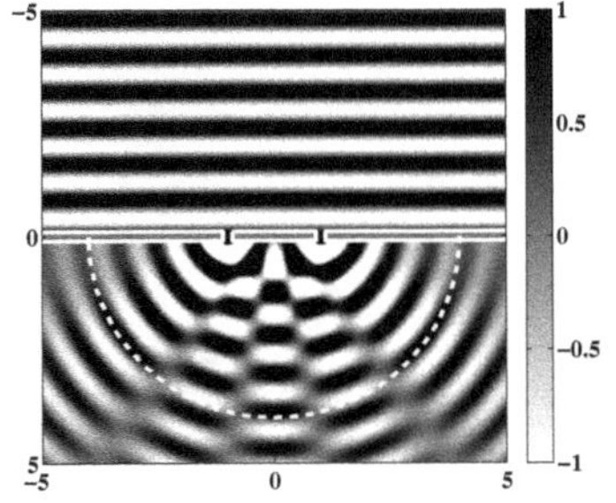

> **Resolve,** . . . 13. a. trans. Science. Originally: (of optical instruments or persons using them) to reveal or perceive (a nebula) as a cluster of distinct stars. Later more widely: to distinguish parts or components of (something) that are close together in space or time; to identify or distinguish individually (peaks in a graph, lines in a spectrum, etc.) . . .
>
> **Resolution,** . . . 6. a. Originally: the effect of a telescope in making the stars of a nebula distinguishable by the eye. Later more widely: the process or capability of rendering distinguishable the component parts of an object or image; a measure of this, expressed as the smallest separation so distinguishable, or as the number of pixels that form an image; the degree of detail with which an image can be reproduced; = resolving power. . . [6]

In this book, resolution refers to the ability to determine the presence of two closely spaced points. As a first attempt at a definition, we will consider two very small test objects. We may consider these as point sources of light, radiating spherical waves. We decrease the space between them to the minimum distance, δ at which we can see two

Resolution is defined by the ability to measure a valley between two closely–spaced points.

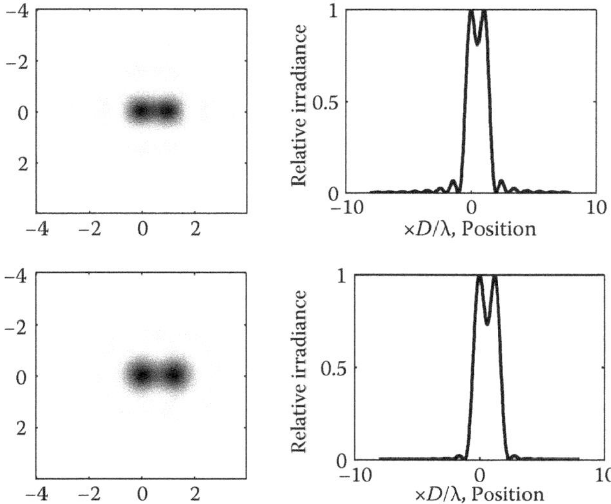

FIGURE 8.20
Resolution in an image. The Rayleigh resolution criterion states that two points can be resolved by an imaging system with a uniform circular aperture if each of them is placed over the first null of the other. For a square aperture, the resolution is λ/d, and the irradiance at the central valley has a height of 81%. For a circular aperture, the resolution is $1.22\lambda/D$ and the valley height is about 73% of the peak.

distinct points. That is, we must be able to see some "background" between them. We call this minimum distance the **resolution** of the system. The objects themselves must be smaller than the resolution, so that no two points within the objects can be resolved. We say that objects smaller than the resolution are **unresolved** objects.

Unfortunately, this attempt at a very precise definition fell short in one important aspect. Suppose that we plot the irradiance or some related quantity as a function of position along a line across the two point objects. We would like to say that they are resolved if there is a valley between two peaks, as shown in Figure 8.20. If the points are very far apart, then the valley is large and it is quite obvious that the two objects are present. If they are very close together, then they may appear as a single object. The deeper the valley, the more likely we are to decide that there are two objects. For example, we recognize the two headlights of an approaching car at night if we can see dark space between them. At a great enough distance they will appear as a single light. However, there is no physical or fundamental basis on which to decide how deep the valley must be to ensure that the points are resolved. If the signal is detected electrically and there is a large amount of noise, even a moderately deep valley can be masked. In contrast, if there is very little noise, a clever signal processor could resolve the two points when no valley is visible.

We will define the **point-spread function** or PSF as the image of an unresolved object or "point object." We can determine the point-spread function by applying diffraction equations to the field in the pupil, just as we computed the field in the pupil from that in the object.

Our actual ability to resolve objects depends on signal and noise statistics and our tolerance for errors. Usually no exact definition is possible.

Noise affects resolution.

Rigorously, we would have to look at the signals at the two peaks and at the valley, each with a random sample of noise, do some statistical analysis of the signal and noise. If there are really two points in the object, the valley could be eliminated by the addition of noise in the region between them, resulting in an incorrect decision that only one point is present. Alternatively, if only one point is present, a valley could be created by the subtraction of noise, incorrectly suggesting the presence of two points. The probability of these errors occurring is related to the statistics of signal and noise as well as the spacing between the two points. In any case, we would have to decide on statistical criteria based on the "cost" of these errors. Resolution also depends on contrast. We are developing equations assuming the objects are small light sources in an otherwise dark background. If the objects are, for example, parts of a person's face illuminated by sunlight, the changes in reflectivity do not vary from 0 to 1, but are much more subtle. Thus changes in the image will be more subtle and the valley will be more difficult to detect. In most cases, we will not have enough information to perform the required statistical analysis. We may not know the noise statistics or the object signal strength and we probably cannot decide what acceptable detection statistics would be. Instead we adopt some arbitrary measure like Rayleigh's criterion.

Contrast affects resolution.

Take-Away

Resolution Defined:
Resolution is a measure of our ability to determine that two closely spaced objects are in fact separate. A rigorous definition of resolution would require detailed knowledge of signal and noise statistics and definition of a suitable metric for success. Normally we do not have sufficient information, so we adopt one of several criteria to be described later.

8.7.2 View from the Pupil Plane

In the previous section, we discussed resolution in the image plane, with the concept of a point-spread function. Sometimes it's useful to think in the pupil plane. Looking at the field from two points in Figure 8.2a again, we remind ourselves that light from one point forms a spherical wave. When there are two points present, can we see a difference between the actual wave and a spherical wave over a given aperture? For example, if the pupil is 0.5 units wide (± 0.25 units at zero) at the bottom of the figure, then the wave appears spherical. With that aperture, the light from the two points would be indistinguishable from the light from a single point at the origin. Making the aperture 5 units wide (± 2.5 units at zero) the wavefront is clearly not spherical; there are changes in phase and amplitude associated with sidelobes. With this larger aperture we can clearly tell, at a minimum, that there is something more complicated than a single point source.

Fig. 8.2a.

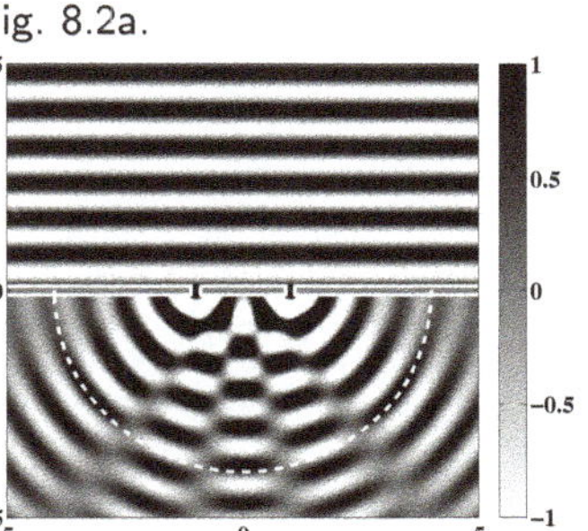

The wave looks spherical over a small aperture, but not over a larger one.

As the points become closer together, a wider pupil is required in order to detect a departure from a spherical wave. Furthermore, the required pupil size is inversely proportional to wavelength. Thus we expect the resolution to be proportional to λ/D, where D is a dimension of the pupil.

Take-Away

Resolution and Aperture:
A larger aperture provides better resolution because it is easier to observe how the actual wave departs from a spherical wave.

8.7.3 Rayleigh Criterion

Lord Rayleigh (John William Strutt) developed a resolution criterion that bears his name. Lord Rayleigh's name is well known in optics for his many contributions, including this criterion, the Rayleigh range which we have already mentioned, the Rayleigh–Jeans law for thermal radiation of light and Rayleigh scattering from small particles (both in Chapter 12). Nevertheless his 1904 Nobel Prize was not for anything in optics but "for his investigations of the densities of the most important gases."

In the Rayleigh criterion, the peak of the second point-spread function is placed over the valley of the first[187]. Thus the **Rayleigh resolution** is the distance from the peak of the point-spread function to its first null, as shown in Figure 8.20. We now call this alignment of peak over valley the **Rayleigh criterion** for resolution. Examples are shown for a square and circular aperture. The plots show a transverse slice of the irradiance along the y axis. The images were obtained by Fourier transform, based on Equation 8.61. For a square aperture, the distance between the centers of the two patterns is

Rayleigh resolution criterion specifies that the peak of one PSF is placed over the first null of the other.

Eq. 8.61: $E(f_x, f_y, z_1) = \frac{j2\pi z_1 e^{jkz_1}}{k} e^{jk\frac{(x_1^2+y_1^2)}{z_1}} \int\int E(x,y,0) e^{-j2\pi(f_x x + f_y y)} dx\, dy.$

$$\delta = z_1\lambda/D \qquad \text{(Rayleigh resolution, square aperture).}$$
$$\delta = \lambda/(2NA) \tag{8.83}$$

The first null of the $\sin \pi x/(\pi x)$ is at $x = 1$, and the maximum value is one, because each peak is over a null from the opposite image point. The height of the valley in the square case is

$$2\left[\frac{\sin(\pi/2)}{\pi/2}\right]^2 = 2\left[\frac{1}{\pi/2}\right]^2 = \frac{8}{\pi^2} = 0.81. \tag{8.84}$$

For a circular aperture the **angular resolution** is

$$\delta\theta = 1.22\frac{\lambda}{D} \qquad \text{Rayleigh resolution, circular aperture.} \tag{8.85}$$

Rayleigh resolution for a circular aperture.

For microscope objectives, numerical apertures are usually large, and it is better to express the Rayleigh criterion in terms of the transverse distance, δ, and numerical aperture

$$\delta = 0.61\frac{\lambda}{NA} \qquad \text{Rayleigh resolution, circular aperture.} \tag{8.86}$$

From the above analysis, one may be tempted to say that a square aperture is "better" than a round one, because the diffraction limit is smaller. However, we must remember that a square of size D occupies more area than a circle of the same diameter. In particular, the "added corners," being farthest from the axis, contribute greatly to improving the diffraction limit.

8.7.4 Alternative Definitions

Considering the uncertainty about noise, signal levels, shape of the point-spread function, our tolerance for making an error in discerning the presence of two points, and indeed the uncertainty of what the underlying question really is when we ask about resolution, it is not surprising that many definitions have been developed over the years. The good news is that the user is free to define resolution in any way that makes sense for the problem at hand. The bad news is that there will never be general agreement. Here we consider some of the approaches taken by others. An excellent review of the subject is available in an article by den Dekker and van den Bos[49].

Rayleigh resolution is probably the most common choice for theoretical predictions.

Rayleigh's definition, in 1879[187] appears to be one of the first, and is probably the most widely used. It provides a way to compare different systems theoretically, but is probably not very practical experimentally, because it involves locating the null of the point-spread function.

In 1903, Wadsworth[239] took the approach of measuring the depth of the valley, which is appropriate for experiments if there is sufficient light, and if the noise is low or can be reduced by averaging. Building on Rayleigh's idea, he suggested, even if the PSF is not a sinc function, that we define the resolution as the spacing of two points that produces an 81% valley/peak ratio. We call this the **Wadsworth criterion**. We can extend it to the circular case and define resolution as the spacing that gives 73%, or we can simply accept the Wadsworth criterion of 81%. We should be careful not to "mix and match," however. It is common, and incorrect, to use the 81% Wadsworth criterion for the valley depth and the $0.61\lambda/NA$ Rayleigh criterion for the spacing of the points.

The Wadsworth resolution is the spacing that produces an 81% valley.

In 1916, Sparrow, recognizing[216] that the 81% (or 73%) valley depth is arbitrary, suggested defining the resolution that just minimally produces a valley, so that the second derivative of irradiance with respect to transverse distance is zero. Wider spacing results in a valley (first derivative positive) and narrower spacing leads to a single peak (first derivative negative). Sparrow's idea is sound theory, but in practice, it is compromised by noise. We need some depth to the valley in order to be sure it exists.

The Sparrow resolution is the smallest spacing that produces a valley.

Ten years later, Houston proposed defining resolution as the full width at half maximum (FWHM)[100] of the PSF as his criterion. Although this criterion does not relate directly to the definition of resolution, it is appealing in that it is often easy to measure.

The Houston resolution is the FWHM of the point-spread function.

Often we may find that a point object does not produce or scatter a sufficient amount of light to produce a reliable measurement of the PSF, and all the criteria in this section can be used to determine theoretical performance of a system, but not to measure it experimentally. For this reason, even more alternatives have been developed. These can be better understood in the context of our discussion of Fourier optics, and will be addressed in Section 11.3.

Take-Away

Resolution Criteria:
Resolution of an imaging system is determined by the same equations used to predict a diffraction-limited spot. Resolution is proportional to λ/D in angle or λ/NA in linear dimension. The proportionality constant depends on the shape of the aperture and our choice of definition of resolution. The definitions discussed here are as follows.

Rayleigh	Max of one PSF over min of Other
Wadsworth	Valley 81% of Peaks
Sparrow	No Valley
Houston	FWHM of PSF

In Practice

As a first approximation, the resolution of a system is λ/D in angle or λ/NA in linear dimension.

8.7.5 Diffraction Examples

Let's consider a couple of examples.

Example

We consider first the example of a camera with a 55 mm, $f/2$ lens. As shown in Figure 8.19, the angle between two points (measured from the entrance pupil) in the object is the same as the angle between them in the image (measured from the exit pupil). Therefore, if we use angular measures of resolution, they are the same for object and image. The resolution on the film is

$$\delta_{camera} = 1.22\frac{\lambda}{D}f = 1.22\frac{\lambda}{D/f} = 1.22\lambda F \quad (8.87)$$

where $F = f/D$ is the f-number.

For our $f/2$ example,

$$\delta_{camera} = 1.22 \times 500\,\text{nm} \times 2 = 1220\,\text{nm}. \quad (8.88)$$

Thus, on the film, the resolution is approximately 1.22 μm. In an electronic camera, the pixels are typically 10 μm so the resolution is better than a pixel until about $f/16$. The angular resolution is

$$\delta\theta_{camera} = 1.22\frac{\lambda}{f}F. \quad (8.89)$$

This limit applies equally in object or image space, because the angles are the same as seen in Figure 8.19. In our $f/2$ example,

$$\alpha_{camera} = 1.22\frac{500\,\text{nm}}{55\,\text{mm}} \times 2 = 22\,\mu\text{rad}. \quad (8.90)$$

If the camera is focused on an object 2 m away, the resolution on the object is 44 μm, which is close to half the width of a human hair, but if it is focused on a landscape scene 10 km away, the resolution becomes 22 cm. In the latter case, even the distance between a person's eyes is not resolvable.

Example

Next, we consider the microscope shown in Figure 8.21. Here we have a separate entrance and exit pupil, as discussed in Chapter 4. In image space, the resolution is given by $0.61\lambda/NA'$ where NA' is the numerical aperture in image space, determined by the triangle in the bottom of the figure. The resolution of the object could be determined by dividing by the magnification. However, it is easy to show that the same result is obtained from

$$\delta_{microscope} = \frac{0.61\lambda}{NA}, \quad (8.91)$$

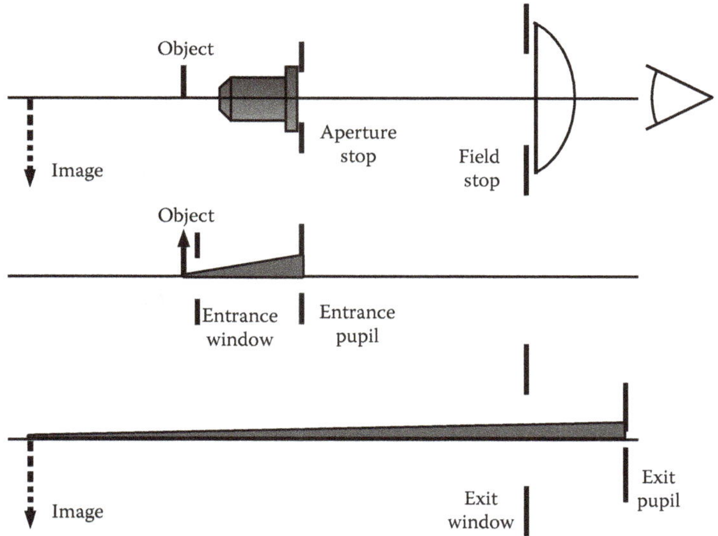

FIGURE 8.21
Microscope. The top shows a two-element microscope. The center shows the microscope in object space, with a cone showing the numerical aperture. The bottom shows the same in image space. The resolution in either space is inversely related to the pupil size.

where NA is the numerical aperture of the objective. One generally defines the diffraction limit of a microscope in object space. In fact, to determine the resolution in the image space, one would most likely compute the resolution in object space, and multiply by the magnification. Such a task would arise if one wanted to determine the relationship between the resolution and the pixel size on a camera.

Finally, we note that as we increase the numerical aperture, $NA = n \sin\theta$, we reach an upper limit of $NA = n$ for $\sin\theta = 1$. Thus, it is often stated that optical imaging systems can never have a resolution less than the **Abbe limit**, at maximum NA,

$$\delta_{Abbe} = \frac{0.61\lambda}{n} \qquad \text{(at Maximum NA).} \tag{8.92}$$

We note that λ is the vacuum wavelength, and the wavelength is λ/n in the medium, so the resolution is a fraction of this shorter wavelength rather than the vacuum wavelength. To consider one example, working in a medium with an index of refraction of 1.4, which might be good for certain types of biological specimens, a vacuum wavelength of 600 nm provides a maximum resolution of 261 nm.

8.7.6 Superresolution

There are at least two notable exceptions to this rule. The first, **near-field microscopy** involves passing light through an aperture much smaller than the wavelength, and scanning the aperture over the area to be measured.[180] Because of diffraction, this must be done at very close distance, and it requires a bright light source. The reason is that when we focus light onto the small aperture, we generate a diffraction-limited spot of diameter d. The aperture passes only a fraction of the photons, $(d_{aperture}/d)^2$, so to pass an average of one photon per measurement through the aperture, we need

$$N = \left(\frac{d}{d_{aperture}}\right)^2 \tag{8.93}$$

photons. The resolution is improved by the factor

$$\frac{d_{aperture}}{d} = \sqrt{N}. \tag{8.94}$$

Secondly, one can always improve resolution by averaging a number of measurements. The results we have derived are applicable to measurements with one photon. In the theoretical limit, resolution can be improved, using N photons, by a factor no larger than $\sqrt{N}$. For example, stimulated-emission-depletion (STED) microscopy[120] can achieve resolution in the tens of nanometers using large numbers of photons. Stochastic techniques such as PALM[10] and STORM[199] as well as structured illumination microscopy (SIM)[88] also lead to so-called "**Superresolution.**" . Eric Betzig, Stefan Hell, and William Moerner won the Nobel Prize in Chemistry in 2014 "for the development of super-resolved fluorescence microscopy."

Nobel Prize in Chemistry 2014

We might ask whether we can achieve resolution improvement in both x and y (and even z) of $\sqrt{N}$ or whether we must divide the N photons among the different directions. Recalling that the uncertainty principle limits the product of uncertainties in position and momentum for each direction separately (Equation 8.3), we see that we can obtain $\sqrt{N}$ improvement in all dimensions with only N photons.

Eq. 8.3: $\delta x\ \delta p_x \ge h$.

Take-Away

Superresolution:
Although the diffraction limit is rooted in fundamental physics, we can always improve the resolution by up to $\sqrt{N}$ by somehow using N photons.

8.7.7 Summary of Resolution

- Resolution is the ability to identify two closely spaced objects as distinct.
- Diffraction provides the ultimate limit on resolution of an imaging system, and the Rayleigh criterion provides a frequently used estimate.
- The diffraction limit expressed as an angle, a small number times λ/D where D is a characteristic dimension of the pupil.
- Expressed as a transverse distance, it is a small number times $\lambda z/D$ or times λ/NA.
- The most commonly used expression is for a circular aperture, $0.61\lambda/NA$.
- The diffraction limit is at best a fraction of the wavelength.
- This limit applies for single photons. Improvement is possible through averaging with many photons. The resolution improves in proportion to the square root of the number of photons used in the measurement.

8.8 Diffraction Grating

A grating is a periodic structure that can be reflective or transmissive.

Diffraction gratings constitute an important class of **dispersive optics**, separating light into its various color components for applications such as spectroscopy. The diffraction grating can be most easily understood through the convolution theorem of Fourier transforms. We begin by looking at a simple analysis of the grating equation, and then build to a mathematical formulation that will allow us to determine the diffraction pattern of any grating.

8.8.1 Grating Equation

Figure 8.22 illustrates a diffraction grating in concept. A periodic structure of reflective facets separated by transmissive slits is placed in the path of a nominally plane wavefront coming from the left at an angle θ_i to the normal. The left panel shows two rays of incident light. Now, assuming that the facets or slits are unresolved, each of them will reflect or transmit light over a wide angle. In some directions, the fields will sum constructively. These angles, θ_d are the ones for which light reflecting from adjacent facets differs by an integer multiple of the wavelength. We choose the notation that both angles are positive in the same direction. In the example, two reflecting paths are shown. The incident path for the bottom ray is longer by an amount $d\sin\theta_i$ and for the diffracted ray, it is longer by $d\sin\theta_d$, where d is the **spatial period** or **pitch** of the grating. Thus, if

The grating equation shows the directions of diffracted light waves.

$$N\lambda = d\left(\sin\theta_i + \sin\theta_d\right), \tag{8.95}$$

where N is an integer, then at every other mirror or opening, the difference will also be an integer multiple of the wavelength. This equation, called the **grating equation**, defines all the angles at which rays will diffract. There are two special angles, defined by $N = 0$, or $\sin\theta_d = -\sin\theta_i$. The first of these, $\theta_d = -\theta_i$ is simply the ray we would expect reflected from a mirror. We note that this is true for all wavelengths. The positive, $N = 1, 2, 3\ldots$, and negative $N = -1, -2, -3\ldots$, reflected orders are on either side of the zero-order reflection.

The second special direction, $\theta_d = 180° + \theta_i$, describes the un-diffracted forward wave. Again, the positive and negative transmitted orders are on opposite sides of this un-diffracted ray. All wavelengths appear in the zeroth order. The center panel shows several reflected and transmitted orders, and the right panel shows a graphical solution of the grating equation for all orders. The diffraction angle, θ_d is plotted on the horizontal axis and its sine,

All wavelengths appear in the zeroth order.

$$\sin\theta_d = -\sin\theta_i + N\frac{\lambda}{d}, \tag{8.96}$$

is plotted on the vertical axis, with lines showing solutions for different values of N. The $N = 0$ line is at $\sin\theta_d = -\sin\theta_i$, and the spacing of the lines separating integer values of N is λ/d. For $\lambda/d > 2$, no diffracted orders exist. For $1 < \lambda/d < 2$, diffraction is possible for only one nonzero order for sufficiently large angles of incidence. As λ/d becomes smaller, the number of orders increases, and the angle between them becomes smaller.

The number of diffracted orders depends on wavelength, grating spacing and the angle of incidence.

Gratings are frequently used in spectroscopy. A **monochromator** is an instrument which selects one wavelength from a source having a broad range of wavelengths. A plane wave diffracts from a grating, in either reflection or transmission, and the output is collected in one particular direction. For a given order, N, only one wavelength will be selected by the monochromator. We want a large dispersion, $d\theta_d/d\lambda$, so that we can select a narrow band:

A monochromator selects one wavelength from the input light.

$$\delta\lambda = \frac{d}{N}\delta\left(\sin\theta_d\right). \tag{8.97}$$

Normally, the slits are in fixed positions, and the mirror is rotated about an axis parallel to the grooves to select the desired wavelengths. The monochromator will be discussed in detail in Section 8.8.7.

Like the monochromator, the **grating spectrometer** disperses incoming collimated light into different directions according to the grating equation, but the grating angle is usually fixed, and the diffracted light is incident on an array of detectors rather than a single slit. The central wavelength detected by each detector element will be determined by its location, and the resulting angle, θ_d, through the grating equation. The range of

A spectrometer separates the light according to wavelength.

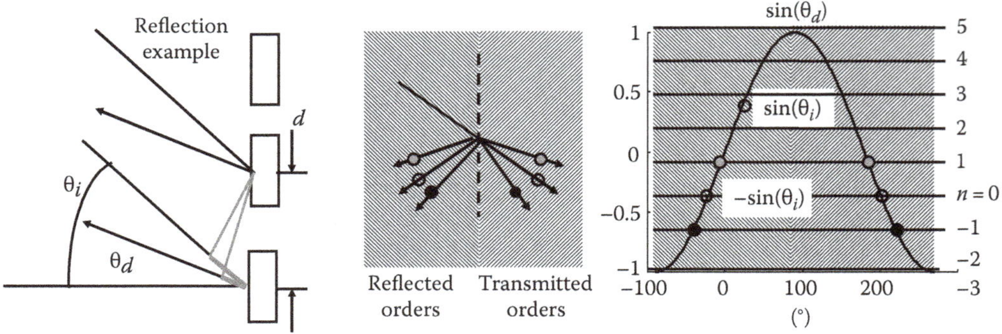

FIGURE 8.22
The diffraction grating. The left side shows the layout for deriving the diffraction grating equation. The center shows reflected and transmitted orders. The right shows how to compute the existing orders from the grating equation. According to the grating equation, the zero order is at $\sin\theta_d = -\sin\theta_i$, shown by the thicker line. The spacing of the lines above and below the zero-order line is λ/d.

wavelengths will be given by the width of the detector, and the angle it subtends, $\delta(\sin\theta_d)$, through Equation 8.97.

Take-Away

Grating Equation:
The grating equation, Equation 8.95,

$$N\lambda = d\left(\sin\theta_i + \sin\theta_d\right),$$

specifies the angles of diffracted light as a function of the input angle and wavelength. The equation is valid for forward and backward diffraction.

8.8.2 Aliasing

Filters can prevent aliasing in the diffraction pattern.

We must also remember that many wavelengths can satisfy the grating equation, Equation 8.95. This unfortunate result, called **aliasing**, is similar to the aliasing that occurs in sampling of electronic signals for similar reasons; the grating is effectively sampling the incident wavefront. If we design a monochromator to pass a wavelength λ in first order ($N = 1$), or if we believe we are measuring this wavelength in a grating spectrometer, the wavelength $\lambda/2$ will satisfy the same equation for $N = 2$, as will $\lambda/3$ for $N = 3$, and so forth. For example, if we are interested in wavelengths around $\lambda \approx 600$ nm, then ultraviolet light around 300 nm will satisfy the grating equation for $N = 2$, and that around 200 nm for $N = 3$. The last is in the portion of the spectrum absorbed by oxygen (see Table 1.2) and would not be of concern unless the equipment is run under vacuum. However, the second-order ($N = 2$) wavelengths would certainly be diffracted and so must be blocked by a filter. Usually a colored glass filter is used to absorb the unwanted wavelengths. For this reason, a spectrometer or monochromator that uses the first order, with an anti-aliasing filter in front of the spectrometer, cannot exceed a wavelength range of a factor of two. To exceed this bandwidth in a monochromator, a motorized filter wheel is frequently used. When the grating is rotated to a new angle for a new wavelength, the filter wheel moves to select an appropriate anti-aliasing filter for that wavelength. In a spectrometer, a special detector array with different filters over different detector elements can be used.

The situation becomes worse at higher orders. If we work at second order, the wavelength of interest, λ is passed with $N = 2$, 2λ is passed with $N = 1$, and $2\lambda/3$ is passed with $N = 3$. Thus the bandwidth that can be accommodated without aliasing is even smaller than for a first-order instrument.

Example

Consider light from an Nd:YAG laser at $\lambda = 1.064\,\mu\text{m}$ incident normally on a grating with 600 lines/mm. What is the angle of the first order?

The grating equation, Equation 8.96, is;

$$\sin\theta_d = -\sin\theta_i + N\frac{\lambda}{d},$$

The angle of incidence is zero and the order, $N = 1$, so

$$\sin\theta_d = 1 \times \frac{1.064 \times 10^{-6}\,\text{m}}{10^{-3}\,\text{m}/600},$$

$$\theta_d = 40°.$$

8.8.3 Fourier Analysis

In Section 8.8.1, Equation 8.95, the grating equation, demonstrated how to determine the relationship among angles, orders, and wavelengths. However, it did not address the issue of strength of the diffraction effect or width of the diffracted orders. Gratings can be fabricated to strengthen or reduce the amount of light diffracted into certain orders. The distribution of light on the grating can be used to determine the widths of the diffracted orders, or the spectral resolution of a monochromator or spectrometer.

Because the far-field pattern of the diffraction grating is described by Fraunhofer diffraction, we can use Fourier analysis to determine specific patterns, as shown by Figure 8.23. We can design gratings using knowledge of a few simple Fourier transforms, and the **convolution theorem**, also called the faltung (German "folding") theorem, which states that the Fourier transform of the convolution of two functions is the product of their Fourier transforms. Using the notation developed for the diffraction integral, with no subscripts on variables in the grating plane, and subscripts 1 on variables in the diffraction pattern,

Grating	Diffraction pattern
$g(x,0) = \int f(x - x', 0)\, h(x', 0)\, dx'$	$g(x_1, z_1) = f(x_1, z_1)\, h(x_1, z_1),$
$g(x,0) = f(x,0)\, h(x,0)$	$g(x_1, z_1) = \int f(x_1 - x_1', z_1)\, h(x_1', z_1)\, dx_1',$

where f and h are arbitrary functions. We have considered the one-dimensional form of the convolution theorem, which is usually sufficient for analysis of gratings, but the analysis can be extended to two dimensions.

Start with the unit pattern in the grating. Take its Fourier transform for the diffraction pattern.

We begin our Fourier analysis of the grating by considering a single element of the grating. For a transmissive grating, this element represents one of the openings in an otherwise non-transmitting material, and in a reflective grating, it is conceptually a mirror. For illustrative purposes, consider that the transmitted or reflected field is uniform across the single element, so the diffraction pattern is a *sinc* pattern, as given by the x part of Equation 8.69. The nulls are located where $\sin\left(kx_1\frac{D_x}{2z_1}\right) = 0$. Next, we want to replicate this pattern with some repetition period. Replication is a convolution with a "comb"

Eq. 8.69:
$$E(x_1, y_1, z_1) = \sqrt{P\frac{D_x D_y}{\lambda^2 z_1^2}}\,\frac{\sin\left(kx_1\frac{D_x}{2z_1}\right)}{kx_1\frac{D_x}{2z_1}}\,\frac{\sin\left(ky_1\frac{D_y}{2z_1}\right)}{ky_1\frac{D_y}{2z_1}}\,e^{jk\frac{\left(x_1^2+y_1^2\right)}{2z_1}}.$$

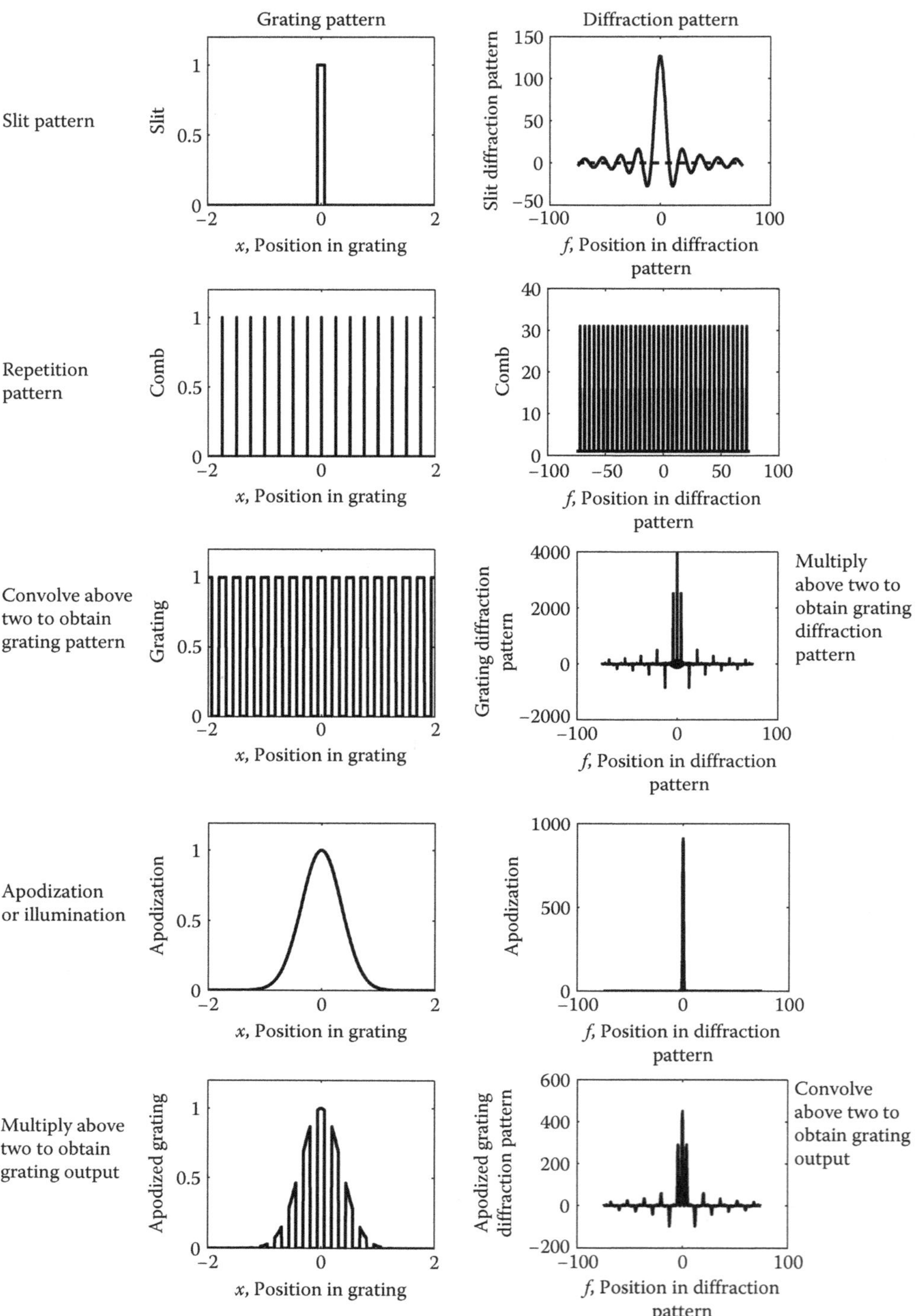

FIGURE 8.23
Diffraction grating. The left side shows patterns in the grating plane, and the right side shows the corresponding diffraction patterns. The grating pattern is represented by convolution of the basic slit pattern with the repetition pattern. The output is obtained by multiplying the grating pattern and the apodization pattern.

TABLE 8.2
Grating Relationships

Item	Grating Pattern	Size	Diffraction Pattern	Size
Basic shape	Uniform	D	sinc	λ/D first nulls
Repetition	Comb	d	Comb	λ/d
Apodization	Gaussian	d_0	Gaussian	$\frac{4}{\pi}\frac{\lambda}{d_0}$

The sizes of the patterns in the example in Figure 8.23 are given.

function or a series of delta functions. The diffraction pattern of such an infinite comb is also an infinite comb, with the angles given by the grating equation, Equation 8.95, with $\theta_i = 0$. Thus, if the period in the grating is d, the period of the diffraction pattern is λ/d. As shown in the third row of Figure 8.23, the grating pattern is a regular repetition of the slit pattern, and the diffraction pattern is a sampled version of the diffraction pattern of the single slit. In particular, if $D = d/2$, the pattern has a 50% duty cycle, and each even-numbered sample is at a null of the sine wave. The odd-numbered peaks are at the positive and negative peaks of the sine wave. For other duty cycles, if one of the sample points, $N\lambda/d$, matches one of the nulls of the *sinc* function, that particular order will be missing.

Convolve with comb in grating. Multiply by comb in diffraction pattern.

At this point, the diffraction pattern is a perfect series of delta functions, with no width, so a monochromator could have perfect resolution. In fact, the resolution will be limited by the illumination, which can be treated through the convolution theorem as well. The word **apodization** is often used here. Apodization is a term used in electronics to describe a modification made to remove Gibbs phenomena or ringing in Fourier transforms. If we use a uniform illumination pattern over a finite aperture, the diffraction pattern of that illumination will have sidelobes (Equation 8.69 or 8.72). Instead, in this example, we illuminate with a Gaussian beam. The diffraction pattern of a Gaussian wave of diameter d_0 is a Gaussian, with width $4\lambda/(\pi d_0)$, and the sidelobes are eliminated. To illustrate the effect, we have deliberately chosen a particularly narrow illumination function so that the widths of the diffracted orders are observable in the plot. To summarize, using the functions we have chosen here, we obtain the general results shown in Table 8.2.

Multiply by apodization in the grating, convolve in the diffraction pattern.

Eq. 8.72: $I(r_1, z_1) = P\frac{\pi D^2}{4\lambda^2 z_1^2}\left(\frac{J_1\left(kr_1\frac{D}{2z_1}\right)}{kr_1\frac{D}{2z_1}}\right)^2.$

We note that the ratio of the spacing of orders to the width of a single order is given by,

$$\frac{1}{F} = \frac{\frac{4}{\pi}\frac{\lambda}{d_0}}{\frac{\lambda}{d}} = \frac{4}{\pi}\frac{d}{d_0}, \tag{8.98}$$

and is approximately the number of grating grooves covered by the illumination pattern. The quantity, F, is a figure of merit, similar to the Finesse of the ètalon in Equation 7.108. As this figure of merit increases the width of the peaks in the diffraction pattern becomes smaller, relative to their spacing.

Eq. 7.108: $F = \frac{4\sqrt{\eta_{rt}}}{1-\eta_{rt}}$.

Take-Away

Fourier Analysis of Grating:
The diffraction pattern of a grating can be analyzed in the grating plane by convolving the slit pattern with a repetition pattern and multiplying by the illumination function. The diffraction pattern can be obtained by a Fourier Transform.

8.8.4 Example: Laser Beam and Grating Spectrometer

Example

As one example, we consider a grating with two overlapping collimated Gaussian laser beams of equal diameter, d_0, but two slightly different wavelengths, $\lambda_1 = \lambda - \delta\lambda/2$ and $\lambda_2 = \lambda + \delta\lambda/2$, incident upon it at $\theta_i = 0$. Can we resolve the two diffracted beams to determine that two wavelengths of light are present, in the same sense as we discussed spatial resolution and the Rayleigh criterion? We measure the diffraction angles by focusing the diffracted light through a Fraunhofer lens onto a detector array. The angular spacing between centers of the two diffracted waves, assuming $N = 1$ is given by

$$\delta\theta = \theta_{d2} - \theta_{d1} \approx \sin\theta_{d2} - \sin\theta_{d1} = \frac{\lambda_2 - \lambda_1}{d} = \frac{\delta\lambda}{d}, \tag{8.99}$$

and their width will be

$$\alpha = \frac{4}{\pi}\frac{\lambda}{d_0}. \tag{8.100}$$

We assume that $\delta\lambda$ is small, so we write this single equation for the width. The irradiance as a function of angle, θ, will be

$$I(\theta) = e^{-2(\theta-\theta_{d1})^2/\alpha^2} + e^{-2(\theta-\theta_{d2})^2/\alpha^2}. \tag{8.101}$$

There are no nulls in Gaussian functions, so we cannot use the same approach we used in Equation 8.84 to determine resolution. Instead we will use the Sparrow resolution defined in Section 8.7.4. Taking the first derivative of Equation 8.101 and setting it equal to zero to locate extrema, we always find a solution at the midpoint between the two angles,

$$\theta = \frac{\theta_{d2} + \theta_{d1}}{2}. \tag{8.102}$$

If $\delta\lambda$ is small, we expect this to be the only solution. There is a peak but no valley in this irradiance. If $\delta\lambda$ is sufficiently large, we expect a valley at the same location with two more solutions for the peaks, but the equations are transcendental and a closed-form solution eludes us.

We can evaluate the second derivative at the midpoint.

$$\left.\frac{d^2 I}{d\theta^2}\right|_{\theta=(\theta_{d1}+\theta_{d2})/2} \qquad \frac{8e^{-2\left(\frac{\delta\theta}{\alpha}\right)^2}\left[\alpha^2 - (\delta\theta)^2\right]}{\alpha^2}. \tag{8.103}$$

We expect a valley if it is positive and a peak if the second derivative is negative. The second derivative is zero at $\alpha = (\delta\theta)$ and negative for

$$\alpha < (\delta\theta),$$

or using Equations 8.100 and 8.99

$$\frac{4}{\pi}\frac{\lambda}{d_0} < \frac{\delta\lambda}{d}. \tag{8.104}$$

Solving for $\delta\lambda$,

$$\frac{\delta\lambda_0}{\lambda} = \frac{4}{\pi}\frac{d}{d_0} = \frac{1}{F}, \qquad (8.105)$$

following Equation 8.98. We use the subscript 0 on this definition of $\delta\lambda_0$ to indicate that it is the $\delta\lambda$ that results in zero height of the valley. Above this value we expect a valley and below it we do not. For a wavelength around $\lambda = 500\,\text{nm}$, with a laser beam diameter on the grating of $d_0 = 5\,\text{mm}$, we consider a grating with 600 lines per millimeter, $d = 1.67\,\mu\text{m}$, and find that

$$\delta\lambda_0 = \frac{4}{\pi}\frac{d}{d_0}\lambda = 0.21\,\text{nm}.$$

8.8.5 Blaze Angle

As seen in Section 8.8.3, the amplitudes of the peaks depend on the product of the Fourier transforms of the basic slit pattern and the repetition pattern, and, as shown in Figure 8.23 for a uniform slit, the $N = 0$ order is strongest. One can shift this peak by introducing a linear phase shift across the grating. In a transmission grating, this means introducing a prism effect into the slit, and in reflection, it means tilting the reflector. The angle chosen is called the **blaze angle**. By selecting the right combination of blaze angle and period, a designer can ensure that most of the incident light is diffracted into a single order. This change in angle simply means tilting the individual mirror facets so that the angle of incidence is equal to the angle of the desired diffraction order; at this angle the facets act like mirrors.

Picking the right blaze angle can improve the performance of a grating on one order at the expense of the others.

Take-Away

Blaze Angle:
Tilting the reflecting facets of a reflective grating can direct the strongest reflection to a particular order. The angle of the tilt is such that the facets satisfy the reflection equation, Equation 2.4.

Eq. 2.4:
$\theta_r = \theta$.

8.8.6 Littrow Grating

A **Littrow grating** is often used as an end mirror in a laser cavity. Many lasers have multiple gain lines and will thus operate on several wavelengths simultaneously or on one unpredictable wavelength among the ones with sufficient gain. Two common examples are the argon ion laser with lines from the violet through the green, and the carbon dioxide laser with numerous lines around 10 µm. By choosing a line spacing and blaze angle such that the first diffracted order in reflection has the same angle as the incident light, $\theta_i = \theta_d$,

$$\lambda = 2d\sin\theta_i, \qquad (8.106)$$

a very high reflectivity can be achieved for a specific wavelength. The peak in the upper right panel of Figure 8.23 is broad enough that the grating can be tilted to change the wavelength over a reasonable range without appreciable loss. At the design wavelength, the shape of the grating is a staircase with steps of $\lambda/2$, leading to constructive interference at the desired wavelength, as shown in Figure 8.24 A Littrow grating is frequently used as the rear reflector in such laser cavities, as shown in Figure 8.25. For reasons which will become clear in Section 9.6, the front mirror is usually curved in such lasers.

A Littrow grating can be used as a laser end mirror to select one of many wavelength bands.

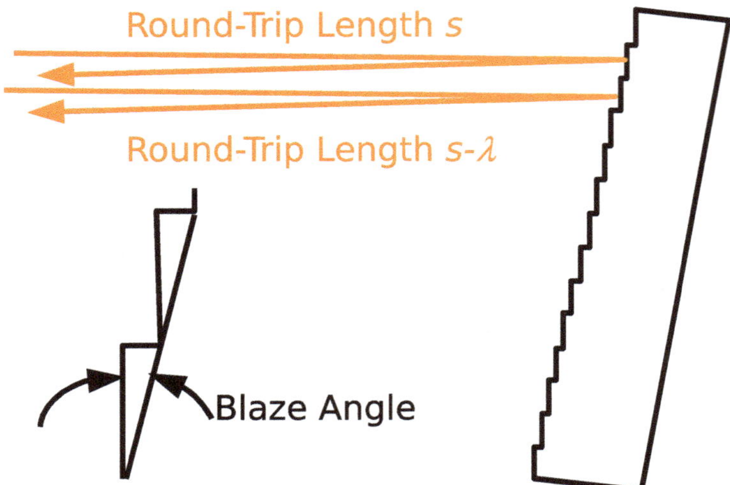

FIGURE 8.24
A Littrow grating has a blaze angle such that a diffracted order is reflected in the direction exactly opposite to the incident light.

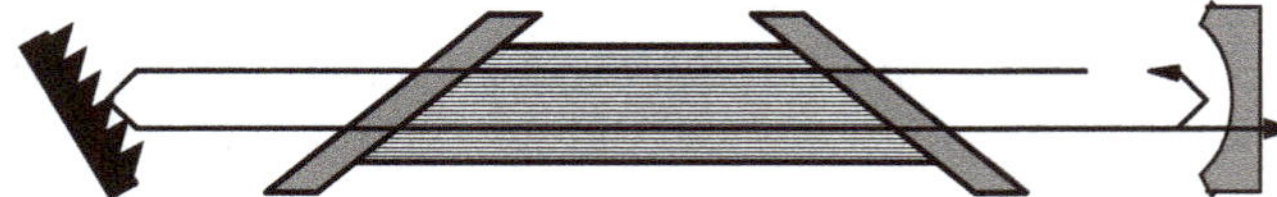

FIGURE 8.25
Littrow grating. A grating in the Littrow configuration can be used to select the operating frequency of a laser from the frequencies of many gain lines.

In Practice

A Littrow grating can be used to select one of several gain bands such as the blue or green lines of the argon-ion laser or the different bands of a carbon-dioxide laser. Generally the resolution is not sufficient to select a single cavity mode. For that purpose, an etalon is usually used (Section 7.5.2). Fine tuning of the frequency of the selected mode is done by adjusting the length of the laser cavity (Section 7.5.3.1).

8.8.7 Monochromators and Spectrometers Again

Next, we consider the monochromator shown in Figure 8.26 or the spectrometer in Figure 8.27. The main distinction between the two is that that the monochromator has a single output slit to pass one wavelength, while the spectrometer has a linear array of detectors, each of which acts as an exit slit for a different wavelength. Although the symmetry of the optical arrangement of the monochromator is slightly compromised in the spectrometer, the aberrations are usually small. While the monochromator can be used as a spectrometer by adjusting the grating and measuring the output one wavelength at a time, the spectrometer shown here measures all wavelengths simultaneously.

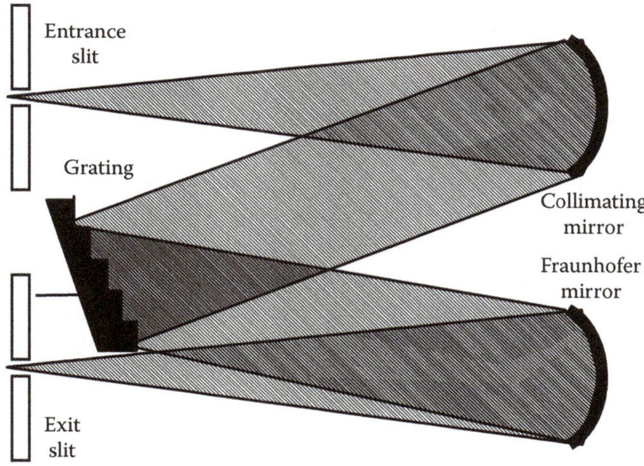

FIGURE 8.26
Grating monochromator. Light from the entrance slit is collimated so that it falls on the grating at a single angle. The diffracted order, which is nearly monochromatic, is focused by the Fraunhofer mirror onto the exit slit.

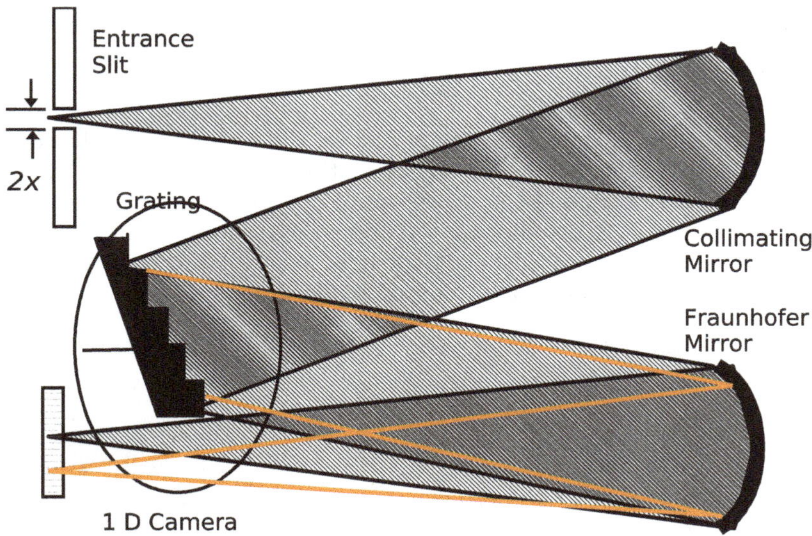

FIGURE 8.27
Grating spectrometer. The slit in the monochromator of Figure 8.26 is replaced by a linear array of detectors (one-dimensional camera). Each pixel receives light at a different wavelength. The black and colored lines show two examples.

> **In Practice**
>
> The scanning monochromator can use a larger, more sensitive, and more expensive detector, since only one measurement is made at a time.

Our goal in this section is to explore the limits of resolution of the monochromator or spectrometer in terms of the optics, the entrance slit, and the exit slit or pixel size.

The entrance slit, of width $d_i = 2w_i$, is placed at the front focal point of the collimating mirror. Thus, light is incident on the grating at angles of $-w_i/(2f) \leq \sin\theta_i \leq w_i/(2f)$. For high efficiency, the grating is blazed for first order at a wavelength in the middle of the band of interest. The diffracted light is then focused onto the back focal plane of an identical second mirror which functions here as a Fraunhofer lens. The exit slit determines the limits of the diffraction angle, in the same way that the entrance slit determines the angles of incidence. If the two slits are very narrow, then the resolution of the monochromator is determined by the apodization function, which is uniform, and has a diameter given by the diameter of the mirrors. Thus, long focal lengths and low f-numbers are useful not only in gathering more light, but also in improving resolution. Using identical mirrors on both sides cancels most aberrations in the plane of incidence of the grating, so that mirror aberrations do not affect the resolution. If the slits are larger, they provide a new function which must be convolved with the resolution limit imposed by the apodization function. Specifically, we can define a transmission function,

Spectral resolution of a grating is determined by the widths of the slits and apodization.

$$T_i(\delta\lambda) = 1 \qquad \left|\delta\lambda - \frac{d}{N}\delta\sin\theta_i\right| < w_i/2, \tag{8.107}$$

and for the exit slit,

$$T_d(\delta\lambda) = 1 \qquad \left|\delta\lambda - \frac{d}{N}\delta\sin\theta_d\right| < w_d/2. \tag{8.108}$$

If at least one of the slits is large, then the resolution will be limited by the convolution of these two functions. In general, the spectral resolution will be limited by the convolution of three functions: these two, and the transform of the apodization function, resulting in an important compromise; if the slits are too large, the resolution is degraded because of this convolution, but if they are too small, the light level is low, the detector signal must be amplified, and the resolution is then limited by noise.

Take-Away

Slit Size:

As the size of the monochromator slit increases, the amount of light collected increases, thereby improving the signal-to-noise ratio, but the resolution becomes worse.

The resolution of a monchromator is limited by the apodization; a wide beam on the grating produces better resolution.

At this point, the astute reader might raise the question, "Isn't the apodization pattern itself just the Fourier transform of the entrance slit?" In fact, if the light source were a laser, the answer would be "yes." However, for an incoherent light source, we can consider the light wave at each point in the slit to be independent of that at any other point. Then each point can be viewed as the source of a spherical wave, but with amplitude and phase independent of its neighbors. This independence prevents interference effects and illuminates the collimating mirror uniformly. This issue will be further discussed in Chapters 10 and 11.

For a grating spectrometer, the analysis is similar, except that the exit slit is replaced by an element of the detector array.

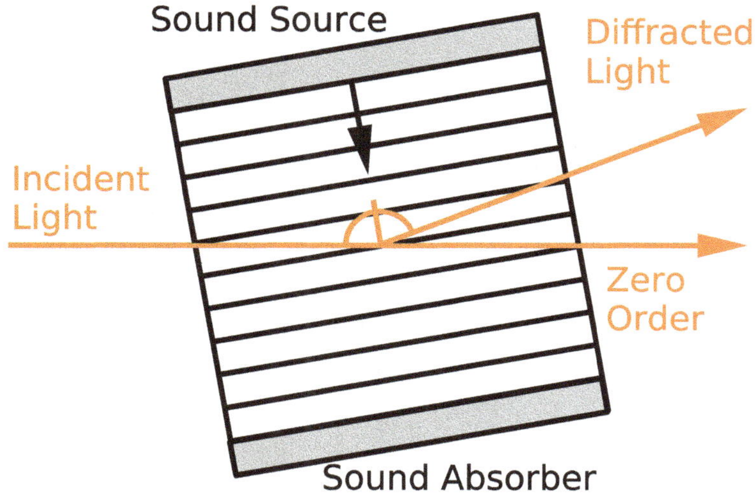

FIGURE 8.28
Bragg Cell. The Bragg Cell can be used to shift the frequency of a laser beam. The black lines represent the acoustic wavefronts traveling approximately from top to bottom. A beam of light from the left is diffracted by the grating with a spacing equal to the acoustic wavelength. The cell is tilted so that in this case the -1 diffracted order is reflected from the wavefronts. The frequency is Doppler shifted by the moving acoustic wavefronts.

In Practice

If we wish to measure the spectrum of a laser, we must be careful. A coherent beam illuminating the input slit will produce a diffraction pattern on the grating given by the diffraction limit. If the slit is too large, the full aperture of the grating will not be used. In fact, if the beam is smaller than the slit, it will pass through the slit completely. It may be useful to use a diffuser to reduce the spatial coherence.

8.8.8 Bragg Cell

In Figure 7.13, we discussed shifting the frequency of a laser beam with a **Bragg Cell.** Here we briefly discuss the operation of the Bragg Cell as a frequency shifter, and provide an intuitive understanding of its operation. The concept is shown in Figure 8.28.

An acoustic source, normally very high frequency ultrasound in the tens of megahertz or more, produces a plane wave which is absorbed at the other end of the cell. The acoustic wavefronts produce a grating with a spacing equal to the acoustic wavelength, Λ,

$$\Lambda = \frac{c_{sound}}{f_{sound}}.$$

The resulting grating diffracts the light according to the grating equation, Equation 8.95,

$$N\lambda = \Lambda\left(\sin\theta_i + \sin\theta_d\right).$$

The grating is tilted so that the acoustic wavefronts act as mirrors, satisfying Equation 2.4 to enhance a particular order.

Eq. 8.95:
$N\lambda = d\left(\sin\theta_i + \sin\theta_d\right)$.

Eq. 2.4:
$\theta_r = \theta$.

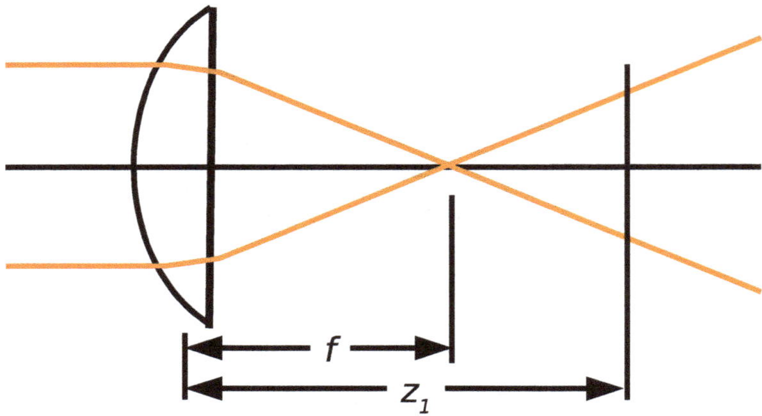

FIGURE 8.29
Geometric-optics calculation of the diameter of a focused beam. Neglecting diffraction, the beam diameter is obtained by similar triangles.

Eq. 7.62:
$f_{DR} = 2\frac{v_{parallel}}{\lambda}$.

Because the wavefronts are moving, the light is frequency shifted according to the Doppler equation, Equation 7.62. It can be shown that th frequency shift will be equal to the acoustic frequency multiplied by the order number. In the example shown, the order is $N = -1$ and the frequency is shifted downward by f_{sound}. The ultrasound absorber is critical. Without it, the acoustic wave would be a standing wave, and no Dopppler shift would occur.

Take-Away

- Diffraction gratings are used to disperse light into its colors or spectral components.
- Diffraction gratings may be reflective or transmissive.
- Diffraction gratings can be analyzed using Fourier transforms.
- The amount of light in each order can be controlled through proper choice of the blaze angle.
- The dispersion angles are determined by the grating pitch or period.
- The resolution is determined by the apodization function.
- Littrow gratings are used to select the wavelength in lasers with multiple gain lines.

8.9 Fresnel Diffraction

Eq. 8.44:
$E(x_1, y_1, z_1) = \frac{jke^{jkz_1}}{2\pi z_1} \int\int E(x, y, 0) e^{jk\frac{(x-x_1)^2}{2z_1} + \frac{(y-y_1)^2}{2z_1}} dx\, dy$.

In the early sections of this chapter, we have considered diffraction in the Fraunhofer zone. We noted that far beyond this zone, the integrand in Equation 8.44 varies rapidly, and numerical integration is difficult. Fortunately, geometric optics works well here, so we can obtain some very good approximations with geometric optics.

For example, if we have $P = 1\,\text{W}$ of light uniformly filling the exit pupil of a beam expander as in Figure 8.29, $D = 30\,\text{cm}$ in diameter, focused to a distance of $f = 100\,\text{m}$, then the irradiance in the pupil is $1\,\text{W}/(\pi D^2/4) \approx 14\,\text{W/m}^2$. At a distance z_1, the diameter is

$$\frac{|z_1 - f|}{f} D,$$

and the irradiance is

$$\frac{P}{\pi (D/2)^2} \frac{f^2}{(z_1 - f)^2}, \tag{8.109}$$

so, for example, the irradiance is $56\,\mathrm{W/m^2}$ at $z_1 = 75\,\mathrm{m}$. This equation provides a correct answer provided that we do not get too close to the edges of the shadow or too close to the Fraunhofer zone.

We have used Fraunhofer diffraction extensively to explore the Fraunhofer zone, and we see above that we can use geometric optics outside the Fraunhofer zone, except at edges. If we could describe the field at edges of shadows, we would have a more complete picture of light propagation. Can we use diffraction theory to explore these edges? Yes, we can, but the approximations of Fraunhofer diffraction, starting with Equation 8.46, do not help us; we must return to Equation 8.44, and find a new approximation that is useful at edges.

Eq. 8.44:
$E(x_1, y_1, z_1) = \frac{jke^{jkz_1}}{2\pi z_1} \int\int E(x, y, 0) e^{jk\frac{(x-x_1)^2}{2z_1} + \frac{(y-y_1)^2}{2z_1}} dx\,dy.$

8.9.1 Fresnel Cosine and Sine Integrals

We begin with the case where the integrals over x and y are separable:

$$E(x_1, y_1, z_1) = \frac{jke^{jkz_1}}{2\pi z_1} \int E_x(x, 0) e^{jk\frac{(x-x_1)^2}{2z_1}} dx \int E_y(y, 0) e^{jk\frac{(y-y_1)^2}{2z_1}} dy. \tag{8.110}$$

Although we mentioned such situations briefly when we discussed the square aperture, we did so in the context of Fraunhofer diffraction. This time, we do not expand the terms $(x - x_1)^2$ and $(y - y_1)^2$. Now the integral becomes more complicated, as we have an exponential of a quadratic, and we do not expect a Fourier transform. Let us consider specifically the case of a uniformly illuminated aperture of size $2a$ by $2b$. The irradiance is $P/(4ab)$, and Equation 8.44 becomes

$$E(x_1, y_1, z_1) \frac{jke^{jkz_1}}{2\pi z_1} \sqrt{\frac{P}{4ab}} I(x_1, a, z_1) I(y_1, b, z_1) \tag{8.111}$$

where

$$I(x_1, a, z_1) = \int_{-a}^{a} e^{jk\frac{(x-x_1)^2}{2z_1}} dx_1$$

$$I(x, a, z_1) = \int_{-a}^{a} \cos\left(k\frac{(x-x_1)^2}{2z_1}\right) dx + j \int_{-a}^{a} \sin\left(k\frac{(x-x_1)^2}{2z_1}\right) dx. \tag{8.112}$$

We can begin to understand Fresnel diffraction by looking at the integrands in this equation. As shown in Figure 8.30, the cosine function starts at $\cos(0) = 1$, after which it also oscillates ever more rapidly. The cosine integral will therefore most likely be dominated by the central region, corresponding to $x \approx x_1$ and $y \approx y_1$. Far from this region, the rapid oscillations will cancel unless the field happens to have matching oscillations. The sine will tend to cancel everywhere. Thus, we can see that for small distances, z_1, such that the integrands go through many cycles across the aperture, the field at (x_1, y_1, z_1) will be approximately the field at $(x_1, y_1, 0)$, as we would have concluded from geometric optics. Near the edges, we would expect fringes, and we will develop equations for these fringes in the next section.

The two integrals in Equation 8.112 are the Fresnel cosine and Fresnel sine integrals respectively[13]. We set

$$u = (x - x_1)\sqrt{\frac{k}{2z_1}} \qquad du = dx_1\sqrt{\frac{k}{2z_1}}$$

with limits

$$u_1 = (x + a)\sqrt{\frac{k}{2z_1}} \qquad u_2 = (x - a)\sqrt{\frac{k}{2z_1}}, \tag{8.113}$$

and Equation 8.112 becomes

$$I(x_1, a, z_1) = \sqrt{\frac{2z_1}{k}}\int_{u_1}^{u_2}\cos^2 u\, du + j\sqrt{\frac{2z_1}{k}}\int_{u_1}^{u_2}\sin^2 u\, du$$

$$I(x_1, a, z_1) = \sqrt{\frac{2z_1}{k}}\left[C(u_2) - C(u_1) + S(u_2) - S(u_1)\right]. \tag{8.114}$$

where

$$C(u_1) = \int_0^{u_1}\cos^2 u\, du \qquad S(u_1) = \int_0^{u_1}\sin^2 u\, du. \tag{8.115}$$

It will simplify the notation if we define

$$F(u_1) = C(u_1) + jS(u_1). \tag{8.116}$$

The Fresnel cosine and sine integrals, and thus F, are odd functions,

$$C(u) = -C(-u) \quad S(u) = -S(-u) \quad F(u) = -F(-u),$$

and have the approximations

$$F(u) \approx u e^{j\pi u^2/2} \qquad u \approx 0$$

$$F(u) \approx \frac{1+j}{2} - \frac{j}{\pi u}e^{j\pi u^2/2} \qquad u \to \infty. \tag{8.117}$$

Finally, returning to Equation 8.111, with the substitutions indicated in Equation 8.114, and the definitions in Equation 8.115 and 8.116, we can write the diffraction field in terms of these integrals:

$$E(x_1, y_1, z_1) = \frac{jke^{jkz_1}}{2\pi z_1}\frac{P}{4ab}\frac{2z_1}{k} \times$$
$$\left\{F\left[(x_1 - a)\sqrt{\frac{k}{2z_1}}\right] - F\left[(x_1 + a)\sqrt{\frac{k}{2z_1}}\right]\right\} \times$$
$$\left\{F\left[(y_1 - b)\sqrt{\frac{k}{2z_1}}\right] - F\left[(y_1 + b)\sqrt{\frac{k}{2z_1}}\right]\right\}. \tag{8.118}$$

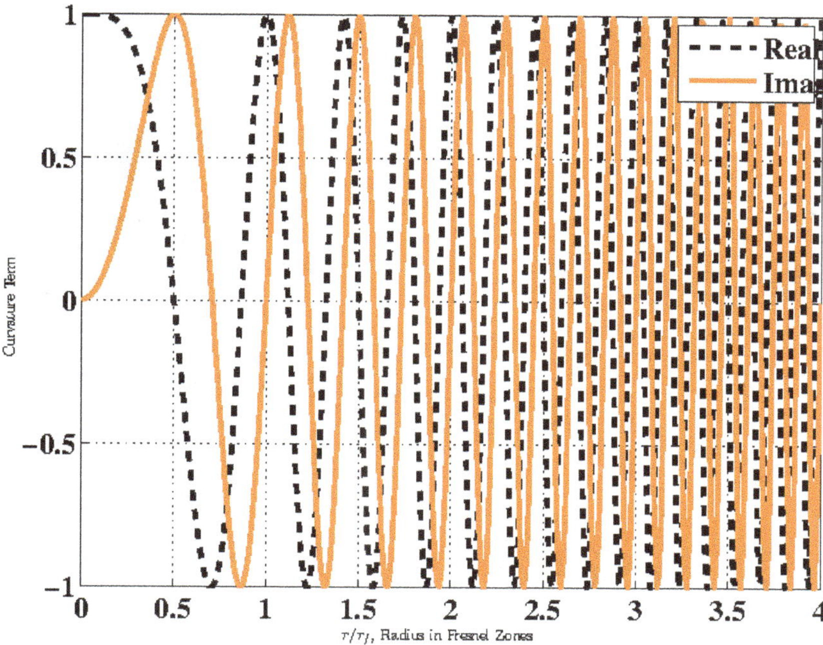

FIGURE 8.30
Integrands of Fresnel integrals. The integrand of the Fresnel cosine integral is shown by the black dashed line, and that of the Fresnel sine integral by the solid colored line. Both integrands oscillate more rapidly further from the origin, because the argument of the trigonometric function is distance squared. Note that this is the same plot as in Figure 8.14.

Take-Away

Fresnel Diffraction Equation:
Diffraction in the near field of a rectangular aperture can be calculated with the equation above.

8.9.2 Cornu Spiral and Diffraction at Edges

Figure 8.31 shows the Fresnel sine integral on the ordinate and Fresnel cosine integral on the abscissa, plotted parametrically. To provide a sense of magnitudes the symbol "o" is shown for integer values of u from -5 to 5, and the symbol "+" is shown for half-integer values.

Near the axis in the near-field diffraction pattern of a large aperture, such that in Equation 8.118,

$$|x_1| \ll a \qquad z_1 \approx 0$$

$$(x_1 - a)\sqrt{\frac{k}{2z_1}} \approx -a\sqrt{\frac{k}{2z_1}} \rightarrow -\infty$$

$$(x_1 + a)\sqrt{\frac{k}{2z_1}} \approx \quad a\sqrt{\frac{k}{2z_1}} \rightarrow \quad \infty, \tag{8.119}$$

In this case, the two values of F are near the asymptotic values, and the field and irradiance

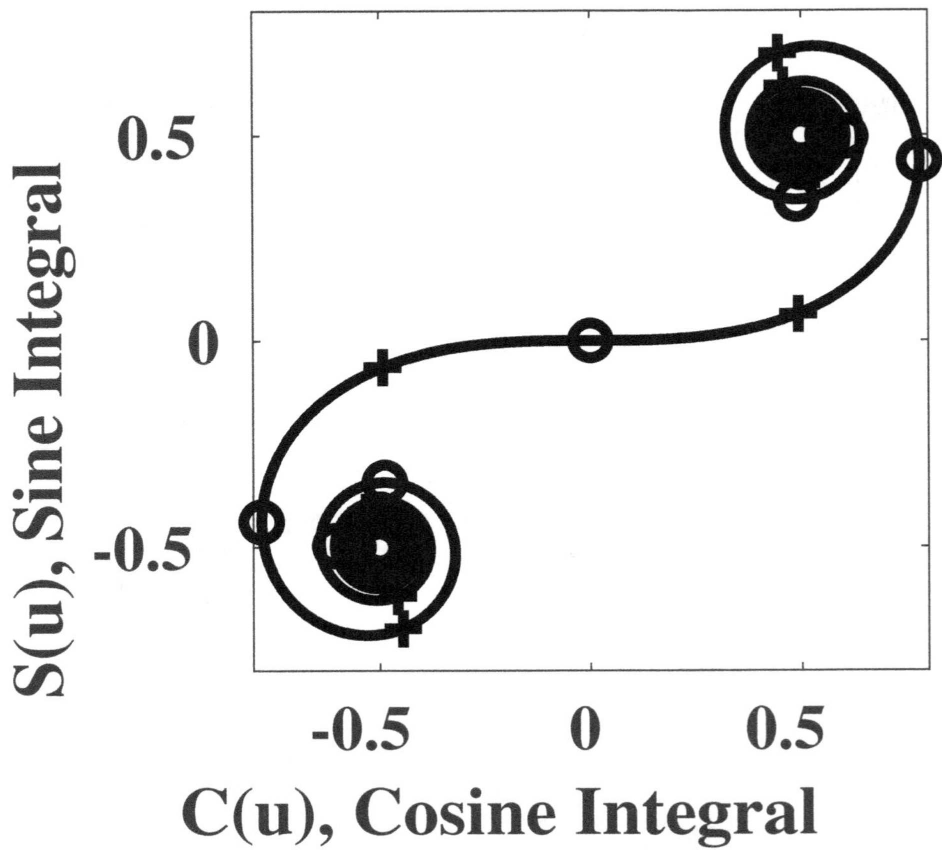

FIGURE 8.31
Cornu spiral. The "o" marks indicate integer values of u, and the "+" marks indicate half-integer values from -5 to 5.

are large, as shown by the **x** in Figure 8.32. In this example, the width, $D = 2a = 2\,\text{cm}$, and the diffraction pattern is examined at $z_1 = 5\,\text{m}$. We use the $\lambda = 514.5\,\text{nm}$ line of the argon ion laser. The far-field distance (Equation 8.50) is $D^2/\lambda = 777\,\text{m}$, so our near-field assumption is valid. As x approaches a or $-a$, one of the end points continues to spiral in toward its asymptote and the other spirals out, leading to larger oscillations in the field and irradiance. When $x = \pm a$, one point is at the origin, the field is about half its maximum value, and the irradiance is thus about one quarter of its maximum, as shown by the *. This point would be the edge of the shadow if we solved the problem entirely by geometric optics. Beyond this point, the field decays rapidly with increasing distance, and the fringes are relatively small, as seen at the symbol, **o**. To illustrate the size of the fringes in the dark region, the figure is plotted in decibels in the lower right panel. The lower left panel shows an image simulating the diffraction pattern for a slit, with $2a = 2\,\text{cm}$ and $b \to \infty$.

What happens if we repeat this analysis with a small aperture? If we choose $D = 2a = 100\,\mu\text{m}$, keeping $z_1 = 5\,\text{m}$ and $\lambda = 514.5\,\text{nm}$, the far-field distance is $D^2/\lambda = 19\,\text{mm}$, and our diffraction pattern is well into the far field. In Figure 8.33, we see that the diffraction pattern is the familiar *sinc* function, and we could in fact, have done the analysis with Fraunhofer diffraction. The endpoints rapidly move from the region around the lower

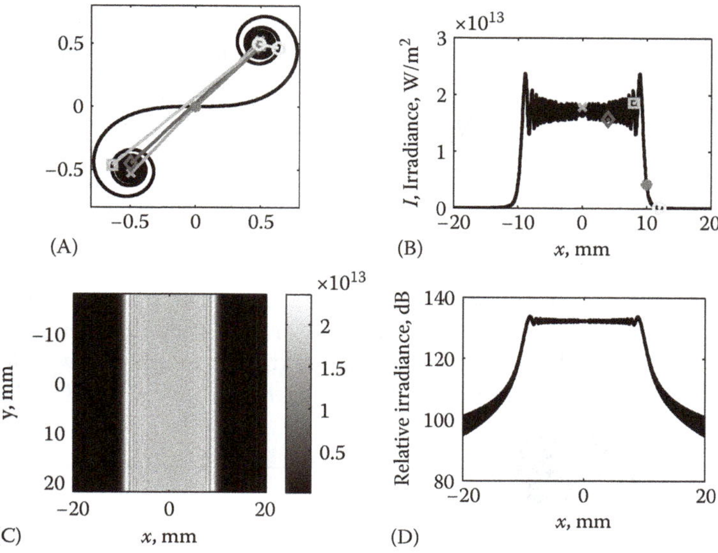

FIGURE 8.32
Diffraction from a large aperture in the cornu spiral. The aperture is a square, with width 2 cm, at a distance of 5 m. The Cornu spiral (A) shows the calculation of the integral for sample values of x. Symbols at endpoints on the lines match the symbols on the irradiance plots (B linear and, D in decibels). The image of a slit with a large width and small height shows the fringes (C).

asymptote through the origin to the upper asymptote, an expansion of which is also shown in the figure.

Eq. 8.44: $E(x_1, y_1, z_1) = \frac{jke^{jkz_1}}{2\pi z_1} \int\int E(x, y, 0) e^{jk\frac{(x-x_1)^2}{2z_1} + \frac{(y-y_1)^2}{2z_1}} dx\,dy.$

8.9.3 Fresnel Diffraction as Convolution

Equation 8.44 can be viewed as a convolution. The field $E(x_1, y_1, z_1)$ is the two-dimensional convolution of the field $E(x, y, 0)$, with

$$\frac{jke^{jkz_1}}{2\pi z_1} e^{jk\frac{x^2}{2z_1} + \frac{y^2}{2z_1}}. \tag{8.120}$$

Given an object with an irregular shape, and a diffraction pattern well into the near field, one could compute the fringe pattern numerically using this convolution approach.

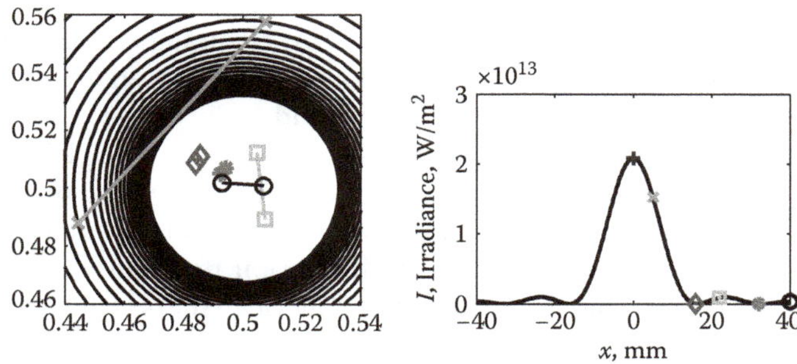

FIGURE 8.33
Diffraction from a small aperture. The aperture is a square, with width 100 µm, at a distance of 5 m.

Eq. 8.49: $N_f = \frac{r_{max}}{r_f}$.

The practicality of doing so is limited by the large number of pixels needed for the two-dimensional convolution in cases where the Fresnel number (Equation 8.49)

$$N_f = \left(\frac{r}{r_f}\right)^2 = \frac{r^2}{2\lambda z_1} \tag{8.121}$$

($r_f = \sqrt{2\lambda z}$ from Equation 8.47) is large. Nevertheless, without completing the whole pattern, we can make some useful guesses. For example, the fringe pattern along relatively straight edges can be estimated by assuming that we are near the edge of a very large rectangular aperture. The irradiance will be a fraction near 1/4 of the maximum at the edge of the shadow, and will decay rapidly thereafter with very weak fringes. Strong fringes will occur on the bright side of the shadow line, and their amplitude will decrease moving into the bright region away from the edge.

In Practice

Although we make the claim that geometric optics can compute the irradiance quite accurately away from the edges, in truth, the fringes can extend quite far from the edges. These effects are particularly noticeable in monochromatic light. This effect can be seen in Figure 8.32C.

8.9.4 Fresnel Zone Plate and Lens

One particularly interesting application of the convolution approach is the **Fresnel zone plate**. Suppose we are interested in the field on the axis, at $(0, 0, z_1)$ somewhere in the near field of a circular aperture of diameter D. Assuming the field is uniform in the aperture, we expect the cosine and sine integrands to look like Figure 8.30. As we have seen, the central lobe of the pattern contributes most to the integral, and because of the rapid alternation of sign, further rings produce contributions that nearly cancel, adding little to the integral. Suppose, however, that we mask the aperture with the pattern shown in Figure 8.34, which is chosen to block the light wherever the cosine is negative. Then as a function of radius, the integrand will be as shown in Figure 8.35. The portions of the sine function allowed through this aperture are almost equally positive and negative, and will integrate to a small value. Without the negative contributions, the integral of the integrand in Figure 8.35 is larger than that of the integrands in Figure 8.30. Therefore, the light on the axis will be brighter than it would be without the zone plate. The zone plate functions as a crude lens of focal length z_1. The real and imaginary parts of the field through the zone plate are shown in Figure 8.36. The performance of the zone plate as a lens will be discussed shortly in Figure 8.38.

A fresnel zone plate can be used in place of a lens. Of course, the quality is not as good.

Even better would be to make the zone plate by changing the thickness of the zone plate rather than its transmission. If the thickness is increased by a half wave where the cosine is negative, then the negative peaks would be inverted, resulting in a further increase. Of course, the best approach would be to taper the thickness to compensate exactly for the phase shifts. This is exactly what is accomplished by a lens of focal length z_1, which provides a quadratic phase shift. As we have already seen, such a lens is called a Fraunhofer lens, and it converts the Fresnel diffraction problem into one of Fraunhofer diffraction. However, we need not use the usual spherical shape for the lens. As the lens diameter increases, the central thickness of a positive lens must also increase.

Instead of increasing the thickness continuously, one can increase it to some multiple of 2π and then step it down by the same amount, producing a **Fresnel lens**, as shown in Figure 8.37. In practice good performance can be obtained if the step size is not a multiple of 2π, although there will be some loss of focusing because of the phase steps at

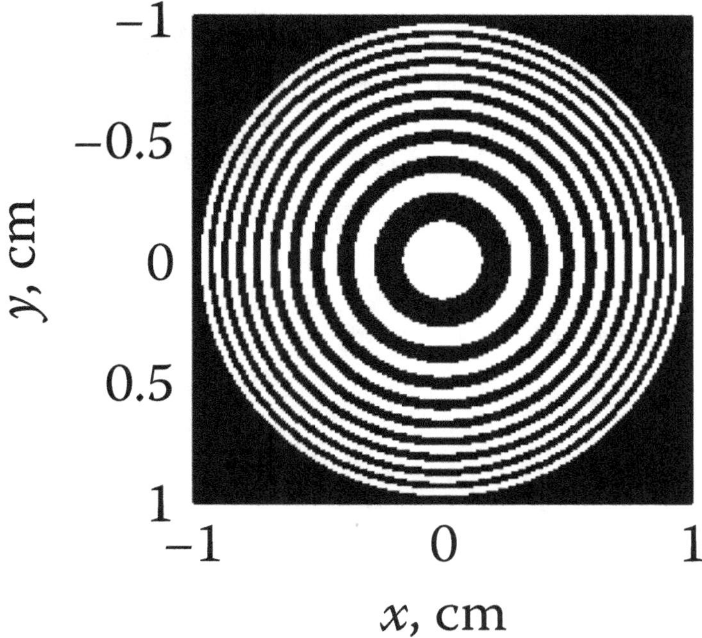

FIGURE 8.34
Fresnel zone plate. This is the zone plate with a diameter of 2 cm at a distance of 10 m.

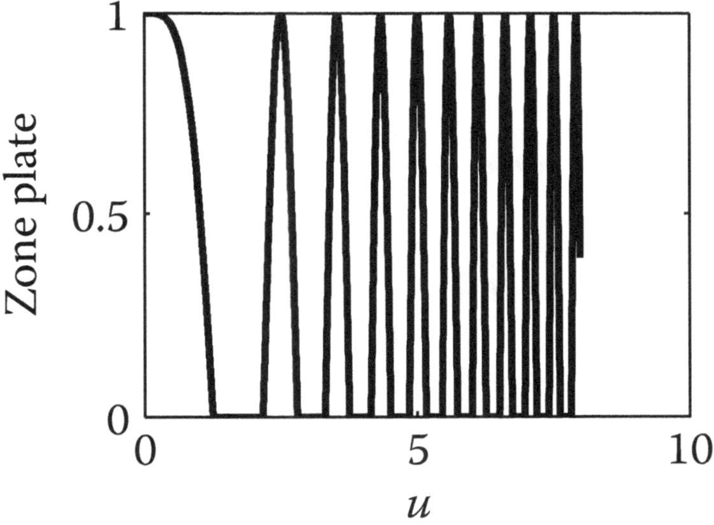

FIGURE 8.35
Real field in Fresnel zone plate. This is the cosine integrand from Figure 8.30. All contributions are positive.

the discontinuities. Fresnel lenses are used for some projectors, as handheld magnifiers, and most notably, in lighthouses.

Figure 8.38 shows the diffraction patterns obtained by Fourier transform. The Fresnel diffraction pattern of a collimated wave is the same size as the aperture, as expected.

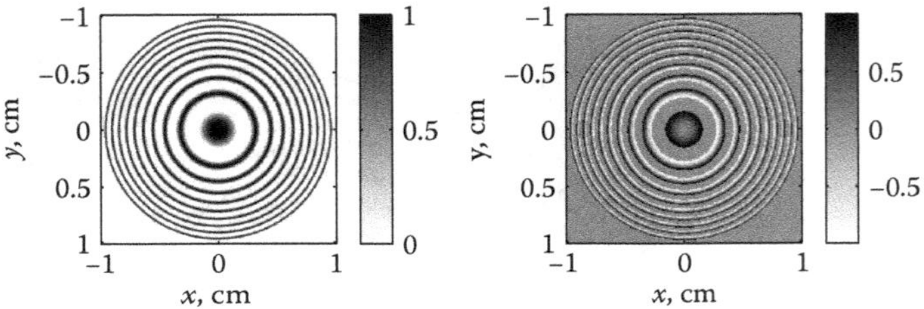

FIGURE 8.36
Fields of Fresnel zone plate. The real part of the field, shown on the left is always positive. The imaginary part alternates sign, and contributes little to the signal.

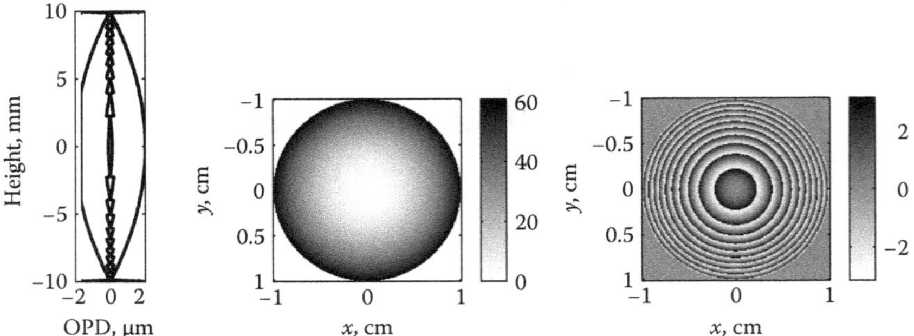

FIGURE 8.37
Fresnel lens. The left panel shows the conventional lens and the reduced thickness of the Fresnel lens. The center panel shows the phase in radians of the conventional lens, and the right panel shows the phase of the Fresnel lens.

Focusing with either a conventional or Fresnel lens produces a diffraction-limited spot. The Fresnel zone plate provides some focusing, but not as much as the lens.

8.9.5 Summary of Fresnel Diffraction

The key points about Fresnel diffraction are

- Fresnel diffraction is useful in the near field when there is no Fraunhofer lens.
- It is also useful even when there is a Fraunhofer lens, if the desired field is away from its focus.
- The key features of Fresnel diffraction are strong fringes on the bright side of a shadow edge. The field far from the edge can be estimated quite well with geometric optics.
- Fresnel diffraction can be treated with Fresnel cosine and sine integrals for rectangular apertures.
- It may be treated as a convolution with a quadratically phase-shifted kernel.
- Numerical convolution may be difficult for high Fresnel numbers.
- A Fresnel zone plate can be used as a crude lens.
- Fresnel lenses make compact lenses of reasonable quality.

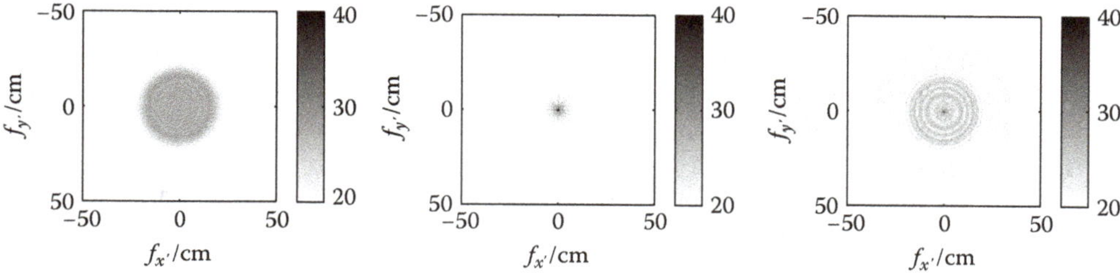

FIGURE 8.38
Fresnel diffraction patterns. The left panel shows the Fresnel diffraction pattern from a uniform collimated wave through the aperture. Because we are in the near field, it is large, with a low irradiance. The center panel shows the focused case, which is equivalent to Fraunhofer diffraction, and could be obtained with either the conventional or Fresnel lens. The right panel shows the diffraction pattern for the zone plate. It has an intermediate size and irradiance.

Take-Away

Fresnel Diffraction:
Fresnel diffraction is more complicated than Fraunhofer diffraction, and is generally less used. It is important in cases where we need to know the fringe pattern at the edges of shadows.

Problems

8.1 Resolution

A car approaches along a long straight road. Initially, it is so far away that the two headlights appear as one.

8.1a Using the Rayleigh criterion for resolution, at what distance do we begin to see that they are two separate lights? Use reasonable numbers for the spacing of the headlights and the diameter of the pupil of the eye.

8.1b Now, consider the fact that we have two eyes, separated by a distance of a few centimeters. Does this change your answer? Why or why not?

8.2 Telescope Diffraction (NS)

Here we look at subtraction of two diffraction patterns to determine the diffraction pattern of a Cassegrain telescope The primary of this telesope is a concave spherical mirror with a circular dimension of $D_2 = 30\,\mathrm{cm}$. The secondary is mounted in front and a hole in the primary allows light to pass through. This configuration obscures a circular region of diameter 5 cm. We use this telescope as a beam expander, putting 10 watts of light light at $\lambda = 10.59\,\mu\mathrm{m}$ into the secondary. Assume uniform illumiation so that you can do the calculation more easily.

Note: You may assume the irradiance is $|E|^2$ so your electric field will have strange units of $\sqrt{\mathrm{Watts}}/\mathrm{m}$.

8.2a Calculate the axial electric field from the whole mirror, assuming no obscuration, at $z = 20\,\mathrm{km}$.

8.2b Calculate the axial electric field at $z = 20\,\text{km}$ that you would obtain from a mirror of just the size of the obscuration. Be careful using the equations in the notes, because those are given in terms of power. Remember that only a fraction of the total power is included in the smaller aperture.

8.2c What is the axial electric field for the telescope?

8.2d What is the axial irradiance?

8.2e As a sanity check, try to estimate the total power in the diffraction pattern. Calculate the width of the diffraction pattern using the larger diameter, D_2. From this and the axial irradiance, **estimate** the total power. You haven't calculated the whole diffraction pattern, so this estimate is going to be only an approximation.

8.3 Resolution (NS)

Over the last few decades, global images of the earth's surface have become available to the public through a number of commercial sources. Let's consider what would be required to image with a resolution of of 10 cm. Such resolution might make it possible to identify a specific vehicle or even a person. Let's use green light in all our examples.

8.3a For a satellite in low–earth orbit at 200 km above the Earth's surface, what aperture would be required on a camera?

8.3b For an aircraft at 12 km what aperture would be required?

8.3c For the human eye, what would be the maximum altitude at which this resolution would be possible? Assume a maximum pupil size of 8 mm. This would be likely at night, but during the day, the pupil would normally be smaller.

9 Gaussian Beams

In Chapter 8, we discussed the Fraunhofer diffraction pattern of a Gaussian beam, and we noted that the Fourier transform of a Gaussian function is also a Gaussian function. We will see in this chapter that the Gaussian beam is a Gaussian function of transverse distances x and y at all locations along its axis (z). The Gaussian beam merits its own chapter for several reasons:

- The output beams of many types of lasers are approximated well by Gaussian functions.
- The Gaussian function is the minimum-uncertainty wave as mentioned in Section 8.1, which makes it a good limiting case against which to compare other waves.
- The equations of Gaussian beam propagation are, as we shall see, very simple. We can write them in closed form and manipulate them encountering nothing more complicated than a quadratic equation.
- More complicated beams can be described in terms of Hermite–Gaussian beams, which have the same propagation properties. Therefore, they can be treated with the same method.

We begin with a brief development of the equations of propagation. We then look at the results in terms of beam propagation, laser cavity design, and other applications. Finally, we consider the Hermite–Gaussian modes. It may be helpful to visualize the shape of a Gaussian beam in Figure 9.1 while reading this section.

9.1 Equations for Gaussian Beams

In this section, we will develop the equations for propagation of Gaussian beams or waves. We use the term "beams" because a Gaussian wave propagating in the z direction is localized in x and y, a feature we will often find highly desirable. We begin with an overview of the derivation of equations for a Gaussian beam, and then discuss the parameters of those equations.

9.1.1 Derivation

The equations for Gaussian beams are described in numerous papers from the early days of lasers. Siegman[211] provides an excellent review. The definitive paper is that of Kogelnik and Li[125] in 1966.

It is interesting to realize that Gaussian beams were "discovered" as solutions to the wave equation in a laser cavity. Simple microwave cavities are modeled as rectangular boxes, often with perfectly reflecting sides. The solutions to the wave equation in these cavities are well known to be sines and cosines. However, "open" resonant cavities are also possible, with reflectors on two ends, but none on the sides. In fact, Kogelnik and Li begin the discussion by looking for eigenvectors of the ABCD matrices we discussed in Chapter 3. By using at least one converging element in the cavity, normally a concave mirror, it is possible to produce a beam which will reproduce itself upon a round trip through the cavity. The convergence caused by this element will exactly offset the divergence caused by diffraction. We will discuss this concept later in this chapter.

Gaussian beams are natural solutions to the wave equation in an open cavity such as a typical laser.

DOI: 10.1201/9781315157047-9

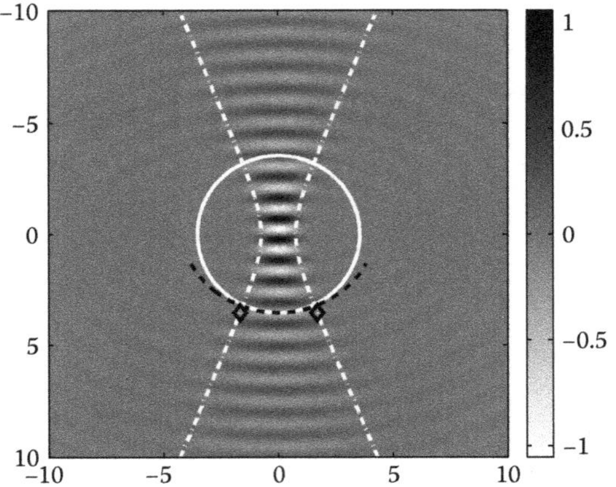

FIGURE 9.1
A spherical Gaussian beam. The imaginary part of the wave, $E(x,0,z)$ is shown in gray-scale. The white circle has a radius of $2b$. The black dashed curve shows ρ, the radius of curvature at $z = 2b$. The white dashed curves show the $1/e$ field or $1/e^2$ irradiance boundaries, $\pm w(z)$.

The equations that explain propagation of Gaussian beams can be derived in at least three different ways. First, we could solve Maxwell's equations directly. Second, assuming that the scalar wave equation is valid, we could use the Fresnel diffraction equation, Equation 8.44. The third derivation[52] is easier than either of the others. We use the scalar field, E, as defined in Section 1.4.3. We know that a spherical wave,

$$E_{sphere} = \sqrt{\frac{P_{sphere}}{4\pi}} \frac{e^{jk\sqrt{x^2+y^2+z^2}}}{\sqrt{x^2+y^2+z^2}}, \tag{9.1}$$

is a solution of the Helmholtz equation (8.28) in a vacuum. More generally, we could substitute nk for k, where n is the index of refraction, and use this equation for any isotropic, homogeneous medium, but we will consider this option later. We have developed a paraxial approximation in Equation 8.39, with the result

$$E_{sphere} \approx \sqrt{\frac{P_{sphere}}{4\pi z}} e^{jkz} e^{jk\frac{x^2+y^2}{2z}}. \tag{9.2}$$

We say that the wavefront has a radius of curvature z. If we substitute a complex term, $q = z + jb$ for z, we obtain another solution of the Helmholtz equation:

$$E = \sqrt{\frac{P_{sphere}}{4\pi}} \frac{e^{jk\sqrt{x^2+y^2+q^2}}}{\sqrt{x^2+y^2+q^2}}, \tag{9.3}$$

and making an equivalent paraxial approximation, namely, taking the zero-power term in the Taylor's series for the square root, $r = q$, in the denominator and the first power in the exponent,

$$E \approx \sqrt{\frac{P_{sphere}}{4\pi}} \frac{1}{q} e^{jkq} e^{jk\frac{x^2+y^2}{2q}}. \tag{9.4}$$

We split the $1/(2q)$ into real and imaginary parts, to obtain

$$E \approx \sqrt{\frac{P_{sphere}}{4\pi}} \frac{1}{q} e^{jkq} e^{jk\left(x^2+y^2\right)\Re\frac{1}{2q}} e^{-k\left(x^2+y^2\right)\Im\frac{1}{2q}}. \tag{9.5}$$

We notice that the first term involving $x^2 + y^2$ looks like a curvature term, so we define the **radius of curvature** as ρ in

$$e^{jk\left(x^2+y^2\right)\Re\frac{1}{2q}} = e^{jk\frac{x^2+y^2}{2\rho}}, \tag{9.6}$$

The last exponential term in Equation 9.5 looks like a Gaussian profile. First by analogy with Equation 9.6, we define

$$e^{-k\left(x^2+y^2\right)\Im\frac{1}{2q}} = e^{-k\frac{x^2+y^2}{2b'}}. \tag{9.7}$$

In Equations 9.6 and 9.7, we have used the definitions

$$q = z + jb \qquad \frac{1}{q} = \frac{1}{\rho} - \frac{j}{b'} \tag{9.8}$$

so that

$$\rho = \frac{z^2 + b^2}{z} \qquad b' = \frac{z^2 + b^2}{b}. \tag{9.9}$$

The inverse of the complex radius of curvature, $1/q$ is called the **complex curvature**. The parameters in Equations 9.8 are defined here, although each will be discussed in more detail later

z	Distance from Waist	
b	Rayleigh Range	Based on waist diameter
ρ	Radius of Curvature	At distance z from waist
b'	Size parameter	Determines beam size at distance z

We note that b' has the same relationship to $w(z)$ as b has to w_0. We will refer to it as the **size parameter**. The Rayleigh range is also called the confocal distance.

If we define the Gaussian beam radius, $w = d/2$ as in Section 8.6.3,

$$w^2 = \frac{2b'}{k} \qquad b' = \frac{\pi w^2}{\lambda} = \frac{\pi d^2}{4\lambda}, \tag{9.10}$$

then Equation 9.7 becomes

$$e^{-k\left(x^2+y^2\right)\Im\frac{1}{2q}} = e^{-\frac{x^2+y^2}{w^2}}. \tag{9.11}$$

If z is negative, then ρ is negative indicating a converging beam. Positive ρ and z denote a diverging beam. When z is negative, the beam diameter, d (solving Equation 9.10), is decreasing with increasing z, while for positive z it is increasing.

We make two more substitutions in Equation 9.4. First, using Equations 9.9 and 9.10,

$$\frac{1}{q} = \frac{1}{z + jb} = \frac{1}{\sqrt{z^2 + b^2}} e^{-\arctan\frac{z}{b}} = \frac{1}{\sqrt{bb'}} e^{-\arctan\frac{z}{b}} = \frac{\lambda}{b} \frac{1}{\sqrt{\pi w^2}} e^{-\arctan\frac{z}{b}}, \tag{9.12}$$

and

$$e^{jkq} = e^{jk(z+jb)} = e^{jkz}e^{-kb}, \qquad (9.13)$$

with the result that

$$E \approx \sqrt{\frac{P_{sphere}\lambda}{4\pi b}} e^{-kb} \sqrt{\frac{1}{\pi w^2}} e^{-\arctan\frac{z}{b}} e^{jkz} e^{jk\frac{x^2+y^2}{2\rho}} e^{-\frac{x^2+y^2}{w^2}}. \qquad (9.14)$$

We note that the first term consists of constants, independent of x, y, z. Because the Helmholtz equation is linear, this solution can be multiplied by any scalar. We choose the scalar so that the integrated irradiance

$$\int\int EE^* dx\, dy = P, \qquad (9.15)$$

with the final result that defines the Gaussian beam, or more correctly, the **spherical Gaussian beam,**

$$E \approx \sqrt{\frac{2P}{\pi w^2}} e^{-\arctan\frac{z}{b}} e^{jkz} e^{jk\frac{x^2+y^2}{2\rho}} e^{-\frac{x^2+y^2}{w^2}}. \qquad (9.16)$$

Take-Away

Spherical Gaussian Beam:
The spherical Gaussian beam is a solution of the Helmholtz equation, or scalar wave equation. It is often a good approximation for a laser beam. It has simple closed-form equations and is frequently used as a first approximation for propagation of other beam profiles.

9.1.2 Gaussian Beam Characteristics

Complete equation for a Gaussian beam.

We next begin to explore the meanings of the parameters in Equation 9.16, which we write substituting $\psi = \arctan z/b$ as

$$E \approx \sqrt{\frac{2P}{\pi w^2}} e^{jkz} e^{jk\frac{x^2+y^2}{2\rho}} e^{-\frac{x^2+y^2}{w^2}} e^{-\psi}. \qquad (9.17)$$

Here we assume cylindrical symmetry, which is most common. However, it is possible to have different beam diameters in x and y.

Because x and y only appear in the form $x^2 + y^2$, the field has cylindrical symmetry about the z axis, and we can explore its features very well by setting $y = 0$. Figure 9.1 shows the imaginary part of the field given by Equation 9.16, or 9.17 as a function of z in the vertical, and x in the horizontal. The minimum beam diameter, d_0, in this example is 1.5λ, and both axes are in wavelengths. To compare this wave field to a spherical wave, we have drawn a white circle of radius $2b$ on the figure.

The Gaussian beam is characterized by four parameters, z, b, ρ, and b', related through the complex radius of curvature and its inverse. Only two of the four variables are independent. If any two are known, the other two may be calculated.

If we define the complex radius of curvature as in Equation 9.8, we note that the **radius of curvature** for any value of q is given by

Eq. 9.8:
$q = z + jb$
$\frac{1}{q} = \frac{1}{\rho} - \frac{j}{b'}$.

$$\frac{1}{\rho} = \Re\frac{1}{q}, \qquad (9.18)$$

The wavefront curvature $1/\rho$ of the Gaussian beam is the real part the complex curvature.

as shown by the dashed black line in Figure 9.1. We note that the radius of curvature of the wavefront is a bit larger than the radius of the corresponding spherical wave. We also note that at large values of x, the wavefronts represented by peaks or valleys in the imaginary part of E have a larger radius of curvature than ρ as indicated by the dashed black curve. This disparity is a result of our decision to take only the first term in the Taylor series to describe the curvature (the paraxial approximation). Practically, the amplitude of the wave is so low outside this region that the approximation works very well. The beam diameter, d, is computed from b', given by

$$\frac{1}{b'} = -\Im\frac{1}{q} \qquad b' = \frac{\pi w^2}{\lambda} = \frac{\pi d^2}{4\lambda}. \qquad (9.19)$$

Beam size is obtained from the imaginary part of the complex curvature.

The diameter of the beam, d, is shown by the distance between the two black diamonds, at the value of q used in our current example, and is shown for all values of z by the distance between the dashed white lines.

The distance from the waist is the real part of the complex radius of curvature.

We also note that

$$z = \Re q \qquad (9.20)$$

is the physical distance along the axis. At $z = 0$, we have $b = b'$, so b defines the **waist diameter**, d_0, which is the minimum diameter along the beam. We call b the **confocal parameter**, the **Rayleigh range**, or sometimes the **far-field distance**. It is given by

The waist diameter is obtained from the imaginary part of the complex radius of curvature.

$$b = \Im q \qquad b = \frac{\pi w_0^2}{\lambda} = \frac{\pi d_0^2}{4\lambda}. \qquad (9.21)$$

Distances far beyond the confocal parameter, $z \gg b$, we call the far field and distances $z \ll b$ are said to be in the near field. We will discuss the significance of these definitions shortly.

Finally, we note the phase of the last term in Equation 9.17:

The Gouy phase goes through π radians as z goes from negative to positive infinity.

$$\psi = \arctan\frac{z}{b}. \qquad (9.22)$$

We see the effect of this phase by looking closely at the white circle, and noticing that the phase of the Gaussian wave has different values at the top and bottom of the circle. This phase difference, ψ, is called the **Gouy phase**.

9.1.3 Summary of Gaussian Beam Equations

- The spherical Gaussian wave, also called a Gaussian wave, or a Gaussian beam, is given by Equation 9.17.

- The axial irradiance is twice what would be expected for a uniformly illuminated circular region of the same diameter, d.
- The Gaussian wave can be characterized by a complex radius of curvature, q.
- The real part of q is the physical distance, z along the axis, with the zero taken at the beam waist location.
- The imaginary part, b, is called the confocal parameter and is related to the beam waist diameter through Equation 9.21.
- At any location, z, the radius of curvature is given by Equation 9.18. We note that $\rho < 0$ indicates a converging wave, and $\rho > 0$ denotes a diverging wave.
- The beam diameter, d, related to b', is given by Equation 9.19.
- Finally, we note that if any two of the four parameters, d, ρ, z, and d_0 are known, then the others may be calculated.

Take-Away

Complex Radius of Curvature:
The complex radius of curvature provides a compact mathematical description of a Gaussian beam, that allows us to determine parameters and deal with propagation.

Example

Let's consider a Gaussian beam with a diameter of 1 cm from a carbon-dioxide laser at a wavelength of 10.59 μm. The beam waist is at the output mirror of the laser. What is the beam size at 1, 10, and 100 m away?

First, we compute the Rayleigh range,

$$b = \frac{\pi d_0^2}{4\lambda} = \frac{\pi \left(10^{-2}\,\mathrm{m}\right)^2}{4 \times 10.59 \times 10^{-6}\,\mathrm{m}} = 7.42\,\mathrm{m}.$$

Next, we compute the complex radius of curvature at the desired distances (we also include b).

$$q = z + jb$$

and the complex curvature, q^{-1}, from which we obtain

$$b' = -\frac{1}{\Im q^{-1}} \qquad d = \sqrt{\frac{4b'\lambda}{\pi}}$$

and

$$\rho = \frac{1}{\Re q^{-1}}$$

z	1 m	$b = 7.42$ m	10 m	100 m
d	1.01 cm	$\sqrt{2}d_0 = 1.41$ cm	1.68 cm	13.52 cm
ρ	56.0 m	$2b = 14.8$ m	15.5 m	100.55 m

9.2 Gaussian Beam Behavior away from the Waist

We begin our discussion of beam propagation by looking at Equation 9.9, which gives us ρ and b' (or d), from z and b. We can put these into a somewhat more useful form by considering Equations 9.21 and 9.19 as well:

Eq. 9.9:
$\rho = \frac{z^2+b^2}{z}$
$b' = \frac{z^2+b^2}{b}$.

$$\rho = z + \frac{b^2}{z} \qquad d = d_0\sqrt{1 + \frac{z^2}{b^2}}. \tag{9.23}$$

These two equations are plotted in Figure 9.2. Both z and ρ are normalized to b, and d is normalized to d_0. The dashed and dash-dot lines show the near- and far-field approximations respectively. In the near field the diameter d_g, is

$$d_g \approx d_0 \qquad \rho \approx b^2/z \to \infty, \tag{9.24}$$

and in the far field, the diameter is d_d,

$$d_d \approx \frac{4}{\pi}\frac{\lambda}{d_0}z \qquad \rho \approx z \to \infty, \tag{9.25}$$

where the subscripts, d and g, will be explained shortly. We note that the far-field approximation for d is the same as we found in Fraunhofer diffraction, in Equation 8.77.

Eq. 8.77:
$d = \left(\frac{4}{\pi}\right)\left(\frac{\lambda}{d_0}\right) z_1$.

The beam diameter in Equation 9.23 has a very simple interpretation. The near-field equation (9.24) predicts that the diameter of the beam will be constant as we would expect from geometric optics. For this reason, we gave it the designation d_g. The far-field diameter in Equation 9.25, is the diffraction limit, d_d. We note that the diameter at any distance, z, is given by

$$d^2 = d_g^2 + d_d^2, \tag{9.26}$$

so the area of the beam is the sum of the area determined geometrically and the area obtained from the diffraction limit. This may be a useful aid to those who use the equations frequently enough to attempt to commit them to memory. We have derived this equation for a collimated Gaussian beam, but we will see in Section 9.3.2 that it is generally valid for Gaussian beams.

Combining the geometric beam diameter and diffraction-limit may be easier to remember than the other equations.

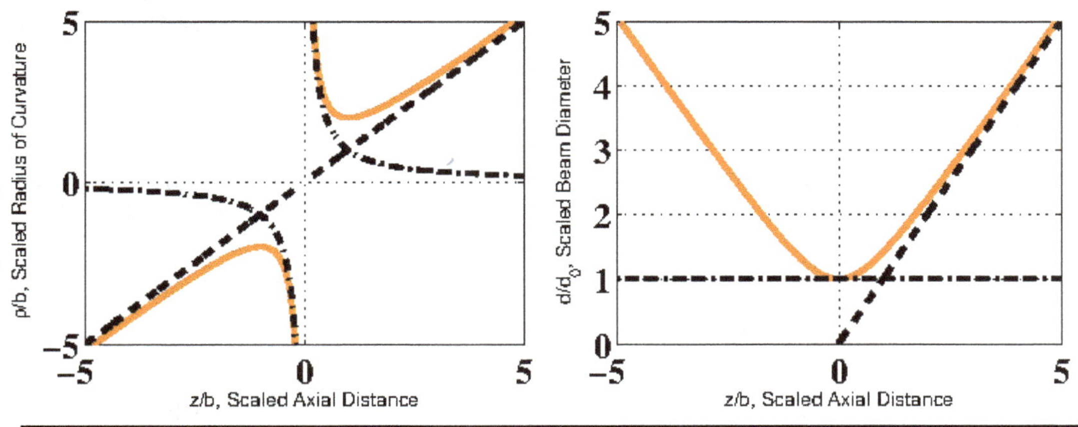

FIGURE 9.2
Gaussian beam propagation. The plot on the left shows the radius of curvature, ρ, scaled by the confocal parameter, b, as a function of the axial distance, z, similarly scaled. On the right, the beam diameter, d, is shown scaled by the waist diameter, d_0.

Take-Away

Propagation Equations:
Gaussian beam propagation from the waist can be described easily using the complex radius of curvature, or equations derived from it.

In Practice

In many problems we can use the simpler near–field or far–field equations, as good approximations. Once we know b, we can decide whether either of these approximations is appropriate.

9.3 Six Questions

As mentioned above, if any two of the four parameters are known, the others can be calculated. We can thus ask any of six questions, depending on which two parameters are known:

z and b in Section 9.3.1, ρ and b' in Section 9.3.2,

z and b' in Section 9.3.3, z and ρ in Section 9.3.5,

ρ and b in Section 9.3.4, b and b' in Section 9.3.6.

Each of these is representative of a class of problems commonly encountered in applications of Gaussian beams. We will explore each of them, along with some numerical examples. There are many more questions that can be asked, and all can be considered using the same basic approach. For example, we might be given the radii of curvature at two locations, ρ_1 and ρ_2 and the distance between these locations, $z_2 - z_1$. However, these six examples are encountered frequently in typical problems, and so we will consider them here.

In each section, we will show a small figure in the margin. The colored lines or markers represent the knowns. For example, in Section 9.3.1, the colored vertical line indicates that z is known and the two small colored arrows indicate that the waist diameter, d_0 (and thus b) are known. The unknowns are shown in black.

9.3.1 Known z and b

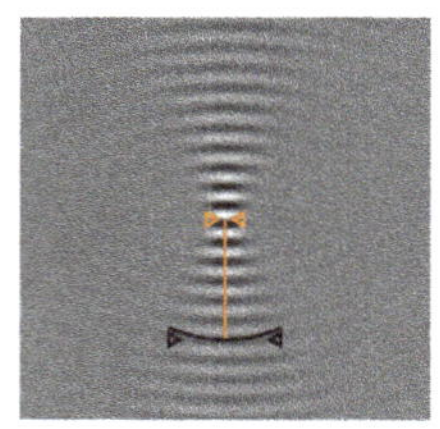

Knowns in color.
Unknowns in black.

First, what are the radius of curvature and beam diameter of a Gaussian beam at distance z from a waist of diameter d_0? Here, z and b are known, and we want to find ρ and b' (and thus d). This is the problem we just addressed with Equation 9.23.

Eq. 9.23:
$\rho = z + \frac{b^2}{z}$
$d = d_0\sqrt{1 + \frac{z^2}{b^2}}$.

Example

As a numerical example, if a green laser pointer (frequency-doubled Nd:YAG laser with $\lambda = 532\,\text{nm}$) has a waist diameter at its output of $2\,\text{mm}$, what are the beam

TABLE 9.1
Diffracction Results from Example. (The last column is the far-field prediction).

Across the room, $z = 3\,\text{m}$:	$\rho = 14.6\,\text{m}$	$d = 2.2\,\text{mm}$	$(1\,\text{mm})$
Across the street, $z = 30\,\text{m}$:	$\rho = 31.2\,\text{m}$	$d = 10.4\,\text{mm}$	$(10.2\,\text{mm})$
To the moon $z = 376{,}000\,\text{km}$:	$\rho = 376{,}000\,\text{km}$	$d = 127\,\text{km}$	$(127\,\text{km})$

diameter and radius of curvature if this laser is pointed across the room, across the street, or at the moon? We first compute

$$b = \frac{\pi d_0^2}{4\lambda} = 5.91\,\text{m}.$$

We then compute

$$\rho = z + \frac{b^2}{z} \qquad d = d_0\sqrt{1 + \frac{z^2}{b^2}}$$

with the results shown in Table 9.1.

The far-field equation is, as expected, almost exact for the longest distance, and is surprisingly good for the intermediate distance, which is about five times b. In both cases, $\rho \approx z$ and $d \approx (4\lambda z)/(\pi d_0)$. Geometric optics predicts that the beam diameter will remain at 2 mm and the radius of curvature will be infinite, which is approximately true across the room. At 3 m, the radius of curvature is 14.6 m, which is nearly planar, and the beam diameter is only 10% larger than it was at the laser. Geometric optics works well in the near field.

9.3.2 Known ρ and b'

Now suppose we know the local beam diameter and radius of curvature. Where is the waist and how large is it? This question arises when we are given a Gaussian beam of a particular diameter, focused by a lens to distance f. According to geometric optics, the light will focus to a point at a distance f away. For the Gaussian beam, considering diffraction, we compute $1/q$ using

$$\rho = -f \tag{9.27}$$

and

$$b' = \pi d^2/(4\lambda). \tag{9.28}$$

Then we take the inverse to obtain q:

$$\frac{1}{q} = \frac{1}{\rho} - \frac{j}{b'}$$

$$q = \frac{\frac{1}{\rho}}{\frac{1}{\rho^2} + \frac{1}{b'^2}} + \frac{\frac{j}{b'}}{\frac{1}{\rho^2} + \frac{1}{b'^2}},$$

Knowns in color.
Unknowns in black.

and

$$z = \Re q = -\frac{f}{1 + \left(\frac{4\lambda f}{\pi d^2}\right)^2}. \tag{9.29}$$

Recalling $\rho = -f$ from Equation 9.27,

$$b = \Im q = \frac{b'}{\left(\frac{b'}{f}\right) + 1} \qquad d_0^2 = \frac{d^2}{1 + \left(\frac{\pi d^2}{4\lambda f}\right)^2}. \tag{9.30}$$

For a beam focused in the near field of the beam diameter, d, so that $f \ll \pi d^2/(4\lambda)$, then $z \approx -f$, which means that the beam is converging toward the focus, almost as predicted by geometric optics, and the spot size is close to $d_0 \approx 4\lambda/(\pi d^2)$ as predicted by Fraunhofer diffraction. In fact, the waist is slightly closer than the geometric focus, f, because $|z| < |\rho|$. If we continue to focus further away, the waist begins to move back toward the source. As the focus becomes longer, the wavefront is becoming more plane, and in the limit of $f \to \infty$ the waist is at the source.

Figure 9.3 shows the results for a laser radar operating on the carbon dioxide laser's P(20) line at $\lambda = 10.59\,\mu\text{m}$. The beam is expanded to a 30 cm diameter, and we neglect truncation by the finite aperture of the expanding telescope. We want this laser radar to focus in order to obtain information from backscattered light at a specific distance, so we focus the beam to a distance f. Figure 9.3A shows that we can achieve a focus near the location predicted by geometric optics in the near field ($\ll b'$) of the 30 cm beam diameter. The vertical and horizontal dash-dot lines in this panel show that we can achieve a waist at a maximum distance of $b'/2$ when we focus at $f = b'$. Beyond that, the outgoing beam is nearly collimated, which means that the waist is close to the source.

Focusing in the near field, the waist will be near the geometric focus and will have a diameter near the diffraction limit.

We note that there are two ways to achieve a waist at a given distance, one with a small beam diameter when we focus as in geometric optics, and another way with a much larger diameter, when the beam is nearly collimated. The waist diameter is close to the diffraction limit in the near field and approaches the initial diameter when the beam is collimated (Figure 9.3B).

We cannot make a waist beyond $b/2$.

There are two ways to generate a waist at a given distance, with different radii of curvature. See Figure 9.3A.

Once we have the waist location and size, we know everything about the beam at every location. For example, if we know for $f = f_1$, that the distance to the waist is $-z = z_1$, and that the waist diameter is $d_0 = d_{01}$, and $b = b_1 = \pi d_{01}^2/(4\lambda)$ then we can compute the beam diameter everywhere as

$$d(z') = d_{01}\sqrt{1 + \left(\frac{z' - z_1}{b_1}\right)^2}. \tag{9.31}$$

One example is shown in Figure 9.3C, for $f = 1\,\text{km}$.

Next, we compute the axial irradiance, which is a measure of our ability to focus the beam. The axial irradiance from Equation 9.17 is given by

A large depth of field is desired for some applications, but a small one is desirable when focusing is used to determine depth.

$$I = \frac{2P}{\pi\frac{d^2}{4}} = \frac{2P\frac{4}{\pi d_{01}^2}}{1 + \left(\frac{z'-z_1}{b_1}\right)^2} \tag{9.32}$$

with the diameter given by Equation 9.31. Equation 9.32 is a Lorentzian function of z' with a full-width at half-maximum (FWHM) of $2b_1$, centered at z_1 is plotted for three values of f in Figure 9.3D. At 200 m, the focusing is quite good. We can describe the focusing ability in terms of a depth of field equal to the FWHM, where the two terms in the denominator are equal, and thus the area of the beam is doubled. If we define the **depth of field** or **range resolution** of a laser radar as the FWHM, then after a moderate amount of rearrangement,

$$z_c = 2b_1 = \frac{8f^2\lambda}{\pi d^2}\frac{1}{\left(\frac{4\lambda f}{\pi d^2}\right)^2 + 1}. \tag{9.33}$$

The range resolution in the near field is proportional to the wavelength and the square of the f-number, as one would expect.

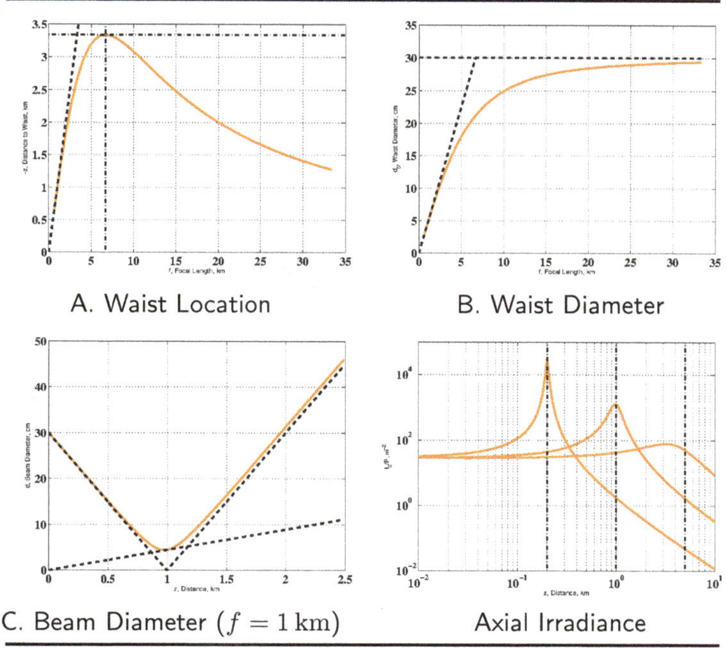

FIGURE 9.3
Focused laser radar beam. The $30\,\mathrm{cm}$ diameter beam is focused to a distance f, and produces a waist at the distance, $-z$ (A). The diagonal dashed line shows $-z = f$, and the dot-dash lines show where the waist distance is a maximum $-z = b/2$ at $f = b$. The beam diameter increases as the waist moves out and then back in again (B). The inclined dashed line shows the diffraction limit, and the horizontal line shows the initial beam diameter. Strong focusing is shown by the rapid increase in beam diameter away from the focal point at $f = 1\,\mathrm{km}$ (C). The axial irradiance for beams focused at $200\,\mathrm{m}$, $1\,\mathrm{km}$, and $5\,\mathrm{km}$ shows a decreasing ability to focus (D). The far-field distance is $6.6\,\mathrm{km}$.

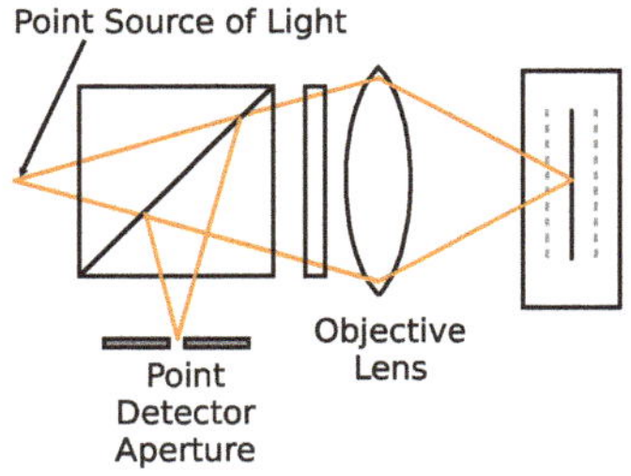

In Practice

For some applications we want a large depth of field. For example in Optical Coherence Tomography which we will discuss in Section 10.3.4, we want to capture signal through as large a depth of field as possible without having to refocus. On the other hand, applications that use focusing for axial resolution require small depth of field. Confocal microscopy (Figure 1.3) and CW laser radar (Section 7.2.4) are two examples. Thus we use a low numerical aperture for optical coherence tomography and a high one for confocal microscopy or CW laser radar.

Fig. 1.3:

Example

How does the depth of field depend on numerical aperture for a confocal microscope and for optical coherence tomography?

For a confocal microscope, we want the smallest possible focused spot and the smallest depth of field. The spot will be hundreds of micrometers from the objective, which will have a pupil many millimeters in diameter, so we are in the near field. Thus we can use the diffraction-limited equations. The spot size from Equation 9.25 with $z = f$ is

$$d_d \approx \frac{4}{\pi}\frac{\lambda}{d_0}f$$

and the depth of field with small f in Equation 9.33 is

$$z_c = \frac{8f^2\lambda}{\pi d^2}.$$

Both decrease as f/d decreases, so we want to make this number as small as possible. If we approximate $NA = d/(2f)$,

$$d_d \approx \frac{2}{\pi}\frac{\lambda}{NA}$$

and the depth of field with small f in Equation 9.33 is

$$z_c = \frac{2\lambda}{\pi NA^2}.$$

With $NA \approx 1$, the Gaussian beam diameter and the depth of field both approach $2\lambda/\pi$. Remember that these are defined differently ($1/e^2$ for the beam diameter, and $FWHM$ for the depth of field), and the results do not account for the size of the aperture.

For optical coherence tomography, depth is resolved by temporal measurements, so we measure all depths without refocusing. Thus we want a large depth of field, but we still want a small spot size for good transverse resolution. These goals are contradictory and we need to compromise. If we let $NA = 0.05$, then

$$d_d \approx \frac{2}{\pi}\lambda \times 20 = 11.5\,\mu\text{m}$$

and the depth of field with small f in Equation 9.33 is

$$z_c = \frac{2\lambda}{\pi} \times 400 = 229\,\mu\text{m}.$$

with the wavelength set to 900 nm. Thus we can image over about a millimeter in depth with only a factor of 2 loss, and even a bit further if we accept a greater loss, and of course a degradation of transverse resolution.

We note that with some effort at rearranging terms, we can express the diameter at any distance, as given by Equation 9.31, according to an equation identical to Equation 9.26,

$$d^2 = d_g^2 + d_d^2.$$

Here we want to put the waist at a given distance.

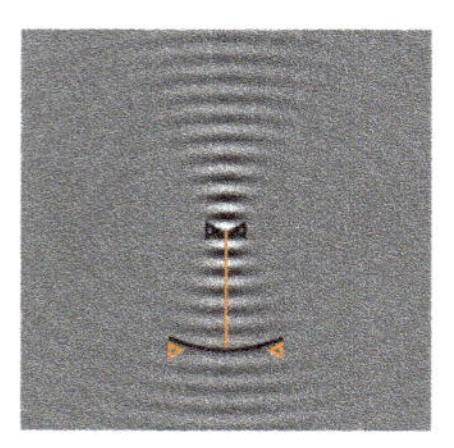

Knowns in color. Unknowns in black.

This time, the geometric-optics beam diameter is computed from similar triangles in Figure 9.4. The equations are

$$d_g = d\frac{|z-f|}{f} \qquad \text{and} \qquad d_d = \frac{4}{\pi}\frac{\lambda}{d}z. \tag{9.34}$$

Thus, once again, the area of the beam is the sum of the areas predicted by geometric optics and the diffraction limit, as shown in Figure 9.3C. In particular, in the far field of a collimated beam ($f \to \infty$, $z \gg b$) or for a beam focused in the near field ($z = f$), the waist is approximately at the focus and has a diameter,

$$d_0 \approx d_d = \frac{4}{\pi}\frac{\lambda}{d}z \qquad \text{(Fraunhofer zone)} \tag{9.35}$$

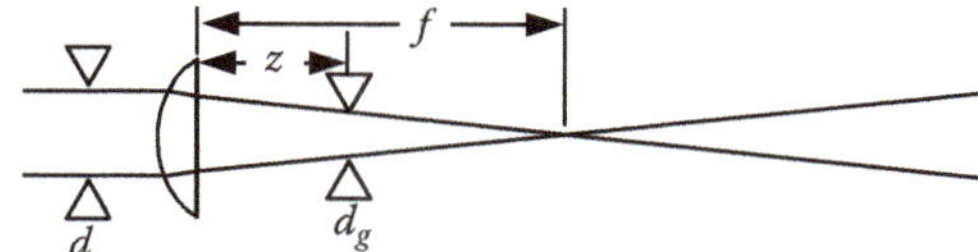

FIGURE 9.4
Geometric-optics beam diameter. We can compute the beam diameter according to geometric optics using similar triangles, where $d_g/|z-f| = d/f$.

9.3.3 Known z and b'

In the previous question, we focused a Gaussian beam of a given diameter to a given geometric focus and asked the location of the waist and its diameter. This time, we still have a known beam diameter, but now our goal is to achieve a waist at a specific distance, and we want to know what lens to use and how large the waist will be. We decide to solve the second part of Equation 9.9 for b:

$$b' = \frac{b^2+z^2}{b} \qquad b = \frac{b' \pm \sqrt{b'^2 - 4z^2}}{2},$$

which will have a solution (or two) provided that $b' > 2z$. In fact, from the previous section, and Figure 9.3, we know that we cannot generate a waist at a distance beyond half the confocal parameter, b' of the beam diameter, d. Once we have solved this equation, we can then determine ρ, which will be negative in this case, and the required focal length, $f = -\rho$.

There will usually be two solutions to this problem, one with a "tight" focus and one with a "weak" focus.

Example

For example, suppose that for an endoscopic application using a needle, we want to keep the beam diameter no larger than 100 μm. We want to focus an Nd:YAG laser beam with wavelength $\lambda = 1.06$ μm to produce a waist at a distance of 1 mm. We compute

$$b' = \frac{\pi d^2}{4\lambda} = 7.4\,\text{mm} \qquad z = -1\,\text{mm}$$

$$b = 3.7047\,\text{mm} \pm 3.5672\,\text{mm}.$$

Thus, we can have either a tightly focused beam or one that is barely focused at all:

$$b = 137\,\mu\text{m} \quad d_0 = 13.6\,\mu\text{m}$$

or

$$b = 7.27\,\text{mm} \quad d_0 = 99\,\mu\text{m}.$$

Almost certainly we will choose the first of these. Noting that we are well within the near field as defined by b', we expect $\rho \approx z = -1$ mm, and calculation shows that the magnitude of ρ is in fact only about 2% higher than that of z. Thus, a lens of focal length $f = -\rho = 1.02$ mm will be appropriate. In all probability, we would not be able to see the difference caused by this 2%.

9.3.4 Known ρ and b

Next, we consider a more unusual case. We know the radius of curvature of the wavefront, and the desired waist diameter. How far away must the waist be located? Later we will see

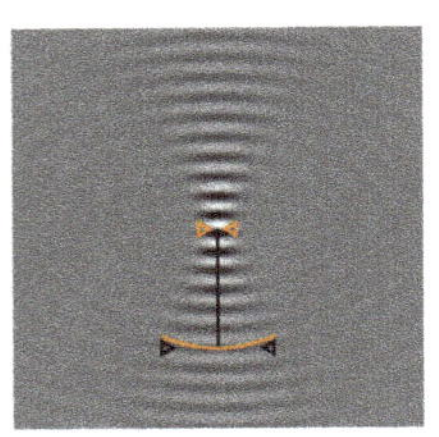

Knowns in color.
Unknowns in black.
Figure 9.2:

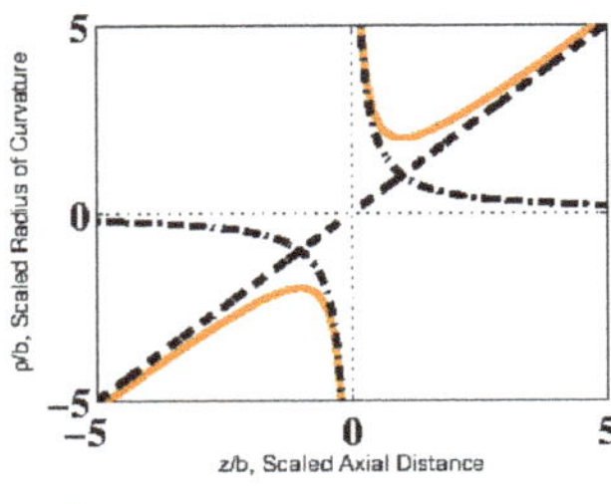

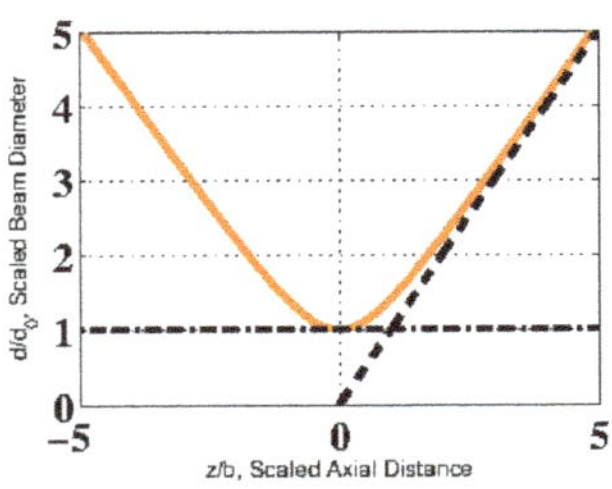

Eq. 9.23:
$\rho = z + \frac{b^2}{z}$
$d = d_0\sqrt{1 + \frac{z^2}{b^2}}$.

that this question is relevant to the design of a laser cavity. If the back mirror in a laser has curvature ρ and the output mirror is flat, placing the mirrors this far apart will cause the laser to produce a Gaussian beam waist of the desired size. As in all these questions, our first goal is to completely define q. We have b, so we need z. Returning to the first part of Equation 9.9,

$$\rho = \frac{z^2 + b^2}{z} \qquad z = \frac{\rho \pm \sqrt{\rho^2 - 4b^2}}{2},$$

which will have solutions provided that $|\rho| > 2b$ which makes sense from Figure 9.2, and Equation 9.23. The radius of curvature can never be less than $2b$. We note that there are two values of z which can produce the same beam diameter. Figure 9.2 shows that indeed, for a given waist diameter, there are two distances that produce the same radius of curvature, although they do so with different beam diameters. We calculate the beam diameter, d in each case:

$$d = d_0\sqrt{1 + \frac{z^2}{b^2}} = d_0\sqrt{1 + \frac{\rho^2 - 2b^2 \pm 2\rho\sqrt{\rho^2 - 4b^2}}{2b^2}}.$$

We will defer a numerical example until the discussion of laser cavities in Section 9.6.

9.3.5 Known z and ρ

The next example is also useful for the design of laser cavities. In this case, we know the distance, z, and the radius of curvature, and the goal is to find the waist diameter. This one is easy:

$$\rho = \frac{z^2 + b^2}{z} \qquad b = \sqrt{\rho z - z^2} \qquad b = |z|\sqrt{\frac{\rho}{z} - 1}.$$

This equation has a solution whenever ρ/z is greater than 1, which means that ρ and z must have the same sign, and that ρ must have the larger magnitude. These requirements, again, are not surprising, in light of Figure 9.2 and Equation 9.23. We only take the positive square root, because b is required by definition to be positive, and there is at most one solution. Again, we defer a numerical example to the discussion of laser cavities.

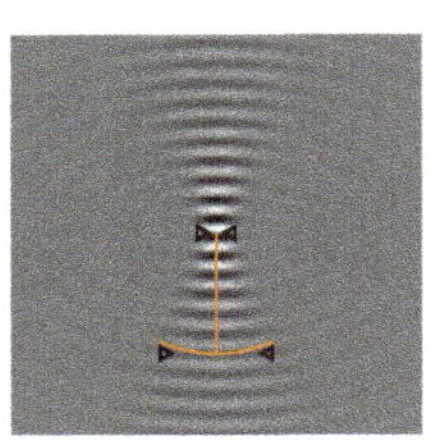

Knowns in color.
Unknowns in black.
There may or may not be a solution to place a waist at a given distance starting with a known radius of curvature.

9.3.6 Known b and b'

Our sixth question is to determine z and ρ when and the diameters at z and at the waist are both known. For example, suppose that we have light from a semiconductor laser with a wavelength of 1.5 μm, with a beam diameter of $d = 15$ μm. We wish to reduce the diameter to $d_0 = 5$ μm to launch into an optical fiber. We solve this problem by first finding z:

$$b' = \frac{z^2 + b^2}{b} \qquad z = \pm\sqrt{bb' - b^2},$$

which has solutions when $b' > b$, or $d > d_0$. This condition is necessary because d_0 is the minimum value of d. There are two solutions, depending on whether we wish to converge the beam toward a waist as in the present example with $z < 0$ or to diverge it with $z > 0$.

Example

In our numerical example, with $\lambda = 1.5$ μm, $d = 15$ μm, and desired $d_0 = 5$ μm,

$$b = \frac{\pi d_0^2}{4\lambda} = 13\,\mu\text{m} \qquad b' = \frac{\pi d^2}{4\lambda} = 118\,\mu\text{m}.$$

Solving for z, knowing we want the negative solution,

$$z = -\sqrt{bb' - b^2} = -37\,\mu\text{m},$$

and the radius of curvature is

$$\rho = z + \frac{b^2}{z} = -41.7\,\mu\text{m}.$$

This means that we would need a lens of focal length, $f = -\rho = 41.7\,\mu\text{m}$ with a diameter greater than $15\,\mu\text{m}$, which is about an $f/3$ lens, so some attention to aberrations would be required.

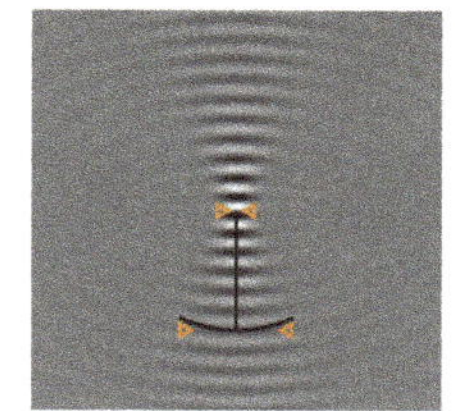

Knowns in color.
Unknowns in black.

In Practice

Astigmatism and aberrations will be challenging because most semiconductor lasers have large numerical apertures that are different in the two transverse directions, x and y. Design of a well-corrected lens is challenging.

Take-Away

Gaussian Beam Characteristics:

- The complex radius of curvature can be manipulated to solve problems of Gaussian beam propagation.
- The complex radius of curvature and its inverse contain four terms, which are related in such a way that only two of them are independent.
- Given two parameters it is possible to characterize the Gaussian beam completely. Six examples have been discussed.
- Solutions are at worst quadratic, producing zero, one, or two solutions.
- Other, more complicated problems can also be posed and solved using this formulation.

9.4 Gaussian Beam Propagation

Next, we examine the changes that occur in Gaussian beams as they propagate through optical systems. We will discuss the propagation problems in terms of the complex radius of curvature.

9.4.1 Free Space Propagation

Propagation through free space is particularly easy. The complex radius of curvature is $q = z + jb$ so as we move along the beam, we simply add the distances, as real numbers, to q:

For translation through a distance δz, add δz to q.

$$q(z_2) = q(z_1) + z_2 - z_1. \tag{9.36}$$

9.4.2 Propagation through a Lens

Eq. 2.51:
$\frac{1}{s} + \frac{1}{s'} = \frac{1}{f}$.

For focusing, subtract the optical power, $1/f$, from the complex curvature.

Focusing a Gaussian beam through a lens is also simple. Considering a simple, perfect lens, the beam diameter (and thus b') stays the same, but the radius of curvature changes. Using the lens equation (Equation 2.51), we relate the curved wavefronts to object and image locations. The main issue is to interpret the signs correctly. The radius of curvature, ρ, is positive for a diverging beam. The object distance is defined as positive to the left of the lens, so this is consistent with a positive radius of curvature, ρ. On the other hand, the image distance is defined as positive to the right, so a positive image distance corresponds to a negative radius of curvature, ρ'.

$$\frac{1}{\rho} + \frac{1}{-\rho'} = \frac{1}{f} \qquad \frac{1}{\rho'} = \frac{1}{\rho} - \frac{1}{f}$$

$$\frac{1}{q'} = \frac{1}{q} - \frac{1}{f}. \tag{9.37}$$

Thus, a lens simply subtracts its optical power from the inverse of q.

9.4.3 Propagation Using Matrix Optics

The ABCD matrix of a system can be used to transform the complex radius of curvature.

We recall that any lens system can be represented by a single $ABCD$ matrix. We can determine the behavior of a Gaussian beam going through any such system by using

$$q_{out} = \frac{Aq_{in} + B}{Cq_{in} + D}. \tag{9.38}$$

Eq. 3.10 (Translation):
$\mathcal{T}_{12} = \begin{pmatrix} 1 & z_{12} \\ 0 & 1 \end{pmatrix}$.

Let us examine some simple cases. We start with translation through a distance z_{12}, given by Equation 3.10. Then Equation 9.38 becomes

$$q_2 = \frac{q_1 + z_{12}}{0 + 1} = q_1 + z_{12}, \tag{9.39}$$

in agreement with Equation 9.36. For a lens, using Equation 3.25 we find

$$q' = \frac{q + 0}{-\frac{P}{n'}q + \frac{n}{n'}}$$
$$\frac{1}{q'} = \frac{n}{qn'} - \frac{P}{n'}. \tag{9.40}$$

Eq. 3.25 (Lens) :
$\mathcal{L} = \begin{pmatrix} 1 & 0 \\ -\frac{P}{n'} & \frac{n}{n'} \end{pmatrix}$.

If the lens is in air, this simplifies to

$$\frac{1}{q'} = \frac{1}{q} - P = \frac{1}{q} - \frac{1}{f},$$

Eq. 3.17 (Refraction):
$\mathcal{R} = \begin{pmatrix} 1 & 0 \\ \frac{n-n'}{n'r} & \frac{n}{n'} \end{pmatrix}$.

as we obtained in Equation 9.37

For refraction at a single surface (see Equation 3.17),

$$q' = \frac{q + 0}{\frac{n-n'}{n'R}q + \frac{n}{n'}}$$
$$\frac{1}{q'} = \frac{n - n'}{n'R} + \frac{n}{qn'}. \tag{9.41}$$

Considering the special case of a plane dielectric interface, $R \to \infty$,

$$q' = q\frac{n'}{n}.$$

Simply, q is scaled by the ratio of the indices of refraction, becoming larger if the second index of refraction is larger. One would expect this, as b is inversely proportional to λ. In a medium other than vacuum, the correct formulation for b, assuming λ is the free-space wavelength, is

$$b = \frac{\pi d_0^2}{4\lambda/n}. \tag{9.42}$$

9.4.4 Propagation Example

Let us consider propagation from the front focal plane to the back focal plane of a lens of focal length f_1. Let us start with a plane Gaussian wave at the front focal plane, which means that $z = 0$. Using subscripts $0, 1, 2$ for the front focal plane, the lens, and the back focal plane respectively, and starting with the waist at the front focal plane,

$$q_0 = jb_0 \qquad b_0 = \frac{\pi d_0^2}{4\lambda},$$

the three steps are

$$q_1 = q_0 + f_1 \qquad \frac{1}{q_1'} = \frac{1}{q_1} - \frac{1}{f_1} \qquad q_2 = q_1' + f_1,$$

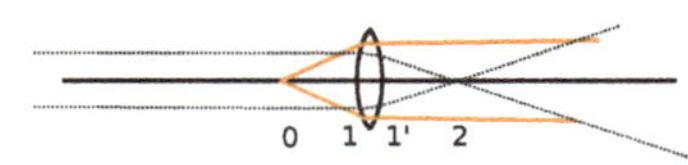

with the solutions

$$q_1 = jb_0 + f_1 \qquad q_1' = -f_1 + j\frac{f_1^2}{b_0} \qquad q_2 = \frac{jf_1^2}{b_0}. \tag{9.43}$$

We note that q_2 is pure imaginary, meaning that this is a beam waist. Furthermore, we note that the beam diameter at this waist is

$$d_2 = \frac{4}{\pi}\frac{\lambda}{d_0}f_1. \tag{9.44}$$

In Chapter 11, we will see that the field at the back focal plane of a lens is the Fourier transform of the field at the front focal plane. The inverse relationship between d and d_0 is consistent with this result. The Gaussian beam diameter is shown as a function of z in Figure 9.5 along with colored geometric rays that illustrate the imaging of a point at the front focal plane. We note that this is not an imaging system. By the lens equation (2.51), the image is at infinity.

Eq. 2.51: $\frac{1}{s} + \frac{1}{s'} = \frac{1}{f}$.

Let us extend this analysis through a second lens. We have used the subscript 2 to refer to the back focal plane of the lens with focal length f_1, so let us call the focal length of the second lens f_3. Then by repeating the analysis above, and changing subscripts we know that the back focal plane of the second lens (subscript 4) will also be a waist, and by analogy with Equation 9.44,

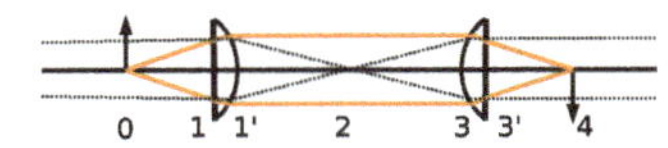

$$d_4 = \frac{4}{\pi}\frac{\lambda}{d_2}f_3 = \frac{f_3}{f_1}d_0. \tag{9.45}$$

This process can be extended any number of times, using pairs of lenses separated by their focal lengths. The Gaussian beam will have a waist at each focal plane, and each waist will

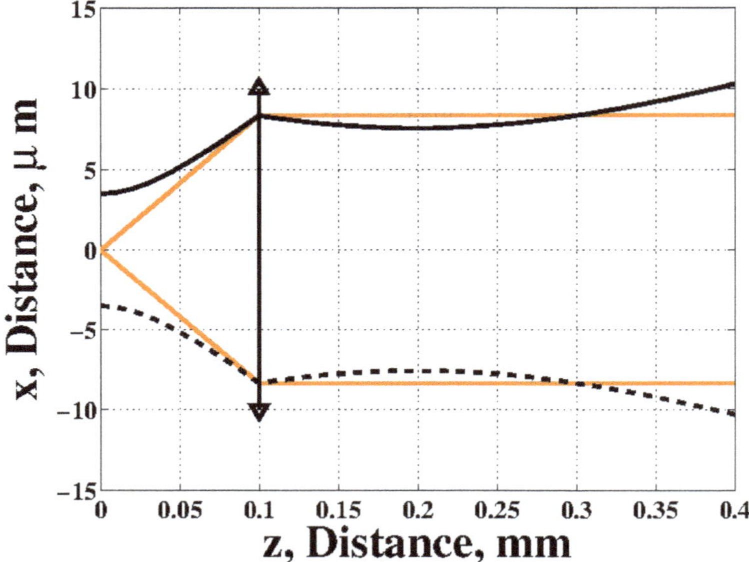

FIGURE 9.5
A Gaussian beam with a waist at the front focal plane of a lens leads to a waist at the back focal plane. The waist diameters are inversely related. This is an example of "weak focusing." With a shorter focal length, we could have achieved diffraction-limited waist diameter. The colored lines show geometric rays and the black lines show the Gaussian beam.

be connected to the next one through a Fourier transform (or inverse Fourier transform). We can treat the waists at locations $0, 4, \ldots$ as image planes and the intervening ones as pupils, $1, 3, \ldots$. Thus Gaussian beams are easily treated in the context of these telecentric optical systems.

To test our results here, let us consider the matrix approach. For the single lens, f_1, we use matrices, $\mathcal{T}_{01}$ (Equation 3.10) for translation from the front focus to the lens, $\mathcal{L}_1$ (Equation 3.25) for the focusing of the lens, and $\mathcal{T}_{12}$ for translation to the back focal plane. The final matrix is

$$\mathcal{M}_{02} = \mathcal{T}_{12}\mathcal{L}_1\mathcal{T}_{01} = \begin{pmatrix} 1 & f_1 \\ 0 & 1 \end{pmatrix} \begin{pmatrix} 1 & 0 \\ -\frac{1}{f_1} & 1 \end{pmatrix} \begin{pmatrix} 1 & f_1 \\ 0 & 1 \end{pmatrix} = \begin{pmatrix} 0 & f_1 \\ -\frac{1}{f_1} & 0 \end{pmatrix}. \tag{9.46}$$

Using Equation 9.38, we find

$$q_2 = \frac{Aq_0 + B}{Cq_0 + q_0} = \frac{0 + f}{-\frac{1}{f}q_0 + 0} = -\frac{f^2}{q_0} \qquad d_2 = \frac{4}{\pi}\frac{\lambda}{d_0}f_1, \tag{9.47}$$

as we obtained in Equation 9.44. The reader may find the negative sign in $q_2 = -\frac{f^2}{q_0}$ troublesome, but it is, in fact, correct. If q_0 has a real component, and is thus not at a waist, the sign of the real part determines the sign of the curvature, and the curvature represented by q_2 has the opposite sign. The imaginary part of q must be always be positive. We note here that the inverse of q_2 has a negative imaginary part, so $\Im q_2 > 0$, as required.

Now considering the complete relay system of two lenses, if we set $f_3 = f_1$,

$$\mathcal{M}_{04} = \mathcal{M}_{24}\mathcal{M}_{02} = \begin{pmatrix} 0 & f_3 \\ -\frac{1}{f_3} & 0 \end{pmatrix} \begin{pmatrix} 0 & f_1 \\ -\frac{1}{f_1} & 0 \end{pmatrix} = \begin{pmatrix} -1 & 0 \\ 0 & -1 \end{pmatrix}, \tag{9.48}$$

and for the pair of lenses acting as an imaging system,

$$q_4 = \frac{Aq_0 + B}{Cq_0 + D} = \frac{-q_0 + 0}{0 - 1} = q_0. \qquad (9.49)$$

Thus this telecentric 1:1 imaging system preserves the Gaussian wavefront perfectly.

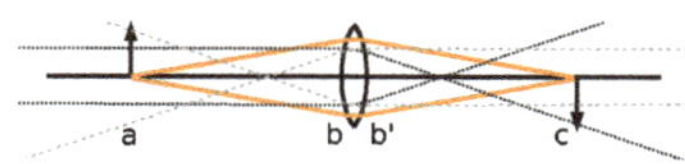

In contrast, let us consider an imaging system consisting of a single lens, $f = f_1/2$, used as an imaging lens, satisfying the lens equation, $1/s + 1/s' = 1/f$. If we consider a 1:1 imaging system, the results should be comparable to the problem above with two lenses, assuming $f_3 = f_1$. Here, we change the subscripts to letters to minimize confusion, and the matrix equation is

$$\mathcal{M}_{ac} = \mathcal{T}_{bc}\mathcal{L}_b\mathcal{T}_{ab} = \begin{pmatrix} 1 & 2f \\ 0 & 1 \end{pmatrix}\begin{pmatrix} 1 & 0 \\ -\frac{1}{f} & 1 \end{pmatrix}\begin{pmatrix} 1 & 2f \\ 0 & 1 \end{pmatrix} = \begin{pmatrix} -1 & 0 \\ -\frac{1}{f} & -1 \end{pmatrix}. \qquad (9.50)$$

The only difference between this equation and Equation 9.48 is the lower left element, representing the optical power. Recall that this element was zero for the pair of lenses working as an afocal relay telescope. The plane a here is equivalent to the plane 0 in the earlier problem, and the plane c is equivalent to 4. Both lenses and the intervening pupil, $1, 2, 3$ are consolidated into a single lens, b. For Gaussian beams, the result is

$$q_c = \frac{Aq_a + B}{Cq_a + D} = \frac{-1q_a + 0}{-\frac{1}{f}q_a - 1} = \frac{q_a}{\frac{q_a}{f} + 1}. \qquad (9.51)$$

We note that although we started with $q_a = q_0$ being pure imaginary and the two-lens relay reproduced it perfectly, the single lens introduces a real term to q_c that did not exist in q_4. Diffraction acting on the Gaussian beam introduces field curvature, and even changes the size of the waist. Figure 9.6 shows the Gaussian beam and the matching geometric ray trace. This single lens imaging system stands in contrast to the non-imaging system in Equation 9.47, where the starting and ending points were at beam waists.

Eq. 9.47: $q_2 = -\frac{f^2}{q_0}$.

Example

Let us consider a numerical example. Suppose we have a waist diameter of 2 μm, at the blue wavelength of the argon ion laser $\lambda = 488$ nm with a pair of lenses of focal length $f_1 = f_3 = 100$ μm, spaced at a distance $f_1 + f_3 = 200$ μm. Then as we have seen,

$$q_4 = q_0 = j\frac{\pi d_0^2}{4\lambda} = j6.438\ \mu\text{m}.$$

For the single lens of $f = f_1/2 = 50$ μm, recalling $q_a = q_0$

$$q_c = \frac{1}{\frac{1}{f} + \frac{1}{q_0}} = (0.815 + j6.333)\ \mu\text{m}, \qquad (9.52)$$

so the waist has been moved 815 nm and the new diameter is $d_4 = 1.984$ μm.

Take-Away

Calculations:

- Once two parameters of a Gaussian beam are known, then its propagation can be analyzed using simple equations for translation, refraction, and focusing.
- Translation changes the beam size and curvature. It is represented by adding the distance to q, as in Equation 9.36.

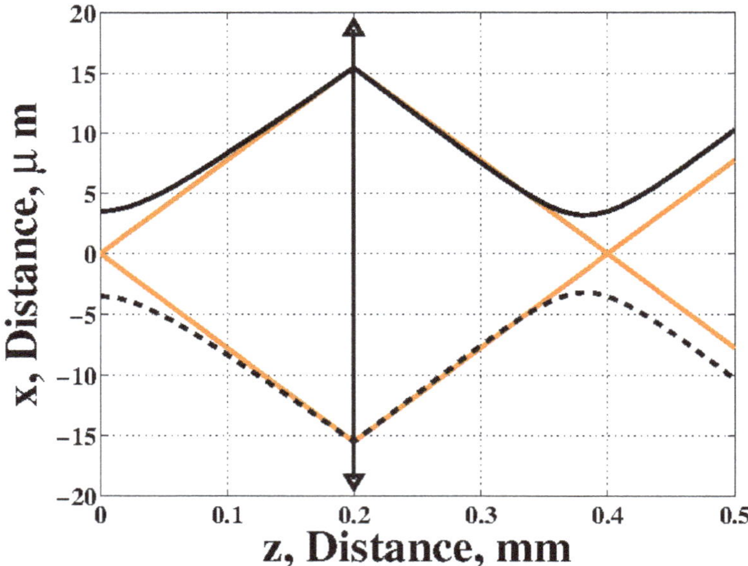

FIGURE 9.6
A Gaussian beam with a waist at the object plane, $s = 2f$ of a lens used as a 1:1 relay leads to a waist slightly inside the image plane, $s' = 2f$. The waist diameters are equal. The beam diameter at the image plane is exactly equal to that in the object plane ($m = -1$, but the wavefront is curved. The colored lines show geometric rays and the black lines show the Gaussian beam.

- Refraction through a lens changes the curvature, but keeps the beam diameter constant. It is represented by subtracting the optical power of the lens from the inverse of q as in Equation 9.37.
- Refraction through a dielectric interface changes the curvature and scale factor of q.
- These equations can be generalized using the $ABCD$ matrices developed for geometric optics, as in Equation 9.38.

9.5 Collins Chart

We have seen that the propagation of Gaussian beams involves mostly adding and subtracting numbers from the complex radius of curvature, $q = z + jb$, and its inverse, $1/q = 1/\rho - j/b'$. Some readers may recognize the mathematical similarity to impedance Z and admittance Y in microwave problems, and perhaps will be familiar with the Smith chart often used in solving such problems. Shortly after the invention of the laser, as interest in Gaussian beams increased, Stuart Collins developed a graphical technique[43] using a chart that bears his name. An example of the Collins chart is shown in Figure 9.7. In principle, Gaussian beam problems can be solved graphically using this chart. In practice, it is normally more convenient now to use computers to do the calculations, but the Collins chart is an excellent visual aid to determine the number of potential solutions to a given problem.

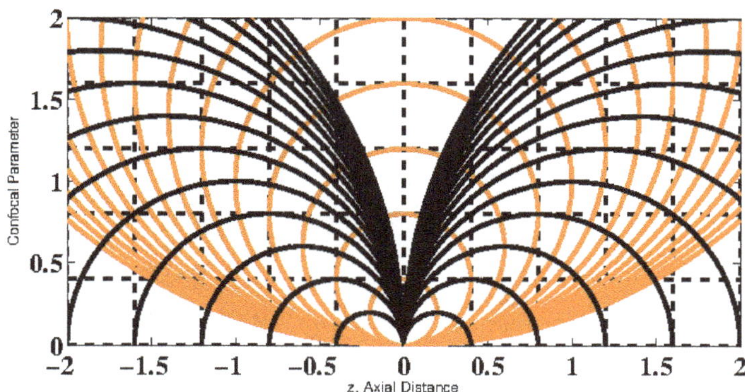

FIGURE 9.7
The Collins chart. The Cartesian grid shows lines of constant z vertically and constant b or d horizontally. The half circles to the left and right are curves of constant ρ. For each curve, the value of ρ is the value of z where the curve intersects the abscissa. The full circles in the center are curves of constant b' or d. The value of b' for each curve is the value of b where the curve intersects the ordinate. Any convenient unit of length can be used.

In Practice

The Collins chart is now seldom used for quantitative design work. It mostly provides a useful way to visualize the number of solutions possible for a given situation.

The plot shows z on the horizontal axis and b on the vertical. Of course, only positive values of b are needed. Moving along the horizontal lines of constant b represents changing the axial position, z, so that the waist diameter d_0 and b do not change. Also shown on the chart are curves of constant b' and constant ρ. The former are circles all passing through the origin. At $z = 0$, $b' = b$, so the curves can be identified easily by the b value at the ordinate. Moving around one of these curves changes ρ while keeping b' and thus d constant. This trajectory represents the focusing power of a lens. Finally, curves of constant ρ are half circles also passing through the origin. At $b' \to \infty$ or $b \to 0$, we know that $\rho \to z$, so the value of ρ for each curve can be identified easily by the value at which it intersects the abscissa. The chart can be generated either showing the real and imaginary parts of q as we have done here, with constant ρ and b' represented by circles, or as Collins did originally, with the coordinates being the real and imaginary parts of $1/q$, with constant z and constant b represented by circles.

Translation through δz is shown on the Collins chart by moving horizontally.

Focusing is represented on the Collins chart by moving along a curve of constant beam diameter (constant b').

Take-Away

Collins Chart:
We can visualize propagation problems on the Collins chart.

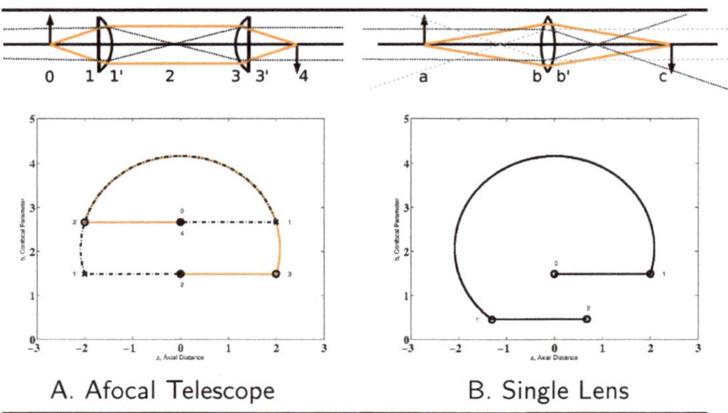

FIGURE 9.8
Lens examples on Collins chart. Propagation through an afocal telescope with unit magnification (A) produces different results from propagation through a single lens (B) with equivalent magnification.

9.5.1 Relay Lenses in the Collins Chart

We consider the examples of our lens systems from the previous section. Figure 9.8A shows our two-lens system, with lenses having a focal length of $f_1 = 2$ units (using arbitrary units here to illustrate the application). We start with a collimated beam with $b = 1.5$, represented by the "o" at Point 0 $(0, 1.5)$ on the plot. We move a distance $f = 2$ to the right in the figure, following the black dashed line to Point 1. Now, just before the lens, we have a radius of curvature of 3.1, and b' has increased to 4.2. Following the curve at $b' = 4.2$, brings us to Point 1' indicated by the circle at $(-2, 2.74)$, with a radius of curvature $\rho = -5.6$. Then, moving a distance $f_1 = 2$ we arrive at Point 2 $(0, 2.74)$. From here, we follow the colored line and points marked with "x," moving to the next lens along the horizontal line at Point 3, then following parts of the same curve as before to Point 3', but then returning to the original point (Point 4 is at Point 0.) at $(0, 1.5)$. The end point is the same as the starting point and the final Gaussian beam is identical to the initial one.

The telecentric relay creates an image beam that is identical to the object one.

We now look at the single lens, shown in the Collins chart in Figure 9.8B. We move to the right exactly as before to Point 1, and start around the circle counterclockwise, but this time, we continue further to Point 1', because the lens has a shorter focal length, $f = f_1/2 = 1$. Now, moving a distance $f_1 = 2$ to the image at Point 2, $(0.68, 0.47)$, we find that the Gaussian beam has a positive curvature, $\rho = 1.00$. The beam size represented by $b' = 1.45$ is not much different from the starting $b = 1.5$, (as can be seen by following the circle of constant b' from this point up to the ordinate) but this is not a waist. In fact, this beam is quite strongly focused, which is indicative of the field curvature problem we discussed in Section 9.4.4. The single-lens relay produces the same Gaussian beam diameter, but a different curvature.

A single 1:1 relay lens preserves the beam diameter but changes the curvature.

9.5.2 Finding Solutions with the Collins Chart

We can use the Collins chart to explore some questions about Gaussian beams. What is the minimum beam size required if a beam is to propagate in free space for a distance, z_{12}? In Figure 9.9A, the length of the line is z_{12}. The smallest b' circle which completely contains it is for $b' = z_{12}$. We could then use this b' to determine the beam diameter. We achieve our goal by placing the waist in the middle of the distance to be traversed. We note that the b' will drop by a factor of 2 at this location, so the minimum diameter is $1/\sqrt{2}$ times the diameter at the ends. This problem is frequently encountered in long

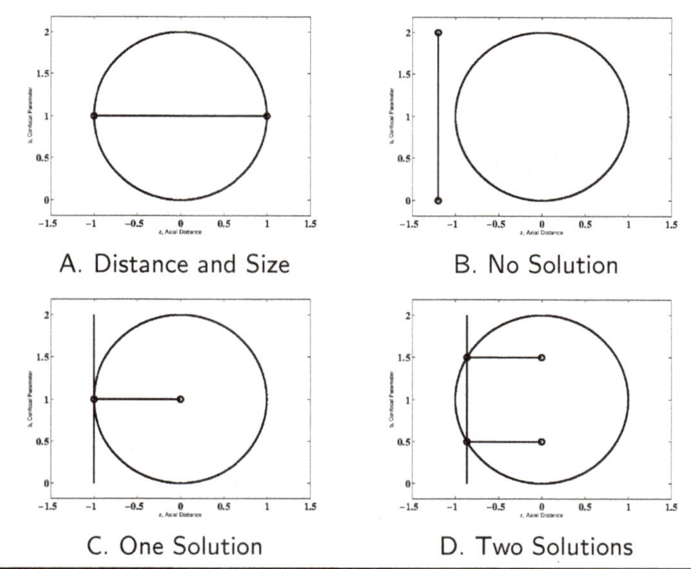

FIGURE 9.9
Collins charts to address common questions. If we wish to ensure that the beam diameter remains less than a prescribed value (e.g., in a laser amplifier tube of a given diameter), then there is an upper limit to the length over which this limit can be maintained (A). The remaining three plots answer the question of how to put a waist at a given distance from a beam of a given diameter. There are, respectively, no solutions (B), one (C), and two (D) solutions, depending on the diameter and distance.

laser amplifiers. We would like to keep the diameter small for cost reasons, but it must be large enough to pass the beam. In practice, we would make the tube at least 1.5 times the largest beam diameter, which occurs at the ends.

Example

We want to build a carbon-dioxide laser amplifier consisting of 6 one-meter glass tubes filled with the appropriate mixture of carbon-dioxide and other gasses. The goal is to amplify a beam of a few watts to a few kilowatts. We know we will need some focusing optics to prepare the beam for the amplifier. What beam diameter and wavefront curvature should we use?

The wavelength is 10.59 µm. If we place the waist at the center of the 6 m path, we want the Rayleigh range to be 3 m. Then the waist diameter is

$$d_0 = \sqrt{4b\lambda}\pi = \sqrt{4 \times 3\,\mathrm{m} \times 10.59 \times 10^{-6}\,\mathrm{m}}\pi = 6.4 \times 10^{-3}\,\mathrm{m}.$$

The beam diameter at the ends of the tube will be

$$d = \sqrt{2}d_0 = 9.0 \times 10^{-3}\,\mathrm{m}.$$

With a Gaussian beam diameter of 9 mm, we would want the glass tube to have an inner diameter at least 1.5 times larger, so perhaps 1.5 cm would be a good choice.

The radius of curvature at $z = b$ is

$$\rho = z + \frac{b^2}{z} = 2b = 6m.$$

Thus we want a very weak focusing lens to focus the beam at the beginning of the amplifier tube.

In Practice

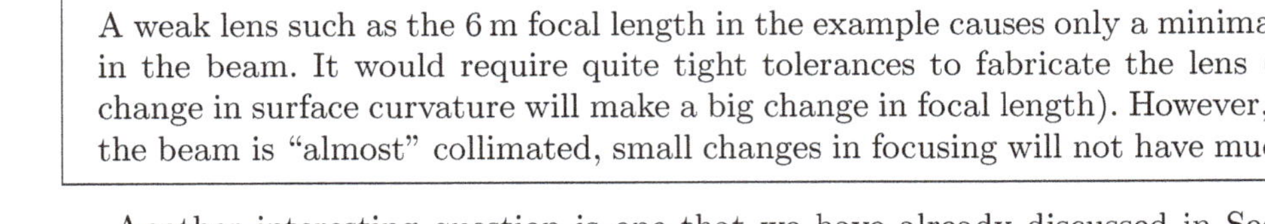

A weak lens such as the 6 m focal length in the example causes only a minimal change in the beam. It would require quite tight tolerances to fabricate the lens (A small change in surface curvature will make a big change in focal length). However, because the beam is "almost" collimated, small changes in focusing will not have much effect.

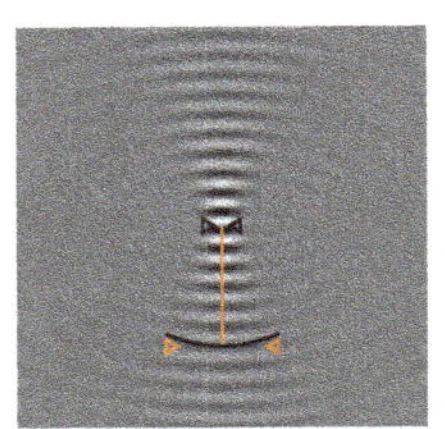

Knowns in color. Unknowns in black.

Another interesting question is one that we have already discussed in Section 9.3.3. Given a beam of known size, and thus b', how do we produce a waist at a given distance? In other words, we want to make the value of z at the current location some given negative number. As we found earlier and as can be seen in the remaining three panels of Figure 9.9, if we are given $z = b'$ there will be exactly one solution (Figure 9.9C). If z is larger than this number, there are no solutions possible (Figure 9.9B). If z is smaller, there are two possible solutions (Figure 9.9D). One is nearly collimated and produces a large waist, the other is tightly focused to a small waist. Looking at the endpoints of one of these lines, one could determine the required radius of curvature. If the wave is initially collimated, one would need to use a lens of this focal length to produce the desired waist. With the computer power available now, one would almost certainly perform the calculations indicated in Section 9.3.3, rather than use a graphical solution. However, the visualization may still be useful to better understand the problem.

9.6 Stable Laser Cavity Design

Fig. 6.11:

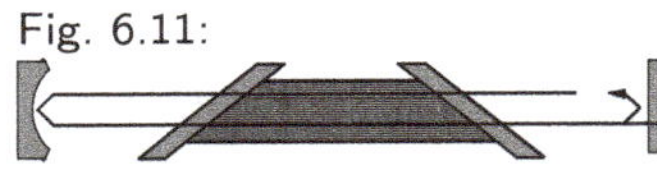

A laser with a flat output coupler was shown in Figure 6.11. At that point, we were concerned with polarization issues in the Brewster plates on the ends of the gain medium. Now we turn our attention to the curvature of the mirrors. We reproduce that laser cavity, along with some others, in Figure 9.10. Using our knowledge of Gaussian beams, we will now be able to design a cavity to produce a specific output beam, or determine the output beam parameters of a given cavity. In steady-state operation of a laser oscillator the field at any position in the cavity is unchanged by one round trip. Thus if the field is influenced by a gain medium having a one-way gain of g, and index of refraction, n, mirrors with reflectivities ρ_1 and ρ_2, and a cavity length of ℓ,

In a stable laser cavity, the round-trip gain is unity.

$$E(x, y, z)\,\rho_1\rho_2 g^2 e^{j2nkz} = E(x, y, z). \tag{9.53}$$

Focusing inside the cavity is required for stability.

This condition had several implications:

- The amplitude after a round trip is unchanged. This means that any loss (including power released as output) must be offset by corresponding gain. A course in laser operation would spend time addressing these topics.
- The phase after a round trip must be unchanged. We discussed, in our study of interference, how this requirement on the axial phase change, e^{jkz}, affects the laser frequency.
- The beam shape must be unchanged, so that the phase and amplitude is the same for all x and y. This is the subject to be considered in this section.

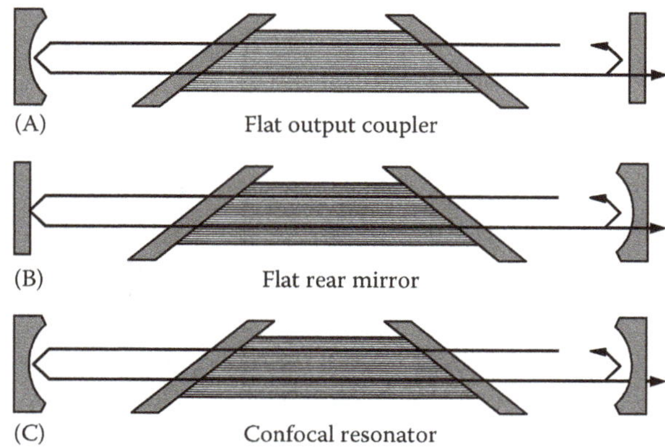

FIGURE 9.10
Some laser cavities. A laser cavity with a flat output coupler (A) produces a collimated beam because the flat mirror enforces a plane wave at this location. A cavity with a flat rear mirror (B) is a good model for one with a Littrow grating (Section 8.8.6) for wavelength selection. A cavity with two concave mirrors (C) has a waist inside the cavity.

In Practice

As a continuous-wave laser begins to operate the field will grow on each round trip. Then the gain will begin to saturate from its "small-signal" value to $1/(\rho_1\rho_2)$, to satisfy the first condition. Of course, if the small-signal gain is less than this value, the laser will not work.

If the laser were built with two flat mirrors then any beam would diverge a little bit on each pass; the beam shape would not match the previous pass, and there would be additional loss as a result. Only a portion of the amplified field would match the shape of the original field. Some focusing power in the cavity is essential to achieve a stable cavity.

If we place a reflector in a Gaussian beam, and we choose the reflector to be everywhere normal to the wavefront, then the wave will be reflected back on itself. This simply requires matching the mirror curvature to that of the wavefront. If we do so at both ends of the laser cavity, then the wave at a particular location will have the same properties, z and b, or ρ and b', after one round trip.

Take-Away

Stable Laser Cavity:
In a stable laser cavity, the beam matches itself after one round trip.

9.6.1 Design Problem: Matching Curvatures

Our goal then in design is to match the mirror curvatures to the wavefront curvatures of a theoretical Gaussian beam. Then when we build the cavity, waves that match that theoretical Gaussian will be amplified to become the laser beam. We first consider a design problem.

Design problem: Find the mirrors that match the desired Gaussian beam.

Example

For this example, we are building a carbon dioxide laser on the P(20) line, with $\lambda = 10.59\,\mu\text{m}$. We desire a 5 mm collimated output, and we need a cavity length of 1 m to achieve sufficient gain. We know immediately that the **output coupler** must be flat because the output is to be collimated. This means that the beam waist is at the output coupler, also called the **front mirror**. This mirror is slightly transmissive to allow some of the power recirculating inside the laser cavity to be released as the output beam. We now need to determine the curvature of the rear mirror. This is just the simple propagation problem of Section 9.3.1 for $z = -1$ m. We calculate the parameters of the beam as $b = 1.85$ m, $\rho = 4.44$ m, and $d = 5.7$ mm. We then need to design the concave back mirror to have a 4.44 m radius of curvature to match ρ. We note that the beam at the back mirror will be slightly larger than at the front.

Examples of matching mirrors are shown in Figure 9.11. The one with the flat output coupler we just discussed is in Figure 9.11A. An alternative configuration, shown in Figures 9.10B and 9.11B, has a flat back mirror and a curved output coupler. The output beam here is diverging, although perhaps only slightly. This configuration is often used when the back mirror is a grating, as discussed in Section 8.8.6. The grating approximates a flat mirror.

Another common configuration for a laser is the confocal cavity, with two curved mirrors, as shown in Figures 9.10C and 9.11C. Designing this cavity follows a similar process to that used for the collimated output. Given a pair of mirrors and a spacing, we can solve for the beam parameters. If there is a unique solution, we refer to the cavity as a **stable cavity** or **stable optical resonator**. Any configuration that matches the wavefronts of a Gaussian beam will be a stable cavity, and will produce the beam for which the cavity is designed. The example Figure 9.11D is still stable. We can use the Collins chart to watch for cavities that are "close to unstable."

The Collins chart can be useful in qualitative analysis of stability.

9.6.2 Analysis of a Laser Cavity

Analysis: Finding the Gaussian beam that fits the cavity as it is built.

Now let us turn to the problem of determining the mode of a given cavity. We will use our example of the carbon dioxide laser above. Figure 9.12 shows this design on the Collins chart. We computed all the relevant parameters above. Now, suppose that we are given the cavity, with a flat output coupler ($z = 0$ at the front mirror, represented by the vertical solid line), and a back mirror with a radius of curvature of 4.44 m, shown by the solid curve, and the distance between them, 1 m, shown by the horizontal solid line with circles at the ends. The three solid lines represent the constraints of the system. We need to find the solution that places the horizontal line so that it "just fits" between the solid curve and the solid vertical line. In this case, we know z and ρ, and we would use the calculations in Section 9.3.5. If $z_{12} > 4.4\ (2\rho)$ in this example, no solution is possible and the cavity is unstable. If $4.4 > z_{12} >> 2.2$, the cavity is stable but the beam diameter is quite small at the flat mirror. The resulting high irradiance leads to excessive saturation and does not take full advantage of the gain medium. In most lasers, We would like the beam diameter to be as constant as possible along the entire length of the gain medium.

In Practice

We want to avoid a tight focus inside a laser cavity. The high irradiance in a small beam would saturate the gain medium at a low gain value. The different cavities shown in Figure 9.11 produce surprisingly similar beams; the beams are nearly collimated in all cases.

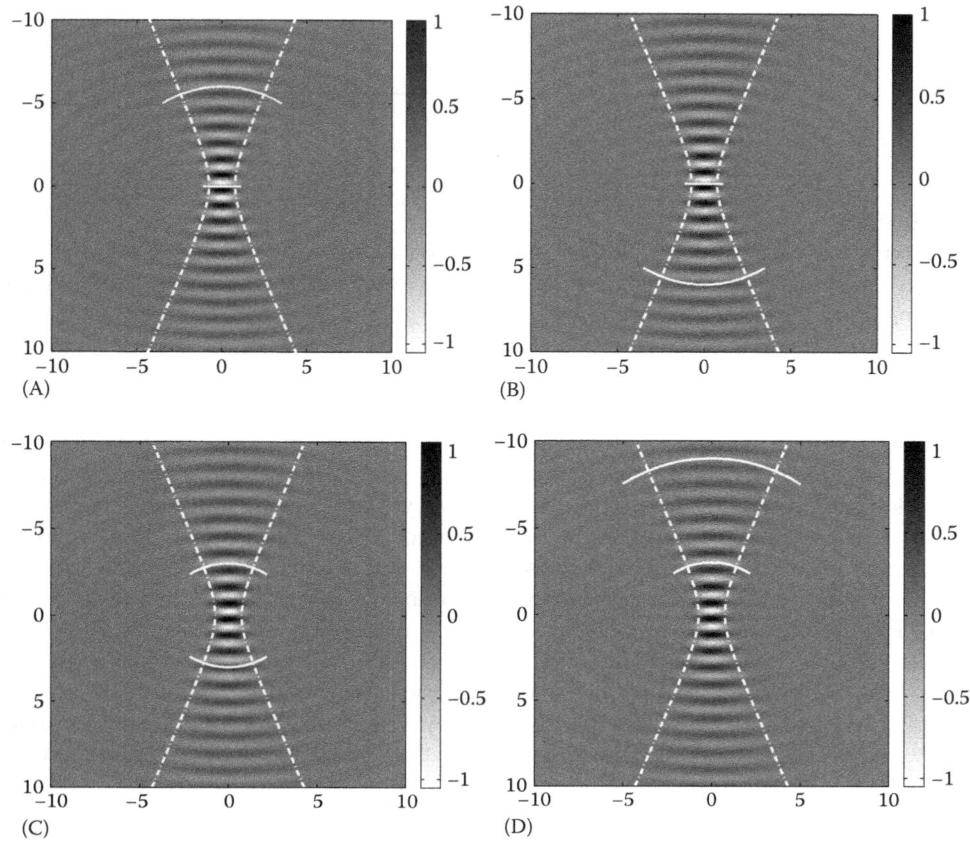

FIGURE 9.11
Laser cavity design. Four cavities are shown. The first three correspond to the cavities in Figure 9.10. A flat output coupler at $z = 0$, and a concave back mirror at $z = -6$ (A) produce a collimated beam. The opposite configuration (B) produces a diverging output beam at $z = +6$. Two identical concave mirrors at $z = \pm 3$ produce a waist in the center (C). With the front mirror at $z = -3$ and the back mirror at $z = -9$ (D), the output beam is converging.

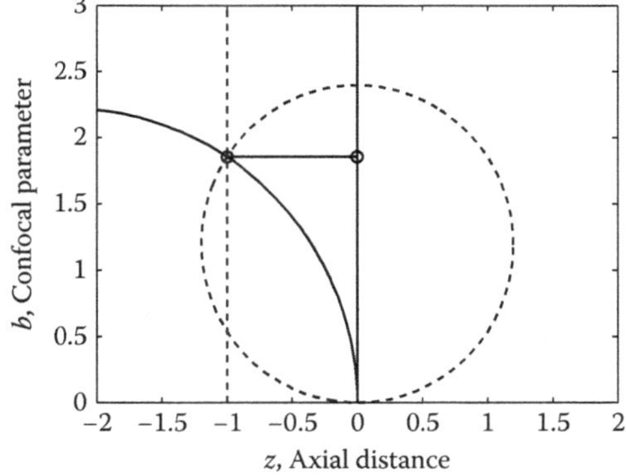

FIGURE 9.12
Collins chart for a laser cavity. The dark solid curves show conditions imposed by the mirrors and their spacing.

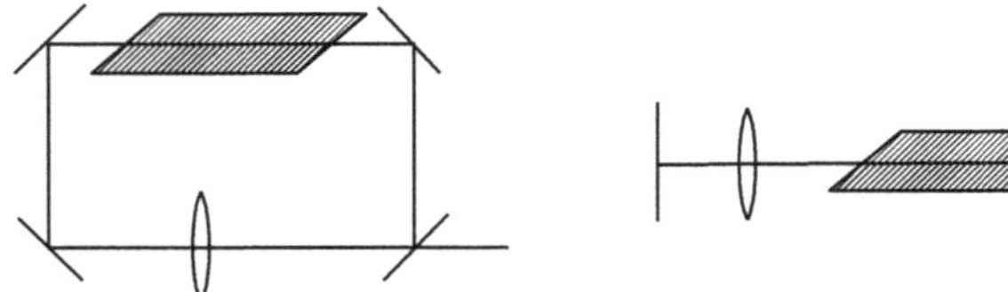

FIGURE 9.13
More complicated laser cavities. The left panel shows a ring cavity with four mirrors, a lens, and a gain medium. The right panel shows a Fabry–Perot cavity with a gain medium and a converging lens.

9.6.3 More Complicated Cavities

A wide variety of laser cavity designs exists. There is always some focusing element within a stable cavity.

We can design more complicated cavities using the basic principles we have developed here. For example, we could design a "ring cavity," with three or more mirrors and some internal focusing elements. In contrast to the Fabry–Perot cavities in which light travels through the same path twice (once in each direction), light travels around a ring cavity and returns to the starting point after passing through each element, and reflecting from each mirror, only once. The analysis would proceed in a very similar manner. The left panel of Figure 9.13 shows a ring cavity with four mirrors, a gain medium, and a lens. We could also include multiple focusing elements in the Fabry–Perot cavity. The right panel in Figure 9.13 shows a Fabry–Perot cavity with two flat mirrors, a gain medium, and a converging lens within the cavity. We can develop a general approach to these problems using the $ABCD$ matrices. Specifically from Equation 9.38, we can have a stable resonator if $q_{out} = q_{in}$, or if we can find a solution, q, to the equation

$$q = \frac{Aq + B}{Cq + D}$$
$$Cq^2 + (D - A)q - B = 0, \tag{9.54}$$

for the matrix elements of a single round trip, and if that solution is a physically realizable beam. Solutions that are physically impossible are those with real q, as this implies a waist diameter of zero. We write the solution of the quadratic equation,

$$q = \frac{A - D \pm \sqrt{(A - D)^2 + 4CB}}{2C}, \tag{9.55}$$

and note that in order to obtain a complex result, the argument of the square root must be negative.

$$(A - Dx)^2 + 4CB < 0. \tag{9.56}$$

The determinant condition of Equation 3.30 requires that the determinant of an $ABCD$ matrix is n/n'. For a round trip, n and n' are the same, so $AD - CB = 1$. Thus

$$(D - A)^2 + 4DA < 4 \qquad (D + A)^2 < 4. \tag{9.57}$$

In Practice

Unstable cavities are sometimes used for high-power applications with high-gain laser materials.

9.6.4 Summary of Stable Cavity Design

- Stable cavities or stable optical resonators have mirrors and possibly other optical elements that cause a Gaussian beam to replicate itself upon one round trip.
- Resonator design can be accomplished using Gaussian beam propagation equations.
- Most cavities will support only one Gaussian beam, and the beam parameters are determined by the curvature of the mirrors and their spacing along with the behavior of any intervening optics.
- Matrix optics can be used to determine the round-trip matrix, which can be used to find the Gaussian beam parameters.
- The Collins chart can help to evaluate the stability of a laser cavity.

9.7 Hermite–Gaussian Modes

We have seen that Gaussian beams are solutions of the wave equation that propagate through free space, constrained in two dimensions by the limits imposed by diffraction. An important aspect is that their Gaussian shape is retained through all values of z. It happens that the Gaussian is not the only such solution. There are two other common sets of solutions, Hermite–Gaussian and Laguerre–Gaussian. Both types of modes can be defined by products of Gaussian waves and orthogonal polynomials. The Hermite–Gaussian modes are particularly useful in problems with rectangular symmetry, and will be discussed in more detail. Intuitively, one would expect the circular symmetry of most laser mirrors to result in Laguerre–Gaussian modes. However, normally the symmetry is imperfect because of slight aberrations caused by the presence of Brewster plates or because of some tilt in the mirrors. The modes are usually Hermite–Gaussian.

Hermite–Gaussian expansion provides a natural description of a laser beam.

9.7.1 Mode Definitions

The Hermite–Gaussian modes are described by products of functions of x and y as

$$h_{mn}(x,y,z) = h_m(x,z)\,h_n(y,z)$$

where the one-dimensional functions are given by

$$h_m(x,z) = \left(\frac{2}{\pi}\right)^{1/4}\sqrt{\frac{1}{2^m m! w}}H_m\left(\frac{x}{w}\right)e^{\frac{x^2}{w^2}}e^{\frac{jkx^2}{2\rho}}e^{j\psi_m},$$

and H_m are the Hermite polynomials[13], given by their first two expressions,

$$H_1(x) = 1 \qquad H_2(x) = 2x$$

and their recurrence relations,

$$H_{m+1}(x) = 2xH_m(x) - 2(m-1)H_{m-1}(x).$$

The Gouy phase is given by

$$\psi_m = \left(\frac{1}{2} + m\right)\arctan\frac{z}{b}. \tag{9.58}$$

All that is required to make these modes correspond to physical fields is to multiply them by an appropriate constant, proportional to the square root of the power in the mode, and having an appropriate phase. For example, we note that

$$\sqrt{P}\,h_{00}(x,y,z) \tag{9.59}$$

The Hermite–Gaussian modes are the transverse modes of the cavity. The Gaussian beam is the lowest mode and is called TEM_{00}.

is the Gaussian mode. We call this the TEM_{00} mode, where TEM means transverse electric and magnetic. We call the Hermite–Gaussian modes the **transverse modes**, or TEM. modes of the cavity. We defined the longitudinal modes in Section 7.5.3.1. Because the wavefront curvature appears mathematically only in the Gaussian term and not in the Hermite polynomials, all the higher order modes have the same curvature and are solutions of the same cavity as the TEM_{00}.

Take-Away

Higher-order Modes:
Just as a laser can operate on multiple longitudinal modes, it can also operate on multiple transverse modes.

The mathematical power of orthogonal functions such as the Hermite–Gaussian functions lies in the fact that any beam can be expanded as a superposition of such functions:

Any beam can be expanded in Hermite–Gaussian functions.

$$E(x,y,z) = \sum_{m=0}^{\infty}\sum_{n=0}^{\infty} C_{mn} h_{mn}(x,y,z). \tag{9.60}$$

The coefficient C_{mn} provides information on the power in the mode, $C_{mn}C_{mn}^*$, and on the phase, $\angle C_{mn}$.

Several of the modes are shown in Figure 9.14. The brightness of the image shows the irradiance. The mode numbers are shown above the individual plots. We note that for the x direction, the number of bright peaks for mode m is $m+1$, and the number of nulls is m, with the same relationship for y and n. For the TEM_{00} mode, there are no nulls. The TEM_{13} mode has two bright spots with one null in the center in x and four bright spots with three nulls separating them in y. All of the modes in the figure are pure Hermite–Gaussian modes, with the exception of the upper right, which shows the **donut mode**. The donut mode is quite common in lasers, and is often thought to be a Laguerre–Gaussian mode. It is, however, a superposition of the TEM_{10} and TEM_{01} modes in quadrature phase,

$$E_{donut}(x,y,z) = \frac{1}{\sqrt{2}} h_{01}(x,y,z) + \frac{j}{\sqrt{2}} h_{10}(x,y,z). \tag{9.61}$$

The donut mode is common in some lasers and is a superposition of the 01 and 10 modes in quadrature.

9.7.2 Expansion in Hermite–Gaussian Modes

The Hermite–Gaussian beam expansion in Equation 9.60 is an example of a very general idea of expanding functions in some type of orthogonal functions. The most common example is the Fourier series, which is one of the staples of modern engineering. Just as a periodic function can be expanded in a Fourier series with a given fundamental frequency, an arbitrary beam in space can be expanded in Hermite–Gaussian functions. Here we show how to obtain the coefficients, C_{mn}, given the beam profile, $E(x,y,z)$. We write Equation 9.60, multiply both sides by the conjugate of one of the Hermite–Gaussian functions, $h_{m',n'}(x,y,z)$, and then integrate over all x and y;

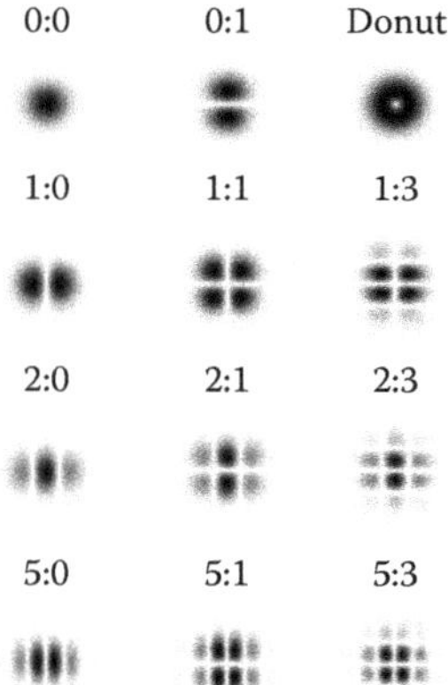

FIGURE 9.14
Some Hermite–Gaussian modes. The modes are identified by numbers above the plots. All are pure modes except the so-called donut mode in the upper right, which is a superposition of the 0:1 and 1:0 modes with a 90° phase difference between them.

$$\int_{-\infty}^{\infty}\int_{-\infty}^{\infty} E(x,y,z)\, h^*_{m'n'}(x,y,z)\, dx\, dy =$$

$$\int_{-\infty}^{\infty}\int_{-\infty}^{\infty} \sum_{m=0}^{\infty}\sum_{n=0}^{\infty} C_{mn} h_{mn}(x,y,z)\, h^*_{m'n'}(x,y,z)\, dx\, dy.$$

Exchanging the order of summation and integration, we obtain

$$\int_{-\infty}^{\infty}\int_{-\infty}^{\infty} E(x,y,z)\, h^*(x,y,z)\, dx\, dy =$$

$$\sum_{m=0}^{\infty}\sum_{n=0}^{\infty} C_{mn} \int_{-\infty}^{\infty}\int_{-\infty}^{\infty} h_{mn}(x,y,z)\, h^*_{m'n'}(x,y,z)\, dx\, dy. \tag{9.62}$$

Now, because the Hermite–Gaussian functions are orthonormal (the ones with different indices are orthogonal to each other, and we have normalized them to unit power),

$$\int_{-\infty}^{\infty}\int_{-\infty}^{\infty} h_{mn}(x,y,z)\, h^*_{m'n'}(x,y,z)\, dx\, dy = \delta_{m,m'}\delta_{n,n'}. \tag{9.63}$$

Therefore,

Given a field, we can compute the coefficients to describe it as a Hermite–Gaussian expansion.

$$C_{m'n'} = \int_{-\infty}^{\infty}\int_{-\infty}^{\infty} E(x,y,z)\, h^*_{m'n'}(x,y,z)\, dx\, dy. \tag{9.64}$$

Thus, just as we obtain the coefficients in a Fourier series by multiplying the desired function by the appropriate sine or cosine term, here we multiply the desired beam profile by the appropriate Hermite–Gaussian function. We need to repeat Equation 9.64 for each m' and n'. Just as Fourier series is useful if coefficients of terms with high order approach zero, the Hermite–Gaussian expansion is useful if

$$C_{m',n'} \to 0 \text{ for } m' \to \infty,\ n' \to \infty.$$

Practically, it is important to find the right w to minimize the number of terms in the expansion.

However, there is one major difference between the Fourier series and this expansion. In expanding a periodic function, we know the period, so we know the fundamental frequency of the sine and cosine waves. In the expansion of a beam profile, we do not have similar guidance to help us choose w; any value we choose will work. However, some choices of w will cause the expansion to converge more slowly, requiring many more terms to achieve a close approximation to the beam. We could simply choose w to be some measure of the beam width, but this depends on our definition of width. We need some method of computing an optimal w, or at least a good one. Usually, we compute C_{00} for different values of w and find a maximum.

$$C_{00} = \max\left[C_{00}(w)\right]$$

$$C_{00}(w) = \int_{-\infty}\int_{\infty} E(x,y,z)\, h_{00}^{*}(x,y,z)\, dx\ dy. \tag{9.65}$$

Then we continue with the computation of the coefficients as defined by Equation 9.64 with the chosen w.

An example of expansion of the field of a uniformly illuminated circular aperture is shown in Figure 9.15. In Panel (A), we see the uniformly illuminated circular aperture of radius 2 units. We begin by computing C_{00} for different values of w, as shown in Panel (B). We find that the best value of w is 1.8. Next, we compute the coefficients using Equation 9.64. The squared amplitude of the coefficients is shown on a dB scale in Panel (C). We see that most of the power (about 86%) is in the TEM_{00} wave. We also notice that the odd orders all have zero coefficients. This is expected, because the original beam has even symmetry, so all the odd contributions to the integral in Equation 9.60 vanish. Now, reconstructing the beam up to 20th order, we obtain the result shown in Panel (D), with several slices through the beam shown in Panel (E), on a dB scale. Each slice is for expansion to a different number of terms. The first term, $C_{00}h_{00}$ is clearly visible. Now, knowing the coefficients at the waist, it is easy to propagate the beam to a new value of z. We note that each mode propagates with the same w and ρ. The only change from one mode to another is the Gouy phase, $\psi = (1+m+n)\arctan(z/b)$. Thus, to go from the waist at $z=0$ to the far field, we replace each coefficient, C_{mn} by

$$C_{mn}e^{(1+m+n)\pi/2}. \tag{9.66}$$

The result is shown in Panel (F) of Figure 9.13. The first null, at $1.22\lambda/D$ is evident, as is the second null, for higher order expansions. Expansions of this sort are useful for propagation of laser beams[55], and for similar propagation problems.

9.7.3 Coupling Equations and Mode Losses

In Equation 9.63, We saw that the Hermite–Gaussian modes are orthonormal. If instead we take the integrals over an aperture,

$$K_{mnm'n'} = \iint_{aperture} h_{mn}(x,y,z)\, h_{m'n'}^{*}(x,y,z)\, dx\ dy = \delta_{m,m'}\delta_{n,n'}, \tag{9.67}$$

An aperture causes loss in each mode and couples modes to each other.

we find that $|K_{mnmn}K_{mnmn}^{*}|$ will generally be less than unity. The exact value depends on the size and shape of the aperture. This term can be considered a transmission efficiency for the TEM_{mn} mode passing through an aperture. Thus, we can think of a **mode loss** $1-|K_{mnmn}K_{mnmn}^{*}|$. Also $|K_{mnm'n'}K_{mnm'n'}^{*}|$ for any two different modes will be greater than zero. Thus, at an aperture, there is a coupling between modes. One can handle

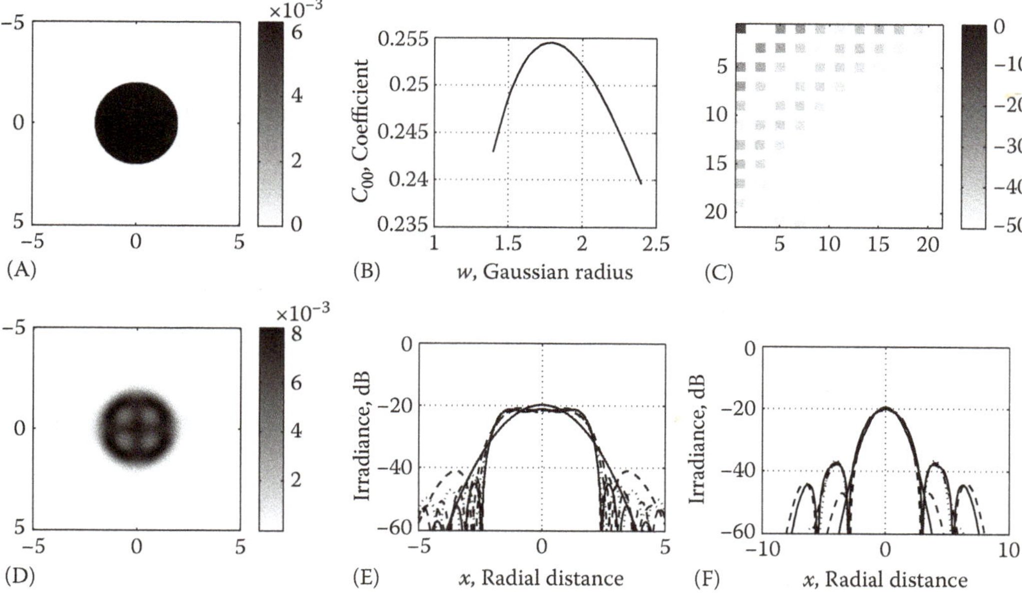

FIGURE 9.15
Expansion in Hermite–Gaussian modes. Panel (A) shows a uniformly illuminated circular aperture, (B) shows the optimization of w, and (C) shows the relative power in each coefficient. Panel (D) shows the reconstruction up to 20th order, and (E) shows a slice though the center with various orders. Finally, (F) shows the diffraction pattern obtained by propagating to the far field.

apertures using a matrix formulation in which the incident beam is described as a vector of its Hermite–Gaussian coefficients, C_{mn}, and the coupling is represented by a matrix of these K values, so that the coefficients C'_{mn} of the output beam are given by

$$\begin{pmatrix} C'_{00} \\ C'_{01} \\ C'_{10} \\ \vdots \end{pmatrix} = \begin{pmatrix} K_{0000} & K_{0001} & K_{0010} & \cdots \\ K_{0100} & K_{0101} & K_{0110} & \cdots \\ K_{1000} & K_{1001} & K_{1010} & \cdots \\ \vdots & \vdots & \vdots & \ddots \end{pmatrix} \begin{pmatrix} C_{00} \\ C_{01} \\ C_{10} \\ \vdots \end{pmatrix}. \tag{9.68}$$

With the double subscripts on the coefficients, one must be careful about the arrangement in the vector.

The diagonals of this matrix are the field transmission values for the individual modes. From Figure 9.14, we see that the size of the modes increases with mode number. Thus the mode losses are expected to become more severe with increasing mode number. To limit a laser to the TEM_{00} mode, we can use a circular aperture inside the cavity, small enough so that the loss on the TEM_{01} and TEM_{10} modes will be sufficient to prevent the laser from operating, but large enough that the loss on the TEM_{00} mode will be minimal.

Take-Away

Limiting Higher Modes:
Beam size increases with mode number. An aperture causes more loss in higher modes.

In Practice

An aperture in a laser cavity can limit operation to TEM_{00}. This may be important to produce a good quality beam profile and/or to avoid beat signals between modes at different frequencies. The aperture can be designed using the analysis in this section, but is often done by trial and error.

Different transverse modes have slightly different frequencies.

Spatial filtering can be quite important, not only for generating a good (Gaussian) spatial distribution, but also for generating a beam with good temporal quality. We recall from our discussion in Section 7.5.4 on the laser frequency modulator, that any change in the round-trip phase changes the frequency of the laser. The exact frequency is determined by the condition that the round-trip phase shift, $2k\ell$ must be an integer, N times 2π. Now we recall that the higher-order modes have a different Gouy phase shift, so the round trip in a cavity of length changes the phase by

$$2k\ell + 2\left(1+m+n\right)\left(\arctan\frac{z_2}{b} - \arctan\frac{z_1}{b}\right) = 2\pi N,$$

and

$$2\frac{f\left(m,n,N\right)}{c}\ell + \frac{1}{\pi}\left(1+m+n\right)\left(\arctan\frac{z_2}{b} - \arctan\frac{z_1}{b}\right) = N.$$

$$f\left(m,n,N\right) = N\frac{c}{2\ell} - \frac{c}{2\ell}\left(1+m+n\right)\frac{\arctan\frac{z_2}{b} - \arctan\frac{z_1}{b}}{\pi}. \tag{9.69}$$

For one increment of mode number, such as changing m from 0 to 1, the frequency difference is

$$f\left(1,n,N\right) - f\left(0,n,N\right) = \frac{c}{2\ell}\frac{\arctan\frac{z_2}{b} - \arctan\frac{z_1}{b}}{\pi}. \tag{9.70}$$

Example

As an example, we consider the carbon dioxide laser we designed in Section 9.6.1. We recall that the output coupler was flat, meaning that it is at $z = 0$, the confocal parameter is $b = 1.85\,\mathrm{m}$, and the back mirror is at $z = 1\,\mathrm{m}$. Then

$$f\left(1,n,N\right) - f\left(0,n,N\right) =$$

$$\frac{c}{2\,\mathrm{m}}\frac{\left(\arctan\frac{1\,\mathrm{m}}{1.85\,\mathrm{m}} - \arctan\frac{0\,\mathrm{m}}{1.85\,\mathrm{m}}\right)}{\pi} = 24\,\mathrm{MHz}.$$

We note that this is only about 16% of the free spectral range, $c/(2\ell) = 150\,\mathrm{MHz}$. Just as two longitudinal modes, N and $N+1$, produce a mixing or beat signal at a frequency equal to the free spectral range, two transverse modes separated any way by one in $m+n$, will produce a mixing signal at this much smaller frequency. In many applications, a signal at this frequency could contaminate a measurement being made.

9.7.4 Summary of Hermite–Gaussian Modes

- Hermite–Gaussian modes (Equation 9.58) form an orthogonal basis set of functions which can be used to describe arbitrary beams.

- The beam can be expanded in a series using Equation 9.60.
- Coefficients of the expansion can be calculated using Equation 9.64.
- Each mode maintains its own profile with a size w and radius of curvature, ρ given by the usual Gaussian beam equations.
- The choice of w in Equation 9.58 is arbitrary, but a value chosen badly can lead to poor convergence of the series expansion.
- Higher-order modes have increased Gouy phase shifts, as shown in Equation 9.58.
- Beam propagation problems can be solved by manipulating the coefficients with appropriate Gouy phase shifts.
- Apertures can be treated using coupling coefficients which can be computed by Equation 9.67.
- Higher-order transverse modes in a laser can mix with each other to produce unwanted mixing, or beat, signals.
- The higher-order modes can be eliminated by placing an appropriate aperture inside the laser cavity.

Problems

9.1 Gaussian Beams

A helium-neon laser (633 nm) produces a collimated output beam with a diameter of 0.7 mm.

9.1a If the laser is 20 cm long, what are the radii of curvature of the two mirrors?

9.1b How big is the beam at the back mirror?

9.1c I now wish to launch the beam into a fiber whose core is best matched to a Gaussian beam with a diameter of 5 µm. Figure out how to do this, using one or more lenses in a reasonable space.

9.2 Diameter of a Laser Amplifier

I wish to build an amplifier tube 6 m long to amplify a carbon dioxide laser beam at a wavelength of 10.59 µm. I can change the beam diameter and radius of curvature entering the amplifier, and I can place lenses at the output so I do not care about the curvature of the wavefront at the output. I need to keep the beam diameter over the entire length as small as possible to keep the cost of the amplifier low. What is the waist diameter? Where is it located? What is the largest size of the Gaussian beam along the tube?

9.3 Gaussian Beam (NS)

In confocal microscopy, we usually illuminate a sample by focusing a laser beam with a microscope objective, where the laser beam is a plane-wave Gaussian beam at what is normally the back focal plane of the objective, and then focuses at the object. Let's consider a $30X$ objective from a vendor that sells microscopes with a 200 mm tube lens. The laser beam is 9 mm in diameter.

9.3a What is the focal length of the objective?

9.3b What is the value of q, the complex radius of curvature as it exits the microscope objective going toward the object?

9.3c What is the location of the waist of the beam, and what is it's diameter?

9.3d We have a laser with a 1 mm plane-wave Gaussian output. How would we couple this to the microscope objective? You don't need to design the coupling optics. Just explain what you would do.

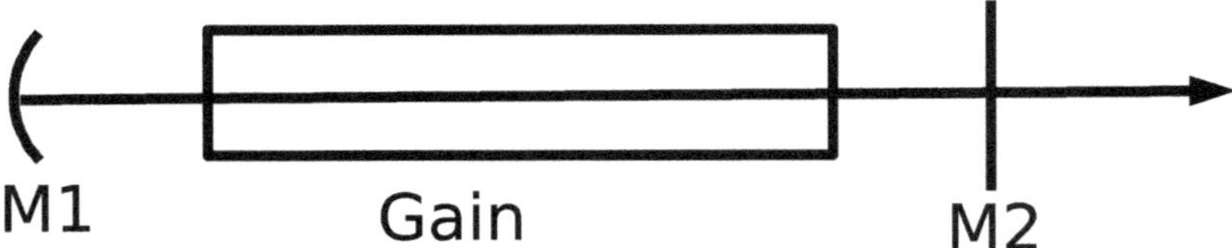

FIGURE 9.4
Figure for Problem 9.4.

9.4 Cavity Stability (NS)

Consider the laser cavity in Figure 9.4. The gain medium is a doped glass rod, with an index of refraction of $n = 1.5$ and a length of 20 cm. The ends would probably be cut and polished at Brewster's angle, but that complicates the problem, so let's ignore that and assume the flat ends shown. The whole cavity is 25 cm long, the radius of curvature of the back mirror, (mirror M1) is $R_1 = 80$ cm and the output coulpler (mirror M2), is flat.

9.4a Write the matrix going through the cavity from right to left.

9.4b Does this matrix make sense? Why or why not?

9.4c Write the matrix going through the cavity from left to right.

9.4d We can think of the curved mirror as a lens of focal length $f = R_1/2$. Plot the stability number, $(D + A)^2$, as a function of $1/f$ in diopters from zero to 15. Also plot the stability number as a function of R_1.

9.4e Why could a Gaussian beam not form if the stability number were to exceed 4? Note that there should be two cases with different answers?

10 Coherence

In this chapter, we address how to describe superposition of two or more light waves. This issue arises in Chapter 7 where we combine two waves at a beamsplitter as in Figure 7.1A, or an infinite number of waves in a Fabry–Perot interferometer in Section 7.5. The question of superposition arises again in Chapter 8 when we integrate contributions to the field in an image over the pupil in Equation 8.44, and implicitly in geometric optics through Chapters 2 through 5, when we combine rays to form an image, the details of which we will address in Chapter 12.

Because Maxwell's equations are linear partial differential equations for the field variables, any sum of any number of solutions is also a solution. Therefore, we added fields in Chapter 7, and predicted effects such as fringe patterns, and variations in constructive or destructive interference with changes in path length. These effects are observed in interferometers, lasers, and reflections from multiple surfaces in a window. In Chapter 8, we integrated fields to predict diffraction effects in which a beam of diameter D diverged at an angle approximately λ/D, and diffraction patterns from slits and gratings. Likewise, all of these phenomena are observed in experiments.

What would happen if we were to sum irradiances instead of fields? The sums in Chapter 7, if done this way, would predict images in multiple surfaces with no interference. In Chapter 8, the integral would predict that light would diverge into a full hemisphere. These phenomena are observed as well. We see our images in the multiple panes of a storm window without interference, and a flashlight beam does indeed diverge at an angle much greater than λ/D. Evidently, in some cases light behaves in a way that requires summation of fields, producing constructive and destructive superposition, and in others superposition can be done with irradiance, and no constructive or destructive effects occur.

We will characterize the first extreme by calling the light **coherent** and we will call the light in the second extreme **incoherent**. We refer to the summation of fields as **coherent addition**, $E = E_1 + E_2$. In contrast, in Chapters 2 and 3, we implicitly assumed that we would sum irradiances or the squares of field magnitudes, in a process we call **incoherent addition**. We will explore the conditions under which light is coherent or incoherent, and we will develop a quantitative measure of coherence, so that we can handle cases between these two extrema. We will see that it is never possible to describe a light source as completely coherent or incoherent, and will explore how the coherence varies as a function of space and time.

Take-Away

Coherent and Incoherent Addition:
For coherent addition we add the fields.

$$I = \left|\sum E_i\right|^2.$$

For incoherent addition, we add irradiances.

$$I = \sum |E_i|^2.$$

Interference effects only occur for coherent addition.

DOI: 10.1201/9781315157047-10

10.1 Definitions

When we wrote Equation 7.3,*

$$I = E_1^* E_1 + E_2^* E_2 + E_1^* E_2 + E_1 E_2^*,$$

Fig. 7.2:

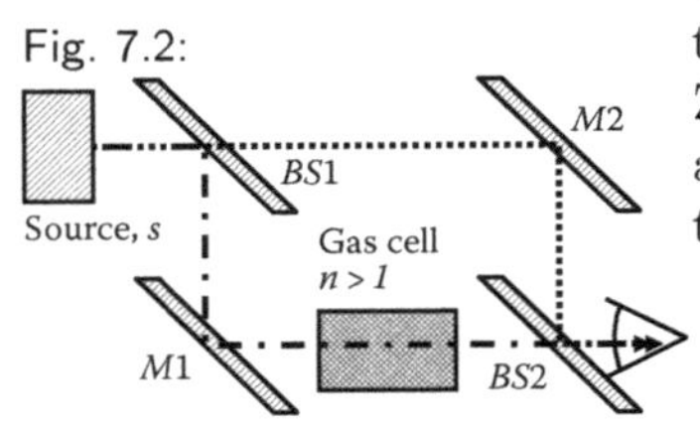

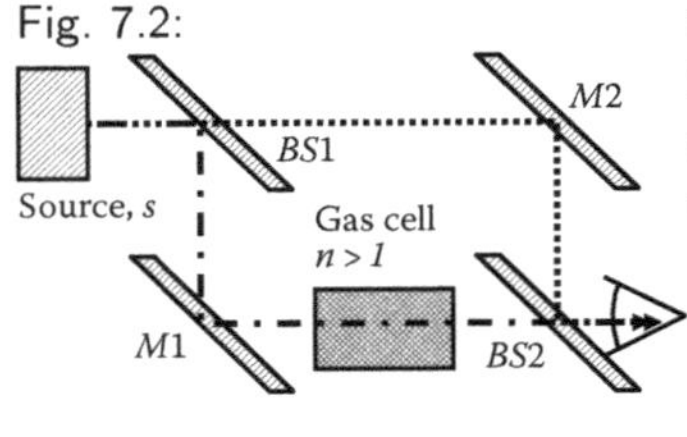

we noted that the first two terms are the result of incoherent addition, and the last two terms, called mixing terms, are required for the coherent sum. For example, in the Mach–Zehnder interferometer (Section 7.1, Figure 7.2), if the path lengths within the two arms are slightly different, the fields that mix at the recombining beamsplitter originated from the source at two different times. Assuming plane waves,

$$E_1(t) = E_s\left(t - \frac{z_1}{c}\right) \qquad E_2(t) = E_s\left(t - \frac{z_2}{c}\right).$$

We explicitly include the $e^{j\omega t}$ time dependence and write

$$E_s(t) = E_0 e^{j\omega t}. \tag{10.1}$$

Then, with $k = 2\pi/\lambda = \omega/c$, the mixing term $E_2 E_1^*$ becomes

$$\begin{aligned} &E_s(t - z_2/c)\, E_s^*(t - z_1/c) = \\ &E_0(t - z_2/c)\, E_0^*(t - z_1/c)\, e^{j(\phi_2 - \phi_1)}, \end{aligned} \tag{10.2}$$

where

$$\phi_1 = kz_1 \qquad \phi_2 = kz_2.$$

If E_0 is independent of time, we obtain Equation 7.6

$$I = I_1 + I_2 + 2\sqrt{I_1 I_2}\cos(\phi_2 - \phi_1).$$

Highly coherent light can produce strong fringes.

If the interferometer splits the beams evenly so that $I_1 = I_2$, then as discussed in Chapter 7, the irradiance varies from $4I_1$ to zero and the fringe visibility or contrast is

$$V = \frac{I_{max} - I_{min}}{I_{max} + I_{min}} = \frac{4I_1 - 0}{4I + 0} = 1. \tag{10.3}$$

The constructive interference then uses all the light from the source, and the destructive interference results in complete cancellation.

In other cases, the result is more complicated. If E_0 is a random variable, then the mixing terms are random and can only be understood statistically. We consider the ensemble average,

$$\langle E_2 E_1^* \rangle = \langle E_0(t - z_2/c)\, E_0^*(t - z_1/c) \rangle\, e^{j(\phi_2 - \phi_1)}, \tag{10.4}$$

where the expression $\langle \cdot \rangle$ denotes the ensemble average, which we discussed in Section 6.1.4. If the source is **ergodic**, the time averages of the source statistics are independent of the time at which we start to measure them, and

$$\langle E_2 E_1^* \rangle = \langle E_0(t)\, E_0^*(t - \tau) \rangle\, e^{j(\phi_2 - \phi_1)}, \tag{10.5}$$

where

$$\tau = \frac{z_2 - z_1}{c} \tag{10.6}$$

is the difference between transit times in the two paths.

*Lest we forget, we are adopting our usual convention here of neglecting to divide by the impedance. See the comments around Equation 1.48 for the original discussion.

Then Equation 7.6 becomes

$$I = I_1 + I_2 + 2\Re\left[\Gamma e^{j(\phi_2 - \phi_1)}\right], \tag{10.7}$$

where

$$\Gamma = \left\langle E_0^*\left(t - \tau\right) E_0\left(t\right)\right\rangle \tag{10.8}$$

is called the temporal **autocorrelation function** of the field. The fringe visibility,

$$V = \frac{I_{max} - I_{min}}{I_{max} + I_{min}} = \frac{2I_1\left(1 + |\Gamma|\right) - 2I_1\left(1 - |\Gamma|\right)}{2I_1\left(1 + |\Gamma|\right) + 2I_1\left(1 - |\Gamma|\right)},$$

even if $I_1 = I_2 = I$, decreases to

$$V = \gamma = \frac{4\,|\Gamma|}{4I}. \tag{10.9}$$

Now, if the interferometer splits the source light evenly, $|E_1| = |E_2|$, then for very short times, $E(t) = E(t - \tau)$,

$$\Gamma = \left\langle E_1^*\left(t\right) E_1\left(t\right)\right\rangle \tag{10.10}$$

and the normalized autocorrelation function,

$$\gamma\left(\tau\right) = \frac{\Gamma\left(\tau\right)}{|E_1 E_1^*|}, \tag{10.11}$$

is

$$\gamma(\tau) \to 1 \qquad \text{as} \qquad \tau \to 0. \tag{10.12}$$

On the other hand, if the time difference is so large that the source field at the second time is unrelated to that at the earlier time we arrive at another simple result. For uncorrelated variables, the mean of the product is the product of the means,

The normalized autocorrelation function characterizes temporal coherence.

$$\Gamma = \left\langle E_1^*\left(t - \tau\right) E_1\left(t\right)\right\rangle = \left\langle E_1^*\left(t - \tau\right)\right\rangle \left\langle E_1\left(t\right)\right\rangle, \tag{10.13}$$

the expectation value of the field, with its random phase, is zero, and the normalized autocorrelation function is

$$\gamma(\tau) \to 0 \qquad \tau \to \infty. \tag{10.14}$$

In Practice

The ensemble average is formally the average over a number of identical systems with different values of the random variable. In a time-invariant system, the ensemble average can be measured as a time average. That is, we average $\left\langle E_1^*\left(t - \tau\right) E_1\left(t\right)\right\rangle$ over t for each value of τ. Often the time scale is a few optical periods and our measurements can only see the average. Such is the case with white light for example. If the system is not time-invariant, it is often the case that the systematic variations are slower than the random ones, and it can be considered time-invariant over short times.

Equations 10.12 and 10.14 show that for small path differences, the interferometer will produce the results anticipated in Chapter 7, while for larger differences, the irradiance will simply be the incoherent sum as we used in geometric optics. A coherent laser radar will work as expected provided that the round-trip distance to the target is small enough so that Equation 10.12 holds, and will not work at all if Equation 10.14 holds.

To summarize, for short times, the light is coherent and we add fields,

$$E = E_1 + E_2 \qquad I = |E|^2 \qquad \text{(Coherent addition)}, \tag{10.15}$$

and for longer times the light is incoherent and we add irradiances

$$\langle I \rangle = \langle I_1 \rangle + \langle I_2 \rangle \qquad \text{(Incoherent addition)}. \tag{10.16}$$

How can we define long and short times quantitatively? We need some measure of how long the field from the source retains its "memory." We will develop a theory in terms of Fourier analysis. Before doing so, we can gain some insight and address a very practical example, by looking at the behavior of a laser which is capable of running on multiple longitudinal modes.

10.2 Discrete Frequencies

Fourier transforms are important in understanding coherence, so it may be useful to first consider a set of discrete frequencies in which the field varies in time according to a sum of frequency contributions rather than an integral. A laser running on multiple longitudinal modes offers a practical example.

Eq. 7.135:
$N = \text{Round}\left(\frac{f_g}{FSR}\right)$
$f_{op} = N \times FSR.$

As discussed in Section 7.5.3, the frequencies of longitudinal modes of a laser are given by Equation 7.135 as those which are integer multiples of the free spectral range, and are under the part of the gain line of the medium that exceeds the loss. In that section, we mostly considered single longitudinal modes. Here we look at the effects associated with multiple modes operating simultaneously. In Figure 10.1, we show the sum of several modes at frequencies spaced equally around f_{center} with spacing equal to the free spectral range:

$$E = \sum_{m=-12}^{12} E_m = \sum_{m=-12}^{12} e^{i2\pi[f_{center}+m\times FSR]t}. \tag{10.17}$$

We note that at $t = 0$ all the contributions to the sum are at the same phase. We say the laser is **mode-locked**. In practice, a modulator is needed inside the cavity to cause the modes to lock, as can be learned from any text on lasers[210]. We have chosen to display Figure 10.1 with $f_{center} = 15\,\text{GHz}$ and $FSR = 80\,\text{MHz}$. The center frequency is not an optical frequency. If we were to choose a realistic frequency, we would have to plot several billion points to illustrate the behavior and could not produce an illustrative plot. However, the equations work equally well for realistic optical frequencies. The free spectral range, however, is quite realistic, being appropriate for a cavity length of 1.875 m, which is typical for titanium-sapphire lasers. The actual wavelengths of operation of these lasers are from near 700 nm to well over 1 μm.

In Figure 10.1A, we plot the real part of the field, and we see that the maximum amplitude is approximately the number of modes, $M = 25$. We have chosen our individual contributions to be unity, so, at time zero, with all contributions in phase, this result is

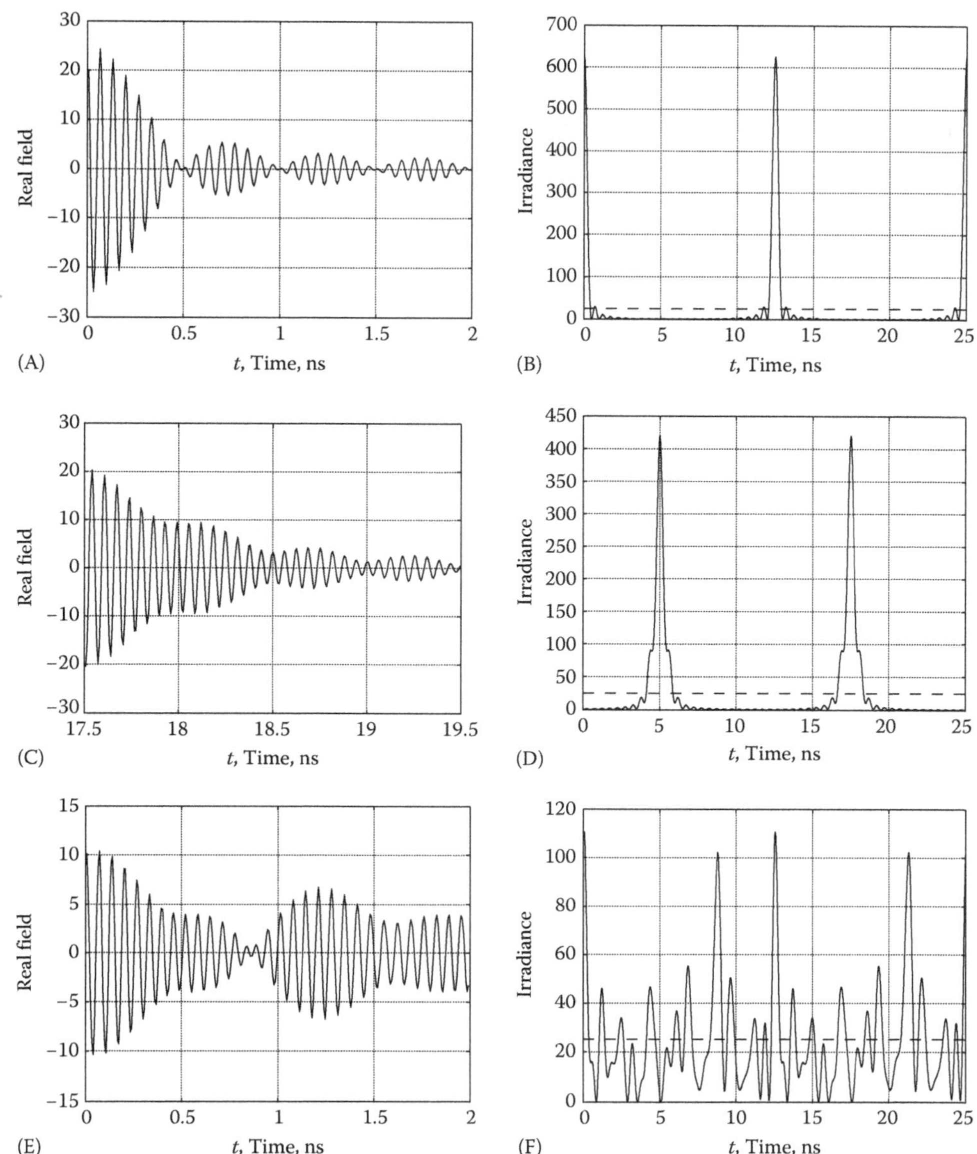

FIGURE 10.1
Mode-locked laser. A mode-locked laser with M modes has a field (A) with a maximum M times that of a single mode. The peak irradiance (B) is M^2 times that of a single mode, and the duty cycle is $1/M$, so the mean irradiance is M times that of a single mode (dashed line). If the pulse is chirped, the height decreases and the width increases (C and D). If the modes are not locked, the irradiance is random (E and F). In all cases, the mean irradiance remains the same. Note the changes in scale on the ordinates.

expected. At slightly later times, the fields lose their phase relationship, and the results are smaller. Figure 10.1B shows the irradiance, calculated as the squared amplitude of the field. The peak irradiance is $M^2 = 625$. After a time, $1/FSR$, the phases are once again all matched and another large pulse is observed. In general, the height of the pulse is the irradiance of a single mode, multiplied by the square of the number of modes. The pulse width is roughly the inverse of the spectral width, $1/(M \times FSR)$, and the average irradiance is thus M times the irradiance of a single mode.

In Figure 10.1C and D, the pulse is slightly "chirped" to model transmission through glass, where different frequencies travel at different speeds. As a result, the separate frequency contributions are never perfectly matched in phase. In this particular example, an incremental time delay

$$t_f = \frac{dt}{df} f,$$

with $\frac{dt}{df} = -(1/3) \times 10^{-18}\,\text{s/Hz}$, as might result from a modest amount of dispersive glass in the path, is introduced. As can be seen in the figure, the amplitude is lower and the pulse width is longer.

In Practice

Mode-locked lasers can produce extremely short pulses with high peak power. The modes are added coherently.

In a multi-mode laser that is not mode-locked, the average power can be computed by incoherent addition, but with random phases, interference will cause the power to vary randomly.

In Figure 10.1E and F, the laser is not mode-locked and the phase of each mode is chosen randomly. The sum in Equation 10.17 becomes

$$E = \sum_{m=-12}^{12} E_m = \sum_{m=-12}^{12} e^{i\{2\pi[f_{center} + m \times FSR]t + \phi_m\}}. \tag{10.18}$$

This situation is the normal one in multimode lasers, unless special efforts are made to lock the modes. The resulting irradiance is random, with a mean value still equal to M times that of a single mode. Occasional "hot" pulses are seen, when by coincidence, several modes have the same phase and sum coherently, but these events are unpredictable, and the output is usually well below the mode-locked pulse. The standard deviation about the mean is quite large, even with only 25 different random parameters, and the pattern repeats at the FSR, 80 MHz or 12.5 ns. This repetition is an artifact of the model. In fact the phases will vary randomly with time and the output will not be periodic. It is interesting to note that in a multimode laser that is not mode-locked, the output power may fluctuate considerably, depending on random alignment of the phases of the different modes. Occasionally, the phases will all be aligned, and there will be a "hot pulse." If the laser mirrors are not able to withstand the high power, this effect can become one of the failure mechanisms of the laser.

If the number of different frequencies is increased, the standard deviation becomes smaller, as shown in Figure 10.2. Here, the FSR has been reduced to 5 MHz and the sum in Equation 10.17 extends from $m = -100$ to 100. The repetition time now extends to $1/FSR = 0.2\,\mu\text{s}$.

As the number of contributions increases and the frequency spacing approaches zero, the spectrum becomes a continuum, and the repetition time approaches infinity. We shall see that as the spectral width increases, fluctuations are reduced by averaging more modes, the time constant of the autocorrelation function decreases, and the light becomes less coherent. In the next section, we will derive the relationship. If the fluctuations are averaged over a time shorter than our measurement time, then we see the source as incoherent.

A transform-limited pulse is the shortest possible pulse for a source with the given bandwidth.

We end this section with a comment on the characteristics of mode-locked lasers. In Figure 10.1, we saw examples of waves consisting of frequency contributions centered around a central frequency or "carrier frequency" with 12 frequencies on each side spaced 80 MHz apart, for a bandwidth of just under 2 GHz. The pulses in Figure 10.1A and B are nearly **transform-limited**, which means all the frequency contributions are in phase, and the resulting pulse is as high and narrow as it can be within that bandwidth. In Figure 10.1C and D, the spectral width is exactly the same, but the contributions have different phases, resulting in a longer pulse with a smaller amplitude. In this case, the pulse is not transform limited. When a pulse is passed through a dispersive medium, such

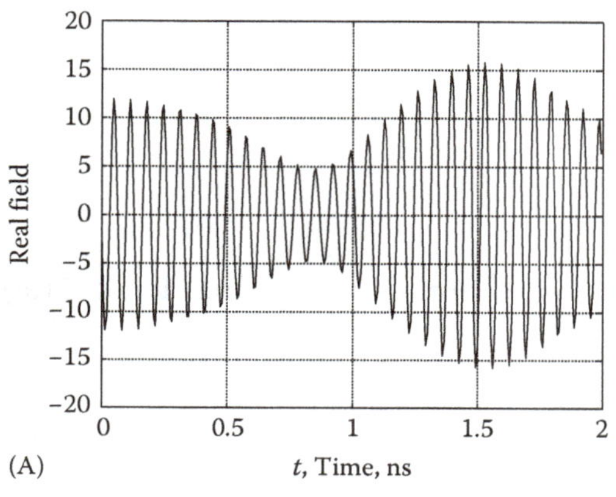

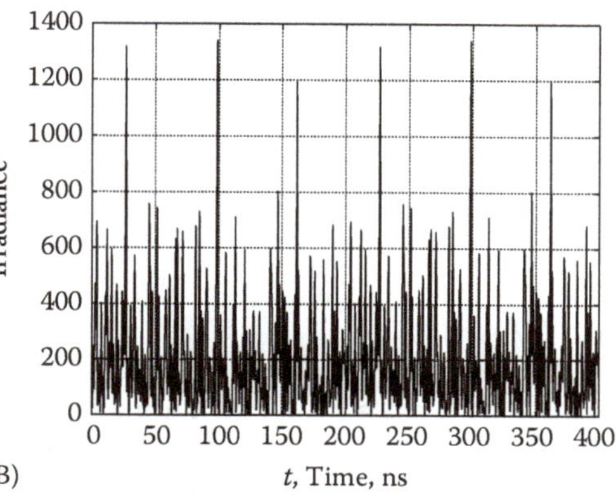

FIGURE 10.2
Incoherent sum. Here 201 modes, spaced 5 MHz apart are added with random phases. The short sample of the field (A) allows an expanded scale so that the optical oscillations are visible, while the longer sample of the irradiance (B) shows the random variations.

as glass, each frequency contribution travels at its own speed, and the resulting pulse is broadened by this "chirp" so that it is no longer transform-limited.

In Practice

If a laser beam with a short mode-locked pulse is passed through a large thickness of glass, the pulse width may be significantly increased by the chirp. Devices are available to create a chirp of equal magnitude but opposite direction. This process is called "pre-chirp," "chirp compensation," or "dispersion compensation."

Finally, in Figure 10.1E and F, the phases are so randomized that the laser does not produce distinct pulses. In all three cases, the irradiance spectral density is exactly the same. As the phase relationships are changed, the temporal behavior changes dramatically, from a transform-limited pulse, to a random signal with an autocorrelation function given by the Wiener–Khintchine theorem, discussed next.

Take-Away

Multi-Mode Laser:
If the different longitudinal modes of a laser are perfectly locked in phase, then the laser will produce a transform-limited pulse, one in which the pulse shape and bandwidth have a perfect Fourier–transform relationship. If the phases are random, the laser output will be a random variable, with a mean power equal to the sum of the powers of the modes. If there is a small departure from perfect mode-locking, then the output will still be a pulse but it will be wider than the transform-limited one.

10.3 Temporal Coherence

The Fourier transform of a field, $E(t)$, is

$$\tilde{E}(\omega) = \int_{-\infty}^{\infty} E(t)\, e^{-j\omega t} dt. \tag{10.19}$$

The units of the field, E, are V/m and the units of the **frequency-domain field**, $\tilde{E}(\omega)$ are V/m/Hz.

In the previous section, we thought of the field as a sum of discrete frequency contributions, each with its own amplitude and phase. This description is an inverse Fourier series. To extend the concept to a continuous wavelength spectrum, we apply the terms of a continuous inverse Fourier transform to the frequency-domain field,

$$E(t) = \frac{1}{2\pi} \int_{-\infty}^{\infty} \tilde{E}(\omega)\, e^{j\omega t} d\omega. \tag{10.20}$$

In this model we let $\tilde{E}(\omega)$ be a different complex random variable for each frequency. This assumption is in contrast to the mode-locked laser in which every mode is locked to every other.

10.3.1 Weiner–Khintchine Theorem

With Equation 10.20 for the field, we can determine the autocorrelation function,

$$\Gamma(\tau) = \langle E^*(t-\tau)\, E(t) \rangle . \tag{10.21}$$

With the field defined by the inverse Fourier transform in Equation 10.20, we can write

$$\begin{aligned} E(t) &= \frac{1}{2\pi} \int_{-\infty}^{\infty} \tilde{E}(\omega)\, e^{j\omega t} d\omega \\ E^*(t-\tau) &= \frac{1}{2\pi} \int_{-\infty}^{\infty} \tilde{E}^*(\omega')\, e^{-j\omega'(t-\tau)} d\omega', \end{aligned} \tag{10.22}$$

where we have chosen to change the name of the variable of integration from ω to ω' to avoid confusion in the next two equations. Then

$$\Gamma(\tau) = \left\langle \frac{1}{2\pi} \int_{-\infty}^{\infty} \tilde{E}^*(\omega')\, e^{-j\omega'(t-\tau)} d\omega' \frac{1}{2\pi} \int_{-\infty}^{\infty} \tilde{E}(\omega)\, e^{j\omega t} d\omega \right\rangle$$

$$\Gamma(\tau) = \left\langle \frac{1}{4\pi^2} \int_{-\infty}^{\infty} \int_{-\infty}^{\infty} \tilde{E}^*(\omega')\, \tilde{E}(\omega)\, e^{j\omega'\tau} e^{j(\omega-\omega')t} d\omega' d\omega \right\rangle$$

Exchanging the order of the linear operations, integration, and expectation value,

$$\Gamma(\tau) = \frac{1}{4\pi^2} \int_{-\infty}^{\infty} \left[\int_{-\infty}^{\infty} \left\langle \tilde{E}^*(\omega')\, \tilde{E}(\omega) \right\rangle e^{j\omega'\tau} e^{j(\omega-\omega')t} d\omega' \right] d\omega. \tag{10.23}$$

Now, if the source is a collection of independent sources, each smaller than the resolution element, all radiating with their own amplitudes and phases, then all $\tilde{E}(\omega)$ are random variables with zero mean and no correlation, so

$$\left\langle \tilde{E}(\omega)\right\rangle = 0. \tag{10.24}$$

What is the ensemble average inside the integrand above,

$$\left\langle \tilde{E}^*(\omega')\tilde{E}(\omega)\right\rangle ? \tag{10.25}$$

Returning for a moment to the sum of discrete modes in Equation 10.18, because the phases, ϕ_m, are random the product of any two terms would be

Eq. 10.18: $E = \sum_{m=-12}^{12} E_m = \sum_{m=-12}^{12} e^{i\{2\pi[f_{center}+m\times FSR]t+\phi_m\}}$.

$$\langle E_m E_n^*\rangle = I_m\delta_{m-n}, \tag{10.26}$$

where δ_{m-n} is the Kronecker delta function, $\delta_0 = 1$ and $\delta_n = 0$ otherwise.

In the continuous case, we replace E_m with $\tilde{E}(\omega_m)$, and obtain

$$\left\langle \tilde{E}^*(\omega')\tilde{E}(\omega)\right\rangle d\omega' = \tilde{I}(\omega')\delta(\omega-\omega')\,d\omega' \tag{10.27}$$

where $\delta(\cdot)$ is the Dirac delta function, the integral in square brackets in Equation 10.18 is

$$\left[\int_{-\infty}^{\infty}\left\langle \tilde{E}^*(\omega')\tilde{E}(\omega)\right\rangle e^{j\omega'\tau}e^{j(\omega-\omega')t}d\omega'\right] = \tilde{I}(\omega)\,e^{j\omega\tau}, \tag{10.28}$$

and the resulting autocorrelation function is

$$\Gamma(\tau) = \frac{1}{4\pi^2}\int_{-\infty}^{\infty}\tilde{I}(\omega)\,e^{j\omega\tau}d\omega \tag{10.29}$$

(Weiner–Khintchine theorem). (10.30)

Eq. 10.20: $E(t) = \frac{1}{2\pi}\int_{-\infty}^{\infty}\tilde{E}(\omega)\,e^{j\omega t}d\omega$.

The autocorrelation function of the field is the inverse Fourier transform of the irradiance spectral density function. Comparing this equation to Equation 10.20, we see that the relationship between these two functions, the autocorrelation function, $\Gamma(\tau)$, and the irradiance spectral density function, $\tilde{I}(\omega)$, is the same as the relationship between the field, $E(t)$ and its spectrum, $\tilde{E}(\omega)$. The Wiener–Khintchine theorem is generally attributed to Wiener[250] and Khintchine[118] (also spelled Khinchine and Khinchin), although Yaglom[253] credits Einstein with the original formulation in 1914[62].

In Figure 10.2, we saw a field composed of 201 discrete frequency contributions separated by 5 MHz each, having a total linewidth of approximately 1 GHz. The Fourier transform of this spectrum will have a width on the order of a nanosecond. In an interferometer, the source will be split into two beams with different transit times based on the difference in path length between the two arms. In Figure 10.3A, a segment of the time history of the field is plotted for the case where the two transit times are separated by

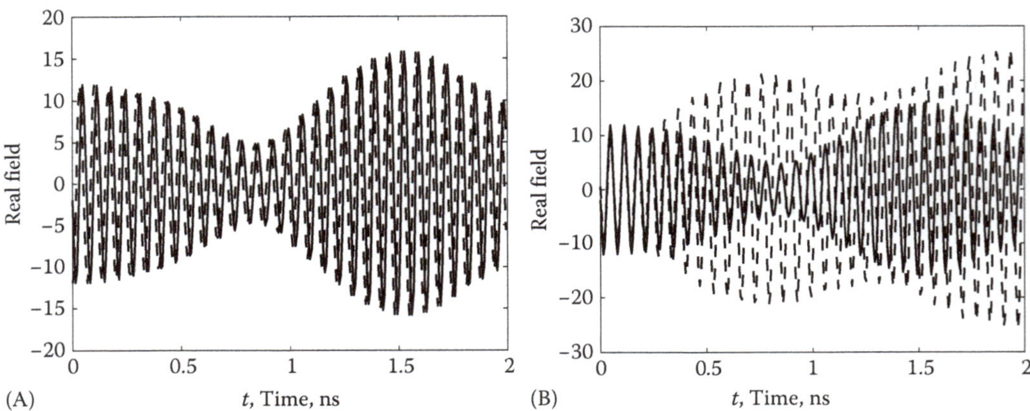

FIGURE 10.3
Coherence in interferometry. The field shown in Figure 10.2 is split into two arms of an interferometer. A short path difference of 10 ps (A) and long path difference of 2 ns (B) are shown. The linewidth of the source is 10 GHz. The sum of the field with short time delay (A) would always exhibit constructive interference, while the sum at a longer delay would produce random interference that would disappear upon averaging.

For small path differences, much less than the coherence time, the fields remain correlated and we have coherent addition of the waves. For large path differences, much more than the coherence time, the fields become random and we have incoherent addition.

10 ps or about 3 mm of optical path length. The fields are close to identical, with a nearly fixed phase difference between the two. We remind ourselves that the carrier frequency here is lower than it would be for optical fields, for purposes of illustration. Because the relationship between the two fields remains constant, the output of the interferometer will be their coherent sum; the mixing term is strong and the fields will add. With a slight change in path length, the interference will change from constructive to destructive with almost complete cancellation, and the fringes will be strong, $V \to 1$.

For comparison, in Figure 10.3B, the path transit difference is 2 ns, or about 60 cm, resulting in incoherent fields. We see that in a short time, the relative amplitudes and phases change. The mixing terms will vary with time, and will be very small when averaged over any reasonable detection time. Thus the mixing terms make no significant contribution to the average, and we have incoherent addition; the irradiances add. The interferometer will produce very weak fringe contrast, $V \to 0$, which will probably not be observable.

To make use of our results, we want to define the terms coherence time, coherence length, and linewidth, and relationships among them. We will examine these with some examples.

Take-Away

Weiner–Khintchine Theorem:

The autocorrelation function in time and the power-density spectrum are a Fourier transform pair. A wide spectrum implies a short correlation time. If light from a single source is split into two paths and recombined, coherent effects are observed if the path difference is very short and not if it is very long.

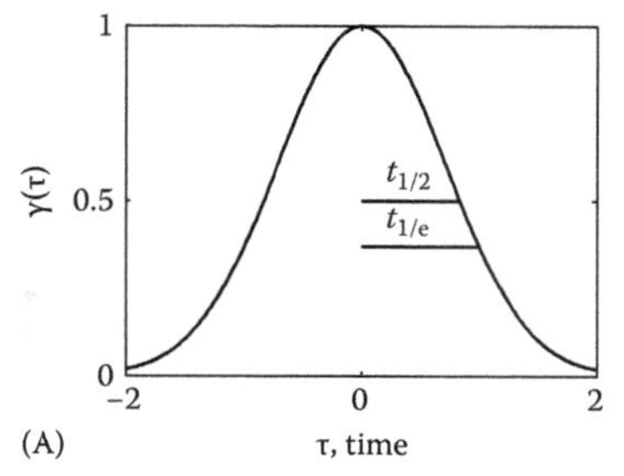

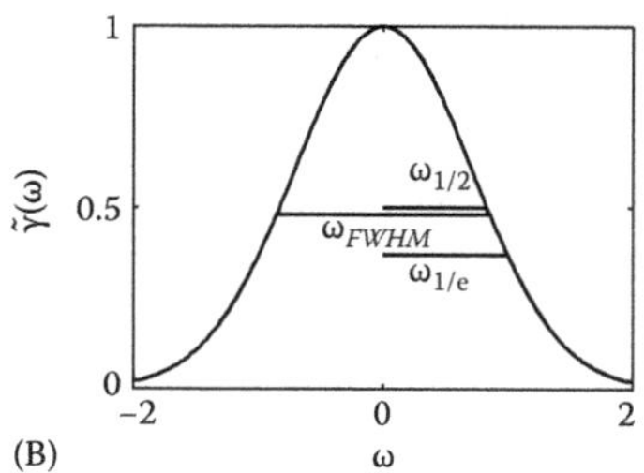

FIGURE 10.4
Gaussian functions. Various definitions of width are shown. We want to define the coherence time in the time domain (A) using the half width at half maximum, $t_{1/2}$, and the linewidth using FWHM, ω_{FWHM} in the frequency domain (B).

10.3.2 Example: LED

As a numerical example, suppose that we wish to build a Mach–Zehnder interferometer using a light-emitting diode (LED) as a source. The LED has a center wavelength of $\lambda_0 = 635\,\text{nm}$ (near that of a helium-neon laser at 633 nm), and a Gaussian spectral shape with a linewidth of $\lambda_{FWHM} = 20\,\text{nm}$. We have chosen to define the linewidth as the full-width at half-maximum (FWHM). How well must the interferometer arms be matched? First, because the linewidth is small compared to the central wavelength, we can use the fractional linewidth approximation we used in Equation 7.127:

$$\lambda f = c \qquad \frac{d\lambda}{\lambda} + \frac{df}{f} = 0,$$
$$\left|\frac{\delta\lambda}{\lambda}\right| = \left|\frac{\delta f}{f}\right|, \tag{10.31}$$

so we can obtain the fractional linewidth in frequency units.

Now we need to take the Fourier transform of a Gaussian. Relationships among widths in time and frequency are confusing because different relationships exist for different shapes, and there are several different definitions of width for different functions. We take some care here to make precise definitions and derive correct constants, but often approximations are sufficient. The Fourier transform pair is

$$\gamma(\tau) = e^{-(t/t_{1/e})^2} \qquad \frac{2\sqrt{\pi}}{\omega_{1/e}}\tilde{\gamma}(\omega) = e^{-(\omega/\omega_{1/e})^2} \tag{10.32}$$

where

$$\omega_{1/e} = \frac{2\sqrt{2}}{t_{1/e}}. \tag{10.33}$$

We note that $t_{1/e}$ and $\omega_{1/e}$ are defined as the half width at the $1/e$ field points ($1/e^2$ irradiance points), so we need to convert to FWHM. The multitude of linewidth terms can become confusing here, so we adopt the convention of using a subscript $1/e$ to denote the half width at $1/e \approx 36.7\%$ of the peak amplitude, $1/2$ to denote the half width at half maximum, and $FWHM$ to denote the full width at half maximum, as shown in Figure 10.4,

$$e^{-(\omega_{1/2}/\omega_{1/e})^2} = \frac{1}{2} \qquad \text{at} \qquad \omega_{1/2} = \sqrt{\ln 2}\,\omega_{1/e} \tag{10.34}$$

and $\omega = 2\pi f$, results in the equations,

$$2\pi f_{1/2} = \omega_{1/2} = \sqrt{\ln 2}\,\omega_{1/e} \qquad \omega_{1/e} = \frac{2\pi f_{1/2}}{\sqrt{\ln 2}}. \tag{10.35}$$

Using the same definitions and subscripts in the time domain as we did in the frequency domain, the autocorrelation falls to half at $t_{1/2}$ defined by

$$\gamma(\tau) = e^{-\left(t_{1/2}/t_{1/e}\right)^2} = 1/2 \tag{10.36}$$

$$t_{1/2} = \sqrt{\ln 2}\,t_{1/e}. \tag{10.37}$$

Using Equations 10.33 and 10.35,

$$t_{1/2} = \sqrt{\ln 2}\frac{2\sqrt{2}}{\omega_{1/e}} = \ln 2\frac{2\sqrt{2}}{2\pi f_{1/2}} \approx \frac{0.31}{f_{1/2}} = \frac{0.62}{f_{FWHM}}. \tag{10.38}$$

If the path lengths differ by

$$z_{1/2} = ct_{1/2}, \tag{10.39}$$

$$\gamma\left(t_{1/2}\right) = e^{\frac{t_{1/2}}{t_{1/e}}} = \frac{1}{2}$$

by Equation 10.35 and the fringe visibility will drop by a factor of two. To evaluate the results numerically, it is convenient to express this distance in wavelengths.

$$\frac{z_{1/2}}{\lambda} = ft_{1/2} = \ln 2 \times \frac{2\sqrt{2}}{\pi}\frac{f}{f_{FWHM}} \approx 0.62\frac{f}{f_{FWHM}}, \tag{10.40}$$

We define $z_{1/2}$ as the **coherence length** of the source, and now we can relate it to the linewidth in wavelengths. The coherence length in wavelengths is proportional to the inverse of the linewidth in wavelengths,

$$\frac{z_{1/2}}{\lambda} \approx 0.62\frac{f}{\delta f_{FWHM}} = 0.62\frac{\lambda}{\lambda_{FWHM}} \qquad \text{(Gaussian spectrum)}, \tag{10.41}$$

In our example, the linewidth is 20/635 wavelengths, and so the coherence length is approximately

$$\frac{z_{1/2}}{\lambda} \approx 0.62 \times \frac{635}{20} \approx 20\,\text{Wavelengths} \qquad z_{1/2} \approx 12.5\,\mu\text{m}, \tag{10.42}$$

If the difference in path lengths exceeds this value, then the fringe visibility will be less than 1/2.

It is important to note that if the shape of the line is not Gaussian, the line is not measured as FWHM, or the desired correlation time is not defined as $t_{1/2}$, the constant, 0.62, will be different, and it is not unusual to approximate Equation 10.41 without the numerical constants as

For an approximate answer we can neglect the constants. The correlation length in wavelengths is the reciprocal of the fractional linewidth. Eq. 10.43 is a good approximation and easier to remember without the numerical constants.

$$\frac{\delta t}{T} = \frac{\delta z}{\lambda} \approx \frac{f}{\delta f} = \frac{\lambda}{\delta\lambda}. \tag{10.43}$$

The correlation length in wavelengths equals the correlation time in cycles. Both are approximately the reciprocal of the fractional linewidth, either in frequency or wavelength units. Generally we are interested in determining whether γ is closer to one or zero, and the exact constant is not important.

Example

For comparison, a typical 633 nm helium-neon laser has a coherence length of about 30 cm or less. What is the linewidth, $\Delta\lambda$?

Using Equation 10.41,

$$\frac{30 \times 10^{-2}\,\mathrm{m}}{633 \times 10^{-9}\,\mathrm{m}} \approx 4.7 \times 10^{5} \approx 0.62\frac{f}{f_{FWHM}} = 0.62\frac{\lambda}{\lambda_{FWHM}} \tag{10.44}$$

$$f_{FWHM} = \frac{0.62f}{4.7 \times 10^{5}} = \frac{0.62c}{47 \times 10^{6}\lambda} = 620 \times 10^{6}\,\mathrm{Hz} \tag{10.45}$$

$$\lambda_{FWHM} = \frac{0.62\lambda}{47 \times 10^{5}} = 830 \times 10^{-15}\,\mathrm{m}. \tag{10.46}$$

The laser has a linewidth of about 620 MHz in frequency or 830 fm in wavelength.

In Practice

Normally we are interested in saying simply that light is coherent or incoherent. Because such a statement requires saying that the measurement time is *much* more or less than the coherence time, we needn't concern ourselves with the constants, as we can see in the example that follows. We simply use Equation 10.43.

Of course, if we are studying partial coherence, then we must pay attention to the details and be very careful about definitions.

Example

Let's consider a white-light interferometer. How closely must the path lengths be matched?

We'll think of white light as having a spectrum extending from about 400 to 800 nm. We could plot the spectrum and compute its Fourier transform to obtain the autocorrelation function, but as we are not asking about small details, the complexity would not be justified. Instead we use Equation 10.43. If we take 600 nm as the center of the spectrum, then the fractional linewidth is

$$\frac{800\,\mathrm{nm} - 400\,\mathrm{nm}}{600\,\mathrm{nm}} = \frac{2}{3},$$

the coherence time is 3/2 cycles and the coherence length is 3/2 wavelengths. Thus the path lengths must be matched to a distance much smaller than 900 nm.

Example

Now Let's pass the white light in the example above through a 6 nm wide bandpass filter, centered at 600 nm, made with the multi-layer coatings discussed in Section 7.7. Now how closely must the path lengths be matched?

Again we use Equation 10.43. The fractional linewidth is

$$\frac{6\,\text{nm}}{600\,\text{nm}} = 0.01,$$

the coherence time is 100 cycles and the coherence length is 100 wavelengths. Thus the path lengths must be matched to a distance much smaller than $60\,\mu\text{m}$.

Take-Away

Coherent and Incoherent Light:
Typically a laser is coherent over a relatively long distance, an LED over a much shorter distance, and a white-light source is incoherent except at path-length differences much less than a wavelength.

10.3.3 Example: Beamsplitters

In Section 7.6, we saw that interference effects can cause undesirable fringes to appear in the output waves of beamsplitters. Now we can see one way to avoid the problem. Let us consider a beamsplitter for visible light built on a glass substrate 5 mm thick ($n = 1.5$). If the round trip through the glass, $\delta z = n \times 2 \times 5\,\text{mm} = 15\,\text{mm}$, is much longer than the coherence length, then no fringes will be observed. We assume normal incidence, so we do not worry about the angle of the path. Let us continue to use the LED and laser sources in Section 10.3.2. Then the wavelength is $\lambda = 633\,\text{nm}$, and

$$\frac{z_{1/2}}{\lambda} = \frac{15 \times 10^{-3}\,\text{m}}{633 \times 10^{-9}\,\text{m}} = 23,700 \approx 0.62\frac{f}{f_{FWHM}} = 0.62\frac{\lambda}{\lambda_{FWHM}}. \tag{10.47}$$

$$\lambda_{FWHM} = \frac{0.62 \times 633\,\text{nm}}{23,700} = 16.6\,\text{pm} \tag{10.48}$$

$$f_{FWHM} = \frac{0.62c}{23,700\lambda} = 12.4\,\text{GHz}. \tag{10.49}$$

The LED ($z_c = 12.5\,\mu\text{m} \ll 15\,\text{mm}$) will not produce fringes in this beamsplitter but the laser ($z_c = 30\,\text{cm} \gg 15\,\text{mm}$) will.

Take-Away

Interference Artifacts in Beamsplitters:
We usually need to be concerned about interference when using beamsplitters with laser beams because of coherent addition of the reflections from the two surfaces.

10.3.4 Example: Optical Coherence Tomography

Here we discuss an optical imaging technique that has proven useful in imaging retina, lung, walls of coronary arteries, and other tissues to depths of near a millimeter. The technique, **optical coherence tomography** (OCT), uses interferometry with a light source having a short coherence length, to select the portion of the light from a biological

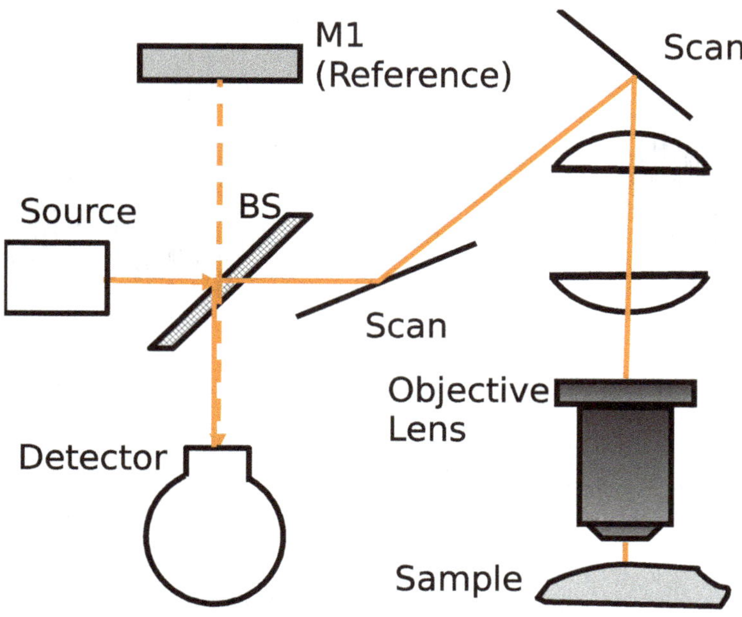

FIGURE 10.5
Optical coherence tomography. A simple OCT system is a Michelson interferometer where the sample replaces the mirror in one path. The goal is to measure backscattered light as a function of depth. "Coherence gating" limits the range of depths from which we detect an interference signal.

imaging measurement that provides the best resolution[101]. The reader may wish to read the material on biological tissue in Appendix D to better understand the need for this technology.

Let us review the Michelson interferometer in Figure 7.17. First, let us call mirror $M2$ the "target" and $M1$ the reference. With these changes, a line drawing of the OCT instrument is shown in Figure 10.5. We start with $M1$ far from the beamsplitter, so that the path lengths differ by more than the coherence length of the source. The signal from our detector will be the result of incoherent summation. Now we move $M1$ toward the beamsplitter. When the path lengths are within a coherence length, we expect to see an oscillating signal. The phase of the mixing term will pass through a full cycle for every half wavelength of displacement, leading to alternating constructive and destructive interference. The signal will appear as shown in Figure 10.6A. The envelope of the interference signal is proportional to the autocorrelation function of the source, and is the impulse response of the OCT system in the z direction.

Fig. 7.17:

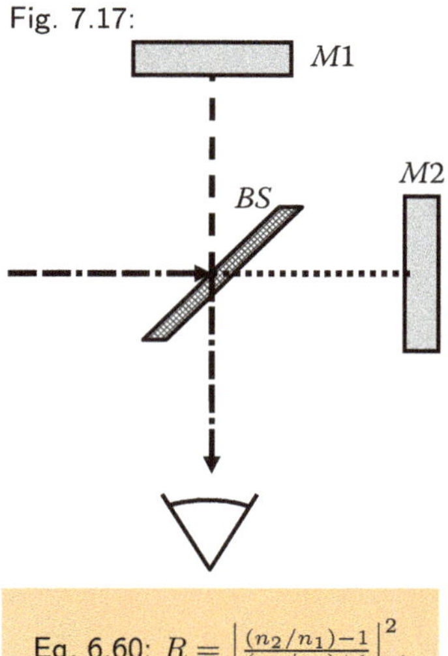

Eq. 6.60: $R = \left|\frac{(n_2/n_1)-1}{(n_2/n_1)+1}\right|^2.$

Next, we replace mirror $M2$ with a target that has two reflective surfaces. We can use a very thin slab of transparent material such as a glass slide. The Fresnel reflection from each surface is about 4% (Equation 6.60). Now when the path length to $M1$ matches to that of the first surface the return from that surface mixes coherently and produces an AC signal. When the path length matches to that of the second surface, then a second AC signal is generated. It is lower than the first one because some light is reflected at the first surface (Fresnel reflection) and possibly because of absorption in the medium. The overall envelope of the total AC signal would be the map of reflected signal as a function of z, if the autocorrelation function were infinitely narrow. Of course, if that were true, the signal would be very small and we could not detect it. The actual signal is the convolution of the autocorrelation function with this desired signal. If we can make the coherence time smaller than the desired resolution, then we can use the signal directly.

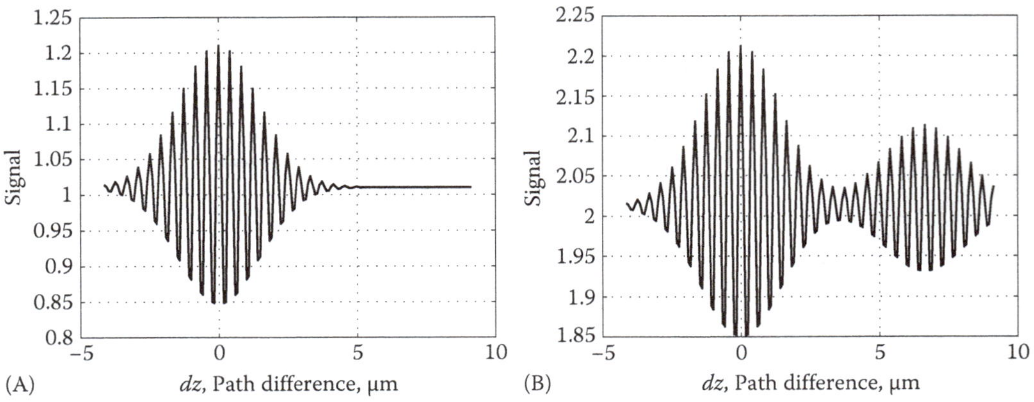

FIGURE 10.6
Simulated OCT signals. The first signal (A) is from mirror $M2$ in Figure 7.17. In B, the mirror is replaced by a target with two surfaces, the first being more reflective than the second.

Often, OCT systems use super-luminescent diodes for moderate resolution, or mode-locked lasers such as titanium-sapphire to achieve resolution limits of a few wavelengths of the optical wave. Actual OCT systems usually use fiber-optical splitters, and a large number of techniques have been developed to optimize performance for various applications. Since the first paper on OCT [101], hundreds of papers on the subject have been published. The interested reader may wish to read a book by Bouma and Tearney on the subject [29].

Take-Away

OCT:
Optical Coherence Tomography is a technique for obtaining axial resolution of a few wavelengths in measurement of backscatter in tissue and other turbid media. Axial resolution is provided by coherence gating. The OCT instrument can be scanned in one or two transverse dimensions to produce an image.

In Practice

In OCT, a complete axial (z) measurement is made at one time while scanning slowly transversely (x and/or y). Thus the depth of field is normally made large, so the Gaussian laser beam must be nearly collimated at the depth of interest (See Section 9.3.3). If the beam is focused with a high-numerical-aperture objective, then the signal will be weak away from focus where the beam is large. The depth of field is approximately λ/NA^2. For a Gaussian beam, the FWHM of the axial signal is twice the Rayleigh range (Equation 9.21).

$$2b = 2\frac{\pi d_0^2}{4\lambda}$$

Therefore, OCT is normally used with a low NA, leading to a poor transverse resolution, λ/NA.

Eq. 9.21:
$b = \frac{\pi w_0^2}{\lambda} = \frac{\pi d_0^2}{4\lambda}$.

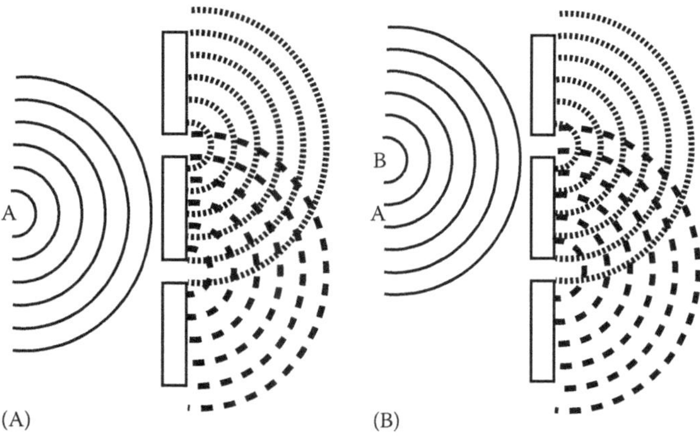

FIGURE 10.7
Spatial coherence. A point source produces a spherical wave (A), which is coherent across the distance between the two points shown. If the source is moved (B), the coherence is maintained. If multiple sources are used, the coherence may be degraded.

10.4 Spatial Coherence

In Section 10.3, we saw that two parts of a light wave from a single source are coherent if their transit times in an interferometer are closely matched, but are incoherent if the disparity in times is long. In this section, we will explore spatial relationships, and we will see that the fields are coherent at two different positions if the positions are sufficiently close together. For example, in Figure 8.2A, a coherent plane wave was incident on two slits to produce a diffraction pattern. If instead the source were "spatially incoherent" so that the fields at the two slits were randomized, then the periodic diffraction pattern might not be observed. As we might suspect by analogy with our study of temporal coherence, if the slits are sufficiently close together, the fields there will be coherent, and if they are far apart, they will not. Our analysis in this section is parallel to that in the previous one.

Fig. 8.2A:

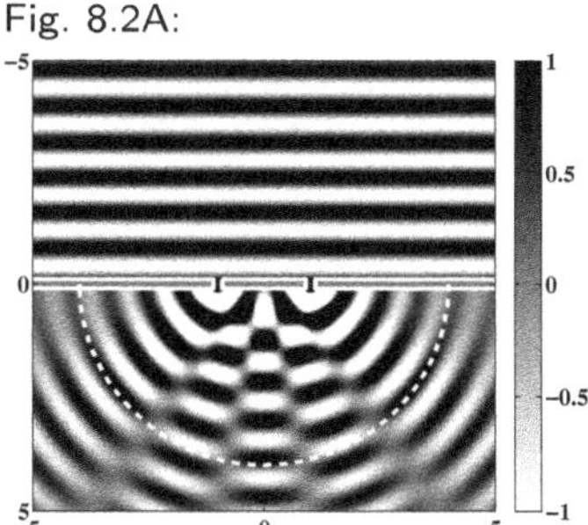

Just as we can characterize the temporal behavior of light through Fourier analysis, we can do the same with the spatial behavior. We recall from Section 8.4.4 that the field in the pupil plane is the Fourier transform of that in the field plane. In the study of temporal coherence, we computed the autocorrelation function, $\gamma(\tau)$ by determining the field at two different times and then the visibility of interference fringes. We usually relate the temporal behavior to differences in the propagation distance, z. Here, we will sample the field at two different transverse distances, in the x, y plane, developing a spatial autocorrelation function, $\gamma(\xi)$ where ξ is a distance between two points in the transverse plane, just as τ was a separation in time for the temporal function, $\gamma(\tau)$. We can understand spatial coherence with the aid of Figure 10.7. A good description of the concepts can be found in an article by Roychoudhuri[198].

In Figure 8.2A, we see the diffraction pattern that results from a plane wave passing through two slits. In the case of Figure 10.7A, the incident wave is spherical, from a point source, but because the slits are small, the diffraction pattern will still look like the one in Figure 8.2A.

If the point source location is shifted up or down the page as in Figure 10.7B, then the phase difference between the fields in the slits will be different and the peaks and valleys of the diffraction pattern will be different, but the fringe contrast will be the same. In order to make a substantial change in the fringe pattern, the source must be moved enough so that the phase difference between the fields in the slits changes significantly. In the case of this spherical wave, the light through the upper slit is coherent with that through the lower one. The phase of γ may change, but the magnitude does not.

What happens to the spatial coherence if we use the two point sources at A and B simultaneously? The answer depends on the temporal coherence of the two sources. If they are mutually coherent,

$$\langle|E_1^* E_2|\rangle = \sqrt{|E_1|^2 |E_2|^2}, \tag{10.50}$$

then the total field at each slit will have a well-defined value. There will be a specific phase relationship between the two. If the amplitudes happen to be equal, the fringe pattern will have a visibility, $V = 1$. On the other hand, if the sources are temporally incoherent (or at least partly so), then at any point, the two fringe patterns will randomly add constructively or destructively as the phase relationship between the sources changes. In this case, averaged over time, the visibility of the resultant fringe pattern will be reduced.

Take-Away

Fringe Visibility:
The ability to produce fringes in the two-slit experiment depends on the spatial coherence width being wider than the space between the slits.

10.4.1 Van–Cittert–Zernike Theorem

The Van–Cittert–Zernike theorem is the spatial equivalent of the Wiener–Khintchine theorem. The derivation is very similar to the Weiner–Khintchine theorem (Section 10.3.1). We write expressions for the fields at two locations as Fourier transforms, combine terms, exchange the order of integration and ensemble average, find a delta function in the inner integral, and simplify. Just as we showed that $\gamma(\tau)$ is the Fourier transform of the irradiance density spectrum, we will show that in a pupil plane, $\gamma(\xi)$ is the Fourier transform of the irradiance distribution in the image, $I(x, y, 0)$. This theorem was first derived by van Cittert in 1934[231], and 4 years later, more simply, by Frits Zernike[261], who went on to win the Nobel Prize in 1953 for the phase contrast microscope.

The geometry for the analysis is shown in Figure 10.8. The coordinates in the object at are $(x, y, 0)$. The pupil coordinates are (x_1, y_1, z_1) at axial location, z_1.

We begin by defining the spatial autocorrelation function between two points (x_1, y_1, z_1), and (x_1', y_1', z_1) in the pupil plane as

$$\Gamma(x_1, y_1, x_1', y_1') = \langle E^*(x_1', y_1', z_1)\, E(x_1, y_1, z_1)\rangle\,. \tag{10.51}$$

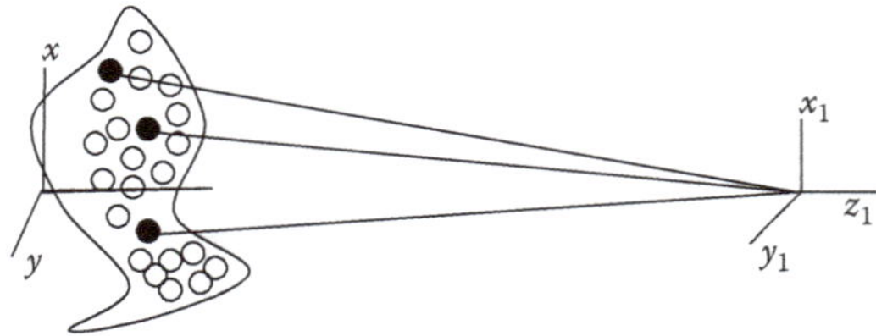

FIGURE 10.8
Model for spatial coherence. The object consists of sources that are all independent with coordinates, $(x, y, 0)$, and the pupil has the coordinates (x_1, y_1, z_1).

Using Equation 8.60 the field is

$$E(x_1, y_1, z_1) = \frac{jke^{jkz_1}}{2\pi z_1} e^{jk\frac{x_1^2+y_1^2}{2z_1}} \iint E(x, y, 0)\, e^{jk\frac{x^2+y^2}{2z_1}} e^{-jk\frac{xx_1+yy_1}{z_1}}\, dx\, dy \tag{10.52}$$

and the conjugate of the field, using new variables, x_1', y_1', x', y' in place of x_1, y_1, x, y, is

$$E^*(x_1', y_1', z_1) = \frac{-jke^{-jkz_1}}{2\pi z_1} e^{-jk\frac{(x_1')^2+(y_1')^2}{2z_1}} \times \iint E^*(x', y', 0)\, e^{-jk\frac{(x')^2+(y')^2}{2z_1}} e^{+jk\frac{x'x_1'+y'y_1'}{z_1}}\, dx'\, dy' \tag{10.53}$$

Now, the autocorrelation function can be written by combining Equations 10.52 and 10.53;

$$\Gamma(x_1, y_1, x_1', y_1') = \frac{-jke^{-jkz_1}}{2\pi z_1} e^{-jk\frac{(x_1')^2+(y_1')^2}{2z_1}} \frac{jke^{jkz_1}}{2\pi z_1} e^{jk\frac{x_1^2+y_1^2}{2z_1}} \times \iint\left[\iint \langle E^*(x', y', 0)\, E(x, y, 0)\rangle\, e^{-jk\frac{(x')^2+(y')^2}{2z_1}} e^{+jk\frac{x^2+y^2}{2z_1}} \times e^{+jk\frac{x'x_1'+y'y_1'}{z_1}} e^{-jk\frac{xx_1+yy_1}{z_1}}\, dx'\, dy'\right] dx\, dy. \tag{10.54}$$

Noting that, for example, $x^2 - (x')^2 = (x + x')(x - x')$, we can combine terms to arrive at

$$\Gamma(x_1, y_1, x_1', y_1') = \frac{-jke^{-jkz_1}}{2\pi z_1} e^{jk\frac{(x_1')^2+(y_1')^2}{2z_1}} \frac{jke^{jkz_1}}{2\pi z_1} e^{-jk\frac{x_1^2+y_1^2}{2z_1}} \times \iint\left[\iint \langle E^*(x', y', 0)\, E(x, y, 0)\rangle\, e^{jk\frac{(x'+x)(x'-x)+(y'+y)(y'-y)}{2z_1}} \times e^{jk\frac{(xx_1-x'x_1')+(yy_1-y'y_1')}{z_1}}\, dx'\, dy'\right] dx\, dy \tag{10.55}$$

Because of the random phases, $\langle E\rangle = 0$. If the source points are all incoherent with respect to each other,

$$\langle E^*(x', y', 0)\, E(x, y, 0)\rangle = 0,$$

for any two distinct points $(x, y, 0)$ and $(x', y', 0)$. Thus, by analogy with Equation 10.27,

$$\langle E^*(x', y', 0)\, E(x, y, 0)\rangle\, dx'dy' = I(x, y, 0)\, \delta(x - x')\, \delta(y - y')\, dx'dy' \tag{10.56}$$

where again, $\delta(\cdot)$ is the Dirac delta function. Then

$$\Gamma(x_1, y_1, x_1', y_1') = \frac{-jke^{-jkz_1}}{2\pi z_1} e^{jk\frac{(x_1')^2+(y_1')^2}{2z_1}} \frac{jke^{jkz_1}}{2\pi z_1} e^{-jk\frac{x_1^2+y_1^2}{2z_1}} \iint \times \left[\iint I(x, y, 0)\, \delta(x - x')\, \delta(y - y')\, e^{jk\frac{(xx_1 - x'x_1') + (yy_1 - y'y_1')}{z_1}} dx'\, dy' \right] dx\, dy, \tag{10.57}$$

and the outer integral reduces to

$$\Gamma(x_1, y_1, x_1', y_1') = \frac{-jke^{-jkz_1}}{2\pi z_1} e^{jk\frac{(x_1')^2+(y_1')^2}{2z_1}} \frac{jke^{jkz_1}}{2\pi z_1} e^{-jk\frac{x_1^2+y_1^2}{2z_1}} \times \iint I(x, y, 0)\, e^{jk\frac{x(x_1 - x_1') + y(y_1 - y_1')}{z_1}} dx\, dy. \tag{10.58}$$

Finally, simplifying this equation,

$$\Gamma(x_1, y_1, x_1', y_1') = \left(\frac{k}{2\pi z_1}\right)^2 e^{jk\frac{(x_1')^2+(y_1')^2}{2z_1}} e^{-jk\frac{x_1^2+y_1^2}{2z_1}} \iint I(x, y, 0)\, e^{jk\frac{x(x_1 - x_1') + y(y_1 - y_1')}{z_1}} dx\, dy. \tag{10.59}$$

We note that, except for the curvature terms, $e^{-jk\frac{x_1^2+y_1^2}{2z_1}}$, this equation depends on its coordinates only in terms of differences. Let us define

$$\xi = x_1' - x_1 \qquad \eta = y_1' - y_1. \tag{10.60}$$

We can ignore the field curvature, which does not affect the magnitude of Γ, or we can correct for it optically by using a telecentric system (see Section 3.4.4). Then we can write the van–Cittert–Zernike theorem as

$$\Gamma(\xi, \eta) = \left(\frac{k}{2\pi z_1}\right)^2 \iint I(x, y, 0)\, e^{jk\frac{x\xi + y\eta}{z_1}} dx\, dy. \tag{10.61}$$

Eq. 8.60: $E(x_1, y_1, z_1) = \frac{jke^{jkz_1}}{2\pi z_1} e^{jk\frac{(x_1^2+y_1^2)}{2z_1}} \int\int E(x, y, 0) \times e^{-jk\frac{(xx_1+yy_1)}{z_1}} dx\, dy.$

Recall that for a coherent source Equation 8.60 applied to the field in the pupil, $E(x, y, 0)$, provides us the field in the image, $E(x_1, y_1, z_1)$:

$$E(x_1, y_1, z_1) = \frac{jke^{jkz_1}}{2\pi z_1} e^{jk\frac{(x_1^2+y_1^2)}{2z_1}} \iint E(x, y, 0)\, e^{jk\frac{(x^2+y^2)}{2z_1}} e^{-jk\frac{(xx_1+yy_1)}{z_1}} dx\, dy.$$

In contrast, Equation 10.61 applied to the irradiance pattern $I(x, y, 0)$ gives us the spatial autocorrelation function, $\Gamma(x_1 - x_1', y_1 - y_1')$ at z_1. The light illuminates the whole

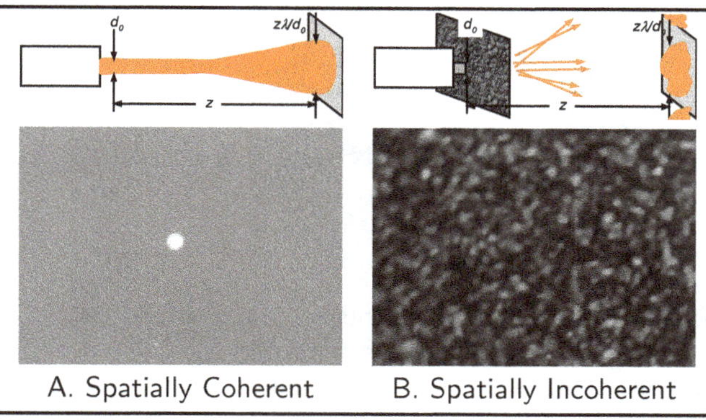

A. Spatially Coherent B. Spatially Incoherent

FIGURE 10.9
Coherent and incoherent illumination. The very coherent laser field (A) propagates to the far field (according to Fraunhofer diffraction), Equation 8.60. The diffraction pattern is Gaussian. With a ground glass placed immediately in front of the laser, the source size is the same, but now it is incoherent (B). In this case, the light is distributed widely across the field, and the spatial autocorrelation obeys the Fraunhofer diffraction equation with different constants, in the form of Equation 10.61. The speckles (B) are approximately the same size as the laser spot (A).

field plane. Let us emphasize this important result. We have two essentially equivalent equations, with different meanings:

$$\begin{array}{lllll} E(x,y,0) & \rightarrow & \text{Eq. } 8.60 & \rightarrow & E(x_1,y_1,z_1) \\ I(x,y,0) & \rightarrow & \text{Eq. } 10.61 & \rightarrow & \Gamma(x_1-x_1',y_1-y_1') \text{ at } z_1 \end{array} \tag{10.62}$$

We can understand this result from the photographs in Figure 10.9. We see a paper illuminated directly by a laser source approximately 1 mm in diameter (A). A spatially incoherent source of the same size is generated (B) by using the same source to illuminate a ground glass . In both cases, the distance from the source to the paper is about 5 m. The coherent laser beam propagates according to the Fresnel–Kirchhoff integral, and illuminates a small spot, while the incoherent light illuminates the whole wall. Speckles in this picture have an average size comparable to the size of the laser beam in the coherent case.

Random Scattering is discussed in Section 10.4.3.

Take-Away

Van–Cittert–Zernike Theorem:
The van–Cittert–Zernike theorem relates the coherence length in x and y to the size of the incoherent source in x_1 and y_1.

10.4.2 Example: Coherent and Incoherent Source

Let us calculate a numerical solution to the problem shown in Figure 10.9. A helium-neon laser beam with a diameter of 1 mm and the same laser beam forward-scattered through a ground glass, each in turn, illuminate a wall 5 m away. If we assume the laser beam has a Gaussian profile, then we can use the Gaussian results in Figure 8.16. The beam diameter at the wall is

$$d = \frac{4}{\pi}\frac{\lambda}{D}z = \frac{4}{\pi}\frac{633\times10^{-9}\,\text{m}}{10^{-3}\,\text{m}}5\,\text{m} = 4\times10^{-3}\,\text{m}, \tag{10.63}$$

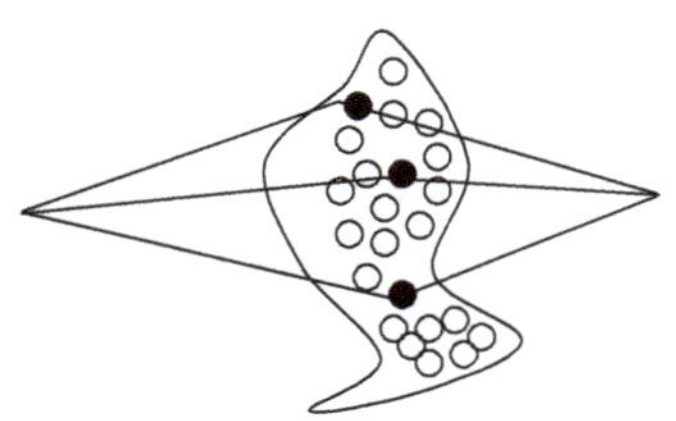

FIGURE 10.10
Speckle. Light from a source on the left is incident on a rough surface such as a ground glass. The surface is modeled as collection of scattering particles. Three paths are shown from a source point to an image point. The scattered field is randomized by variations in the particle location.

at $1/e^2$ irradiance. Now, for the incoherent source, the light is spread over the entire wall as in Figure 10.9B. The diameter of the autocorrelation function is the same as the diameter of the laser beam, 4 mm. If we look at the field at any point on the wall, then at distances less than the radius, $4\,\text{mm}/2 = 2\,\text{mm}$, away, the field will be correlated, and if we were to place pinholes at these locations, the light passing through would create a diffraction pattern. If the spacing of the pinholes were greater, interference would not be produced.

If we perform the incoherent experiment with white light from a tungsten source, filtered to pass the visible spectrum, from about 400 to 800 nm, then in Equation 10.31,

$$\frac{\delta\lambda}{\lambda} = \frac{\delta f}{f} \to 1. \tag{10.64}$$

Eq. 10.43:
$\frac{\delta t}{T} = \frac{\delta z}{\lambda} \approx \frac{f}{\delta f} = \frac{\lambda}{\delta\lambda}$.

The coherence length from Equation 10.43 becomes comparable to the "carrier wavelength" to the extent that a meaningful carrier wavelength can even be defined over such a broad band. The correlation time is on the order of a cycle, and any reasonable detection system will average so many independent samples that the measurement of irradiance will yield a uniform result equal to the expectation value. As we know from everyday observation of white light, no speckles will be observed.

10.4.3 Speckle in Scattered Light

Fig. 10.8:

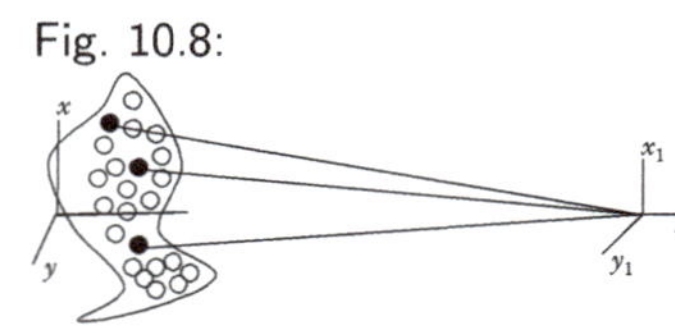

The scattering of light from a rough surface provides an interesting insight into spatial coherence, and is important in many applications. Our model of the rough surface is shown in Figure 10.10. The difference between this figure and Figure 10.8 is that the points in Figure 10.10 represent a collection of randomly located scattering particles while in Figure 10.8 they represented random sources. Each particle here scatters a portion of the incident light as a spherical wave, with a random amplitude and phase so the result is very similar except that the irradiance is determined by the incident light. A physical realization of such an object could be the ground glass we have already discussed. Incident light will be scattered forward and backward. In the figure, we consider the light that is scattered forward, but our analysis would hold equally well for light that is scattered backward.

Eq. 10.61:
$\Gamma(\xi,\eta) = \left(\frac{k}{2\pi z_1}\right)^2 \iint I(x,y,0)\, e^{jk\frac{x\xi+y\eta}{z_1}}\, dx\, dy.$

Now, suppose that the light source is a laser, which we can consider to be monochromatic. Each point in the surface is illuminated by a portion of the wave, and scatters its own contribution to the field at the detector. If the positions are sufficiently random, then each point will act as a random source, with its own amplitude and phase, and the expectation value of the irradiance will be the sum of all the irradiance contributions. However, the local irradiance will fluctuate with the various mixing terms, and the spatial autocorrelation function will be given by Equation 10.61. The light will be spread over the wall at the detector plane, with the mean irradiance uniform, but with bright and dark

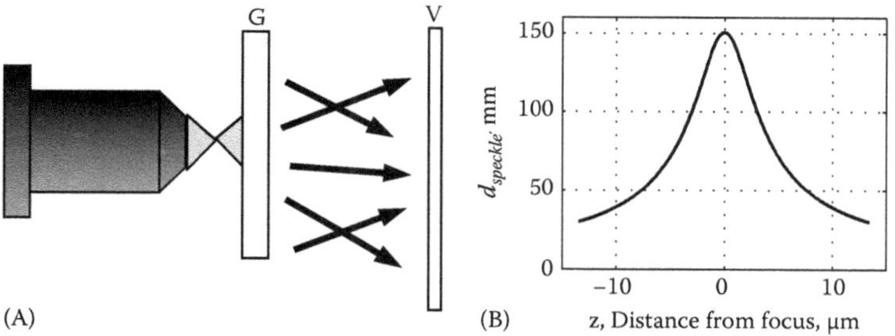

FIGURE 10.11
Finding the focal point. The experiment (A) includes a focused laser beam, a ground glass (G), and a viewing screen (V). The speckle size from a numerical example (B) becomes very large when the ground glass is placed at the focus (≈ 0).

spots associated with constructive and destructive interference. The size of the spots will be determined by the size of the spatial autocorrelation function.

10.4.3.1 Example: Finding a Focal Point

The exact focal point of a beam can be found by watching the speckle pattern in light scattered from a ground glass plate as it is translated through the focus. The speckles will be largest when the rough surface is closest to the focus. As an example, let us consider a Gaussian beam from the frequency-doubled output of an Nd:YAG laser focused through a microscope objective as shown in Figure 10.11A, toward a piece of ground glass. The scattered light from the ground glass, viewed on a screen some distance beyond the ground glass, behaves as an incoherent source, with an irradiance profile determined by the profile of the incident light. As shown, the focus is before the ground glass, the spot is large, and the speckles are small. As the ground glass is moved toward the focus, the speckle pattern becomes larger, and after the focus, it becomes smaller again. If the numerical aperture is 0.25, and the objective is slightly underfilled, the waist of the Gaussian beam will be as shown by Equation 9.35 approximately

$$d_0 = \frac{4}{\pi}\frac{\lambda}{d}z \qquad d_0 = \frac{2}{\pi}\frac{\lambda}{NA}. \tag{10.65}$$

For the present case,

$$d_0 = \frac{2}{\pi}\frac{532\times 10^{-9}\text{m}}{0.25} = 1.4\times 10^{-6}\,\text{m}. \tag{10.66}$$

Now, as we move the ground glass axially through the focus, the size of the focused spot on the ground surface of the glass will vary according to Equation 9.23

$$d_{glass} = d_0\sqrt{1+\frac{z^2}{b^2}}, \tag{10.67}$$

where, z is the distance from focus, and, according to Equation 9.21,

$$b = \frac{\pi d_0^2}{4\lambda}.$$

The spatial autocorrelation function of the scattered light can be obtained by Equation 10.61, but we do not really need to do the integral. Equation 10.61 states that the autocorrelation function can be obtained from the irradiance using the same integral as is used for coherent light to obtain the field. We can thus imagine a coherent beam of the same diameter as the illuminated spot on the ground glass surface, and propagate that beam to the screen. Fortunately we have a closed-form expression for that field from Equation 9.25:

$$d_{field} = \frac{4}{\pi}\frac{\lambda}{d_{glass}}z$$

The spatial autocorrelation function will be Gaussian with a diameter,

$$d_{speckle} = \frac{4}{\pi}\frac{\lambda}{d_{glass}}z, \tag{10.68}$$

using Equation 10.65. The speckles will, on average, be approximately this size, although they will have random shapes. The diameter is plotted in Figure 10.11B. We can use this technique to find the location of a focused spot.

In summary, if we know the equation for propagation of the electric field from the object to the pupil, we can apply the same equation to the object irradiance to determine the spatial autocorrelation function in the pupil.

In Practice

Often we need to find the focal point of a laser beam. For example, using a laser as a light source for a microscope, we need to know where to place the sample. By moving a ground glass axially along the beam and observing the speckle pattern of the scattered light, it is easy to find the focal point. At that point, the speckles will be largest.

10.4.3.2 Example: Speckle in Laser Radar

Next, we consider the effects of speckle in a laser radar. As shown in Figure 10.12, the light starts from a collimated source, typically a Gaussian beam expanded to a large diameter. Just as in Section 10.4.3.1, the scattered light behaves like an incoherent source with a diameter determined by the laser footprint on the target. The equation is the same as Equation 10.67, but the values are quite different. Let us assume that the beam is Gaussian with a diameter $d_0 = 30\,\text{cm}$, from a carbon dioxide laser with $\lambda = 10.59\,\mu\text{m}$. Then, at the target, using Equation 9.23,

$$d_{target} = d_0\sqrt{1 + \frac{z_t^2}{b^2}}, \tag{10.69}$$

where z_t is the distance from the laser transmitter to the target. The Rayleigh range is approximately $b = \pi d_0^2/(4\lambda) = 6.7\,\text{km}$. In the far field, $z \gg b$, the equation simplifies to

$$d_{target} = \frac{4}{\pi}\frac{\lambda}{d_0}z_t, \tag{10.70}$$

and at $z_t = 20\,\text{km}$, the spot size is about $90\,\text{cm}$. The speckle size is about

$$\frac{4}{\pi}\frac{\lambda}{d_{target}}z_1 = 15 \times 10^{-6}\,\text{rad} \times z_1. \tag{10.71}$$

Now, let us consider the speckle pattern of the backscattered light, as shown in Figure 10.13. The characteristic size of the speckle pattern at a receiver colocated with the transmitter, will be

$$d_{speckle} = \frac{4}{\pi}\frac{\lambda}{d_{target}}z_r,$$

where z_r is the distance from the target to the receiver. Combining this equation with Equation 10.70, for the very common case of a monostatic or nearly monostatic laser radar where $z_t = z_r = z$,

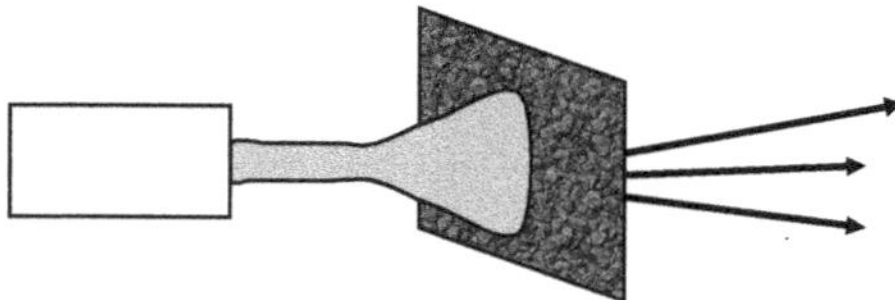

FIGURE 10.12
Speckle size in laser radar. A laser radar system with a diverging Gaussian beam produces a footprint on a distant target that increases linearly with distance.

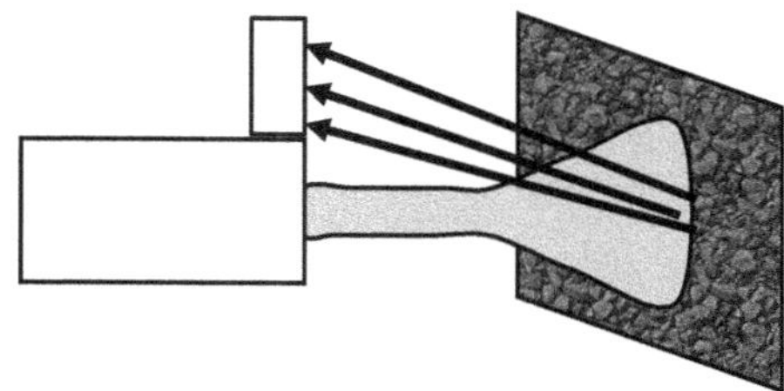

FIGURE 10.13
Speckle in a laser radar. The characteristic size of the speckles at the receiver is equal to the beam diameter of the transmitter, provided that the target is in the far field of the transmitter, and that the receiver is located at the same z location as the transmitter. The analysis is also correct at the focal plane of a focused system, and would be valid for the microscope shown in Figure 10.11.

$$d_{speckle} = \frac{4}{\pi} \frac{\lambda}{\frac{4}{\pi}\frac{\lambda}{d_0}z} z = d_0 \qquad \text{for} \qquad z_t = z_r = z. \tag{10.72}$$

In a monostatic laser radar viewing a diffuse (scattering) target, the speckle size at the receiver is equal to the transmitter beam diameter.

This simple result is very useful in the design of a laser radar. It is valid for a collimated beam when the receiver is located with the transmitter, when the target is in the far field. It is also valid at the focal plane of a focused system. In these cases, the characteristic size of a speckle is equal to the size of the transmitted spot.

Now if the receiver's aperture is small, it may lie within a bright speckle, producing a signal stronger than expected, or it may lie within a dark speckle, producing little or no

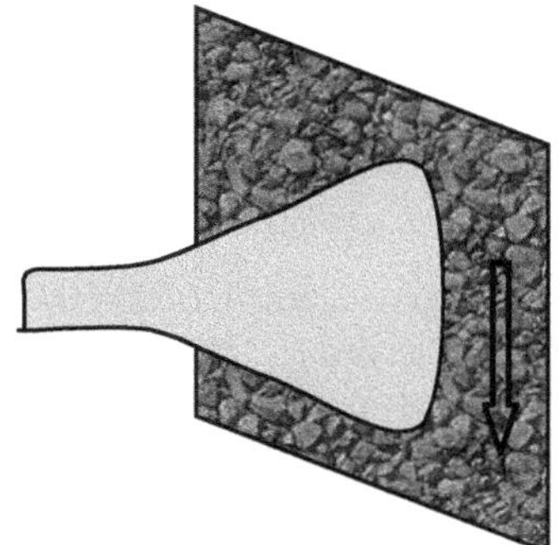

FIGURE 10.14
Speckle speed. As the target moves (arrow), new scatterers enter the beam and old ones leave. The speckle pattern changes, and the correlation time is the dwell time of a particle in the beam.

signal. However, if the target moves, as shown in Figure 10.14, new scattering particles will enter the laser footprint and old ones will leave. As a new target configuration with the same statistics is realized, the mean irradiance at the receiver will remain the same, but the mixing terms from the new particles will result in a new speckle pattern. The speckle pattern will change on a timescale comparable to the dwell time or the "time on target" of a particle in the beam because, in that time, the original scattering particles are mostly replaced with new ones. For a given position, $\gamma(\tau)$ has a decay time equal to the time on target.

If the transverse velocity of the target is $v_\perp$, the dwell time or **time on target** is

$$t_{target} = \frac{d_{target}}{v_\perp}. \tag{10.73}$$

For a Doppler laser radar, we recall from Equation 7.62 that the Doppler frequency is related to the parallel (axial) component of the target's velocity, by

$$f_{DR} = \frac{2v_\parallel}{\lambda}.$$

The spectral shape of that signal will be given by the Fourier transform of the autocorrelation function. Because the characteristic time of the autocorrelation function is given by Equation 10.73, the bandwidth associated with the time on target, B_{tot}, is proportional to the transverse component of the target's velocity:

The bandwidth contribution from transverse motion is the reciprocal of the time the beam spends on a point in the target.

$$B_{tot} = \frac{v_\perp}{d_{target}}. \tag{10.74}$$

We will be careful not to refer to this bandwidth at "the" bandwidth of the Doppler signal, because it is just one contribution to the bandwidth. The short pulse of a pulsed laser radar is one other contribution. The pulse bandwidth is at least as wide as the inverse of the pulse, as discussed in Section 10.2.

If we make a measurement of the strength of the Doppler signal in a time short compared to the width of the autocorrelation function, we will sample one value of the speckle pattern, and as a result, the measurement will be unreliable. This signal variation about the mean is called **speckle fluctuation**, and is present to some degree in *all* optical devices using coherent sources.

Take-Away

Speckle Size:
The speckle size of a fully random scattered light field is inversely proportional to the size of the illuminated spot on the scattering surface.

10.4.3.3 Speckle Reduction: Temporal Averaging

We can improve our measurement of the strength of the Doppler signal by averaging several samples, for example, by integrating over a longer time. In general, the improvement will be proportional to the square root of the number of samples, so if we integrate the signal for a time $T_{integration}$, we expect the standard deviation, σ of our measurement to improve according to

$$\sigma = \sqrt{\frac{t_{target}}{T_{integration}}}. \tag{10.75}$$

However, integration over long times is problematic for varying targets. In OCT, coherent laser radar, or any similar application, if we integrate too long, we lose temporal resolution.

10.4.3.4 Aperture Averaging

For an incoherent laser radar, we can further improve the estimate of the Doppler signal strength by **aperture averaging**. If the speckle sizes are matched to the diameter of the transmitter, then the use of a receiver larger than the transmitter will reduce the fluctuations associated with speckle. As is the case with integrating over a longer time, the improvement is proportional to the square root of the number of samples, and the number of samples is proportional to the area of the receiver, so for circular apertures the standard deviation reduces to

$$\sigma = \sqrt{\frac{A_{tx}}{A_{rx}}} = \frac{D_{tx}}{D_{rx}}. \tag{10.76}$$

where

- A_{tx} is the transmitter area
- A_{rx} is that of the receiver
- the A variables are the diameter

Frequently, a coaxial geometry is used with a small circular transmitter inside a larger circular receiver.

It is important to note that aperture averaging does not work for coherent detection. The mixing term in Section 7.2 is obtained by summing the scattered fields, rather than irradiances, and as new speckles contribute, both the mean and standard deviation of the signal are increased. Thus, instruments such as Doppler laser radars and OCT imagers suffer from speckle fluctuations, and one must make multiple measurements with incoherent summation to reduce them.

Aperture averaging does not occur with coherent detection.

Eq.10.72: $d_{speckle} = \frac{4}{\pi}\frac{\lambda}{\frac{4}{\pi}\frac{\lambda}{d_0}z}z = d_0$.

Example

A monostatic downward-looking Nd:YAG laser radar has a transmitter wavelength of 1.06 μm and a beam 3 cm in diameter and a coaxial receiver aperture 20 cm in diameter. What variability is expected in the signal returned from a diffuse surface in the far field? If the laser radar is mounted in an aircraft with a speed of 500 knots at an elevation of 10000 ft , what is the minimum bandwidth of the received signal?

Equation 10.72 states that the received speckle size is equal to the transmitter size. Thus, the speckle diameter is 3/20 of the receiver diameter, and there are 20/3 speckles in each direction. Over the area of the receiver there are about $N = (20/3)^2 \approx$ 45 speckles. The fluctuation will be reduced by $1/\sqrt{N} \approx 0.15$.

The far–field spot size on target is given by Equation 10.70, and 10000 ft is 3049 m.

$$d_{target} = \frac{4}{\pi}\frac{\lambda}{d_0} = 13.7\,\text{cm}.$$

This diameter is significantly larger than d_0 so our guess of being in the far field is valid. The speed of 500 nautical miles per hour is

$$500\,\text{nauticalmiles/hour}\times$$

$$1852\,\text{meters/nauticalmile/}$$

$$(3600\,\text{seconds/hour}) = 257\,\text{m/s}.$$

Using Equation10.74.,

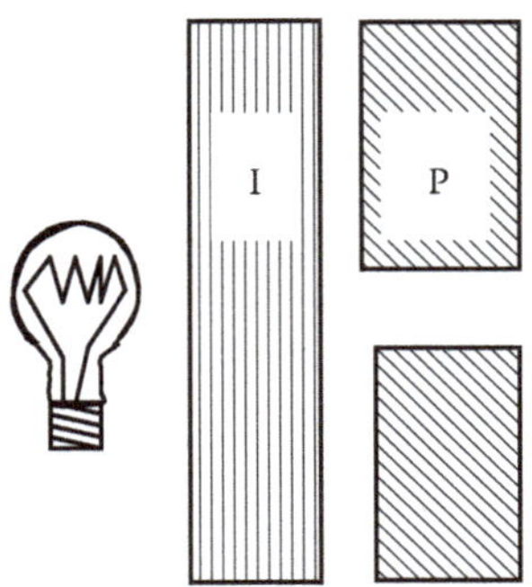

FIGURE 10.15
Making light coherent. We can improve the coherence of a source with an interference filter (I), and a pinhole (P).

$$B_{tot} = \frac{v_\perp}{d_{target}} = \frac{257\,\text{m/s}}{0.0137\,\text{m}} = 1874\,\text{Hz}.$$

The bandwidth is about 2 kHz.

10.4.3.5 Speckle Reduction: Other Diversity

When we average to reduce speckle, it is important that we average over diverse and independent samples. We can use temporal or spatial diversity as discussed above. We can also, in some cases, use angular diversity, wavelength diversity, or even, in some cases polarization diversity.

10.5 Controlling Coherence

We conclude this chapter by looking at techniques for controlling the coherence of light. Many applications of optics require highly coherent sources. Prior to the invention of the laser, it was difficult to obtain high brightness sources with good spatial and temporal coherence. It was necessary to begin with an incoherent source and modify the light to make it coherent, as shown in Figure 10.15, which was extremely inefficient. Now that lasers exist, sometimes we want to make light less coherent, to perform a particular experiment or to reduce undesirable speckle artifacts.

10.5.1 Increasing Coherence

Eq. 10.43:
$\frac{\delta t}{T} = \frac{\delta z}{\lambda} \approx \frac{f}{\delta f} = \frac{\lambda}{\delta \lambda}$.

Eq. 10.64:
$\frac{\delta \lambda}{\lambda} = \frac{\delta f}{f} \to 1$.

A narrow–band filter makes the coherence time much longer at the expense of greatly reduced irradiance.

A small spatial filter increases the spatial coherence of light, at the expense of greatly reduced irradiance.

First, because the coherence time is inversely related to the bandwidth, we can improve the temporal coherence by using a narrow interference filter such as we saw in Section 7.7, to pass only a narrow linewidth. For example, if we use a filter with a width of $\delta\lambda = 1\,\text{nm}$ at $\lambda = 633\,\text{nm}$, using Equation 10.43, we can achieve a coherence length of $\delta z = 633\lambda = 400\,\mu\text{m}$, which is a considerable improvement over a natural white light source (approximately one wavelength using Equation 10.64), but very small compared to many lasers. Furthermore, we accomplish this goal at the expense of removing most of the light from the source. We could achieve a "head start" on the spectral narrowing, by using a narrow-band source such as a sodium lamp, mercury lamp, or other source of narrow-band light. The low-pressure mercury lamp emits a set of discrete narrow lines, and an interference filter can be used to transmit only one of these lines. Light-emitting diodes typically have linewidths of 20–40 μm, and may have sufficient light for many applications.

The van–Cittert–Zernike theorem tells us that we can make light spatially coherent over an angle of approximately λ/D by focusing through an aperture of diameter D. A

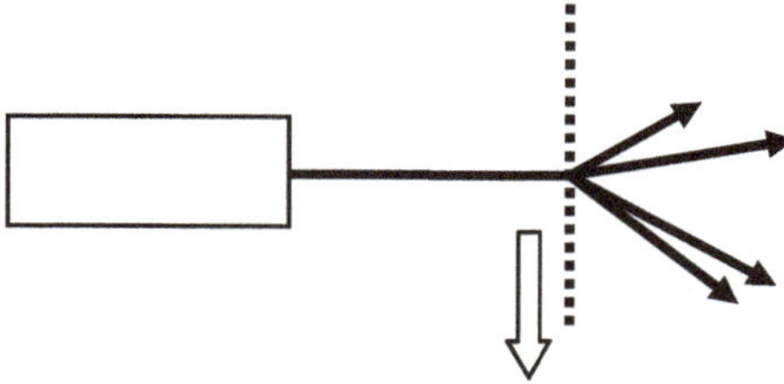

FIGURE 10.16
Making light incoherent. We can make light less coherent by scattering it from a moving ground glass.

small pinhole will improve spatial coherence, but this goal is only achieved at the expense of lost light.

We will see in Chapter 12 that if we want to retain a large amount of light by spatially and temporally filtering an incoherent source, the key parameter is the spectral radiance. Even the most intense sources of light, such as the sun, prove to be quite weak indeed when filtered through a narrow filter and a small pinhole. The invention of the laser revolutionized optics by providing a high-brightness source in a small spatial spot and a narrow spectral bandwidth.

In Practice

For highest temporal and spatial coherence, we normally choose to use a laser. Other narrow–band sources exist and can be made more coherent by the use of temporal and spatial filters.

10.5.2 Decreasing Coherence

To make light less coherent, we wish to reduce both the spatial coherence width, and the temporal coherence time or coherence length. The spatial coherence can be reduced by scattering the light from a ground glass or other rough surface, as shown in Figure 10.16. The larger the spot on the ground glass, the smaller the speckle size or spatial coherence width. Finally, moving the ground glass transversely will generate temporal variations, reducing the coherence time to the time on target, given by Equation 10.73. Degrading the coherence is often useful to eliminate the speckle fluctuations that are so troubling to many coherent sensing and imaging techniques.

Eq. 10.73: $t_{target} = \frac{d_{target}}{v_\perp}$.

Temporal coherence can also be reduced by using a number of lasers at different wavelengths. If the wavelengths are spaced sufficiently far apart, then each will produce its own speckle pattern, and if we sum the signals from several equally bright sources, the signal will increase by the number of sources, while the standard deviation will only increase by the square root of this number, thereby reducing the speckle fluctuations. This approach is simply another way of increasing the bandwidth of the source. Using lasers spaced widely in wavelength we can achieve a wider spectrum than we can by moving a ground glass rapidly to take advantage of the time on target.

Example

Consider the experiment in Figure 10.17. Coherent laser light is scattered by a moving ground glass surface and then passed through a two-slit experiment. Will the experiment produce fringes?

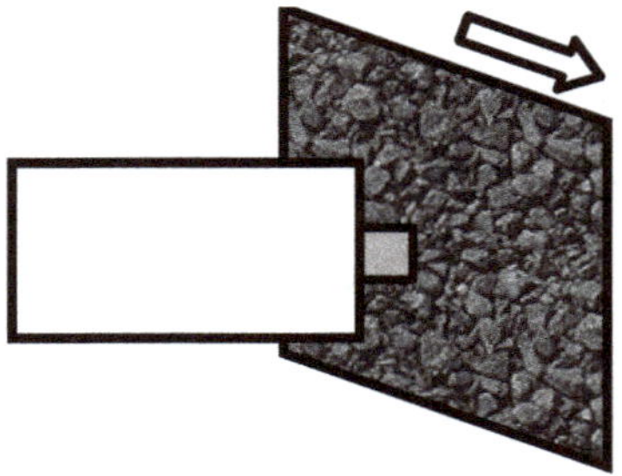

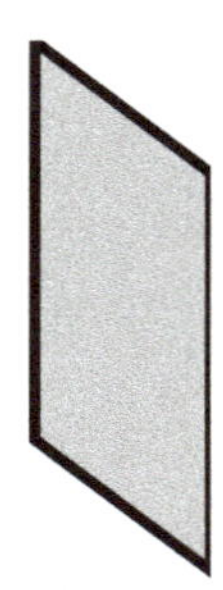

FIGURE 10.17
Ground-glass experiment. A moving ground glass is placed after a laser, and then followed by a double slit. Will light passing through this system produce fringes on the screen behind the slits?

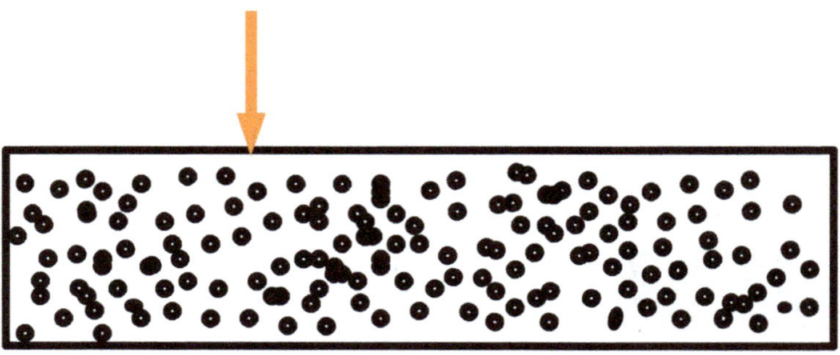

FIGURE 10.18
Light propagation in a scattering medium. If some of the particles are moving, the correlation time is reduced. If the detector integration time is long enough, the spatial contrast is reduced.

Let's first consider the ground glass to be stationary. There will be a random speckle pattern on the slits, but it will be stationary. If the slits are closer than the coherence width

$$\delta x = \frac{\lambda}{d_0} z_1,$$

then the fields will be the same at both slits and we expect the usual fringe pattern. Now if the ground glass moves, the pattern will change in brightness as bright and dark speckles move over the slit, but the shape of the pattern will not change.

If the spacing of the slits is larger than δx, then there will be a different random field on each slit. For the stationary case, there will be a fringe pattern but the locations and brightness of the bright fringes will be unpredictable. Now if the ground glass moves, the pattern will change. If it moves fast enough compared to the speed of our detector, we will detect only the average signal and not see any fringes. For example, if the detector is a camera with an integration time of $1/30 sec$, then fringes will be seen if the speed of the glass is much slower than

$$v = \frac{d_0}{1/30\,\text{sec}}.$$

For a 2 mm laser beam, the velocity of the ground glass must be much slower than 60 mm/sec to observe the fringes. If the velocity is much larger, the fringes will not be observed. Around this velocity, the fringes will appear with reduced contrast.

10.6 Summary

- Coherence of light is characterized by the autocorrelation function in space and time.
- The temporal autocorrelation function is described by its shape and a characteristic time called the coherence time, which is inversely proportional to the linewidth.
- The temporal autocorrelation function is the Fourier transform of the irradiance spectral density.
- Light is temporally coherent if the temporal autocorrelation function is high over the time of a measurement.
- The spatial autocorrelation function in a field plane is the Fourier transform of the irradiance distribution in the pupil plane.
- The coherence length, in the propagation direction, is the coherence time multiplied by the speed of light.
- The coherence length in wavelengths is inversely proportional to the linewidth of the source in wavelengths.
- The spatial autocorrelation function is described by its shape and a characteristic distance called the coherence distance, which is inversely proportional to the width of the irradiance distribution in the pupil.
- Light is spatially coherent if the spatial autocorrelation function is high over the area of a measurement.
- Speckle fluctuations limit the performance of sensing and imaging systems using coherent light.
- Speckle fluctuations can be reduced by temporal diversity, spectral diversity, spatial diversity, angle diversity, or polarization diversity.
- Light can be made more coherent by filtering through a narrow band filter (temporal), and by spatial filtering with a pinhole (spatial).
- Light can be made less coherent by scattering from a moving rough surface.

Imaging in partially coherent light is treated in a number of articles and books. A small sample includes an article by Hopkins[99], and texts by Goodman[81] and O'Neill[5] (recently reprinted[9]), and of course Born and Wolf[28].

FURTHER READING: Imaging and Coherence:[99;81;9;28], among many others.

Problems

10.1 Köhler Illumination (G)

A common way of illuminating an object for a microscope is to focus the filament of a tungsten lamp on the pupil of the system (which means focusing it in the back focal plane of the objective lens). The image plane will then contain the two-dimensional (2-D) Fourier transform of the filament. The goal is to make illumination very uniform. Start with the following code:

```
\% Koehler Illumination
\%
\% By Chuck DiMarzio, Northeastern University, April 2002
\%
\% coherent source
\%
[x,y]=meshgrid(1:256,1:256);
coh=zeros(size(x));
coh(60:69,50:200)=1;
```

```
coh(80:89,50:200)=1;
coh(100:109,50:200)=1;
coh(120:129,50:200)=1;
coh(140:149,50:200)=1;
coh(160:169,50:200)=1;
coh(180:189,50:200)=1;
fig1=figure;
subplot(2,2,1);imagesc(coh);
colormap(flipud(bone));colorbar;
title('Coherent Source');
```

10.1a Now, take the 2-D Fourier transform and image the magnitude of the result. A couple of hints are in order here. Try "help fft2," and "help fftshift." If you have not used subplots, try "help subplot" to see how to display all the answers in one figure.

10.1b The aforementioned result is not very uniform, but the prediction was that it would be. See if you can figure out why it does not work. You have two more subplots to use. On the third one, show the corrected source.

10.1c On the fourth one, show the correct image (Fourier transform).

There are different degrees of correctness possible in Part (b), which will affect the results somewhat. You only need to develop a model that is "correct enough" to show the important features.

10.2 LED and Laser Coherence (NS)

A certain red LED has a linewidth of 20 nm centered at 635 nm. We wish to use it as a light source for a Michelson interferometer.

10.2a How precisely do we need to match the distances from the mirrors to the beam-splitter?

10.2b Now we want to test the surface shape of a large telescope and we need unequal paths with a difference of 5 meters. The LED will not work. What linewidth (in frequency is needed to make this interferometer work?

10.3 Short Pulse in Water (NS)

We have a laser that produces pulses about 100 fs wide at a wavelength of 1.1 μm.

10.3a What is the approximate linewidth of this transform-limited pulse in frequency?

10.3b Look up the dispersion of water, $\frac{dn}{d\lambda}$ possibly using https://refractiveindex.info/ (hint: Water is an oxide, H_2O.

10.3c How far can this pulse go through water before it is increased in width by 20%?

10.4 Speckle Reduction (NS)

Consider a laser radar with incoherent detection. We will use a small transmitter beam of 4 cm diameter and a nearly co-located receiver telescope. The target is a large, flat, uniform area such as sand on the desert.

10.4a What is the size of a speckle at the receiver?

10.4b If the receiver has an aperture diamter of 20 cm diameter, the variations in signal caused by speckle will be reduced. By how much?

11 Fourier Optics

We noted in Chapter 8 that the Fresnel–Kirchhoff integral is mathematically very similar to a Fourier transform, and in fact, in the right situations, becomes exactly a Fourier transform. In this chapter, we explore some of the applications of this relationship to certain optical systems. Using the tools of Fourier analysis, following the techniques used to analyze linear time-invariant filters in electronics, we can describe the point-spread function (PSF) (see Section 8.7.1) as the image of an unresolvable object. By analogy with the study of temporal signals in electronics, the PSF is the spatial impulse response. In this case, it is a two-dimensional function. The Fourier transform of the PSF can be called the transfer function. These functions are analogous to the impulse response and transfer function of an electronic filter, but are two-dimensional. We may recall from the study of electronics that the output of a filter is the convolution of the input with the impulse response, and the Fourier transform of the output is the Fourier transform of the input multiplied by the transfer function. In optics, the input is the object, and the output is the image, and the same rules apply. Our discussion here must be limited to a first taste of the field of Fourier optics. The reader will find a number of excellent textbooks available on the subject, among which one of the most famous is that of Goodman[82].

When we discuss the PSF and transfer function, we may apply them to objects and images which are either field amplitudes or irradiances. Fortunately the functions for these two different types of imaging problems, coherent and incoherent, are connected. We will find it useful to develop our theory for coherent imaging first, defining the PSF in terms of its effect on the amplitude. We may call this function the amplitude point-spread function or APSF, or alternatively the coherent point-spread function or CPSF. Then we turn our attention to incoherent imaging. In the end, we will develop coherent and incoherent PSFs, and corresponding transfer functions, which are consistent with the facts that the irradiance must be a nonnegative quantity and that amplitude is a complex quantity. There is some overlap of terminology, so the reader may wish to become familiar with the terms in Table 11.1 before they are formally defined. In this table, we try to cover all the terminology that is commonly in use, which involves having multiple names for certain quantities, such as the APSF or CPSF. The term "PSF" is often used to mean either the coherent or incoherent PSF. We will use this term when the meaning is obvious from context, and we will use "coherent" or "incoherent" whenever necessary to distinguish one from the other.

This chapter is organized into three sections covering coherent imaging, incoherent imaging, and application of Fourier optics to characterization of the performance optical systems.

11.1 Coherent Imaging

We have really started to address the basics of Fourier optics in our study of diffraction. Specifically, let us now take another look at Equation 8.46,

$$E\left(x_1, y_1, z_1\right) = \frac{jke^{jkz_1}}{2\pi z_1} e^{jk\frac{\left(x_1^2+y_1^2\right)}{2z_1}} \quad \times$$

DOI: 10.1201/9781315157047-11

TABLE 11.1
Fourier Optics Terminology

	Field Plane	Fourier Plane
C	Field amplitude, $E(x,y)$	$\tilde{E}(f_x,f_y)$
	Amplitude point-spread function, $h(x,y)$ Coherent point-spread function Point-spread function PSF, APSF, CPSF	Amplitude transfer function, $\tilde{h}(x,y)$ Coherent transfer function Transfer Function ATF, CTF
	$E_{image} = E_{object} \otimes h$	$\tilde{E}_{image} = \tilde{E}_{object} \times \tilde{h}$
I	Irradiance, $I(x,y)$	$\tilde{I}(f_x,f_y)$
	Incoherent point-spread function, $H(x,y)$ Point-spread function PSF, IPSF	Optical transfer function, $\tilde{H}(f_x,f_y)$ OTF Modulation transfer function, $\left\lvert\tilde{H}\right\rvert$ MTF Phase transfer function, $\angle\tilde{H}$ PTF
	$I_{image} = I_{object} \otimes H$	$\tilde{I}_{image} = \tilde{I}_{object} \times \tilde{H}$

The names of PSFs and transfer functions for coherent (C) and incoherent (I) imaging are summarized. In some cases, there are multiple names for the same quantity.

$$\iint E(x,y,0)\, e^{jk\frac{\left(x^2+y^2\right)}{2z_1}} e^{-jk\frac{(xx_1+yy_1)}{z_1}} dx\, dy.$$

Eq. 8.60:
$E(x_1,y_1,z_1) = \frac{jke^{jkz_1}}{2\pi z_1} e^{jk\frac{\left(x_1^2+y_1^2\right)}{2z_1}} \int\int E(x,y,0)\, e^{-jk\frac{(xx_1+yy_1)}{z_1}} dx\, dy.$

There are several different optical configurations that produce a Fourier transform of an image.

We recall that we said this equation described (1) a curvature term at the source plane, (2) a Fourier transform, (3) a curvature term at the final plane, and (4) scaling. In Section 8.4.4, we saw that the curvature term at the source could be canceled by the addition of a Fraunhofer lens of focal length, $f = z_1$. With this lens the term Q in Equation 8.56 became zero, and Equation 8.46 simplified to Fraunhofer diffraction (Equation 8.60). In the same way, the curvature term in x_1 and y_1 can be canceled by a second lens with the same focal length, as shown in Figure 11.1A. Likewise, if we make z_1 sufficiently large, we can neglect both curvature terms over the range of transverse coordinates for which the field is not close to zero, as in Figure 11.1B where the source is at the origin and the measurement plane is at infinity. Noting that plane 1 in Figure 11.1C is conjugate to plane 1 in Figure 11.1B, propagation from the front focal plane of a lens (plane 0) to the back focal plane (plane 1) in Figure 11.1C is also described by Equation 8.60.

As we noted in Section 8.5.1, we can always convert Equation 8.60 to Equation 8.61

$$E(f_x,f_y,z_1) = \frac{j2\pi z_1 e^{jkz_1}}{k} e^{jk\frac{\left(x_1^2+y_1^2\right)}{z_1}}$$

$$\iint E(x,y,0)\, e^{-j2\pi(f_x x+f_y y)} dx\, dy.$$

In the three examples in Figure 11.1, we can remove the remaining curvature term to obtain

$$E(f_x,f_y,z_1) = \frac{j2\pi z_1 e^{jkz_1}}{k}$$

$$\iint E(x,y,0)\, e^{-j2\pi(f_x x+f_y y)} dx\, dy.$$

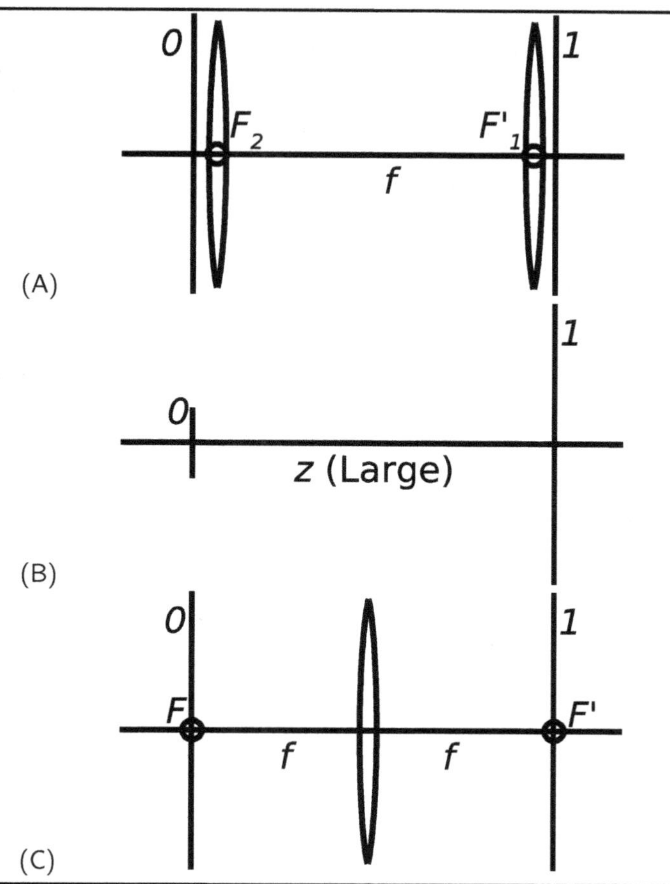

FIGURE 11.1
Fourier optics. Equation 8.46 describes propagation by a curvature term, a Fourier transform, and another curvature term. Lenses can be used to cancel the curvature terms (A). If the lenses are placed at each others' focal points, then the field at plane 1 is the Fourier transform of the field at plane 0. The lenses function similarly to Fraunhofer lenses. Alternatively, Equation 8.60 has the curvature term only outside the integral. The curvature term disappears as $z \to \infty$ (B). The Fourier relationship holds in this example as well. The plane at infinity is conjugate to the back focal plane of the lens (C), and the Fourier relationship still holds.

Then defining

$$\tilde{E}\left(f_x, f_y\right) = j\; z_1 \lambda e^{jkz_1} E\left(f_x, f_y, z_1\right), \tag{11.1}$$

we obtain the equations of Fourier optics, the Fourier transform,

$$\tilde{E}\left(f_x, f_y\right) = \iint E\left(x, y, 0\right) e^{-j2\pi\left(f_x x + f_y y\right)} dx\, dy, \qquad \text{(FT)} \tag{11.2}$$

and its inverse,

$$E\left(x, y\right) = \iint \tilde{E}\left(f_x, f_y\right) e^{-j2\pi\left(f_x x + f_y y\right)} df_x\, df_y. \qquad \text{(IFT)} \tag{11.3}$$

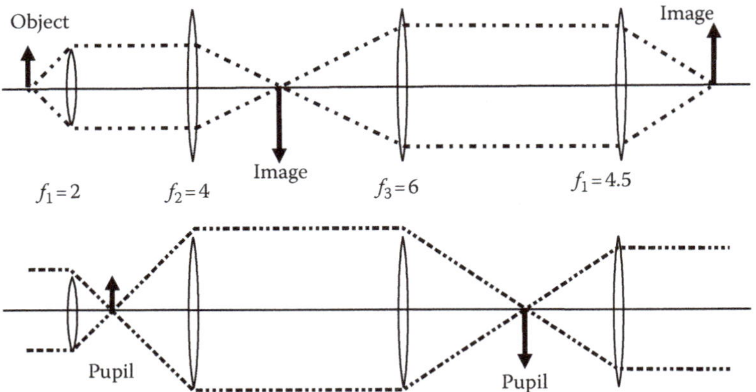

FIGURE 11.2
Fourier optics system. The field planes are labeled in the upper drawing, with edge rays traced from a point in the object through each image. Recall that these rays define the numerical aperture. In the lower drawing, the pupil planes are labeled, and edge rays from a point in the pupil are shown to define the field of view.

Eq. 8.62:
$f_x = \frac{k}{2\pi z_1}x_1 = \frac{x_1}{\lambda z_1}$
$f_y = \frac{k}{2\pi z_1}y_1 = \frac{y_1}{\lambda z_1}.$

We recall that the relationship between transverse distances in the pupil plane (x_1 and y_1) and spatial frequencies (f_x and f_y) are given by Equation 8.62 and Table 8.1 as

$$f_x = \frac{x_1}{\lambda z} \qquad f_y = \frac{y_1}{\lambda z}.$$

We can now consider an optical path such as that shown in Figure 11.2, where we have labeled field planes (object and image) in the upper drawing and pupil planes in the lower one. This configuration is an extension of Figure 11.1C, which is used in many applications. In this telecentric system, we find alternating field and pupil planes. The basic unit of such a system can be repeated as many times as desired. We recognize that we can make a **spatial filter** by placing a mask with a desired transmission pattern in one or more of the pupil planes.

11.1.1 Fourier Analysis

Eq. 8.46: $E(x_1, y_1, z_1) = \frac{jke^{jkz_1}}{2\pi z_1}e^{jk\frac{(x_1^2+y_1^2)}{2z_1}}$
$\int\int E(x, y, 0)\, e^{jk\frac{(x^2+y^2)}{2z_1}}$
$e^{-jk\frac{(xx_1+yy_1)}{z_1}} dx\, dy.$

To develop our Fourier transform and inverse Fourier transform theory, we can begin either in a field plane or a pupil plane. Let us, somewhat arbitrarily, begin in one of the field planes, for example, the object plane in Figure 11.2, and writing the results in terms of $x_1 = \lambda z f_x$, and $y_1 = \lambda z f_y$. We can define propagation from the object plane to the first pupil plane as a Fourier transform, using Equation 11.2. How do we get from here to the next pupil plane? The physics of the problem that led to Equation 11.2 was described by Equation 8.46. Thus, the physics would tell us to use the same equation with x, y in place of x_1, y_1, in the starting plane, and some new variables, x_2, y_2 in place of x, y in the ending plane:

$$E(x_2, y_2, z_2) \stackrel{?}{=} \frac{jke^{jkz_2}}{2\pi z_2}\int\int E(x, y, 0)\, e^{-jk\frac{(xx_2+yy_2)}{z_2}} dx\, dy. \tag{11.4}$$

However, if we use a Fourier transform from a field plane to a pupil plane, should we not use an inverse Fourier transform (Equation 11.3)

$$E(x, y) = \int\int \tilde{E}(f_x, f_y, 0)\, e^{j2\pi(f_x x + f_y y)} df_x\, df_y.$$

from a pupil plane to a field plane? Examining the connection between Equations 11.3 and 11.2, we should write the physical equation for the latter as

$$E\left(x_2, y_2, z_2\right) \stackrel{?}{=} \frac{jke^{-jkz_1}}{2\pi z_1} \iint E\left(x, y, 0\right) e^{jk\frac{(xx_2+yy_2)}{z_1}} dx\, dy. \tag{11.5}$$

Which of Equations 11.4 and 11.5 is correct? The only distinctions are the scaling and introduction of a negative sign, through the use of a new distance $-z_2$ in the former and z_1 in the latter. However, both the negative sign and the scaling would better be associated with the coordinates, because

$$x_2 = -\frac{f_2}{f_1} x_1 \qquad y_2 = -\frac{f_2}{f_1} y_1. \tag{11.6}$$

In both equations, we conceptually take a Fourier transform to go from one field plane to the following pupil plane. In the former equation, we use another Fourier transform to go to the next pupil plane, resulting in a scaling and inversion of the coordinates. In the latter equation, we take the inverse Fourier transform to go from the pupil plane back to the original field plane, and then use geometric optics embodied in Equation 11.6 to go from this field plane to the next.

This result is particularly useful because we can now move easily from any one chosen field plane (say the object plane) to a pupil plane with a Fourier transform (Equation 11.2) back to the same field plane using an inverse Fourier transform (Equation 11.3), accounting for modifications of the field amplitude in each plane, and end by using geometric optics to obtain the field in any other field plane, accounting only for the magnification.

Field to Pupil: FT
Pupil to Field: IFT

In Figure 11.2, we can take the Fourier transform of the object, multiply by a "spatial filtering" function in the pupil, inverse Fourier transform to the intermediate image plane, multiply by another filtering function to account for the field stop of the system, and continue in this fashion until we reach the image.

Take-Away

Using Fourier Transforms:
We use the Fourier transform from a field plane to the pupil and the inverse Fourier transform from pupil back to the same field plane. For other field planes we scale according to the magnification. This process avoids confusion about scale factors in the Fourier transform.

11.1.2 Computation

We can simplify the process with two observations: (1) The Fourier transform and its inverse are linear operations on the amplitude, and (2) most optical devices are also represented by linear operations, specifically multiplication by functions of position that are independent of the field amplitude. Therefore, we can perform all the pupil operations in one pupil plane, provided that we use f_x, f_y coordinates rather than the physical (x_1, y_1) ones in the functions. We will call this product of pupil functions the transfer function, or specifically the **amplitude transfer function** or ATF, for reasons that will soon become apparent. We can also combine all the field-plane operations (e.g., field stops) into a single operation. Thus, we can reduce any optical system that satisfies certain constraints to Figure 11.3, and compute the expected image through five steps:

1. Multiply the object by a binary mask to account for the entrance window, and by any other functions needed to account for nonuniform illumination, transmission effects in field planes, etc.

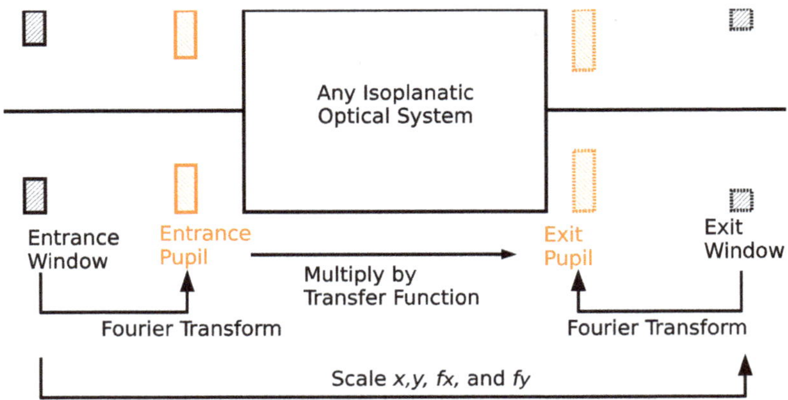

FIGURE 11.3
Fourier optics concept. Any isoplanatic system can be evaluated in terms of functions in the field plane and pupil plane.

2. Fourier transform
3. Multiply by the ATF, which normally includes a binary mask to account for the pupil, and any other functions that multiply the field amplitude in the pupil planes (e.g., aberrations)
4. Inverse Fourier transform
5. Scale by the magnification of the system

All pupil functions are converted to coordinates of spatial frequency, f_x, f_y, before they are multiplied to generate the ATF. In addition to the simplicity and insight that Fourier analysis provides, formulating the problem in this particular way has computational advantages as well. While different authors and computer programmers define the Fourier transform in slightly different ways, e.g., using different constants, all maintain the relationship between the transform variable pairs (f_x, f_y), and (x, y) on which Fourier analysis is based. Every convention is self-consistent in that the inverse Fourier transform of the Fourier transform of a function is the function itself, unchanged in either magnitude or phase. Thus, if we develop equations for Fourier optics using one convention, and then implement them using a different convention for the Fourier transform, the frequency-domain description of the fields may disagree by a constant multiplier, but the imaging results will still be correct.

We have chosen the term transfer function by analogy to the transfer function of an electronic filter. One way to analyze the effect of such a filter on a time-domain signal is to take the Fourier transform of the input, multiply by the transfer function, and inverse Fourier transform, a process analogous to the five steps above.

Example

Let's consider the image plane of a camera that has an $f/4$ lens, looking at a distant object. What is the highest spatial frequency that can be seen by this camera?

The image plane is at the focus, a distance f from the back principal plane of the lens, so The pupil radius is $r = D/2 = f/4/2 = f/8$ so the angle is approximately

1/8. Then the maximum spatial frequency is

$$f_{max}\frac{1}{8\lambda} \approx \frac{1}{4}\,\mu\mathrm{m}^{-1}$$

for green light at $\lambda = 500\,\mathrm{nm}$, or one cycle per four micrometers. With a lens of $f = 25\,\mathrm{mm}$ and the object at $s = 1\,\mathrm{km}$, $m \approx -f/s = 2.5 \times 1-^{-5}$ and the maximum spatial frequency at the target is $f_{max} * m = 6.25\,\mathrm{m}^{-1}$ or one cycle per 16 cm; it would be impossible to resolve a person's hand.

11.1.3 Isoplanatic Systems

We must now take a brief look under the metaphorical rug, where we have swept an important detail. We have developed this theory with lenses and propagation through space, using the paraxial approximation, but we would like to generalize it to any optical system. Just as the Fourier analysis of electronic filters holds for linear time-invariant systems where the contribution to the output at time t' depends on the input at time t only as a function of $t' - t$, the present analysis holds for systems in which the field at an image plane, (x', y', z') depends on that at the object $(x, y, 0)$ only as a function of $x' - x$ and $y' - y$. We call such a system **shift-invariant** or **isoplanatic**. It is easy to envision systems that are not isoplanatic. Imagine twisting a bundle of optical fibers together, and placing one end of the bundle on an object. At the other end of the bundle, the image will probably be "scrambled" so that it is no longer recognizable. Such a fiber bundle is frequently used to transfer light from one location to another. There is still a mapping from (x_1, y_1) to (x, y) but it is not isoplanatic. A more subtle variation arises with departures from the paraxial approximation at high numerical apertures, or in systems with barrel or pincushion distortion. These systems can be considered isoplanatic over small regions of the field plane and the Fourier analysis of an optical system as a spatial filter can still prove useful.

Fourier optics works for isoplanatic or shift-invariant systems, just as Fourier analysis works for time-invariant systems.

In Practice

Systems with aberrations are usually not isoplanatic. For example, coma and astigmatism result in different point-spread functions for different parts of an image. We can still evaluate the image quality with point spread functions for a few different positions in the field of view. In designing a circularly symmetric system, we might choose the center (normally the best PSF), the edge of the field of view and one point in between.

Experimentally, we would want to measure the field at all the edges because imperfect alignment could lead to loss of circular symmetry.

11.1.4 Sampling: Aperture as Anti-Aliasing Filter

We have already seen an example of spatial filtering; the finite aperture stop we considered in Section 8.5.2 could be considered a low-pass spatial filter. We recall the relationship between the size of the mask and the cutoff frequency of the filter from Table 8.1.

$$f_x = \frac{u}{\lambda},$$

so that the cutoff frequency is

$$f_{cutoff} = \frac{NA}{\lambda}. \tag{11.7}$$

For example, with $\lambda = 500$ nm and $NA = 0.5$, the cutoff frequency is 1 cycle/μm. This result is of great practical importance. If we sample at twice the cutoff frequency, to satisfy the Nyquist criterion, we are assured that we can reconstruct the image without aliasing; the aperture stop acts as the anti-aliasing filter. In the present case, the minimum sampling frequency is $f_{sample} = 2$ cycles/μm. Of course, these frequencies are to be measured in the object plane. In Figure 11.2, the magnification of the first stage to the intermediate image is $-f_2/f_1 = -2$, and the magnification of the second stage, from the intermediate image to the final one, is $-f_4/f_3 = -3/4$, so the total magnification is $2/4 = 1.5$. Placing a camera at the intermediate image, we would need 1 pixel/μm and at the final image we would need 1.33 pixels/μm. The pixels in a typical electronic camera (CCD or CMOS) are several micrometers wide, so in order to satisfy the Nyquist sampling theorem, we would need a much larger magnification.

Example

Considering an example using a 100 × oil-immersion microscope objective with $NA = 1.4$, at the same wavelength, $\lambda = 500$ nm. What pixel pitch (spacing) is required on the camera to satisfy the sampling theorem for imaging? The sampling theorem discussed in Section 8.5.2, states that we can recover signals having a frequency spectrum that is zero beyond some $fmax$ if we sample at the Nyquist frequency, $f2max$.

The cutoff frequency is determined from the numerical aperture;

$$f_{cutoff} = \frac{NA}{\lambda} = 2.8\,\text{cycles}/\mu\text{m}, \tag{11.8}$$

in the object plane, or

$$f_{cutoff} = \frac{NA}{\lambda}/m = 0.028\,\text{cycles}/\mu\text{m}, \tag{11.9}$$

in the image plane, where $m = 100$ is the magnification. Thus, the image must be sampled at a frequency of at least

$$f_{sample} = 2f_{cutoff} = 0.056\,\text{pixels}/\mu\text{m}, \tag{11.10}$$

which requires a pixel spacing ≤ 4.5 μm. This spacing is easily achieved in solid-state imaging chips.

Example

As another example, consider an ordinary camera with an imaging chip having square pixels spaced at 5 μm. We want to know the range of f-numbers for which the sampling will satisfy the Nyquist criterion. We recall Equation 4.10 relating the f-number, F, to the numerical aperture.

$$NA_{image} = n'\frac{1}{|m-1|\,2F} \qquad (\text{Small NA, large } F), \tag{11.11}$$

Normally, a camera lens is used in air, so $n' = 1$. With the exception of "macro" lenses, the magnification of a camera is usually quite small ($m \to 0$), and

$$NA_{image} \approx \frac{1}{2F} \qquad \text{(Small NA, large } F\text{, and small } m\text{, air).} \tag{11.12}$$

In contrast to the microscope in the example above, the f-number or NA of the camera is measured in image space, so we can use the pixel spacing, without having to address the magnification. Now the minimum sampling frequency in terms of the minimum f-number (associated with the maximum NA) is

$$f_{sample} = 2f_{cutoff} = 2 \times \frac{NA}{\lambda} = \frac{1}{\lambda F_{min}}, \tag{11.13}$$

Finally, solving for the f-number,

$$F_{min} = \frac{1}{\lambda f_{sample}} = \frac{x_{pixel}}{\lambda}, \tag{11.14}$$

where $x_{pixel} = 1/f_{pixel}$ is the spacing between pixels. For our 5 μm pixel spacing, and green light with $\lambda = 500\,\text{nm}$,

$$F_{min} = 10, \tag{11.15}$$

and the Nyquist criterion will be satisfied only for $f/10$ and smaller apertures. This aperture is quite small for a camera, and frequently the camera will not satisfy the Nyquist limit.

Take-Away

Anti-Aliasing:
The aperture stop of an optical system can be used as an anti-aliasing filter to avoid aliasing.

In Practice

Many cameras, microscopes, and other optical instruments do not satisfy the Nyquist criterion. In theory aliasing may produce unwanted artifacts in the resulting image, and sometimes such artifacts do degrade an image. It is quite common to see some very strange images of striped shirts on a television screen. However, in many cases the image quality appears very good. The extra light received through a large NA may offset the damage done to image quality by aliasing.

Failure to satisfy the Nyquist limit does not necessarily imply a picture of poor quality. However, if the limit is satisfied, then we can be sure that higher spatial frequencies will not affect the quality of the image because the exit pupil acts as an anti-aliasing filter, removing these frequencies before sampling. We can interpolate between pixels to obtain an image with any desired number of pixels, and the resolution of the image will be limited by the resolution of the lens. Interpolation is best done with a sinc function, often called the Shannon interpolation function[203], or the Whittaker–Shannon interpolation function. Shannon's article gives credit to Whittaker[248] for the concept. It is important to recall that aberrations become more significant at low f-numbers. Also if the lens is not diffraction-limited, the highest spatial frequency may be limited by aberrations rather than by diffraction. Thus, the Nyquist limit may be satisfied at larger apertures than we would expect through analysis of diffraction effects alone.

11.1.5 Amplitude Transfer Function

The Amplitude Transfer Function (ATF) is the Coherent Transfer Function (CTF).

We have considered examples where the transmission of light through the pupil is limited by an aperture. However, in general, we can multiply by any function in the pupil, for example, placing the surface of a sheet of glass at the pupil, and modifying the transmission of the glass. The modification can be a complex number; the amplitude can be controlled by an absorbing or reflecting material such as ink to limit light transmission and the phase can be controlled by altering the thickness or refractive index of the glass. The position, x_1, y_1, in the pupil can be related to the spatial frequencies using Table 8.1. The complex function of spatial frequency is the ATF or amplitude transfer function, $\tilde{h}\,(f_x, f_y)$. It may also be called the **coherent transfer function**, or **CTF**, because it acts on the field of the object to produce the field of the image. In contrast, we will shortly define these functions for incoherent imaging.

11.1.6 Point-Spread Function

Fig. 8.16 (Part): Circular Aperture

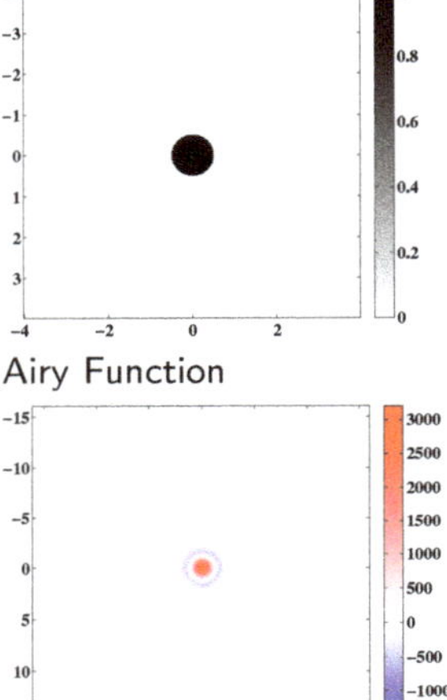

Airy Function

The most common Fourier transform pair in optics is shown in the middle row of Figure 8.16. The pupil function is a uniformly illuminated aperture. Converting the x_1 and y_1 axes to f_x and f_y using Table 8.1, and scaling, we can equate this to the amplitude transfer function. If in Figure 11.2 the object is a point, and the aperture is located in one of the pupils, then the image will be the Fourier transform shown as the Fraunhofer pattern in Figure 8.16. This function is called the **point-spread function** or **PSF**. The Fourier transform of the uniformly illuminated aperture is the Airy function. When we discuss incoherent imaging, we will develop another PSF, so to avoid ambiguity, we may call this one the **coherent point-spread function** or **CPSF** or equivalently the **amplitude point-spread function** or **APSF**.

Given any object, we know that the field entering the pupil will be the scaled Fourier transform of the object field, and the field leaving the pupil will be that transform multiplied by the ATF, provided that we are careful to convert the x_1, y_1 coordinates in the pupil to the spatial frequencies f_x and f_y. If the object is a point, then the Fourier transform of the object is uniform across the pupil, and the Fourier transform of the image is just the ATF, $\tilde{h}\,(f_x, f_y)$. Thus, the PSF is

The Coherent point-spread function (CPSF) is the Amplitude point-spread function (APSF).
The ATF is the Fourier transform of the APSF.

$$h\,(x, y) = \iint \tilde{h}\,(f_x, f_y, 0)\, e^{-j2\pi(f_x x + f_y y)} df_x\, df_y, \tag{11.16}$$

or equivalently, the ATF is given by

$$\tilde{h}\,(f_x, f_y) = \iint h\,(x, y, 0)\, e^{j2\pi(f_x x + f_y y)} dx\, dy. \tag{11.17}$$

11.1.6.1 Convolution

Multiplication of the pupil field (Fourier transform of the object) by the ATF to obtain the Fourier transform of the image,

$$\tilde{E}_{image}\,(f_x, f_y) = \tilde{E}_{object}\,(f_x, f_y)\, \tilde{h}\,(f_x, f_y)\,, \tag{11.18}$$

is equivalent to convolution of the object field with the PSF to obtain the image field.

$$E_{image}(x, y) = E_{object}(x, y) \otimes h(x, y), \qquad (11.19)$$

where $\otimes$ denotes convolution. The concept is analogous to electronic filtering of a temporal signal, where multiplication of the input signal spectrum by the filter's transfer function produces the output signal spectrum, and equivalently convolution of the input signal with the impulse response of the filter produces the output signal. The PSF may be understood as the two-dimensional spatial impulse response of the spatial filter formed by the pupil function.

The PSF is the two-dimensional impulse response.

Take-Away

PSF and Transfer Function:
The Fourier transform of the point-spread function, h, is the transfer function, $\tilde{h}$. Convolution of h with the image is equivalent to multiplication of $\tilde{h}$ with the Fourier transform of the image.

11.1.6.2 Example: Gaussian Apodization

The finite aperture of an imaging system results in a limited resolution, expressed by an equation like Equation 8.91. We can now understand this limitation as a result of the object field being convolved with the PSF to produce the image. However, the PSF that is the diffraction pattern of a uniform circular aperture, as we saw in Table 8.1, is an Airy function, with first sidelobes down 17 dB from the peak. Depending on our criteria for a "good" image, these sidelobes, shown in Figure 11.4A and C, may do more damage to the image quality than the limited resolution.

Eq. 8.91: $\delta_{microscope} = \frac{0.61\lambda}{NA}$.

Examining Table 8.1, we note that the Fourier transform of a Gaussian is a Gaussian, so if we could use a Gaussian pupil function (transfer function) we would have a Gaussian PSF. Of course, a Gaussian function is nonzero for all locations, and strictly requires an infinite aperture. Practically, however, if we make the diameter of the Gaussian small enough compared to the aperture diameter, the field beyond the aperture is so small that its contribution to the Fourier transform can be neglected. The Gaussian transfer function in Figure 11.4B has a diameter (at $1/e$ of its peak amplitude or $1/e^2$ of its peak irradiance) of one half the aperture diameter. If we were to consider a Gaussian laser beam passing through an aperture with these diameters, the fraction of the power transmitted through the aperture would be $1 - e^{-4}$ or over 98%. As shown in Figure 11.4D, the PSF is nearly Gaussian, with no sidelobes observable.

Gaussian apodization removes sidelobes but degrades resolution.

Figure 11.5 shows the effects of these two PSFs on the image of a knife-edge. In the experiment, the edge of a knife blade could be illuminated with a coherent light source. The transmitted light would be detected over half the plane, but would be blocked by the knife blade in the other half, as shown in Figure 11.5A. Alternatively, metal may be deposited in a pattern on a glass substrate. Many such metal-on-glass targets are available from vendors, and arbitrary patterns can be made to order. Results are shown for an imaging system with a finite aperture in Figure 11.4A. The PSF shown in Figure 11.4B is convolved with the object, resulting in the image shown in Figure 11.5C. A slice of this image is shown as a dashed line in Figure 11.5B. The effects of the limited resolution and the fringes are readily apparent. If the pupil of the system is modified with the Gaussian apodization shown in Figure 11.4C the PSF in Figure 11.4D, convolved with the knife edge, produces the image in Figure 11.5D and the solid line in (B). It is readily apparent that the resolution is further degraded by this PSF, but the fringes are

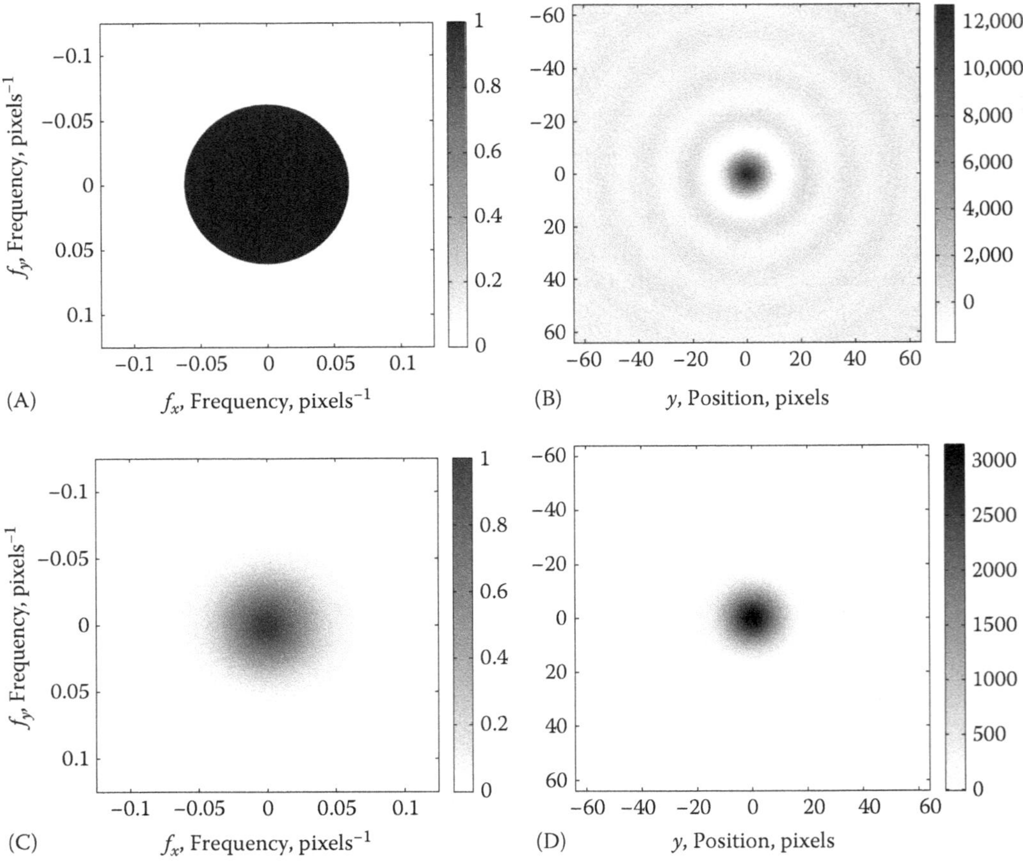

FIGURE 11.4
Point-spread functions. The finite aperture (A) in the pupil generates an Airy PSF (B), which limits the resolution and results in sidelobes. With Gaussian apodization (C), the resolution is further degraded, but the sidelobes are almost completely eliminated (D).

eliminated. If the purpose of the experiment is to measure the location of the knife edge with the best possible resolution, we would prefer to use the full aperture, and accept the fringes. If the purpose is to make a "pleasing" image, we may prefer to use the Gaussian apodization.

The Gaussian apodization in this example has a small diameter, equal to half the aperture diameter, and results in no measurable fringes with the dynamic range of the gray-scale display available here. In fact, a close examination of the fringe pattern has shown that the amplitude of the fringes is less than 6×10^{-4}. The irradiance is down more than 60 dB from the peak irradiance.

11.1.7 General Optical System

We can see now that we have built a very powerful methodology for analysis of any optical system. Having constructed a system by the rules of Chapters 2 and 3, we can find the entrance and exit windows and pupils following Chapter 4. The details of the optical system can all be captured as shown in Figure 11.3 by taking the Fourier transform of the object field with suitable scaling to obtain the field at the entrance pupil. We can then multiply this field by the amplitude transfer function to obtain the field at the exit pupil. We then scale this field to account for magnification. If the magnification of the

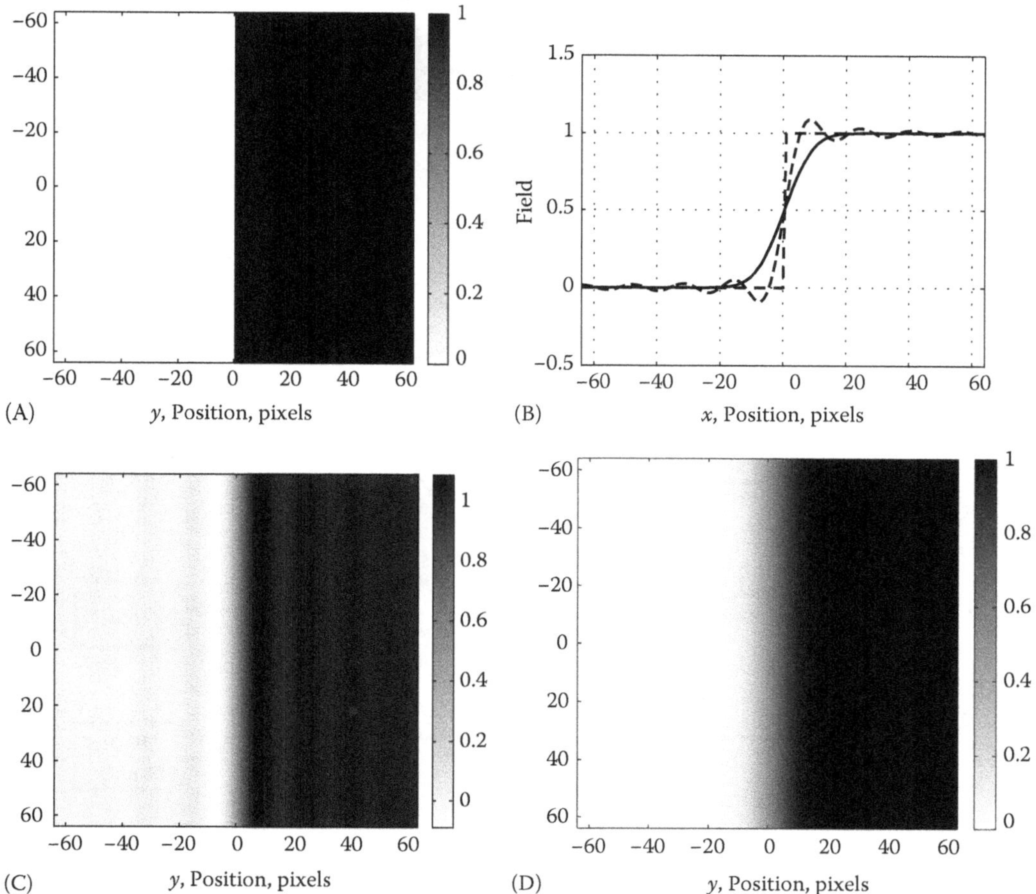

FIGURE 11.5
(A) Imaging a knife-edge. The image is the convolution of the object with the PSF. The transfer and PSFs are shown in Figure 11.4. The finite aperture limits the spatial frequency, and its diffraction pattern introduces fringes into the image (C and dashed line in B). The Gaussian apodization further degrades the resolution, but nearly eliminates the diffraction fringes (D and solid line in B).

system is m, then the exit pupil size is the entrance pupil size divided by m. Finally, we take the inverse Fourier transform to obtain the image field in the exit window. Obviously, the OTF will include the limiting aperture of the system, but it can also include a phase description of the aberrations, as developed Section 5.3.

In fact, we can do the scaling to account for the magnification at any time during this process, and we can describe the transfer function in the entrance pupil or exit pupil, and PSF in either the object or image plane. Figure 11.6 illustrates alternative approaches to the calculation. Figure 11.6A shows the process most directly, as described in the paragraph above (Fourier transform, multiplication and scaling, and inverse Fourier transform). Note that the PSF can be obtained from the OTF using the inverse Fourier transform. In practice, it is important and difficult to keep track of all the multiplicative constants and scaling factors. Often the result is an image which is qualitatively correct, but with amplitudes scaled by some number.

Because the process is linear, we can avoid most of these problems by using the process shown in Figure 11.6B. Here we assume that we know the transfer function in the entrance pupil. We take the Fourier transform of the object field, multiply by the transfer function,

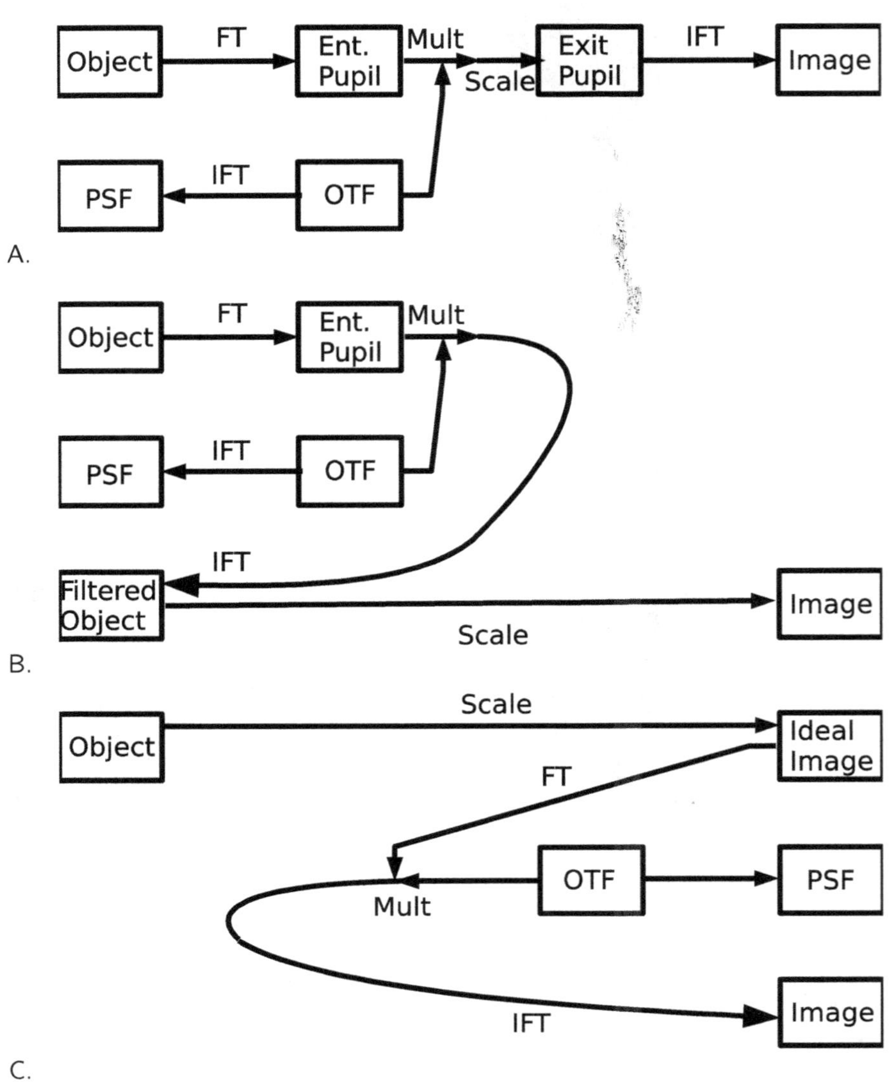

FIGURE 11.6
Computational approaches to Fourier optics. The fundamental steps for analyzing a system by Fourier techniques are shown in A. In practice it is almost always easier to perform the Fourier analysis in either the object or image space, and relate the two through the magnification.

and take the inverse Fourier transform to obtain a "filtered object" which looks like the image, but in object coordinates. If we wish, we can obtain the PSF in the object plane by inverse Fourier transform. We then scale the filtered object to obtain the image. Here scaling means multiplying the x, y coordinates by the magnification, m, and multiplying the irradiance by $1/m^2$ (because the power in an object area A appears in the image area $A' = m^2A$.

If we know the transfer function in the exit pupil instead of the entrance pupil, we can change the order of operations as in Figure 11.6C. First we obtain the "ideal image" which is a scaled version of the object. Then we can perform the Fourier analysis as before to obtain the actual image. The result, of course, is the same in either case.

Take-Away

Fourier Analysis:
Fourier analysis can be used to predict and analyze the performance of an optical system. The transmission in the pupil (including any phase shifts and amplitude changes) is the coherent optical transfer function. The Fourier transform of the object, multiplied by this transfer function and appropriately scaled is the Fourier transform of the image.

In Practice

Because different implementations of the Fourier transform may have different scale factors, it is always preferable to use one of the two processes in Figure 11.6B or C, rather than A.

The processes described here may seem complicated, so let us consider a simple example of the telescope in Figure 4.18. We'll consider the object space process in Figure 11.6B. For an object, we'll use the data in Figure 11.7A. This is a bit artificial as we are assuming here that the data represents the electric field, E, rather than the irradiance, I, but it is good for illustration.

Fig. 4.18

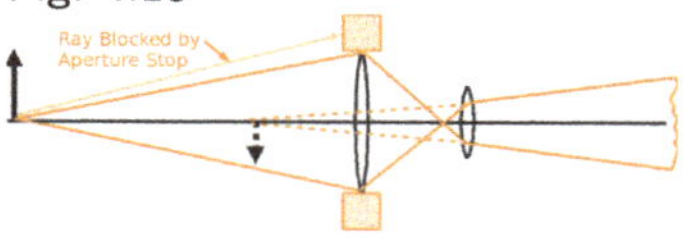

Example

Fourier analysis of Lidar Telescope in Object Space

Let's assume the telescope primary is a lens of focal length, $f = 75\,\text{mm}$ and 10 mm diameter, and the magnification is 1/5, and that the system is well-designed so that the primary is the aperture stop. The application is to a lidar that is imaging an object at $s = 300\,\text{m}$, illuminated by a laser with wavelength, 905 nm. Suppose that we know that the optical system has spherical aberration amounting to two wavelengths at the edge of the aperture stop. We would like to compute the image with and without aberration.

First, the relationship between spatial frequency, f_x, and position in the pupil, x_1, for an object at distance $z = s$ is given from Table 8.1 as

$$f_x = \frac{x_1}{\lambda z} = 3680\,\text{m}^{-2} \times x_1. \tag{11.20}$$

Thus the aperture is a low–pass filter having unit transmission inside the radius

$$f_x = 3680\,\text{m}^{-2} \times 0.0015\,\text{m}/2 = 27.6\,\text{m}^{-1} \tag{11.21}$$

The Fourier transform of the object is shown in Figure 11.7B, in decibels (10 times the common logarithm of the number). After being limited by the aperture of the lens to frequencies less than $27.6\,\text{m}^{-1}$, it is shown in Figure 11.7C. Taking the inverse Fourier transform, we obtain the modified object in Figure 11.7D. The degradation is evident, as expected; the maximum spatial frequency corresponds to a distance of $1/f_x = 36\,\text{mm}$, so features of the object much smaller than this distance will not be well resolved.

Returning to Figure 11.7C, we multiply by

$$e^{j(f/f_{max})^4 \times 2 \times 2\pi}, \tag{11.22}$$

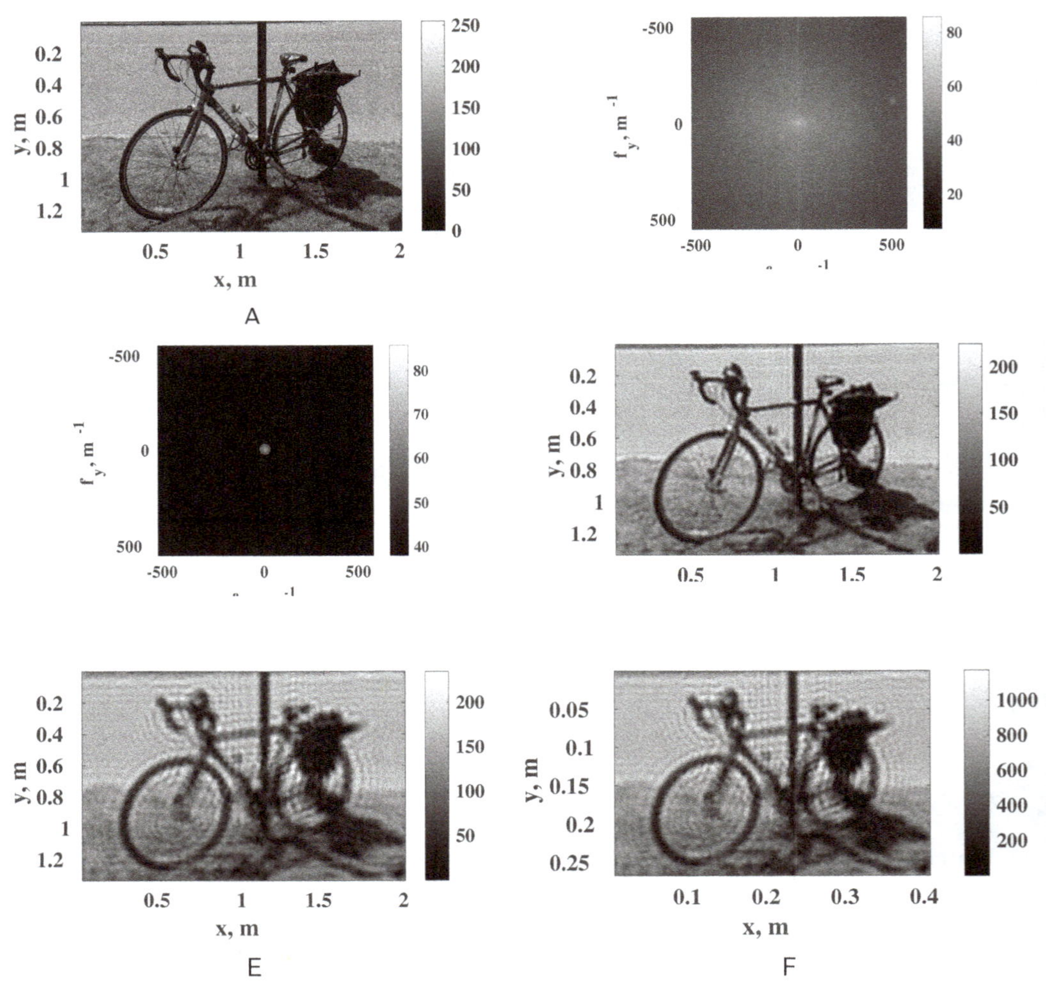

FIGURE 11.7
Fourier optics demonstration. Panel A represents the object electric field. We take the Fourier transform (B) and multiply by the transmission function of the pupil (C) and then inverse Fourier transform to obtain the modified object (D). When we include spherical aberration, the modified object is shown in E. Finally, we scale E to obtain the image (F).

which introduces a quartic phase variation rising to 2 cycles at $f = f_{max}$, in accordance with Equation 5.33. Taking the inverse Fourier transform of the result, we obtain a worse modified object, as expected (Figure 11.7E). Finally, for the actual image, we scale this figure appropriately, using

$$x' = mx \qquad y' = my \qquad E_{image} = E_{object}/m, \tag{11.23}$$

being careful to note that scaling the irradiance by $1/m^2$ means scaling the Electric field by $\sqrt{1/m^2} = 1/m$. The final image is in Figure 11.7F.

Eq. 5.33
$\Delta_{sa} = c_1 \rho^4$.

In some sense, we have reduced the study of optical systems to the mathematical description of electronic filters that is used with linear, time-invariant systems. The OTF plays the role of the filter's transfer function, and the PSF plays the role of the impulse response. One of the differences lies in the way we determine the transfer function.

Instead of examining the effects of resistors, capacitors, and inductors, we need to consider numerical aperture, apodization, and aberrations. Another difference is that in optics, the transforms are two-dimensional.

Take-Away

Summary of Coherent Imaging:

- In an isoplanatic imaging system, we can locate pairs of planes such that the field in one is a scaled Fourier transform of that in the other.
- One such pair consists of the planes immediately in front of and in back of a matched pair of lenses as in Figure 11.1A.
- Another consists of the front and back focal planes of a lens as in Figure 11.1C.
- It is often useful to place the pupil at one of these planes and the image at another.
- Then the aperture stop acts as a low-pass filter on the Fourier transform of the image.
- This filter can be used as an anti-aliasing filter for a subsequent sampling process.
- Other issues in an optical system can be addressed in this plane, all combined to produce the transfer function.
- The PSF is a scaled version of the inverse Fourier transform of the transfer function.
- The transfer function is the Fourier transform of the PSF.
- The image can be viewed as a convolution of the object with the PSF.

11.2 Incoherent Imaging Systems

With our understanding of optics rooted in Maxwell's equations, it is often easier for us to begin any study with the assumption of coherent light, as we have done in this chapter. However, in reality, most light sources are incoherent, and we must extend our theory to deal with such sources. In fact, even if the source is coherent, the detection process usually is not, so we need to understand both coherent and incoherent imaging. One approach is to begin with a coherent description, and then average over different members of an ensemble to obtain expectation values. Let us consider a nearly monochromatic, but incoherent light source, and assume that we have the ability to collect data only by averaging over a time long compared to the coherence time of the source. This assumption is often valid. A light-emitting diode may have a spectral linewidth of about 20 nm, with a center wavelength of 600 nm, and thus a fractional linewidth of $2/60 \approx 0.033$. Using Equation 10.43, the coherence length is about 30 wavelengths or equivalently the coherence time is about 30 cycles or

Fractional linewidth was discussed first in Section 7.5 in Equation 7.127.

Eq. 7.127: $\left|\frac{\delta\lambda}{\lambda}\right| = \left|\frac{\delta f}{f}\right|.$

$$\tau_c \approx \frac{30}{\nu} = 30\frac{\lambda}{c} \approx 60\,\text{fs}. \tag{11.24}$$

Nearly all detectors are too slow to respond to variations at this timescale, and so we can sum or integrate incoherently over the wavelengths (Equation 10.16). Thus, the expectation value of the PSF, given random amplitudes and phases, is not particularly useful,

Eq. 10.16: $\langle I\rangle = \langle I_1\rangle + \langle I_2\rangle.$

$$\langle h(x, y)\rangle = 0. \tag{11.25}$$

We need a more useful statistical description.

11.2.1 Incoherent Point-Spread Function

We can define a nonzero function,

$$H(x, y) = \langle h(x, y) h^*(x, y) \rangle, \tag{11.26}$$

The incoherent point-spread function is the squared magnitude of the coherent one.

which we will call the **incoherent point-spread function** or **IPSF**. The irradiance of the image will be

$$I_{image}(x, y) = \left[h(x, y) \otimes E_{object}(x, y)\right] \left[h^*(x, y) \otimes E^*_{object}(x, y)\right] \tag{11.27}$$

The product of the two terms in brackets defies our assumption of linearity, but our assumption of incoherence allows us to average all of the cross-terms to zero, so that

$$I_{image}(x, y) = \left[h(x, y) h^*(x, y)\right] \otimes \left[E_{object}(x, y) E^*_{object}(x, y)\right] \tag{11.28}$$

$$I_{image}(x, y) = H(x, y) \otimes I_{object}(x, y). \tag{11.29}$$

11.2.2 Incoherent Transfer Function

Incoherent OTF is the convolution of the ATF with its complex conjugate.

Can we now determine the **incoherent transfer function** or **optical transfer function** (OTF)? Using Equation 11.26, we note that the incoherent PSF is the product of the coherent PSF with its own complex conjugate. We recall the convolution theorem which states that if two quantities are multiplied in the image domain, they are convolved in the spatial frequency domain, and *vice versa*. Thus, the incoherent OTF is simply

$$\tilde{H}(f_x, f_y) = \tilde{h}(f_x, f_y) \otimes \tilde{h}^*(f_x, f_y). \tag{11.30}$$

Then

$$\tilde{h}^*(f_x, f_y) = \tilde{h}(-f_x, -f_y), \tag{11.31}$$

The MTF is the amplitude of the OTF and the PTF is it's phase.

and the OTF is the autocorrelation of the ATF. Its amplitude, $\left|\tilde{H}(f_x, f_y)\right|$, is called the **modulation transfer function** or **MTF** and its phase, $\angle\left[\tilde{H}(f_x, f_y)\right]$, is called the **phase transfer function** or **PTF**.

Take-Away

Incoherent and Coherent Fourier Optics:
The incoherent point-spread function is the squared magnitude of the coherent point-spread function. It is non-negative.

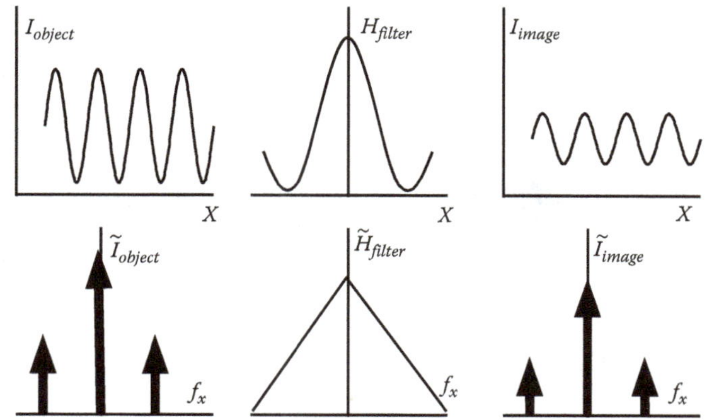

FIGURE 11.8
Fourier optics example. The object (left), system (center), and image (right) are shown in the spatial domain (top) and the frequency domain (bottom).

The incoherent optical transfer function is the autocorrelation of the coherent optical transfer function. It has a maximum at zero spatial frequency.

Figure 11.8 shows the operations we have been discussing. The top left panel shows an irradiance which is a cosinusoidal function of x. Although the field amplitude can be a negative or even complex quantity, the irradiance must be real and nonnegative. Shown here is a high-contrast image, in which the minimum irradiance is close to zero. The Fourier transform of this function has peaks at the positive and negative values of the frequency of the cosine function. The transfer function at zero frequency measures the amount of light transmitted through the system so the "DC" term is lower in the image than in the object. At higher spatial frequencies, the OTF is reduced both because of losses in the optical system and the "smoothing" of the sinusoidal signal by convolution with the PSF. Thus, the higher frequencies are attenuated more than the lower ones. The result is that the image has not only lower brightness, but also lower contrast.

A numerical example is shown in Figure 11.9. Transfer functions are shown on the left and PSFs are shown on the right. This figure describes a simple diffraction-limited optical imaging system, in which the numerical aperture is determined by the ATF in (A). The coherent PSF in (B) is the inverse Fourier transform, and the incoherent PSF is obtained by squaring according to Equation 11.26. Then, the OTF (C) is obtained by Fourier transform. Slices of the functions are shown in the lower row. It is interesting to note that the incoherent PSF is narrower than the coherent PSF, but we must not draw too many conclusions from this fact, which is simply the result of the squaring operation. In the case of the coherent PSF, we are considering the field and in the case of the incoherent PSF, the irradiance. Equation 11.31 is always valid and useful, regardless of whether the light source is coherent or incoherent. The maximum spatial frequencies transmitted are now twice those of the incoherent case, requiring that we revisit our imaging examples.

Eq. 11.26:
$H(x,y) = \langle h(x,y)\, h^*(x,y)\rangle$.

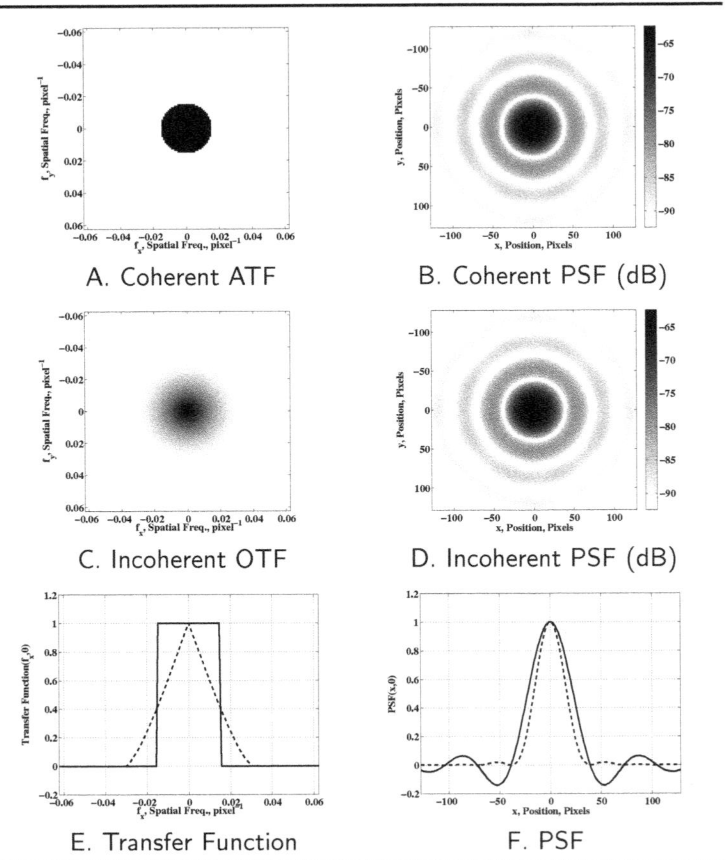

FIGURE 11.9
Transfer functions and PSFs. The PSF is the inverse Fourier transform of the transfer function. The IPSF is the squared magnitude of the CPSF. Therefore, the optical transfer function is the autocorrelation of the amplitude transfer function. The solid lines are for coherent imaging and the dashed for incoherent. (A) Coherent ATF. (B) Coherent PSF (dB). (C) Incoherent OTF. (D) Incoherent PSF (dB). (E) Transfer function. (F) PSF.

As an example, we consider a shifting operation in the x direction, as seen in Figure 11.10. The transfer function consists of a linear phase ramp,

$$\tilde{h}\left(f_x, f_y\right) = \begin{array}{ll} \exp\left[i2\pi\dfrac{f_x/4}{1/128}\right] & \text{for } f_x^2 + f_y^2 \le f_{max}^2 \\ 0 & \text{Otherwise} \end{array} \tag{11.32}$$

so that the phase shift is $\pi/2$ or one quarter cycle at a frequency of $1/128$ cycles per pixel. with $f_{max} = 1/64$ per pixel.

The linear phase ramp is shown in Figure 11.10A. Transforming to the image plane we obtain the coherent PSF, a slice of which is shown in Figure 11.10C by the solid line. The squared magnitude of the coherent PSF is computed as the incoherent PSF, and a slice of that is plotted with the dashed line. The incoherent PSF is then transformed back to the pupil plane, to obtain the incoherent OTF. The incoherent PSF is always positive, as expected, and the incoherent OTF varies in phase twice as fast as the coherent OTF.

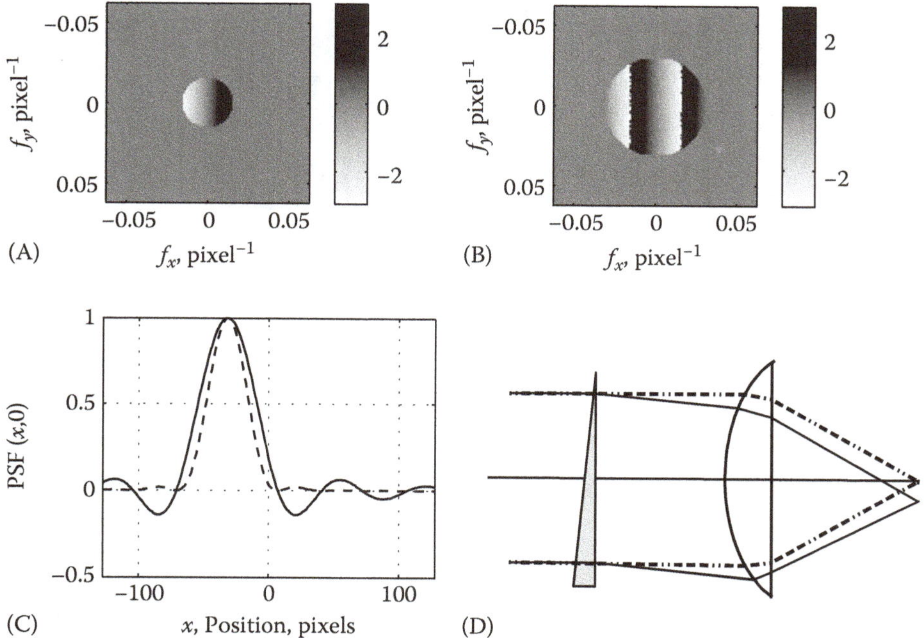

FIGURE 11.10
A shifting operation. The coherent OTF (A, Phase in radians) consists of a linear phase ramp within the boundaries of the numerical aperture. The phase shift is by a quarter cycle or $\pi/2$ radians at a frequency of $1/128$ per pixel. The incoherent transfer function has a phase with double the slope of the coherent one (B) so we see two cycles across the aperture instead of one that we see in (A). The resulting PSFs are shifted in position, but otherwise equivalent to those of Figure 11.9.

In Figure 11.11, an object with illumination that varies sinusoidally at a frequency of $1/128$ per pixel is shown. This coherent object is mainly for instructional purposes. This object is imaged through the system described in Figure 11.10. When the coherent object is imaged with the system using coherent imaging equations, the image is shifted by $1/4$ cycle as expected. The incoherent object is obtained by squaring the magnitude of the coherent one to obtain a sine wave at twice the spatial frequency of the coherent one, along with a DC term, as expected from trigonometric identities,

$$[\sin(2\pi f_x x)]^2 = \frac{1}{2} - \frac{1}{2}\cos(2\pi \times 2f_x x). \tag{11.33}$$

The sinusoidal coherent object is difficult (but not impossible) to manufacture, but the incoherent sinusoidal chart is somewhat easier. It can be produced, in principle, by using a printer to produce sinusoidally varying amounts of ink on a page. Now the object is shifted by a half cycle rather than a quarter cycle, but because the frequency is doubled, this shift amounts to the same physical distance as in the coherent image.

We should note that the linear phase shift can be implemented by a prism placed in the pupil plane, as shown in Figure 11.10D. The thickness of the prism, t, introduces a phase shift, relative to that present in the absence of the prism (dashed lines) of $\phi = 2\pi(n-1)t/\lambda$. The angle of the prism is given by $\sin\alpha = dt/dx$. We can develop the displacement of the focused spot using Snell's Law, Equation 2.2, in Section 2.1. The results will be the same as we have developed here. However, the Fourier analysis allows us to address much more complicated problems using a consistent, relatively simple, formulation.

Eq. 2.2:
$n\sin\theta = n'\sin\theta'$.

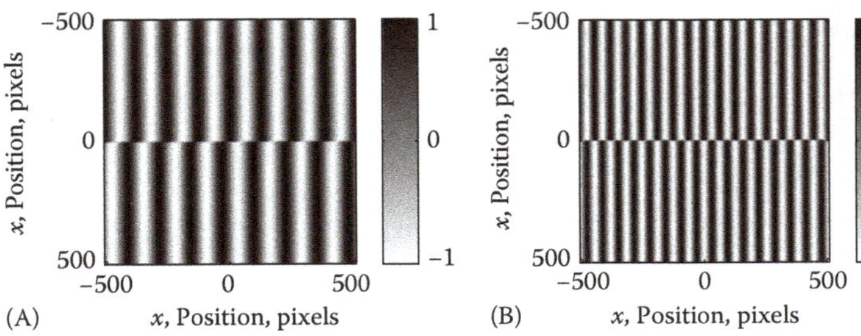

FIGURE 11.11
An object with sinusoidal illumination. The upper part of the figure shows the object and the lower part shows the image so that their positions may be compared, In the coherent case (A), the pattern is shifted by one quarter cycle. Note that the scale includes negative and positive fields. The incoherent object (B) has positive irradiance, and a sinusoidal component at twice the frequency of the coherent one. The phase shift is now one half cycle, amounting to the same physical distance as in the coherent case.

> *Take-Away*
>
> **Comment on Resolution:**
> We note that the incoherent transfer function is wider than the coherent one, and we might be tempted to say that incoherent imaging results in better resolution. However, it's important to remember that the incoherent image has frequencies twice as high as those of the coherent image.

11.2.3 Camera

Let us consider the detection step in view of our understanding of Fourier optics. Now our assumptions of linearity can no longer serve us; the contribution to the detector signal from an increment of area, dA on the camera is the product of the irradiance (squared magnitude of the field) and the responsivity of the detector material, typically in Amperes per Watt (see Equation 13.7 which will be discussed in detail in Chapter 13). This product is then integrated over the area of each pixel to obtain the signal from that pixel:

The detector ideally produces a current consisting of one electron, e for each incident photon, $h\nu$. In practice only a fraction, η of the photons lead to an electron.

> Eq. 13.7:
> $\rho_i = \frac{\eta e}{h\nu}$.

$$i_{mn} = \iint_{pixel} \rho_i \left(x - m\delta x, y - n\delta y\right) E(x,y)E^*(x,y)\; dx\; dy, \tag{11.34}$$

where $\rho_i \left(x - m\delta x, y - n\delta y\right)$ is the current responsivity of the pixel as a function of position in coordinates $(x - m\delta x, y - n\delta y)$ referenced to the center of the pixel, assumed here to be the same for every pixel. This process can be described by a convolution of the irradiance with the "pixel function," followed by sampling.

$$i_{mn} = \left\{\left[E(x,y)E^*(x,y)\right] \otimes \rho_i\right\} \times \delta\left(x - x_m\right) \delta\left(y - y_n\right). \tag{11.35}$$

The convolution step may be incorporated into the ATF of the microscope, by writing the transfer function as

$$\tilde{i} = \tilde{I}\tilde{H}\tilde{\rho}_i. \tag{11.36}$$

Normally the current from the camera is converted to a serial signal, one horizontal line after another. Therefore, adjacent pixels in a horizontal row are read out one after another. This serial signal is passed through an electronic system, which will have a finite bandwidth, so the line of data will be convolved with the impulse response, $h(t)$ of this electronic system, resulting in a convolution of the image in the horizontal direction with a PSF, given by $h(x/v)$, where x is the position coordinate and v is the scan rate at which signals are collected from the row in the array.

Thus, the complete imaging process can be described by (1) a window function describing the field stop and illumination pattern and (2) an MTF describing (a) the pupil aperture, (b) any apodization, (c) aberrations, and (d) the pixel. The MTF may also include the electronic filter, but we must be careful here because the effect is only in the horizontal direction.

Take-Away

Summary of Incoherent Imaging:

- The incoherent PSF is the squared magnitude of the coherent one.
- The optical transfer function (incoherent) is the autocorrelation of the amplitude transfer function (coherent).
- The OTF is the Fourier transform of the IPSF.
- The DC term in the OTF measures transmission.
- The OTF at higher frequencies is usually reduced both by transmission and by the width of the PSF, leading to less contrast in the image than the object.
- The MTF is the magnitude of the OTF. It is often normalized to unity at DC. The PTF is the phase of the OTF. It describes a displacement of a sinusoidal pattern.

11.3 Characterizing an Optical System

In Section 8.7, we discussed how to compute performance metrics for optical systems using resolution as a criterion. We included the Rayleigh criterion and others in that section, but noted that a single number does not adequately describe system performance. Here we present an alternative approach which provides more information. In discussing the performance of an electronic filter, we can specify some measure of the width of the impulse response in the time domain or the width of the spectrum in the frequency domain, or we could plot the complete impulse response or spectrum. The single numbers present us with two problems, just as was the case in terms of the PSF. First, we must decide what we mean by "width." We must decide if we want the full width and half maximum, the $1/e$ or $1/e^2$ width, or some other value. Second, this single number does not tell us anything about the shape of the curve. Figure 11.4 illustrates the problem of relying on a single number. The Gaussian PSF (lower right in the figure) is "wider" than the Airy function (upper right) that uses the full aperture, but in many cases, we may find the Gaussian produces a "better" image.

Fig. 11.4:

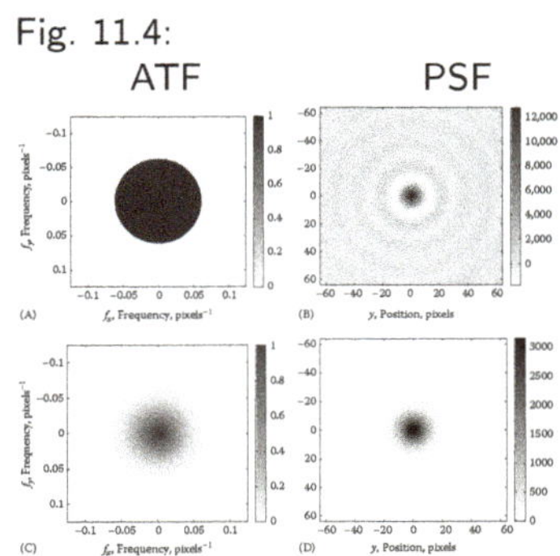

We can completely define the incoherent performance of an optical system by presenting either the IPSF or the OTF, including magnitude and phase. Because either of these two can be obtained from the other, only one is required. Just as the spectral response of an electronic filter is easily interpreted by an electrical engineer, the OTF, or even the MTF provides useful information to the optical engineer. For coherent detection, we can make the same statements about the coherent PSF and the ATF. In general, if we are interested in understanding what information we can recover from an optical system, we may ask about:

- Some measure of the overall light transmission through the optical system. This information may be the DC component of the OTF, or equivalently the integral under the incoherent PSF, but it is not uncommon in both prediction and measurement, to normalize the OTF to a peak value of 1. If normalization is done, a separate number is required to describe overall transmission of light through the system.
- The 3 dB (or other) bandwidth or the maximum frequency at which the transmission exceeds half (or other fraction) of that at the peak, This quantity provides information about how small an object we can resolve.
- The maximum frequency that where the MTF is above some very low value which can be considered zero. To recover all the information in the image, the sampling spatial frequency must be twice this value.
- Height, phase, and location of sidelobes. Often consecutive sidelobes will alternate in sign. Large sidelobes can result in images that are hard to interpret without signal processing.
- The number and location of zeros in the spectrum. Spatial frequencies at zeros of the MTF will not be transmitted through the optical system, and thus cannot be recovered, even with signal processing.
- The spatial distribution of the answers to any of these questions in the case that the system is not isoplanatic.

Often the MTF is normalized so the DC value is 1. This is useful for predicting contrast.

In Practice

Often the MTF is normalized. In this case, it offers no information about the transmission of the system. For example, loss caused by absorption or reflection from lens surfaces is not included. This normalized MTF is good for predicting contrast.

Fig. 11.9:

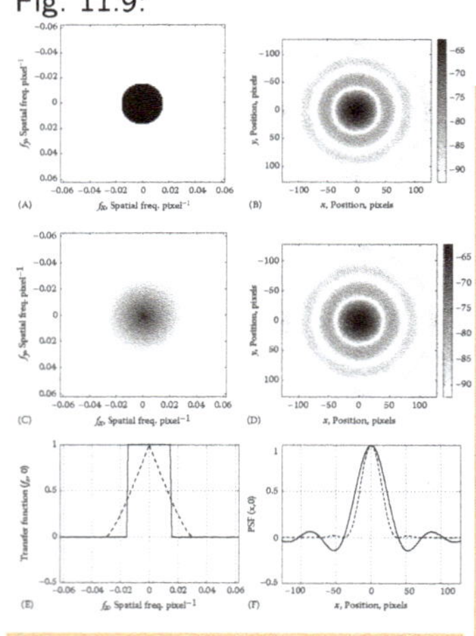

Example

Consider the normalized MTF in Figure 11.9E. Suppose that the object irradiance is sinusoidally modulated;

$$I = \frac{I_0}{2} + \frac{I_0}{2} \sin 2\pi x$$

where I_0 is some constant irradiance and w is the period of the sinusoidal modulation. What is the contrast in the object and image?

For our definition of contrast we'll use the same one we used for fringe contrast in Equation 7.30.

Then for the object, $I_{max} = I_0$ and $I_{min} = 0$.

$$V = 1.$$

In the image, for $w = 50$ pixels, the spatial frequency is 0.02 cyclesperpixel and the transmission is about 0.25. Thus,

$$I_{max} = \frac{I_0}{2} + 0.25\frac{I_0}{2} = 0.625 I_0$$

Eq. 7.30:
$V = \frac{I_{max} - I_{min}}{I_{max} + I_{min}}$.

and

$$I_{min} = \frac{I_0}{2} - 0.25\frac{I_0}{2} = 0.375 I_0,$$

so

$$V = 0.25.$$

The contrast is reduced to 25%.

Example

In the previous example, suppose the object is replaced by a bar chart with a period of $w = 50$ pixels. What is the image?

The bar chart consists of a square wave, which is expanded in a Fourier series of odd harmonics. The first term, a sinusoid at 0.02 cycles/pixel is transmitted with transmission of 0.25. The second harmonic is zero. The third harmonic is at 0.06 cycles/pixel, which is beyond the cutoff of the transfer function. Thus the image is the same as in the above example; a sine wave with contrast of 25%.

In Practice

It is not easy to make a light source or object with sinusoidal modulation. A good way to produce such a pattern is to use a bar chart which can be made with metal on glass, and then pass the light through a carefully chosen pupil to remove the third and higher harmonics.

We need metrics for system performance that can be computed from the design, using mathematical techniques, and that can be measured experimentally. For example, given a particular optical system design, we may ask, using the above criteria,

- What is the diffraction-limited system performance, given the available aperture?
- What is the predicted performance of the system as designed?
- What are the tolerances on system parameters to stay within specified performance limits?
- What is the actual performance of a specific one of the systems as built?

The diffraction-limited performance is usually quite easy to calculate, and we have seen numerous examples.

The OTF and PSF contain all the information about the system. They can be computed from the design parameters of the system, and studied as design tolerances are evaluated. However, they are not particularly amenable to direct measurement in the system once it is built. Measurement of OTF requires objects with brightness that varies sinusoidally in space at all desired spatial frequencies. Electrical engineers use network analyzers to sweep the spectra of filters. In optics, because spatial frequency is a two-dimensional variable, the number of test objects often becomes unmanageable, and good sinusoidal objects, particularly coherent ones, are not easily manufactured. Direct measurement of the PSF requires an object small enough to be unresolvable, yet bright enough so that the PSF can be measured to the desired level of accuracy. In many cases, such small and bright objects are not easily manufactured.

Many different test objects and patterns have been developed for measuring some parameters of the OTF or PSF. Some are shown in Figure 11.12. A point target (A) can,

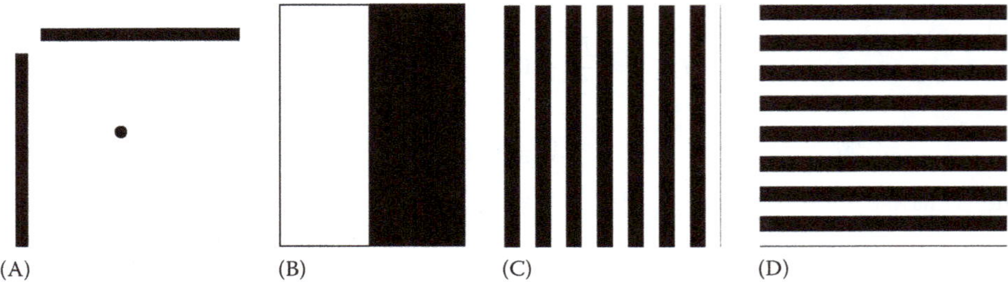

FIGURE 11.12
Some test targets. The most fundamental test target is the point or pinhole. For one-dimensional measurements a line is useful (A). For practical reasons, it is often useful to use a knife edge (B) or a bar chart (C and D).

in principle, be used to measure the PSF directly. The image will be the convolution of the PSF and the object, so if the actual point object is much smaller than the width of the PSF, the image will be a faithful reproduction of the PSF. For transmission systems, a small pinhole in a metal film is a potential point object. For reflective systems, a small mirror or highly scattering object may be used. However, for a high-NA system, the dimensions of the object must be less than a wavelength of light, and Rayleigh's scattering law (see Section 12.5.3) tells us that the scattering from a sphere varies like the inverse sixth power of the radius, so the resulting image may be too dim to be useful.

If a point object is too faint to measure, we could use a larger object, but then we would measure the convolution of the object and the PSF, so we would need to perform some mathematical deconvolution. In principle, we could compute the Fourier transform of the image, divide by the Fourier transform of the object, and inverse transform to obtain the OTF. If the object's transform has any zeros (a likely prospect), then we would need to find a way to "work around" them. Near these zeros we would be dividing by a small number, so we would amplify any noise in the measurement. More importantly, however, the deconvolution process always involves noise-enhancing operations. Ultimately, there is "no free lunch." If we use a very small pixel size to measure a PSF, so that we collect 100 photons in some pixel, then as we will see later, the noise would be 10 photons, by Equation 13.23, or 10%. If we collect photons from a larger object and then deconvolve, we can never reconstruct the signal in the smaller region with better SNR. The computational process can never alter the fact that we only detect 100 photons from that region of the image.

eq:13.23: $\sqrt{\overline{n^2} - \bar{n}^2} = \sqrt{\bar{n}}.$

When the number of photons is insufficient to answer a question, the most pragmatic approach is to ask a different question. For example, we can change our point object to a line object. For a transmissive system, this will be a slit, for example, etched into metal on a glass substrate, or constructed by bringing two knife edges close together. For a reflective imaging system, a line target could consist of a fine wire, of diameter well below the theoretical resolution limit of the system. The image of this line object will be the **line-spread function** or **LSF**,

The line-spread function gives 1-D information about the PSF, but the signal-to-noise ratio is higher because it includes more light.

$$h(x) = \int_{-\infty}^{\infty} h(x, y)\, dy. \tag{11.37}$$

We can predict the width of the LSF analytically, and compare it to experimental results. Because it only samples one direction, it provides less information, but with a higher SNR. We will want to measure the line-spread function in different directions such as x and y, as shown in Figure 11.12A, because we cannot be sure that the PSF will be rotationally symmetric.

If the line-spread function cannot be measured for some reason, we can obtain measurements with even more light, using the image of a knife edge, as shown in Figure 11.12B. The resulting function,

$$ESF(x) = \int_{-\infty}^{x} h(x'), \tag{11.38}$$

called the **edge-spread function** or **ESF**, can be predicted theoretically and measured experimentally. We saw an example of a computed ESF in Figure 11.5. One way we can compare experimental results to predictions is to plot the ESF, find the minimum and maximum values corresponding to the dark and light sides of the knife edge, respectively, and compute the distance between points where the signal has increased from minimum by 10% and 90% of the difference[59]. This so-called 90-10 width is a useful single number to characterize the system. Levels other than 10% and 90% may be used as well. However, examination of the complete ESF offers further information such as the presence of sidelobes and other details that may be of interest. Again, the ESF can be measured in different directions to ensure that the system meets the desired specifications. The ESF can, in principle, be differentiated to obtain the LSF. We must be careful, however, because taking the derivative increases the effect of noise, just as we saw in using deconvolution to obtain the PSF. Effectively, we are subtracting all the photons measured in the interval $\infty < x' < x + dx$ from those measured in the interval $\infty < x' < x$, each of which is large and therefore has a good signal-to-noise ratio. However, the difference is the number of photons in the interval $x < x' < x + dx$, which is small, and thus subject to larger fractional noise. In general, it is best to work directly with the ESF if the measurement of LSF is not practical.

The edge-spread function provides less information, but includes even more light than the line-spread function.

Take-Away

Testing Performance:
Although the ideal test object to characterize an optical system is a point source, it often does not produce enough light, and we need to use less perfect objects such as slits or edges.

A variety of test patterns are used in different applications.

It is important to test different parts of the field of view.

One problem frequently encountered in characterizing an optical system, particularly with an electronic camera, is that the pixel density may not be sufficient to satisfy the Nyquist criterion. One may argue that this failure can introduce artifacts into images through aliasing, and is characteristic of a "bad design," and that the final resolution of such a system is limited by pixel density. However, it is often the cases that systems are designed so that they fail to satisfy the Nyquist criterion at least over part of their operating range, and we may still desire to know the performance of the optical system alone. In some cases, we may use a camera to characterize an optical system that will eventually be used by the human eye or a detector of greater resolution than the camera in question. Thus, we may want to consider techniques for sub-pixel resolution measurements. In the case of the ESF, we can attempt to extract information with sub-pixel resolution. One approach is to use a tilted knife edge. Each row of pixels samples a shifted image of the knife edge, so that each row becomes a convolution of the ESF and the pixel function, sampled at the pixel spacing. After data collection, the images can be processed to deconvolve the pixel function, to shift the lines appropriately, and

A tilted knife edge can provide a measure of resolution at a sub-pixel level.

to interpolate the ESF[227]. Such a computation depends, of course, on there being sufficient SNR.

Resolution may not be the same across the field of view. Optical systems are rarely space-invariant.

These test objects may, if we like, be replicated in several positions throughout the field of view in order to sample multiple areas if we are not sure that the system is sufficiently close to isoplanatic that we can trust a single measurement. The spacing must be sufficiently sparse that the widest expected PSFs or LSFs from the closest pair of targets will not overlap.

Bar charts are good for full-field tests of resolution.

One common object is the bar chart, which consists of parallel bars, normally of equal spacing. Typically they consist of metal deposited on glass substrates, and to be useful they must have very sharp, straight edges. Ronchi rulings, originally designed by Vasco Ronchi for testing reflective optics[191], can be used here. If a large enough spacing is used, any edge can be used to measure a local ESF in a specific region. We can thus obtain the ESF over a large field of view, without assuming an isoplanatic system. The lines must be spaced further apart than the widest expected PSF at any point in the system.

In Practice

Aberrations such as coma, field curvature, and astigmatism prevent an optical system from being perfectly space invariant. Generally we need to measure the PSF or transfer function at several points in the field of view. Often we choose the center (which usually produces the best result, the edges of the field of view, the corners, and perhaps a few points in between.

On the other hand, we can use a bar chart with narrow bars to measure the OTF. Ideally we would like a set of test objects having sinusoidal transmission, reflection, or illumination. As we have already discussed, such charts may or may not be easy to manufacture, and a different target would be needed for each spatial frequency. We may find that a bar chart can be more easily produced, and although the bar chart suffers the same problem of probing only a single frequency (or at most a few harmonics), there are situations in which a bar chart may be useful. For example, we might want to see if a system meets its design specifications at certain set spatial frequencies. In such a case, we might require, for example, that the contrast, $(max - min)/(max + min)$, be greater than 80% at 150 **line pairs** per millimeter. By this, we mean that the spatial frequency is $150\,\text{mm}^{-1}$, so there are 150 dark lines and 150 bright lines per millimeter.

The bar chart works well when we are asking about the performance at a specific frequency, but it does not address the problem of sampling multiple spatial frequencies. An alternative chart, which has become a well-known standard is the **Air-Force resolution chart**[228] shown in Figure 11.13. Despite the fact that the military standard that established this chart was canceled in 2006, it remains in frequent use. The chart consists of sets of three bars with equal widths and spacing equal to the width. The length of each bar is five times its width. Two bar charts in orthogonal configurations are placed beside each other with a spacing between them equal to twice the width of the bars in an "element." The elements of different size are clustered into "groups." Within each group are six "elements" or pairs of patterns, each pair having a different size. The patterns are arranged in a spiral pattern with the largest at the outside. The elements of any group are twice the size of those in the next smaller group. The width of a single bright or dark line (not the width of a line pair which some authors use) is

$$x = \frac{1\,\text{mm}}{2^{G+1+(E-1)/6}}, \tag{11.39}$$

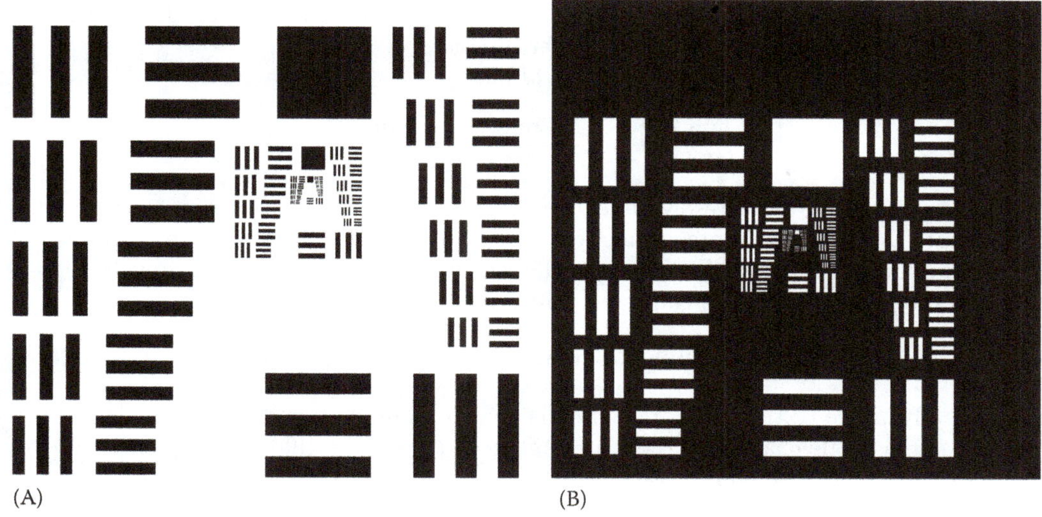

FIGURE 11.13
Air-Force resolution charts. The first element of the first group is in the lower right. The remaining five elements are on the left side from top to bottom. Each element is $2^{-1/6}$ the size of the one before it. Each group is smaller than the previous by a factor of 2. The elements of the next group are from top to bottom along the right side. The chart can be used in positive (A) or negative (B) form, with bright lines or dark lines, respectively.

where

- G is the group number (possibly negative)
- E is the element number

The Air Force resolution chart is widely used.

For example, the first element in group -1 has a line width of 1 mm and a line length of 5 mm. The third element of group 1 has a line width of 198.4 µm and a line length of 992.1 µm. The chart in Figure 11.13 consists of six groups and was generated by computer. The Air-Force resolution chart is available from a variety of vendors in several forms, and often includes numbers as text identifying the groups. It may be printed with ink on paper, or may be produced in metal on glass, in either positive or negative form. The inherent contrast of the ink-on-paper targets will be determined by the qualities of the ink and the paper. For the metal-on-glass target, the metal region will be bright in reflection because the metal reflection may be as high as 96% or more, and glass regions will reflect about 4%, for contrast of almost 25:1. In a transmissive mode, the glass transmission through two surfaces will be about 92%, and the transmission of the metal will be low enough to be neglected. Researchers may be clever in designing their own resolution charts. For example, someone designing a system to characterize two different materials might construct a resolution chart made completely of the two materials. In another interesting variation, someone designing a system for measuring height profiles might use the thickness of the metal on glass target rather than the reflectivity.

Example

Suppose we look at an image of the Air Force resolution chart and determine that we can resolve the fourth element in group 3. That is, we can see the space between the lines in this element to our satisfaction. How do we interpret this statement in terms of point-spread function or transfer function?

The line width is given by Equation 11.39.

$$x = \frac{1\,\text{mm}}{2^{G+1+(E-1)/6}} = \frac{1\,\text{mm}}{2^{3+1+(4-1)/6}} = \frac{1\,\text{mm}}{2^{4.5}} = 0.044\,\text{mm}.$$

This means that it can resolve lines which are approximately 44 μm wide. We can estimate that the point-spread function has a width somewhat close to this distance.

Thinking of the transfer function, the width of a line pair (bright and dark) is twice the line width, so we could say that this system resolves

$$1\,\text{line pair}/(2 \times 0.044\,\text{mm}) = 11.3\,\text{Line pairs per mm}. \tag{11.40}$$

The advantage of the Air Force chart is that it presents a wide range of spatial frequencies on a single chart, and the disadvantage is that these different frequencies are, of necessity, located at different positions in the field of view, so we must assume that our system is isoplanatic, or else move the chart around through the field of view. We must guard against overly optimistic estimates of performance, because the smallest parts of the chart are in the center. If we place these near the center of the field of view, they will be exactly where we expect the least aberrations and the best resolution. A good approach is to find the resolution at the center, and then move the center of the chart to other regions of the field of view.

In Practice

Using the Air Force chart, it is important to test different portions of the field of view. Often we choose the center, points at the edges, points at the corners, and some points in between. Even if we expect the system to be circularly symmetric, alignment will not be perfect and there may be differences between left and right or top and bottom, so it is important to cover the complete field of view.

Another test object is the **radial bar chart** shown in Figure 11.14, and often called the Siemens Star.[44] Like the Air-Force chart, this one provides different spatial frequencies at different locations. It has the advantages of being continuous, and providing a large number of orientations. It has the disadvantage of being harder to use to determine quantitative resolution limits, and is less commonly used than the Air-Force chart. It does allow us a different view of the OTF. For example, in Figure 11.15 the radial bar chart has been convolved with a uniform circular PSF. The OTF of this PSF is an Airy function. This

FIGURE 11.14
Radial test target. This target can help visualize performance of an optical system at multiple spatial frequencies.

FIGURE 11.15
Image of radial test target. The radial test target in Figure 11.14, 1000 units in diameter, has been convolved with a point-spread top-hat function 50 units in diameter.

PSF might be found in a scanning optical system with a detector aperture in an image plane having a diameter greater than the diffraction limit. For example, in a confocal microscope, a pinhole about three times the diffraction limit is often used. In the outer regions of the image we can see a slight blurring of the edges as expected. As we approach the center, we reach a frequency at which the MTF goes to the first null, and we see a gray circle, with no evidence of light and dark regions; this spatial frequency is completely blocked. Inside this gray ring, we once again see the pattern, indicating that at these higher spatial frequencies, the MTF has increased. However, if we look closely, we notice that the bright and dark regions are reversed. Now that the OTF has passed through zero, it has increased in magnitude again, but this time with a negative sign. Looking further into the pattern, the reversal occurs several times. Our ability to observe this process on the printed page is ultimately limited by the number of pixels we used to generate the data, or the resolution of the printing process.

One interesting application of this chart is in looking at the effect of pixelation. The bar chart in Figure 11.14 was produced with an array of 1200 elements in each direction. Let these pixels be some arbitrary unit of length. Then to illustrate the effect of large pixels we can simulate pixels that are $n \times m$ units in size, by simply averaging each set of $n \times m$ elements in the original data, and reporting that as the value in a single pixel. Samples are shown in Figure 11.16. For example, if the pixels are 30 units square, the first

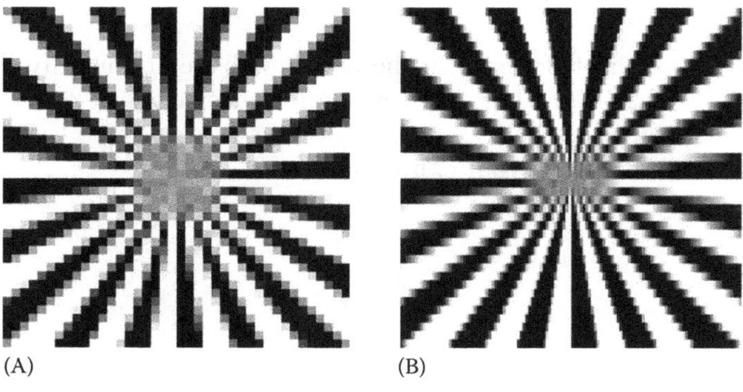

FIGURE 11.16
Effect of pixels. The original image is 1200 by 1200 units. In these figures, they have been imaged with pixels 30 units square (A), and 30 units high by 10 units wide.

pixel in the first row is the average of the data in elements 1–30 in each direction. The next pixel in the row is the average of elements 31–60 in the horizontal direction, and 1–30 in the vertical. We continue the process until all pixels have been computed, and display the image in the original configuration.

Figure 11.16 shows results for pixels that are 30 units square (A), and 30 units high by 10 units wide (B). Near the edges, we can see the pixels directly, and near the center we can see the effects of averaging. In the first case (A), the center of the image is quite uniformly smeared to a gray value, and we cannot see the pattern. At lower spatial frequencies in the outer regions, we can see the pattern, but with pixelation artifacts. In the asymmetric case, the pixelation is obviously asymmetric, and the smearing is preferentially in the vertical direction. Thus, the vertical bars are better reproduced than the horizontal ones, even at the lowest spatial frequencies.

Take-Away

Summary of System Characterization:

- Some systems may be characterized by measuring the PSF directly.
- Often there is insufficient light to do this.
- Alternatives include measurement of LSF or ESF.
- The OTF can be measured directly with a sinusoidal chart.
- Often it is too tedious to use the number of experiments required to characterize a system this way.
- A variety of resolution charts exist to characterize a system. All of them provide limited information.

Problems

11.1 Point-Spread Function

Generate bar charts with one bar pair (light and dark) per 8 μm. The bars are in the x direction for one and y for the other. Then generate an elliptical Gaussian point-spread function with diameters of $d_x = 4\,\mu\text{m}$ and $d_y = 2\,\mu\text{m}$. Show the images of the bar charts viewed through a system having this point-spread function.

11.2 Resolution (NS)

Use the Fourier optics script, fo.m, from the book's website and modify it as you see fit.

Make a pattern of 2 bright bars with a width and spacing of δx, and compute the image with a numerical aperture, $NA = 0.2$. Vary δx to determine the resolution of the system. Compare your result to your expectation.

12 Radiometry and Photometry

We have discussed light quantitatively in a few cases, giving consideration in Sections 1.4.4 and 1.5.3 to the quantities irradiance, power and energy. We have also considered the transmission of irradiance and power in the study of polarization in Chapter 6, and thin films in Section 7.7. We have briefly considered the concept of radiometry in the discussion of f-number and numerical aperture in Section 4.1. In this chapter, we describe the fundamentals of the quantitative measurement of light. In one approach, we can think of light as an electromagnetic wave, and use the units developed in that field, Watts, Joules, and their derivatives in what we call the subject of **radiometry**. Alternatively, we can talk in terms of the spectral response of the human eye, using some measure of the visual perception of "brightness." This approach is the subject of **photometry**, and it uses units such as lumens, and lux.

There has been great confusion regarding the terminology and units discussed in this chapter, and recently, international organizations have made an effort to establish consistent standards. A very precise vocabulary has been developed with well-defined names, symbols, and equations, and we will use the results of these efforts in this chapter. Occasionally, this notation will conflict with that used elsewhere, such as in the use of E for irradiance, which can be confused with the scalar electric field. We make an effort to minimize such conflict but will use the international radiometric conventions. We begin with a discussion of five important radiometric quantities, radiant flux or power (in Watts = W), radiant exitance (W/m^2), radiant intensity (W/sr), radiance (W/m^2/sr), and irradiance (W/m^2), and we establish equations connecting each of these to the others in Section 12.1.1. For each of these five quantities, there is a corresponding photometric quantity, luminous flux (in lumens, abbreviated lm), luminous exitance (lm/m^2), luminous intensity (lm/sr), luminance (lm/m^2/sr), and illuminance (lm/m^2). In Section 12.2, we will briefly discuss spectral radiometry which treats the distribution of radiometric quantities in frequency or wavelength and then will address photometry and color in Section 12.3. We will discuss some instrumentation in Section 12.4, and conclude with blackbody radiation in Section 12.5. Definitions, units, and abbreviations shown here conform to a guide[224], published by the National Institute of Standards and Technology (NIST), of the U.S. Department of Commerce, and are generally in agreement with international standards.

As we proceed through these sections, it may be useful to refer to the chart in Figure 12.1 and Table 12.1 which summarize all of the nomenclature and equations of radiometry and photometry presented herein. The new reader will probably not understand all these concepts at first but will find it useful to refer back to them as we develop the concepts in more detail throughout the chapter. The four main panels in the second and third columns of Figure 12.1 describe four quantities used to measure light emitted from a source. The total emitted light power or "flux" may be considered to be the fundamental quantity, and derivatives can be taken with respect to solid angle going down the page or with respect to area going to the right. Radiometric and photometric names of these quantities are given at the top of each panel, and a cartoon of an arrangement of apertures to approximate the desired derivatives are shown in the centers of the panels. Apertures are shown to limit the area of the source, the solid angle subtended at the source, and the area of the receiver.

DOI: 10.1201/9781315157047-12

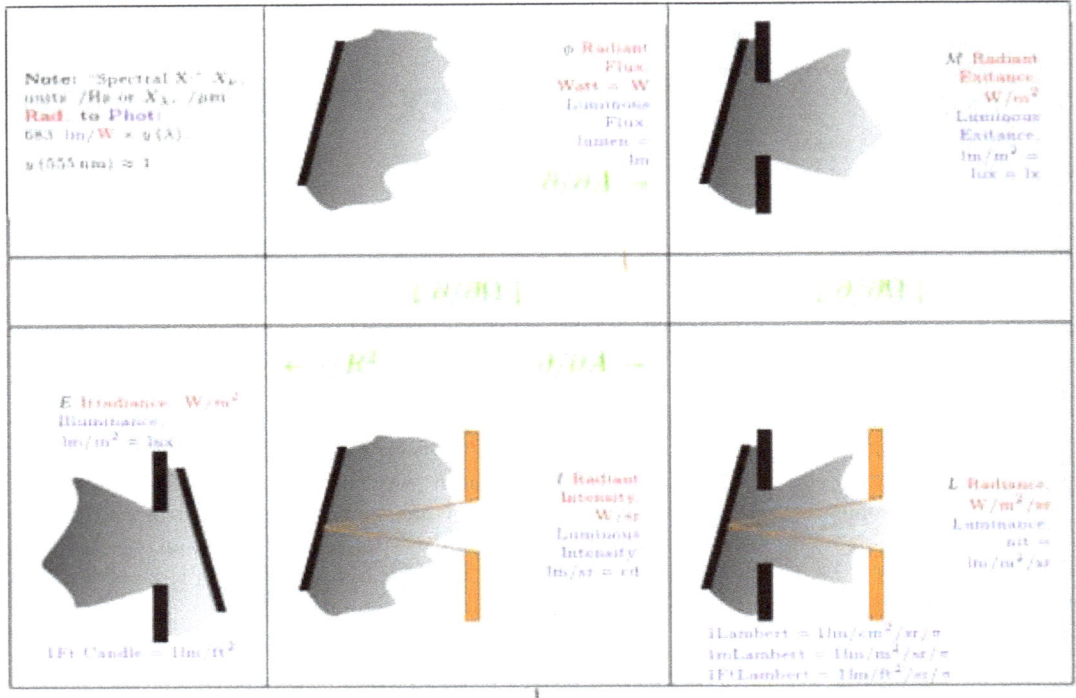

FIGURE 12.1
Radiometry and photometry. The fundamental quantities of radiometry and photometry and their relationships are shown. Abbreviations of units are lm for lumens, cd for candelas.

> *Take-Away*
>
> **SI Units:**
> To bring some order to a confusing discipline, a rigorous set of international standards has been established for radiometry and photometry[224]. However, it is not unusual to see non-standard units based on a variety of historical precedents.

> Eq. 1.45:
> $\mathbf{S} = \mathbf{E} \times \mathbf{H}$.
> Note the use of $\mathbf{E}$ for field here while we use E for irradiance in this chapter to conform to international standards.

12.1 Basic Radiometry

We will begin our discussion of radiometry with a detailed discussion of the quantities in Figure 12.1, their units, definitions, and relationships. We will then discuss the radiance theorem, one of the fundamental conservation laws of optics, and finally discuss some examples.

12.1.1 Quantities, Units, and Definitions

We have defined the power per unit area transverse to a plane wave in terms of the Poynting vector in Equation 1.45. However, very few light sources are adequately defined by a single wave with a well-defined direction. Consider an extended source in an area A,

TABLE 12.1
Some Radiometric and Photometric Quantities

Quantity	Symbol	Equation	SI Units
Radiant energy	J		J
Radiant energy density	w	$w = \frac{d^3 J}{dV^3}$	$\mathrm{J/m^3}$
Radiant flux or power	P or Φ	$\Phi = \frac{dJ}{dt}$	W
Radiant exitance	M	$M = \frac{d\Phi}{dA}$	$\mathrm{W/m^2}$
Irradiance	E	$E = \frac{d\Phi}{dA}$	$\mathrm{W/m^2}$
Radiant intensity	I	$I = \frac{d\Phi}{d\Omega}$	W/sr
Radiance	L	$L = \frac{d^2\Phi}{dA\cos\theta\, d\Omega}$	$\mathrm{W/m^2/sr}$
Fluence	Ψ	$\Psi = \frac{dJ}{dA}$	$\mathrm{J/m^2}$
Fluence rate	F	$F = \frac{d\Psi}{dt}$	$\mathrm{J/m^2/s}$
Emissivity	ϵ	$\epsilon = \frac{M}{M_{bb}}$	Dimensionless
Spectral ()	$()_\nu$	$\frac{d()}{d\nu}$	()/Hz
	or $()_\lambda$	$\frac{d()}{d\lambda}$	()/μm
Luminous flux or Power	P or Φ		lm
Luminous exitance	M	$M = \frac{d\Phi}{dA}$	$\mathrm{lm/m^2}$
Illuminance	E	$E = \frac{d\Phi}{dA}$	$\mathrm{lm/m^2}$
Luminous intensity	I	$I = \frac{d\Phi}{d\Omega}$	lm/sr
Luminance	L	$L = \frac{d^2\Phi}{dA\cos\theta\, d\Omega}$	$\mathrm{lm/m^2}$
Spectral luminous efficiency	$V(\lambda)$		Dimensionless
Color matching functions	$\bar{x}(\lambda)$		
	$\bar{y}(\lambda)$		
	$\bar{z}(\lambda)$		Dimensionless
Spectral luminous	$\bar{y}(\lambda)$		
efficiency function	Photometric conversion	$y \times 683\,\mathrm{lm/W}$	
	$V(\lambda) = \bar{y}(\lambda) = 1$ at	$\approx 555\,\mathrm{nm}$	
	1 nit	$1\,\mathrm{lumen/m^2/sr/}$	
	1 $\mathrm{cd/m^2}$	$1\,\mathrm{lumen/m^2/sr/}$	
	1 Lambert	$1\,\mathrm{lm/cm^2/sr/}\pi$	
	1 Meterlambert	$1\,\mathrm{lm/m^2/sr/}\pi$	
	1 Footlambert	$1\,\mathrm{lm/ft^2/sr/}\pi$	
	1 Footcandle	$1\,\mathrm{lm/ft^2}$	
	1 candela	$1\,\mathrm{cd} = 1\,\mathrm{lm/sr}$	

Each quantity is presented with its SI notation, and units[224]. Other appropriate units are often used.

consisting of multiple point sources as shown in Figure 12.2. Each source patch, of area dA, contributes a power dP to a spherical wave. At a receiver with area A', far enough away that the transverse distances x and y within the source area A can be neglected (or equivalently, at the focus of a Fraunhofer lens as discussed in Section 8.5), the Poynting vector will be

$$d\mathbf{S} \approx \frac{dP}{4\pi r^2}\left(\sin\theta\cos\zeta\hat{\mathbf{x}} + \sin\theta\sin\zeta\hat{\mathbf{y}} + \cos\theta\hat{\mathbf{z}}\right), \tag{12.1}$$

where

- $r^2 = x^2 + y^2 + z^2$
- θ is the angle from the z axis

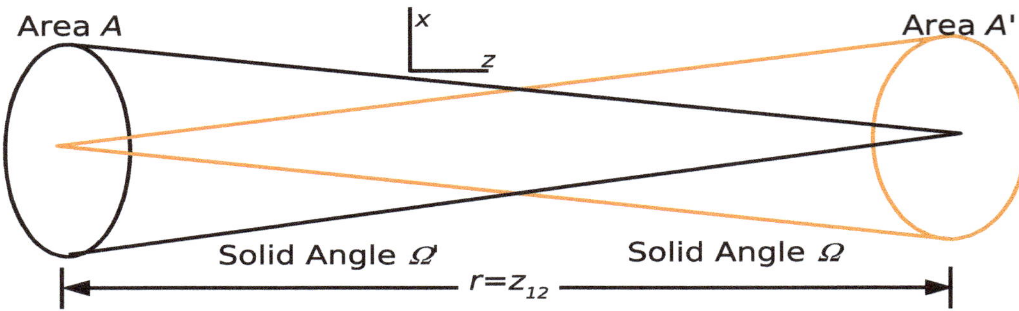

FIGURE 12.2
Source and receiver. The solid angle seen from the source, Ω, is defined by the area of the receiver, A', divided by the square of the distance r between source and receiver. Likewise, Ω' seen from the receiver, is defined by A/r^2.

Irradiance, E: $\mathrm{W/m^2}$

- ζ is the angle of rotation in the x, y plane about that axis

The irradiance* will be the magnitude of the Poynting vector, in units of power per unit area ($\mathrm{W/m^2}$);

$$E = \frac{dP}{dA'} = \frac{P}{(4\pi r^2)}, \tag{12.2}$$

assuming that the light distribution is isotropic (uniform in angle).

12.1.1.1 Irradiance, Radiant Intensity, and Radiance

We may often find it useful to avoid discussing the distance to the receiver. The solid angle is defined by

$$\Omega = \frac{A'}{r^2}. \tag{12.3}$$

Radiant Intensity, I: W/sr

so if we know the irradiance in units of power per unit area ($\mathrm{W/m^2}$), we can define the **radiant intensity**† in units of power per solid angle (W/sr) by

$$I = Er^2 \qquad E = \frac{I}{r^2}. \tag{12.4}$$

Then the increment of intensity of the source described in Equation 12.2 for this unresolved source of size, dA, is

$$dI = \frac{dP}{d\Omega} = \frac{dP}{d\frac{A'}{r^2}} = \frac{dP}{(4\pi)}, \tag{12.5}$$

again assuming an isotropic source.

Radiant intensity or "intensity" is a property of the source alone. The intensity of the sun is the same whether measured on the planet Mercury, or on Pluto (but unfortunately not always as reported in the literature on Earth). It is quite common for writers to use the term "intensity" to mean power per unit area. However, the correct term for power

*Here we use E for irradiance in keeping with the SI standards. Elsewhere in the book we have used I to avoid confusion with the electric field. This inconsistency is discussed in Section 1.4.4.

†The symbol, I, is used for intensity to comply with the SI units, and should not be confused with irradiance. Elsewhere in the book, I is used for irradiance. This inconsistency is discussed in Section 1.4.4.

per unit area is "irradiance." The irradiance from the sun decreases with the square of the distance at which it is measured.

If we think of the unresolved source as, for example, an atom of hot tungsten in a lamp filament, then dP and dA are small, but significant power can be obtained by summing over many such atoms, or in practice, integrating over a whole tungsten filament.

$$I = \int dI = \int \frac{\partial I}{\partial A} dA$$

In this equation, the quantity $\partial I/\partial A$ is sufficiently important to merit a name, and we will call it the **radiance**, L, given by

Radiance, L, $W/m^3/sr$

$$L(x, y, \theta, \zeta) = \frac{\partial I(\theta, \zeta)}{\partial A} \tag{12.6}$$

in $\mathrm{W/m^2/sr}$. Then, given L, we can obtain I by integration. When we perform the integrals in radiometry, the contributions add incoherently. If the contributions were coherent with respect to each other we would need to compute electric fields instead of irradiances and integrate those with consideration of their phases. In the present case, each point can be considered, for example, as an individual tungsten molecule radiating with its own phase, and the random phases cancel. Thus,

$$I(\theta, \zeta) = \int_A L(x, y, \theta, \zeta)\, dx\, dy. \tag{12.7}$$

On the axis,

$$L(x, y, \theta, \zeta) = \frac{\partial I(\theta, \zeta)}{\partial A} = \frac{\partial^2 P}{\partial A \partial \Omega}, \qquad x' = y' = 0.$$

However, from a point on the receiver far from the axis, the source will appear to have a smaller area because we only see the **projected area**, $A_{proj} = A\cos\theta$, and we would report the radiance to be the derivative of I with respect to projected area

$$L = \frac{\partial I}{\partial A \cos\theta} = \frac{\partial^2 P}{\partial A \partial \Omega \cos\theta}. \tag{12.8}$$

We have completed the connections among the lower three panels in Figure 12.1:

- The radiant intensity from a source is the irradiance of that source at some receiver, multiplied by the square of the distance between source and receiver. Alternatively, if we know the intensity of a source, we divide it by the squared distance to the receiver to obtain the irradiance on the receiver. These relationships are in Equation 12.4.
- The radiance of a source is the derivative of its radiant intensity with respect to projected source area, $A\cos\theta$, according to Equation 12.6. The radiant intensity is the integral of the radiance times the cosine of θ over the area of the source, as in Equation 12.7.

12.1.1.2 Radiant Intensity and Radiant Flux

If we integrate the radiance over both area and solid angle, we will obtain the total power received from the source. There are two ways to accomplish this double integration. First, we can use Equation 12.7 and then integrate the intensity over solid angle.

Details of the integration are shown in Appendix B. The integral to be performed is

$$P = \Phi = \iint I(\theta, \phi) \sin\theta \, d\theta \, d\zeta. \tag{12.9}$$

We use the notation, P, here for power, as well as the notation of **radiant flux**, Φ, commonly used in radiometry.

In general, the integrand may have a complicated angular dependence, but if it is uniform then one of the special cases in Appendix B can be used, including

Equation B.6,

$$\Omega = 2\pi \left(1 - \sqrt{1 - \left(\frac{NA}{n}\right)^2}\right) \qquad \text{(Cylindrical symmetry)},$$

Equation B.7

$$\Omega \approx \pi \left(\frac{NA}{n}\right)^2 \qquad \text{(Small NA)},$$

Equation B.8

$$\Omega = 2\pi \qquad \text{(Hemisphere)},$$

or Equation B.9

$$\Omega = 4\pi \qquad \text{(Sphere)}.$$

We have now addressed the center column of Figure 12.1; power or radiant flux is the integral of radiant intensity over solid angle as in Equation 12.9, and radiant intensity is the partial derivative of power with respect to solid angle:

$$I = \frac{\partial P}{\partial \Omega} \tag{12.10}$$

12.1.1.3 Radiance, Radiant Exitance, and Radiant Flux

As the second route to computing the radiant flux, we could start with radiance, and integrate first with respect to solid angle, and then with respect to area. This integral can be accomplished, like Equation 12.9 as shown in Figure B.1. We define M, the **radiant exitance** as

$$M(x, y) = \iint \left[\frac{\partial^2 P}{\partial A \partial \Omega}\right] \sin\theta \, d\theta \, d\zeta$$

$$M(x, y) = \iint L(x, y, \theta, \zeta) \cos\theta \sin\theta \, d\theta \, d\zeta. \tag{12.11}$$

Equivalently, the radiance can be obtained from the radiant exitance as

Radiant Exitance, M: $\mathrm{W/m^2}$

$$L(x, y, \theta, \zeta) = \frac{\partial M(x, y)}{\partial \Omega} \frac{1}{\cos\theta}. \tag{12.12}$$

We note that the radiant exitance has the same units as irradiance, and we will explore this similarity in Section 12.1.3, where we estimate the radiant exitance reflected from a diffuse surface as a function of the incident irradiance.

On one hand, if L is independent of direction over the solid angle of interest, and this solid angle is small enough that the cosine can be considered unity, then the integral becomes

$$M(x, y) = L\Omega \cos\theta,$$

where the solid angle is given by Equations B.4, B.6, or B.7.

On the other hand, if the solid angle subtends a whole hemisphere, we detect all the light emitted by an increment dA of the source, and Equation 12.11 becomes

$$M(x, y) = \int_0^{2\pi} \int_0^{\pi/2} L \cos\theta \sin\theta \, d\theta \, d\zeta = 2\pi L \frac{\sin^2 \frac{\pi}{2}}{2}.$$

$$M(x, y) = \pi L, \qquad \text{(Lambertian source)} \tag{12.13}$$

This simple equation holds only when L is constant. In this case, the source is said to be **Lambertian**, with a radiance $L = M/\pi$. Many sources are in fact very close to Lambertian, and the assumption is often used when no other information is available. A Lambertian source appears equally bright from any direction. That is, the power per unit projected angle is a constant.

Lambertian Source

In Practice

Many sources are nearly Lambertian. A light-emitting diode without any optics is likely to have a radiance approximated by Equation 12.13. Light from an external source, scattered by a diffuse surface such as a painted wall, may also be well described by this equation. Sometimes a source distribution may be very complicated, but it is challenging to make enough measurements to characterize it. Often we use the Lambertian assumption as a first guess. This assumption is a reasonable one for light scattered from diffuse surfaces as discussed in Section 1.7 and Figure 1.10B.

Fig. 1.10B:

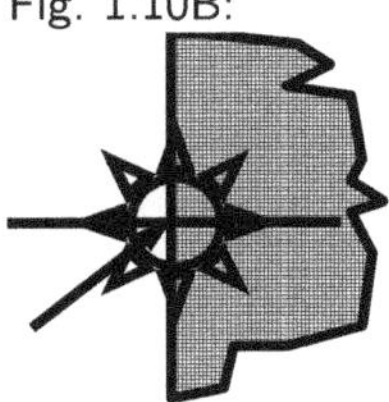

In Figure 12.1, we have just established the relationships indicated in the right column; radiance is the derivative of radiant exitance with respect to solid angle, and radiant exitance is the integral of radiance over solid angle. In our thought experiment of a hot tungsten filament, we now have the power per unit area of the filament. Now we move from the top right to the top center in the figure, integrating over area:

$$P = \Phi = \iint M(x, y) \, dx \, dy, \tag{12.14}$$

or

$$M(x,y) = \frac{\partial P}{\partial A}. \tag{12.15}$$

This section completes the radiometric parts of the five panels of Figure 12.1, P, M, I, L, and E, and the relationships among them. We will return to this figure to discuss its photometric parts in Section 12.3.

It may be useful to conclude the section with an example to illustrate these five quantities. For illustration, we'll consider a uniform "bare" light-emitting diode (LED), exposed to air with no lens.

Example

Radiometric Properties of an LED Lets consider a square LED, 200 μm wide, emitting 16 mW. We would like to know the radiant exitance, radiance and intensity of this source as well as the irradiance on a paper held 1 meter away.

Lacking any information, we will guess that the light is emitted uniformly over the surface and follows a Lambertian function. Given the power,

$$\Phi = 16\,\text{mW},$$

the radiant exitance from Equation 12.15 is

$$M = \frac{P}{A} = \frac{16\times 10^{-3}\,\text{W}}{(200\times 10^{-6}\,\text{m})^2} = 4\times 10^5\,\text{W/m}^2,$$

or 400 kW/m^2. Under the Lambertian assumption, from Equation 12.13, the radiance is

$$L = M/\pi = 1.27\times 10^5\,\text{W}/\left(\text{m}^2\text{sr}\right) \tag{12.16}$$

The intensity is obtained by integrating the radiance over the area according to Equation 12.7. Because of the assumption of uniformity,

$$I = LA = \frac{MA}{\pi\,\text{sr}} = \frac{P}{\pi\,\text{sr}} = \frac{16\times 10^{-3}\,\text{W}}{\pi} = 5.09\times 10^{-3}\,\text{W/sr}.$$

Finally, on a surface $r = 1$ m away, the irradiance, from Equation 12.4, is

$$E = \frac{I}{r^2} = \frac{5.09\times 10^{-3}\,\text{W/sr}}{(1\,\text{m})^2} = 5.09\times 10^{-3}\,\text{W/m}^2$$

or a bit over 5 mW per square meter.

12.1.2 Radiance Theorem

Recalling Equation 12.3 and Figure 12.2, the increment of power in an increment of solid angle and an increment of projected area is

$$dP = L\,dA\,d\Omega = L\,dA\frac{dA'}{r^2}\cos\theta, \tag{12.17}$$

where $d\Omega = dA'/r^2$. It is equally correct to write

$$dP = L\,dA'\,d\Omega' = L\,dA'\,\frac{dA}{r^2}\cos\theta, \tag{12.18}$$

where,

$$d\Omega' = \frac{dA}{r^2}, \tag{12.19}$$

and θ is the angle between the normals to the two patches of area, A and A'.

Equations 12.17 and 12.18 provide two different interpretations of the radiance. In the first, the source is the center of attention and the radiance is the power per unit projected area at the source, per unit solid angle subtended by the receiver as seen from the source. The second interpretation centers on the receiver, and states that the radiance is the power per unit area of the receiver per unit solid angle subtended by the source as seen from the detector. This result suggests that the quantity $A_{proj}\Omega$ is invariant, at least with respect to translation from the source to the receiver. In fact, we recall from the Abbe sine invariant (Equation 3.31) that transverse and angular magnification are related by

Eq. 3.31:
$n'x'd\alpha' = nxd\alpha.$

$$n'x'd\alpha' = nx\,d\alpha,$$

where x and x' are the object and image heights so that the magnification is $m = x'/x$, while α and α' are the angles so that the angular magnification is $d\alpha'/d\alpha = n/(n'm)$. Therefore,

$$n^2 A\Omega = (n')^2 A'\Omega'. \tag{12.20}$$

Because the product of A, Ω, and n^2 is preserved in any optical system, we can define $L = d^2P/(dA\,d\Omega)$ and $L' = d^2P/(dA'\,d\Omega')$, and recognize that conservation of energy demands that

$$\iint L\,dA\,d\Omega = \iint L'dA'd\Omega'$$

$$\iint \frac{L}{n^2} n^2 dA\,d\Omega = \iint \frac{L'}{n^2}(n')^2\,dA'\,d\Omega'$$

so that

$$\frac{L}{n^2} = \frac{L'}{(n')^2}, \tag{12.21}$$

or the **basic radiance**, L/n^2, is a conserved quantity, regardless of the complexity of the optical system. Of course, this is true only for a lossless system. However, if there is loss to absorption, reflection, or transmission, this loss can be treated simply as a multiplicative factor. Equation 12.21 is called the **radiance theorem**, and, as we shall see, provides a very powerful tool for treating complicated radiometric problems.

Eq. 6.60:
$R = \left|\frac{(n_2/n_1)-1}{(n_2/n_1)+1}\right|^2.$

Take-Away

Conservation Law:
Basic radiance is a conserved quantity, except for multiplicative losses such as transmission, reflection, and absorption. This fact greatly simplifies radiometric calculations in complicated optical systems.

Example

Radiance of an LED

Let's consider the 16 MW LED in the earlier example. We computed the radiance of the "bare" LED in Equation 12.16 as $127\,\text{Mw}/\left(\text{m}^2\text{sr}\right)$. Now this LED is encased in a plastic housing which ends in a lens. What is the radiance of the output?

The ease with which we obtain the answer is a great example of the importance of the radiance theorem, Equation 12.21. Assuming the plastic has an index of refraction of about 1.5, the radiance in the plastic is

$$L' = \left(\frac{n'}{n}\right)^2 L = \left(\frac{1.5}{1}\right)^2 \times 127\,\text{MW}/\left(\text{m}^2\text{sr}\right) =$$

$$286\,\text{MW}/\left(\text{m}^2\text{sr}\right).$$

However, we didn't need to do this calculation because upon exiting the plastic, the radiance returns to the original value. The radiance is still $127\,\text{MW}/\left(\text{m}^2\text{sr}\right)$. This result is true, even despite the presence of the lens, and we do not need to know anything about the lens.

Finally, we should remember that the lens probably does not have an anti-reflection coating so we should include Fresnel reflection at the surface. If the angles are not too large, the reflection is about 4%, (Equation 6.60), so 96% of the light is transmitted and the final radiance is

$$127\,\text{MW}/\left(\text{m}^2\text{sr}\right) \times 0.96 = 122\,\text{MW}/\left(\text{m}^2\text{sr}\right).$$

12.1.2.1 Examples

The radiance theorem is true for any lossless optical system. We shall see in Section 12.5 that it follows from fundamental principles of thermodynamics. The radiance theorem is a fundamental conservation law that enables us to solve many complicated problems easily. We will consider four common examples to illustrate its broad validity. Using matrix optics we will consider translation, imaging, a dielectric interface, and propagation from image to pupil in a telecentric system. Translation was already discussed in the steps leading to Equation 12.2. In terms of matrix optics, using Equation 3.10, an increment of length, dx, and an increment of angle, $d\alpha$, before the translation, are related to those afterward by

Eq. 3.10:
$$\mathcal{T}_{12} = \begin{pmatrix} 1 & z_{12} \\ 0 & 1 \end{pmatrix}.$$

Fig. 12.2:

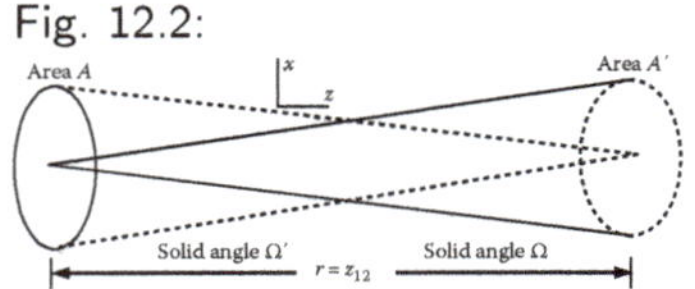

$$\begin{pmatrix} dx_2 \\ d\alpha_2 \end{pmatrix} = \begin{pmatrix} 1 & z_{12} \\ 0 & 1 \end{pmatrix} \left[\begin{pmatrix} x_1 + dx_1 \\ \alpha_1 + d\alpha_1 \end{pmatrix} - \begin{pmatrix} x_1 \\ \alpha_1 \end{pmatrix} \right] =$$
$$\begin{pmatrix} 1 & z_{12} \\ 0 & 1 \end{pmatrix} \begin{pmatrix} dx_1 \\ d\alpha_1 \end{pmatrix}. \quad (12.22)$$

The determinant of the matrix is unity, and $dx_2 d\alpha_2 = dx_1 d\alpha_1$. Figure 12.2 illustrates this relationship, with z_{12} here being the distance from source to receiver in the figure.

Eq. 9.50:
$$\begin{pmatrix} 1 & 2f \\ 0 & 1 \end{pmatrix} = \begin{pmatrix} -1 & 0 \\ -\frac{1}{f} & -1 \end{pmatrix}.$$

Second, in an imaging system, as seen in Equation 9.50, the magnification, m, of the transverse distances is exactly offset by the magnification, $n/n'm$, of the angles, so that Equation 12.20 is true. This situation is illustrated in Figure 12.3, for $n' = n$.

We made use of this result in the example in Section 4.1.2 where we found that we did not need to know the distance or size of the object to compute the power on a camera pixel. This simplification was possible because the basic radiance at the image plane is the same as that at the object plane (except of course for multiplicative loss factors such as the transmission of the glass).

Third, at a dielectric interface, as in Figure 12.4, the same conservation law applies. The corresponding ABCD matrix is

$$\begin{pmatrix} 1 & 0 \\ 0 & \frac{n}{n'} \end{pmatrix}, \tag{12.23}$$

and again, the fact that the determinant of the matrix is n/n' leads directly to Equation 12.20.

Physically, we can understand this because the flux through the two cones is equal:

$$d^2\Phi = L_1 dA \cos\theta_1 d\Omega_1 = L_2 dA \cos\theta_2 d\Omega_2$$

$$L_1 dA \cos\theta_1 \sin\theta_1 d\theta_1 d\zeta_1 = L_2 dA \cos\theta_2 \sin\theta_2 d\theta_2 d\zeta_2. \tag{12.24}$$

Applying Snell's law and its derivative,

$$n_1 \sin\theta_1 = n_2 \sin\theta_2 \qquad n_1 \cos\theta_1 d\theta_1 = n_2 \cos\theta_2 d\theta_2,$$

we obtain

$$\frac{\sin\theta_1}{\sin\theta_2} = \frac{n_2}{n_1},$$

and

$$\frac{d\theta_1}{d\theta_2} = \frac{n_2 \cos\theta_2}{n_1 \cos\theta_1}.$$

Substituting these into Equation 12.24, we obtain Equation 12.21, as expected.

As a fourth and final example, we consider the telecentric system which we have seen in Figure 4.34. A segment of that system is reproduced in Figure 12.5, where $n' = n$. We see that the height in the back focal plane x is related to the angle in the front focal plane, α_1 by $x = f\alpha_1$, where f is the focal length of the lens. Likewise, the angle in the back focal plane is related to the height in the front focal plane by $\alpha = x_1/f$, so that $x\alpha = x_1\alpha_1$. Again, for the ABCD matrix from the front focal point to the back focal point which we

Fig. 4.34:

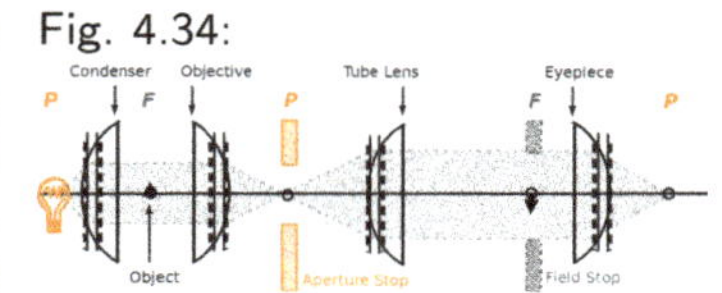

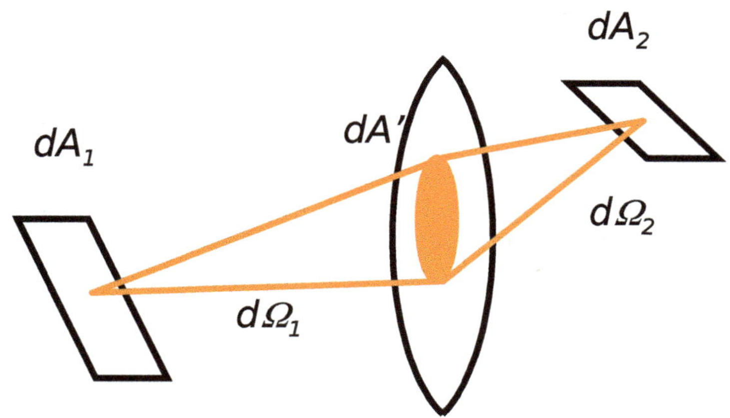

FIGURE 12.3
Radiance in an image. The image area is related to the object area by $A_2 = mA_1$, and the solid angles are related by $\Omega_2 = \Omega_1/m^2\,(n/n')^2$.

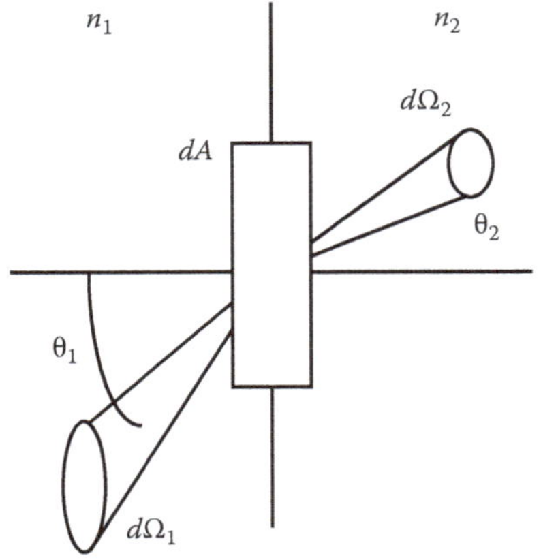

FIGURE 12.4
Radiance at an interface. The solid angles are related by $\Omega_2 = \Omega_1 (n/n')^2$, and the projected areas are related by $A_{2proj} = \cos\theta_2 / \cos\theta_1$.

have calculated in Equation 9.46:

$$\mathcal{M}_{FF'} = \begin{pmatrix} 0 & f \\ -\frac{1}{f} & 0 \end{pmatrix},$$

$\det \mathcal{M}_{FF'} = 1 = n'/n$ leads to Equation 12.20, and thus to the radiance theorem, Equation 12.21.

The key points in this section bear repeating:

Take-Away

Radiance Theorem:

- The radiance theorem (Equation 12.21) is valid for any lossless optical system
- In such a system, basic radiance, L/n^2, is conserved
- Loss can be incorporated through a multiplicative constant

12.1.3 Radiometric Analysis: An Idealized Example

We have introduced five precise definitions of radiometric terms, radiant flux or power, radiant exitance, radiance, radiant intensity, and irradiance. We developed the concepts starting from the radiant intensity of an unresolved point source, using a single molecule of hot tungsten in a thought experiment, and integrating to macroscopic quantities of

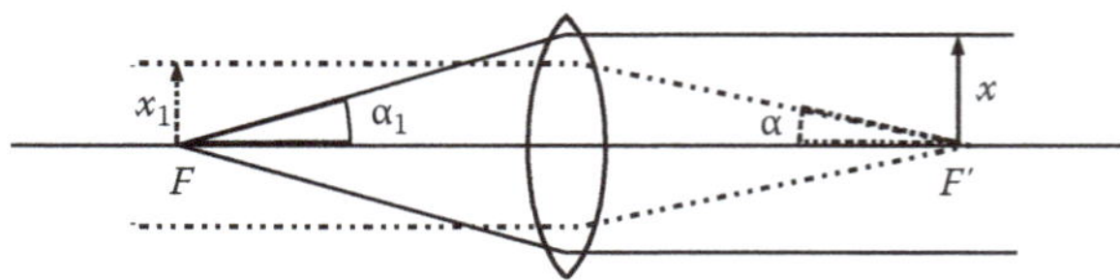

FIGURE 12.5
Radiance in telecentric system. The concept here is the same as in Figure 12.2, because the plane at F' in this figure is conjugate to the plane at $r \to \infty$ in Figure 12.2.

radiance, intensity, and power or flux. To further illustrate these concepts, we start with the power in a large source and take derivatives to obtain the other radiometric quantities. Consider a hypothetical ideal 100 W light bulb radiating light uniformly in all directions. Although this example is completely impractical, it will allow us to review the concepts of radiometry. The filament is a piece of hot tungsten of area A, and we view it at a distance r, using a receiver of area A'. Let us first take the derivative of the power with respect to the area of the filament. Let us suppose that the filament is a square, 5 mm on a side. Then the radiant exitance is $M = 100\,\text{W}/\left(5\,\text{mm}\right)^2 = 4\,\text{W}/\text{mm}^2$. If the light radiates uniformly, an admittedly impossible situation but useful conceptually, then the radiance is $L = M/\left(4\pi\right) \approx 0.32\,\text{W}/\text{mm}^2/\text{sr}$. Now, if we reverse the order of our derivatives, the intensity of the light is $I = 100\,\text{W}/\left(4\pi\right) \approx 7.9\,\text{W}/\text{sr}$, and the radiance is again $L = I/A \approx 0.32\,\text{W}/\text{mm}^2/\text{sr}$.

At the receiver a distance r from the source, the power in an area, A', will be $PA'/\left(4\pi\right) = IA'/r^2$. Thus, the irradiance is

$$E = \frac{I}{r^2}. \tag{12.25}$$

At 10 m from the source, the irradiance will be $E = I/r^2 \approx 8\,\text{W}/\text{sr}/\left(10\,\text{m}\right)^2 = 0.08\,\text{W}/\text{m}^2$.

We could also obtain the irradiance using the radiance theorem, Equation 12.21, to say that the radiance, L, is the same at the receiver as it is at the source. Thus, the irradiance can be obtained by integrating over the solid angle of the source as seen from the receiver,

$$E = L\frac{A}{r^2} = L\Omega', \tag{12.26}$$

obtaining the same result;

$$E \approx 0.32\,\text{W}/\text{mm}^2/\text{sr} \times \frac{\left(5\,\text{mm}\right)^2}{\left(10\,\text{m}\right)^2} \approx 0.08\,\text{W}/\text{m}^2.$$

This form is particularly useful when we do not know r but can measure Ω'.

12.1.4 Practical Example

Next, we will see how the radiance theorem can simplify a complex problem. Suppose that we have an optical system consisting of perhaps a dozen lenses, along with mirrors, beamsplitters, and polarizing elements, imaging the surface of the moon, which has a radiance of about $L_{moon} = 13\,\text{W}/\text{m}^2/\text{sr}$ onto a CCD camera. The task of accounting for light propagation through a complicated instrument could be quite daunting indeed. However, making use of the radiance theorem, we know that the radiance of the image is equal to the radiance of the object, multiplied by the transmission of the elements along the way. We can now simplify the task with the following steps:

1. Use Jones matrices (or Mueller matrices) to determine the transmission of the polarizing elements. Include here any scalar transmission factors, such as reflectance and transmission of beamsplitters, absorption in optical materials, and other multiplicative losses.
2. Find the exit pupil and exit window of the instrument (Section 4.3). If it is designed well, the exit window will be close to the CCD, and the CCD will determine the field of view. Compute the numerical aperture or f-number from these results.
3. Compute the solid angle from the numerical aperture or f-number, using Equation B.6, or one of the approximations.
4. Compute the irradiance on a pixel by integrating L over the solid angle. If L is nearly constant, then the result is $E = L\Omega$. Some pixels will image the moon and others will image the dark sky. We consider only the former here.

Eq. B.6: $\Omega = 2\pi\left(1 - \sqrt{1 - \left(\frac{NA}{n}\right)^2}\right)$.

5. Compute the power on a pixel by multiplying the irradiance by the pixel area; $P = EA$. In Chapter 13, we will learn how to convert this power to a current or voltage in our detection circuitry.
6. If desired, multiply the power by the collection time to obtain energy.

Example

For example, let us assume that the polarization analysis in the first step produces a transmission of 0.25, and that each of the 12 lenses has a transmission of 92%, including surface reflections and absorptions. This value is typical for two glass surfaces with no anti-reflection coatings. If there are no other losses, then the radiance at the CCD is

$$L = 13\,\text{W/m}^2/\text{sr} \times 0.25 \times 0.92^{12} =$$
$$13\,\text{W/m}^2/\text{sr} \times 0.25 \times 0.367 = 1.2\,\text{W/m}^2/\text{sr}. \tag{12.27}$$

If our system has $NA = 0.1$, then using Equation B.6 and assuming a square pixel $10\,\mu\text{m}$ on a side,

$$P = 1.2\,\text{W/m}^2/\text{sr} \times 2\pi\left(1 - \sqrt{1 - NA^2}\right)\,\text{sr}\times$$
$$\left(10 \times 10^{-6}\,\text{m}\right)^2$$
$$P = 1.2\,\text{W/m}^2/\text{sr} \times 0.315\,\text{sr} \times 10^{-10}\,\text{m}^2 = 3.8 \times 10^{-12}\,\text{W}. \tag{12.28}$$

Thus, we expect about 4 pW on a pixel that images part of the moon. While this number sounds small, recall that the photon energy, $h\nu$, is on the order of 10^{-19} J, so in a time of 1/30 s, we expect about a million photons.

In Practice

While we've seen that compound lenses can reduce aberrations we notice in the above example that they come at a cost. The transmission of 12 uncoated lenses, with 0.04 reflection at each surface is $0.92^{12} = 0.37$. Even using an anti-reflection coating of one percent, the transmission is only 80%. Furthermore the multiple surfaces can present unwanted reflections that can degrade an imaging system.

We could just as well have computed the same number by considering the area of the pupil and the solid angle subtended by a pixel on the moon. In this example, we used $NA = 0.1$. If the lens is a long telephoto camera lens with $f = 0.5$ m, then the aperture must have a radius of about 5 cm. The $10\,\mu\text{m}$ pixel subtends an angle of $10\,\mu\text{m}/0.5$ m, or $20\,\mu\text{rad}$, so

$$P = LA\Omega = 1.2\,\text{W/m}^2/\text{sr} \times \pi\,(0.05\,\text{m})^2 \times \left(20 \times 10^{-6}\right)^2\,\text{sr} =$$
$$3.8 \times 10^{-12}\,\text{W},$$

in agreement with Equation 12.28. In a third approach, we could have used the area of the $10\,\mu\text{m}$ pixel projected on the moon, and the solid angle of the lens, as seen from the moon, with the same result. However, to do that, we would have needed to know the distance to the moon. Our choice simply depends on what numbers we have available for the calculation.

The point of this exercise is that we know nothing about the optical system except the magnification, numerical aperture, field of view, and transmission losses, and yet we can make a good estimate of the power detected.

Take-Away

Applications of the Radiance Theorem:
The radiance theorem says that we can do the radiometric calculation at any point along the path, provided that we keep track of the multiplicative losses. The calculation is easiest where we know A and Ω. Usually this location is at the entrance or exit window, or the entrance or exit pupil.

In Practice

In a camera, we usually know the size of the imaging chip and the f-number of the lens, so we work in image space. We don't need to know the object distance or object size.

In a microscope, we usually know the NA of the objective. We can compute the entrance window by dividing the camera size by the magnification. We normally work in object space. We don't need to know the actual aperture diameter or internal details of the microscope.

Take-Away

Summary of Basic Radiometry:

- There are at least five important radiometric quantities, radiant flux Φ or power P, radiant exitance, M, radiant intensity, I, radiance, L, and irradiance, E. They are related by derivatives with respect to projected area, $A\cos\theta$, and solid angle, Ω.
- Basic radiance, L/n^2, is conserved throughout an optical system by the radiance theorem, with the exception of multiplicative factors. This fact can be used to reduce complicated systems to simple ones.
- To determine the power collected in a lossless optical system, we need only know the numerical aperture and field of view in image (or object) space, and the radiance.
- Losses can be included by simple multiplicative factors. We will raise one qualification to this statement which we will discuss in Section 12.2. The statement is true for monochromatic light, or for systems in which the loss is independent of wavelength over the bandwidth of the source.
- Finally, although the word "intensity" has a very precise radiometric definition as power per unit solid angle, it is often used in place of the word "irradiance," even in the technical literature.

12.2 Spectral Radiometry

The results in the previous section are valid when all system properties are independent of wavelength over the bandwidth of the source. They are equally valid in any case, if we apply them to each wavelength individually. Often, the interesting information from an optical system lies in the spectral characteristics of the source and the system. In this section, we develop methods for spectral radiometry, the characterization of sources and optical systems in radiometric terms as functions of wavelength.

12.2.1 Some Definitions

In Section 12.1.1, we discussed the fact that plane waves are unusual in typical light sources, and learned how to handle waves that are distributed in angle. Likewise, our model of light as a monochromatic wave is rather exceptional. Indeed, even a source that follows a harmonic function of time for a whole day, has a spectral width at least as large as the Fourier transform of a 1-day pulse, and is thus not strictly monochromatic. Most light sources consist of waves that are distributed across a much wider band of frequencies. Provided that the waves at different frequencies are incoherent with respect to each other, we can treat them with incoherent integration. Specifically, for each radiometric quantity, we define a corresponding spectral radiometric quantity as its derivative with respect to frequency or wavelength. For example, if the radiance of a source is L, then we define the spectral radiance as either

$$L_\nu(\nu) = \left.\frac{dL}{d\nu}\right|_{c/\lambda} \qquad \text{or} \qquad L_\lambda(\lambda) = \left.\frac{dL}{d\lambda}\right|_{\lambda}. \tag{12.29}$$

Spectral Radiometry:
For any radiometric quantity, add the word "spectral" in front of the name, add the subscript λ to the variable label, and append nm^{-1} or $\mu\mathrm{m}^{-1}$ to the units.
or ... add the word "spectral" in front of the name, add the subscript ν to the variable label, and append Hz^{-1} to the units.

Usually in optics, we think in terms of wavelength, rather than frequency, and the second expression in Equation 12.29 is more often used. However, in applications such as coherent detection, particularly where we are interested in very narrow (e.g., RF) bandwidths and in some theoretical work, the first is used. No special names are used to distinguish between the two, and if there is any ambiguity, we must explicitly make clear which we are using. The relationship between the two expressions in Equation 12.29 is

$$L_\lambda d\lambda = L_\nu d\nu \qquad L_\lambda = L_\nu \frac{d\nu}{d\lambda}$$

$$L_\lambda = L_\nu \frac{\nu}{\lambda} = L_\nu \frac{c}{\lambda^2} \tag{12.30}$$

We use the subscript ν or λ on any radiometric quantity to denote the spectral derivative, and we modify the vocabulary by using the word "spectral" in front of the name of the quantity. The units are the units of the original quantity divided by a wavelength or frequency unit. Some common examples are shown in Table 12.2. It is common to find mixed units, such as $\mathrm{w/cm^2/\mu m}$ for spectral irradiance. These units are appropriate because they fit the problem well, and they avoid misleading us into thinking of this quantity as a power per unit volume as could occur if consistent units such as $\mathrm{W/m^3}$ were used. The price we pay for the convenience these units afford is the need to be ever vigilant for errors in order of magnitude. It is helpful to learn some common radiometric values to perform frequent "sanity checks" on results. Some of these values will be discussed later in Table 12.3.

Spectral Fraction, $f_\lambda(\lambda)$b: nm^{-1} or $\mu\mathrm{m}^{-1}$

Sometimes, the spectral distribution of light is nearly uncoupled from the spatial distribution, and it is useful to think of a generic **spectral fraction**,

$$f_\lambda(\lambda) = \frac{X_\lambda(\lambda)}{X}, \tag{12.31}$$

TABLE 12.2
Spectral Radiometric Quantities

Radiometric Quantity	Units	Frequency Derivative	Units	Wavelength Derivative	Units
Radiant flux, Φ Power, P	W	Spectral radiant flux, Φ_ν	W/Hz	Spectral radiant flux, Φ_λ	W/μm
Radiant exitance, M	W/m^2	Spectral radiant exitance, M_ν	$W/m^2/Hz$	Spectral radiant exitance, M_λ	$W/m^2/\mu m$
Radiant intensity, I	W/sr	Spectral radiant intensity, I_ν	W/sr/Hz	Spectral radiant intensity, I_λ	W/sr/μm
Radiance, L	$W/m^2/sr$	Spectral radiance, L_ν	$W/m^2/sr/Hz$	Spectral radiance, L_λ	$W/m^2/sr/\mu m$

For each radiometric quantity, there are corresponding spectral radiometric quantities, obtained by differentiating with respect to wavelength or frequency. Typical units are shown.

TABLE 12.3
Typical Radiance and Luminance Values

	Radiance		Photometric Units		
Object	$W/m^2/sr$		Units = $lm/m^2/sr$	Footlamberts	lm/W
Minimum visible	7×10^{-10}	Green	5×10^{-7}	1.5×10^{-7}	683
Dark clouds	0.2	Vis	40	12	190
Lunar disk	13	Vis	2500	730	190
Clear sky	27	Vis	8000	2300	300
Bright clouds	130	Vis	2.4×10^4	7×10^3	190
	300	All		82	
Solar disk	4.8×10^6	Vis	7×10^8	2.6×10^7	190
	1.1×10^7	All		$\times 10^7$	82

Estimates of radiance and luminance are shown for some typical scenes. It is sometimes useful, when comparing radiance to luminance, to consider only the radiance in the visible spectrum, and at other times it is useful to consider the entire spectrum. In the radiance column, the wavelength band is also reported. The final column shows the average luminous efficiency.

where X is any of Φ, E, M, I, or L. One example where the spectral fraction is useful is in determining the spectral irradiance at a receiver, given the spectral distribution of a source and its total irradiance, as in the sunlight example that follows.

Take-Away

Spectral Definitions:

- The derivative of any quantity with respect to wavelength is called the spectral quantity.
- If the quantity in question is identified as X, the spectral quantity is identified as X_λ.
- The units of the spectral quantity are those of the quantity divided by the units of wavelength.
- Sometimes spectral quantities are defined as derivatives with respect to frequency instead of wavelength. Conversion between the two is given by Equation 12.30.
- The spectral fraction, f_λ is useful when we know the spectral behavior of one

radiometric quantity and need to evaluate that of another. An example is shown in the next section.

12.2.2 Examples

Here we consider two examples, one of sunlight incident on Earth, and one involving a molecular tag in microscopy.

12.2.2.1 Sunlight on Earth

Later we will compute the spectral radiant exitance $M_\lambda(\lambda)$, of an ideal source at a temperature of 5000 K, a reasonably good representation of the light emitted from the sun, measured at sea level, after some is absorbed by the Earth's atmosphere. We consider one question.

Example

What is the solar irradiance in the visible spectrum? There is no reason to expect significant spectral changes in different directions, so one would expect $f_\lambda(\lambda)$ as defined for M to be equally valid for L and I. Now, without knowing the size of the sun, or its distance to the Earth, if we know that the total irradiance at the surface of the Earth is $E = 1000\,\text{W/m}^2$, then the spectral irradiance could be obtained as

$$E_\lambda(\lambda) = E f_\lambda(\lambda),$$

with E being the measured value, and $f_\lambda(\lambda)$ being computed for the 5000 K blackbody,

$$f_\lambda(\lambda) = \frac{M_\lambda(\lambda)}{M} = \frac{M_\lambda(\lambda)}{\int_0^\infty M_\lambda(\lambda)\,d\lambda}. \tag{12.32}$$

At worst, the integral could be done numerically. We will discuss sunlight in greater detail in Section 12.5, but for now we will note that the radiant exitance of the sun, corrected for atmospheric transmission is about $35\,\text{MW/m}^2$, and about $15\,\text{MW/m}^2$ of that is in the visible spectrum. Thus

$$\int_{380\,\text{nm}}^{780\,\text{nm}} f_\lambda(\lambda)\,d\lambda = \frac{15\,\text{MW/m}^2}{35\,\text{MW/m}^2} = 0.42. \tag{12.33}$$

Therefore about 42% of the incident $1\,\text{kW/m}^2$ or $420\,\text{W/m}^2$ is in the visible spectrum.

12.2.2.2 Molecular Tag

As another example of spectral radiometry, we consider Figure 12.6A. Here we see the excitation and emission spectrum of **fluorescein**, a commonly used fluorophore in biomedical imaging. An optical **molecular tag** consists of a fluorophore attached to another molecule that in turn binds to some specific biological material of interest. The fluorophore absorbs light of one color (blue in this case) and emits at another (green). We call this process **fluorescence**. Two curves are shown. The one to the left is the relative absorption spectrum which indicates the best wavelengths at which to excite its fluorescence. We see that the absorption spectrum has a strong peak in the ultraviolet, at a wavelength too short to be of interest to us in this example. Such short wavelengths tend to be more harmful

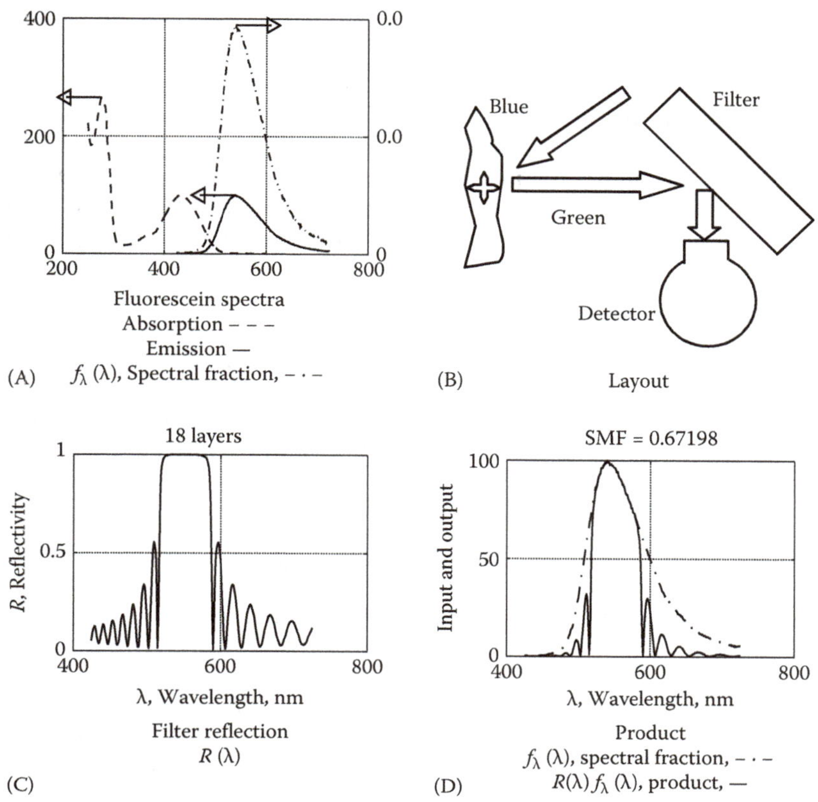

FIGURE 12.6
Spectral matching factor. Relative absorption and emission spectra of fluorescein are shown (A). The absorption spectrum is the dashed curve, and the emission spectrum is the small solid curve. Units are arbitrary, and scaled to a maximum value of 100% at the blue peak for absorption and the green peak for emission. The large dash-dot curve shows the spectral emission fraction, for which the integral is unity. Light is reflected from a band-pass filter (B) with the spectrum shown (C). The reflected spectrum is the product of the two (D). The spectral fraction (a dash-dot curve) is repeated for comparison. The spectral matching factor here is 0.672.

to living material than longer ones, and do not penetrate far into water. There is a more useful peak around 430 nm in the blue, and the blue wavelength of the argon ion laser 488 nm, is commonly used as an excitation source, despite being to one side of the peak. Of more interest here is the other curve, the **emission spectrum.** This spectrum, $S_\lambda(\lambda)$ is in arbitrary units. We do not know the number of molecules, the brightness of the light incident on them, or anything about the collection geometry. Let us assume that none of these factors affects the shape of the detected spectrum. Then if we somehow compute the total power or radiant flux, Φ, of the green light, we can compute the spectral radiant flux from this curve as

$$\Phi_\lambda(\lambda) = \Phi f_\lambda(\lambda),$$

where we use the estimated or measured power, Φ, and compute $f_\lambda(\lambda)$ from the given spectrum, $S_\lambda(\lambda)$,

$$f_\lambda(\lambda) = \frac{S_\lambda(\lambda)}{\int_0^\infty S_\lambda(\lambda)\, d\lambda}. \tag{12.34}$$

Given a relative spectrum, we do not have any reason to pick a particular set of units. Suppose that we decide that this curve represents "something" per nanometer. Integrating over wavelength numerically, we obtain 1.03×10^4 "somethings." Thus, if we divide the curve shown by 1.03×10^4, we have f_λ in nm^{-1}. The large dash-dot curve in the figure, with the axis on the right, shows this function. As an example of its use, if an analysis of the fluorescence process predicts Φ as the total power in Watts, we multiply Φ by f_λ, to obtain the spectral radiant flux in W/nm. We return to this example shortly in discussing the spectral matching factor, but we can see that if we use the filter in Figure 12.6B and C, most of the emission light will be detected while most of the excitation light will be blocked. The spectrum of this filter was generated using the equations of Section 7.7.

12.2.3 Spectral Matching Factor

In several earlier chapters, we have considered multiplicative effects on light propagation. We considered the Fresnel reflection and transmission coefficients in Section 6.4, and the transmission and reflection of multilayer dielectric stacks in Section 7.7. In the latter, we considered the transmission and reflection to be functions of wavelength. If the light has a wide bandwidth and the transmission or reflection varies with wavelength, such definitions may become problematic.

12.2.3.1 Equations for Spectral Matching Factor

Consider, for example, a filter with a reflection for irradiance of

$$R(\lambda) = |\rho(\lambda)|^2 .$$

Note that R is dimensionless, and although it is a function of λ, it does not have the subscript λ because it is not a wavelength derivative. Assuming the incident light has a spectral radiance, $L_\lambda(\lambda)$, the spectral radiance of the output will be

$$L'_\lambda(\lambda) = L_\lambda(\lambda) R(\lambda) . \tag{12.35}$$

The total radiance out is

$$L' = \int_0^\infty L'_\lambda(\lambda)\, d\lambda = \int_0^\infty L_\lambda(\lambda) R(\lambda)\, d\lambda, \tag{12.36}$$

and $R(\lambda)$ cannot, in general, be taken outside the integral. Nevertheless, it would be useful to define a reflection, R such that

$$R = \frac{L'}{L} = \frac{\int_0^\infty L'_\lambda(\lambda)\, d\lambda}{\int_0^\infty L_\lambda(\lambda)\, d\lambda}. \tag{12.37}$$

We can write R as

$$R = R_{max} \int_0^\infty \frac{R(\lambda)}{R_{max}} f_\lambda(\lambda)\, d\lambda = R_{max} SMF, \tag{12.38}$$

where R_{max} is the transmission of the filter at its peak, and we have made use of the fact that $\int_0^\infty f(\lambda)\, d\lambda = 1$. Then for narrow-band light at the appropriate wavelength, $R = R_{max}$.

For light with a broader bandwidth, the **spectral matching factor**, SMF, must be computed. The SMF is

$$SMF \approx 1 \tag{12.39}$$

if the filter holds its maximum reflection over the entire wavelength band of the source

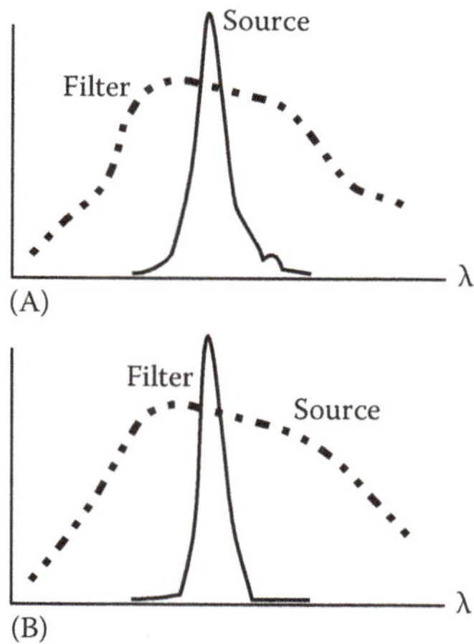

FIGURE 12.7
Spectral matching factor approximations. (A) If the bandwidth of the signal is much narrower than that of the filter, and the signal is located near the peak of the filter's spectrum, then $SMF \approx 1$. (B) If the filter's width is much narrower than the signal's, then the SMF is approximately the ratio of the widths.

as in Figure 12.7A. At the other extreme, if the filter bandwidth, $\delta\lambda_f$, is much narrower than the source bandwidth $\delta\lambda_s$, of the source, then

$$SMF \approx \frac{\delta\lambda_f}{\delta\lambda_s} \qquad \delta\lambda_f \ll \delta\lambda_s, \tag{12.40}$$

as in Figure 12.7B.

Take-Away

Spectral Matching Factor:
To describe the net transmission through a filter or other component, we need to know both the spectrum of the filter and the spectrum of the light.

Example

Figure 12.6C shows the reflectance spectrum for a multilayer dielectric stack consisting of 18 layers, using the equations in Section 7.7. If the fluorescence discussed in Section 12.2.2 is reflected from this stack, then the spectrum of the reflection is as shown by the solid line in Figure 12.6D, where the original spectrum is shown by the dash-dot line. Integrating the reflected spectrum (solid line) over all wavelengths we obtain the total reflected fluorescence and integrating the incident spectrum (dash-dot line) we obtain the incident fluorescence before reflection from the filter. The spectral matching factor is the ratio of these two, which in this case is 0.672, from numerical calculation.

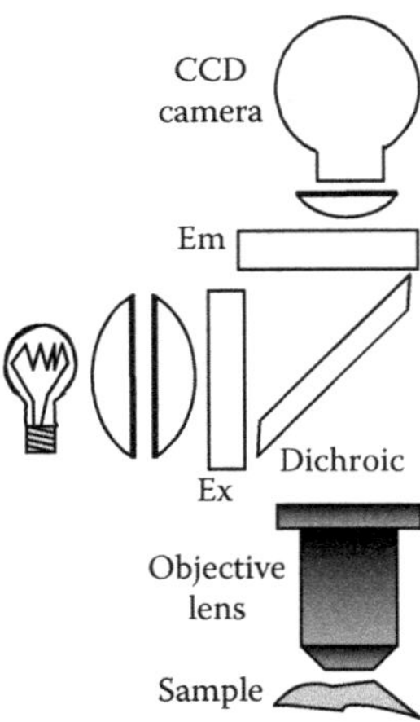

FIGURE 12.8
Epifluorescence measurement. One line of the microscope's light source, a mercury lamp, is selected by the excitation filter (Ex) and reflected by the dichroic beamsplitter onto the sample. Fluorescent light is returned through the dichroic beamsplitter and an emission filter (Em) to the camera.

We could have produced a larger spectral matching factor by using a wider filter. In fact we could make $SMF \approx 1$ by making the filter bandwidth wider than the whole fluorescence spectrum. One disadvantage of wider bandwidth is that on the short wavelength side, it would reflect some scattered blue light from the excitation source. Another disadvantage is that a wider filter would admit more stray light, which would increase the noise level. This issue will be discussed in more detail in Chapter 13.

Do not multiply spectral matching factors.

Finally, it is important to note that we cannot combine spectral matching factors. If we use one filter following another, the reflection or transmission at each wavelength is given by the product, $R(\lambda) = R_1(\lambda) R_2(\lambda)$ but the spectral matching factor is not the product of the two individually, because the integral of a product in Equation 12.38 is not the product of the integrals:

$$\int_0^\infty \frac{R_1(\lambda)}{R_{1max}} \frac{R_2(\lambda)}{R_{2max}} f_\lambda(\lambda)\, d\lambda \neq$$

$$\int_0^\infty \frac{R_1(\lambda)}{R_{1max}} f_\lambda(\lambda)\, d\lambda \int_0^\infty \frac{R_2(\lambda)}{R_{2max}} f_\lambda(\lambda)\, d\lambda. \tag{12.41}$$

Let us briefly consider a more complicated example using the epifluorescence microscope shown in Figure 12.8. The light source is a mercury arc lamp. An excitation filter (Ex) transmits only one line of the mercury spectrum (e.g., 365 nm in the ultraviolet), which is then reflected by the dichroic mirror through the objective, to the sample. The dichroic is designed to reflect the excitation wavelength and transmit the emission. Returning fluorescent light passes through the dichroic and an emission filter (Em), which rejects unwanted light. We can approximate the detected irradiance using

$$E_{det} = L_{source}\Omega_{ex}T_{ex}R_{dichroic}T_{obj}SMF_{ex}[F/(4\pi)]$$
$$T_{obj}T_{dichroic}T_{em}\Omega_{em}SMF_{em}, \tag{12.42}$$

where

- the subscripts Ex, Em, and $dichroic$, are related to the component names
- F is the fraction of incident irradiance that is converted to fluorescent radiance

In this example, SMF_{ex} is calculated to account for all the components in the path from the source to the sample, and SMF_{em} for all those between the sample and the camera. From the detected irradiance, we could multiply by the pixel area and the detection time to obtain the number of Joules of energy detected. A similar numerical example for a different measurement is shown in Section 13.6.2.

Take-Away

Summary of Spectral Radiometry:

- For every radiometric quantity, there exists a spectral radiometric quantity which is its derivative with respect to frequency or wavelength. The name of this derivative is formed by placing the word "spectral" before the name of the original quantity. In the name, no distinction is made between frequency and wavelength derivatives. Wavelength is used more commonly on optics, except in certain equations involving coherent detection and theoretical analysis.
- The notation for the derivative is formed by placing a subscript ν for frequency or λ for wavelength after the notation used for the original quantity.
- The units are the units of the original quantity divided by frequency or wavelength units.
- It is useful to define a spectral fraction, f_λ as in Equation 12.31, which can be applied to any of the radiometric quantities.
- The spatial derivatives in Figure 12.1, and the equations defined with reference to that chart, are valid for the spectral quantities, wavelength-by-wavelength.
- The behavior of filters is more complicated, and is usually treated with the spectral matching factor defined in Equation 12.38.

12.3 Photometry and Colorimetry

Before beginning our discussion of photometry, we will discuss the conversion from radiometric to photometric units because it has a fundamental similarity to the spectral matching functions we have just been discussing in Section 12.2. We will then discuss photometry in the context of Figure 12.1, and conclude with some examples.

12.3.1 Spectral Luminous Efficiency Function

One spectral function deserves special attention, the **CIE spectral luminous efficiency function**, often called $V(\lambda)$, but which we will call $\bar{y}(\lambda)$[3]. We have chosen to use this terminology because $\bar{y}$, along with two more functions, $\bar{x}$ and $\bar{z}$, defines the xyz color space, to be discussed in Section 12.3.3. This function describes the response of the human eye to light[4], based on a study with a small number of observers. This function is a measure of the visual perception of a source as detected by human observers comparing sources of different wavelengths. Shown in Figure 12.9, $\bar{y}$ peaks at 555 nm, and goes to zero at the edges of the visible spectrum which is defined for this purpose as 380–780 nm. The dotted curve shows the actual $\bar{y}$ data points, and the solid line shows a rough functional approximation. The approximation is useful in computations because it can be defined for any wavelength, while the data are given in increments of 5 nm. The approximation is

Spectral Luminous Efficiency, $\bar{y}(\lambda)$: Dimensionless.

$$\bar{y} = \exp\frac{-833.202209 + 2.995009\lambda - 0.002691\lambda^2}{1 + 0.011476\lambda + 0.000006\lambda^2}. \tag{12.43}$$

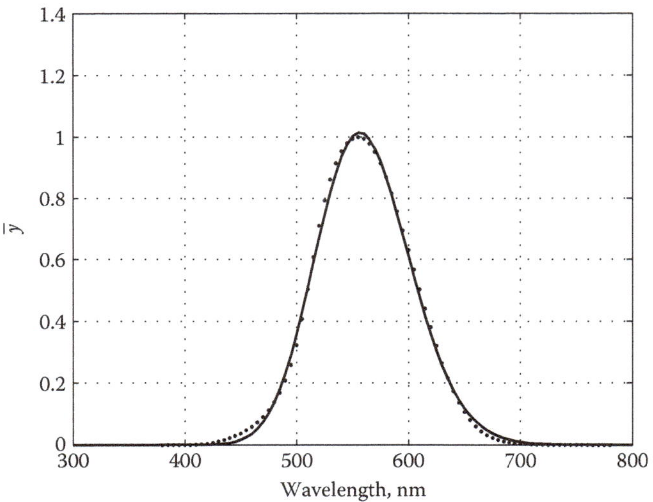

FIGURE 12.9
Photometric conversion. The $\bar{y}$ curve shows the response of the standard human observer. The dotted line shows the data, and the solid line shows the fit provided by Equation 12.43.

Eq. 12.38:
$R = R_{max} \int_0^\infty \frac{R(\lambda)}{R_{max}} f_\lambda(\lambda)\, d\lambda = R_{max} SMF.$

Eq. 13.7:
$i = \rho_i P$
$\rho_i = \frac{\eta e}{h\nu}.$

Fig. 13.2:

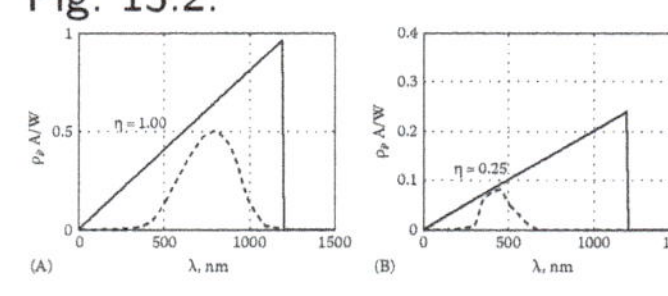

We can think of $\bar{y}$ as defining the transmission of a filter with the output determining the visual appearance of "brightness." If an object, described by its spectral radiance as a function of wavelength and position, $L_\lambda(\lambda, x, y)$, is imaged through an optical system that includes such a filter, the resulting image will be

$$Y(x, y) = \int_0^\infty \bar{y}(\lambda)\, L(\lambda, x, y)\, d\lambda, \qquad (12.44)$$

where Y has the same units as L. This equation is analogous to Equation 12.38 for the spectral matching factor. In some sense, it quantifies the spectral matching factor of the human eye.

Suppose that a scene includes four light-emitting diodes (LEDs), blue, green, red, and near infrared, all with equal radiance. We will see in Equation 13.7 and Figure 13.2 that a silicon CCD will usually be more sensitive to longer wavelengths, making an image from a monochrome camera (single-channel black and white) more likely to show the infrared diode to be the brightest and the blue to be the weakest. If the CCD is corrected for this spectral effect by putting filters in front of it, all diodes will appear equally bright in the image. An observer viewing the diodes directly instead of using the camera will see the green diode as brightest because it is closest to the peak of the $\bar{y}$ curve (the spectral luminous efficiency), and the infrared one will be invisible. Placing a filter that transmits the $\bar{y}$ spectrum in front of the camera will produce a monochrome image which agrees with the human observer's visual impression of brightness. In fact, monochrome cameras frequently contain a filter to, at a minimum, reduce the infrared radiance. Such filters are not perfect; one can sometimes observe some leakage through the filter when watching an infrared remote control device through a mobile phone camera. Newer cameras use a strong infrared blocking filter to reduce infrared light and thus improve contrast in pictures, so observing the infrared remote control light may not be possible.

Equation 12.44 can be used not only with L, but with any radiometric quantity. Suppose that we have a power meter that provides a correct reading of power regardless of wavelength. Looking at a source with a spectral radiant flux, $P_\lambda(\lambda)$, and recording measurements first and without the filter we would measure

$$P = \int_0^\infty P_\lambda (\lambda) .$$

Then with the $\bar{y}$ filter between the source and the power meter the measured power would be

$$Y = \int_0^\infty \bar{y} (\lambda) P_\lambda (\lambda) \, d\lambda,$$

where in this case, Y is in Watts. While P is a measure of the radiant flux, Y is a measure of the luminous flux in Watts. The conversion to photometric units requires only a scale factor:

$$P_{(V)} = \frac{683 \, \mathrm{lm/W}}{\max (\bar{y})} \int_0^\infty \bar{y} (\lambda) P_\lambda (\lambda) \, d\lambda \qquad (12.45)$$

We have used the notation $P_{(V)}$ to denote "luminous" radiant exitance. Normally we do not use any subscript on the photometric quantities, but in this case, with photometric and radiometric quantities in the same equation, we find it convenient. The conversion factor, in lumens per Watt, varies with the spectrum of the source, analogously to Equation 12.38:

$$\frac{P_{(V)}}{P} = 683 \, \frac{\mathrm{lm}}{\mathrm{W}} \int_0^\infty \frac{\bar{y} (\lambda)}{\max (\bar{y})} \frac{P_\lambda (\lambda)}{P} d\lambda. \qquad (12.46)$$

As one example of the use of Equation 12.44, at each of the two wavelengths, 508 and 610 nm, $\bar{y} \approx 0.5$, and twice as much light is required at either of these wavelengths to produce the same apparent power as is needed at 555 nm.

12.3.2 Photometric Quantities

The difficulty of converting between radiometric and photometric units arises because the photometric units are related to the visual response. Therefore, the conversion factor depends on wavelength according to the spectral luminous efficiency curve, $\bar{y}$, in Figure 12.9, or approximated by Equation 12.43. The unit of lumens is usually used for photometric work. Because the photometric units are based on experiments with human subjects, providing no fundamental tie for photometric units to other SI basic units such as meters, kilograms, and seconds, it is necessary that one photometric unit be established as a one of the basic SI units. Despite the fact that the lumen holds the same fundamental status in photometry that the Watt does in radiometry, the unit which has been chosen as the basic SI unit is the unit of luminous intensity, I, the **candela**. The candela is defined as the luminous intensity of a source emitting (1/683) W/sr at a frequency of 450 THz (approximately 555 nm)[1]. From luminous intensity, following Figure 12.1 and Table 12.1, we can obtain **luminous flux** by integrating over solid angle. The unit of luminous flux, the **lumen**, is defined through its connection to the candela;

Candela: Defined as an SI Unit

$$I = \frac{d\Phi}{d\Omega}, \qquad (12.47)$$

where I is in lumens per steradian, or candelas.

The remaining photometric units are defined by analogy to the radiometric units. Figure 12.1 shows the relationships. Integrating the luminous intensity over solid angle, we obtain the **luminous flux**, Φ. Differentiating with respect to area yields M, the **luminous exitance**, and differentiating M with respect to solid angle, yields L, the **luminance**. However, some special units are in sufficiently common use that they merit mention. By analogy with radiance, the unit of luminance would be the nit or $\mathrm{lm/m^2/sr}$. However, units such as Lamberts are also frequently used. The curious unit "nit" may be from the Latin word "niter" meaning "to shine"[30]. Because the assumption of a Lambertian source is often used, the Lambert is particularly convenient. If a Lambertian source has a luminous exitance of $1\,\text{lm/m}^2 = 1\,\text{lux}$, then its luminance is $L = M/\pi$ or $1/\pi$ nits. It is convenient to use units which include a factor of π as do Lamberts. This source has a luminance of 1 meterlambert. As shown in Figure 12.1, similar units are in use, based on centimeters and feet.

One old unit which is still used occasionally, and has been the subject of numerous jokes is the **footcandle**, or lumen per square foot, which is a unit of illuminance. The footcandle defies numerous conventions; units of "feet" for length, are not accepted as SI units, and the footcandle is not a product of "feet" and "candles." The meterlambert and footlambert also defy conventions. Some typical values of luminance are shown in two units in the third and fourth columns of Table 12.3. For each entry, the radiance is also shown in the second column. In the fifth and final column, the average luminous efficiency is shown.

Take-Away

Photometry:
Photometric analysis follows the same approach as radiometric analysis, with an appropriately modified vocabulary. The fundamental difference is that in radiometry the flux unit is the Watt, which is a measure of power, while in photometry it is the Lumen, which is a measure of the response of the eye.

The coefficient relating photometric to radiometric units is therefore a function of wavelength.

In Practice

Although SI units have been established for photometry, older units such as footlamberts are still in wide use.

12.3.3 Color

The normal human eye is capable of detecting three different colors via its rod and cone cells. In the 1930s, experiments were conducted to determine the spectral response of the eye's sensors[3;66]. Since then, considerable increases in experimental capability have added to this field of research, and what we will study here is one small part of this subject.

Equation 12.44:
$Y = \int_0^\infty \bar{y}(\lambda)\, L(\lambda)\, d\lambda.$

12.3.3.1 Tristimulus Values

By analogy with Equation 12.44, we can define two more equations:

$$X = \int_0^\infty \bar{x}(\lambda)\, L(\lambda)\, d\lambda, \tag{12.48}$$

and

$$Z = \int_0^\infty \bar{z}(\lambda)\, L(\lambda)\, d\lambda. \tag{12.49}$$

The **color-matching functions**, $\bar{y}$ (already presented in Figure 12.9), $\bar{x}$, and $\bar{z}$, are shown in Figure 12.10. These functions are experimental results of color matches performed by a small sample of observers. There is no physical basis for these functions. Again the dots show the actual values, and the solid curves show the functional approximations,

$$\bar{x} = \exp\frac{-9712.539062+32.536427\lambda-0.027238\lambda^2}{1+0.270274\lambda-0.00028\lambda^2} + \exp\frac{-225.256866+1.011834\lambda-0.001141\lambda^2}{1-0.004699\lambda+0.00001\lambda^2}. \quad (12.50)$$

and

$$\bar{z} = \exp\frac{0.620450+0.000582\,(\lambda-450\,\text{nm})-0.000973\,(\lambda-450\,\text{nm})^2}{1+.008606\,(\lambda-450\,\text{nm})}. \quad (12.51)$$

$$\text{for } 380\,\text{nm} \le \lambda \le 780\,\text{nm},$$

$$\text{and } \bar{x}=\bar{y}=\bar{z}=0 \text{ otherwise.}$$

The integrals, X, Y, and Z, are called the **tristimulus values** of the light. As they are presented here, they have units of radiance, but we can also define tristimulus values in units of Φ, M, E or I in place of L.

Filters with the spectra $\bar{x}$, $\bar{y}$, and $\bar{z}$, are available, and can be used with monochrome cameras to produce data for color images. The $\bar{x}$ filter is difficult to implement physically because of its two peaks. In some cases, the blue peak is neglected, and in other cases four filters and cameras are used, combining two to generate the $\bar{x}$ signal. Many modern cameras use colored filters over individual pixels (See the paragraph below on the Bayer Matrix), and compute the color components after the data has been collected electronically. It is interesting to note that if one observes a variety of natural scenes with a spectrometer, and performs a principal component analysis on the spectra, most of the information will be in the first three principle components, which will be similar to these spectra. The choice of a limited number of spectra was explored through measurements of natural scenes in the early days of color television[95] but the recent development of hyperspectral imagers,

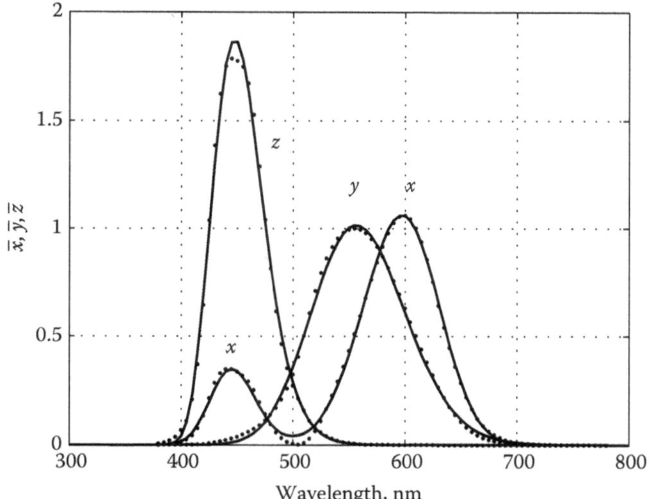

FIGURE 12.10
Tristimulus values. These three curves are used to define color in the xyz system. Generally, $\bar{x}$ is mostly red, $\bar{y}$ is mostly green, and $\bar{z}$ is mostly blue. The solid curves show the fits in Equations 12.43, 12.50, and 12.51, while the dotted curves show the actual data. (From *Commission Internationale de l'Eclairage Proceedings*, Cambridge University Press, Cambridge, U.K., 1931.)

which collect images at a large number of closely spaced wavelengths, has greatly increased the availability of data, and among the many papers published, we mention two[238;230].

Take-Away

Tristimulus:
The X, Y, and Z tristimulus values describe the response of the normal human eye to light. There can be many spectra that produce the same three integrals, and the eye cannot distinguish among them.

In Practice

Digital cameras often use the Bayer Matrix[20] to produce a color image. The Bayer Matrix consists of a two-by-two-pixel pattern repeated across the surface of the device. Two diagonal pixels are green, and the remaining two are red and blue. A pattern of four pixels is easier to repeat in two dimensions and it is appropriate to use two green pixels because green is the "luminance channel" which provides most of the brightness information.

12.3.3.2 Discrete Spectral Sources

Example

As an example, consider three laser lines, the red and green lines of the Krypton ion laser at $\lambda_{red} = 647.1$ nm and $\lambda_{green} = 530.9$ nm, and a blue line of the argon ion laser at $\lambda_{blue} = 457.9$ nm. Because these are single wavelengths, the integrals in Equations 12.44, 12.48, and 12.49 are easily accomplished. If for a moment, we assume that the lasers all have unit power, then for the red laser,

$$X_R = \int_0^\infty \bar{x}(\lambda)\,\delta(\lambda - 647.1\,\text{nm})\,d\lambda = \bar{x}(647.1\,\text{nm}) = 0.337$$

and likewise

$$Y_R = 0.134 \qquad Z_R = 0.000.$$

For the green and blue lasers, respectively,

$$X_G = 0.171 \qquad Y_G = 0.831 \qquad Z_G = 0.035$$

$$X_B = 0.289 \qquad Y_B = 0.031 \qquad Z_B = 1.696,$$

as shown in Figure 12.11. We note that the red laser has a value of X greater than Y or Z, indicating that the $\bar{x}$ sensor is sensitive to red light, and in the same way, we see that the $\bar{y}$ sensor is sensitive to green, and the $\bar{z}$ sensor is sensitive to blue.

12.3.3.3 Chromaticity Coordinates

Now let's control the power of the laser sources to generate different colors. Letting the laser powers be R, G, and B in units of Watts, we can compute a color vector,

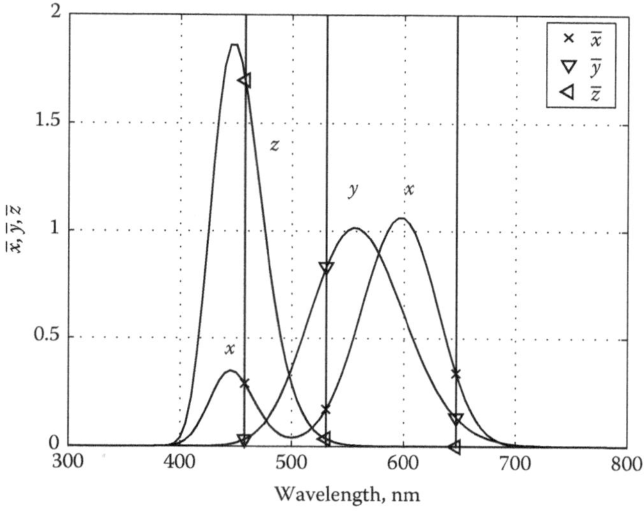

FIGURE 12.11
Tristimulus values of three lasers. The vertical lines show the wavelengths of the three lasers. Each laser's coordinates on the tristimulus curves are shown.

$$\begin{pmatrix} X \\ Y \\ Z \end{pmatrix} = \begin{pmatrix} X_R & X_G & X_B \\ Y_R & Y_G & Y_B \\ Z_R & Z_G & Z_B \end{pmatrix} \begin{pmatrix} R \\ G \\ B \end{pmatrix}. \tag{12.52}$$

The color vector defined here contains all the information required to describe how the light will appear to an observer, or at least to the "standard" CIE observer. The luminous flux will be $(683\,\text{lm/W}) \times Y$, and the color will be completely described by ratios of the three numbers. Because these ratios reduce to two independent quantities, we define the **chromaticity coordinates**

Chromaticity Coordinates, x and y, define a color.

$$x = \frac{X}{X+Y+Z} \qquad y = \frac{Y}{X+Y+Z}, \tag{12.53}$$

to describe color. It is unfortunate that these variable names are the same as we use for position coordinates in many cases, but the notation is so widely accepted that we will continue to use it. The meaning of x and y will be clear from the context, and no confusion is expected. For monochromatic light at a wavelength, λ,

$$x = \frac{\bar{x}(\lambda)}{\bar{x}(\lambda) + \bar{y}(\lambda) + \bar{z}(\lambda)}$$

Monochromatic Light

$$y = \frac{\bar{y}(\lambda)}{\bar{x}(\lambda) + \bar{y}(\lambda) + \bar{z}(\lambda)}. \tag{12.54}$$

Figure 12.12 shows the horseshoe-shaped plot of y as a function of x. All of the colors the normal human eye can observe are superpositions of combinations of the colors along this curve, with varying positive weights according to Equations 12.44, 12.48, and 12.49, such that all possible colors lie inside of this curve. This region of the curve is called the **gamut**

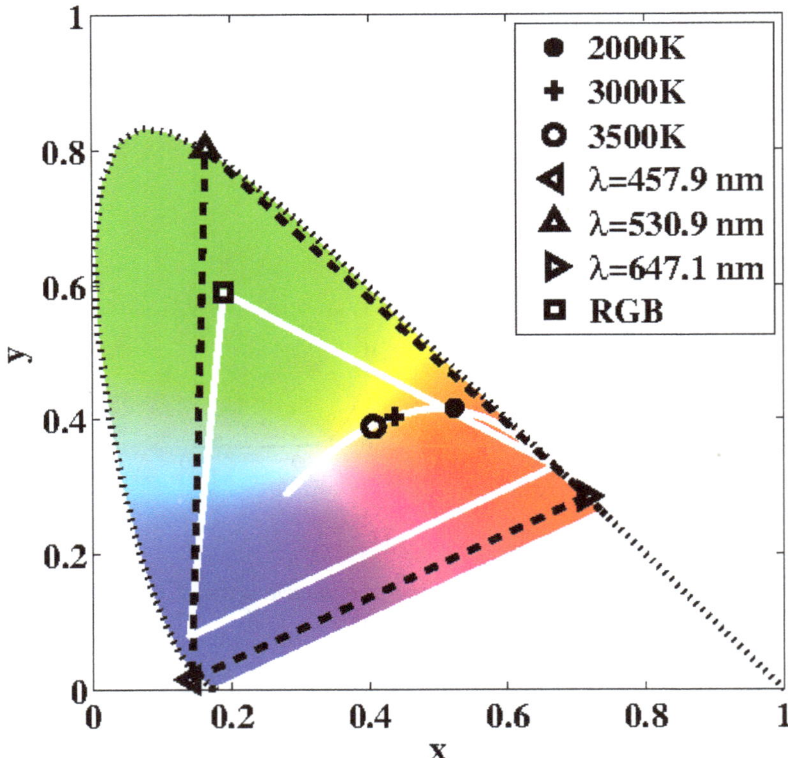

FIGURE 12.12
CIE (1931) chromaticity diagram. The coordinates x and y are the chromaticity coordinates. The horseshoe-shaped boundary shows all the monochromatic colors. The region inside this curve is the gamut of human vision. Three lasers define a color space indicated by the black dashed triangle, and three fluorophores that might be used on an old television monitor define the color space in the white triangle. The white curve shows the locus of blackbody radiation. The specific colors of three blackbody objects at elevated temperatures are also shown. These will be discussed in Section 12.5.

of human vision. It is important in generating this curve to use the exact values for $\bar{x}$, $\bar{y}$, and $\bar{z}$, rather than the approximations in Equations 12.43, 12.50, and 12.51, because the small errors at low values will distort the horseshoe shape.

The points associated with the three lasers discussed above are widely spaced on the curve, as shown in the Figure 12.12. Any combination of powers of these three lasers will result in a point inside the triangle defined by their three color points. We refer to this region of the xy plot as the **color space** of this combination of lasers. Equations 12.52 and 12.53 predict the point in the color space and Equation 12.52 with $P_V = 683\,\text{lm/W} \times Y$ predicts the luminous flux.

12.3.3.4 Generating Colored Light

In the previous section, we determined the color of a set of three sources with known power. Of course, we'd like to invert this problem and determine what powers to use to generate a color of our choice.

How do we generate light that appears to have a given luminous flux (or luminance or any of the other photometric quantities) and color, using these three sources? If we are given the photometric and colorimetric quantities, $P_{(V)}$, x, and y, that we would like to produce, we can write

$$Y = \frac{yP_V}{y \times 683\,\text{lm/W}} \qquad X = \frac{xP_V}{y \times 683\,\text{lm/W}}$$
$$Z = \frac{(1-x-y)\,P_V}{y \times 683\,\text{lm/W}}, \qquad (12.55)$$

and determine the powers using

$$\begin{pmatrix} R \\ G \\ B \end{pmatrix} = \begin{pmatrix} X_R & X_G & X_B \\ Y_R & Y_G & Y_B \\ Z_R & Z_G & Z_B \end{pmatrix}^{-1} \begin{pmatrix} X \\ Y \\ Z \end{pmatrix}. \qquad (12.56)$$

In Practice

The above matrix is only valid for these specific light sources. Any other combination of sources will have a different matrix. It is also possible to have more than three sources, in which case the matrix is not square. Generating colors in these cases is more complicated.

Example

As an example, suppose our goal is to produce an illuminance of 100 lux of "white" light, $x = y = 1/3$ on a screen 2 m^2, using these lasers. We need 400 lm of luminous flux from the source to provide this illuminance over $(2\,\text{m})^2 = 4\,\text{m}^2$, so

$$X = Y = Z = 400\,\text{lm}/(1/3)/683\,\text{lm/W}.$$

Inverting the matrix, in Equation 12.56, we find that we need 1.2 W of red light, 0.50 W of green, and 0.33 W of blue. These equations, like all involving X, Y, and Z, can be used with any of the radiometric and photometric quantities. Adding the three sources, we have used 2.03 W to produce 400 lm, so the overall luminous efficiency is 197 lm/W. We may find this disappointing, given that we can achieve over three times this value at 555 nm. Of course, the price we would pay for such a high efficiency would be that everything would look green.

The word "white" in the first paragraph of this example is in quotation marks, because we need to pause and think about the meaning of the word. In one sense, "**white**" light has a flat spectrum, with all wavelengths equally bright. On the other hand, we could define white light as having $X = Y = Z$. In the present case, we are going to make the "white" light from just three discrete colors. Although a standard observer could not see the difference between this white light, and light which is white in the first sense, the difference would be readily noted on a spectrometer.

12.3.3.5 Color outside the Color Space

We consider one more example.

Example

Suppose we want to reproduce the color of the strong blue line of the argon ion laser at $\lambda = 488$ nm, using these sources. From Equations 12.44, 12.48, and 12.49, the three color coordinates of this wavelength are $X = 0.0605$, $Y = 0.2112$, and $Z = 0.5630$, for unit power. The results are $R = -0.2429$, $G = 0.2811$, and $B = 0.3261$. This result is problematic. The negative power for the red source indicates that the desired color is out of the color space of these sources. The best one can do is use $R = 0.0000$, $G = 0.2811$, and $B = 0.3261$, but the resulting color will not look quite like the desired one. It will have the right **hue**, but not the right saturation. We refer to the monochromatic colors as highly **saturated** and this color can be considered to be less saturated, or to be closer to the white point in Figure 12.10.

Take-Away

Color Vision:

The human eye has three color sensors. Light from two different sources can produce the same appearance even though the spectra are different. All that is necessary is that the three integrals that determine X, Y, and Z be the same.

Monochromatic sources have chromaticity coordinates along the horseshoe curve in Figure 12.10. All visible colors are on or inside that curve. Three nearly monochromatic sources can be mixed in different proportions to produce any color inside the triangle formed by their chromaticity coordinates.

12.3.4 Other Color Sources

Next, we consider sources having wider bandwidth. Such sources could include, for example, a tungsten light source with colored glass filters, a set of LEDs or the phosphors in a cathode-ray tube (CRT) as in an old television. These sources lie inside the curve on Figure 12.12 because they have wide spectra. Chromaticity coordinates of typical phosphors for a CRT are shown by the white triangle in Figure 12.12.

We would prefer to have the sources as monochromatic as possible so that we can span the largest possible color space. Typical visible LEDs have spectra as narrow as 20 nm, and are reasonably good approximations to monochromatic sources. Phosphors are considerably worse, as is evidenced by their smaller color space in Figure 12.12. If tungsten light is used, the filters are often even wider than these phosphors because narrow spectra can be obtained from the tungsten light only at the expense of power. Their spectral width makes them closer to the white point and decreases the size of their color space.

12.3.5 Reflected Color

Most readers will have observed a situation in which a colored material looks different under different lighting conditions. Here we consider an example in Figure 12.13. In Figure 12.13A, we see the reflectance spectrum, $R(\lambda)$, of some material. Figure 12.13B shows the three different spectral fractions for the incident light, $f_{inc}(\lambda)$, in order from top to bottom: natural sunlight, light from a quartz-halogen tungsten (QHT) lamp, and light from an ordinary tungsten lamp. We will see in Section 12.5 that the colors of these

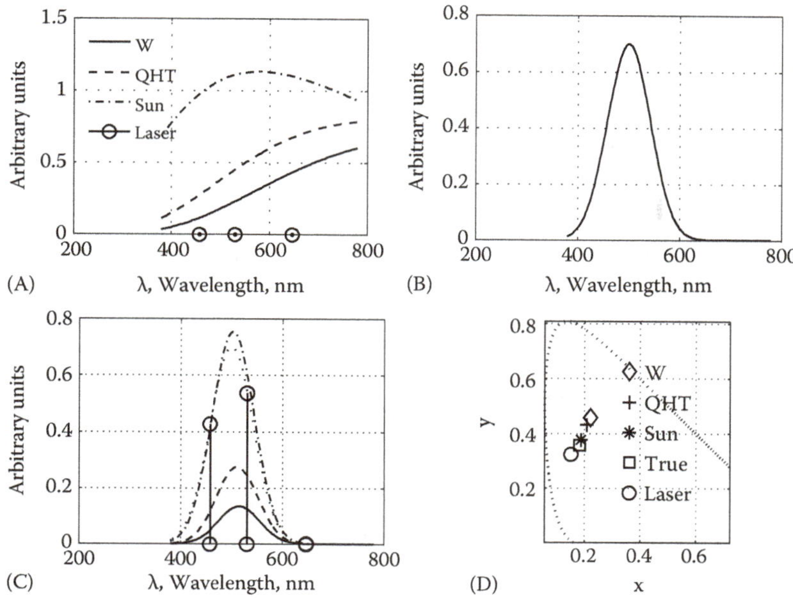

FIGURE 12.13
Reflected color. Several source spectra are shown (A), including the sun (5000 K), standard tungsten (W, 3000 K), and quartz-halogen tungsten (QHT, 3500 K). Multiplying by the reflectance spectrum (B) of a material, we obtain the reflected spectra (C) for these three sources, and our three-laser source from Section 12.3.3.2. The chromaticity coordinates (D) show points for the "true" reflectance and each of the reflected spectra.

sources change toward the red in this order. Figure 12.13C shows the spectral radiant exitance of the reflected light,

$$M_{reflected} = E_{inc} f_{inc}(\lambda) R(\lambda), \tag{12.57}$$

where E_{inc} is the incident irradiance. It is evident that the spectra shift to the right and the object looks redder when it is illuminated by the tungsten lamps.

We also consider illumination with the three laser lines discussed in Section 12.3.3.2 and Figure 12.12. Assuming equal incident irradiances from the three lasers, the reflected spectrum consists of three discrete contributions represented by the circles in Figure 12.13C.

The chromaticity coordinates of the reflected light are computed in all four cases using Equations 12.44, 12.48, and 12.49 to obtain X, Y, and Z, and then the pair of Equations 12.53 to obtain x and y, which are plotted in Figure 12.13D. The sun, QHT, and tungsten show increasing departures from the "true" spectrum of the material, with the changes being toward the red. The source composed of three lasers produces a spectrum which is shifted toward the blue. Thus the apparent color of the material depends on the illumination. The exact details would vary with different material spectra, and thus even the contrast in a scene could be different under different conditions of illumination.

Take-Away

Reflected Colors:
The color of a reflective object depends on the spectrum of the illumination source. Tungsten lights generally appear slightly red.

Take-Away

Summary of Color:

- Three color-matching functions $\bar{x}$, $\bar{y}$, and $\bar{z}$, in Figure 12.10 are used as spectral matching functions, to be multiplied by a spectrum and integrated over wavelength to determine how the color will be perceived by the human eye. Approximate functional representations in Equations 12.43, 12.50, and 12.51 may be used in some situations.
- The three integrals X, Y, and Z, (the tristimulus values), in Equations 12.44, 12.48, and 12.49, describe everything required to determine the apparent brightness of the light and the apparent color. Tristimulus values can be developed for any of the spectral radiometric quantities.
- The Y tristimulus value is related to the brightness, and with a conversion factor, 683 lm/W, provides the link from the radiometric units to the photometric units.
- Two normalized tristimulus values, x and y, defined by Equation 12.53, called chromaticity coordinates, describe the color. The gamut of human vision lies inside the horseshoe curve in Figure 12.12.
- Given three light sources with different spectra, called R, G, and B, having different radiometric spectra (spectral radiance, spectral radiant flux, or any of the others), one can compute the tristimulus values using Equation 12.52.
- Sources with different spectra may appear to be the same color.
- Light reflected or scattered from materials may appear different under different sources of illumination, even if the sources have the same tristimulus values.
- To generate light with a desired set of tristimulus values, one can compute the powers, R, G, and B, using Equation 12.56.
- Only a limited color space can be reproduced from any three sources. More than three sources can be used, but the matrix in Equation 12.56 is not square and the solutions are not unique.

12.4 Instrumentation

Having discussed computation with the radiometric and photometric quantities in Figure 12.1, we now turn our attention to the instruments used to measure them. Optical detectors measure power or energy (see Chapter 13). To determine radiance, one must pass the light through two small apertures, a field stop and an aperture stop, measure the power, and divide by the window area, A, and the pupil solid angle, Ω, both in the same space, which may be either object or image space. The reader may wish to review some of the material in Chapter 4 before reading the rest of this section.

12.4.1 Radiometer or Photometer

A radiometer measures power, Φ, and computes $L = \frac{\Phi}{A\Omega}$ for small A and Ω.

Measurements of radiance are made with a **radiometer**. The basic principle is to collect light from a source, limiting the solid angle and area, measure the radiant flux or power, Φ, and compute the radiance. The concept is shown in Figure 12.14. The choice of aperture sizes depends on the object being measured. We need to provide a field stop limiting the field of view to a patch area, δA, and aperture stop limiting the solid angle to an increment, $\delta\Omega$, both of which must be small enough so that the radiance is nearly uniform

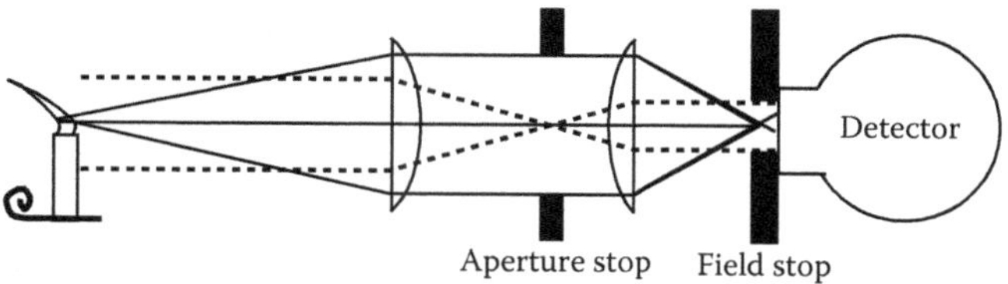

FIGURE 12.14
Radiometer concept. An aperture stop is used to limit the solid angle (solid lines), and a field stop is used to limit the area of the detector (dashed lines). The field stop is chosen too large for the source in this example.

across them, in order that we can compute the partial derivatives

$$\frac{\partial \frac{\partial \Phi}{\partial A}}{\partial \Omega} \approx \frac{\delta \Phi}{\delta A \delta \Omega} \tag{12.58}$$

Usually a number of different user-selectable field stops are available to limit the area with some sort of visual aid for selecting the appropriate one. For example, many radiometers are equipped with a sighting telescope, and have a black dot placed in an image plane of the telescope to match the size of the aperture selected. The user will choose a dot that is small enough so that it subtends a small portion of the image in the telescope. The chosen field of view shown in Figure 12.14 is a bad example, because it is larger than the source. Thus, the instrument will measure the power of the source and divide by this too-large solid angle, to arrive at an incorrect low radiance. The aperture stop can be chosen large enough to gather sufficient light to provide a strong signal, but small enough so that the radiance does not vary over the collected solid angle. Then the detected power is converted to a voltage signal, digitized, and divided by the product of δA and $\delta \Omega$, to compute the radiance.

Normally the detector will not have a constant response across wavelength. The nonuniform response can be corrected by a carefully designed optical filter to "flatten" the spectrum, so that a given power input corresponds to the same voltage at all wavelengths. The filter will reduce the sensitivity at some wavelengths, but is essential to give a correct reading of a wide-bandwidth source.

In Practice

For monochromatic light, on some instruments the user can enter the wavelength, which will be used as a calibration factor. This allows the instrument to have the maximum sensitivity at every wavelength, but opens the door for human error in entering the data. Such an instrument is not suitable for wide bandwidth sources.

A **spectral radiometer** has the same optical path for selecting the area and solid angle, but is followed by a spectrometer of some sort. The power at each wavelength is divided by the $A\Omega$ product to obtain the spectral radiance, L_λ. In a spectral radiometer, the nonuniform response can be corrected computationally by storing calibration values in the instrument and correcting the digitized voltage through division by the calibration factor.

A **photometer** can be implemented in the same way, computing the spectral luminance, $683\,\text{lm/W} \times \bar{y} \times L_\lambda$, and then computing the integral over wavelength to obtain the luminance. Alternatively, a photometer can be made with a single detector and a filter that accounts for both the $\bar{y}$ spectrum and the spectral response of the detector, so that the voltage reading is directly proportional to luminous flux.

In Practice

Many radiometers (and power or energy meters) have silicon detectors. They can be used across the visible wavelengths and a bit into the near infrared, but normally must be calibrated at each wavelength. Such detectors are not suitable for broad-band light sources.

A silicon detector can be covered with a correcting filter to produce uniform response, or with a photometric filter to produce a response comparable to the eye in photometric units. Such filters will result in loss, but are suitable for broadband visible light.

Thermal detectors can often be used across a much wider bandwidth, even into the far infrared.

Detectors are discussed in more detail in Chapter 13.

12.4.2 Power Measurement: The Integrating Sphere

In Section 12.1, we discussed integrating the radiance over solid angle and area to obtain radiant flux (e.g., using Equations 12.7 and 12.9 in succession). In some cases, physical implementation of the integration is easy. For example, given a Gaussian laser beam with a diameter, d_0, the beam divergence is $(4/\pi)(\lambda/d_0)$, and the product of $A\Omega$ is of the order of λ^2. If the beam is small enough, all of the light can easily be captured by a **power meter**. A power meter is simply a detector which produces a voltage or current proportional to the incident power (Chapter 13). If the beam is too large for the area of the detector, a lens can be used to reduce the beam diameter, and the resulting increase in divergence will likely not prevent measurement of the total power.

However, in other cases, capturing all of the light is more difficult. A tungsten lamp or a LED radiates its intensity over a large angle. Another example is a semiconductor laser, without any optics to collimate it. Such a laser has a beam diameter of a few micrometers and thus a large beam divergence. Although the $A\Omega$ product is small, the beam may diverge so fast that it is difficult to get a power meter close enough to it. Finally, measurement of the transmission or reflection of a thin scattering sample such as paper is often challenging, because the scattered light subtends a wide angle.

One technique for performing the integration is **goniometry**. We measure the intensity by making a power measurement on a small detector far from the source, perform this measurement at all angles, and integrate over solid angle. This process provides a wealth of information, but it is very tedious.

The **integrating sphere**, shown in Figure 12.15, can often be used to provide a direct measurement of the total radiant flux in a single measurement. The integrating sphere consists of a sphere coated with a highly reflecting diffuse material. Such spheres are available commercially from a variety of vendors. A crude integrating sphere may be made by painting the inside of a sphere with white paint. A small hole is available for the input light, and another for the detector. Regardless of the location or angle of a ray of light at the input, the ray will be scattered by the interior wall of the sphere multiple times until it is eventually absorbed. If the wall is highly reflective, the light will eventually reach the detector without being absorbed by the wall.

When the source is first turned on, the scattered radiance inside the sphere will increase, until equilibrium is reached. The time this increase takes depends on the path

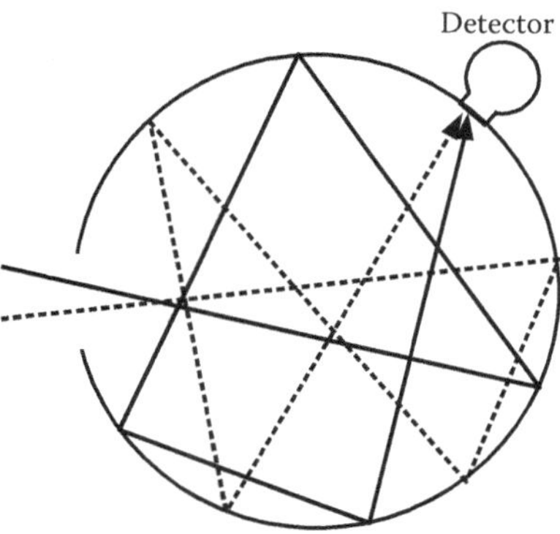

FIGURE 12.15
Integrating sphere. The interior of the sphere is coated with a diffusely reflective material. Light incident on this surface is scattered multiple times until it is eventually detected.

of the many multiple reflections, but is typically nanoseconds. The radiance is ideally uniform on the surface of the sphere, and the amount of power detected is equal to the amount incident. The detector and the input opening must both be small compared to the total area of the sphere, so that they do not contribute to unaccounted losses. The power measured by the detector is ideally equal to the total incident power or radiant flux, and in fact can be calibrated to account for a nonideal sphere. The theory of the integrating sphere is described by Jacquez[109] and some results for nonideal spheres were developed a decade later[79].

Because the power falling on the detector is equal to the incident power, the total power falling on the interior wall of the sphere is many times larger than the incident power. The reader may wonder why it would not be possible to make the detector larger, capture more of the light on the surface, use the detected power to drive the source, and create a perpetual motion machine. The fallacy of this argument is that increasing the detector size would decrease the overall radiance.

Integrating spheres can be used in pairs to measure both transmission and reflection, and have found application in such tasks as determining the optical properties of biological tissue[176].

In Practice

In using an integrating sphere with pulsed light sources, one must be mindful of equilibrium issues. The user must allow sufficient time for the light to achieve a large number of reflections before measuring the power. For example, if the diameter of the sphere is 10 cm and the surface reflects 99.9% of the light, the integrating sphere takes more than 1000 passes to come to equilibrium. Light will travel $s = 1000 \times 0.01$ m in $t = 10\,\text{m}/c = 3.33 \times 10^{-8}$ s so equilibrium will require many tens of nanoseconds.

12.5 Blackbody Radiation

Next, we explore the subject of blackbody radiation, which is fundamental to many common light sources. The important equation is the Planck equation. First, we will discuss the background that leads to this equation, and then illustrate some applications.

12.5.1 Background

We begin with a background discussion and derivation of the equations. The reader may find this discussion useful, but may prefer to proceed directly to the Planck Equation, Equation 12.81.

We can understand the basic principles of blackbody radiation by considering a cubic cavity, as shown in Figure 12.16, with perfectly reflecting walls. Using the Helmholtz equation, Equation 1.39, for the electric field, E,

$$\frac{\partial^2 E}{\partial x^2} + \frac{\partial^2 E}{\partial y^2} + \frac{\partial^2 E}{\partial z^2} = \frac{1}{c^2}\frac{\partial^2 E}{\partial t^2},$$

with one corner of the cavity at the origin of coordinates, the boundary conditions that the field be zero at the walls dictate that the solutions be sine waves,

$$E(x, y, z, t) = \sin\frac{2\pi\nu_x x}{c}\sin\frac{2\pi\nu_y y}{c}\sin\frac{2\pi\nu_z z}{c}\sin 2\pi\nu t, \qquad (12.59)$$

where, for a square cavity of dimensions ℓ,

$$\frac{2\pi\nu_x\ell}{c} = N_x\pi \qquad \frac{2\pi\nu_x\ell}{c} = N_y\pi \qquad \frac{2\pi\nu_x\ell}{c} = N_z\pi, \qquad (12.60)$$

$$\nu = \sqrt{\nu_x^2 + \nu_y^2 + \nu_z^2}, \qquad (12.61)$$

and N_x, N_y, N_z are integers. The resonant frequencies of the cavity are given by

$$\nu = \nu_0\sqrt{N_x^2 + N_y^2 + N_z^2} \qquad \nu_0 = \frac{c}{2\ell}. \qquad (12.62)$$

The fundamental frequency, ν_0, is already familiar from the one-dimensional laser cavity in Section 7.5.3.

Next, we compute the mode density, or the number of modes per unit frequency. If we envision a three-dimensional space, defined by frequency coordinates ν_x, ν_y, and ν_z, then the spacing between modes is $c/(2\ell)$, and the "volume occupied" by a mode is $c^3/(8\ell^3)$. The volume in a shell from radius ν to radius $\nu + d\nu$ is $4\pi\nu^2 d\nu$. For only the positive frequencies, the volume is $1/8$ of the total volume, so the **mode density** is given by

$$dN = \frac{4\pi\nu^2 d\nu/8}{c^3/(8\ell^3)}$$

$$N_\nu = \frac{dN}{d\nu} = \frac{4\pi\nu^2\ell^3}{c^3}.$$

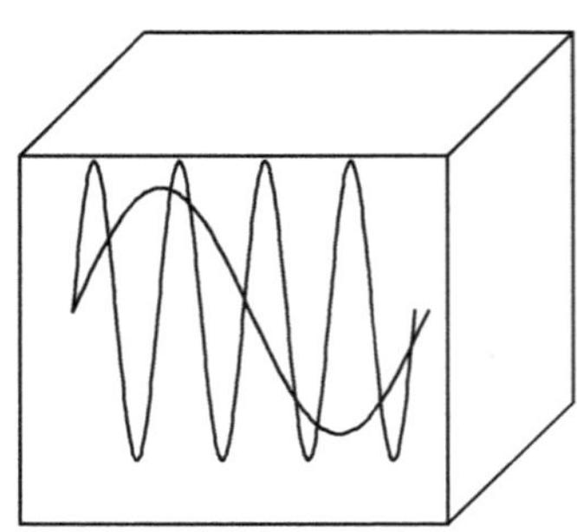

FIGURE 12.16
Blackbody cavity. Because the walls are conductors, the field must be zero at the walls. Two modes are shown.

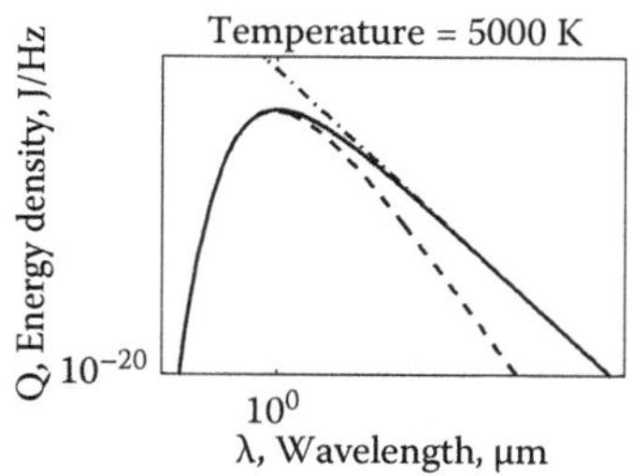

FIGURE 12.17
Energy density in a 1 L cavity at 5000 K. The Rayleigh-Jeans law fails in the ultraviolet (dash-dot line), and the Wien law in the infrared (dashed line).

Now, because we started with the scalar wave equation, we have considered only the modes in one polarization, so the total mode density will be twice this number or

$$N_\nu = \frac{8\pi\nu^2\ell^3}{c^3}. \tag{12.63}$$

Next, we need to assign a distribution of energy to the modes. Near the turn of the twentieth century the distribution of energy was an exciting topic of conversation in the physics community, a topic that in fact contributed to the foundations of quantum mechanics. One proposal, the Rayleigh-Jeans law[111], suggested assigning equal energy, k_BT, to each mode, where k_B is the Boltzmann constant, and T is the absolute temperature. The resulting equation,

$$J_\nu = k_BT\frac{8\pi\nu^2\ell^3}{c^3} \qquad \text{(Rayleigh–Jeans)} \tag{12.64}$$

is plotted in Figure 12.17 as the highest curve (dash-dot), a straight line going toward infinity for short wavelengths (high frequencies). The integral over frequency does not converge, so the energy in the cavity is infinite. This **ultraviolet catastrophe** was one of the disturbing aspects of a subject that many scientists in the late 1800s thought was quite nearly solved. A second approach by Wien[249] assigned an energy $h\nu\exp(h\nu/kT)$ to each mode. This equation,

$$J_\nu = h\nu\exp(h\nu/kT)\frac{8\pi\nu^2\ell^3}{c^3} \qquad \text{(Wien)} \tag{12.65}$$

also plotted in Figure 12.17, as the lowest curve (dashed line), predicted a spectrum which disagreed with experiment at longer wavelengths. Of even greater concern was the result that the Wien distribution predicted energy densities at some wavelengths that decreased with increasing temperature.

Max Planck[178] developed the idea that the energy distribution is quantized, leading to his being awarded the Nobel Prize in 1918. The energy, J, of each mode, is an integer multiple of some basic quantum, proportional to the mode's frequency, and designated as $J_N = Nh\nu$, where h is **Planck's constant**, 6.6261×10^{-34} J/Hz. After this first significant step recognizing the quantization of energy, Planck took an additional step and recognized that the number N would be a random number with some probability distribution. The average energy density per unit frequency is

$$\bar{J} = \frac{\sum_{N=0}^{\infty} J_N P(J_N)}{\sum_{N=0}^{\infty} P(J_N)}, \tag{12.66}$$

where $P(J_N)$ is the probability distribution, and the denominator ensures normalization. Choosing the **Boltzmann distribution**

$$P(J_N) = e^{\frac{-J_N}{k_B T}}, \tag{12.67}$$

the average energy is

$$\bar{J} = \frac{\sum_{N=0}^{\infty} Nh\nu e^{\frac{-Nh\nu}{k_B T}}}{\sum_{N=0}^{\infty} e^{\frac{-Nh\nu}{k_B T}}} = h\nu \frac{\sum_{N=0}^{\infty} N e^{\frac{-Nh\nu}{k_B T}}}{\sum_{N=0}^{\infty} e^{\frac{-Nh\nu}{k_B T}}}. \tag{12.68}$$

The sum of a geometric progression

$$\sum_{N=0}^{\infty} a^N = \frac{1}{1-a} \qquad \text{where} \qquad a = e^{\frac{-Nh\nu}{k_B T}}, \tag{12.69}$$

provides an expression for the denominator, and the derivative,

$$\frac{d}{da} \sum_{N=0}^{\infty} a^N = \frac{d}{da} \frac{1}{1-a} = \sum_{N=0}^{\infty} N a^N \tag{12.70}$$

provides the numerator in Equation 12.68:

$$\bar{J} = h\nu \frac{\frac{e^{h\nu/k_B T}}{\left(1-e^{h\nu/k_B T}\right)^2}}{\frac{1}{\left(1-e^{h\nu/k_B T}\right)}} = \frac{h\nu}{e^{h\nu/k_B T} - 1}. \tag{12.71}$$

Now to obtain Planck's expression for the energy in the cavity, we need to multiply this average energy density per unit frequency by the mode density in Equation 12.63, just as we did for the Rayleigh-Jeans and Wien distributions. The total average energy density is

$$N_\nu \bar{J} = \frac{8\pi\nu^2 \ell^3}{c^3} \frac{h\nu}{e^{h\nu/k_B T} - 1}. \tag{12.72}$$

This equation is plotted in Figure 12.17 as the solid line. This result increases with temperature for all wavelengths, integrates to a finite value, and agrees with spectra observed experimentally.

We obtain the average energy density per unit spatial volume, $\bar{w}$, per unit frequency as

$$\frac{d\bar{w}}{d\nu} = \frac{N_\nu \bar{J}}{\ell^3} = \frac{8\pi\nu^2}{c^3} \frac{h\nu}{e^{h\nu/k_B T} - 1}, \tag{12.73}$$

and note that this is a function of only ν, T, and some constants. It admits fluctuations about the average value, and thus one expects to see random fluctuations in the amount of energy, which are experimentally observed.

This energy constitutes a collection of random electromagnetic waves moving inside the cavity always at the speed of light, c. As we have seen for elevated temperatures, the wavelengths at which the energy is concentrated are within the band we claim as belonging to the field of optics. Thus light is continually absorbed and radiated by each wall of the cavity, bouncing from wall to wall, maintaining the average spectral energy density satisfying Equation 12.73. This notion will be familiar to anyone who has learned to build a good fire. A single burning log in cool surroundings radiates its energy, heating its surroundings, but receives little energy in return. Thus it cools and soon stops burning. Several logs with burning sides facing each other, forming a "box of heat," and exchange radiant energy among themselves, maintaining near thermal equilibrium at a high temperature. A small opening will allow some of the energy to escape, providing a modest amount of heat to the surroundings, but for a longer time (Figure 12.18). From this description of the cavity, we can draw several important conclusions.

First, to accommodate the different orientations of the walls, it is necessary that the radiance be independent of direction. Thus, the walls act as Lambertian sources. If they did not, then walls oriented favorably could heat to a higher temperature and become an energy source for a perpetual motion machine.

Second, the light emitted by a surface must, on average, be balanced by light absorbed, so that the temperature remains constant. Thus, if the average irradiance incident on a wall in a cavity at some temperature is $\bar{E}(T)_{incident}$, and the amount absorbed is $\epsilon_a \bar{E}(T)_{incident}$ (the rest being reflected), the radiant exitance must be the same:

$$\bar{M}(T)_{emitted} = \epsilon_a \bar{E}(T)_{incident}. \tag{12.74}$$

Otherwise, the wall would change its temperature. In this equation, ϵ_a is the absorption coefficient. The rest of the light is reflected either specularly or diffusely, and the total reflectance is $1 - \epsilon_a$. We state that "the absorption is equal to the emission," and refer to this statement as the principle of **detailed balance**.

Third, detailed balance must be true wavelength-by-wavelength. If it were not, we could use a clever arrangement of filters to heat one wall while cooling another and use the thermal difference to make a perpetual motion machine.

Fourth, because the light moves at speed of light, c, in all directions equally, the integral of the radiance over a full sphere divided by c is the energy density

$$w_\lambda = \frac{1}{c} \int_{sphere} L_\lambda \, d\Omega = \frac{4\pi l}{c}. \tag{12.75}$$

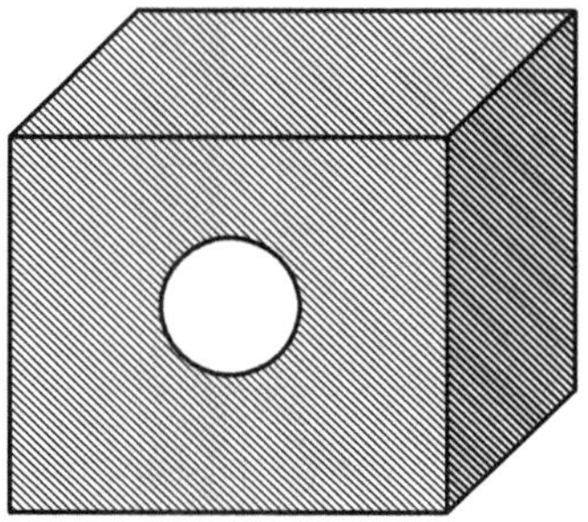

FIGURE 12.18
Cavity with a hole. At the hole, the energy in a volume one wavelength thick is allowed to exit during each cycle of the optical frequency.

The irradiance is always given by an equation identical to Equation 12.11, and in this specific case, the result is the same as Equation 12.13

$$E_\lambda = \iint L_\lambda \cos\theta \sin\theta \, d\theta \, d\zeta = \pi L_\lambda. \tag{12.76}$$

From these two equations,

$$E_\lambda = 4cw_\lambda. \tag{12.77}$$

From Equation 12.74 for a blackbody, which is a perfect absorber, ($\epsilon_a = 1$), $\bar{M}(T)_{bb} = \bar{E}(T)_{incident}$. Thus Equation 12.73 leads to

$$M_{\nu(bb)}(\lambda, T) = h\nu c \frac{2\pi\nu^2}{c^3} \frac{1}{e^{h\nu/k_B T} - 1} = \frac{2\pi h\nu^3}{c^2} \frac{1}{e^{h\nu/k_B T} - 1}; \tag{12.78}$$

the mean radiant exitance is equal to the mean incident irradiance. For any wall in thermal equilibrium at temperature, T, we can define the **emissivity** such that

$$M_\nu(\lambda, T) = \epsilon M_{\nu(bb)}(\lambda, T) \tag{12.79}$$

The principle of detailed balance requires that the emissivity is equivalent to the absorption coefficient,

$$\epsilon(\lambda) = \epsilon_a(\lambda). \tag{12.80}$$

This principle is important in the analysis of infrared imaging systems where reflection, transmission, absorption, and emission all play important roles in the equations.

With Equation 12.79 and this statement of the principle of detailed balance as given by Equation 12.74, we have a precise definition of a **black** surface. A black surface is one that has a spectral radiant exitance given by Equation 12.78. According to Equation 12.80 it is also a perfect absorber. This definition fits nicely with our intuitive definition of the word "black."

Wien conceived the idea that a heated cavity with a small opening, shown in Figure 12.18, would well represent a blackbody source, and for this idea he was awarded the Nobel Prize in 1911. If we allow the energy to exit the hole at a rate corresponding to one wavelength per period, so that it is traveling at the speed of light, then the resulting spectral radiant exitance is given by Equation 12.78.

More commonly, we want to express this important result in terms of wavelength as opposed to frequency. We use the fractional linewidth equation, Equation 7.127, to obtain $|d\nu/\nu| = |d\lambda/\lambda|$,

$$M_\lambda = \frac{dM}{d\lambda} = \frac{d\nu}{d\lambda}\frac{dM}{d\nu} = \frac{\nu}{\lambda}\frac{dM}{d\nu} = \frac{c}{\lambda^2}\frac{dM}{d\nu},$$

and the result is the **Planck Equation**,

$$M_\lambda(\lambda, T) = \frac{2\pi hc^2}{\lambda^5} \frac{1}{e^{h\nu/k_B T} - 1}. \tag{12.81}$$

Although this equation depends only on temperature and wavelength, it is customary to leave the frequency in $h\nu$ in the exponential when writing this equation. For computational purposes, $h\nu = hc/\lambda$.

Take-Away

Planck Radiation Equation:
The Planck equation, Equation 12.81, is the basis for describing all thermal sources. The equation computes a spectral radiant exitance that is a function of only temperature and wavelength.

In Practice

Blackbody radiation is more than a mathematical curiosity. A blackbody can be made using a heated chamber with a small hole in it. Such devices can be purchased as calibration sources and the light out of this hole is an excellent approximation to the Planck equation. The light is also nearly Lambertian, so

$$L_\lambda(\lambda, T) = M_\lambda(\lambda, T)/\pi.$$

12.5.2 Useful Blackbody-Radiation Equations and Approximations

Here we find the most commonly used blackbody radiation equations, and some approximations.

12.5.2.1 Equations

As discussed above, a heated cavity with a small opening provides a good physical approximation to a blackbody. The Planck equation, Equation 12.81, is plotted in Figure 12.19 for several temperatures. As discussed in Section 12.5.1, the spectral radiant exitance increases with temperature at all wavelengths. The peak shifts to the shorter wavelengths with increasing temperature. Setting the derivative of Equation 12.81 to zero in order

Wien Displacement Law: Hotter objects emit light with a spectral peak shifted toward the blue. Cooler objects have a peak in the infrared and thus are not luminous (Their Y values are low).

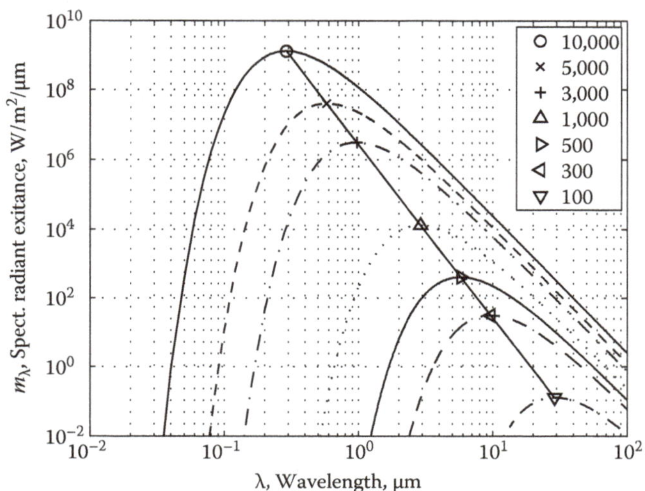

FIGURE 12.19
Blackbody spectra. Temperatures are shown in Kelvin. The diagonal line passes through the spectral peaks, and is given by the Wien displacement law, Equation 12.82.

to obtain the location of the peak (a task which requires numerical computation[127]), produces the **Wien displacement law**,

$$\lambda_{peak} T = 2898\,\mu\text{m} \cdot \text{K}. \tag{12.82}$$

Thus, the wavelength of the peak is inversely related to temperature.

The area under the curve, M, is the radiant exitance, given by the **Stefan–Boltzmann law**,

Stefan–Boltzmann Law: Hotter objects emit more light (a LOT more light). In fact, the spectral radiant exitance goes up with temperature at all wavelengths.

$$M(T) = \frac{2\pi^5 k^4 T^4}{15 h^3 c^2} = \sigma_e T^4, \tag{12.83}$$

where

$$\sigma_e = 5.67032 \times 10^{-12}\,\text{W/cm}^2/\text{K}^4 = 5.67032 \times 10^{-8}\,\text{W/m}^2/\text{K}^4$$

is the **Stefan–Boltzmann constant.**

12.5.2.2 Examples

In Section 12.5.3, we will look in detail at some examples, but before that, we briefly examine the Planck equation, Equation 12.81, and the Wien displacement law, Equation 12.82, both plotted in Figure 12.19, along with the Stefan–Boltzmann law, Equation 12.83.

These two examples are very simple, but provide some surprisingly reasonable results.

Example

Estimate the temperature of the Earth

First, we recall that the typical daytime solar irradiance on Earth is $1000\,\text{W/m}^2$, and that only half the Earth is illuminated at one time, so the temporal average is $500\,\text{W/m}^2$. If we were to consider the Earth as a uniform body with a well-defined equilibrium temperature, T, it would radiate the same amount of power, or

$$500\,\text{W/m}^2 = \sigma_e T^4,$$

resulting in $T = 306\,\text{K}$. Considering that we have made very unjustified assumptions about uniformity, this answer is surprisingly reasonable.

Example

How many calories per day are required to maintain body temperature?

At human body temperature, 310 K, the peak wavelength is about 9.34 μm, which is in the 8-to-14 μm far-infrared band transmitted by the atmosphere, and the radiant exitance is $524\,\text{W/m}^2$. Enclosed in a room at 295 K, the body receives about

429 W/m^2, so the net power radiated out per unit area is about 95 W/m^2. An adult human has about 2 m^2 of surface area. Some of that surface area is facing other parts of the body, as when the arms are held at the sides. Furthermore, $\epsilon < 1$, so let us assume that the effective area times the average ϵ is 1 m^2. Then the radiant flux is 95 W × (3600 × 24) s/day/4.18 J/cal = 2 kcal/day, which is quite surprisingly close to the amount of energy in the average daily diet. Thus, to a first approximation, in a comfortable room, we radiate as much energy as we consume. This power is also similar to the allowance of 100 W per person often invoked in determining building cooling requirements.

The net power radiated out is

$$P_{net} = \epsilon A\sigma_e \left(T_{body}^4 - T_{ambient}^4\right) = \epsilon A\sigma_e \left(T_{body}^2 - T_{ambient}^2\right)$$

$$\left(T_{body} + T_{ambient}\right)\left(T_{body} - T_{ambient}\right)$$

$$P_{net} \approx \epsilon A\sigma_e T^4 \frac{4\left(T_{body} - T_{ambient}\right)}{T}. \tag{12.84}$$

In our example,

$$P_{net} \approx 500\,\mathrm{W} \times \frac{4\left(T_{body} - T_{ambient}\right)}{300\,\mathrm{K}}. \tag{12.85}$$

If the temperature difference changes by only 3 K because the room cools or the person has a fever, this slight change in temperature results in 20 W of increased cooling, which is over 20% of the net power of 95 W radiated by the human body as computed above. Thus, we humans are very sensitive to small temperature changes.

12.5.2.3 Temperature and Color

As the temperature of an object increases, its radiant exitance increases and its peak wavelength decreases. Figure 12.20 shows

$M_{(R)}$ is Radiant Exitance

$M_{(V)}$ is Photometric Exitance

$$M_{(R)} = \int_0^\infty M_{(R)\lambda} d\lambda$$

and

$$M_{(V)} = \int_0^\infty M_{(V)\lambda} d\lambda$$

$$M_{(V)} = 683\,\mathrm{lm/W} \times \int_0^\infty \bar{y} M_{(R)\lambda} d\lambda$$

as functions of temperature. We use the subscripts (R) and (V) for radiometric and photometric quantities because we are using both in the same equations. Both curves increase monotonically with temperature, although their ratio has a maximum around 6600 K, as shown in Figure 12.21. At a temperature somewhat above 600 K objects appear faintly visible, and after 1000 K they become as bright as the lunar disk. Although an object around 1000 K has a spectral peak in the infrared, it has significant light in the red region of the spectrum and appears red to an observer. Around 3000 K, a tungsten filament in an incandescent light bulb has a peak in the middle of the visible spectrum. Thus objects that are "white hot" are hotter than those that are merely "red hot." The chromaticity coordinates are shown as a function of temperature in Figure 12.22. The locus of these coordinates is shown as the white curve in Figure 12.12. The computation of tristimulus values is shown in Figure 12.23 for three temperatures, the first at 2000 K,

Tristimulus Values from Eq. 12.48, 12.44, 12.49: $X = \int_0^\infty \bar{x}(\lambda) L(\lambda) d\lambda$, *etc.*

Chromaticity Coordinates from Eq.12.53 $x = \frac{X}{X+Y+Z}$ $y = \frac{Y}{X+Y+Z}$.

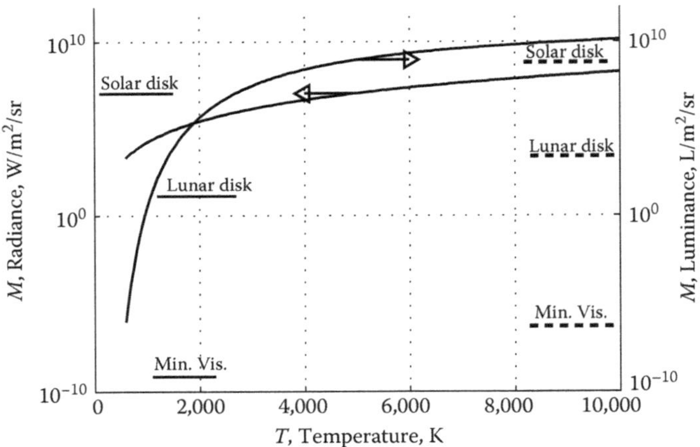

FIGURE 12.20
Radiant and luminous exitance of hot objects. The radiant exitance is plotted as a curve with an arrow pointing to the left axis which is in radiometric units. The luminous exitance curve has an arrow toward the right axis which has photometric units. Both increase monotonically with temperature but their ratio does not (see Figure 12.21). Horizontal lines on the left show the radiance of a minimally visible object, the lunar disk and the solar disk for comparison. Horizontal lines on the right show the luminance for the same three objects.

A quartz–halogen lamp is brighter and whiter than an ordinary tungsten lamp because the filament is hotter.

a second at 3000 K to simulate a tungsten filament, and the third at 3500 K to simulate the hotter tungsten filament of a **quartz-halogen lamp**. The halogen is a catalyst so that evaporated tungsten will condense back onto the filament rather than on the glass envelope. It thus allows the filament to operate at a higher temperature. The quartz envelope is required because a glass one would melt upon absorbing the higher power of the hotter filament. As we can see, the hotter filament produces a color much closer to white than the cooler one.

12.5.3 Some Examples

Here we consider examples of scenes under natural and manufactured light.

12.5.3.1 Outdoor Scene

After this analysis, it may be useful to consider some specific examples. First, we look at the solar spectrum. The sun is not really a blackbody, and its spectrum, measured on Earth, is further complicated by absorption in the atmosphere. Nevertheless, to a good approximation, the sun behaves like a blackbody over a portion of the spectrum so that the Planck equation is at least somewhat useful as a means of describing it. The dashed lines in Figure 12.24 show measured solar spectra[8] outside the atmosphere and at the surface. The lower line, of course, is the spectrum at the surface. We approximate the exo-atmospheric spectral radiance by computing the spectral radiance of a 6000 K blackbody, and scaling it to an irradiance of 1480 W/m^2;

$$E_\lambda(\lambda) = 1560\,\mathrm{W/m^2} f_\lambda(\lambda)\,(6000\,\mathrm{K}). \tag{12.86}$$

The total irradiance obtained by integrating the measured curve is 1480 W/m^2. We need to use a higher normalization in our approximation because we are not considering that a portion of the spectrum is absorbed. At sea level, we use the approximation,

$$E_\lambda(\lambda) = 1250\,\mathrm{W/m^2} f_\lambda(\lambda)\,(5000\,\mathrm{K}). \tag{12.87}$$

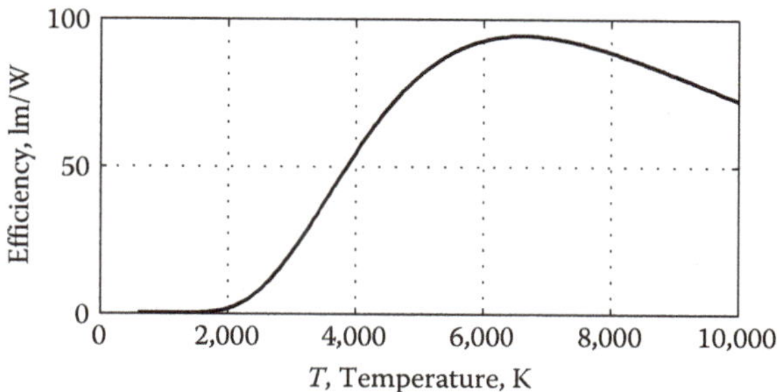

FIGURE 12.21
Luminous efficiency of a hot object. The luminous efficiency increases sharply from about 2000 K, as the radiant exitance increases and the spectral peak shifts from the infrared into the visible spectrum. The efficiency reaches a maximum at 6600 K, and then falls. Although the radiant exitance continues to increase with temperature, the peak shifts into the ultraviolet, reducing the efficiency.

Again, the scale factor is higher than the $1000\,\mathrm{W/m^2/sr}$ obtained by integrating the measured curve. Typical actual spectra are shown in Figure 12.24,

> **Take-Away**
>
> **Solar Spectrum:**
> The exo-atmospheric solar spectrum is reasonably approximated by a 6000 K blackbody. At sea level it is better matched by a 5000 K blackbody. These approximations must be used carefully. For example, there are strong absorption bands in the infrared that attenuate the light reaching sea level.

With a few additional facts, we can make some approximations for outdoor scenes. Attenuation and scattering by the atmosphere add a high level of complexity to the radiometric picture of the world, but these approximations are a first step toward better models. Looking at the solar disk from Earth, we use a 5000 K blackbody, which has a radiance of $10\,\mathrm{MW/m^2}$ or a luminance of about 1 giganit. Taking a cloud to be highly reflective, with an incident irradiance of $E = 1000\,\mathrm{W/m^2}$, we assume that it scatters a radiant exitance, $M = E$, with the solar spectrum so that the spectral radiant exitance is $M_\lambda(\lambda) = E_\lambda(\lambda)$, where $E_\lambda(\lambda)$ is given by Equation 12.87. Scattered light from the ground could be reduced to account for a diffuse reflection. A better model would consider the reflectance to be a function of wavelength, so that

$$M_\lambda(\lambda) = R(\lambda)\, E_\lambda(\lambda)\,. \tag{12.88}$$

We can treat a diffuse reflector illuminated by an irradiance E as a source with radiant exitance, $M = RE$, where R is the fraction of the incident light that is returned. The remaining fraction, $(1 - R)$ is absorbed or transmitted.

These values are plotted in Figure 12.25.

We next consider sunlight scattered from molecules (primarily nitrogen) in the atmosphere to obtain a spectrum of the blue sky. We know that the molecular scatter will follow Rayleigh's scattering law for small particles of radius, a, which predicts it to be proportional to a^6/λ^4. Reportedly, Lord Rayleigh derived this scattering law after watching a sunset, while floating down the Nile on his honeymoon, presumably to the distress of the new Mrs. Rayleigh. With this result, we expect the sky's radiant exitance to be

Rayleigh scattering theory is used for particles much smaller than the wavelength. The amount of scattering is proportional to a^6/λ^4, where a is the radius of the particle.

$$M_\lambda = M_{sky} f_\lambda(5000\,\mathrm{K}) \times \lambda^{-4}. \tag{12.89}$$

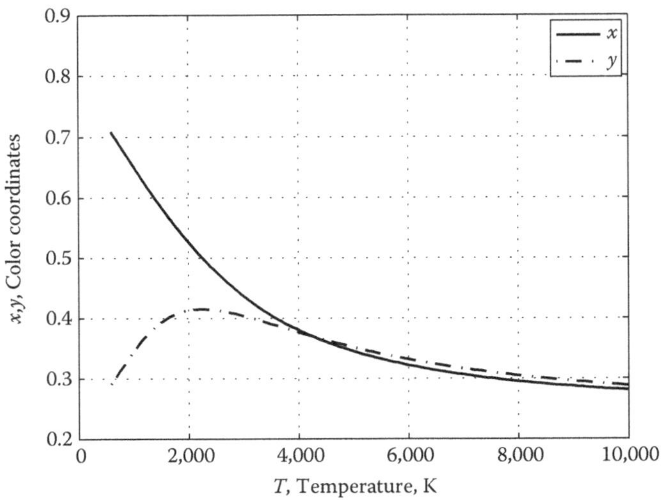

FIGURE 12.22
Chromaticity coordinates of a blackbody. The x and y chromaticity coordinates are plotted as functions of temperature. The luminous exitance is in Figure 12.20, and the locus of these points is in Figure 12.12. At temperatures where the object is barely visible, the color is strongly red (high x). As the temperature continues to increase, the chromaticity coordinates approach a constant slightly below $x = y = 0.3$, which results in a slightly blue color.

From Table 12.3, we choose $M_{sky} = 27\pi\,\mathrm{W/m^2}$, and obtain the result shown in Figure 12.25. The radiant exitance of an object at ambient temperature, about 300 K, is also shown in Figure 12.25. This could represent an object viewed at night. To some extent, it could also represent the night sky, but we must be careful here. The sky is quite transmissive in some bands, so that sunlight can reach the surface of the earth. Thus the night sky will follow this spectrum weighted with the emissivity spectrum, which, by the principle of detailed balance, will also be the absorption spectrum.

Fig. 1.1B:

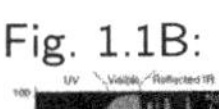

In fact, we must be mindful of the fact that the atmospheric absorption is quite strong with the exception of the visible spectrum and some wavelengths in the infrared, as shown in Figure 1.1B. Nevertheless, we can use these approximations as a starting point for understanding outdoor scenes.

In Practice

The Planck equation and scaling terms above can be useful as a first approximation to scene parameters for thermal imagers. Because of the importance of this field, a large number of computational models have been developed. These models often include atmospheric gasses in different concentrations.

As one example, let us consider a daytime outdoor scene. The emission spectrum of objects will be approximated by the "night" curve with slight variations in radiant exitance to account for variations in temperature, and spectral variations to account for emissivity. The spectra of reflected sunlight will be given by the "cloud" curve, reduced slightly because most objects reflect less than clouds, and modified slightly to account for reflectance variations. From the curves, it is evident that in the visible spectrum, 380–780 nm, we will see only reflected sunlight. The "night" curve is below the limits of visibility. In the far-infrared band from 8 to 14 μm, we will see only emission from the objects, the reflected sunlight as represented by the curve for the cloud, being too low to be detected. In the mid-infrared band, from 3 to 5 μm, the two curves are nearly equal, and we expect to see some of each. For example, if we view a shadow on the side of a

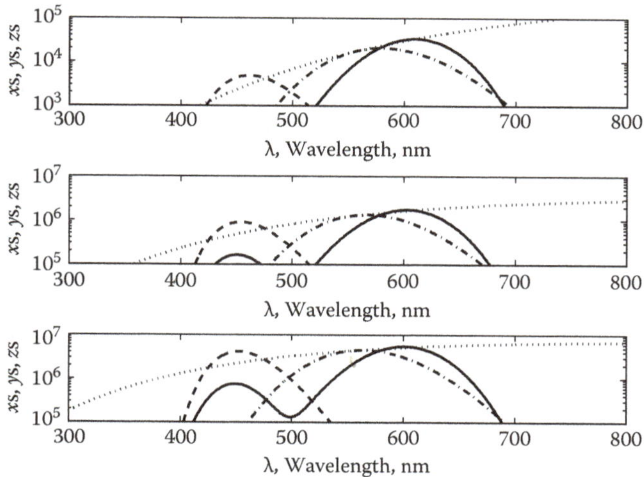

FIGURE 12.23
Tristimulus computation for three hot objects. The integrands, $M_\lambda \cdots$, $M_\lambda \bar{x}$ —, $M_\lambda \bar{y}$ $- \cdot -$, and $M_\lambda \bar{z}$ $--$ are shown for three different temperatures: $2000\,\mathrm{K}$ (top), $3000\,\mathrm{K}$ (center), typical of an ordinary tungsten filament, and $3500\,\mathrm{K}$ (bottom) typical of a tungsten filament in a quartz-halogen lamp. The luminous and radiant exitance, $683\,\mathrm{lm/W} \times Y = 683\,\mathrm{lm/W} \times \int_0^\infty M_\lambda \bar{y} d\lambda$, and $\int_0^\infty M_\lambda d\lambda$ are shown in Figure 12.20, and the chromaticity coordinates, x and y are shown in Figure 12.22.

building with cameras sensitive to the three bands, then if the object casting the shadow moves to a new location, the shadow in the visible image will move with it, and in the far infrared the old shadow will slowly fade and a new one will appear as the formerly shaded region is heated by the sun, and the newly shaded region cools by radiating to its surroundings. The image in the mid-infrared will show both behaviors; part of the shadow will move immediately, but the remainder will fade slowly with time as the new shadow strengthens in contrast as the new heating pattern appears. At wavelengths between these bands, the atmosphere absorbs most of the light, and the camera would see only the atmosphere at very close range.

Take-Away

Natural Light:
For natural daytime scenes reflect sunlight or other visible light sources dominate in the visible spectrum. In the far-IR from about 8 to 14 micrometers, emitted light dominates. In the mid-IR from about 3 to 5 micrometers, the contributions are more equal.

At night, the far-IR and mid-IR are often used for viewing emitted light.

12.5.3.2 Illumination

Illumination is one of the oldest and most common applications of optics, improving over time from fire through candles and gas lamps, to electric lighting based on tungsten filaments, ions of various gases, and more recently LEDs. The ability to light interior spaces has allowed us to remain active past sundown, and inside areas without exposure

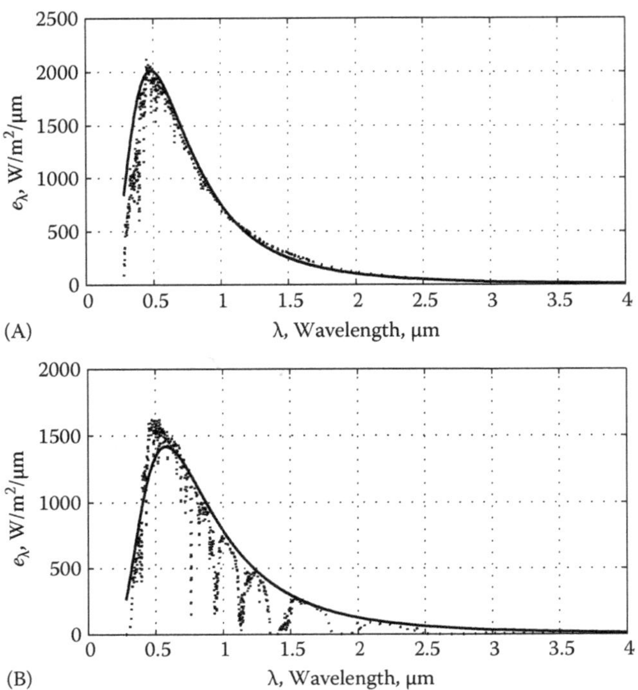

FIGURE 12.24
Solar spectral irradiance. The exo-atmospheric measured spectral irradiance (A, dotted) matches a $6000\,\mathrm{K}$ blackbody spectrum (A, solid), normalized to $1480\,\mathrm{W/m^2}$. The sea-level measured spectral irradiance (B, dotted) matches a $5000\,\mathrm{K}$ blackbody spectrum, normalized to $1000\,\mathrm{W/m^2}$ (B, solid). The surface irradiance varies considerably with latitude, time of day, season, and atmospheric conditions. (Data from ASTM Standard g173-03, Standard tables for reference solar spectral irradiance at air mass 1.5: Direct normal and hemispherical for a 37 degree tilted surface, ASTM International, West Conshohocken, PA, 2003.)

to daylight. Outdoor lighting has similarly allowed us to move from place to place with increased safety and speed. We have seen in Figure 12.22 that the peak luminous efficiency of a blackbody ($94\,\mathrm{lm/W}$), is at $7000\,\mathrm{K}$. This is fortuitous, because the spectrum at this color is quite white, with $x = 0.31$ and $y = 0.32$. Our enthusiasm for this maximal efficiency is reduced by the myriad problems such as melting of the filament and of any ordinary enclosure in which we try to contain it. More realistically, we must be satisfied with temperatures around $3000\,\mathrm{K}$, with a luminous efficiency of only $20.7\,\mathrm{lm/W}$. Figure 12.26 shows these efficiencies, along with data points for several common light sources. We see that some sources (not blackbodies) can achieve efficiency that meets and even exceeds the maximum efficiency of the blackbody.

Nevertheless, those of us who find our glass half empty must complain that even this efficiency falls far short of the fundamental limit of $683\,\mathrm{lm/W}$. As we pointed out in Section 12.3.3.4, we can only achieve this value at the expense of seeing everything in green. In that section, we discussed a light source of three lasers with an efficiency of $197\,\mathrm{lm/W}$, which is still much more than a factor of three below the limit. How far can we go toward improving efficiency? We could use LEDs as a more practical everyday substitute for the lasers in Section 12.3.3.4, with very little loss of luminous efficiency, and indeed a great gain in overall efficiency, because the LEDs are usually much more electrically efficient than the lasers. We can choose red and blue LEDs that are closer to the peak of the $\bar{y}$ curve. In Figure 12.12, this choice would move the two triangular data points at the bottom of the triangular color space up toward the center of the visible gamut. Thus, less light will be

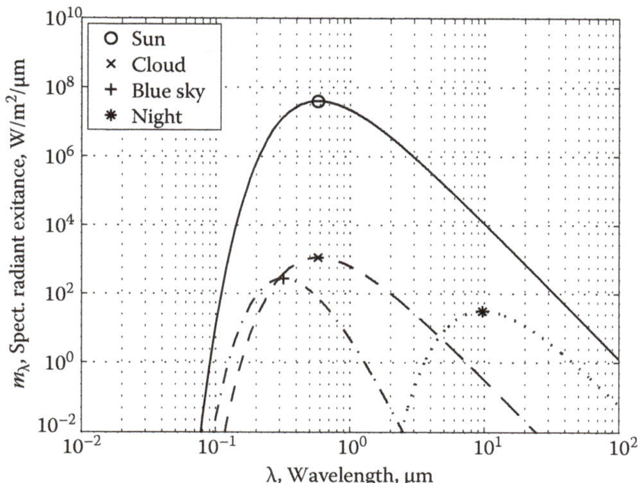

FIGURE 12.25
Outdoor radiance approximations. Simple approximations can be made using blackbody equations and a few facts about the atmosphere. The solar spectrum is assumed to be a $5000\,\mathrm{K}$ blackbody (o). A cloud is assumed to have the same spectral shape with a radiant exitance of $1000\,\mathrm{W/m^2}$ (x). Blue sky has that spectrum weighted with λ^{-4} (+). The light emitted at $300\,\mathrm{K}$ describes the night sky, or an object at ambient temperature (*). It is important to note that these approximations are not valid at wavelengths where atmospheric attenuation is strong. See Figures 1.1 and 12.24.

required from these sources to make the light source appear white. In fact, we can achieve "white" light with only two sources. However, this approach tightens the requirement to maintain the appropriate power ratios, and more importantly, will result in drastic changes to the appearance of any colored material viewed under this source, as discussed in Section 12.3.5. For example, a material that reflects only long red wavelengths would appear black under this illumination.

As efficient use of energy becomes more important to society, the search for higher luminous efficiencies will continue. How close can we come to the fundamental limit? There is no quantitative answer to this question, because the criteria for success involve qualitative factors such as the appearance of various materials under different types of illumination. Color researchers use a **color rendering index** to try to establish a numerical rating, but this is based on a small sample of different materials[7;11]. Along with other measures of performance, this one can be understood through the use of equations in this chapter.

Take-Away

Optimizing Illumination:
White light is inherently inefficient. The highest luminous efficiency occurs with green light. However, we gain so much useful information from color that it is usually advantageous to trade some efficiency for a broader spectrum. Sometimes the color rendering index is used along with the efficiency to characterize a light source.

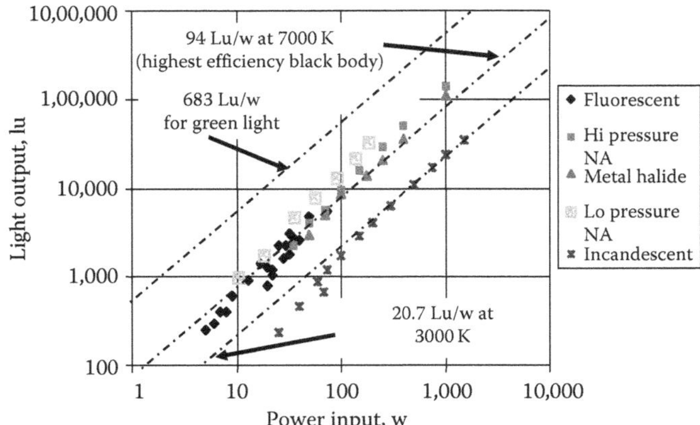

FIGURE 12.26
Lighting efficiency. The luminous output is plotted as a function of the electrical input for a variety of sources. Added straight lines show various luminous efficiencies. The highest black body efficiency of $94\,\text{lm/W}$ occurs at $7000\,\text{K}$. The upper limit of $683\,\text{lm/W}$ occurs at the green wavelength of $555\,\text{nm}$, and a typical tungsten filament at $3000\,\text{K}$ produces only $20.7\,\text{lm/W}$. (Reproduced with permission of Joseph F. Hetherington.)

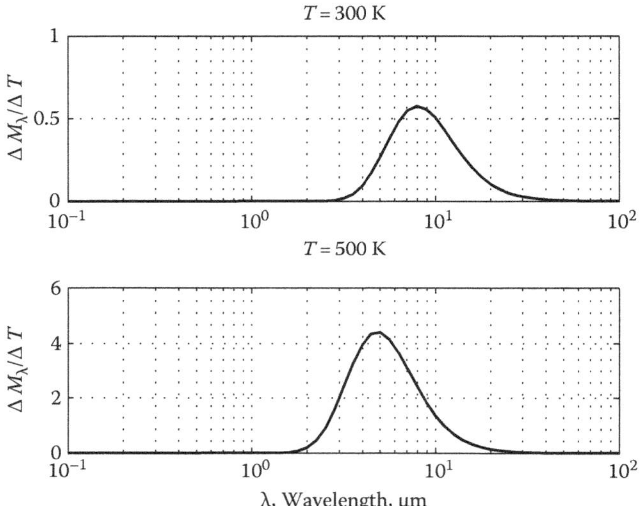

FIGURE 12.27
Infrared imaging bands. The change in spectral radiant exitance for a 1 K change in temperature occurs mostly in the far-infrared band for a temperature change from 300 to 301 K. To measure small temperature changes around ambient temperatures, this band is most useful. For a temperature change from 500 to 501 K, the change is mostly in the mid-infrared band. There is considerable overlap.

12.5.3.3 Thermal Imaging

Infrared imaging can be used to measure the temperature of an object. More precisely, it measures

$$\int_{\lambda_1}^{\lambda_2} \rho(\lambda)\, \epsilon(\lambda)\, M_\lambda(\lambda, T)\, d\lambda, \tag{12.90}$$

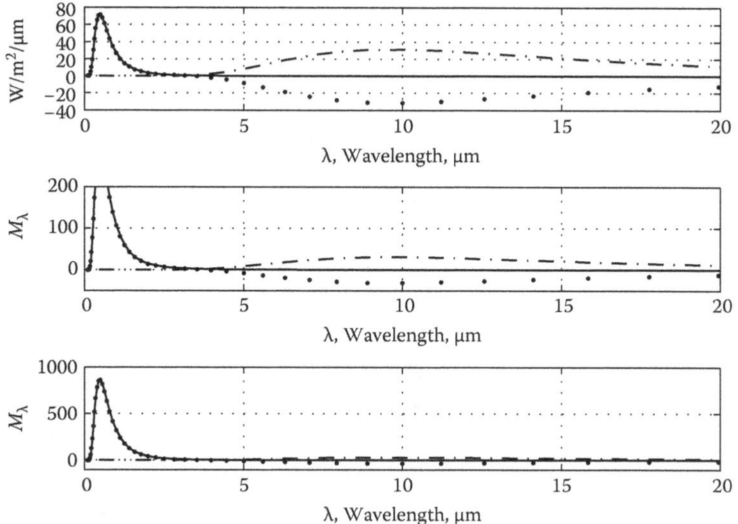

FIGURE 12.28
Solar heating. An object is heated by sunlight at an effective temperature of about 5000 K (solid line) mostly in the visible spectrum, but subtending a small portion of the sky. The object at body temperature radiates a much lower radiance, but into a full hemisphere (dash-dot). The net spectral radiant exitance (dots) is both positive and negative, but the negative portion is at longer wavelengths. If the integral of the net spectral radiant exitance is positive, heating occurs. Otherwise, cooling occurs. The incident irradiance is $50\,\mathrm{W/m^2}$ in the top plot, $200\,\mathrm{W/m^2}$ in the center, and $600\,\mathrm{W/m^2}$ in the bottom.

where

- $\rho(\lambda)$ is the response of the camera at a particular wavelength
- ϵ is the emissivity
- $M_\lambda(\lambda, T)$ is the blackbody spectral radiant exitance given by the Planck equation (12.81)

We would like the camera to be sensitive to small changes in temperature, and the dependence on temperature is only in M_λ. Figure 12.27 shows the derivative of M_λ with respect to temperature at 300 K, a typical temperature for objects in a natural scene such as a person or animal at 310 K walking through a forest with temperatures around 300 K. The peak derivative lies near 8 μm in the far-infrared band, although high values extend to either side by several micrometers. For hotter objects around 500 K, the peak shifts down toward the mid-infrared band, although again, the useful spectrum extends some distance in both directions.

Thus for natural scenes, one would prefer a far-infrared camera while for hotter objects one would choose a mid-infrared one. Because of the overlap, decisions are often made on the basis of available technology and cost considerations. As we noted earlier, mid-infrared imagers are more sensitive to scattered sunlight, while far-infrared ones more directly measure the blackbody radiance.

12.5.3.4 Glass, Greenhouses, and Polar Bears

We conclude this chapter with an example that illuminates a number of points raised in the chapter. Figure 12.28 shows the spectral irradiance of the sun with various scale factors, representing situations where the sun is very low in the sky ($50\,\mathrm{W/m^2}$ in the top plot), and higher in the sky ($200\,\mathrm{W/m^2}$ and $600\,\mathrm{W/m^2}$). The spectral shape is always that of the solar spectrum at sea level, approximated by a 5000 K source as in Figure 12.24. We

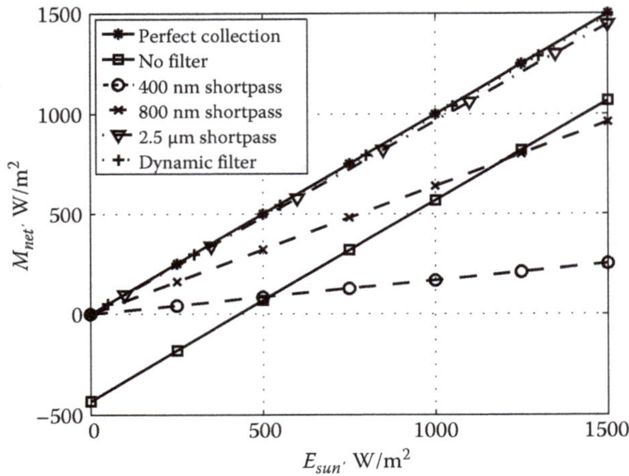

FIGURE 12.29
Keeping warm. The lower solid line (representing a black body in sunlight) shows cooling for $E_{sun} < 432\,\text{W/m}^2$. Because the radiation from the body is strongest at wavelengths much shorter than that from the sun, short-pass filters can be used between the sun and the object, to reduce the loss of energy caused by radiative cooling. A UV-transmission filter blocks most of the radiative cooling, but also blocks most of the heating. Filters with longer wavelengths perform better, and one with a cutoff wavelength of $2.5\,\mu\text{m}$ is almost as good as a dynamic filter that varies with changes in solar irradiance. The top solid line is the "45°" line representing the limit in which all the incident light is collected and none is emitted.

Eq. 12.83:
$M(T) = \sigma_e T^4$.

will neglect the atmospheric attenuation for this example. The object being illuminated is itself a blackbody, but at ambient temperature, having a constant radiant exitance of $\sigma_e T^4$ from Equation 12.83 into a full hemisphere of sky. Because of the low temperature, this radiant exitance is concentrated at far-infrared wavelengths. Subtracting the spectral radiant exitance emitted by the object from the spectral irradiance incident upon it, we obtain a net spectral power absorbed per unit area. If the net absorbed power per unit area, given by the integral of this function, is positive, the object is warmed, and if negative, it is cooled.

If we construct these curves for each different amount of solar irradiance, and plot the integral over wavelength,

$$M_{net}(T) = \int_0^\infty \left[E_{sun} f_\lambda(\lambda, 5000\,\text{K}) - M_\lambda(\lambda, 310\,\text{K})\right] d\lambda \tag{12.91}$$

as a function of E_{sun}, we see that at high values of E_{sun} the net result is positive, but it becomes negative below $E_{sun} \approx 432\,\text{W/m}^2$, as seen in the solid curve with square markers in Figure 12.29.

Because the emission is at a longer wavelength than the excitation, we can improve the heating by using some type of wavelength selective filter as in Figure 12.30. In Figure 12.29, several idealized filters are shown. The dotted line, highest in the plot, labeled "perfect collection," is the "45° line," $M_{net} = E_{sun}$, that we could obtain if we could magically keep all the power in. The solid line at the bottom shows the result if the object is exposed directly to the elements with no filter. Moving up, we consider a short-pass filter with a sharp cutoff at 400 nm, which passes the ultraviolet light. This filter prevents radiative cooling in the infrared, but also prevents most of the solar spectrum in the visible from reaching the object. A filter with a cutoff at 800 nm, provides considerable

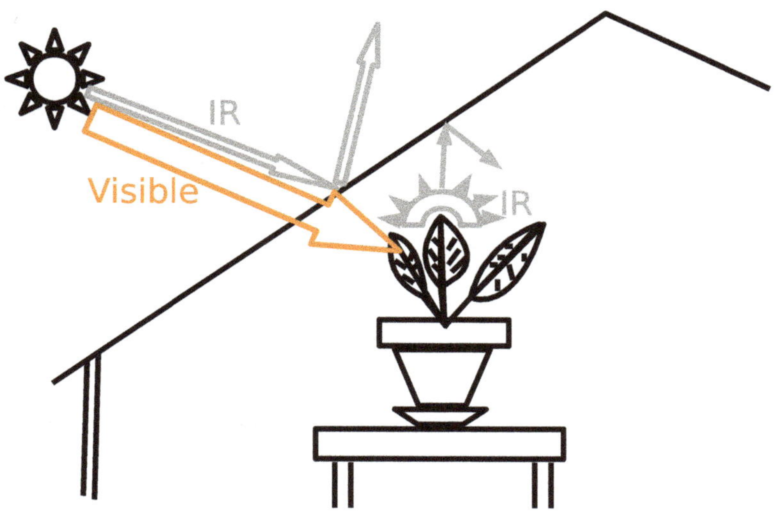

FIGURE 12.30
Greenhouse effect. Strong incident light from the small solid angle subtended by the sun is weighted toward the visible spectrum and so passes through the glass. Weaker light emitted in all directions by objects inside the greenhouse is in the infrared, and thus blocked by the glass.

warming because, with this filter, Equation 12.91 becomes

$$M_{net}(T) = \int_0^{\lambda_f} \left[E_{sun} f_\lambda(\lambda, 5000\,\text{K}) - M_\lambda(\lambda, 310\,\text{K})\right] d\lambda, \tag{12.92}$$

where $\lambda_f = 800\,\text{nm}$. A long-pass filter at $\lambda_f = 2.5\,\mu\text{m}$ is even better, and in fact is so close to the "perfect" line that it is difficult to see any difference. The final curve shown is one in which the filter is adaptively chosen for each E_{Sun}, so that it blocks all the wavelengths for which the integrand is negative. This curve is labeled "dynamic filter." Because these curves are all so close to each other, we plot the amount of the input that is lost,

$$M_{lost} = E_{sun} - M_{net},$$

in Figure 12.31. The "dynamic filter" loses only about $9\,\text{W/m}^2$ of $1000\,\text{W/m}^2$ input, or 1%, while the $2.5\,\mu\text{m}$ filter loses some $35\,\text{W/m}^2$, which is still a small 3%. These losses scale approximately linearly with decreasing E_{sun}, so they are not significant. The simple analysis presented here is to illustrate the process of solving radiometric problems.

Fortuitously, it happens that glass blocks light at wavelengths longer than about $2.5\,\mu\text{m}$, so the ordinary window glass of a greenhouse is quite effective in allowing the sun to heat the interior, while preventing radiative cooling. We have neglected several points here which would be needed in a more complete analysis. Some light is reflected from the glass through Fresnel reflection (Section 6.4). Although the absorption coefficient of glass is quite low, it is not zero, and some light is absorbed, as is evident to anyone touching a window on a hot day. We could characterize the glass in terms of absorptivity and emissivity, and develop equations for its heating and cooling. The direct heating of the object by sunlight would be reduced. The re-radiated light from the glass would contribute in the outward direction to a loss, and in the inward direction, to an additional heat source. From a practical point of view, this task is challenging, in part because of conductive and convective cooling.

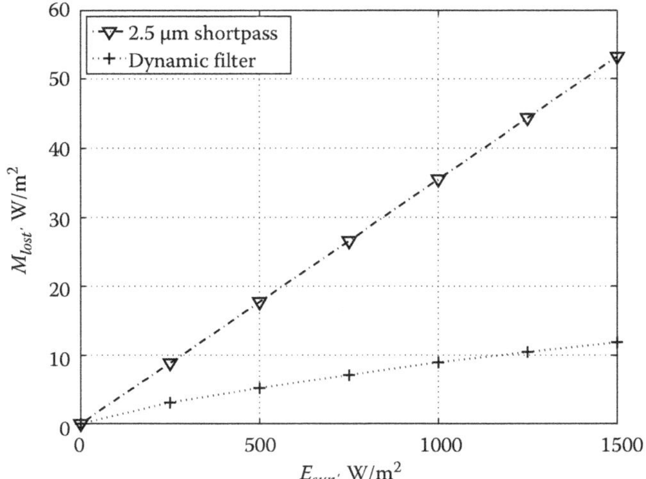

FIGURE 12.31
Keeping warmer. The two best filters considered in Figure 12.29 only fail to capture about 1% and 3% of the available light.

The phenomenon discussed here is called the "greenhouse effect." In the atmosphere, carbon dioxide blocks many infrared wavelengths, while transmitting visible light quite strongly. It is in part for this reason that researchers study the effects of carbon dioxide on the Earth's climate. We could repeat the analysis of the Earth's temperature in Section 12.5.2.2 using the absorption spectrum of carbon dioxide to see this effect. Of course, a quantitative analysis would require attention to many details that have been omitted here, but the discussion in this section should provide an understanding of the computations required.

Finally, animal pelts may also have spectral filtering properties. The animal's pelt serves multiple and often conflicting purposes of camouflage, thermal insulation, and control of heating and cooling. The sum of powers transmitted to the skin, reflected to the environment, and absorbed in the pelt is equal to unity. The relationship between reflectance and heating is more complicated than one might expect. Some animals may reflect and absorb heat quite efficiently[240]. A polar bear needs to make use of every available incoming Watt. The polar bear's skin is black, and the pelt consists of hairs that are transparent and relatively devoid of pigment. The white color of the bear indicates that a large amount of the incident light is reflected in the visible. The bear is often difficult to detect in infrared imagery although it appears dark in ultraviolet imagery[138;137]. The lack of an infrared signature suggests that the outer temperature of the pelt is close to ambient. As seen in Figure 12.29, transmission at wavelengths shorter than about 2.5 μm contributes to heating the bear, while for longer wavelengths, any transmission will more likely contribute to cooling. Measurements show that the transmission increases across the visible spectrum to 700 nm[226], decreases gradually toward 2.3 nm[184], at which wavelength it falls precipitously because of water in the hair, and is very low around 10 μm. The transmission in the visible through near-infrared ensures that some solar energy contributes to warming. The low value at the far-infrared peak of the black body spectrum demonstrates that the pelt reduces radiative cooling. The mechanisms and spectra of light transfer in the pelt of the polar bear was the subject of a number of papers in the early 1980s[240;87;26].

Problems

12.1 Radiation from a Human

For this problem, suppose that the surface area of a human being is about 1 m^2.

12.1a What is the power radiated in an infrared wavelength band from 10.0 to 10.1 µm? How many photons per second is this?

12.1b What is the power radiated in a green wavelength band from 500 to 600 nm? In what time is the average number of photons equal to one?

13 Optical Detection

Most of our study so far has been about the manipulation of light, including reflecting, refracting, focusing, splitting, and recombining. In Section 12.1, we discussed quantifying the amount of light. In this chapter, we address the detection process. Normally, detection means conversion of light into an electrical signal for display or further processing. Two issues are of greatest importance: sensitivity and noise.

The first is a conversion coefficient that determines the amount of electrical signal produced per unit of light. Typically, at the fundamental level, this is the ratio of the detected electrical charge to optical energy that produced it. For a continuous source, we can measure both the optical energy and the charge over some period of time. We define the **current responsivity** as

$$\rho_i = \frac{q}{J} = \frac{i}{P}, \tag{13.1}$$

where

- q is the electrical charge
- J is optical energy
- i is the electrical current
- P is the optical power

We may characterize the conversion process in other ways as well. For example, in some detectors, the output may be a voltage and we can define a **voltage responsivity** as

$$\rho_v = \frac{V}{P}, \tag{13.2}$$

where V is the measured voltage.

The second issue is the noise that is superimposed on the signal. Regardless of the detection mechanism, quantum mechanics imposes a fundamental uncertainty on any measurement, and the measurement of light is no exception. In addition to this so-called quantum noise, a detector will introduce other noise. We say that the detection process is **quantum-limited**, if the ratio of the signal to the noise is limited by the fundamental quantum-mechanical noise, and all other sources of noise are smaller.

At optical frequencies, all detectors are **square-law detectors**; they respond to the power, rather than to the electric field amplitude. We also call the optical detection process **incoherent detection** or **direct detection**. Coherent detection can only be accomplished indirectly, using techniques from interferometry as discussed in connection with Table 7.1, and in Section 7.2. As discussed in those sections, the approach to solving these problems is to compute the irradiance that results from interference, and then recognize that the phase is related to this irradiance.

Two types of detectors; photon and thermal.

Detectors can be divided roughly into two types, photon detectors and thermal detectors, as shown in Figure 13.1. In a photon detector, each detected photon changes the state of an electron in some way, and the electron is detected. We will see that the responsivity of a photon detector includes a function that is proportional to wavelength.

DOI: 10.1201/9781315157047-13

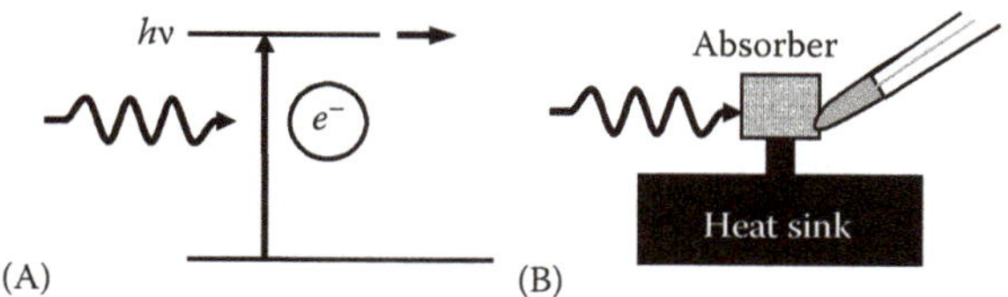

FIGURE 13.1
Optical detection. In a photon detector, an incident photon is absorbed, changing the state of an electron that is subsequently detected as part of an electrical charge (A). The current (charge per unit time) is proportional to the number of photons per unit time. In a thermal detector, light is absorbed, and the resulting rise in temperature is measured (B). Temperature is proportional to energy or power.

In a thermal detector, the light is absorbed by a material, and produces a temperature increase. The temperature is detected by some sensor. Assuming perfect absorption, or at least absorption independent of wavelength, the responsivity of a thermal detector is therefore constant across all wavelengths.

We will discuss the photon detectors first, beginning with a discussion of the quantization of light (Section 13.1), followed by a discussion of photon statistics (Section 13.2), a noise model (Section 13.3), and development of the equations for detection (Section 13.4). We then discuss thermal detectors in Section 13.5 and conclude with a short discussion of detector arrays (Section 13.6).

Detection alone is the subject of many texts, and this chapter is only a brief overview of the subject. Two outstanding books may be useful for further information[65;119]. More recently, in the field of infrared imaging detectors, a more specific book is available[190]. However, particularly in the area of array detectors, the technology is advancing rapidly, and the reader is encouraged to seek the latest information in papers and vendor's catalogs. The purpose of this chapter is to provide a framework for understanding the deeper concepts discussed elsewhere.

FURTHER READING:
Two references on optical detection are books by Dereniak[65] and Kingtson[119].

Take-Away

Detector Basics:

Optical detectors are square-law devices; they respond to the irradiance or square of the field.

Detectors can be divided into photon detectors that generate electrons when photons are absorbed and thermal detectors that increase their temperature when photons are absorbed.

There is a fundamental limit on signal-to-noise ratio for a given amount of light, rooted in the discrete nature of light.

13.1 Photons

The hypothesis that light is somehow composed of indivisible "particles" was put forward by Einstein in 1905[60], in one of his "five papers that changed the face of physics" gathered into a single collection in 2005[217]. He provided a simple but revolutionary explanation of an experiment, which is still a staple of undergraduate physics laboratories. Light is incident on a metal photocathode in a vacuum. As a result of the energy absorbed from the light, electrons are emitted into the vacuum. Some of them find their way to an **anode** and are collected as a photocurrent. The anode is another metallic component, held at a positive voltage in normal operation. Not surprisingly, as the amount of light increases, the photocurrent increases. The goal of the experiment is to measure the energies of these so-called **photoelectrons**. For the experiment, a negative voltage, V, is applied to the anode and increased until the current just stops. The energy of the most energetic electron must have been eV, where e is the charge on an electron. The surprising result of this experiment is that the required nulling voltage is independent of the amount of light, but is linearly dependent on its frequency.

$$eV = h\nu - \Phi_c, \tag{13.3}$$

where

- ν is the frequency of the light c/λ
- h is Planck's constant
- Φ_c is a constant called the **work function** of the photocathode

Einstein's hypothesis to explain this result was that light is composed of individual quanta, each having energy, $J_1 = h\nu$, rather than being composed of something infinitely divisible. In 1916, Millikan called Einstein's hypothesis

> bold, not to say ... reckless first because an electromagnetic disturbance which remains localized in space seems a violation of the very conception of an electromagnetic disturbance, and second because it flies in the face of the thoroughly established facts of interference. The hypothesis was apparently made solely because it furnished a ready explanation of one of the most remarkable facts brought to light by recent investigations, vis., that the energy with which an electron is thrown out of a metal by ultraviolet light or X-rays is independent of the intensity of the light while it depends on its frequency[158].

The concept that a wave could somehow be quantized is now readily accepted in quantum mechanics, but still remains a source of puzzlement and confusion to many.

Despite his other contributions to science, Einstein's Nobel Prize in 1921 was for his explanation of the photoelectric effect in Equation 13.3. It can truthfully be said that Einstein won the Nobel Prize for the equation of a straight line. This hypothesis was in agreement with Planck's earlier conclusion about quantization of energy in a cavity[177], with the same constant, h, which now bears Planck's name.

13.1.1 Photocurrent

If an electron is detected for each of n photons incident on the photocathode, then the total charge, q detected will be related to the total energy, J, by

$$q = ne \quad n = \frac{J}{h\nu} \quad q = \frac{e}{h\nu}J, \tag{13.4}$$

or if only a fraction, η, of the photons result in electrons, the number of electrons is

$$n_e = \frac{\eta}{h\nu}J. \tag{13.5}$$

and the charge is

$$q = \frac{\eta e}{h\nu}J. \tag{13.6}$$

Photocurrent in a detector is proportional the the incident optical power.

Differentiating both sides with respect to time, we relate the electrical current $i = dq/dt$ to the optical power, $P = dJ/dt$:

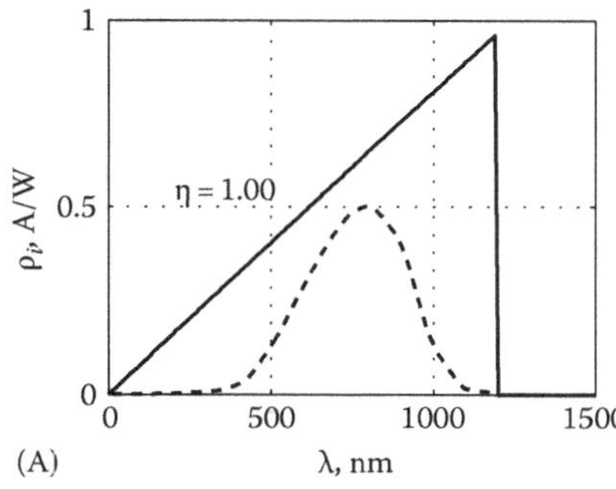

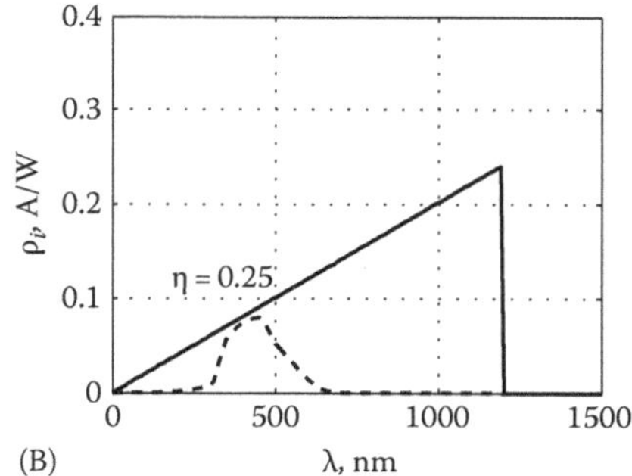

FIGURE 13.2
Current responsivity. Spectral shapes are influenced by quantum efficiency, window materials, and other considerations. The silicon photodiode (A) has a relatively high quantum efficiency at its peak (over 80% is common). The photocathode in a photomultiplier tube (B) usually has a much lower quantum efficiency. The solid line in (A) is for $\eta = 1$ and that in (B) is for $\eta = 0.25$.

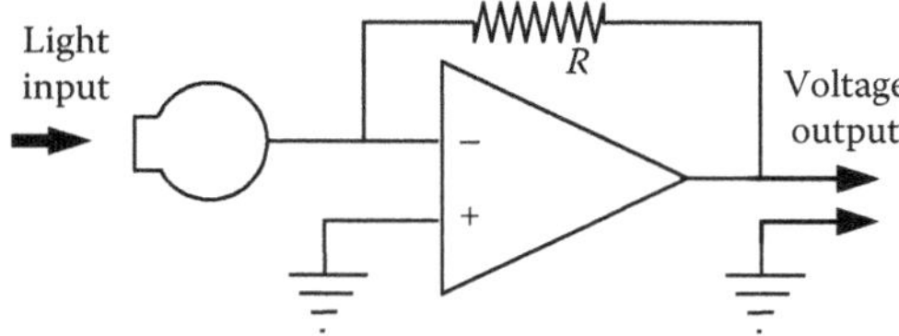

FIGURE 13.3
Detector circuit. A transimpedance amplifier converts the detector current to a voltage. The conversion factor is the resistance of the feedback resistor, R.

$$i = \rho_i P \quad \rho_i = \frac{\eta e}{h\nu} \quad \text{(Photon detector).} \tag{13.7}$$

The constant, η, is called the **quantum efficiency**. Assuming constant quantum efficiency, the responsivity of a photon detector is inversely proportional to photon energy (and frequency), and thus proportional to wavelength. Figure 13.2 shows two examples of detector current responsivity for typical laboratory detectors.

Quantum efficiency is the probability that absorption of a photon will result in an electron being detected.

Generally, we want to convert the current to a voltage. Often a circuit such as Figure 13.3 is used. Recalling the rules for operational amplifiers from any electronics text[89], (1) there is no current into either input of the amplifier and (2) there is no voltage across the inputs; the amplifier will, if possible, produce an output voltage to satisfy these two conditions. Thus, (by 1) all the detector current passes through the resistor, R, and (by 2) the voltage across the detector is zero, in what is called a **virtual ground**. The output then is

$$V = iR = \rho_i RP. \tag{13.8}$$

If the amplifier and detector are packaged as a single unit, the vendor may specify the voltage responsivity

$$\rho_v = R\rho_i. \tag{13.9}$$

13.1.2 Examples

Example

Detector responsivity at two wavelengths

One microwatt of green light includes, on average, about $P/(h\nu) = 2.5 \times 10^{12}$ photons per second. A microwatt of carbon dioxide laser light at $\lambda = 10.59\ \mu\text{m}$ contains about 20 times as many photons of correspondingly lower energy. The variation in this rate determines the limiting noise associated with a measurement, and will be discussed shortly. For now, let us look at the mean values.

The maximum current responsivity for any detector without gain is

$$\rho_i = \frac{e}{h\nu} = 0.40\ \text{A/W (green)} \quad \text{or} \quad 8.5\ \text{A/W}\ (\text{CO}_2) \tag{13.10}$$

and the maximum photocurrent for $1\ \mu\text{W}$ is

$$\bar{i} = \rho_i P = 400\ \text{nA (green)} \quad \text{or} \quad 8.5\ \mu\text{W}\ (\text{CO}_2). \tag{13.11}$$

If the current is amplified by the transimpedance amplifier in Figure 13.3 with a transimpedance (determined by the feedback resistor) of $R = 10\ \text{k}\Omega$, then the voltage is

$$\bar{v} = \bar{i}R = 4.0\ \text{mV (green)} \quad \text{or} \quad 85\ \text{mV}\ (\text{CO}_2). \tag{13.12}$$

Take-Away

Wavelength Dependence:

For longer wavelengths, the photon energy is lower, so a given amount of optical energy includes more photons.

Because a photon detector produces electrons in quantity proportional to the quantity of photons, longer wavelengths result in higher current responsivity for the same optical power.

13.2 Photon Statistics

We want to develop a basic understanding of photon statistics so that we can better understand noise issues in detection. Much of the fundamental research in this field was developed by Roy J. Glauber, who was awarded the Nobel Prize in 2005 "for his contribution to the quantum theory of optical coherence." He shared the prize, with the other half going to John L. Hall and Theodor W. Hänsch, "for their contributions to the development of laser-based precision spectroscopy."

Consider the photon counting experiment shown in Figure 13.4. Light from the source is detected by some type of detector. Knowing that the light arrives in quanta called photons, we perform a counting experiment. A clock controls a switch that connects the detector to a counter for a prescribed length of time. At the end of the time interval, the counter reports its result, the clock resets it, and we repeat the experiment. Suppose, for example, that the laser power is such that we expect one photon per microsecond. For green light, the power level is about 400 fW, so in practice, the laser beam would be strongly attenuated before making this measurement. Now if we close the switch for 10 μs, we expect on average, 10 photons to be counted. We repeat the experiment a number of times, as shown on the left in Figure 13.5. At the end of each clock interval, we record the number of photons counted. We then count how many times each number occurs, and plot the probability distribution, $P(n)$. Will we always count exactly 10, or will we sometimes count 9 or 11? The answer, as we shall see, is considerably worse than that.

In Practice

It is possible to build or buy photon counters which consist of fast detectors and electronics to implement the experiment shown in Figure 13.4. They are often used to detect light at low levels, and to study photon statistics.

13.2.1 Mean and Standard Deviation

Here we derive the Poisson distribution for photons. Coherent light sources such as lasers follow this distribution very well. The reader simply looking for an equation to describe the fluctuations in such a light source may skip directly to Equation 13.16.

We will consider a light source that we would think of as "constant" in the classical sense, with a power, P. If we consider a very short time interval, dt, there are only two possibilities. First there is a small probability that a photon will be detected. We assume that this probability is

- Independent of time (the source is constant)
- Independent of previous photon detections

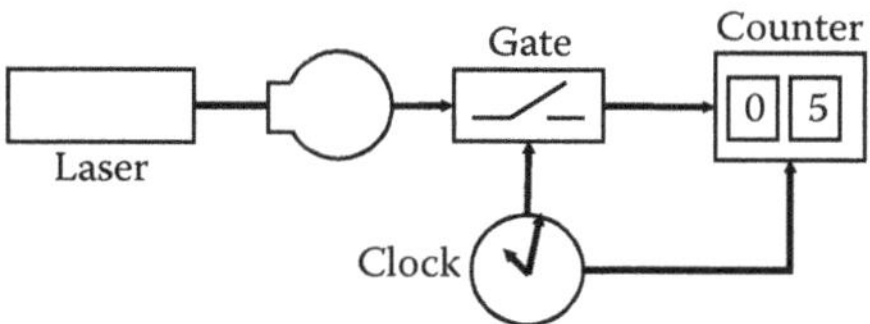

FIGURE 13.4
Photon counting experiment. Photons entering the detector result in electrons in the circuit. The electrons are counted for a period of time, the count is reported, and the experiment is repeated many times to develop statistics.

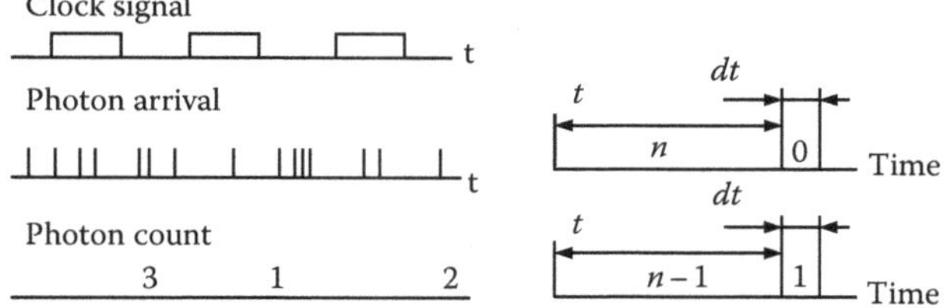

FIGURE 13.5
Timing for photon counting. In the example on the left, the clock signals and photon arrivals are shown. At the end of each clock interval, the count is reported. On the right, dt is a short time interval so that the number of photons is 1 or 0. There are two ways to count n photons in time $t + dt$.

- Proportional to the source power
- Proportional to the duration of the interval, dt

Therefore, we call this probability $\alpha\, dt$ where α is a constant. The only other possible result is that no photon is detected. The probability of two or more photons arriving in this time is vanishingly small. Unless the distribution of photons proved to be extremely unusual, it should be possible to make dt small enough to satisfy this assumption. Now, if we append this time interval to a longer time t, and ask the probability of counting n photons in the extended time interval, $t + dt$, we must consider two cases, as shown on the right in Figure 13.5. Either we detect n photons in time t and none in time dt, or we detect $n - 1$ photons in time t and one in time dt. Thus,

$$P(n|t + dt) = P(n|t)(1 - \alpha dt) + P(n - 1|t)\, \alpha dt \tag{13.13}$$

or

$$\frac{P(n|t + dt) - P(n|t)}{dt} = -P(n|t)\, \alpha + P(n - 1|t)\, \alpha.$$

We thus have a recursive differential equation

$$\frac{dP(n)}{dt} = P(n)\, \alpha - P(n - 1)\, \alpha, \tag{13.14}$$

except of course that

$$\frac{dP(0)}{dt} = -P(0)\, \alpha. \tag{13.15}$$

Photons from what we consider a "constant" source such as a laser have a Poisson distribution.

The recursive differential equation in Equation 13.14 has the solution

$$P(n) = \frac{e^{-\bar{n}}\bar{n}^n}{n!}, \tag{13.16}$$

with a mean value, $\bar{n}$. Given that the energy of a photon is $h\nu$, we must have

$$\bar{n} = \frac{Pt}{h\nu}, \tag{13.17}$$

for a power P during a time t. More usefully, we consider just those photons that are actually detected:

$$\bar{n} = \eta\frac{Pt}{h\nu}, \tag{13.18}$$

where η is the probability that an incident photon is actually detected, meaning that it is absorbed and excited an electron that can be measured. We refer to η as the **quantum efficiency**, .

Figure 13.6 shows several Poisson distributions. For a low mean number, $\bar{n} = 0.1$, the most probable number is 0, and the probability of any number greater than one is quite small. The probability of exactly one photon is very close to the mean number, 0.1. For $\bar{n} = 1$, the probabilities of 0 and 1 are equal, and there is a decreasing probability of larger numbers.

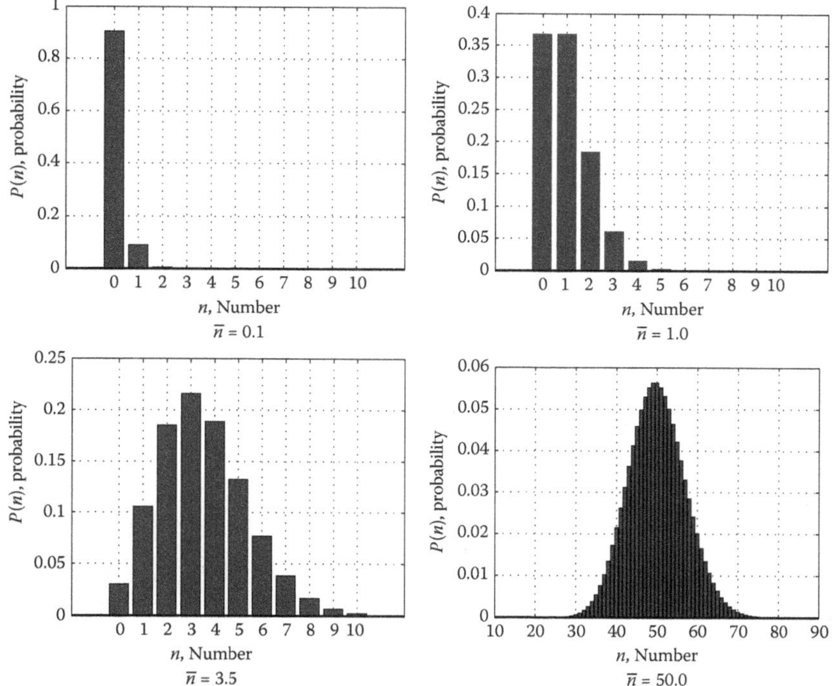

FIGURE 13.6
Poisson distribution. For $\bar{n} \ll 1$, $P(n) \approx \bar{n}$. For $n \gg 1$ the distribution approaches a Gaussian. Note the horizontal scale change in the last panel.

For a mean number of $\bar{n} = 3.5$, there is a wide range of numbers for which the probability is large. The probability of a measurement much larger than the mean, or much smaller, including 0, is quite large. Thus the photons appear to bunch into certain random intervals, rather than to space themselves out with three or four photons in every interval. Photons are much like city busses. When the stated interval between busses is 5 min, we may stand on the corner for 15 min and watch three of them arrive nearly together.

For larger numbers, the distribution, although discrete, begins to look like a Gaussian. In fact, the central-limit theorem states that the distribution of a sum of a large number of random variables is Gaussianly distributed. The variable, n, for a large time interval, because of the independence on previous history, is merely a sum of the many smaller random numbers of photons collected in smaller time intervals.

Figure 13.7 shows the probability of detecting at least one photon (solid curve):

$$\sum_{n=1}^{\infty} P(n) = 1 - P(0) = 1 - e^{-\bar{n}} \tag{13.19}$$

or at least two photons (dashed curve):

$$\sum_{n=2}^{\infty} P(n) = 1 - P(0) - P(1) = 1 - e^{-\bar{n}}\bar{n} - e^{-\bar{n}}. \tag{13.20}$$

For small mean numbers, the probability of at least one photon is $\bar{n}$, proportional to power:

$$1 - P(0) \approx \bar{n} \quad \bar{n} \ll 1 \tag{13.21}$$

and the probability of at least two is proportional to $\bar{n}^2$ and the square of the power

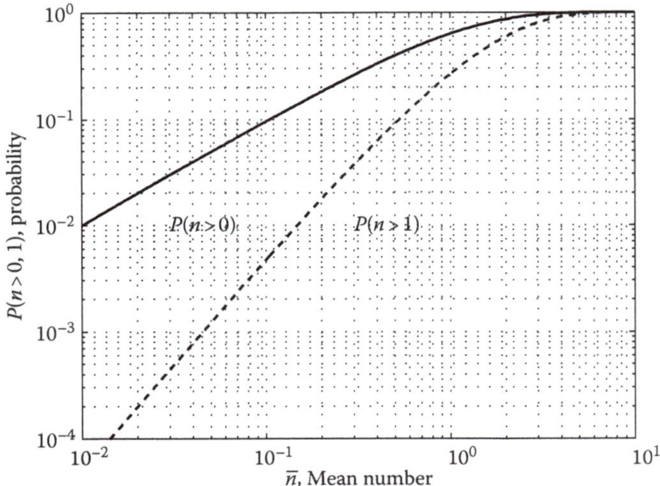

FIGURE 13.7
Probability of at least one or two photons. For low mean numbers, $\bar{n} \ll 1$ the probability of detecting at least one photon, $P(n > 0)$ (solid line), is equal to the mean number. The probability of at least two photons, $P(n > 1)$ (dashed line), is equal to the square of the mean number.

according to

$$1 - P(0) - P(1) \approx \frac{\bar{n}^2}{2} \quad \bar{n} \ll 1 \tag{13.22}$$

where the approximations are computed using $e^{-\bar{n}} \approx 1 - \bar{n} + \bar{n}^2/2$.

We can think of the mean number, $\bar{n}$, as a signal and the fluctuations about that mean as noise. These fluctuations can be characterized by the standard deviation:

$$\sqrt{\overline{n^2} - \bar{n}^2} = \sum_{m=0}^{\infty} n^2 P(n) - \left[\sum_{m=0}^{\infty} nP(n)\right]^2.$$

The standard deviation of the Poisson distribution is $\sqrt{\bar{n}}$.

Using the Poisson distribution,

$$\sqrt{\overline{n^2} - \bar{n}^2} = \sqrt{\bar{n}} \quad \text{(Poisson distribution)}. \tag{13.23}$$

Take-Away

Poisson Distribution:

Because photons from a classically constant source are Poisson-distributed, there is an uncertainty in the measurement of $\bar{n}$ photons with a standard deviation $\sqrt{\bar{n}}$. Every measurement will result in a different number with a mean, $\bar{n}$ and standard deviation, $\sqrt{\bar{n}}$. The variation is a source of noise that limits our ability to make an accurate measurement.

13.2.2 Example

Example

Quantum noise current

Let us return to the example from Section 13.1.2, where we computed mean voltages that would be measured for two light sources in a hypothetical detector circuit. Now let us compute the standard deviation of these voltages. We need to take the square root of the mean number of photons, $\bar{n}$, and we know that $\bar{n} = \eta Pt/(h\nu)$, but what do we use for t? Let us assume that the amplifier has a bandwidth, B, and use $t = 1/B$. Then

$$\bar{n} = \eta \frac{P}{h\nu B} \qquad \sqrt{\bar{n}} = \sqrt{\eta \frac{P}{h\nu B}}. \tag{13.24}$$

The fluctuations in the number of photons result in a fluctuation in the current of

$$i_n = \frac{\sqrt{\bar{n}}}{t} e = \sqrt{\bar{n}}\ Be = \sqrt{\frac{\eta P}{h\nu}} Be \tag{13.25}$$

or as a voltage:

$$v_n = \sqrt{\eta \frac{P}{h\nu} B}\ eR. \tag{13.26}$$

We refer to this noise, caused by quantum fluctuations in the signal, as **quantum noise**. We say that the detection process is **quantum-limited** if all other noise sources have lower currents than the one given by Equation 13.25.

Table 13.1 summarizes the results.

The **signal-to-noise ratio, (SNR)** is computed here as the ratio of electrical signal power to electrical noise power. Thus the SNR is given by

$$SNR = \frac{\bar{v}^2}{v_n^2} = \left(\frac{\bar{n}}{\sqrt{\bar{n}}}\right)^2 = \bar{n} \quad \text{(Quantum-limited)}. \tag{13.27}$$

The SNR is better at the longer wavelength by a factor of 20 (13 dB), because more photons are required to produce the same amount of power. If the mean number increases by 20, the standard deviation increases by $\sqrt{20}$, so the SNR increases by

$$\left(20/\sqrt{20}\right)^2 = 20.$$

It is important to note that some authors define SNR as a ratio of optical powers, while we, as engineers, have chosen to use the ratio of electrical powers, which would be measured on typical electronic test equipment.

In this book SNR means the ratio of electrical signal to electrical power.

Take-Away

Photon Statistics:
The standard deviation of the number of photons in a time interval is equal to the square root of the mean.

TABLE 13.1
Signals and Noise

Wavelength	λ	500 nm	10.59 μm
Example detector		Si	HgCdTe
Photons per second	$P/(h\nu)$	2.5×10^{12}	5.3×10^{13}
Current responsivity	ρ_i	0.40 A/W	8.5 A/W
Mean current	$\bar{i}$	400 nA	8.5 μA
Mean voltage	$\bar{v}$	4 mV	85 mV
Mean photon count	$\bar{n}$	2.5×10^5	5.3×10^6
Standard deviation	$\sqrt{\bar{n}}$	500	2300
Quantum noise current	i_n	800 pA	3.6 nA
Quantum noise voltage	v_n	8.0 μV	37 μV
SNR	$10\log_{10}\left(\frac{\bar{v}}{v_n}\right)^2$	54 dB	67 dB

Signal and noise values are shown for 1 μW of light at two wavelengths. The bandwidth is 10 MHz, $\eta = 1$ is assumed, and the transimpedance is $R = 10$ kOhms.

Thus the quantum-limited signal-to-noise ratio is equal to the mean number. We define SNR as the square of the number of photons, because that is related to the electrical power.

In Practice

The quantum noise current here is a result of the quantum nature of light, and stands as the theoretical limit for SNR. We call it the quantum limit or photon limit. Actual SNR will be lower because of (1) photons that are not detected ($\eta < 1$) and (2) other sources of noise.

Thus the quantum limit is often used as a "target." It may be possible to come very close to the limit in photon-counting devices, coherent detection systems, and others. We might compare a measured SNR to this theoretical limit, and then improve the system to better approach it.

However, it is usually challenging to approach the limit, and many useful optical devices are not quantum-limited.

13.3 Detector Noise

Inevitably, there will be some other noise in the detection process. In fact, if the light source is turned off, it is likely that we will measure some current in the detector circuit. More importantly, this current will fluctuate randomly. We can subtract the mean value, but the fluctuations represent a limit on the smallest signal that we can detect with confidence. We will not undertake an exhaustive discussion of noise sources, but it is useful to consider a few.

13.3.1 Thermal Noise

First, let us consider thermal noise in the feedback resistor. Electrons will move randomly within the resistor because of thermal agitation. This motion will result in a random current called **Johnson noise**[114], with a mean value of zero and an RMS value of

$$i_n^2 = \frac{4k_BTB}{R} \quad \text{(Johnson noise)}, \tag{13.28}$$

where

- $k_B = 1.3807 \times 10^{-23}\,\mathrm{J/K}$ is the Boltzmann constant
- T is the absolute temperature

As shown in Figure 13.8, this current is added to the signal current. For the example detector considered in Table 13.1, the noise at $T = 300\,\mathrm{K}$ in the $10\,\mathrm{k\Omega}$ resistor with a bandwidth of 10 MHz is

$$i_n = \sqrt{\frac{4k_BTB}{R}} = 4.1\,\mathrm{nA} \quad v_n = \sqrt{4k_BTBR} = 41\,\mu\mathrm{V}, \tag{13.29}$$

which is considerably higher than the photon noise for the visible detector in Table 13.1; this detector will not be quantum-limited.

13.3.2 Noise-Equivalent Power

It is often useful to ask how much power would have been required at the input to the detector, to produce a signal equivalent to this noise. This power is called the **noise-equivalent power**, or **NEP**. In general,

$$NEP = \frac{i_n}{\rho_i}, \tag{13.30}$$

and specifically for Johnson noise,

$$NEP = \sqrt{\frac{4k_BTB}{R}}\frac{h\nu}{\eta e} \quad \text{(Johnson noise)}. \tag{13.31}$$

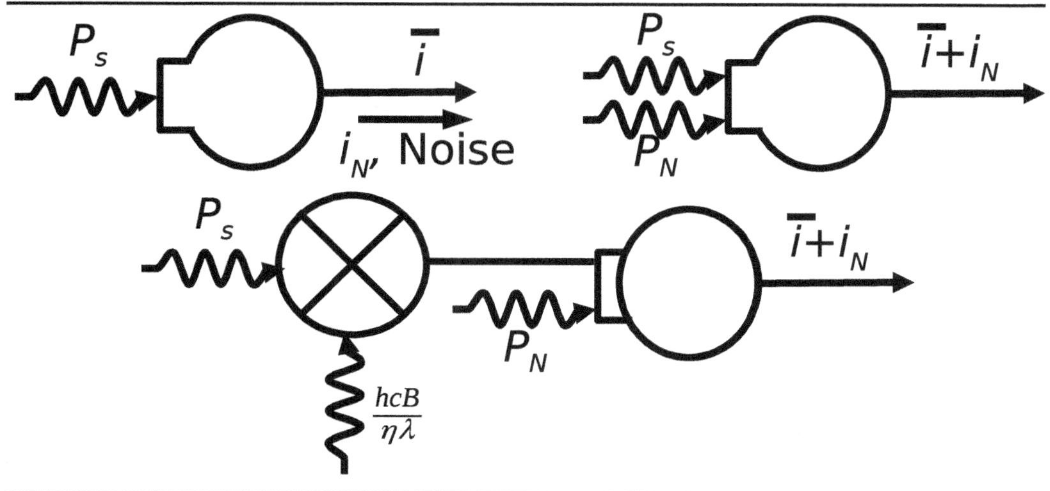

FIGURE 13.8
Noise models. Top: Noise currents are added to the signal. On the right, a "noise-equivalent power" is added to simulate the noise current. Bottom: Photon noise is simulated by adding one photoelectron per reciprocal bandwidth coherently.

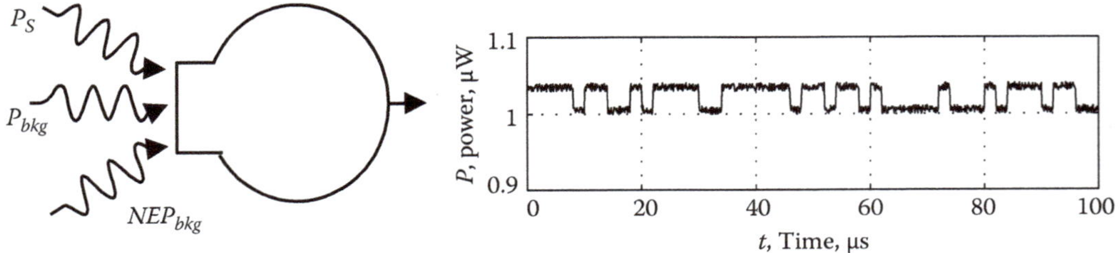

FIGURE 13.9
Switched signal. The small signal is shown on top of a DC background, along with quantum fluctuations of that background.

Several other sources of noise may also arise in detection. A typical detector specification will include the NEP, or more often the NEP per square root of bandwidth, so that the user can design a circuit with a bandwidth suitable for the information being detected, and then compute NEP in that bandwidth.

Eq. 13.25: $i_n = \frac{\sqrt{\bar{n}}}{t}e = \sqrt{\bar{n}}\ Be = \sqrt{\frac{\eta P}{h\nu}}Be.$

Applying the same approach to Equation 13.25,

$$NEP = \sqrt{\frac{\eta P}{h\nu}B}\ e\ \frac{h\nu}{\eta e} = \sqrt{P\frac{h\nu B}{\eta}} \quad \text{(Quantum noise).} \qquad (13.32)$$

This equation looks much like our mixing terms in Equation 7.5:

$$|I_{mix}| = |I^*_{mix}| = \sqrt{I_1 I_2},$$

using $I = P/A$. The NEP can be calculated by coherently mixing the signal power, $P_1 = P$, with a power $P_2 = h\nu B/\eta$, sufficient to detect one photon per reciprocal bandwidth. This model of quantum noise may make the equations easier to remember, and it is a model similar to the one obtained through full quantum-mechanical analysis. The bottom part of Figure 13.8 illustrates this idea conceptually.

13.3.3 Example

Returning to the example in Section 13.1.2, let us assume that the 1 μW incident on the detector is a background signal. Perhaps we have an outdoor optical communication system, in which we have placed a narrow-band interference filter (Section 7.7) over the detector to pass just the source's laser line. If the source is switched on and off to encode information, we will then measure a signal like that shown in Figure 13.9. The small signal, of 2 nW in this example is added to the larger background of 1 μW, along with the fluctuations in the background. Because the signal switches between two discrete values, we may be able to recover it, even in the presence of the large background. The question is whether the small signal power is larger than the combined NEP associated with fluctuations in the background and Johnson noise. The results are shown in Table 13.2, for the two wavelengths we considered in Section 13.1.2. At the green wavelength, the NEP is dominated by Johnson noise, and the signal is smaller than the noise with $SNR = -14$ dB. For the CO_2 laser line, with the same powers, the two contributions to the NEP are nearly equal, and the signal is larger, with $SNR = 9.9$ dB. We could improve the situation a little for the IR detector by reducing the temperature to 77 K, the temperature of liquid nitrogen, bringing the Johnson noise down by a factor of almost four.

If the background NEP were the largest contribution to the noise, we would say that the detector is **background-limited**. The limit is the result of fluctuations in the background,

TABLE 13.2
Switched Input

Wavelength	λ	500 nm	10.59 μm
Example detector		Si	HgCdTe
Background power	P_{bkg}	1 μW	1 μW
Mean voltage	$\bar{v}$	4 mV	85 mV
Background noise voltage	v_n	8.0 μV	37 μV
Background NEP	NEP_{bkg}	2.0 nW	430 pW
Johnson noise voltage	v_n	41 μV	41 μV
Johnson NEP	NEP_J	10 nW	480 pW
Total NEP	NEP	10 nW	640 pW
Signal power	P_{sig}	2 nW	2 nW
SNR		−14 dB	9.9 dB

The signal now is 2 nW, but it is superimposed on a background of 1 μW. The bandwidth is 10 MHz, and $\eta = 1$ is assumed.

and not its DC level. However, if the background DC level is high, it may saturate the detector or the electronics. For example, if the background were 1 mW instead of 1 μW, then the background mean voltage would be 85 V for the CO_2 wavelength and the current would be 8.5 mA. In the unlikely event that this current does not destroy the detector, it is still likely to saturate the amplifier, and the small signal will not be observed. Even with smaller backgrounds, if we digitize the signal before processing, we may need a large number of bits to measure the small fluctuation in the presence of a large mean. In some cases, we can use a capacitor to **AC couple** the detector to the circuit, and avoid saturating the amplifier.

In Practice

Background light is often a serious problem. If we are interested in light within a narrow band of wavelengths we can filter the remaining light with a good interference filter (Section 7.7) to prevent saturating the detector bias circuit. If the signal varies at some known frequency, we may then be able to use capacitive coupling to block the electrical signal from the remaining DC background, so that the amplifier is not saturated.

For the infrared detector, one limit of performance is the background power from thermal radiation. Viewing the laser communication source in our example, the detector will see thermal radiation from the surroundings as discussed in Section 12.5. To reduce this power, we can place a mask over the detector to reduce the field of view. However, we will then see background thermal radiation from the mask. To achieve background-limited operation, we must enclose the mask with the detector inside a cooled dewar, to reduce its radiance. We say that the detector is background-limited if the noise is equal to the quantum fluctuations of the background radiance integrated over the detector area, A, and collection solid angle, Ω,

$$P_{bkg} = \iiint L_{\lambda:T}\, dA\, d\Omega\, d\lambda. \tag{13.33}$$

Following the rule of mixing this with one detected photon per reciprocal bandwidth in Equation 13.32,

$$NEP_{bkg} = \sqrt{\iiint \frac{hcB}{\eta\lambda} L_{\lambda:T}\, dA\, d\Omega\, d\lambda}. \tag{13.34}$$

It is common, particularly in infrared detection, where this noise associated with background is frequently the limiting noise source, to describe a detector in terms of a figure of merit, D^*, or **detectivity**, according to

Detectivity, D^*, is a common figure of merit for infrared detectors.

$$NEP_{bkg} = \frac{\sqrt{AB}}{D^*}. \tag{13.35}$$

The goal of detector design is to make D^* as large as possible while satisfying the other requirements for the detector. The upper limit, making full use of the numerical aperture in a wavelength band $\delta\lambda$, is

$$\left(D^*_{max}\right)^2 = \frac{1}{\pi \frac{hcB}{\eta\lambda} L_{\lambda:T} \delta\lambda}, \tag{13.36}$$

where the $d\Omega$ integral is completed using Equation 12.11.

Eq. 12.11:
$M(x,y) = \int\int L(x,y,\theta,\zeta) \cos\theta \sin\theta \, d\theta \, d\zeta$.

Take-Away

Noise:
Noise sources include photon noise, thermal noise, background noise, and more.

Fig 13.1A

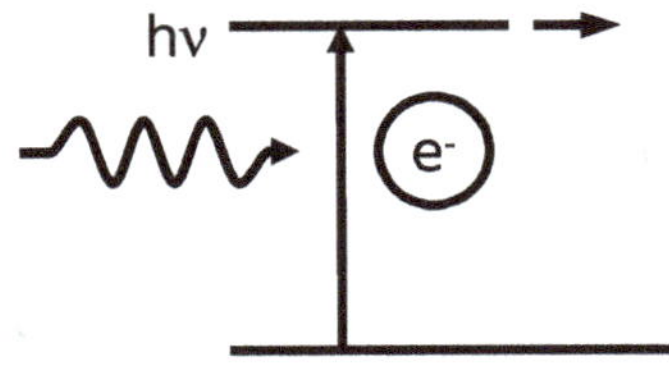

Fig. 13.2

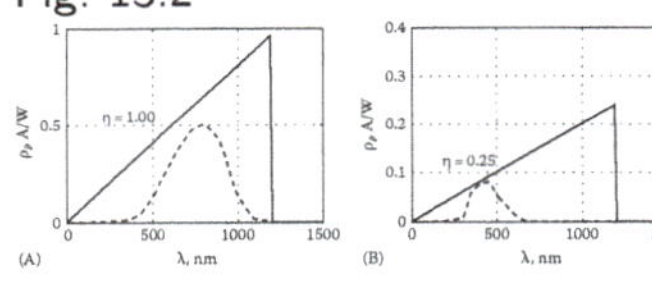

13.4 Photon Detectors

There are many different types of detectors that fall under the heading of photon detectors. We saw their general principle of operation in Figure 13.1A, and showed some examples of their responsivity in Figure 13.2. An electron is somehow altered by the presence of a photon, and the electron is detected. In a **vacuum photodiode**, the electron is moved from the photocathode into the vacuum, where it is accelerated toward an anode and detected. The modern photomultiplier in Section 13.4.1 uses this principle, followed by multiple stages of amplification. In the semiconductor detector, the electron is transferred from the valence band to the conduction band and is either detected as a current under the action of an externally applied voltage in a photoconductive detector or modifies the I–V curves of a semiconductor junction in a photodiode.

13.4.1 Photomultiplier

The photoelectric effect described conceptually in Section 13.1 describes the operation of the vacuum photodiode or the first stage of the photomultiplier. In the photomultiplier, multiple dynodes are used to provide gain. The dynode voltages are provided by a network of electronic components. The photomultiplier is very sensitive and can be used to count individual photons.

13.4.1.1 Photocathode

Light is incident on a **photocathode** inside an evacuated enclosure. If the photon energy, $h\nu$, is sufficiently large, then with a probability, η, the photon will be absorbed, and an electron will be ejected into the vacuum. Thus, the quantum efficiency, η (Equation 13.7) characterizes the combined probability of three events. First the photon is absorbed as

Eq. 13.7:
$i = \rho_i P \qquad \rho_i = \frac{\eta e}{h\nu}$.

opposed to reflected. The quantum efficiency cannot exceed $(1 - R)$ where R is the reflectivity at the surface. Second, an electron is emitted into the vacuum and third, it is subsequently detected. The emission of an electron requires a low work function as defined in Equation 13.3. Detection requires careful design of the tube so that the electron reaches the anode. The **cutoff wavelength** is defined as the wavelength at which the emitted electron energy in Equation 13.3 goes to zero.

Eq. 13.3:
$eV = h\nu - \Phi_c$.

The work function determines the cutoff wavelength. Only wavelengths shorter than the cutoff wavelength will be detected.

$$0 = h\nu_{cutoff} - \Phi_c \qquad \frac{hc}{\lambda_{cutoff}} = \Phi_c \qquad \lambda_{cutoff} = \frac{hc}{\Phi_c}. \tag{13.37}$$

There are many materials used for photocathodes to achieve both low reflectivity and low work function. The problem is complicated by the fact that good conductors tend to be highly reflective, and that work functions of metals tend to correspond to wavelengths in the visible spectrum. Alkali metals are common materials for photocathodes, and semiconductor materials such as gallium arsenide are common for infrared wavelengths. Considering the potential for noise, it is advantageous to pick a photocathode with a cutoff wavelength only slightly longer than the longest wavelength of interest. For example, if one is using a 1 μm wavelength laser for second-harmonic generation (see Section 14.4.2) then the emission will be at 500 nm, and the photomultiplier tube with responsivity shown in Figure 13.2B will detect the emitted light while being insensitive to the excitation and stray light at long wavelengths.

Fig. 13.2B:

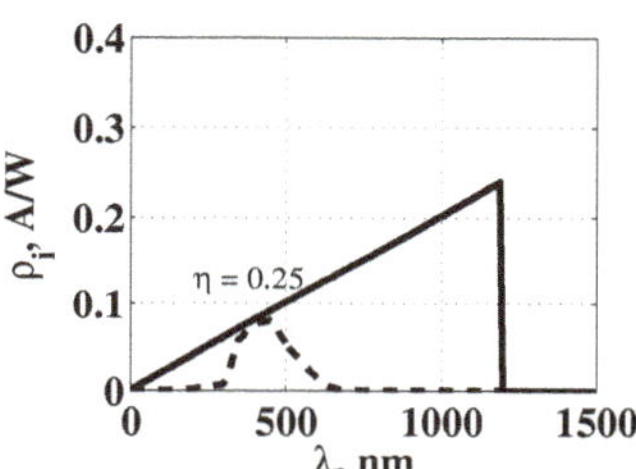

13.4.1.2 Dynode Chain

In a vacuum photodiode, we would simply apply a positive voltage to another element inside the vacuum envelope, called the anode, and measure the current. However, the photoelectric effect can be used at light levels corresponding to a few photons per second, and the resulting currents are extremely small and difficult to measure. Photomultipliers employ multiple **dynodes** to provide gain. A dynode is a conductor that produces a number of **secondary electrons** when struck by an incident electron of sufficient energy. Figure 13.10 shows the concept. The electrons emitted from the photocathode are accelerated by the electric field from the photocathode at voltage V_0 toward the first dynode at V_1 with a force

$$ma = \frac{V_1 - V_0}{z_{01}} e, \tag{13.38}$$

until they reach the dynode at a distance z_{01}:

$$a = \frac{V_1 - V_0}{z_{01}} \frac{e}{m} \qquad z_{01} = \frac{1}{2} a t^2 = \frac{V_1 - V_0}{2 z_{01}} \frac{e}{m} t^2, \tag{13.39}$$

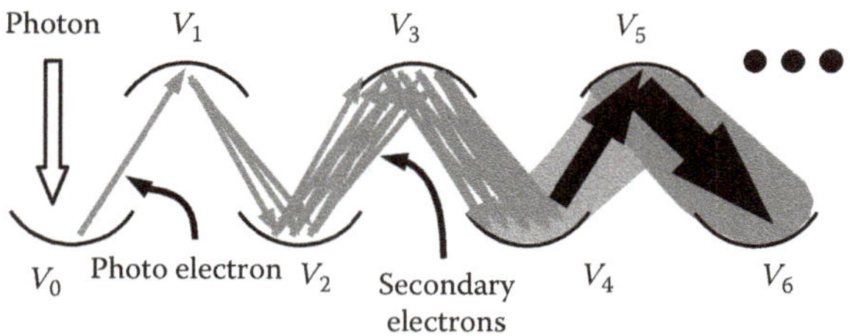

FIGURE 13.10
Photomultiplier. The photomultiplier tube consists of a photocathode, a number of dynodes, and an anode. At each dynode, electron multiplication provides gain.

arriving at a time

$$t = z_{01}\sqrt{\frac{2}{V_1 - V_0}\frac{m}{e}} = z_{01}\sqrt{\frac{2}{V_1 - V_0}\frac{9.11 \times 10^{-31}\,\text{kg}}{1.60 \times 10^{-19}\,\text{C}}}.$$

$$= \frac{z_{01}}{2.9655 \times 10^5\,\text{m/s}}\sqrt{\frac{1\,\text{V}}{V_1 - V_0}}. \tag{13.40}$$

where

- m is the electron mass
- a is the acceleration

At several tens of volts, distances of millimeters will be traversed in nanoseconds. At higher voltages, the speed approaches c, and relativistic analysis is required. Incidentally, with electrons moving at such high velocities, the effects of magnetic fields cannot be neglected, and photomultipliers may require special shielding in high magnetic fields.

The voltage difference between dynodes adds kinetic energy to the electrons. The resulting collision emits many secondary electrons.

The voltage difference defines the change in energy of the charge, so the electrons will arrive at the dynode with an energy $(V_1 - V_0)\,e$. One of these electrons, arriving with this energy, can cause **secondary emission** of electrons from the dynode. We would expect, on average N_1 secondary electrons if

$$N_1 = \frac{V_1 - V_0}{\Phi_c}e, \tag{13.41}$$

where Φ_c is the work function of the dynode material. In practice the process is more complicated, and above some optimal voltage, additional voltage increase can actually decrease the number of secondary electrons. At the voltages normally used, this functional relationship is, however, a good approximation. From the first dynode, the secondary electrons are accelerated toward a second dynode, and so forth, until the **anode**, where they are collected.

Voltages are applied to the dynodes using a chain of resistors and other circuit components called the **dynode chain**. A typical example is shown in Figure 13.11. The physics of the photomultiplier requires only that each successive dynode be more positive than the one before it by an appropriate amount. In most cases, the anode is kept at a low voltage, simplifying the task of coupling to the rest of the detector circuit. The cathode is then kept at a high negative voltage, and a resistor chain or other circuit is used to set the voltage at each dynode. Often, but not always, this **dynode chain** consists of resistors with identical values. If the currents arising from electron flow in the PMT, shown as current sources in the figure, are small, then the voltages between adjacent dynodes will all be the same. To maintain this condition, the resistors must be small enough so that the resistor current exceeds the current in the photomultiplier. The smaller resistors, however, result in greater power dissipation in each stage, V_{mn}^2/R_{mn}, and this power is all dissipated as heat. Alternative bias circuits including active ones that provide more constant voltage with lower power consumption are also used.

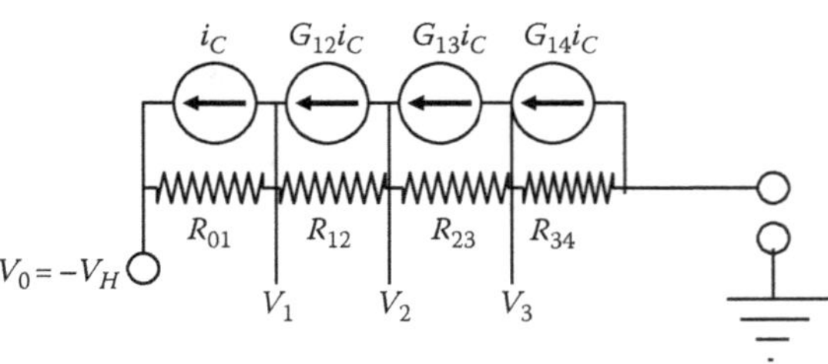

FIGURE 13.11
Dynode chain. The photocurrent in each stage is larger than in the one before.

If there are M dynodes with equal voltages, the PMT gain is

$$i_{anode} = G_{PMT} \times i_{cathode} \quad G_{PMT} = N_1^M. \tag{13.42}$$

With 10–14 dynodes, gains from 10^6 to 10^8 are possible. The output is best measured by placing a transimpedance amplifier between the anode and ground, as in Figure 13.3. The voltage is then

$$V = Ri_{anode} = RG_{pmt}\frac{\eta e}{h\nu}P_{optical}, \tag{13.43}$$

where R is the transimpedance, or the resistance of the feedback resistor.

Alternatively, a small resistor may connect the anode to ground, and the anode current may be computed from the voltage measured across this resistor. The value of the resistor must be kept small to avoid disturbing the performance of the dynode chain in setting correct voltages. In addition, the added resistor, in combination with capacitance of the tube may limit the bandwidth of the detector. Thus a transimpedance amplifier is usually the preferred method of coupling.

Take-Away

Photomultiplier:

The photomultiplier tube is a vacuum photodiode with gain. The dynodes with increasing positive voltages can provide large gain. It is a very sensitive detector.

In Practice

Some photomultipliers usually look like old-fashioned vacuum tubes. Some are designed to receive light from the side and some from the end.

It is possible to buy a package containing a photomultiplier tube with a power supply and transimpedance amplifier built in.

Photomultipliers are sensitive to magnetic fields and often some magnetic shielding is required.

Small arrays of miniaturized photomultipliers are now available.

13.4.1.3 Photon Counting

Ideally the anode current will be proportional to the rate at which photons are incident on the photocathode. Let us consider some things that can go wrong along the way. Figure 13.12 in the upper left shows a sequence of photons arriving at a detector. For any detector, there is a probability, η (the quantum efficiency) that a photon will be detected. Other photons may be reflected by the detector, or may be absorbed without producing the appropriate signal. Thus, some photons will be missed (1 in the figure). On the other hand, an electron may escape the photocathode in the absence of a photon (2). This is called a **dark count**, and contributes to the **dark current**. After a photon is detected, there is a **dead time** during which another photon cannot be detected. If the photon rate is sufficiently high, some photons will be missed (3) as a result. The gain process is not perfect, and there will be variations in the number of electrons reaching the anode as a result of a single cathode electron. Of course at this point, electrons caused by photons and dark counts are indistinguishable. If we were to deliver this

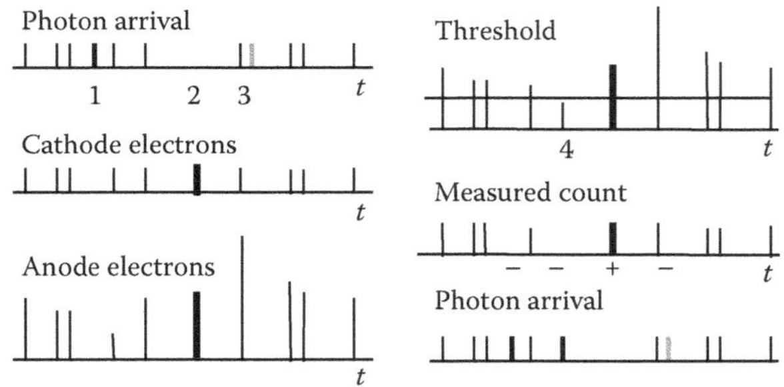

FIGURE 13.12
Photon counting errors. Wide lines show missed detections, and hatched lines show false alarms. The quantum efficiency is the probability of a photon producing an electron. Many photons will be missed (1). Dark counts are electrons generated in the absence of photons (2). After a photon is detected, there is an interval of "dead time" during which photons will be missed (3). Random variations in gain mean that the anode current will not be strictly related to the cathode current. Thresholding the anode signal results in discrete counts, but some pulses may be missed (4) if the threshold is set too high. The measured counts still carry useful information about the photon arrivals. At the bottom right, we repeat the original photon arrival plot and compare it to the measured counts, indicating missed detections (−) and false alarms (+).

signal to an amplifier and a low-pass filter, we would have a current with a mean proportional to the optical power, but with some noise. The direct measurement of the anode current is sufficient for many applications. However, we can improve the detection process by thresholding the anode signal and counting the pulses. Of course if we choose too high a threshold, some pulses will be missed (4). Finally, we compare the measured count to the original history of photon arrivals, which is reproduced at the lower right in the figure. In this case, the detection process has three missed detections (the wide lines), and one false alarm (the dashed line). In practice, at low light levels, the majority of the errors will be missed detections. With a quantum efficiency of 25%, at least three out of four photons will be missed. At higher light levels, the photons lost in the dead time become significant. To some extent, corrections for dead time can be made. Good choices of photocathodes, operating voltages, and thresholds can minimize the dark counts.

13.4.2 Semiconductor Photodiode

In a semiconductor, absorption of a photon can result in an electron being excited from the valence band to the conduction band, provided that the photon energy exceeds the energy gap between the valence and conduction bands. The cutoff wavelength that was related to the work function in the photomultiplier in Equation 13.37 is now related to the energy gap:

The bandgap of the semiconductor determines the cutoff wavelength, just as the work function did for the photomultiplier

$$\lambda_{cutoff} = \frac{hc}{E_g}. \tag{13.44}$$

At wavelengths below cutoff, the responsivity is given by Equation 13.7:

$$i = \frac{\eta e}{h\nu} P = \rho_i P.$$

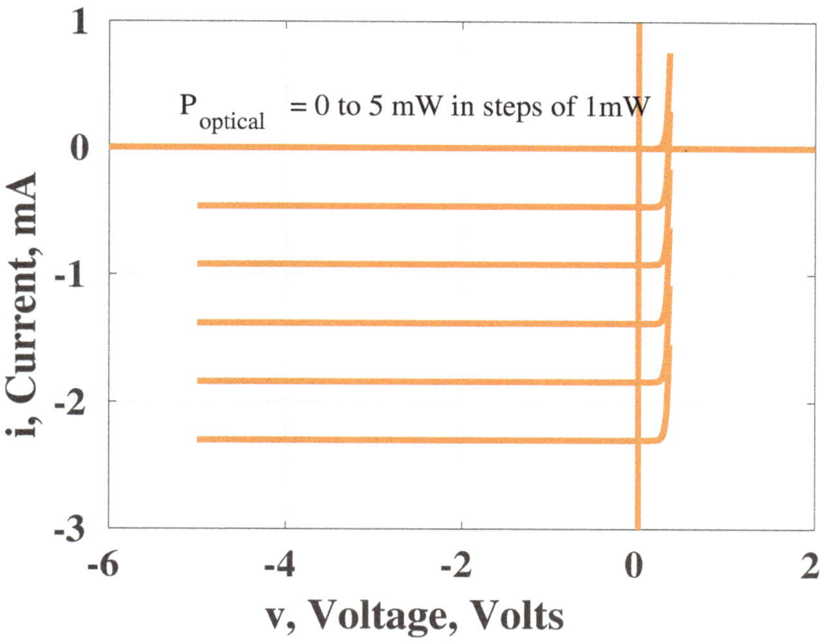

FIGURE 13.13
Photodiode characteristic curves. This detector has a quantum efficiency of $\eta = 0.90$, at the HeNe wavelength of $632.8\,\text{nm}$, and a saturation current of $I_s = 1\,\text{nA}$.

In contrast to the photomultiplier, the semiconductor detector has

- High quantum efficiency, easily reaching 90%
- Potentially longer cutoff wavelength for infrared operation
- Lower gain
- Generally poorer noise characteristics

In Practice

Semiconductor detectors can also be purchased in a package with power supply and amplifier.

Photon-counting packages are available as well. Some of these are time-resolved to report the arrival time of a photon, for example to measure fluorescence lifetime after an excitation pulse.

In its simplest form, the semiconductor material can be used in an external circuit that measures its conductivity. The device is called **photoconductor**. The current will flow as long as the electron remains in the conduction band. Let us visualize the following thought experiment. One photon excites an electron. A current begins to flow, and electrons in the external circuit are detected until the photoelectron returns to the valence band, at which time the current stops. We have thus detected multiple electrons for one photoelectron. There is a photoconductive gain equal to the ratio of the photoelectron lifetime to the transit time across the detector's active region.

Photoconductors act like light-controlled variable resistors. They can also have gain, although usually not as large as a PMT.

Alternatively, a semiconductor junction is used, resulting in a **photovoltaic** device, or **photodiode** in which the photocurrent adds to the drift current, which is in the diode's reverse direction. The current–voltage (I–V) characteristic of an ideal diode is given by the **Shockley equation** named for W. B. Shockley, co-inventor of the transistor.

$$i = I_s \left(-1 + e^{V/V_T}\right), \tag{13.45}$$

where

- $V_T = k_B T/e$ is the **thermal voltage**
- k_B is the Boltzmann constant
- T is the absolute temperature
- e is the charge on an electron
- I_s is the saturation current

A photovoltaic detector behaves as an ordinary diode with the addition of a reverse photocurrent. There is no gain.

For a nonideal diode, a factor usually between 1 and 2 is included in the exponent. With the addition of the photocurrent,

$$i = I_s \left(-1 + e^{V/V_T}\right) - \frac{\eta e}{h\nu} P_{optical}, \tag{13.46}$$

as shown in the example in Figure 13.13. In this figure, the quantum efficiency is $\eta = 0.90$, at the HeNe wavelength of 632.8 nm, and the saturation current is $I_s = 1$ nA. At these optical powers, the saturation current makes very little difference in the reverse current. Curves are shown for five different powers, from 0 to 5 mW in 1−mW steps.

13.4.3 Detector Circuit

Fig. 13.3:

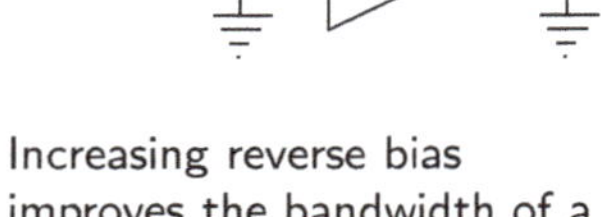

Many different detector circuits are possible. Three simple ones will be considered here. The first is the transimpedance amplifier we discussed in Figure 13.3. The voltage across the detector remains at zero because of the virtual ground of the amplifier, the current is exactly equal to the photocurrent, and the voltage at the output of the amplifier is

$$V = Ri = R\frac{\eta e}{h\nu} P_{optical}. \tag{13.47}$$

Increasing reverse bias improves the bandwidth of a photovoltaic detector.

However, the bandwidth of a photodiode can be improved by using a reverse bias to increase the width of the depletion layer[89] and thus decrease the capacitance. The circuit in Figure 13.14A allows for a reverse bias.

Take-Away

Photovoltaic Circuits:
Different photovoltaic-detector bias circuits are needed for different applications. Reverse bias improves the bandwidth of the detector.

13.4.3.1 Example: Photodetector Circuit

In Figure 13.14A, suppose that the optical power incident on the detector is nominally 3 mW. Assuming that the detector is operating in the third quadrant (i.e., the lower left), the current is well approximated by the photocurrent

$$i_{DC} \approx -\frac{\eta e}{h\nu} P_{optical} = -1.4\,\text{mA}, \tag{13.48}$$

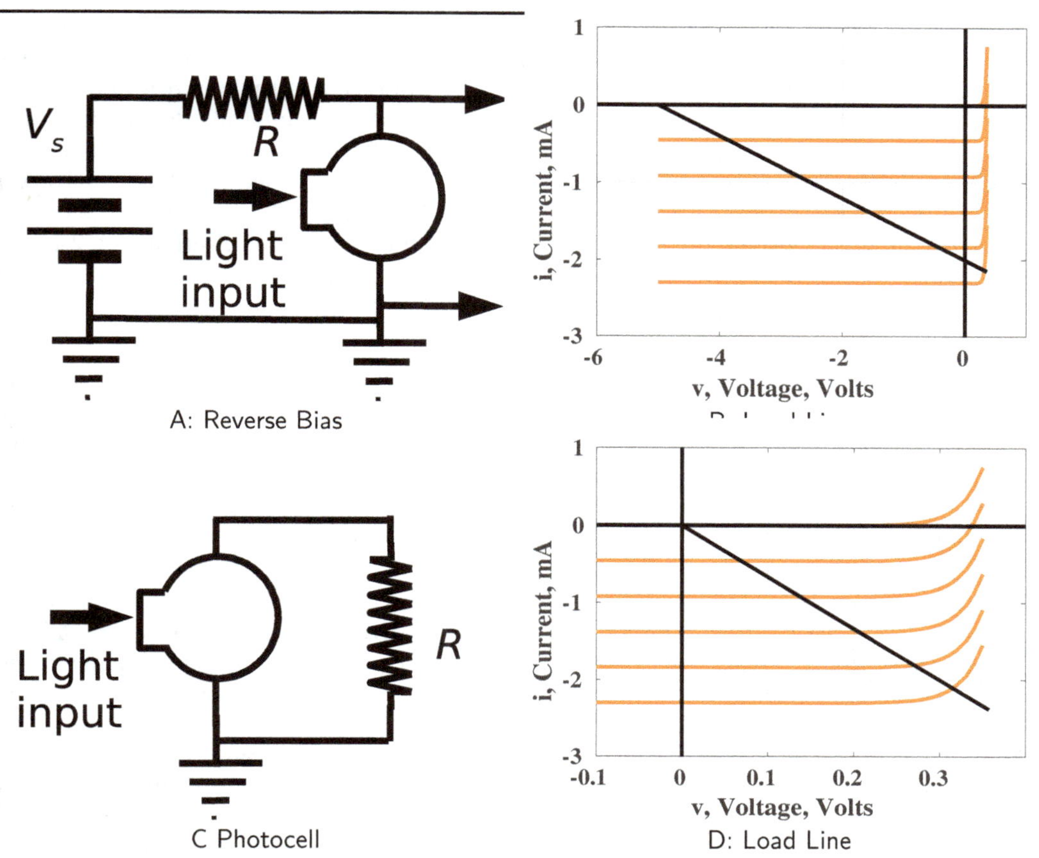

FIGURE 13.14
Circuits for a photodiode. Normally a detector is used in reverse bias (A) to maintain a good bandwidth. Here (B) the supply voltage is $V_s = 5\,\text{V}$ and the load resistor is $2.5\,\text{k}\Omega$. With no bias voltage (C), the diode operates in the fourth quadrant (D), where it produces power.

at $\eta = 0.90$ and $\lambda = 632.8\,\text{nm}$. With a supply voltage of $V_S = 5\,\text{V}$ and a load resistor of $R = 2.5\,\text{k}\Omega$, the detector voltage is $V_{DC} = V_s - i_0 R = -1.56\,\text{V}$. If the power varies because of a signal imposed on it, then the instantaneous voltage and current will vary about these DC values. How much can the power vary and still achieve linear operation? Looking at the plot in Figure 13.14B, the voltage across the detector will be $-5\,\text{V}$ if the power drops to zero, and will increase linearly until the power reaches a little more than 4 mA, at which point the voltage falls to zero. If we wish to measure higher powers, we can decrease the load resistor, and thereby increase the maximum reverse current. If we want greater sensitivity $dv/dP_{optical}$, we can increase the supply voltage and increase the load resistor. Of course, we must be careful that the supply voltage does not exceed the maximum reverse bias rating of the photodiode, in case the optical power is turned off while the electrical circuit is turned on.

We must also be aware of the total power being consumed by the photodiode, which is ultimately dissipated as heat. In our example, at the nominal operating point, $P_{electrical} = iV = 2\,\text{mW}$. Adding the optical power, the total power to be dissipated is 5 mW.

13.4.3.2 Example: Solar Cell

Next, let us consider the circuit in Figure 13.14C. In this case, the photodiode is connected directly to a $150\,\Omega$ resistor. The operating point is now moved into the fourth quadrant,

A photovoltaic detector can be used as a source of electrical power.

where V is positive, but i remains negative. Thus, the electrical power consumed by the photodiode is negative; it has become a source of electricity! A solar cell is a photodiode used in the fourth quadrant. In Figure 13.14D, we can estimate the power graphically as $P_{electrical} = 0.206\,\text{V} \times (-1.375\,\text{mA}) = 0.28\,\text{mW}$. This particular photodiode circuit is only about 10% efficient, producing this power from 3 mW of optical power. On one hand, we could design the circuit for better efficiency. On the other hand, much of the incident power may lie in regions where η is low (e.g., wavelengths beyond cutoff). Therefore, the overall efficiency may be considerably lower. For example, we have noted in Section 12.2.2.1 that only about 42% of the solar spectrum at sea level is in the visible band.

13.4.4 Typical Semiconductor Detectors

Silicon is the most common semiconductor detector material.

Certainly the most common semiconductor detector material is silicon (Si), which is useful over the visible wavelengths and into the near infrared (NIR), with a cutoff wavelength of 1.2 μm. However, because of the indirect bandgap of silicon, the quantum efficiency begins to fall at shorter wavelengths. For NIR applications, indium gallium arsenide (InGaAs) detectors are preferred. A variety of other materials including germanium (Ge), gallium arsenide (GaAs), and indium phosphide (InP) are also in use.

Avalanche photodiodes have gain. It is usually much less than that of a photomultiplier.

Avalanche photodiodes (APDs) operate in a reverse bias mode near breakdown, and can achieve gains of up to a few hundred, but generally have more noise than other photodiodes. A relatively new device is the **single-photon avalanche diode (SPAD)**[45] can be fabricated in large arrays and can be used for precise timing on the nanosecond scale. In the early 2000s, large arrays of SPAD detectors were fabricated and connected in parallel in a device called a **silicon photomultiplier (SiPM)**[38]. These devices are finding new applications and are becoming quite inexpensive. The parallel nature of many elements means that each element has its own dead time and the probability of another photon falling on a single element during the dead time of a previous one is thus lower. Fewer photons are usually lost to dead time. Generally SiPMs are easier to use and less susceptible to damage than traditional photomultipliers.

The mid- and far-IR bands used for thermal imaging require different detector materials.

In the MIR and FIR, mercury cadmium telluride (HgCdTe) is a common material. The amount of mercury can be varied to control the cutoff wavelength. Other common materials are indium antimonide (InSb) and platinum silicide (PtSi). Newer devices using quantum wells and bandgap engineering are also becoming popular. Most infrared photon detectors, and some visible ones, are cooled to reduce noise.

Take-Away

Semiconductor Detectors:

Semiconductor detectors are mostly made from silicon which works well in the visible wavelengths, and high quantum efficiency is common. Other materials are used in the infrared.

The cutoff wavelength depends on the band gap. Wavelengths below the cutoff wavelengths are detected.

Both photoconductive and photovoltaic detectors are used. Photoconductors and avalanche photodiodes have gain, but usually less than that of PMTs.

Take-Away

Summary of Photon Detectors:

- Photomultipliers and semiconductor diodes are the common photon detectors.
- Photon detectors have a cutoff wavelength that depends on the choice of material. Longer wavelengths are not detected.
- Below the cutoff wavelength, the current responsivity is ideally a linear function of wavelength, multiplied by the quantum efficiency, η, but η is a function of wavelength.
- Quantum efficiencies of semiconductor detectors are usually higher than those of the photocathodes in photomultipliers.
- The photomultiplier has a gain that can be greater than 10^6.
- Photomultipliers do not perform well in high magnetic fields.
- Semiconductors usually have lower gains than PMTs. Avalanche photodiodes may have gains up to a few hundred.
- Reverse bias of a semiconductor diode detector improves its frequency response.
- Transimpedance amplifiers are often used as the first electronic stage of a detector circuit.
- The most sensitive detectors are cooled to improve SNR.

13.5 Thermal Detectors

In thermal detectors, light is absorbed, generating heat, and the rise in temperature is measured by some means to determine the optical energy or power. Thermal detectors are often used as power meters or energy meters.

Fig. 13.1A:

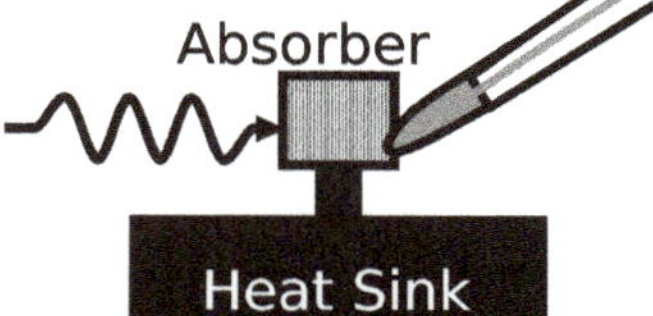

13.5.1 Energy Detection Concept

The concept of the thermal detector is quite different from that of the photon detector. Figure 13.1A shows the concept of a thermal detector or **bolometer**. A photon is absorbed, and the energy is converted to heat. The heat causes the sensing element to increase its temperature, and the rise in temperature is measured. Assuming a small absorber, the temperature rises according to

$$\left.\frac{dT}{dt}\right|_{heating} = CP \tag{13.49}$$

where C is a constant that includes the fraction of light that is absorbed, the material's specific heat, and its volume. If nothing else happens, the temperature rise is proportional to the energy.

$$T(t_1) - T(0) = \int_0^{t_1} CP(t)\,dt = CJ(t_1),$$

where $J(t_1)$ is the total energy between time 0 and t_1. With the proper calibration, this detector can be used to measure pulse energy:

For a short pulse, the temperature rise is proportional to energy

$$T(t_1) - T(0) = CJ, \tag{13.50}$$

Thus, the thermal detector can be used as an **energy meter**.

13.5.2 Power Measurement

If the pulse is sufficiently short, the above approach will work well. However, as the temperature rises, the material will begin to cool. If we assume cooling exclusively by conduction to a heat sink at temperature, T_s, then

$$\left.\frac{dT}{dt}\right|_{cooling} = -K\left(T - T_s\right), \tag{13.51}$$

where K is a constant that includes the thermal conductivity of the elements connecting the sensor to the heat sink, along with material and geometric properties. The final differential equation for temperature is

$$\frac{dT}{dt} = CP\left(t\right) - K\left[T\left(t\right) - T_s\right]. \tag{13.52}$$

If no more light enters the sensor after a time t_1, then cooling occurs according to

$$T\left(t\right) = T\left(t_1\right)e^{-K(t-t_1)}. \tag{13.53}$$

The $1/e$ cooling time is thus $1/K$. If the duration of the pulse is very small compared to this time,

$$t_1 \ll 1/K, \tag{13.54}$$

cooling can be neglected during the heating time, and the optical pulse energy measurement described in Equation 13.50 will be valid.

At the other extreme, if $P(t)$ is constant over a longer period of time

$$t_2 \gg 1/K, \tag{13.55}$$

For a long pulse the detector comes to thermal equilibrium and the temperature is proportional to power.

then by this time, the sensor will approach steady state, and can be used as a **power meter**. The steady state equation is

$$\frac{dT}{dt} = CP - K\left[T\left(t\right) - T_s\right] = 0,$$

$$T\left(t_2\right) - T_s = \frac{C}{K}P. \tag{13.56}$$

13.5.3 Thermal Detector Types

Thermopiles

Thermal detectors are often characterized by the way in which temperature is measured. In the original bolometer, invented by Langley[135], the sensor element was used as a **thermistor**, a resistor with a resistance that varies with temperature. Alternatives are numerous, and include in particular thermocouples and banks of thermocouples called **thermopiles**. Thermopiles can be used relatively easily to measure power levels accurately in the milliwatts.

Pyroelectrics

Another type of thermal detector is the pyroelectric detector. The temperature of this device affects its electrical polarization, resulting in a change in voltage. It is sensitive not to temperature, but to the time derivative of temperature, and therefore can only be used with an AC light source.

Thermal detectors are often used for energy and power meters because of their broad range of wavelengths. Because all of the absorbed energy is converted into heat, the responsivity, ρ_i, or more commonly ρ_v, does not depend explicitly on photon energy. The only wavelength dependence is related to the reflectivity of the interface to the sensor. Laser power meters that are useful from tens of milliwatts to several watts over most of the UV, visible, and IR spectrum are quite common. Higher and lower power ranges are achievable as well. The devices made for higher power, in particular, tend to have long time constants and are less useful for fast measurements, where the temporal behavior of the wave must be resolved.

Microbolometers

However, micro-bolometers are small devices that are amenable to arrays and are in fact quite sensitive and fast. The operation of these devices is dependent on a well-defined heat sink at ambient temperature. Air currents on the face of the detector would produce a variable heat sink, and thus the detectors are normally used in a vacuum, and the light must pass through a window. Although the detectors respond quite uniformly across the spectrum, the actual detector spectrum may be limited by the window material. Only very recently, the possibility of using micro-bolometers at atmospheric pressure has been considered[166].

Take-Away

Thermal Detectors:

- Thermal detectors are often better than photon detectors for measurements where accuracy is important.
- However, they are often slower and less sensitive than photon detectors.
- Thermal detectors are usually less sensitive to changes in wavelength than photon detectors are.
- Thermal detectors are often used for energy and power meters.

13.6 Array Detectors

It is often desirable to have an array of detectors, placed in the image plane of an imaging system. In this way, it is possible to capture the entire scene "in parallel," rather than scanning a single detector over the image. Parallel detection is particularly important for scenes that are changing rapidly, or where the light level is so low that it is important to make use of all the available light.

13.6.1 Types of Arrays

Here we briefly discuss different types of arrays. Each of these small sections could easily be a chapter, or even a book, but our goal here is simply to introduce the different types that are available.

13.6.1.1 Photomultipliers and Micro-Channel Plates

Photomultipliers, in view of their size, and their associated dynode chains, transimpedance amplifiers, and other circuitry, are not particularly well suited to be used in arrays. Nevertheless, some small arrays are available, with several photomultiplier tubes and associated electronics in a single package. A **micro-channel plate** consists of a transparent photocathode followed by an array of hollow tubes with a voltage gradient to accelerate electrons and create a multiplication effect similar to that of a photomultiplier, and a phosphor that glows when struck by electrons. The micro-channel plate serves as an amplifier for imaging, in that one photoelectron at the cathode end results in multiple electrons striking the phosphor at the anode end. The brighter image can be detected by an array of less sensitive semiconductor detectors.

In Practice

Photomultipliers are available in arrays, but the pixels are large and few in number. More recently SPAD and SiPM devices are becoming available with reasonable performance, and provide increasingly attractive alternatives. See Section 13.4.4.

13.6.1.2 Silicon CCD Cameras

Silicon semiconductor detectors are strongly favored for array technology, as conventional semiconductor manufacturing processes can be used to make both the charge transfer structure and the detectors. A **charge-coupled device** or **CCD** is an arrangement of silicon picture cells or **pixels**, with associated circuitry to transfer electrons from one to another. The charge-transfer pixels can be combined with silicon detector pixels to make a CCD camera. In early CCD cameras, the same pixels were used for both detection and charge transfer, leading to artifacts in the images. For example, electrons generated by light in one pixel would be increased in number as the charge passed through brightly illuminated pixels en-route to the output, leading to a "streaking" effect. In more modern devices, the two functions are kept separate. Typical pixels are squares from a few micrometers to a few tens of micrometers on a side, and arrays of millions of them are now practical.

13.6.1.3 Performance Specifications

Some performance indicators for CCD cameras are shown in Table 13.3.

They include, in addition to the size and number of pixels, the "full well capacity" and some measure of noise. The **full-well capacity** is the maximum charge that can be accumulated on a single pixel. usually given in electrons. The noise may also be given in electrons. Typically there is a base number of electrons present when no light is incident, and a variation about this number. The variation represents the lower limit of detection. Other parameters of importance include the transfer rate, or the rate at which pixel data can be read and the **frame rate**, or the rate at which a frame of data can be read out of the camera. Many cameras have an electronic shutter that can be used to vary the **exposure time** or **integration time** independently of the frame rate. Fixed video rates of 30 Hz, with a 1/30 s integration time are common in inexpensive cameras, but slower frame rates and longer exposures are possible for detecting low light levels, and faster times are possible if sufficient light is available. For imaging in bright light, short exposure times are required. If we do not need the corresponding high frame rate, we can

TABLE 13.3
Camera Specifications

Specification	Notes	Units
Pixel size	x and y	μm
Pixel number	x and y	
Full well		Electrons
Noise floor		Electrons
Noise fluctuations		Electrons
Transfer speed		MHz
Frame rate		Hz
Shutter speed	(Variable?)	ms
Sensitivity	Analog	V/J V/electron
Sensitivity	Digital	Counts/J Counts/electron
Uniformity	Responsivity	
Uniformity	Noise	
Spectral response	(Per channel for color)	
Digitization		Bits
Gain	(Value)	
Gain control	Automatic/manual	
Linearity	Gamma	
Modes	Binning, etc.	
Lens mount	(If applicable)	

Some parameters that are of particular importance in camera specifications are shown here. In addition, the usual specifications of physical dimensions, power, weight, and temperature, etc., may be important for many applications.

set the exposure time to achieve the desired level, and a slower frame rate to minimize data storage issues. Some research cameras can achieve frame speeds of several kilohertz. Some other important issues are the conversion factor from charge to voltage, or charge to counts if a digital output is used, uniformity of both dark level and responsivity, and fixed-pattern noise, which describes a noise level that varies with position, but is constant in time. Camera chips can be manufactured with different colored filters arranged in a pattern over the array, typically using three or four different filters, and combining the data electronically to produce red, green, and blue channels. More information can be found in Section 12.3.3.

Gain control and linearity are also important parameters. For example, an 8-bit camera can be used over a wide variety of lighting conditions if automatic gain control (AGC) is used. A variable electronic gain is controlled by a feedback loop to keep the voltage output within the useful range of the digitizer. However, we cannot then trust the data from successive frames to be consistent. For quantitative measurements, manual gain is almost always used.

Another way to improve dynamic range is the choice of gamma. The output voltage, V, and the resulting digital value, are related to the incident irradiance, I, according to

$$\frac{V}{V_{max}} = \left(\frac{I}{I_{max}}\right)^{\gamma}, \tag{13.57}$$

where at the maximum irradiance, I_{max}, the output voltage V_{max} is chosen to correspond to the maximum value from the digitizer (e.g., 255 for an 8-bit digitizer). The gamma can be used to "distort" the available dynamic range to enhance contrast in brighter (high gamma) or darker (low gamma) features of the image, as shown in Figure 13.15. Usually

FIGURE 13.15
Gamma. Increasing γ makes the image darker and increases the contrast in the bright regions because the slope is steeper at the high end of the range; a small range of values in the input maps to a larger range of values in the output. Decreasing γ makes the image brighter and increases the contrast in the dark regions (A). A photograph as digitized (B) has been converted using $\gamma = 2.0$ (C). Note the increased contrast in the snow in the foreground, and the corresponding loss of contrast in the tree at the upper right. The same photograph is modified with $\gamma = 0.5$ (D), increasing the contrast in the tree and decreasing it in the snow. These images are from a single digitized photograph. The results will be somewhat different if the gamma is changed in the camera, prior to digitization. (Photograph used with permission of Christopher Carr.)

for research purposes, we choose $\gamma = 1$ to avoid dealing with complicated calibration issues. However for certain applications, a different gamma may be useful.

In Practice

Many low–cost cameras are designed to produce a "pleasing" image, with automatic gain control and different values of gamma. For most research applications, these make it difficult or impossible to make quantitative measurements. It is usually best to turn off AGC and choose $\gamma = 1$.

Example

Let's consider an illustrative example, in which we illuminate an outdoor scene with laser light, and try to detect the backscattered light with a camera. We'll go through a full radiometric analysis to demonstrate the process.

We use a laser 50 mW of power in a collimated beam of diameter $d_0 = 1\,\text{mm}$ at $\lambda = 905\,\text{nm}$. The scene is assumed to be a 10% Lambertian reflector, $z = 100\,\text{m}$ away. The camera has $w = 7\,\mu\text{m}$ square pixels with a quantum efficiency of $\eta = 0.8$, exposure time of $t = 1/30\,\text{s}$, full-well capacity of 50,000 electrons, and dark noise of 10 electrons. It uses a 10-bit digitizer and the maximum digital output, $2^{10} - 1$ corresponds to a full well of 50,000 electrons. The digitizer reads $DARK = 50$ counts when the camera is in complete darkness.

We'd like to compute the signal-to-noise ratio seen by the camera, and typical digital values of the signal and noise.

First, we use Gaussian beam equations to determine the beam size at the target. Equation 9.25 gives $d = 0.115\,\text{m}$. The maximum irradiance is given by Equation 9.32. The reflected radiant exitance is given by Equation 12.88 and the Lambertian nature of the target dictates that

$$L = M/\pi = 0.305\,\text{W/m}^2/\text{ster}.$$

Now, neglecting any loss in the atmosphere and camera lens, the radiance theorem tells us that the radiance on the camera chip will be the same. The irradiance there can be computed using Equation 12.76. Because the numerical aperture is small, we can just multiply the radiance by the solid angle to get the irradiance where Equation 4.4 ($s = z$ tells us the solid angle, so

$$E = L\Omega = L\frac{\pi}{4}\left(\frac{D}{s}\right)^2.$$

The total optical energy on a pixel during the exposure time is $J = EAt = Ew^2t$ and Equation 13.5 predicts the number of electrons;

$$n_e = 17,800.$$

If we had used $\eta = 1$, then this would have been the number of photons, about $nn_p = 22,300$. The fundamental limit on SNR is

$$SNR_{max} = \left(\frac{n_p}{\sigma_p}\right),$$

where for the Poisson distribution, $\sigma = \sqrt{n^2 - \bar{n}^2}$ is given by Equation 13.23 as $\sqrt{\bar{n}_p}$, so

$$SNR_{max} = 22,300 \qquad (43.5\,\text{dB}).$$

By comparison, using the number of electrons, we obtain

$$SNR_{max} = 17,800 \qquad (42.5\,\text{dB}),$$

so we have only lost 1 dB for the quantum efficiency.

Eq. 9.25:
$d_d \approx \frac{4}{\pi}\frac{\lambda}{d_0}z$

Eq. 9.32
$I = \frac{2P}{\pi\frac{d^2}{4}}$.

Eq. 12.88:
$M_\lambda(\lambda) = R(\lambda)\,E_\lambda(\lambda)$.

Eq. 12.76:
$E_\lambda = \iint L_\lambda \cos\theta \sin\theta\, d\theta\, d\zeta = \pi L_\lambda$.

Eq. 4.4:
$\Omega = \frac{\pi}{4}\left(\frac{D}{s}\right)^2$.

Eq. 13.5:
$n_e = \frac{\eta}{h\nu}J$.

Eq. 13.23:
$\sqrt{n^2 - \bar{n}^2} = \sqrt{\bar{n}}$.

Eq. 12.81:
$M_\lambda(\lambda, T) = \frac{2\pi hc^2}{\lambda^5}\frac{1}{e^{h\nu/k_BT}-1}$.

Eq. 12.83:
$M(T) = \frac{2\pi^5 k^4 T^4}{15h^3c^2} = \sigma_e T^4$.

However, we want to use this system outdoors so we must contend with ambient sunlight. To minimize the effect of the sunlight, we'll put a filter in front of the detector to pass only a small portion of the scattered sunlight. between 395 nm and 405 nm. Using the Planck equation, Equation 12.81, and the Stefan–Boltzmann Equation 12.83 the spectral fraction for a 5000 K black body is $f_\lambda = M_\lambda d\lambda / M = 0.008$. Assuming a solar irradiance of $E_{solar} = 1000\,\text{w/m}^2$, the scattered solar radiance is

$$L_{ss} = E_{solar} \times 0.1/pi = 31.8\,\text{W/m}^2/\text{ster}.$$

Running this number through the same calculation as we used above for the signal radiance, we obtain an electron count

$$n_{e-solar} = 1.86 \times 10^6.$$

This greatly exceeds the full-well capacity, so we are not going to be able to work with direct sunlight on the target. Let's assume that we can keep the target in shade, and reduce the incident solar irradiance by two orders of magnitude. Then

$$n_{e-solar} = 18,600.$$

To compute the total RMS noise, we use

$$n_{noise} = \sqrt{sigma^2_{signal} + sigma^2_{solar} + darknoise^2}$$

$$n_{noise} = \sqrt{n_e + n_{solar} + 10^2} = 171.$$

Now with $CPE = \frac{2^{10}\,\text{counts}}{50{,}000\,\text{electrons}}$, we obtain a total count of

$$C_1 = CPE \times (n_e + n_{solar}) + DARK = 1000.$$

The noise in counts is

$$CPE \times (n_{noise}) = 3.5.$$

Without the laser turned on,

$$C_2 = CPE \times (n_{solar}) + DARK = 354,$$

with a noise of

$$\sqrt{n_{solar} + 10^2} = 122.4$$

Finally, we make a measurement with the laser on, and another with the laser on. We subtract the two to obtain the signal count of 646.

To compute the SNR, we use electron counts

$$SNR = \frac{n_e + n_{solar} - n_{solar}}{\sqrt{171^2 + 122^2}} = \left(\frac{646}{\sqrt{171^2 + 122^2}}\right)$$

$$SNR = 7210 \qquad (38.5\,\text{dB}).$$

To summarize this long example, our photon-limited SNR was 43.5 dB. We lost only 1 dB to the quantum efficiency, but we lost an additional 4 dB because of background noise, and even at that we had to give up the idea of placing the target in direct sunlight.

13.6.1.4 Other Materials

Silicon is useful in the visible spectrum, and to some extent in the shorter wavelengths of the NIR. However, it does not work for the infrared cameras or **thermal imagers** that typically image in the 3-to-5 or 8-to-14-μm band. Detector materials for longer wavelengths, such as mercury cadmium telluride (HgCdTe) can be attached to silicon CCDs to make cameras, but the process is difficult, and array size is limited by the difference in thermal expansion properties of the materials. However, progress is being made in this area, and the interested reader is encouraged to consult recent literature for information.

More recently, arrays of micro-bolometers have been developed into arrays that are quite sensitive for thermal imaging, and are considerably less expensive than CCD cameras at the longer wavelengths of the MIR and FIR bands. Early micro-bolometers required cooling to achieve the sensitivity required for thermal imaging, but more recently, room-temperature micro-bolometer arrays have become available.

13.6.1.5 CMOS Cameras

Recently, **complementary metal-oxide-semiconductor (CMOS)** imagers have become popular. The major difference in the principles of operation is that in CMOS, both the detection and conversion to voltage take place on the pixel itself. Frequently, the CMOS camera has a built-in analog-to-digital converter, so that a digital signal is delivered at the output. Generally CMOS detectors are more versatile. For example, it is possible to read out only a portion of the array. On the other hand, CMOS is generally described as having less uniformity than CCD technology. Because of current rapid development in CMOS technology, a comparison with CCD technology is not really possible, and the reader would do well to study the literature from vendors carefully.

The CMOS technology enables integration of electronic circuits into each pixel, enabling cameras with "intelligent pixels" that can measure arrival time of a pulse, or changes with time. Such tasks have typically required single detectors and scanning optics.

13.6.2 Example

Let us conclude this overview of array detectors by considering one numerical example. We return to the problem of imaging the moon in Section 12.1.4. In that section, we computed a power of about 4 pW on a 10 μm square pixel. The moon is illuminated by reflected sunlight, and we will disregard the spectral variations in the moon's surface, but include absorption of light by the Earth's atmosphere. Therefore, we assume that the spectral distribution of the light follows the 5000 K spectrum as suggested by Figure 12.24. Figure 13.16 shows $f_\lambda = M_\lambda/M$, the fractional spectral radiant exitance with a solid line, and $f_{P\lambda} = M_{P\lambda}/M_P$, the fractional spectral distribution of photons as a dotted line. The photon distribution is weighted toward longer wavelengths, and the ratio $M_p/M = 5.4 \times 10^{18}$ photons/J, averaged over all wavelengths. If the camera has a short-pass filter that only passes wavelengths below $\lambda_{sp} = 750$ nm, then

Fig. 12.24:

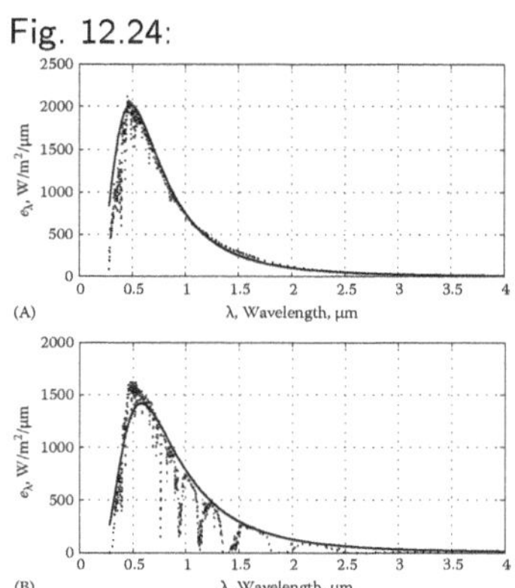

$$\frac{\int_0^{\lambda_{sp}} M_{p\lambda} d\lambda}{\int_0^{\infty} M_{p\lambda} d\lambda} = 0.22, \tag{13.58}$$

and only 22% of these photons reach the detector surface. In an exposure time of 1/30 s, this number is 1.5×10^5. If we assume a quantum efficiency of 0.8 then we expect 1.2×10^5 electrons. This number is still probably optimistic, as the quantum efficiency of the camera is not constant over all wavelengths. However, for a detector with a full-well capacity of 30,000 electrons, it will be necessary to reduce the numerical aperture from the assumed value of 0.1 or to reduce the exposure time. With a camera having an RMS noise level of 20 electrons, the signal could be reduced by almost four orders of magnitude and still be detected.

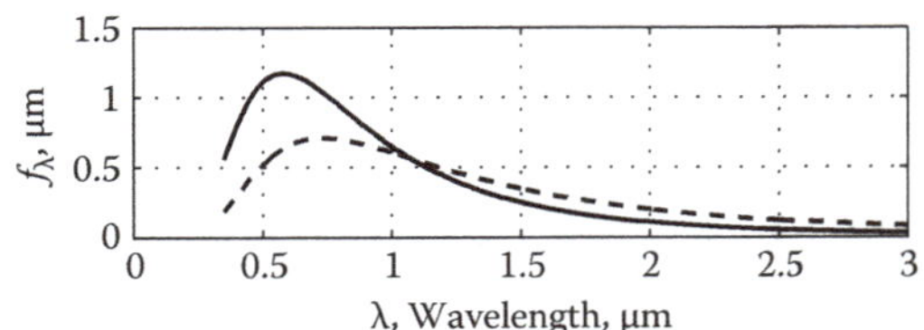

FIGURE 13.16
Moonlight spectrum. The solid line shows $f_\lambda = M_\lambda/M$, the fractional spectral radiant exitance. The dashed line shows $f_{P\lambda} = M_{P\lambda}/M_P$, the fractional spectral distribution of photons. The photon distribution is weighted toward longer wavelengths, and the ratio $M_p/M = 5.4 \times 10^{18}$ photons/J.

Take-Away

Array Detectors:

- Array detectors allow parallel detection of all the pixels in a scene.
- Silicon arrays in charge-coupled devices (CCDs) are commonly used in visible and very NIR applications.
- Pixel sizes are typically a few micrometers to a few tens of micrometers square.
- The CMOS camera is an alternative to the CCD, also using silicon.
- Color filters can be fabricated over individual pixels to make a color camera.
- Other detector materials are used, particularly for sensitivity in the infrared.
- Micro-bolometer arrays are becoming increasingly common for low-cost thermal imagers.

14 Nonlinear Optics

In Chapter 6, we discussed the interaction of light with materials, using the Lorentz model which treats the electron as a mass on a spring, as shown in Figure 6.7. The restoring force, and the resulting acceleration in the x direction were linear as in Equation 6.26:

$$\frac{d^2x}{dt^2} = -\mathbf{E}_r e/m - \kappa_x x/m,$$

where

- e is the charge on the electron
- m is its mass

Fig. 6.7:

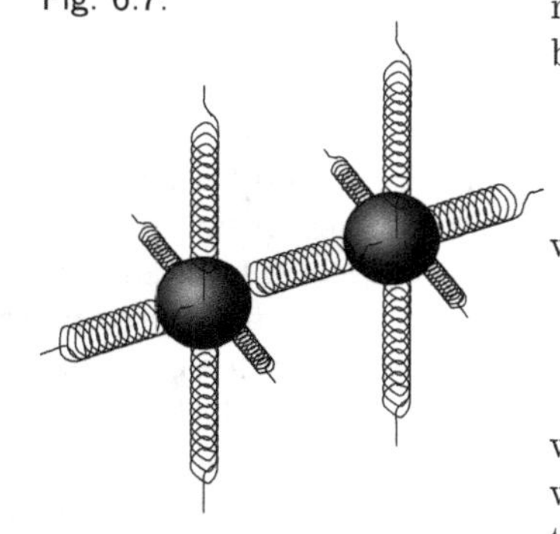

The restoring force, $-\kappa_x x$, is the result of a change in potential energy as the electron moves away from the equilibrium position. The force is in fact the derivative of a potential function, which to lowest order is parabolic. However, we know that if we stretch a spring too far, we can observe departures from the parabolic potential; the restoring force becomes nonlinear. We expect linear behavior in the case of small displacements, but nonlinear behavior for larger ones. If we replace Equation 6.26

$$\frac{d^2x}{dt^2} = -\mathbf{E}_r e/m - \kappa_x x/m,$$

with

$$\frac{d^2x}{dt^2} = -\mathbf{E}_r e/m - \kappa_x x/m - \kappa_x^{[2]} x^2/m \ldots, \tag{14.1}$$

where $\kappa_x^{[2]}$ is a constant defining the second-order nonlinear behavior of the spring, then we expect a component of acceleration proportional to x^2. If x is proportional to $\cos\omega t$ then the acceleration will have a component proportional to

$$\cos^2\omega t = \frac{1}{2} + \frac{1}{2}\cos 2\omega t. \tag{14.2}$$

FURTHER READING: See references[31;21;19;221]

Now in connection with Equation 6.26, we discussed anisotropy; if a displacement in one direction results in an acceleration in another, then that acceleration produces a component of the polarization vector, $\mathbf{P}_r$, in that direction. The polarization acts as a source for the electric displacement vector $\mathbf{D}_r$. In nonlinear optics, the polarization term at frequency 2ω becomes a source term in a wave equation at that frequency. The result is second-harmonic generation.

Fig. 1.9A:

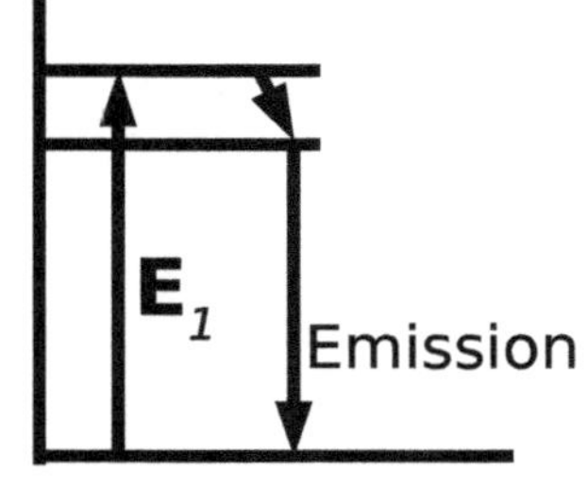

We may readily see that optical inputs at two different frequencies will result in mixing to produce sum and difference frequencies. All that is needed is to replace the trigonometric identity in Equation 14.2 with the appropriate one for a sum of cosines. We may extend this approach to an arbitrary number of harmonics and mixing signals, based on sums of cosine waves raised to the third, fourth, or higher power.

In this chapter, we briefly discuss some of these nonlinear optical phenomena. Non-Linear optics is worthy of a book by itself, and in fact is the exclusive subject of many books including these four[31;21;19;221]. Without attempting to be inclusive, We present some very brief discussions of a few selected topics to show the range of concepts and applications that are part of non-linear optics.

DOI: 10.1201/9781315157047-14

First let us mention one effect that might be viewed as nonlinear, but is not. Nonlinear processes often produce output light at new frequencies not present in the input, but not all such processes are nonlinear. In fluorescence (see Section 1.6 and Figure 1.9A), a material absorbs light at one wavelength and emits light at a longer wavelength, or one having lower energy. This phenomenon is distinct from the nonlinear processes discussed here, and in fact, the output is proportional to the input, at least at low incident irradiance. Fluorescence is indeed considered a linear optical process.

Before proceeding, we make one observation. We have used the complex notation for the frequency-domain description of a field, as defined in Equation 1.58:

$$\mathbf{E}_r = \mathbf{E}e^{j\omega t} + \mathbf{E}^* e^{-j\omega t}.$$

It is important to consider positive and negative frequency components when working with non-linear equations.

We noted in Chapter 1 that it is normally sufficient to use the positive frequency portion of a wave in our calculations because of linearity. For the same reason, we could also often be careless as to whether we were dealing with a time-domain or frequency-domain representation of the wave. Because we are now dealing with nonlinear behavior, we must remember to be careful to make the distinction between time domain and frequency domain, and to include the negative-frequency component in all our calculations.

We will begin this chapter with a quick overview of the non-linear processes we have chosen to consider. Then we proceed to a short discussion of the nonlinear susceptibility. We will discuss practical issues such as phase matching, and then consider a number of applications. The interested reader may want to consult any of a number of texts on the subject as mentioned above, or recent literature for specific topics.

14.1 Some Non-Linear Processes

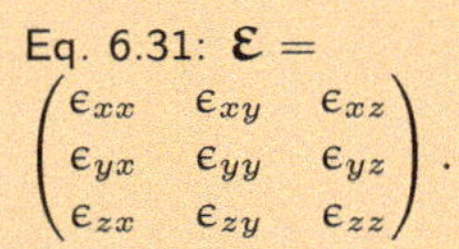

Eq. 6.31: $\boldsymbol{\mathcal{E}} = \begin{pmatrix} \epsilon_{xx} & \epsilon_{xy} & \epsilon_{xz} \\ \epsilon_{yx} & \epsilon_{yy} & \epsilon_{yz} \\ \epsilon_{zx} & \epsilon_{zy} & \epsilon_{zz} \end{pmatrix}.$

Non-linear optics is a particularly rich and diverse discipline. Many non-linear processes are best described in terms of non-linear susceptibility, the relationship between $\mathbf{D}$ and $\mathbf{E}$. In linear optics as in Chapter 6, three vector directions resulted in nine possible relationships among components of these two vectors, and we used a three-by-three matrix in Equation 6.31. There are six quadratic combinations of the three vector components, so second-order susceptibility is described by a three-by-six matrix, and there are ten cubic combinations, leading to a three-by-ten matrix. Even higher orders may be interesting. Other types of non-linearity combine with these to provide a multitude of different processes and applications.

In this chapter we will discuss

- In Section 14.4.2
  - second-harmonic generation to produce green laser light for laser pointers and as alternatives to Argon ion lasers.
  - second-harmonic generation microscopy, specifically in biological material.
  - Optical frequency sum and difference generation including optical parametric oscillators.
  - Raman spectroscopy.
- In Section 14.4.3
  - third-harmonic generation
  - third-harmonic microscopy
  - irradiance-controlled index of refraction
  - dynamic holography for amplifying optical waves
  - dynamic holography for phase conjugation
  - self focusing
  - coherent anti-Stokes Raman spectroscopy

- In Section 14.4.4
  - multiphoton fluorescence
  - stepwise multiphoton fluorescence
- In Section 14.4.5
  - thermal lensing
  - photorefractive mixing

14.2 Wave Equations

We return for a moment to a point we made in Chapter 1, in which we invoked Maxwell's equations and the constitutive relations in a loop shown in Figure 1.7 to arrive at the wave equation. The change we are going to make in this section is to the constitutive relation between **D** and **E**. Before we had Equation 1.5,

Fig. 1.7:

$$\begin{array}{ccccc} & & \mathbf{D}=\epsilon\mathbf{E} & & \\ & \mathbf{D} & \leftarrow\ \ \rightarrow & \mathbf{E} & \\ \nabla\times\mathbf{H}=\frac{\partial\mathbf{D}}{\partial t} & \updownarrow & & \updownarrow & \nabla\times\mathbf{E}=-\frac{\partial\mathbf{B}}{\partial t} \\ & \mathbf{H} & \leftarrow\ \ \rightarrow & \mathbf{B} & \\ & & \mathbf{B}=\mu_0\mathbf{H} & & \end{array}$$

$$\mathbf{D}=\epsilon\mathbf{E},$$

which was valid in either the time domain or frequency domain as a result of linearity. We now change this equation to

$$\mathbf{D}_r=\epsilon\mathbf{E}_r+\epsilon^{[2]}\mathbf{E}_r^2, \tag{14.3}$$

where

$$\epsilon=\epsilon_0\left(1+\chi\right), \tag{14.4}$$

$$\epsilon^{[2]}=\epsilon_0\chi^{[2]} \tag{14.5}$$

and $\mathbf{E}_r^2$ is a vector, which contains all second-order combinations of real fields. We call $\chi^{[2]}$ the second-order susceptibility, but there are six possible second-order combinations of the three input field components and three components of the output, so Equation 14.3 must take on a more complicated form than we might expect. We can write it as

$$\begin{pmatrix} D_{rx}\\ D_{ry}\\ D_{rz}\end{pmatrix}=\epsilon_0\begin{pmatrix} E_{rx}\\ E_{ry}\\ E_{rz}\end{pmatrix}+\epsilon_0\begin{pmatrix}\chi_{11} & \chi_{12} & \chi_{13}\\ \chi_{21} & \chi_{22} & \chi_{13}\\ \chi_{31} & \chi_{32} & \chi_{33}\end{pmatrix}\begin{pmatrix} E_{rx}\\ E_{ry}\\ E_{rz}\end{pmatrix}$$

$$+\epsilon_0\begin{pmatrix}\chi_{11}^{[2]} & \chi_{12}^{[2]} & \chi_{13}^{[2]} & \chi_{14}^{[2]} & \chi_{15}^{[2]} & \chi_{16}^{[2]}\\ \chi_{21}^{[2]} & \chi_{22}^{[2]} & \chi_{23}^{[2]} & \chi_{24}^{[2]} & \chi_{25}^{[2]} & \chi_{26}^{[2]}\\ \chi_{31}^{[2]} & \chi_{32}^{[2]} & \chi_{33}^{[2]} & \chi_{34}^{[2]} & \chi_{35}^{[2]} & \chi_{36}^{[2]}\end{pmatrix}\begin{pmatrix} E_{rx}E_{rx}\\ E_{ry}E_{ry}\\ E_{rz}E_{rz}\\ E_{rx}E_{ry}\\ E_{rx}E_{rz}\\ E_{ry}E_{rz}\end{pmatrix} \tag{14.6}$$

or

$$\mathbf{D}_r=\epsilon_0\mathbf{E}_r+\epsilon_0\chi\mathbf{E}_r+\epsilon_0\chi^{[2]}\mathbf{E}_r{}^2, \tag{14.7}$$

with the understanding that we have defined the product $\chi^{[2]}\mathbf{E}_r{}^2$ as the product of the 3×6 matrix with the 6×1 column vector. We have defined the **second-order susceptibility**, $\chi^{[2]}$, as a tensor acting on all possible second-order products of the field[31]. Many authors prefer to write the three equations rather than use this matrix form. We have chosen the matrix notation used here to conform to that most often used.

Following Equation 1.58, and noting that we now have two angular frequencies, ω and 2ω, we can write

Eq. 1.58:
$\mathbf{E}_r=\mathbf{E}e^{j\omega t}+\mathbf{E}^*e^{-j\omega t}.$

$$\begin{aligned}\mathbf{E}_r=&\mathbf{E}\left(\omega\right)e^{j\omega t}+\mathbf{E}\left(2\omega\right)e^{j2\omega t}+\\ &\mathbf{E}^*\left(\omega\right)e^{-j\omega t}+\mathbf{E}^*\left(2\omega\right)e^{-j2\omega t},\end{aligned} \tag{14.8}$$

where the two fields $\mathbf{E}\left(\omega\right)$ and $\mathbf{E}\left(2\omega\right)$ may vary with position.

The first two terms in Equation 14.7 are exactly what we saw in Figure 1.7, and have a wave solution at the fundamental frequency. We should be careful to write this part of the equation explicitly showing the frequency dependence of $\mathcal{E}(\omega)$ as

$$\mathbf{D}(\omega) = \mathcal{E}(\omega)\,\mathbf{E}(\omega), \tag{14.9}$$

because the linear susceptibility may depend on wavelength. Our model in Figure 1.7 is insufficient to describe the nonlinear effects, and we must modify it as shown in Figure 14.1. The last term in Equation 14.7 results in a polarization vector, $\mathbf{P}_2$, which is a source term for a new wave equation described by the right-hand loop. This wave equation has a solution that is harmonic at twice the fundamental frequency. Once again, in the right-hand loop of Figure 14.1, we must be careful to allow for a different value of ϵ for this frequency, so in this loop,

$$\mathbf{D}(2\omega) = \epsilon(2\omega)\,\mathbf{E}(2\omega) + \epsilon_0\chi^{[2]}\mathbf{E}^2(\omega), \tag{14.10}$$

where the latter term acts as a source.

The concept shown in Figure 14.1 can be extended to address mixing of different waves, or higher-order harmonics, and provides a simple basis for understanding nonlinear phenomena.

Figure 14.2 illustrates the process. Excitation light at ω is incident on a crystal from the left. For simplicity, assume a plane wave in the positive z direction. At the beginning of the crystal, some of the incident light creates a second-harmonic displacement vector, $\mathbf{D}$ according to Equation 14.10. That light propagates from left to right. Along the crystal, the excitation light generates more second harmonic which adds coherently to the light generated, so the irradiance of the second harmonic increases with z. Many second-harmonic processes are inefficient so we can neglect the depletion of the excitation, but others are highly efficient and result in significant depletion of the excitation.

14.3 Phase Matching

Dispersion can limit the effective length of a SHG crystal. Polarization tricks can be used to compensate for this effect.

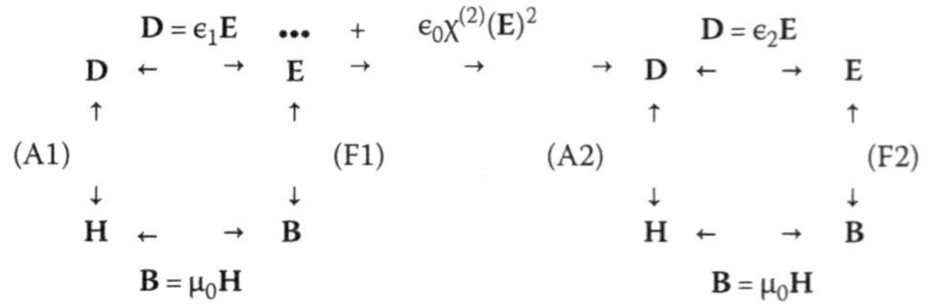

If the material is not dispersive then $\epsilon(\omega) = \epsilon(2\omega)$, and the waves will be synchronized in phase. Thus, if the waves are traveling in the z direction, the second harmonic $\mathbf{D}(2\omega)$ at any distance, z, will be in phase with the source term $\mathbf{P}(2\omega)$ at that location. The result is that the second harmonic will increase in strength with increasing z as shown in Figure 14.2. We can thus use a long crystal for efficient harmonic generation. However, nature is seldom so kind. Most materials are sufficiently dispersive that the dielectric properties are quite different at frequencies that differ by a factor of two. In this case, after a certain distance, the waves will lose their phase relationship, and the additional

FIGURE 14.1
Second-harmonic generation. The left loop results in a wave equation for $\mathbf{E}$ (or any other field variable) as in Figure 1.7, at the fundamental frequency. Ampere's law for the fundamental frequency (A1) is $\nabla \times \mathbf{H} = j\omega\mathbf{D}$ and Faraday's law (F1) $\nabla \times \mathbf{E} = -j\omega\mathbf{B}$. Ampere's law for the second harmonic (A2) is $\nabla \times \mathbf{H} = j2\omega\mathbf{D}$ and Faraday's law (F2) $\nabla \times \mathbf{E} = -j2\omega\mathbf{B}$.

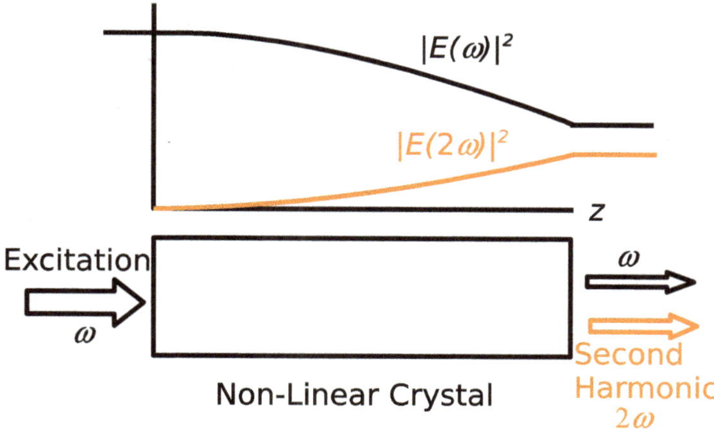

FIGURE 14.2
Second-harmonic generation. The excitation light field, multiplied by $\chi^{[2]}$, generates a second harmonic. The second harmonic grows and propagates to the right.

second-harmonic being generated will combine destructively with that which has been generated at smaller z distances. Therefore, in practice, there is an upper limit on the length of the crystal we can use.

We can often achieve phase matching through the use of polarization. If the nonlinear material is birefringent, and if we are willing to use two different states of polarization for the waves, then we may be able to match their speeds if the index of refraction at 2ω for one polarization is equal to that for the other polarization at ω. This seemingly improbable result is in fact quite possible, and is frequently used. A uniaxial crystal (See Equation 6.73) has two different indices of refraction, one along the axis, and the other in all orthogonal directions. By cutting and orienting the crystal correctly, it is possible to achieve any index of refraction between these extrema. Thus with careful choices, phase matching can often be achieved. The reader may wish to consult any one of a number of sources for a discussion of the index ellipsoid[17;254;147] to clarify this point.

Eq. 6.73:
$$\underline{\epsilon} = \begin{pmatrix} \epsilon_{xx} & 0 & 0 \\ 0 & \epsilon_{yy} & 0 \\ 0 & 0 & \epsilon_{yy} \end{pmatrix}.$$

Phase matching becomes considerably more complicated with multiple waves in such processes as sum-and-difference-frequency generation, where the phases of three or more waves must be considered.

In Practice

Dispersion will often limit the useful length of a non-linear crystal because the waves of different frequencies travel at different speeds, limiting the amount of constructive interference. Crystals can be designed to take advantage of the different polarizations of the waves to mitigate this effect.

14.4 Nonlinear Processes

Next, we will consider some of the important nonlinear processes. We will begin with a discussion of energy-level diagrams, which will often add clarity to our analysis, and we will then consider several important processes, based on second-order and third-order susceptibilities and on other processes.

14.4.1 Energy-Level Diagrams

Fig. 1.9:

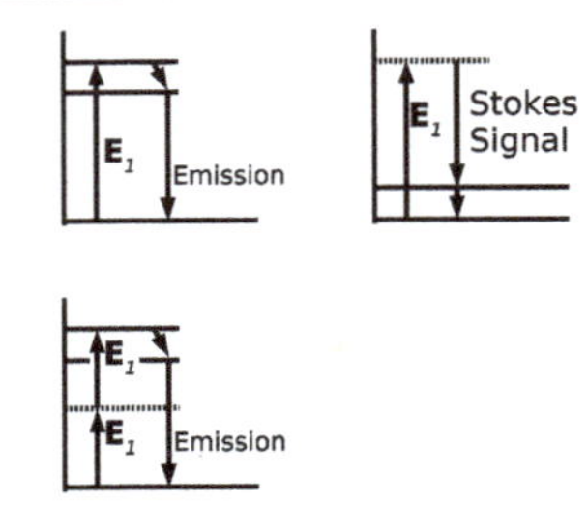

Figure 1.9 described some energy-level diagrams and their meaning. We can describe the generation of light at the second harmonic using Figure 1.9C. Two photons of equal energy, $h\nu$, are absorbed, and one of energy $2h\nu$ is emitted. Our result in Equation 14.7, based on a classical Lorentz model, provides the right frequency relationships but fails to address conservation of energy. We could introduce a conservation law empirically, by including an imaginary component in the index of refraction of the fundamental wave. If the second-harmonic process has a low efficiency, the amount of energy removed from the fundamental is small and may often be neglected. However, in many cases, the efficiency can be substantial, and we must consider the depletion of the fundamental. Doing so is beyond the scope of this book.

Eq. 14.7:
$\mathbf{D}_r = \epsilon_0 \mathbf{E}_r + \epsilon_0 \chi\ \mathbf{E}_r + \epsilon_0 \chi^{[2]} \mathbf{E}_r{}^2$.

In Section 1.6, we discussed the fact that some of the excited states in Figure 1.9 are virtual states with lifetimes on the order of an optical cycle. In Figure 1.9A, the states were all real. In Figure 1.9B, the upper state was virtual, but the intermediate state was real. In Figure 1.9C, the upper state was real, but the intermediate state was virtual. This two-photon-absorption process can be described by a complex second-order susceptibility.

We have already commented that the fluorescence in Figure 1.9A is considered a linear process; the amount of light absorbed is proportional to the amount of light incident (unless the population of the upper state becomes large), as is the amount of fluorescent light emitted (again unless the middle level becomes highly populated).

Eq. 13.16:
$P(n) = \frac{e^{-\bar{n}}\bar{n}^n}{n!}$.

From a quantum-mechanical perspective, we can say that the probability of absorption of a photon is proportional to the number of incident photons. We recall the Poisson distribution in Equation 13.16, and specifically Equation 13.19, for the probability of a photon count greater than or equal to one,

Eq. 13.19:
$\sum_{n=1}^{\infty} P(n) = 1 - P(0) = 1 - e^{-\bar{n}}$.

$$\sum_{n=1}^{\infty} P(n) = 1 - P(0) = 1 - e^{-\bar{n}}.$$

The mean number of photons absorbed is

$$\bar{n} = E\sigma_a, \tag{14.11}$$

where

- E is the irradiance
- σ_a is the **absorption cross section** of the molecule

The probability of a single photon being absorbed at low irradiance is proportional to the irradiance,

$$1 - e^{-E\sigma_a} \approx E\sigma_a. \tag{14.12}$$

Eq. 13.22:
$1 - P(0) - P(1) \approx \frac{\bar{n}^2}{2}$.

Likewise, the probability of absorbing two photons from Equation 13.22 is

$$\frac{(-E\sigma_a)^2}{2}, \tag{14.13}$$

which is proportional to the irradiance squared. This simple analysis is for illustrative purposes only. The actual nonlinear susceptibilities are best obtained from measurements.

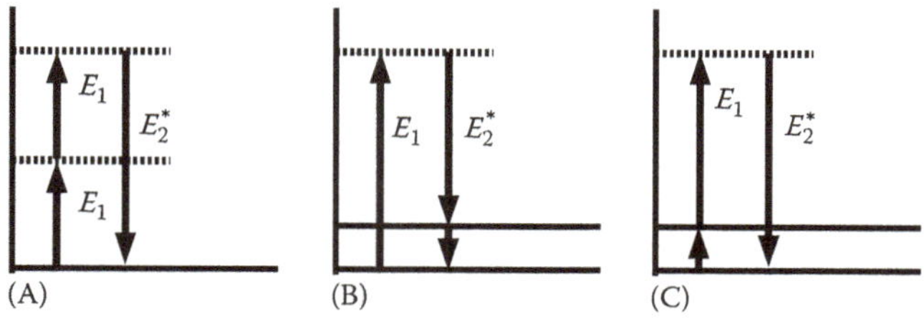

FIGURE 14.3
Some second-order processes. Probably the most common second-order process is second-harmonic generation (A). Raman scattering is a first-order process, but stimulated Raman scattering (B and C) is second-order.

> *Take-Away*
>
> **Quadratic Dependence:**
> Because the Poisson distribution predicts a probability of two photons proportional to the square incident irradiance, most nonlinear processes require high input power.

14.4.2 $\chi^{[2]}$ Processes

We have already discussed the generation of second-harmonic light, shown in Figure 14.3A. One of the greatest practical successes of this process is the frequency doubling of the Nd:YAG laser from its fundamental wavelength of 1064 to 532 nm. This process can be achieved with high efficiency (greater than 80% with good phase matching), and the resulting light is coherent, so the beam has the quality of a laser beam. In fact, such devices have been called "KTP lasers," after the SHG medium potassium titanyl phosphate, although strictly the term is incorrect as the KTP is not lasing but simply doubling the frequency of the laser. These light sources are often used as replacements for the green 514 nm line of argon ion lasers that require gas-filled tubes, high voltages, and often water cooling. The Nd:YAG laser can be efficiently pumped with near-infrared diode lasers, which in turn can be excited with low voltages. On one hand, green "laser" light can now be produced at levels sufficient for a good laser pointer using a small battery as a power source. On the other hand, starting with high-power laser diodes, it is possible to produce several watts of this light. Many other wavelengths can be generated as second and higher harmonics of other lasers.

> *Take-Away*
>
> **Frequency-Doubled Nd:YAG:**
> One of the great success stories of non-linear optics is the green laser pointer. A battery-powered diode at 780 nm excites an Nd:YAG laser generating 1064 nm laser light. This light excites a KTP crystal to generate coherent second-harmonic light at 532 nm.
>
> The concept can be scaled to produce bench-top green lasers at hundreds of milliwatts of power.

Second harmonic generation also occurs in natural materials, including biological ones[68], and is useful in microscopy[72]. In particular, collagen, with its triple helix, generates a strong second harmonic when excited with near-infrared light.

Take-Away

Second Harmonic Microscopy:
Second-harmonic generation is used in microscopy to visualize certain biological materials that have a high $\chi^{[2]}$. Most notable among these materials is collagen. Because the process is coherent, the second harmonic is generated in the "forward" direction, parallel to the incident light.

In Practice

Second-harmonic generation is inherently a weak process, so in practice, high-power, short-pulse laser sources are required. Although the average power may be low, instantaneous power must be high. Mode-locked lasers (see Section 10.2) such as titanium-sapphire are usually used.

Backward second-harmonic light is often detected. This light is mostly forward-generated second harmonic that has been scattered by the tissue.

Another common second-order process is the mixing of two waves at different frequencies to generate sum and difference frequencies. In electronics, a mixer is employed to multiply two signals at different frequencies. The product of two cosine waves is

$$\cos\omega_1 t\cos\omega_2 t = \frac{1}{2}\cos(\omega_1+\omega_2)t + \frac{1}{2}\cos(\omega_1-\omega_2)t, \tag{14.14}$$

so the output of an ideal mixer will contain two signals, one at the sum frequency and one at the difference. An alternative formulation is to write the fields in the frequency domain, being careful to include the negative-frequency contributions. Multiplication in the time domain is convolution in the frequency domain, as shown in Figure 14.4.

In electronics, the mixing is often achieved by adding the two signals, imposing the sum on a nonlinear element (e.g., a diode), and filtering out (or ignoring) the unwanted frequencies. The same approach is used in optics. If two fields with different frequencies, ω_1 and ω_2, are incident on a material with a nonzero $\chi^{[2]}$, the polarization will include terms at the sum and difference frequencies:

$$\begin{aligned}\mathbf{D}_r = {} & \epsilon_0(1+\chi)\left[\mathbf{E}(\omega_1)e^{j\omega_1} + \mathbf{E}(\omega_2)e^{j\omega_2}\right.\\ & \left. + \mathbf{E}^*(\omega_1)e^{-j\omega_1} + \mathbf{E}^*(\omega_2)e^{-j\omega_2}\right]\\ & + \epsilon_0\chi^{[2]}\left[\mathbf{E}(\omega_1)e^{j\omega_1} + \mathbf{E}(\omega_2)e^{j\omega_2}\right.\\ & \left. + \mathbf{E}^*(\omega_1)e^{-j\omega_1} + \mathbf{E}^*(\omega_2)e^{-j\omega_2}\right]^2.\end{aligned} \tag{14.15}$$

We can separate this equation into complex-conjugate pairs such as the one at the difference frequency:

$$\mathbf{D}(\omega_2-\omega_1) + \mathbf{D}^*(\omega_2-\omega_1)$$

where

$$\mathbf{D}(\omega_2-\omega_1) = \epsilon_0\chi^{[2]}\mathbf{E}(\omega_2)\,e^{j\omega_2}\mathbf{E}^*(\omega_1)\,e^{-j\omega_1}$$
$$\mathbf{D}^*(\omega_2-\omega_1) = \epsilon_0\chi^{[2]}\mathbf{E}(\omega_1)\,e^{j\omega_1}\mathbf{E}^*(\omega_2)\,e^{-j\omega_2}, \tag{14.16}$$

the second harmonic of ω_1

$$\mathbf{D}(2\omega_1) = \epsilon_0\chi^{[2]}\mathbf{E}^2(\omega_1)\,e^{j\omega_1}, \tag{14.17}$$

and similar terms at $\omega_1+\omega_2$ and $2\omega_2$. The fields, $\mathbf{E}$, may, of course, be functions of position, arising, for example, from plane waves in different directions. The waves generated by the various frequency-domain components of $\mathbf{D}_r$ will have also spatial variations. It can be shown that for each component, photon energy and momentum, $\mathbf{p} = h\mathbf{k}$ (Equation 1.73), will both be conserved. Thus, for the sum frequency term, recalling that energy is proportional to frequency (Equation 1.70)

Eq. 1.73: $|p_1| = \frac{h}{\lambda}$.

$$\omega_{sum} = \omega_1+\omega_2, \tag{14.18}$$

Eq. 1.70: $J_1 = h\nu$.

and

$$\mathbf{k}_{sum} = \mathbf{k}_1+\mathbf{k}_2, \tag{14.19}$$

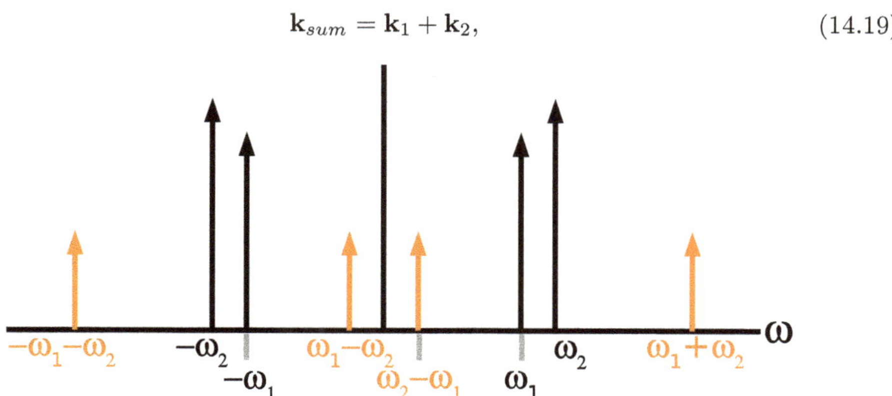

FIGURE 14.4
Mixing in the frequency domain. Multiplication in the time domain is convolution in the frequency domain. The sum and difference frequencies arising from two inputs (in black) are shown in the output (color). Also present, but now shown are the second harmonics of both inputs.

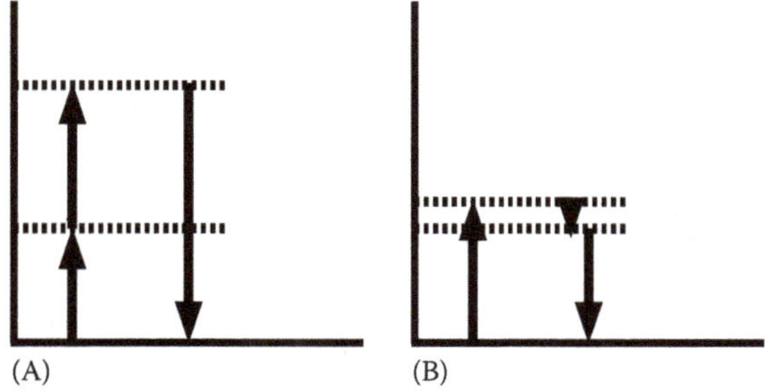

FIGURE 14.5
Generation of (A) sum and (B) difference frequencies. These two processes are second order. Generation of a sum frequency involves absorption of two photons and emission of one. Generation of a difference frequency involves the absorption of the higher-energy input photon and emission of a photon at both the lower input frequency and the difference frequency.

while for the difference term

$$\omega_{dif} = \omega_1 - \omega_2, \tag{14.20}$$

and

$$\mathbf{k}_{dif} = \mathbf{k}_1 - \mathbf{k}_2. \tag{14.21}$$

The energy-level diagrams are shown in Figure 14.5. In sum generation, photons are absorbed from both inputs, and a single photon with energy equal to the sum of the two energies is emitted.

Although a modification of the simple model in Figure 14.1 will explain the generation of both sum and difference frequencies, we need to examine the energy-level diagrams in Figure 14.5 to see that the process that generates a sum depletes each of the incident waves, whereas the process that generates a difference amplifies the wave at the lower frequency. A photon is absorbed from the signal with the higher frequency, but an extra photon is emitted at the lower frequency, along with the expected difference frequency. This gain, used in a cavity, is the basis of operation of an **optical parametric oscillator** or OPO[78]. The OPO can be used as a tunable laser source.

The equations for conservation of momentum (**k**) follow the same rules. The sum frequency will propagate in the direction of the sum of the input **k** vectors, and the difference will propagate in the direction of their difference.

Take-Away

Directionality of Harmonics:
Generation of sum and difference frequencies requires conservation of both energy and momentum. As a result, the directions of propagation of the waves are constrained.

In Practice

Despite the complexity of an excitation laser and a frequency-mixing crystal, the OPO is often used when a high-power tunable source is required. While some lasers are tunable, the available wavelengths often are not suitable for the application at hand.

Another second-order process is stimulated Raman scattering. We discussed Raman scattering briefly in Section 1.6. In Raman scattering, a photon is absorbed, exciting the material to a virtual state, and a photon is emitted at a lower energy or longer wavelength, with the material ending in an excited state, which typically has energy that is characteristic of an infrared frequency. Raman scattering is a linear process. However, the signal can be enhanced by introducing light at the Stokes-shifted frequency. If the second wavelength is tuned across the spectrum, it will be amplified just as in a parametric oscillator. This process, shown in Figure 14.3B can be used to generate gain in a Raman laser. The anti-Stokes-shifted frequency will suffer a loss. The energy-level diagram is shown in Figure 14.3C. For further information on stimulated Raman scattering, the reader may find a recent article useful[179].

Fig. 14.3:

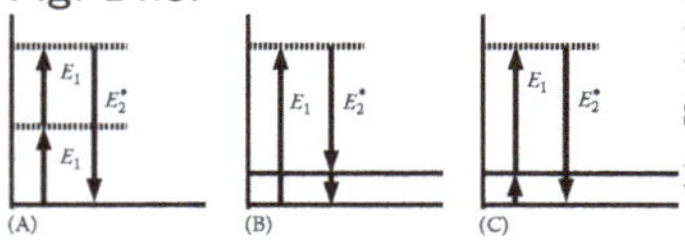

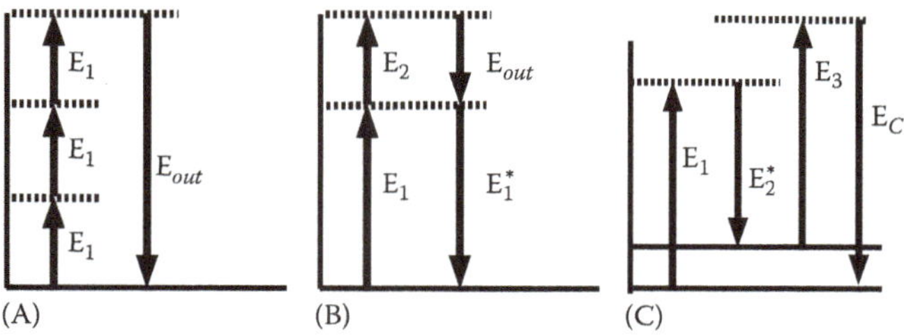

FIGURE 14.6
Some third-order processes. Processes involving $\chi^{[3]}$ include (A) third-harmonic generation, (B) irradiance-dependent index of refraction, and (C) coherent anti-Stokes Raman scattering (CARS). In CARS, often the pump $\mathbf{E}_1$ and probe $\mathbf{E}_3$ are the same.

> *Take-Away*
>
> **Raman Spectroscopy:**
> Raman spectroscopy is a powerful tool for identifying composition of materials. Stimulated Raman spectroscopy is a non-linear extension of the concept. These techniques are used in biology, material science, and more.

14.4.3 $\chi^{[3]}$ Processes

If we combine multiple fields in second-order processes, we expect to generate signals related to the product

$$\left(\mathbf{E}_1 + \mathbf{E}_2 + \cdots + \mathbf{E}_1^* + \mathbf{E}_2^* + \cdots\right)^2, \tag{14.22}$$

which consists of all possible products of fields taken two at a time. For third-order processes, there is a richer collection of possibilities, because the signals generated are related to the product

$$\left(\mathbf{E}_1 + \mathbf{E}_2 + \cdots + \mathbf{E}_1^* + \mathbf{E}_2^* + \cdots\right)^3, \tag{14.23}$$

which includes all possible products of fields taken three at a time. The second-order susceptibility tensor, Equation 14.7, is three-by-six. To consider all possibilities in third order, the tensor, $\chi^{[3]}$ is three-by-ten. The full list of possibilities is too long to consider here so we will look at two types of applications.

> Eq. 14.7: $\mathbf{D}_r = \epsilon_0\mathbf{E}_r + \epsilon_0\chi\ \mathbf{E}_r + \epsilon_0\chi^{[2]}\mathbf{E}_r{}^2$.

The first application is generation of the third harmonic, which proceeds using the same approach as second-harmonic generation using an equation analogous to Equation 14.17:

$$\mathbf{D}(3\omega) = \epsilon_0\chi^{[3]}\mathbf{E}^3(\omega). \tag{14.24}$$

In the energy-level diagram, we have three photons being absorbed and one being emitted, as shown in Figure 14.6A.

> **In Practice**
>
> Generally, because the simultaneous arrival of three photons is even more rare than the arrival of two, third-harmonic generation is often weaker than second-harmonic. However, in centro-symmetric crystals the symmetry ensures that there will be no

second-harmonic. In amorphous materials, the random orientation has the same effect. In these cases, the first non-zero harmonic is 3ω.

Another important process is called **irradiance-dependent index of refraction**, which arises from terms in the product (Equation 14.23) like

$$\mathbf{E}_1\mathbf{E}_1^*\mathbf{E}_2. \tag{14.25}$$

The electric displacement vector associated with $\mathbf{E}_2$ becomes

$$\mathbf{D}_2 = \epsilon_0\left(1+\chi\right)\mathbf{E}_2 + \epsilon_0\chi^{[3]}\left(\mathbf{E}_1\mathbf{E}_1^*\right)\mathbf{E}_2, \tag{14.26}$$

consisting of the usual linear term and the third-order term. We can regroup terms and say

$$\mathbf{D}_2 = \epsilon_0\left[1+\chi+\chi^{[3]}\left(\mathbf{E}_1\mathbf{E}_1^*\right)\right]\mathbf{E}_2 = \epsilon_{eff}\mathbf{E}_2. \tag{14.27}$$

We can therefore say that the material is described by

$$\epsilon_{eff} = \epsilon_0 n_{eff}^2, \tag{14.28}$$

where n_{eff} is a function of $\mathbf{E}_1\mathbf{E}_1^*$:

$$n_{eff}^2 = 1+\chi+\chi^{[3]}E_1E_1^*. \tag{14.29}$$

Take-Away

Irradiance Modulation:
Irradiance modulation of the index of refraction is a $\chi^{[3]}$ process. The irradiance of one light wave modulates the index of refraction for another.

The concept of irradiance modulating the index of refraction is often useful in holography, illustrated conceptually in Figure 14.7. We will digress for a few paragraphs to discuss the subject of generating and reading holograms. We will see that a hologram is generated by modulating the transmission of a medium with the irradiance that results from the mixing of two fields. Because we have just shown that the third-order susceptibility results in an irradiance-dependent index of refraction, we will then return to show how processes similar to holography can be implemented in such a medium.

14.4.3.1 Dynamic Holography

A hologram is generated by mixing two waves, $\mathbf{E}_{ref}$ and $\mathbf{E}_{object}$. The resulting hologram has a transmission H given by

$$H = \left|\mathbf{E}_{ref}+\mathbf{E}_{object}\right|^2 = \\ \left|\mathbf{E}_{ref}\right|^2 + \left|\mathbf{E}_{object}\right|^2 + \mathbf{E}_{ref}\cdot\mathbf{E}_{object}^* + \mathbf{E}_{ref}^*\cdot\mathbf{E}_{object}, \tag{14.30}$$

where we have neglected scale factors. These scale factors are critical in the implementation, but are best ignored in developing the concept. The hologram can be played back using a playback wave $\mathbf{E}_{playback}$, to produce $\mathbf{E}_{image}$.

$$\mathbf{E}_{image} = H\mathbf{E}_{playback}. \tag{14.31}$$

If the playback wave is the same as the reference, then the fourth term of the four in Equation 14.30, using Equation 14.31, is the image

$$\mathbf{E}_{image} = \cdots + \mathbf{E}_{playback}\mathbf{E}_{ref}^*\cdot\mathbf{E}_{object} = \cdots + \left|\mathbf{E}_{ref}\right|^2\cdot\mathbf{E}_{object}. \tag{14.32}$$

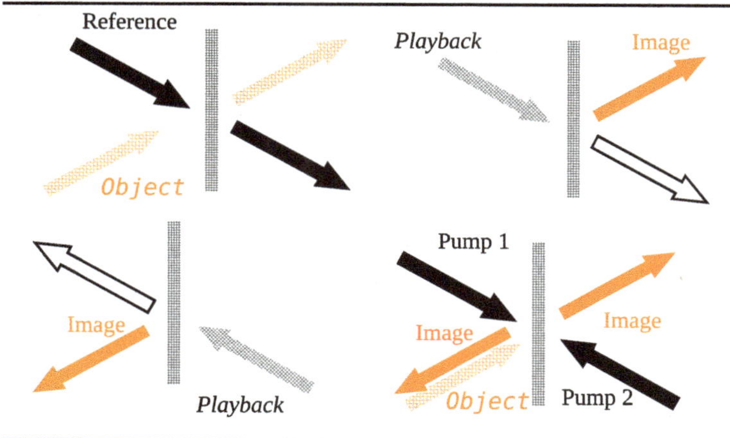

FIGURE 14.7
Holography and phase conjugation. (A) A hologram is recorded by mixing the signal with a reference, and recording on a nonlinear medium, which responds to the irradiance. Often film is used as the nonlinear medium. (B) The hologram can be "played back" using a beam just like the reference. (C) Alternatively, it can be played back with a beam that is conjugate to the reference to generate a conjugate image. (D) If the nonlinear medium responds fast enough, we can make a dynamic hologram, and perform phase conjugation.

The image wave is a reproduction of the object wave, scaled by the irradiance of the reference and playback waves and the constants we have omitted.

In Equation 14.32, we can increase the contrast of the hologram by increasing the reference wave, up to the point where we cause saturation of the film (a nonlinear effect beyond the desired one). We can increase the brightness of the image by simply increasing the amplitude of the playback wave. Thus the holographic technique can amplify a weak object wave.

If we play the hologram back with a beam in the opposite direction from the reference beam,

$$\mathbf{E}_{playback} = \mathbf{E}^*_{ref}$$

the third term forms an image wave, which is the conjugate of the object wave:

$$\mathbf{E}_{image} = \cdots + \mathbf{E}_{playback}\mathbf{E}_{ref}\cdots$$
$$\mathbf{E}^*_{object} + \cdots = \cdots + \left|\mathbf{E}_{ref}\right|^2 \cdots \mathbf{E}^*_{object} + \cdots . \tag{14.33}$$

We can obtain gain in Equation 14.33 to make a conjugate wave stronger than the original object wave, just as we did for the image. Later we will see that we can replace the film with a faster nonlinear medium, and amplify waves, for a variety of applications.

14.4.3.2 Phase Conjugation

We will briefly discuss one of the important third-order processes associated with irradiance-dependent index of refraction. These processes have the name **four-wave mixing**, because they combine three input and one output wave. In Figure 14.7C, we showed that a hologram can be "played back" with a wave conjugate to the reference wave, producing a wave conjugate to the object wave. Figure 14.8A shows a schematic diagram of a system for **phase conjugation**. The medium here is a nonlinear crystal, which has a $\chi^{[3]}$. The mixing process here includes the object beam and two pump beams. One of the pump beams serves the function of the reference wave and the other that of the playback

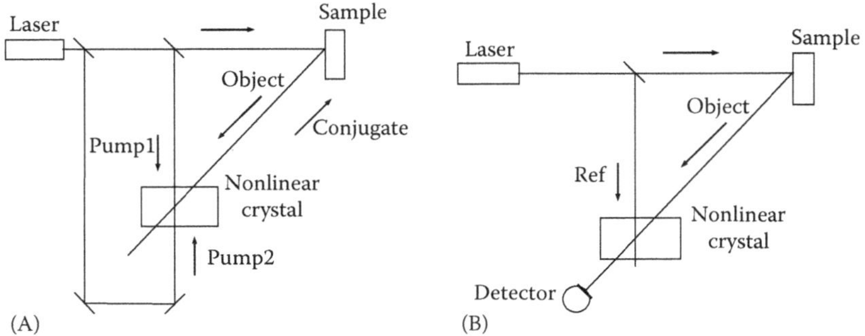

FIGURE 14.8
Applications of third-order processes. A nonlinear crystal can be used to conjugate a wave, or to amplify it. (A) Phase conjugation. (B) Wave amplification.

wave. The roles are in fact arbitrary. The source term for the conjugate wave is

$$\epsilon_0 \chi^{[3]} \left(\mathbf{E}_{object} + \mathbf{E}^*_{object}\right) \left(\mathbf{E}_{pump1} + \mathbf{E}_{pump1}\right) \left(\mathbf{E}_{pump2} + \mathbf{E}_{pump2}\right) \tag{14.34}$$

$$\mathbf{D}_{conj} = \epsilon_0 \chi^{[3]} \mathbf{E}^*_{object} \mathbf{E}_{pump1} \mathbf{E}_{pump2}, \tag{14.35}$$

where out of all the terms in Equation 14.34 we have chosen the one that produces the conjugate wave in Equation 14.35. If the second pump has the same wavefront shape as the first, but is traveling in the opposite direction

$$\mathbf{E}_{pump2} = C\mathbf{E}^*_{pump1}, \tag{14.36}$$

where C is a constant, then

$$\mathbf{D}_{conj} = \epsilon_0 \chi^{[3]} \mathbf{E}^*_{object} C \left|\mathbf{E}_{pump1}\right|^2. \tag{14.37}$$

This source will result in a propagating wave conjugate to E_{object} with irradiance proportional to the geometric mean of the irradiances of the two pump fields:

$$C \left|\mathbf{E}_{pump1}\right|^2 = C \left|\mathbf{E}_{pump1}\right| \left|\mathbf{E}_{pump2}\right| = C \frac{\sqrt{P_{pump1} P_{pump2}}}{\sqrt{A_{pump1} A_{pump2}}},$$

where A_{pump1} and A_{pump2} are the beam areas, and we have assumed uniform irradiance. If we reverse the roles of the two pump beams, we simply change subscripts and arrive at the same result. Thus we can view the first pump beam as reference and the second as playback, or vice versa.

The terms reference and playback are arbitrary here. What is important is that there are two counter–propagating pump waves.

Suppose that the object consists of a small scattering medium embedded in a material with a spatially varying index of refraction. Regions where the spatial variation is large will be scattering regions, but let us assume that the spatial variations are not that strong, except for one point, which we will call the object. One example could be a missile illuminated by a laser. Atmospheric turbulence causes variations in the index of refraction. The object scatters a spherical wave, but the wave is distorted by this turbulence, so that the transmitted laser beam is not focused well on the object, and the received light cannot produce a good image. The conjugate wave will return through the same turbulence, and will ultimately become a converging spherical wave focused on the object. We can call this system a **phase conjugate mirror**. Unlike a conventional mirror that would result in further distortion of the wavefronts, this one returns exactly the desired wave. Also unlike a conventional mirror, the "reflectivity" is controlled by $\sqrt{P_{pump1} P_{pump2}}$, so we can make a phase conjugate mirror with a reflectivity much greater than unity, with the extra power coming from the pump waves. The first work in phase conjugation was done in the mid-1970s[255;22;94;113], and a particularly good overview of the early work was written by David Pepper in 1986[174].

In Figure 14.8B, we see an example of amplification of a signal. The mathematical analysis follows the same approach as phase conjugation, except for the lack of a second pump beam. The interesting term in Equation 14.23 is

$$\mathbf{D}_{image} = \epsilon_0 \chi^{[3]} \left(\mathbf{E}_{object} \mathbf{E}^*_{ref} \right) \mathbf{E}_{ref} = \epsilon_0 \chi^{[3]} \mathbf{E}_{object} \left| \mathbf{E}_{ref} \right|^2 . \tag{14.38}$$

instead of Equation 14.37. The output wave will be a scaled copy of the input object wave, with a scale factor that can exceed unity if the reference wave is sufficiently strong.

Take-Away

Dynamic Hologram:
We can generate dynamic holograms to amplify a signal wave or to phase conjugate it. Phase conjugation is achieved with two counter–propagating pump waves.

14.4.3.3 Self-Focusing

Another application of third-order effects we will mention is that of **self-focusing**, in which a light beam can focus (or defocus) itself. We briefly mentioned gradient index materials in Chapter 1. Generally light refracts toward regions of high index of refraction. The important term here is

$$\mathbf{D}_1 = \epsilon_0 \chi^{[3]} \mathbf{E}_1 \mathbf{E}_1^* \mathbf{E}_1 = \left[\epsilon_0 \chi^{[3]} \left| \mathbf{E}_1 \right|^2 \right] \mathbf{E}_1, \tag{14.39}$$

in which the index of refraction is

$$n = \sqrt{\epsilon_0 \left(1 + \chi + \chi^{[3]} \left| \mathbf{E}_1 \right|^2 \right)}. \tag{14.40}$$

Thus, for positive $\chi^{[3]}$, the index of refraction will be highest at the strongest part of the field. The field will then be focused by the resulting gradient index lens.

Take-Away

Self-Focusing:
Self-focusing is a $\chi^{[3]}$ process to be expected in beams with high power. It may be offset by thermal lensing to be discussed in Section 14.4.5.

14.4.3.4 Coherent Anti-Stokes Raman Scattering

Our final third-order process is **coherent anti-Stokes Raman scattering** (CARS)[57]. In this process, as shown in Figure 14.6C, a pump photon is absorbed, and a Stokes-shifted photon is emitted through a stimulated-emission process, bringing the material to a real intermediate state. A second pump, often at the same frequency as the first, excites another virtual state, leading to emission of the "CARS photon" as the process returns the material to the ground state. Keeping fields $\mathbf{E}_1$ and $\mathbf{E}_3$ at fixed frequencies (often the same) and tuning $\mathbf{E}_2$, the Raman signals can be used to distinguish among different materials.

14.4.4 Nonparametric Processes: Multiphoton Fluorescence

A parametric process is one in which photon energy is conserved.

Before leaving the subject of nonlinear susceptibility, we take a moment to discuss the term **parametric**. We define a parametric process as one in which the initial and final energy states of the material are the same[31]. Thus, when all the states in our energy-level diagrams are virtual, the process is always parametric. We can still use the idea of a nonlinear susceptibility in nonparametric cases, but the susceptibility must be complex to account for the loss or gain of photon energy. Parametric processes include harmonic generation, and generation of sum and difference frequencies. The optical parametric oscillator derives its name from this term. Nonparametric processes include fluorescence, two-photon excited fluorescence[50], and Raman scattering.

Using two-photon excited fluorescence (Figure 14.9), we can excite the same fluorescence that we can in conventional fluorescence. Two photons are absorbed to raise the material's energy level to an excited state. Then the material decays nonradiatively to a lower state with a longer lifetime. From there, it decays to the ground state, emitting a photon. Two-photon excited fluorescence is a second-order nonparametric process, resulting in light at a wavelength longer than half the excitation wavelength. In comparison, second-harmonic generation is a second-order parametric process resulting in light at exactly half the wavelength of the input. The fluorescent emission occurs over the decay time of the intermediate state, typically nanoseconds. The second harmonic is generated instantaneously. Third-order (three-photon) and higher-order excitation of fluorescence is also common. In addition, stepwise multiphoton fluorescence can occur when real intermediate states are available. To mention one example, stepwise two-photon fluorescence has been observed in melanin[223]. This same process is also used to fabricate devices that convert near–infrared energy to visible.

In Practice

Viewing a near–infrared laser beam is a challenge during system alignment. Viewing cards can be purchased that are treated with a stepwise multi-photon fluorescent material that absorbs the near–infrared light and emit visible light.

Take-Away

SHG *vs.* Two-Photon:

Many readers are confused by second-harmonic generation (SHG) and two-photon fluorescence, $2h\nu$. Often both of these occur in biological materials. There are at least four fundamental differences;

- Temporal: SHG is instantaneous and $2h\nu$ fluorescence decays over a time that is typically measured in nanoseconds.

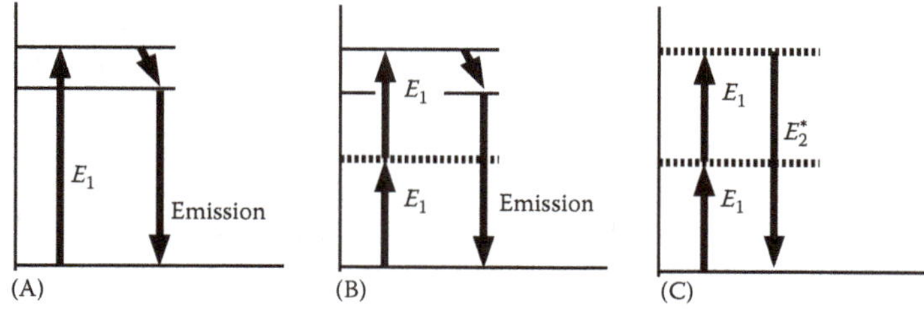

FIGURE 14.9
Two-photon excited fluorescence. In conventional fluorescence (A), light is absorbed at one wavelength and emitted at a longer wavelength. In two-photon excited fluorescence, it takes two photons to excite the material (B). This process is nonparametric, and occurs more slowly than the parametric process of second-harmonic generation (C).

- Spectral: SHG light is emitted at exactly half the excitation wavelength and $2h\nu$ fluorescence covers a broader band at wavelengths longer than half the excitation wavelength.
- SHG is coherent and forward-directed, while $2h\nu$ fluorescence is incoherent and random in orientation.
- SHG emission is inherently polarized. Fluorescence is usually unpolarized.

14.4.5 Other Processes

Some important nonlinear processes are not explained by the nonlinear susceptibilities. For example, most optical materials are at least slightly absorbing. As light is absorbed, the energy is converted to heat, and this heating changes the index of refraction, normally lowering it. The result then is an index of refraction that depends not only on the irradiance, but on its time history and on the thermal properties of the medium. The local heating from an irradiance, E, is given by

$$\frac{dT}{dt} = \frac{E\mu_a}{C}, \tag{14.41}$$

where

- μ_a is the absorption coefficient
- C is the specific heat

This source term is offset by the diffusion of heat from hotter to cooler regions. During a pulse of light, the material may heat, and after the pulse it will cool. If a continuous light source is turned on, the temperature will rise to some steady state distribution. If the change in index of refraction with temperature, dn/dT, is negative, as is usually the case, then the wavefront will be diverged as if by a negative lens. The divergence will reduce the heating effect, until an equilibrium is established. The time to reach equilibrium can be very long, lasting more than a few seconds, leading to complicated variations in beam profile with time. Various models have been used to predict the effect using closed-form equations[124;207] and computation[126;245]. Effects can be significant in laser materials[124], biological materials[162;236], and the atmosphere[69], where the term **thermal blooming** is used. The few references mentioned here are drawn from an extremely large body of literature on the subject. It is interesting to note that this **thermal lensing** process usually produces a result opposite that produced by the third-order process of self-focusing.

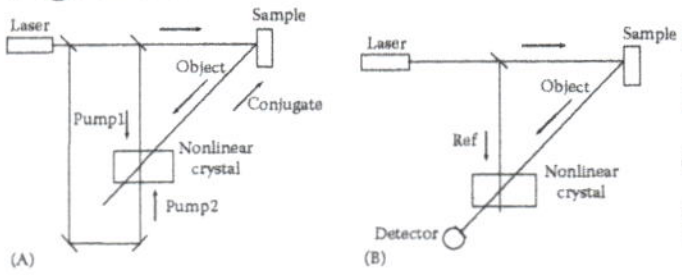

Another important nonlinear effect is the **photorefractive effect**. In a semiconductor material, bright light can excite some electrons to the conduction band as discussed in Section 13.4. The electrons diffuse in time, so there is a distribution of charge with more positive charge in the bright regions, and more negative charge in the dark ones. The resulting field produces a change in the index of refraction via the electro-optical effect. This process is slower than the $\chi^{[3]}$ effects, but, with time constants in the tens to hundreds of milliseconds, may be faster than the thermal effect. One example application is in the coherent detection of signals from highly scattering media. In Section 7.2.4, we discussed interferometric techniques such as laser radars, and saw that it was important for the reference wavefront to be well-matched to the signal one in order to achieve efficient mixing. If the scattering object is very diffuse, then the resulting signal wavefront can be complicated, and impossible to match effectively with a plane or spherical wavefront. Figure 14.8B shows a way around this problem. The wave from the object is mixed with a reference in a photorefractive crystal, resulting in amplification of the signal. However, if the object is moving, the photorefractive crystal will respond to the signal integrated over earlier times, and the amplified output of the photorefractive crystal as a function of time will be

Fig. 14.8:

$$\mathbf{E}_{prc}(x,y,t) = C\frac{1}{\tau}\int_{t-\tau}^{t} \mathbf{E}_{object}(x,y,t')\, e^{j\omega t'} \times$$
$$\mathbf{E}^*_{ref}(x,y)\, e^{-j\omega t'} \times \mathbf{E}_{ref}(x,y)\, e^{j\omega t} dt', \tag{14.42}$$

$$\mathbf{E}_{prc}(x,y,t) = C\frac{1}{\tau}\int_{t-\tau}^{t} \mathbf{E}_{object}(x,y,t')\, \mathbf{E}^*_{ref}(x,y) \times$$
$$\mathbf{E}_{ref}(x,y)\, dt' \times e^{j\omega t}, \tag{14.43}$$

where C is a constant. If the surface is vibrating, then

$$\mathbf{E}_{object}(x,y,t) = \mathbf{E}_0(x,y)\, e^{jk\delta e^{j2\pi f_v t}} e^{j\omega t}. \tag{14.44}$$

Then,

$$\mathbf{E}_{prc}(x,y,t) = C\frac{1}{\tau}\mathbf{E}_0(x,y)\, \mathbf{E}^*_{ref}(x,y)\, \mathbf{E}_{ref}(x,y)\, e^{j\omega t}. \tag{14.45}$$

The shape of the wavefront stays constant at the average value, but the position and the resulting phase of the optical signal vary at the vibration frequency, f_v. Thus, the averaged wavefront, obtained by the system in Figure 14.8B, is exactly the right wavefront to mix with the object wave for high mixing efficiency (Equation 7.59). The mixing signal is

Eq. 7.59: $\eta_{mix} = \int \frac{E_{sig}(x_d,y_d)E^*_{LO}(x_d,y_d)}{\sqrt{P_{sig}P_{LO}}} dA.$

$$\mathbf{E}_{object}(x,y,t)\, \mathbf{E}^*_{prc}(x,y,t) =$$
$$\mathbf{E}_0(x,y)\, e^{jk\delta e^{j2\pi f_v t}} C\frac{1}{\tau}\mathbf{E}^*_0(x,y)\, \mathbf{E}^*_{ref}(x,y)\, \mathbf{E}_{ref}(x,y). \tag{14.46}$$

Provided that the reference wave is uniform, this result is exactly what we want to maximize the mixing efficiency.

Take-Away

Photorefractive Crystal:

A photorefractive crystal provides another method for non-linear optical mixing. It is amenable to applications such as wavefront amplification and phase conjugation.

14.5 Summary

Non-linear optics is an important tool to produce coherent waves with desired characteristics. It also forms the basis of many different measurement techniques to determine material characteristics and morphology. We have presented a small subset of the many processes and applications along with some of the underlying theory. The interested reader will find detailed discussion of these and other topics in the literature.

15 Optical Breadboarding

In the previous chapters we've discussed the important concepts of optics from an engineering perspective, including the equations needed to understand the behavior of optical components, devices, and systems. We conclude the book with a chapter discussing how to build expertise in experimental work. Because so much laboratory work is done on optical breadboards or tables, we will focus the chapter on designing, building, and testing prototypes on such boards. This chapter will be useful for beginning optics researchers, but also for those using optics as a tool for research in biology, chemistry, physics, engineering, and other applications. For this reason, we attempt to make this chapter stand alone as much as possible.

For the researcher faced with the design of a new laboratory, we begin with a general discussion of some of the design considerations for such a facility, whether it be a small corner of a room with a small table for a single user or a large facility supporting multiple researchers. For the researcher moving into an existing group, we will then discuss some of the issues encountered in shared spaces.

In Section 15.3, we will focus on a few experiments that will introduce the reader to some of the basic components, instruments, and techniques used in prototypes. In each experiment, there is a background section with pointers to the relevant material in the earlier chapters and helpful hints, followed by suggested activities before, during, and after the experiment. The reader may be working in a laboratory with different equipment from different vendors, and may modify the experimental configurations to make use of what is available.

Finally, we include sections on cleaning optics and on laser safety. These sections are placed at the end to avoid interrupting the continuity of topics, but Section 15.5 should be read before starting any experiments involving lasers or entering a space where lasers are used.

15.1 Optics Research Laboratory

Here we discuss some of the issues one may encounter in designing a new laboratory for optics research. Ideally we want an environment with adequate space, with low vibrations and the ability to eliminate stray light. Of course, many compromises are necessary, particularly when building such a space within an existing building. This section may be of some interest even to those working in an existing laboratory.

15.1.1 Location and Facilities

It is likely that most readers will begin working in an existing laboratory with established facilities and an existing supply of equipment. However, we will begin with a discussion of some ideas on how to build a new laboratory from the beginning. Obviously such a task requires compromises and will vary depending on the type of work to be done, but some general guidelines are useful.

As we have seen in earlier chapters, optical experiments often involve measurements on a size scale of a fraction of the wavelength (sub-micrometer) and very low levels of light. Experiments at these levels will be very sensitive to vibration, temperature changes, and

DOI: 10.1201/9781315157047-15

stray light. Vibration of the laboratory floor can cause vibration of the optical components, varying path lengths and orientations. In some cases, optical instruments are sensitive to motion smaller than one nanometer. Temperature changes can cause changes in the size of mechanical components, and can change the index of refraction of air by a small amount, which can lead to changes in optical path length. These issues are particularly important for high-NA microscopy and for interferometry. Temperature changes can also change the length of optical fibers the thickness of coatings, and the index of refraction of all optical components, which can lead to changes in spectral properties of fiber Bragg gratings and dielectric filters. Almost any amount of stray light can be a problem in a system designed to detect a few photons per second. For these reasons, an ideal laboratory location is in the basement of a very stable building far from sources of vibration. The basement is usually the most stable location, provided that the room is not close to large machinery. Often basement laboratories do not have windows, and thus reduce the amount of stray light.

Windows pose two problems. First stray sunlight entering the room can corrupt measurements of low light levels. Second, if lasers are used in the laboratory, light exiting from the windows could pose a hazard to people outside. For this reason, usually windows should be covered with light-absorbing material. Sheets of cardboard may be sufficient to block stray light if careful attention is given to the edges. If high-power lasers are in use, a potential fire hazard exists and special laser curtains will be needed at the windows.

Stray light may also enter through cracks in walls and the space around and under doors. Special doors are available to block light entering, for example, using a shield that drops down to the floor level when the door is closed. It is a good idea to have all walls and ceiling painted flat black to absorb as much light as possible.

Entrance doors to the laboratory also require careful consideration. When Class-3b or Class-4 lasers are to be used in the laboratory, government or institutional regulations may require power interlocks on the doors so that power is turned off when someone enters the laboratory. A person entering the laboratory may thus interrupt the power to an experiment that requires hours of warm-up time to stabilize, and furthermore may allow light levels into the room that will contaminate signals at detectors, and even damage the detectors in the case of sensitive photomultipliers. Some laboratories have double doors or other configurations to prevent stray light, and an override switch for the interlocks that can be activated by authorized users of the laboratory.

The sensitivity of optical experiments to temperature suggests a need for careful attention to the heating, ventilation, and air-conditioning (HVAC) of the space. Calculations of HVAC requirements depend on many variables, including equipment, occupancy, outside temperature, insulation, and more. If the reader is fortunate enough to be involved with the laboratory design it would be useful to work with the HVAC company to address temperature stability as well as other needs such as air flow and filtration. Temperature control to less than one Kelvin (about 1.8 degrees Fahrenheit) may be needed in some cases, and it is often important to avoid air currents near the tables. To illustrate the sensitivity, a temperature change of one Kelvin (one degree Celsius) causes about 10^{-6} change in the index of refraction. Over ten centimeters of path length, this results in an optical path length change of about ten degrees of phase at the 632.8 nm wavelength of a helium–neon laser, which is much worse than the resolution of an interferometer. Enclosures for sensitive optical experiments may be required, even in a carefully controlled room.

Dust and air contaminants can degrade optical devices. If possible, the HVAC system should include HEPA filters. Depending on the work being done, a clean-room environment may be needed. In existing facilities an enclosure may be provided to meet the requirements for clean air.

Electrical, internet, and other utilities should be available from the ceiling or floor so that cables are not placed across the floor. Users working in the dark should not have to worry about where the cables are located. At least two elecrical drops per table, on

FIGURE 15.1
Optical table. This table is in the author's research lab. Note that the location against two walls is not ideal.

separate circuit breakers are recommended. Compressed air may be needed for optical tables, and other utilities may be needed for specific experiments (*e.g.* gas, vacuum, *etc.*). Because needs will change with time, having the utilities come from above is easier than having them built into the floor.

For some devices such as sensitive detectors, electro-static discharge should be considered. Grounding of the table, along with ESD straps and mats may be needed.

It is useful to have switches for the room lights near the location where the researchers will be working, as it is frequently necessary to turn the lights on to observe something and turn them off when experiments are being done. Benchtop task-level lights offer another alternative to allow the researcher flexibility in lighting. In very low-light-level experiments, the researcher will take up to 20 minutes to become adapted to working in a dark room, and it will be very inconvenient to turn on lights frequently. Some low-power red light-emitting diodes may be used to provide extra light with minimal impact on the user's night vision, because night vision is mediated by rhodopsin, which is less bleached by red light than by other colors.

15.1.2 Tables

The most sensitive optical work is done on "floating tables." The table surface is often a complicated structure with an aluminum or steel surface having threaded holes for mounting optical components. Most manufacturers provide different sizes and thicknesses of tables. Generally the thicker the table the more vibration is reduced. The table is generally mounted on four legs that use compressed air for vibration isolation. Frequently the legs are self-leveling, maintaining the table in a level configuration regardless of the load. If alignment tolerances are not severe, rigid legs can be used instead. A typical table with vibration isolation is shown in Figure 15.1.

The table height is an important consideration. Usually the ideal height is a bit higher than a desk, as most of the work is done while standing. A typical height is 35 inches or about 90 cm. from the floor. Keeping the table height well below eye level provides some extra safety when lasers are used.

15.1.3 Support Equipment

Most optical experiments require a variety of support equipment such as power supplies, signal generators, oscilloscopes, computers, and more. However many of these may have an adverse impact on the optical system. A frequent problem is the vibration introduced by fans in electrical equipment. Any instrument containing a fan should be located off the optical table. Any instrument that generates heat or light should be placed away from optical paths with which it could interfere. For example, indicator lights are a frequent source of stray light. Power cables, signal wires, and other connections between the table and the rest of the laboratory can couple vibrations to the table. When necessary, wires should be loose so that the transfer of vibration is minimized.

In recent decades, computers have become an essential tool of researchers in every discipline. In optics they are essential for data collection, image acquisition, and often experiment control. However, they provide new problems as sources of light, heat, and vibration. Computers and other electronic devices should be placed off the table because of potential vibration from fans and from typing on the keyboard. Of course they should also be located where the heat will not affect experiments. Perhaps the major problem is the light produced by the screen. This light can reach detectors and corrupt measurements and can also impair the user's night vision. If the experiment is being performed with narrow–band light from a laser or even an LED, a filter on the detector may provide some help but often not enough, depending on the wavelengths of the experiment and the computer screen.

Some tools that are useful to carry in a pocket or have easily at hand include

- A flashlight, possibly with a red filter, like the bench-top light discussed above.
- some white cards for viewing the light path (old business cards are good for this purpose),
- a small inspection mirror such as a dentist uses.

15.1.4 Basic Equipment

It will be helpful in starting a new lab to purchase some basic components with the understanding that later purchases will be needed for specific projects. The reader can look through the experiments in Section 15.3 to see representative equipment and supplies that will be useful as these experiments were designed to use many of the most common components. At a minimum an assortment of the following will be useful:

- Assorted posts, post holders, baseplates and lens mounts and an assortment of lenses. Often these are available in starter kits.
- Lens-cleaning paper.
- Mirrors, beamsplitters, filters, and iris apertures.
- Hardware kits including commonly used screws, washers, setscrews, etc.
- Polarizers, waveplates, polarizing and non-polarizing beamsplitters.
- Power meter, camera, spectrometer, computer.
- Tungsten light source with fiber coupler.
- Assorted LEDs with different wavelengths.
- Kinematic mounts.

A low-magnification microscope objective may be useful for epi-fluorescence measurements, and an Air-Force resolution chart is useful for almost all quantitative resolution experiments.

15.2 Starting in an Existing Lab

For the reader unfamiliar with an optics lab, there are particular challenges relating to such topics as the handling and care of optics, as well as new issues when working in shared space. it may be helpful to list some common rules

Don't touch the optics unless you know what you are doing. Fingerprints are detrimental to almost any optics. Expensive coatings can sometimes be damaged by a simple touch. Fingerprints on optical surfaces can become permanent sources of scattered light.

Be gentle with front-surface mirrors. At home, the reader probably has a mirror with a rear reflecting surface. The glass on the front protects the fragile reflective coating. The 4% reflection from the glass is normally not a problem, and the glass surface can be cleaned with household cleaners, provided that there are no hard particles such as sand present. In many optics experiments multiple reflections are a problem and so front-surface mirrors are used. These are more susceptible to damage.

Don't clean anything unless you know what you are doing. Again, many coatings can be damaged by cleaning. Special procedures can minimize the damage.

Don't turn on a laser unless you have been trained and are authorized to do so by your institution.

Be mindful of the beam path for your own safety and that of others in the room. Use beam blocks in unused outputs of beamsplitters. Be especially cautious during alignment as the beam may move off the path to a mirror and go in an unexpected direction.

Don't bump anything. Some alignment procedures can take hours or days and the slightest bump can spoil this work. If you do touch inadvertently something in another person's project, say something.

If you borrow test equipment or tools, leave a note. The other user may trust the settings on the instrument upon returning. Indicate that you borrowed it and the other user should check the settings before using it. Some groups have a policy of leaving the equipment unplugged when returning it so that the original user knows that it was borrowed. In other groups there may be a strict policy or custom against borrowing. Ask first. Have your own basic tools such as ball drivers, screwdrivers, and the like to minimize borrowing.

Be mindful of dust accumulation. If you leave anything uncovered, dust will accumulate over time. This is particularly important for cameras, gratings and optics with expensive coatings. Use a lens cap for a camera and cover expensive optics when leaving the lab at the end of a day.

Use washers. Reduce damage to mounting hardware by using washers as discussed later in Section 15.3.1, and shown in Figure 15.6.

Cables: When using support equipment, it is a good idea to attach cables to the table to avoid pulling on the equipment. See Figure 15.6.

Announce "lights on" and "lights off." Many optics experiments require very low background light levels, but in a shared space turning on or off the lights can present problems for other users. For example, a sensitive photomultiplier can be destroyed if room lights are turned on while it has high voltage applied. Use small local lights when possible rather than turning on and off the main room lights. Warn the group before changing the light level in the room.

Return unused components. It is easier to put things away in the right place when they are no longer needed than to try to find them later.

Overall, in shared space, start by talking to others, learn what they are doing and learn the rules and customs of the group.

15.3 Some Experiments

Here we include some experiments that introduce a variety of components, instruments, and procedures that are commonly used in optics laboratories. After completing these

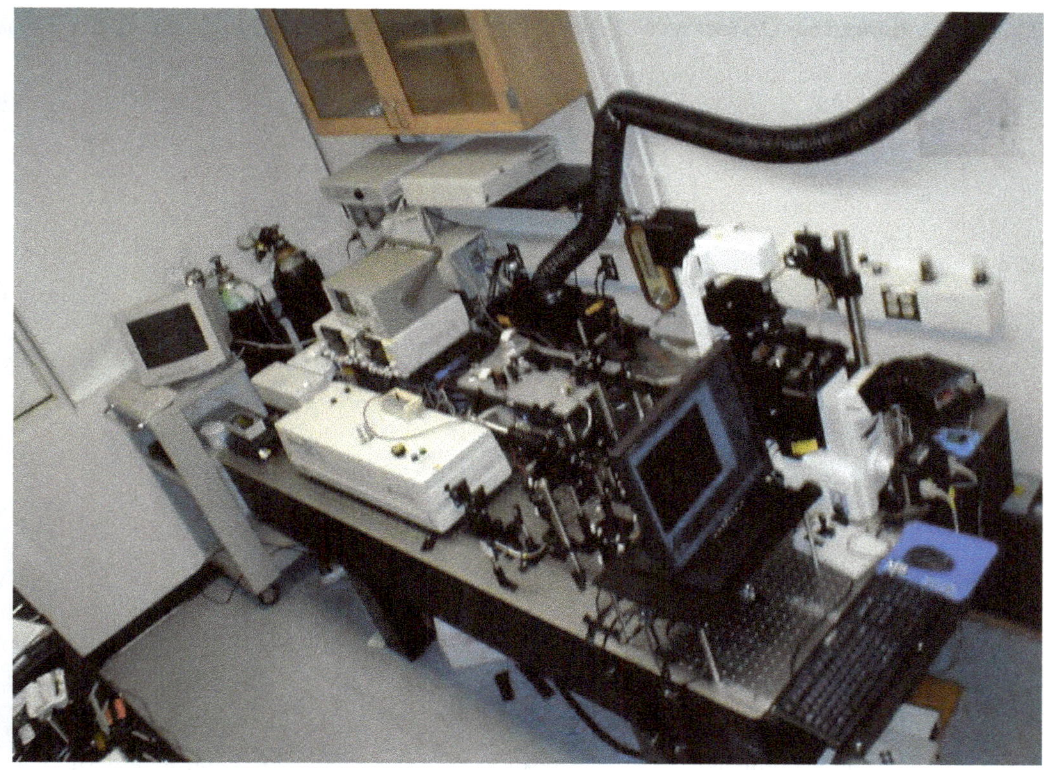

FIGURE 15.2
Multi-modal microscope. This instrument was built using the procedures that reader will learn in this chapter.

experiments, the reader will have some understanding of how to build a complicated optical breadboard system such as shown in Figure 15.2.

Each experiment includes a set of learning objectives, introducing new light sources, measurement devices, mounting hardware, and alignment techniques. Next, some background material discusses the underlying concepts as they relate to the experiment with pointers back into earlier chapters where the interested reader can find detailed discussion of them. Each experiment includes suggestions for pre-lab planning, implementation of the experiment, and post-lab reflections on what was learned. The first and last of these could be written homework assignments for a course.

Because users will have different equipment available the instructions are incomplete and may need modification. The skills to be gained by understanding the issues and making these modifications are an important part of the learning experience as these skills will be needed when the reader encounters more complicated real-life projects.

15.3.1 Basics: An Imaging System

We begin by building a simple imaging system with two lenses and an aperture. We have chosen to start with such a simple system because there are many basic concepts to be learned that will find use in the other projects. The reader may be surprised to learn how much time is required to build even a simple system that can be sketched on paper in a few seconds. Simply mounting one lens can take over a half hour, use several components costing over seventy US Dollars, and require more time for alignment. As the reader gains experience, each assembly will be easier, and in later sections we will be able to tackle more challenging experiments.

15.3.1.1 Learning Objectives

Some of the main learning objectives of this experiment are...

- Plano-convex lenses
- Working on the Bench
  - Basic mounting hardware
  - Alignment
- Incoherent Sources: Tungsten, LED
- Imaging: Lens Equation, Magnification, Telecentric Lenses
  - Numerical Aperture, Diffraction
  - Aberrations
- Linear Micropositioners
- The Air Force Resolution Chart
- Cameras
  - Pixels, Bit Depth and Dynamic Range
  - Integration Time, Gamma
  - Saturation, Clipping Blacks, Clipping Whites, Blooming.

15.3.1.2 Basic Setup

We will use an incoherent light source, a resolution target, two lenses and a camera. The numerical aperture of the system will be controlled by an iris. All components will be mounted on a breadboard or optical table.

The light source may be a fiber-coupled tungsten lamp, an LED, or almost any other light source that is readily available.

Fig. 3.10

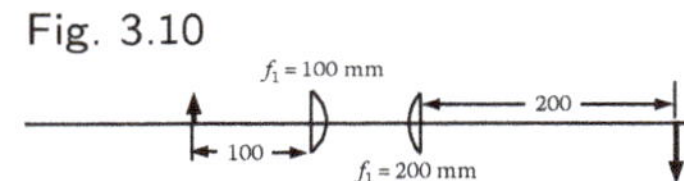

We will design a telecentric magnifier using two lenses like the one in Figure 3.10. The figure shows a 2× magnifier, but we can choose any magnification, recalling Equation 3.91.

Eq. 3.91:
$m = -f_2/f_1$.

We choose to mount the lenses using the fixed-diameter lens mounts, posts, post holders, and post holder bases shown in Figure 15.3.

In particular, for this experiment, we will mount the lens with a simple fixed-diameter lens mount as shown in Figure 15.4. We consider two options.

In the first, we have an unmounted lens. We first ensure that the retaining ring (B in the figure) is removed from the lens mount (A). Then we carefully remove the lens from its packaging holding it by the edges and using lens tissue or gloves to prevent oil from our fingers from reaching the surface. We carefully drop the lens into the mount and secure it in place gently tightening the retaining ring with a spanner wrench (C). See also Figure 15.7A for other spanner wrenches. Look carefully in B for the two slots in the ring that will hold the spanner wrench.

In the second case, we have a lens already mounted in a threaded mount. Figure 15.4D shows the mounted lens on the left and the lens mount on the right. Assuming both parts are from the same vendor, the threaded mount containing the lens will screw into the lens mount that we will put on the post. Before doing so, we should ensure that the retaining ring in the empty lens mount is removed or screwed in far enough to make room for the one containing the lens. Many researchers like to leave the ring in the mount so that it is not lost. The completed assembly is shown in D. Note that one option is to remove the lens from the original mount and use the first procedure above, but the original mount

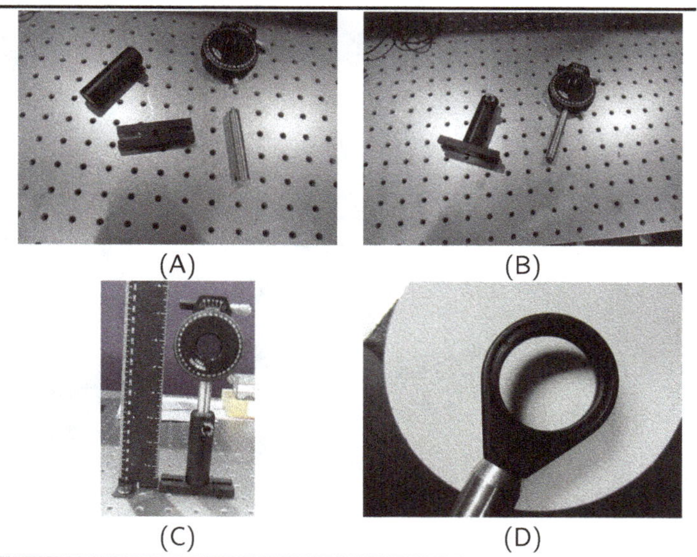

FIGURE 15.3
Optical components. The lenses, posts, post holders, and post holder bases are used to make the imaging system. The major components are shown in A. In this example, the optical mount is a rotational mount suitable for a polarizer. The assembly is the same for most other mounts. In B, the post holder and baseplate are assembled with a short 1/4–20 or M6 Screw, and the optical mount is attached to the post with an 8–32 screw. In C the whole assembly is shown with a ruler used to set the height. A simple lens mount assembled to a post is shown in D.

normally has a label indicating the lens's focal length and other information. When optical components are removed from the labelled mount, it is easy to forget which is which.

In Practice

If one side of the lens is to be placed close to another optical element, it may be important to think about the way the lens is mounted. For example, if there is an object close to the flat side of the lens, it is best to mount the lens with the flat side deepest in the mount. In Figure 15.4E the flat side would then be toward the bottom.

The bases will be secured to the table with 1/4–20 screws (Or M6, if the table was built to metric specifications).

Screw Sizes and Threads: It is helpful to understand the basic terminology of screw thread specifications. In the imperial system, screws are defined by their diameter and number of threads per inch. Diameter is denoted by a number up to Number 12, and larger diameters are indicated in inches. For example, an 8–32 thread specification indicates a "Number 8 diameter" with 32 threads per inch and a 1/4–20 has a quarter–inch diameter and 20 threads per inch. Metric thread specifications are normally given by the letter "M" and a number for the size, occasionally followed by the thread pitch (M4 or M4 x 0.7). Both the diameter and the pitch are in millimeters. M4 is close to 8–32 and M6 is close to 1/4–20. Commonly used sizes are shown in Table 15.1, and common screw types are shown in Figure 15.5. Often the length will be expressed in the appropriate units. For example, a 1/4–20 X 3/4 screw is 3/4 inch long, from below the head to the tip.

FIGURE 15.4
Mounting a lens. (A, B, C) Mounting an unmounted lens. (D, E) Mounting a lens in a lens holder.

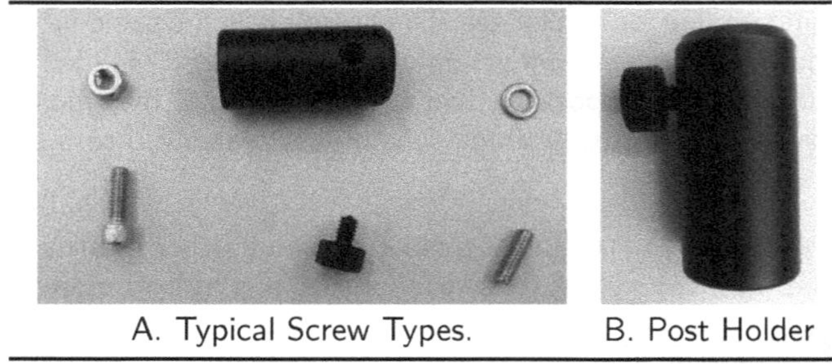

FIGURE 15.5
Common screw types. Cap screws (A, lower left) are generally used for assembling components where one side is accessible. Thumbscrews (lower center) are used for frequent adjustments such as post holders (upper center and B) Setscrews (lower right) are used in assembling parts such as posts and optics mounts, as well as some adjustments that are not done frequently. Nuts (upper left) and washers (upper right) are also used.

In Practice

Because both imperial and metric units are in wide use, it often happens that a laboratory will, over time, acquire both types of components. The 1/4–20 imperial and M6 metric screws are very nearly, but not quite, the same. It is possible to damage the threads by forcing one type of screw into the other type of threaded hole. It is good practice to try to keep the two separated, but inevitably errors will occur so it is important to be careful when mounting components. When in doubt, it is best to try an unknown screw with a nut known to be the correct type. Damaging a screw or nut is much less serious than damaging a table or expensive mount.

Avoid confusing 1/4–20 and M6 screws, nuts, and threaded holes. It is easy to damage the threads.

Usually these screws have an Allen or Hex head and it is useful to keep the appropriate Allen wrench or ball driver handy in the laboratory. In some cases it may be desirable to

TABLE 15.1
Common Screw Threads
A wide variety of threads are used for different purposes, but in bench-top optical assembly these are the most important. Metric threads are normally specified by the diameter in millimeters The thread pitch is often included, but in contrast to the imperial system, it is specified in millimeters per thread.

Imperial
2–56
4–40
6–32
8–32
10–32
1/4–20
Metric
M3
M4
M5
M6

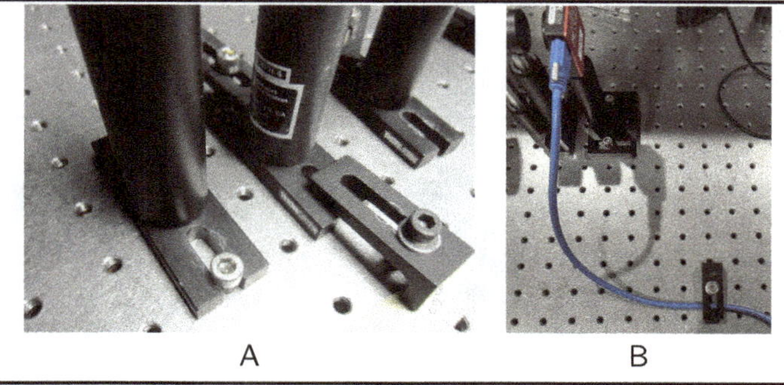

FIGURE 15.6
Angle brackets. (A) Holes in the table may not be located where we need them. In this case we can use angle brackets to secure the base plate to the table. Damage to the mounting hardware is visible and was caused by tightening mounting screws without washers. (B) Angle brackets can also be used to secure cables to the table.

use angle brackets to secure the baseplates to the table as in Figure 15.6 if the locations of the holes are not useful.

As mentioned earlier, it is best to use the mounting screws with a washer. Figure 15.6 shows one angle bracket with a washer and a baseplate without one. On each, there is visible damage to the mounting hardware from previous use without a washer. Such damage makes it more difficult to make small adjustments in position because the component will try to move so that the screw is in the same position as the one that caused the damage. See https://www.thorlabs.com/newgrouppage9.cfm?objectgroup_id=12211#Washers for further information.

Using washers avoids damage to components when tightening the screws.

We need to decide on a height for the optical axis of the imaging system. Too great a height will lead to vibration of the posts, which may impair the quality of the image. Too small a height will make it difficult to position our light source, object, and camera. It is best to agree on a height for a particular experiment and use a vertical ruler such as shown in Figure 15.7B. One can easily build something, such as a ruler taped to a block

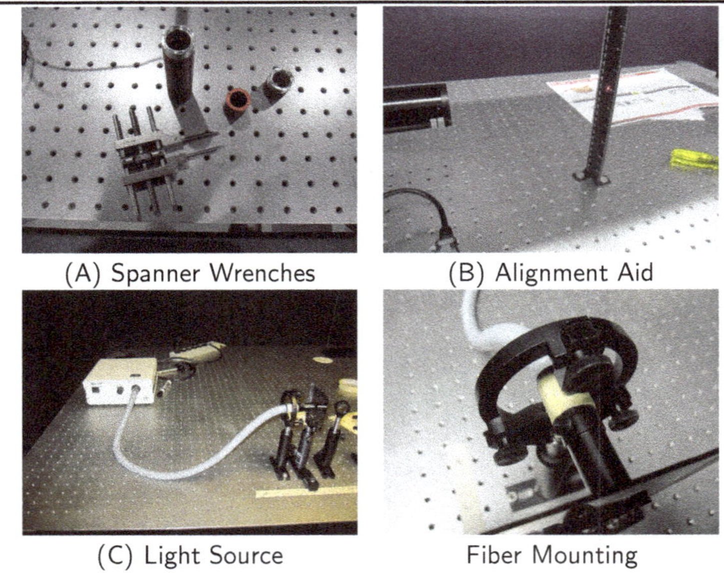

FIGURE 15.7
Some useful tools. These are some tools and components that will be useful in this and other experiments.

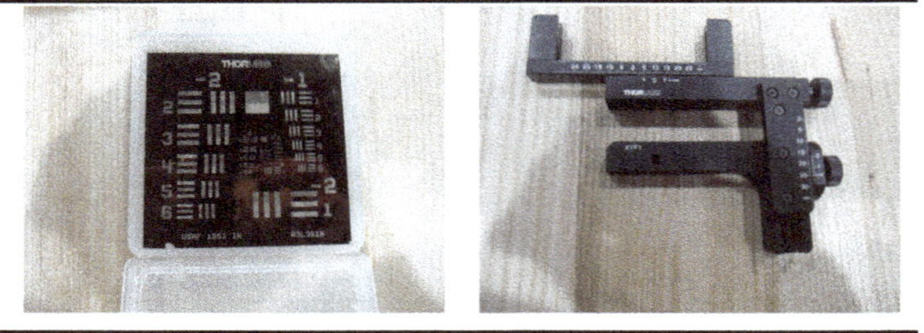

FIGURE 15.8
Air-Force resolution target. This large metal-on-glass target (A) can be mounted in the mount shown in B. The transverse position can be adjusted easily and precisely.

of wood, with a clip at the correct height instead. However, the ruler in the figure has a magnetic base and has small holes at selected heights which we will see later are useful for aligning a laser beam. Different alignment aids can be purchased, and it is not difficult to make your own.

We begin by determining locations for the object, the two lenses, and the camera. By using the slotted holes to secure the bases to the table, we will allow for more precise alignment later.

Fig. 11.13

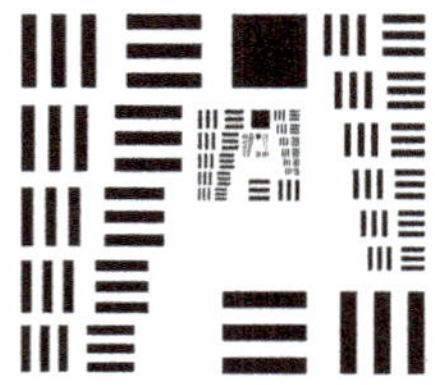

For the light source we can use a fiber–coupled white light source such as shown in Figure 15.7C. This source uses a quartz–halogen–tungsten light source with a fiber bundle to deliver the light. One method of mounting the fiber is shown in Figure 15.7D. We can also use a small V–block to mount the fiber or LED.

For an object, we use the Air–Force resolution chart shown in Figure 11.13. We want to be able to move the object transversely to select the right portion of the chart for our measurements. A variety of mounting hardware can be used for the object, but it is helpful to use something specifically designed for the task as shown in Figure 15.8.

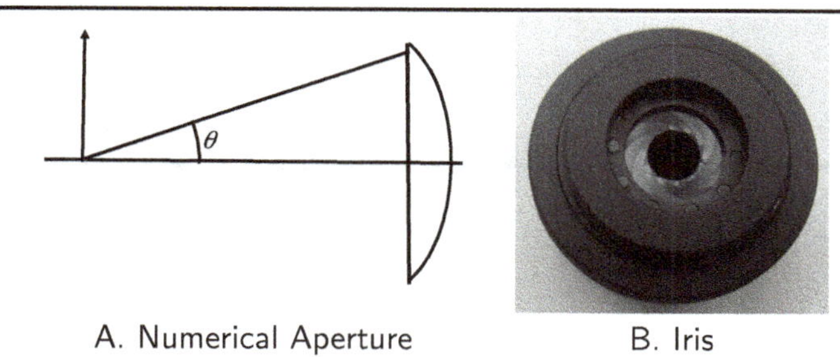

FIGURE 15.9
Numerical aperture. The numerical aperture (NA) of an optical system is defined as $n \sin\theta$ where n is the index of the ambient medium. An iris is useful for setting a variable aperture.

Next, we position the iris at the back focal plane of the first lens (front focal plane of the second), using the same mounting components we used for the lenses. The axial position of this iris is not too critical so we can place it by hand.

15.3.1.3 Resolution

The diffraction limit for resolution of the imaging system is determined by the numerical aperture and often given by the Rayleigh resolution according to Equation 8.86. However, the numerical value in the definition of resolution depends on many parameters as discussed in Section 8.7. For simplicity, we often use the approximation,

Eq. 8.86.
$\delta = 0.61\frac{\lambda}{NA}$.

Resolution: Sec. 8.7
At high numerical aperture resolution is better because of diffraction, but worse because of aberrations. The diffraction limit is fundamental, but aberrations can be reduced by using more complicated and expensive lenses.

$$\delta \approx \frac{\lambda}{NA}. \tag{15.1}$$

The numerical aperture is measured from the object or image as shown in Figure 15.9A. It can be controlled with a fixed aperture of diameter, d, or an iris as shown in Figure 15.9B.

The diffraction limited resolution decreases with increasing NA, but unavoidably aberrations in the optical system will cause the resolution to become worse at higher NA. We hope that the resolution will follow the diffraction limit for small NA and that aberrations will only become significant at NA beyond what we plan to use. If the aberrations are larger than we want, we will need to use a different lens design. The subject of aberrations is discussed in Chapter 5.

Aberrations: Ch. 5

We need to be careful to use the correct numerical aperture. Figure 15.10 shows that the numerical aperture in image space is different from that in object space. Specifically, $NA' = NA/m$. If we are interested in the resolution at the camera, to compare it to pixel size, we use

$$NA' = \sin\frac{D/2}{f_2}, \tag{15.2}$$

while if we are interested in the resolution at the object, to determine what features we can resolve, we use

$$NA = \sin\frac{D/2}{f_1}. \tag{15.3}$$

In practice we don't need to compute both of these because

$$\delta' = \delta \times m. \tag{15.4}$$

15.3.1.4 Camera

The camera will be placed at the back focal plane of the second lens, using a mounting post, post holder, and base. In this case a simple monochrome camera is the best choice.

CCD: Section 13.6.1.2

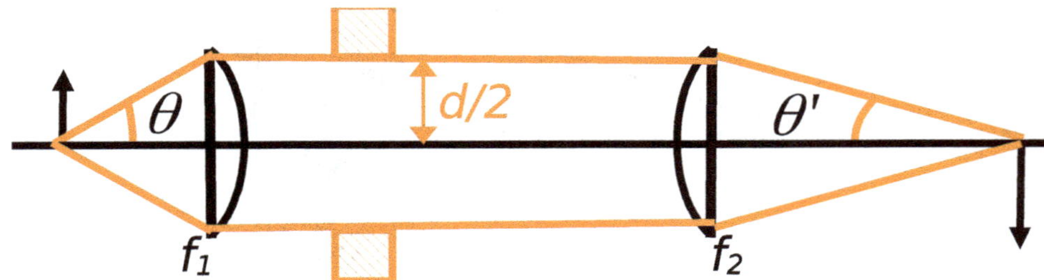

FIGURE 15.10
NA in object and image space. We can calculate the numerical aperture in either space; $\theta = \frac{d}{2f_1}$ or $\theta' = \frac{d}{2f_2}$.

Section 13.6.1.2 and Table 13.3 provide some information about cameras that may be useful here. However, in this simple demonstration of an imaging system, we only need to concern ourselves with pixel size and image size.

Fig. 4.30

The camera will form the field stop for this system and the iris in the pupil plane will form the aperture stop, as shown in Figure 4.30. The colored lines show the rays that pass through the aperture stop, and the black dashed lines show the limits of the field of view. It is useful to think of the field of view on the camera in two different ways. In the first, we consider the boundaries of a pixel to define the field stop.

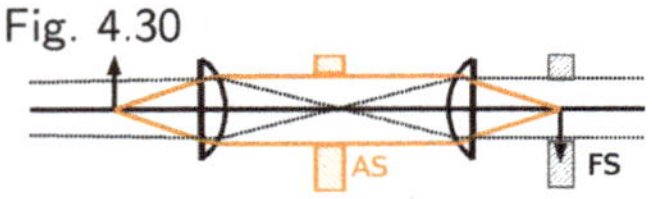

Example

For example, if the pixel size on the camera is $dx = dy = 5\,\mu\text{m}$ and the magnification of the system is

$$m = -\frac{f_2}{f_1} = 0.1,$$

then the light collected in a single camera pixel is from an area of size $5\,\mu\text{m}/0.1 = 50\,\mu\text{m}$ on the object. Any object smaller than this will not be resolved by the camera.

Although the size of a single pixel is important, we are also concerned with how big an object we can see. If the camera includes N_x by N_y pixels, then the total horizontal field of view is $FOV_x = dx \times N_x$ Likewise for height, $FOV_y = dy \times N_y$.

Example

Consider a microscope with a magnification of $m = -40$, with the same camera above, having $N_x = 1280$ and $N_y = 960$ pixels. Remember $dx = dy = 5\,\mu\text{m}$. The field of view of the camera (image space) is given by the full size of the sensor;

$$FOV_x = N_x dx = 1280 \times 5\,\mu\text{m} = 6.4\,\text{mm}$$

and

$$FOV_y = N_y dy = 960 \times 5\,\mu\text{m} = 4.8\,\text{mm}.$$

The field of view in object space is

$$FOV_{object} = FOV_{image}/m,$$

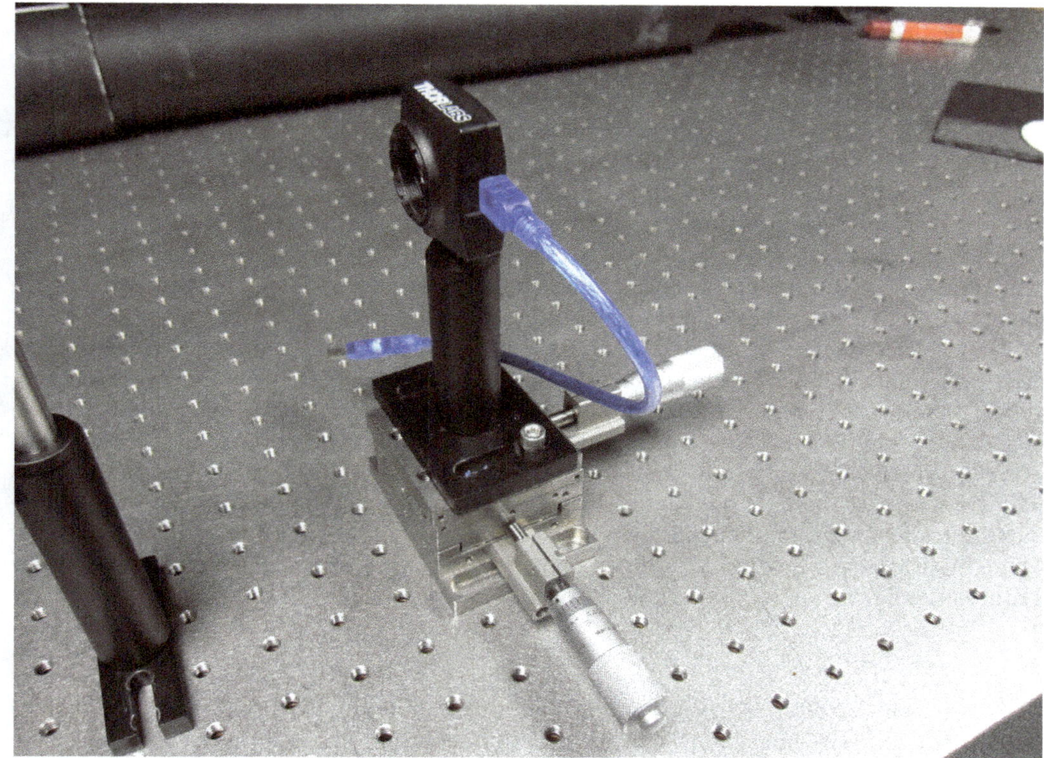

FIGURE 15.11
Camera. The camera is mounted on an axial micropositioning stage. The transverse micropositioner is not really necessary.

or 160 μm horizontally and 120 μm vertically, while the single-pixel field of view is 5 μm/40 = 125 nm. The resolution of the digital image is limited by the pixel size even if the resolution of the optical system is smaller than the pixel. A biological cell of diameter 15 μm will be imaged by 15 μm/0.125 μm = 120 pixels in each direction. A mouse embryo of 100 μm will almost completely fill the field of view, and anything much larger cannot be shown in a single image with this microscope.

As the focusing of the system is critical to producing a good image, we may want to mount the camera on an axial translation stage for micropositioning as shown in Figure 15.11. For our purposes, the actual distance between the lenses is not critical, so the intermediate image may not be exactly at infinity, but at least one degree of freedom is needed in the axial direction. For this reason we only microposition the camera. We recall that the depth of field is given by Equation 8.82, so even for relatively low numerical apertures, the depth of field is smaller than a millimeter. We need to position the camera to a small fraction of this depth. Here we use the image-space numerical aperture,

Eq. 8.82
$DOF \approx \frac{\lambda}{NA^2}$.

$$NA_{image} \approx \frac{D/2}{f_2}, \tag{15.5}$$

because we are interested in moving the camera. If we were to microposition the object, we would use the object-space numerical aperture,

$$NA_{object} \approx \frac{D/2}{f_1}, \tag{15.6}$$

where in both cases, D is the diameter of the aperture.

15.3.1.5 Alignment

Fig. 2.20

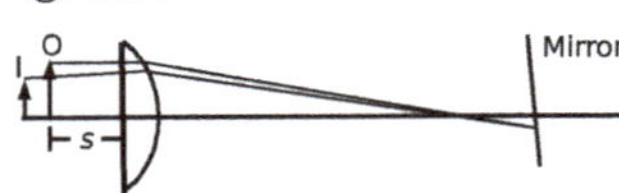

Autocollimation can be used to ensure that the object is at the front focal plane of a lens.

Our first step is to place the object at the front focal plane of the first lens. To do this we use autocollimation. We illuminate the object and place a flat mirror somewhere after the first lens. If the object is at the front focal plane of the first lens, the image will be at infinity. That image, reflected in the mirror will focus back onto the object. Thus we adjust the object position until the returning image is well focused, and we are sure that the object is properly positioned. Figure 2.20 illustrates the concept.

Now our task is simply to adjust the camera position to achieve the best image. Usually the camera will include software for different types of acquisition. We begin with a live image displayed in real time on the computer screen. Once we have achieved proper focus, we can collect images for analysis. We next need to check the irradiance of the image. The camera has a limited range of image brightness, producing a low number representing darkness and a high number (255 for an 8-bit camera) for the maximum irradiance. If the image has a higher irradiance the number will still indicate 255, and we will not be able to compare the brightness of one pixel to that of another. Therefore we may need to reduce the irradiance of the light source on the target. We can do so by moving the light source further away, by adding neutral density filters, partially–crossed polarizers, or maybe by changing the power supply if the light source has a variable one. A neutral density filter reduces the light source by transmitting a fraction of the light,

$$T = 10^{-ND}. \tag{15.7}$$

Thus an ND 0.3 filter reduces the light by a factor of 2. On the other hand, if the light level is too low, the image will have a reduced dynamic range, and may be degraded by camera noise. Many light sources have a "brightness" control to change the voltage on the lamp, but it is usually not calibrated. Furthermore changing the voltage changes the temperature of the filament and thus the spectrum according to Equation 12.81. In some cases spectral changes could affect the experiment.

Eq. 12.81:
$M_{\nu(bb)}(\lambda, T) = h\nu c\frac{2\pi\nu^2}{c^3}\frac{1}{e^{h\nu/k_BT}-1} = \frac{2\pi h\nu^3}{c^2}\frac{1}{e^{h\nu/k_BT}-1}.$

The brightness control on a tungsten light source is not calibrated. The spectrum of the light will vary with brightness as well. It is better to control the brightness with neutral density filters or crossed polarizers.

15.3.1.6 Pre-Lab Homework: Imaging System

Design a telecentric imaging system. Decide on a "reasonable" magnification so that you can test the diffraction limit by imaging the Air–Force resolution chart. Also be sure that you can measure changes in the diffraction-limited spot using your camera. In other words, make sure that the smallest aperture will result in a point-spread function that subtends many pixels on the camera. Use plano-convex lenses and put them in "the right way." See Section 5.4.2.

Lens Orientation: Sec. 5.4.2

Calculate the numerical aperture using different diameters for an iris placed in the pupil plane. Calculate the diffraction limit using these numerical apertures. Recall that the numerical aperture is defined as shown in Figure 15.9.

Calculate the transverse spherical aberration. This will take a bit of work. Remember that the axial aberrations (in diopters) add. More specifically, calculate the total axial aberration using Equation 5.43, and then compute the transverse aberration using Equation 5.46. Do the same with both lenses installed "the wrong way."

Eq: 5.43:
$L_s = \frac{1}{s'(x_1)} - \frac{1}{s'(0)}.$

Eq: 5.46:
$\Delta x(x_1) = x_1\frac{\Delta s'(x_1)}{s'(0)}.$

Optionally if ray–tracing software is available, plot spot diagrams and compare the results to the aberration calculations above.

15.3.1.7 Lab Experiment: Imaging System

Set up the imaging system with a camera at the image plane and the Air–Force chart at the object plane. You'll probably need to microposition the axial direction of at least one element of the system.

Use autocollimation (Section 15.3.1.5 to assure that the object is at the front focal plane of the first lens.

Autocollimation (Sec. 15.3.1.5)

Collect an image. Is the magnification what you would expect? You will need to know the pixel size on your camera, and the size of different elements of the AF chart (Equation 11.39) according to Equation 11.39. Look at the "brightness" of the image. Adjust the light source to different values and look for the phenomenon of "clipping whites" where the brightest part of the image saturates the digitizer in the camera (or the full-well capacity of the pixel (See Section 13.6.1.2). In extreme cases, electrons beyond the full-well capacity may spill into adjacent pixels in a phenomenon called blooming. At the other extreme, if the brightness is too low, dim pixels may be reported as zero (clipping blacks). Remove the AF chart and just look at the light source. Adjust it so the brightness is near but not too near the upper limit. Call this image *White*. Also block the light source and collect an image called *Dark*. Put the AF chart back without moving anything else and take an image called *Sample*.

Eq: 11.39:
$x = \frac{1\,\mathrm{mm}}{2^{G+1+(E-1)/6}}.$

In Practice

When we perform mathematical operations on images it is important that the setup is unchanged between image collections. We need to avoid moving anything but also we need to avoid changing camera settings. In particular, many cameras offer an "automatic exposure" option, which will change the exposure time depending on the amount of light in the image. While this produces some very pleasing pictures, any mathematical operations we perform are likely to be incorrect. The camera must be set for manual operation, and it is often necessary to experiment a few times to find a setting that will work for the brightest and darkest image.

It is important to keep the original raw image for any future calculations.

If the differences among pictures are too great, it may be necessary to use neutral density filters on the brightest images and correct for these in the equations.

It is also important that the camera be in a linear mode of operation, meaning that the "gamma" is $\gamma = 1$. See Section 13.6.1.3 for discussion of gamma.

The glass areas of the AF chart should transmit about $T_{Fresnel} = 0.96^2$ because about 4% of the light is reflected at each surface using Equation 6.61, with $n_{glass} \approx 1.5$.

Eq. 6.61.
$R = \left|\frac{n-1}{n+1}\right|^2.$

Create an image of the transmission,

$$T_{experiment} = \frac{Sample - Dark}{White - Dark}. \tag{15.8}$$

Choose a good color axis and see how closely the transmission matches the theory.

A photo of an experiment is shown in Figure 15.12. In this implementation, rotating polarizers were used to control the brightness.

15.3.1.8 Post-Lab Reflections: Imaging System

How close does the resolution match your expectation for different iris sizes? Answer this question first at the center of the field of view and then at the edges and corners. You may get different answers in the vertical and horizontal directions. Remember that the equations derived from spherical aberration are for the ray at the edge of the aperture stop. Most rays are considerably closer to the center and the equation will give a very pessimistic estimate of the resolution. On the other hand, imperfect alignment will result in larger aberrations.

How do your answers change if you install the lenses "the wrong way?"

Discuss changes in resolution and changes in contrast among your images.

Is the transmission of the glass in the air–force chart about what you would expect?

FIGURE 15.12
Imager experiment. A green LED was used and the brightness was controlled with two polarizers. (Photo by Sam Koblensky, Northeastern University.)

15.3.2 Telescope

We started with a simple two-lens telecentric system with the object at the front focal plane of the first lens and the image at the back focal plane of the second one. The common intermediate focal plane was the pupil. In this experiment we will still use the two-lens telecentric configuration, but we will place the object and image at infinity, making a Keplerian telescope. In this case there will be a real image at the common internal focal plane. The telescope is discussed in detail in Section 4.4.1.

Telescope: Sec. 4.4.1

15.3.2.1 Learning Objectives

In this section we will learn more about telecentric systems and will pay more attention to stops and windows that were discussed in Chapter 4. Specific learning objectives include

Stops: Ch. 4

- Telescopes: More Telecentric Optics
- Transverse and Axial Magnification
- Aperture Stops and Pupils
  - Real Pupils
  - Matching Pupils for Cascaded Systems
  - Scanning in the Pupil
- Field Stops and Windows
- Focusing
- Stray Light

- Lens Tubes, Cage Assemblies, *etc.*
- Scanning

15.3.2.2 Background

We'll design and build the telescope, and compare our experimental results to our calculations. Looking at the telescope in Figure 4.18 we see that the primary lens (the left-hand one in the figure) is the aperture stop and the secondary lens, or eyepiece, is the field stop. Furthermore we note that because $f_1 \gg f_2$, the exit pupil is very near the back focal plane of the secondary and the entrance window is very near infinity, as we noted in Section 4.4.1. We recall from that section that when we combine two optical systems, it is important to match the exit pupil of the first (here the telescope) with the entrance pupil of the other (the eye). The distance between the secondary lens and the exit pupil is called the eye-relief.

We recall that the magnification of any two-lens telecentric system is given by the ratio of the focal lengths of the two lenses, $m = -f_2/f_1$, but we recall that we are using the telescope for angular magnification. With air at the input and output of the telescope, Equation 3.31 is $m_\alpha = 1/m$. Thus in the telescope, the image will appear bigger in angle than the object. This makes sense because although the object is smaller by the ratio, m, it is closer by the ratio m^2, so the angle subtended is larger.

Fig. 4.18

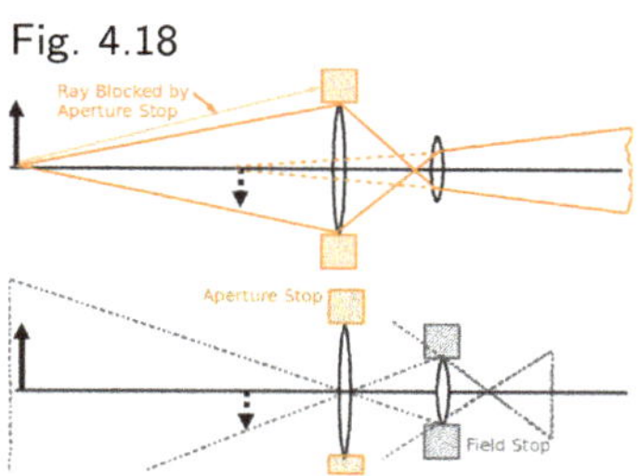

Eye relief is the distance between the secondary and the exit pupil.

Magnification of an astronomical telescope is less than 1.

Eq. 3.31
$n'x'd\alpha' = nxd\alpha.$

Scanning. Telescopes are frequently used as beam expanders. In cases where an image is created by scanning a focused beam it is often desirable to scan before the telescope. The beam at the output of a laser radar may be quite large and would require large mirrors to scan. There is also no space to locate scanning mirrors after the objective of a confocal microscope. In such cases we need to place the scanning mirrors very close to a pupil as discussed in Section 4.4.2. The angular displacement of the beam will be magnified by $1/m$, as discussed in Section 4.4.2, and the angular displacement of the beam at the scanner will be twice the angular displacement of the scanning mirror.

Scanning in a Pupil: Sec. 4.4.2
A scanning mirror must be placed close to a pupil.

Tubes and Cages. Although our emphasis in this chapter is on building systems on a breadboard, we may want to build a telescope that we can hold in our hands and take out of the lab. Depending on available equipment, lens tubes or cage assemblies may be used for this purpose. Figure 15.13 shows examples.

15.3.2.3 Pre-Lab Homework: Telescope

Design a $5X$ telescope with at least 2 cm of eye relief. Specify the lenses (focal length and diameter).

Use plano-convex lenses and decide which way to mount them (curved surface toward the object or away from it?). Explain your decision in view of what you learned in Homework Assignment 15.3.1.6.

Determine the size of the pupil (mm) and field of view (radians). Note that the magnification as we usually define it is actually 1/5. Why does it appear to magnify the object?

What is the diffraction limited resolution (radians) assuming visible light.

15.3.2.4 Lab Experiment: Telescope

Implement your telescope design in the laboratory as closely as possible with available components. Measure the exit pupil and the field of view. Hint: If you illuminate the telescope from the object side with a flashlight, you should see the real image of the aperture stop in the appropriate place.

Use the telescope as a beam expander to expand the beam from a low–power laser with a mirror on a rotation stage or simply rotating it in its post holder as shown in Figure 15.14. Place the scanner in the pupil of the telescope and adjust it through a known angle. Measure the angular change at the output. What is the angular magnification?

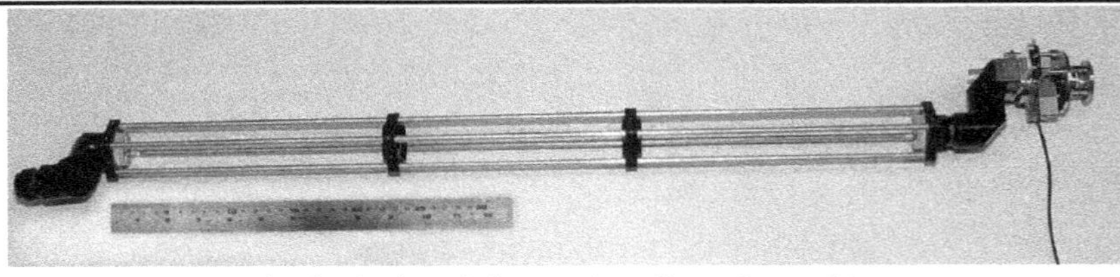

A. Articulated Arm using Cage Assembly

B. Some Tube System Components

FIGURE 15.13
Cages and tubes. On the left, an articulated arm relay system is used in a confocal microscope. On the right a telescope is assembled using lens tubes. Both systems can be used on or off the optical bench.

What happens if the point of rotation is not in the pupil plane? Notice that if we use a scanning mirror it is important to place the mirror close to a pupil plane.

Implement your telescope using the "tube" or "cage" components if these are available. Try taking a photo with a camera (mobile phone camera for example). Make sure you can adjust the focus. Look at different targets to see what you can and cannot resolve. Try moving the telescope toward and away from your eye to see the effect of the exit pupil.

Take the telescope outdoors. Pick a distant object that you can just barely resolve (eg. a brick in a wall). Estimate the angular extent of the object.

Some results of this experiment are shown in Figure 15.15. The two lenses were mounted on a rail as shown in Figure 15.15A, so that the telescope could be carried away from the bench. A piece of paper was placed in front of the primary (entrance pupil) and illuminated by the fiber-coupled white-light source on the left. The student on the right is holding a piece of paper used as an imaging screen to see the exit pupil. In Figure 15.15B, the exit pupil is the image of the primary, seen on the paper behind the secondary. It is a real image. The telescope was then moved into a long hallway and placed on a cart in Figure 15.15C. With a camera's entrance pupil placed at the exit pupil of the telescope to produce the image in Figure 15.15D.

A telescope built using tubes is shown in Figure 15.16. An additional tube section is included on the output end (right) to control the eye relief. An image of the Air-Force resolution chart was collected with this telescope using the camera of a mobile phone.

15.3.2.5 Post-Lab Reflections: Telescope

How well did your telescope match your design? Did you get the expected magnification? Is the image as good as you would have expected?

How easy was the tube or cage system to use? What advantages does it provide? What are some of the possible problems?

Explain why it is important to position your eye correctly to match the eye relief.

FIGURE 15.14
Rotating a mirror. In place of a rotating stage, it is possible to rotate a mirror on a post within a post holder. A small clip (left) can be locked on the post to keep the height from changing. Not that the surface of the mirror may not be perfectly aligned with the axis of rotation.

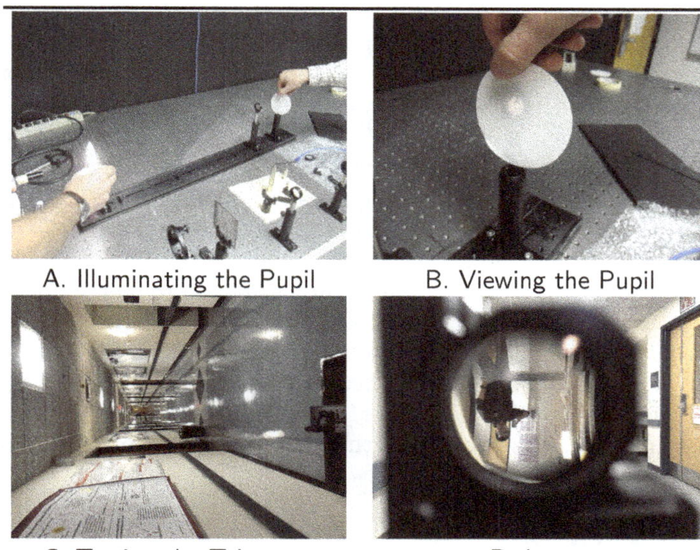

FIGURE 15.15
Telescope experiment. This telescope was assembled by students as part of a laboratory course in the author's lab. (Photos C and D by Fernando Ivich, Northeastern University.)

How well does the resolution compare with your expectation?

Sketch a Galilean telescope and locate the exit pupil. What problem occurs when looking through this telescope by eye? Why would this be a bad design to use with a scanner?

15.3.3 Spectroscopy

In spectroscopy we wish to measure some parameter as a function of wavelength. Spectral information can be used to study absorption, scattering, fluorescence and other phenomena that are often indicative of specific sizes and shapes (scattering) or compositions (absorption and fluorescence) of materials. Spectroscopy is a common technique for identifying materials. Fluorescence microscopy has become a critical tool in biology, where fluorophores are bound to cellular components. The tools available include dispersive prisms (Section 2.6) gratings, (Section 8.8), and filters (Section 7.7). Prisms and grat-

Prisms: Sec. 2.6
Gratings: Sec. 8.8
Thin-Film Filters: Sec. 7.7

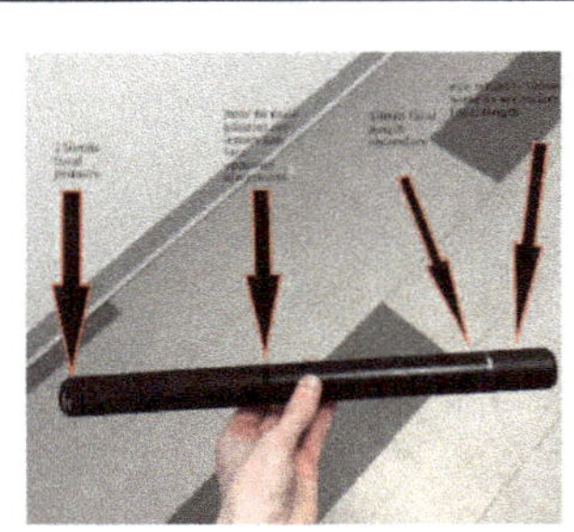

FIGURE 15.16
Telescope using tubes. This telescope was used to image the Air-Force resolution target at up to 20 yards. The photo shown on the right is from 5 yards, or approximately 5 meters. (Photos by Eric Hall (left) and Jane Lee (right), Northeastern University.)

ings disperse light to different angles depending on wavelength, and are useful to separate contributions from multiple wavelengths. Filters normally select a band of wavelengths (often narrow). Therefore, they can be used in imaging systems to select light of a specific wavelength for the image. In this set of experiments we will first measure transmission and reflection spectra using a compact spectrometer covering visible wavelengths along with some near–ultraviolet and near–infrared wavelengths.

Next, we will image a fluorescent object using a filter to select the emission wavelengths and reject other light, including the incident light at a different wavelength.

In recent decades, hyperspectral imaging has begun to merge these techniques producing images that are functions of position and wavelength using spectroscopic cameras and tunable filters. Although these are beyond the scope of this chapter, the experiments here will help to explain the underlying concepts.

After listing the learning objectives, we will discuss the concepts of spectroscopy and fluorescence imaging separately, and then consider some of the common issues such as light sources and detectors.

15.3.3.1 Learning Objectives

- Spectroscopy: Transmission, Reflection, Diffuse and Specular
- Colors and Wavelengths
- Compact Spectrometer Operation
  - Principles, Resolution
  - Calibration: White and Dark Measurements
  - Absolute Calibration Issues
- Fluorescence Imaging

Eq. 8.95 $N\lambda = d\left(\sin\theta_i + \sin\theta_d\right).$

Fig. 8.27

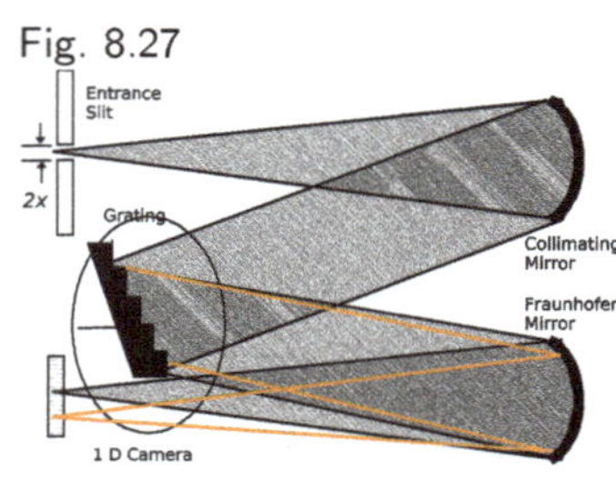

15.3.3.2 Spectroscopy

In this section, we'll take a look at spectroscopy using a small spectrometer. We discussed the grating spectrometer in Section 8.8. Grating spectrometers are more common than prism spectrometers because the dispersion angle is larger and it is easier to control the amount of dispersion by controlling the period of the grating according to Equation 8.95. The concept was shown in Figure 8.27. The input slit and mirror or lens (focal length, f_i produce a beam of collimated light. The slit must be narrow enough to ensure collimation. If the half-width of the slit is x, then light at the edge of the slit will be at an angle of $\theta_i = x/f_i$, which will change the angle of diffraction in Equation 8.95. The light is then

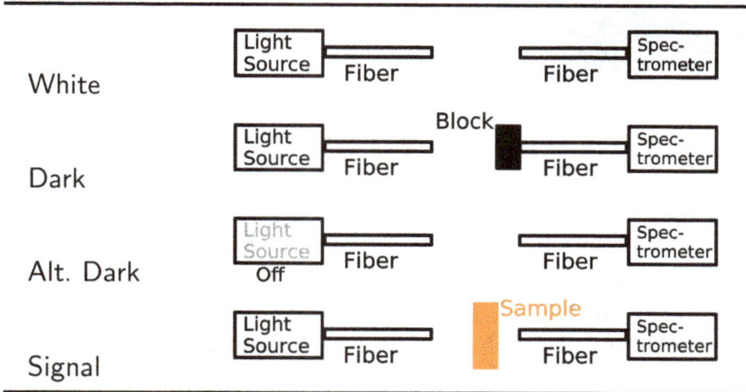

FIGURE 15.17
Transmission measurements. Transmission is compared to a white reference.

dispersed by the grating, and then refocused by the output mirror or lens onto a one-dimensional detector array, where the position is determined by the angle of the dispersed light beam. Thus each pixel in the array collects light in a narrow band of wavelengths. The spectrometer is calibrated so that each pixel is characterized by its center wavelength. The fibers are included in the calibration because different fibers have different transmission spectra.

In Practice

The resolution of the spectrometer will depend on the grating and the pattern with which it is illuminated, as discussed in Section 8.8, and will also be limited by the size of the entrance slit and the size of the detection pixel. The grating equation, 8.95, can be used to determine the degradation of resolution by these size effects. While smaller slits and pixels permit better resolution, larger ones gather more light, reducing noise.

In Practice

The spectrometer calibration can be affected by how well the light source fills the modes of the multi-mode fiber. Some caution is needed with coherent light sources.

There are multiple implementations of the spectrometer concept, but many compact spectrometers include a fiber to collect the input light and a USB connection from the detector array to a computer. All alignment is done internally and the user simply aims the fiber toward the light source.

Because of variations in the efficiency of gratings at different wavelengths and the quantum efficiency of detector arrays, some calibration is necessary. Here we look at a few calibrated measurements that can be done easily in the lab.

Transmission. Figure 15.17 shows a basic transmission measurement that we can use, for example, a dielectric-coated interference filter .

We can use any white light source but we will assume the Tungsten source shown in Figure 15.7C. We first make a "white" measurement by pointing the spectrometer's fiber toward that of the source. It is important to mount these so that they will not move during the experiment. We then make a dark measurement. We may choose to block the spectrometer fiber or to turn off the light source. Ideally the two will be equivalent, but the latter may be a better choice if background light is present. Comparing the two types of dark measurements may be a good idea to ensure that background light is minimal.

In Practice

To some extent, background light can be subtracted from a measurement. However, the noise associated with it cannot. The limit of noise is given by the Poisson distribution of photon arrival. For example, if a signal consists of 8000 photons in the measurement time and the background consists of 100, the result will be $8100 \pm \sqrt{8100}$ and subtracting the background will produce $8000 \pm \sqrt{8100}$. However, if the signal is 20 and the background is 80, the result, after background subtraction, will be $20 \pm \sqrt{100}$. It is best to reduce the background as much as possible before making a measurement.

Finally, we insert the sample between the two fibers and measure the signal. Then the transmission is given by Equation 15.8,

$$T_{experiment} = \frac{Sample - Dark}{White - Dark}.$$

One source of error may arise if the sample is not planar or if it is tilted at a sufficient angle to move the apparent position of the light source.

Reflection. In principle, measurements of specular reflection can be done in the same way as transmission, as shown in Figure 15.18A, but it is not possible to avoid moving something between measurements of white and sample, so some careful planning is required. A rapidly diverging and uniform light source will allow more tolerance of alignment errors. Assuming that the geometry is correct, the reflectivity is then given by an equation similar to Equation 15.8;

$$R_{experiment} = \frac{Sample - Dark}{White - Dark}. \tag{15.9}$$

Extinction. Having completed a measurement of transmission and reflection, we can determine extinction (absorption plus scattering) using

$$T + R + A + S = 1, \tag{15.10}$$

where T and R are the transmission and reflection, and A and S are absorption and scattering. Unfortunately, this calculation often involves subtraction of large numbers, leading to considerable error, and much greater care is needed in making the measurements. More complicated experiments can separate the absorption from the scattering.[183]

Diffuse Reflection. In principle the reflection from a diffuse surface such as a painted wall can be measured by comparison to a known standard. Calibrated white Lambertian diffuse standards are readily available, and one can simply make two measurements, one with the white standard and one with the unknown sample, along with a dark measurement, and then use Equation 15.9 to calculate the diffuse reflectance. If the sample is not Lambertian, this experiment could become time consuming as different angles of incidence and measurement would yield different results.

15.3.3.3 Fluorescence Imaging

Fluorescence: Sec. 12.2.2.2

In this section, we'll complete a simple fluorescence imaging experiment following the ideas in Section 12.2.2.2. Epi-fluorescence microscopy is a commonly used technique in biology and medicine. A material is stained with a molecular tag that binds to some specific material of interest. Light at an appropriate wavelength is absorbed by the tag (emission) and light of a longer wavelength is emitted (emission). Normally the illumination and detection are done on the same side of the sample, in a configuration we refer to as epi-illumination. A typical configuration is shown in Figure 15.19. This

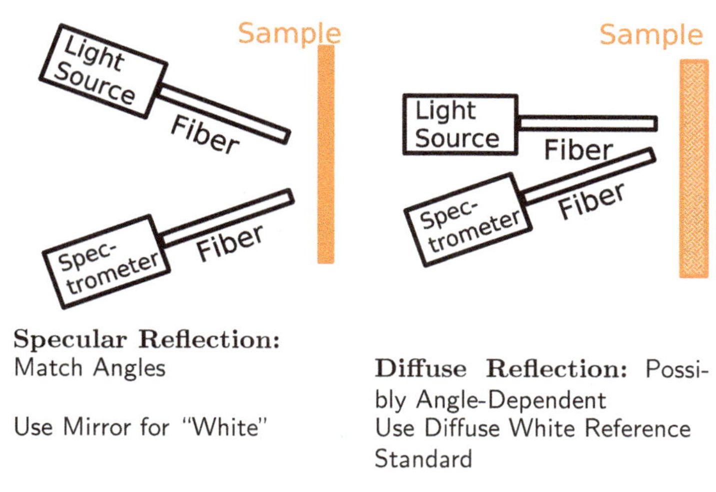

FIGURE 15.18
Reflection measurements. Specular reflection is challenging because of the need to move the receiver fiber. Diffuse reflection is compared to a white diffuse standard.

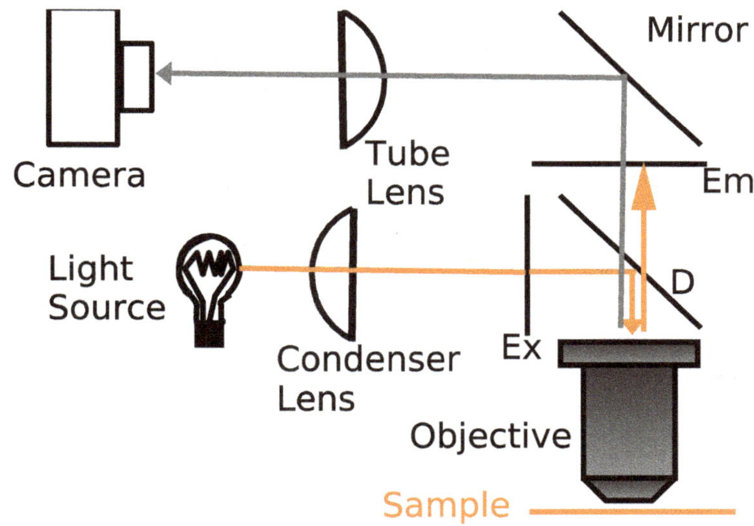

FIGURE 15.19
Epi-fluorescence microscopy layout. The light path contains a filter cube that controls the path of excitation and emission light. The filter cube consists of the dichroic (D), excitation filter (Ex) and emission filter (Em). The excitation filter and dichroic reflection ensure the correct excitation spectrum and the dichroic transmission and the emission filter ensure that only the fluorescence spectrum reaches the camera.

configuration includes a filter cube consisting of three elements, an excitation filter, a dichroic beamsplitter, and an emission filter.

A dichroic beamsplitter by itself is usually insufficient to reject unwanted light. Emission and maybe excitation filters are usually needed.

Dichroic Beamsplitter: In principle, only a dichroic beamsplitter is required. Dichroic refers to "two colors;" the excitation light is reflected by this beamsplitter from the source to the sample, and the emission light is transmitted from the sample to the camera. Ideally, the beamsplitter should have 100% reflectance for the excitation wavelength and zero for all others, and it should also have 100% transmission for the emission wavelength range and zero for all others. In practice this requirement is impossible to meet, and the result is that some light from the source, scattered by the sample, would pass to the camera. This "leakage" would likely be much larger than the fluorescence signal. For this reason, separate emission and excitation filters are used to reject the unwanted wavelengths.

In Practice

An excitation filter may be needed even with a narrow-band light source. Some sources have multiple wavelengths, even though each one is very narrow. Some lasers may produce visible spontaneous emission at unwanted wavelengths.

Filters select a portion of the spectrum. In the first part of this experiment, we measure the properties of filters. In the second we use filters for spectral imaging.

Filters: Filters are used to select a portion of the spectrum. For example, a colored glass window transmits one particular band of wavelengths. A variety of colored-glass and gelatin filters are available to filter the spectrum of light. Typically these filters have large bandwidths, and are useful, for example, in selecting red, green, and blue light in an image. For more narrow wavelength selection, multiple layer dielectric coatings are used as discussed in Section 7.7. In the visible spectrum these coatings are typically applied to a glass substrate. Multiple layers can be used to achieve a variety of effects, including short-pass filters (those that transmit short wavelengths and reflect long ones), bandpass filters and bandstop filters. A short-pass filter with a cutoff wavelength between visible and near–infrared is called a cold mirror, and the opposite is called a hot mirror.

Bandpass filters are usually used for excitation and emission.

For the excitation and emission filters in our experiment here, we want bandpass filters. Generally these are specified in terms of their center wavelength and full width, using terminology such as "532/25" which means the passband is 532 nanometers plus or minus 35/2 nanometers at half-maximum. Another important characteristic of these filters is the out-of-band leakage. For example, if we are using the 488 nm line of an Argon laser for excitation, the emission filter should transmit as little as possible at 488 nm. Manufacturers can produce multi-layer coatings with desired characteristics.

15.3.3.4 Light Sources

Black body: Sec. 12.5.3

Eq. 12.81
$$M_\lambda(\lambda, T) = \frac{2\pi hc^2}{\lambda^5}\frac{1}{e^{h\nu/k_BT}-1}.$$

We have a great variety of light sources from which to choose. Direct or scattered sunlight is often used for simple imaging applications. We can have more control of the light source by using a tungsten or quartz–halogen–tungsten light source. These and other examples of near-black-body sources are discussed in Section 12.5.3. Their distinguishing characteristic is a broad spectrum with a spectral radiant exitance having a peak value and peak location given by the temperature using the Planck equation (Equation 12.81). Tungsten is often used as a light source for spectroscopy because of its broad and relatively smooth spectrum.

Arc lamps often have narrow spectra, useful for spectroscopic imaging.

Various arc lamps can produce higher spectral radiant exitance at selected wavelengths. For example, a mercury arc lamp produces a variety of strong lines in the UV and visible spectrum, and is frequently used as a light source.

Lasers normally have very narrow spectra and high radiance.

Mode-Locked Laser: Sec. 10.2

Even narrower spectral lines can be obtained from lasers and lasers are available across the spectrum from UV to far–infrared. Lasers can be continuous-wave (CW) in which case the power output is a constant over time, or they can be modulated. In particular, it is possible to obtain very short pulses with techniques such as mode-locking (Section 10.2). Such lasers may be nearly transform–limited meaning that their time-bandwidth product is limited by the uncertainty principle. Furthermore lasers have a high degree of spatial coherence, meaning that the product of diameter and angular divergence is diffraction limited. The result of these two conditions is an extremely high spectral radiance. Arc lamps and lasers are narrow–band sources useful for spectral imaging where a single excitation wavelength is desired. Lasers have high radiance and are also used for non-linear spectroscopy such as multi-photon and harmonic spectroscopy.

LEDs have moderate bandwidth and can be good sources for spectral imaging.

Light-emitting diodes (LEDs) have bandwidths typically 20 to 40 nm, much larger than those of most lasers, but considerably smaller than tungsten. Spatially, the emission is usually near Lambertian ($\cos\theta$), so the spatial coherence is closer to that of tungsten sources. Many LEDs can be modulated at moderately high frequencies. Now LEDs are available across the spectrum from UV through near infrared wavelengths. Although high-power LEDs (Several Watts) are becoming more available, these have a larger area and the spectral radiance is usually less than that of tungsten sources. LEDs are finding more use in fluorescence microscopy as higher radiances become available.

There are two types of white-light LEDs. In the first, three different LEDs, red, green, and blue, are combined with appropriate drive currents to produce what looks like white light. However, it must be noted that the spectrum is a superposition of three relatively narrow lines and is not white in the sense of being a uniform function of wavelength across the visible spectrum. In the second type, the LED is blue, and coated with a phosphor that absorbs the blue light and emits a broad spectrum across the green and red. These LEDs have a spectrum peaked strongly in the blue, with a valley between the blue LED and the phosphore emission. Many people dislike them for room illumination because of the resulting blue appearance. Furthermore, the spectral radiance outside the narrow blue band is considerably lower.

Most white LEDs have spectra with a narrow blue peak and a wide band at longer wavelengths, that may be less useful for spectroscopy.

15.3.3.5 Detectors

Compact spectrometers are often built with one-dimensional arrays. Almost all such arrays are built with semiconductor detectors. Sensitivity of the detector (and maybe more importantly efficiency of the grating, vary with wavelength so calibration is needed. Furthermore, the array is usually protected by a thin coating which can produce interference fringes that will also be removed by calibration. It is important to remember that small signals will result in greater noise and large signals may saturate the digitizer. Usually the integration time (also called acquisition time, sample time, or exposure time) can be varied to ensure that the data fall within a useful range.

Photomultipliers are often used for better sensitivity, but usually the grating must scan in angle across the exit slit, because large photomultiplier arrays do not exist. These detectors are very sensitive to light and have a high current amplification. This amplification can be controlled by the high voltage, as discussed in Section 13.4.1. Photomultipliers are easily damaged by even low light levels when the high-voltage power supply is turned on. It is important to avoid accidental illumination. A photomultiplier in the author's lab was destroyed by the flashing light from a fire alarm.

Photomultiplier: Sec. 13.4.1

15.3.3.6 Pre-Lab Homework: Spectroscopy

Download software for the spectrometer, and spectral data for the light-source.

Think about absorbing gelatin filters and interference filters. How do you expect the transmission spectra to differ?

Interference filters often have a very wide dynamic range to block unwanted wavelengths. Find a way to measure transmission spectra over a wide range of magnitudes

What difference do you expect between specular and diffuse reflection from an object?

Look up the spectra of hemoglobin and oxyhemoglobin. What impact do these materials have on skin color?

Prepare an image that has red, green, blue, and white quadrants.

Think about safe use of the ultraviolet LED.

15.3.3.7 Lab Experiment: Spectroscopy

Measure the spectra of the room lights, white LED, UV LED, laser, and the colors on your computer screen *etc.* If you have access to a Wood lamp, that is an excellent UV source. If you have external windows, look at the spectrum of the sun.

Observe fluorescence of white paper, various markers, skin, maybe minerals, under UV illumination *etc.*

Set up the white light source and spectrometer for convenient use.

Measure transmission spectra of some filters, your glasses, gel or colored glass filters, lenses, two Polaroid sheets parallel and crossed, *etc.*

Measure specular reflectance on some of the components and compute the extinction $(A + S)$. This may be interesting for the interference filters and blue blockers among others.

Measure the diffuse reflectance spectra of some materials; Plastic, clothing, your finger, *etc.* See if you can see a specular reflection and measure it.

Collect a spectrum of the white light source (without the fiber. Why?)

15.3.3.8 Post-Lab Reflections: Spectroscopy

How well do your spectral measurements agree with your visual observations such as color?

How do the bandwidths of different filters compare?

What type of lights do you think are used to light the room? Why do you think this?

Were you able to see the hemoglobin spectrum in skin?

What problems would you expect in measuring the (specular) reflection spectrum of a dichroic filter? How well did you do?

Why might it be a bad idea to use paper as a calibration standard for diffuse reflectance?

In measuring the spectrum of a light source discuss the issues of knowing or being able to perform an absolute *vs.* relative calibration with a reference source.

15.3.3.9 Pre-Lab Homework: Epi-Fluorescence

For this experiment you will need a fluorescent object to image. Prepared microscope slides with fluorescent samples are available. Some everyday objects also exhibit fluorescence, including some olive oil and laundry detergents. Fibers in paper often exhibit broad blue fluorescence when excited with ultraviolet light. After choosing a material, learn about the excitation and emission spectra.

Select a light source and appropriate dichroic, excitation, and emission filters, and plot the spectra of all of these along with the spectra of the chosen material.

Plot the product of transmission along the excitation path (normally excitation source, excitation filter transmission, and dichroic reflection, and fluorophore excitation) and the emission path (normally emission spectrum, dichroic transmission, emission filter

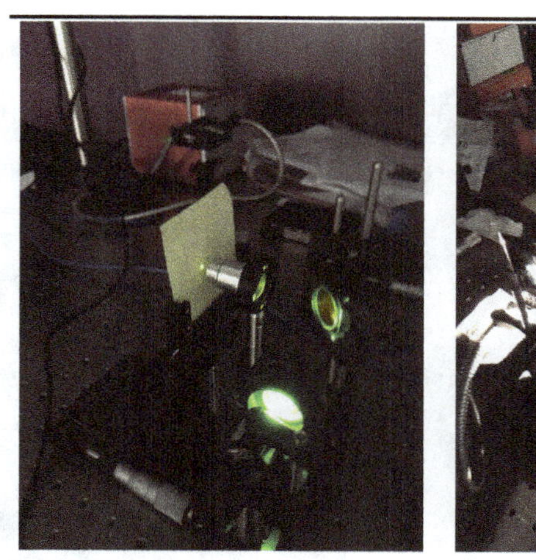

FIGURE 15.20
Epi-fluorescence experiment. The left side shows a close view of the LED source, dichroic, objective, and the spot on a paper target. The right side shows the whole layout, in which focusing was tested using the Air-Force resolution chart with white light transmission. The image of the chart is seen in the distance on the camera. (Setup and photos by Eric Hall, Jane Lee, Sophia Bauer, and Mohammadrasoul Taghavi, Northestern University.)

transmission, and detector response). If everything is chosen correctly, the excitation light will peak near the peak of the excitation spectrum and most of the emission spectrum will be detected.

Determine what fraction of scattered source light will pass through the excitation and emission path. This light will compete with the fluorescence signal, and should be very small.

15.3.3.10 Lab Experiment: Epi-Fluorescence

Fig. 15.19:

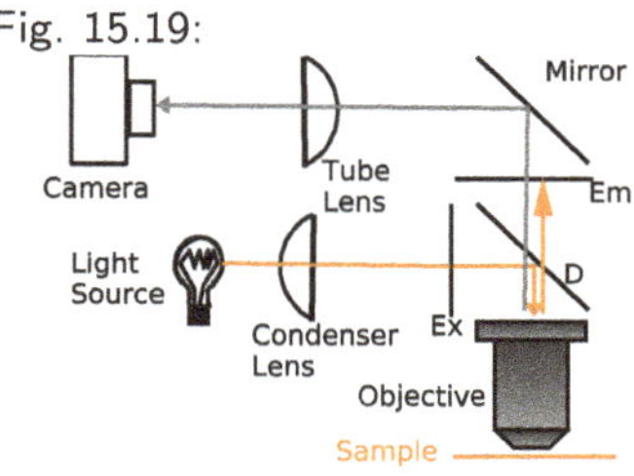

Set up the epi-fluorescence experiment as shown in Figure 15.19. Variations may be needed to accommodate your fluorophore with available equipment. A typical experiment is shown in Figure 15.20.

Try to align the system by removing the emission filter. The leakage of excitation light through the dichroic will probably be strong enough to produce a measurable signal. Alignment tolerances are relaxed because a broad illumination is required to cover the whole sample region, so slight changes in angle do not have effect. It is a good idea to keep the illumination spot only slightly larger than the field of view to maximize incident irradiance.

Collect images with different excitation levels. Measure the irradiance from your source with a power meter.

See if the fluorescence decays with time. Most materials will photo-bleach under strong illumination.

An example of fluorescent cells collected with this laboratory experiment is shown in Figure 15.21. These cells would be nearly invisible in reflection or transmission microscopy because their index of refraction is close to that of the culture medium.

15.3.3.11 Post-Lab Reflections: Epi-Fluorescence

Was the fluorescence image stronger than the background image caused by backscattered excitation light?

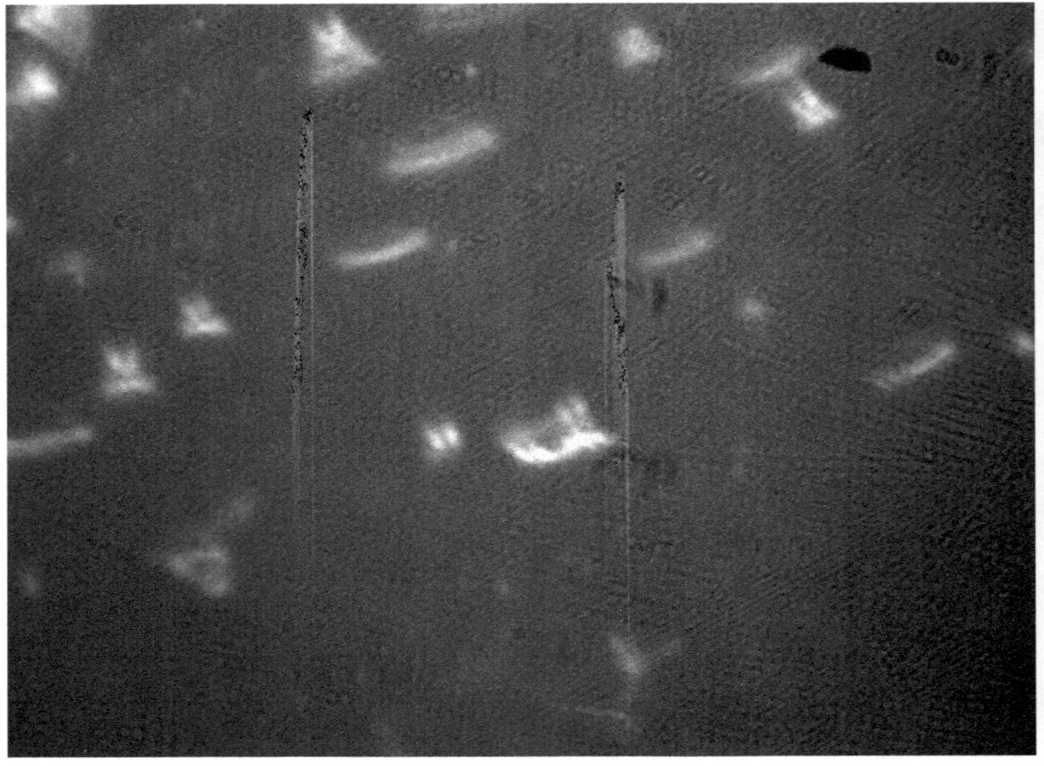

FIGURE 15.21
Epi-fluorescence. Fluorescent cells were imaged with the experiment shown in Figure 15.20. The cells are easily observed against a dark background. (Photo by Jane Lee, Northeastern University.)

What could you do to improve the image?
Was the fluorescence linearly related to the incident irradiance?

15.3.4 Polarization

Polarization: Ch. 6

T/R Switch: Sec. 6.6.5.2

In Chapter 6, we discussed the control, propagation, and measurement of polarization. We considered components such as polarizers, waveplates, rotators, and polarizing beamsplitters. In this experiment, we will build the transmit-receive switch discussed in Section 6.6.5.2, which will provide us with practical experience using several of these components, and will also introduce lasers and power meters.

15.3.4.1 Learning Objectives

- Lasers
  - Polarization
  - Alignment
- Linear Polarization, Circular Polarization
- Polarizers:
  - Rotational Mounts: Alignment
  - Insertion Loss, Extinction
- Waveplates: Changing the Polarization
- Power Meter: Calibration and Measurement

15.3.4.2 Background

The T/R Switch Before we begin with this experiment, it is useful to remind ourselves of the definitions of "P" and "S" polarization from Section 6.1.2, and particularly in Figure 6.1. Parallel, or "P" polarization has the electric field parallel to the plane of polarization, which is the plane of the page in the figure. Perpendicular (German Senkrecht) or "S" polarization has the electric field perpendicular to the plane of polarization. These definitions are only meaningful when light is reflected from a surface, as the plane of incidence is defined as containing the incident ray and the normal to the surface.

Polarization Labels: Sec. 6.1.2
Fig.6.1

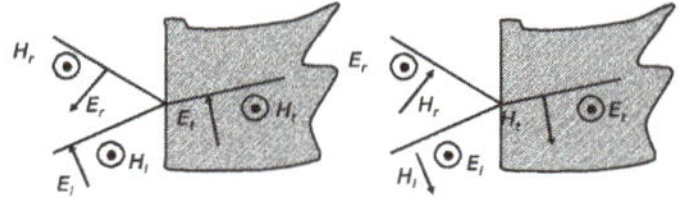

In some optical systems, we want to use common optics for transmission and reception of light. Doing so is important when the optics are expensive or when it is important that the light is received along exactly the transmission path. Some such systems include confocal laser radars and epi-illumination microscopes (including confocal). In such systems we may be limited by the reciprocity of beamsplitters, which dictates that the power transmission at an interface is the same in either direction, and the conservation law given by Equation 6.59. Then if we use a beamsplitter to separate transmission and reception paths, Equation 6.59 leads to the optimal choice of $T = R = 1/2$ with the product $TR = 1/4$ as explained in Section 6.6.5.2, and shown in the left side of Figure 6.32. This product is important because the signal is proportional to TR. As shown in that section the right side of the figure, we can eliminate the constraint, $T + R = 1$, by transmitting with one polarization and receiving with the other. We need a quarter-wave plate as shown on the right side of the figure to control the polarization, and we can only do this if the target retains the polarization on reflection. We recall that "S" polarization is more strongly reflected than "P" so we choose to use "P" polarization for transmission and "S" for reflection, and we need a beamsplitter for which both T_P and R_s are as close to 100% as possible. Fortunately such beamsplitters are available using appropriate dielectric coatings. An example is shown in Figure 15.22A.

Eq. 6.59:
$T + R = 1.$

T/R Switch: Sec. 6.6.5.2
Fig. 6.32

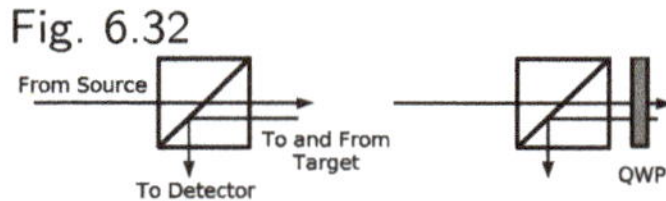

> **In Practice**
>
> One would expect the transmission of the beamsplitter to be the same from either direction. However, the optical adhesive holding the two prisms together to make a beamsplitter cube may be slightly birefringent. Therefore, when possible, it is recommended to have the light enter from the side away from the adhesive. The direction is usually marked on the beamsplitter. Of course, in the present experiment, transmitted light is incident in one direction and reflected light in the other. Because the birefringence is small, the efficiency of the T/R switch should not be affected.

The other key component is the quarter-wave plate. Precise rotation of the waveplate to 45 degrees from the incident linear polarization is important, so a precision rotation mount such as in Figure 15.22B is important.

Waveplates: Sec. 6.5.2.1

> **In Practice**
>
> The performance of the quarter-wave plate depends on wavelength. Broadband waveplates are available over certain ranges of wavelength as discussed in Section 6.5.2.1.

Laser Polarization: For this experiment, the transmitted light must be linearly "P" polarized. For this reason a polarized laser is highly desirable. A quick survey of vendors will show that a polarized helium–neon laser is more expensive than an "unpolarized" one, because of the Brewster plates needed to maintain polarization. The reader may be tempted to place a polarizer in front of an unpolarized laser. However, an "unpolarized" helium–neon laser actually runs on two longitudinal modes (see Section 7.5.3.1) in orthog-

An "unpolarized" helium neon laser actually runs with two polarizations in adjacent cavity modes. After passing through a polarizer, there may be mode-beating and power instability.

Laser Cavity Length: Sec. 7.5.3.1

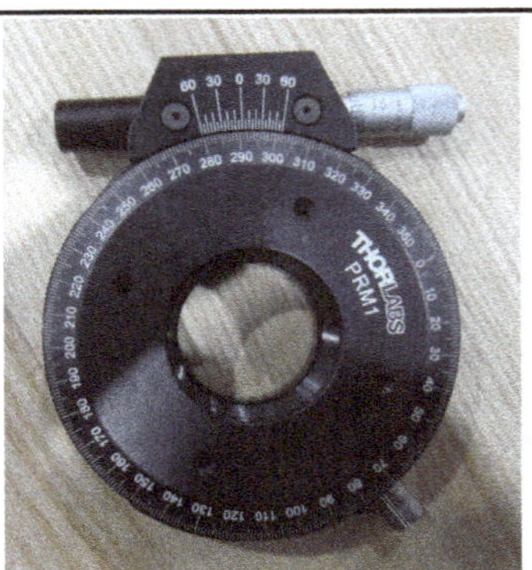

FIGURE 15.22
Key components for polarization experiment. The key element of the T/R switch is the polarizing beamsplitter (A). A precision rotation mount is useful for mounting the waveplate (B).

onal (not necessarily linear) polarizations. As the laser temperature changes, the modes will shift, changing the relative power in each polarization, making the polarized output very unstable. Such a configuration, or even a bright LED with a polarizer can be used in this experiment if necessary, but a polarized laser is preferred.

Power meters may be based on photon detectors or thermal detectors. Photon detectors must be calibrated for the wavelength being measured.
Detectors: Ch. 13

Power Meters A power meter is essential for this experiment. Power meters include a calibrated detector which may be either a photon or thermal detector as discussed in Chapter 13. Either detector will have a defined active area which detects the light. The detector produces a signal proportional to power, and measures the power on the active area. Thus, if a beam smaller than the active area is measured, the signal is proportional to power, and if the beam is much larger, it is proportional to irradiance, the power per unit area. Many meters will include a switch to select between the two measurements, or at least provide the user with the area to enable manual calculation.

In Practice

To measure power, ensure that all the power falls on the active area of the power meter. Doing so may require a lens and the loss in the lens may be measured to correct the reading. To measure irradiance the beam being measured should overfill the active area.

Generally thermal detectors have a broader and more uniform response across wavelength, and are thus suitable for measurement of white light or other sources with broad bandwidth or unknown wavelength. In contrast, a photodiode's response is more strongly wavelength dependent, and usually the meter must be calibrated for each wavelength. The user must set the wavelength before reading the meter.

In Practice

It is important to remember that if the wrong wavelength is selected, the power meter will still produce a reading, but that reading will be wrong. This type of meter cannot be used to measure the power of broadband sources or sources of unknown wavelength.

Some power meters have the option to upload the light spectrum so that broadband sources can be measured quantitatively.

15.3.4.3 Pre-Lab Homework: T/R Switch

This experiment can be done with a polarized laser or another light source. If you choose to use an unpolarized light source, you will need one polarizer after the light source to begin with the appropriate polarization.

If you choose to do this experiment with a laser source, begin by reading Section 15.5.

Also for a laser, watch this about aligning a laser to the table https://www.youtube.com/watch?v=qzxILY6nOmA

Watch this about aligning a polarizer to the table https://www.youtube.com/watch?v=W9pALZ5Z8ms

Look at the graphs for your polarizing beamsplitter cube. Calculate the product TR at the source wavelength.

What fraction of the laser light will be transmitted if the laser's polarization angle is θ with respect to the axis of a polarizer?

Design your experiment to have the HeNe laser as a source, a polarizer oriented to pass horizontally–polarized light, followed by the polarizing beamsplitter and quarter wave-plate. Use a mirror as the target. Leave room to insert the power meter between every pair of optical elements. Look at the mechanical components available and think about how to mount the beamsplitter cube and how to align the quarter-wave plate.

15.3.4.4 Lab Experiment: T/R Switch

Assemble your system. Align the polarizer to transmit horizontally-polarized light. Then rotate the laser (or add a half-wave plate between the laser and polarizer) to convert the laser's polarization to horizontal.

Measure the power at each location and determine the "insertion loss" of each component.

The quarter-wave plate should produce circular polarization. Rotate a polarizer after it and see how constant the power is during rotation.

Compute the product TR from your experiment.

15.3.4.5 Post-Lab Reflections: T/R Switch

How well does your experiment compare to your prediction?

What are the other losses in your experiment?

What fraction of the laser light reaches the output? What would this fraction have been with an optimized non-polarizing beamsplitter? To answer this, think about what components you would not have needed in that case.

How would this experiment be different if the target were a scattering surface instead of a mirror?

15.3.5 Interferometry

Interferometry was discussed in detail in Chapter 7, and again when we studied coherence in Chapter 10. Interferometry is useful for measuring phase with high precision for

Interference: Ch. 7
Coherence: Ch. 10

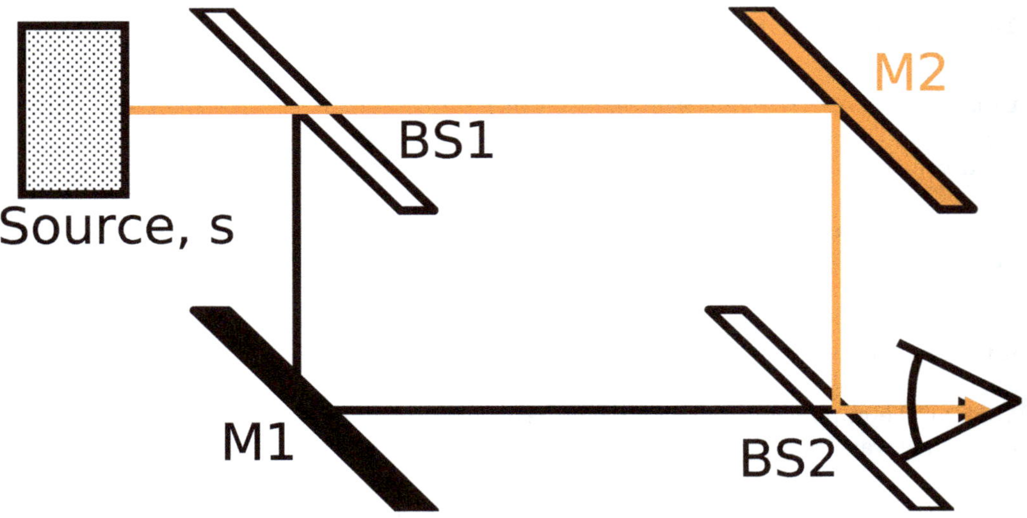

FIGURE 15.23
Mach–Zehnder interferometer. Light from the source is split into two paths by BS1, and recombined at BS2.

applications such as surface shape measurements for optical fabrication (Section 7.4.3), measurements of Doppler velocity in coherent lidar (Section 7.2), vibration measurement, quantitative phase microscopy (Sections 7.3.1 and 7.3.3), optical coherence tomography (Section 10.3.4), and many more.

Optics Testing: Sec. 7.4.3
Coherent Lidar: Sec. 7.2
Phase Shifting: Sec. 7.3.1
Optical Quadrature: Sec. 7.3.3
OCT: Sec. 10.3.4

The sensitivity of interferometry requires precise attention to alignment, and probably presents the greatest alignment challenge of all the experiments in this chapter.

15.3.5.1 Learning Objectives

- Interferometry Principles
- Interferometer Design
- Micropositioning Mirrors
- Precision Alignment
- Coherent Light Artifacts
- Phase Measurement

15.3.5.2 Background

Mach–Zehnder: Sec. 7.1

We will use the Mach–Zehnder interferometer discussed in Section 7.1 and shown in Figure 7.2 in this experiment, because it is conceptually simple, and meets a number of different learning objectives. For convenience the figure is reproduced in Figure 15.23. The interferometer consists of two mirrors and two beamsplitters. Note that if the aternate output of the interferometer, in the downward direction, is used as in Section 7.1.5 on balanced mixing , compensator plates (Section 7.4.2) may be required if the coherence length of the source is short.

Interference: Ch. 7

Equations: We briefly review the equations for interference developed in Chapter 7. The electric field at the output of the interferometer in Figure 15.23 is given by

$$|E|^2 =$$

$$(E_1 + E_2)(E_1 + E_2)^* =$$

$$(E_1 + E_2)(E_1^* + E_2^*) =$$

FIGURE 15.24
Nonpolarizing beamsplitter. This beamsplitter is 50% reflecting and 50% transmitting, at any polarization.

$$|E_1|^2 + |E_2|^2 + E_1E_2^* + E_1^*E_2 =$$

$$|E_1|^2 + |E_2|^2 + 2\,|E_1E_2|\cos\frac{2\pi\Delta s}{\lambda}.$$

where Δs is the difference in optical path length between the two paths, λ is the wavelength, E_1 is the field of the light that followed the path BS_1, M_1, BS_2, and E_2 is the field of the light that followed the path BS_1, M_2, BS_2. We can compute these fields knowing the incident field and the properties of the beamsplitters. Recall that the field is related to the irradiance by $I = EE^*$) so we can write equations such as

$$E_1 = \sqrt{I_{in}}\sqrt{R_{BS1}}\sqrt{R_{M1}}\sqrt{T_{BS2}}.$$

Remember that if we wanted the electric field in the correct units, it would be $\sqrt{IZ}$ where Z is the impedance. Because we don't care about the units, we normally neglect the impedance as we did in Chapter 7.

Fig.7.6

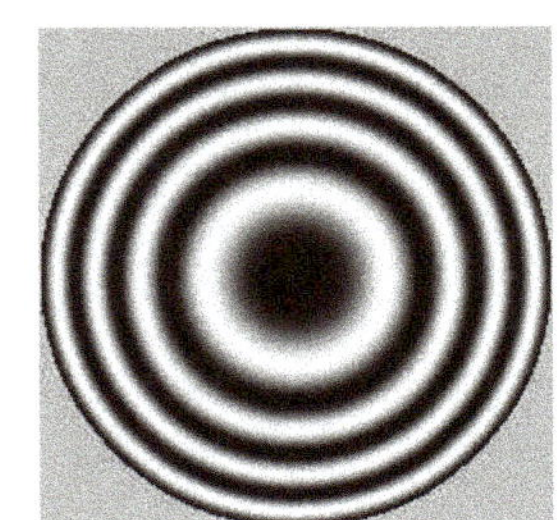

With a slight difference in path lengths, Δs, if the interferometer is otherwise well-aligned, we expect fringes as we saw in Figure 7.6.

Ideally the beamsplitters will have the same reflectivity, which we have assumed here is 50%. We avoid polarizing beamsplitters because those would have near-zero transmission or reflection for certain polarizations. We recall that we achieve the highest fringe contrast, V when the reflectivities are equal, as a result of Equation 7.31. If both reflectivities are 50%, then both the right and bottom outputs of the interferometer will be equal, allowing for balanced detection.

The beamsplitters used in this experiment are cage-mounted as shown in Figure 15.24. The correct input direction is shown by a drawing on the top of the cage. This direction is chosen because the optical adhesive holding the prisms of the beamsplitter is on the output side, minimizing concerns about birefringence in the adhesive.

Eq. 7.31: $V = 2\frac{\sqrt{R_1T_2T_1R_2}}{R_1T_2+T_1R_2}$.

Without the cage assembly, a non-polarizing beamsplitter cube looks just like the polarizing beamsplitter in Figure 15.22, and if they are not identified, they must be tested by using crossed polarizers. Non-polarizing beamsplitters are also available as flat glass components with a dielectric coating for the proper reflection on one side and often an anti-reflection coating on the other. The flat beamsplitters may be easier to use in some cases as the cubes produce at least weak reflections from the cube faces. An anti-reflection coating on the back surface of a flat beamsplitter or on the cube faces of a cube beamsplitter will reduce these unwanted reflections.

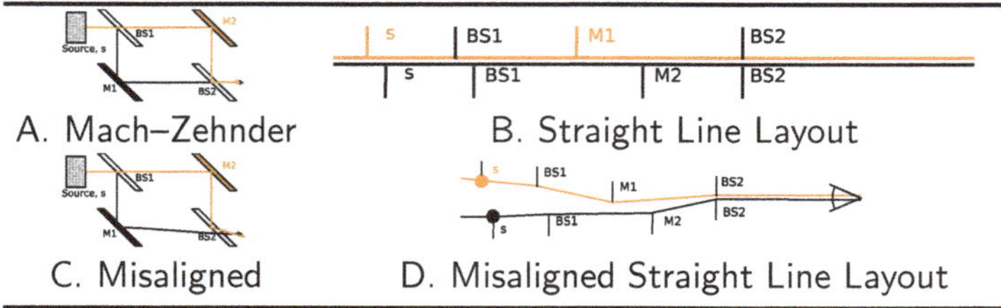

FIGURE 15.25
Interferometer straight-line layout. The top figures show the two paths as we want them . The bottom figure shows them in a state of misalignment. The goal is to adjust the mirrors to make the output beams coaxial.

In Practice

When purchasing a beamsplitter, it is necessary to specify the reflection and/or transmission as functions of wavelength and of polarization, along with the angle of incidence.

Alignment: Probably the greatest challenge in interferometry experiments is alignment. Even with careful measurements, the interferometer will be misaligned. It is useful to think of looking back into the interferometer and drawing the straight-line layout that we would see. It might look something like Figure 15.25.

A slight change of angle between the two beams will cause the beams to eventually separate so that no interference occurs, but even where they do overlap a change of path length of one half wavelength will change the output from constructive to destructive interference. To see the fringe pattern in Figure 7.6, the angular displacement of the beams must be much less than a wavelength over the width of the beam. For this, we use kinematic mounts as shown in Figure 15.26.

Example

For example, if the beam is 1 cm in diameter and we want a half wavelength change across the beam at $\lambda = 632.8\,\text{nm}$, the angular tolerance is about

$$\delta\theta = \frac{632.8 \times 10^{-9}\,\text{m}/2}{0.01\,\text{m}} = 32\,\text{microradians}. \tag{15.11}$$

If the screw in Figure 15.26C is 5 cm from the pivot point, then the screw must move

$$\delta z = \delta\theta \times 0.05\,\text{m} = 1.6\,\mu\text{m}. \tag{15.12}$$

A screw with 80 threads per inch (318 μm per thread) must be turned 0.005 turns, or about 1.8 degrees.

Successful alignment depends on careful attention to alignment during the design process, being sure to have the necessary adjustments with sufficient precision, and on developing an alignment procedure to use after arranging the components. We want the two beams at the output to be co-linear, and we achieve this when they overlap at two different points (two points determine a line). Thus we need two adjustable mounts with two degrees of freedom, not counting path length, and we have already chosen to use kinematic mounts on the two mirrors.

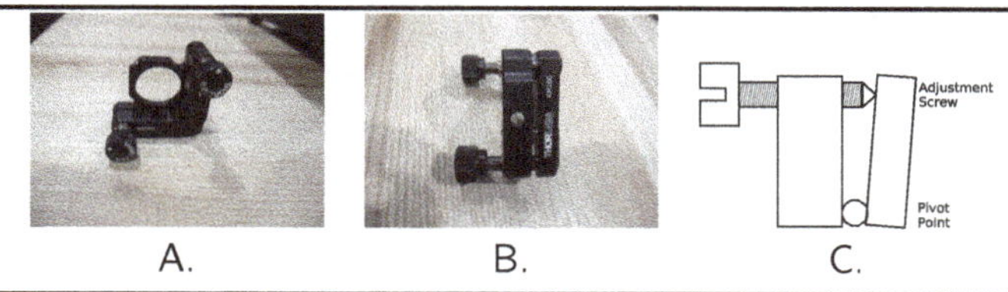

FIGURE 15.26
Angle adjustment.

FIGURE 15.27
Iterative alignment. Align the mirror further from the output on a screen near the output. Align the nearer mirror at a longer distance.

One of the common mistakes made by students in this experiment is to make the interferometer in Figure 15.23 square. In that case, the two mirrors are at the same location in the straight-line layout so their adjustments are redundant, and we will never achieve alignment. Making the interferometer rectangular is essential for alignment with these particular adjustments. Another common mistake is adding every imaginable degree of freedom to the optical components. Not only does this add extra expense in kinematic mounts, but it also introduces interactions among the different adjustents, making the alignment procedure complicated.

We note that in the figure, M_2 is closer to the output than M_1. Thus, if we observe the two laser beams on a viewing screen close to the interferometer, adjusting M_1 will have more effect than adjusting M_2. We thus iterate on the following two steps in either order; (1) adjust M_2 while viewing the output as far as possible from the interferometer. We then (2) adjust M_1 viewing the output as close as possible. Each step will undo some of the alignment from the previous step, so it is necessary to continue to iterate until some fringes are visible. The process is shown in Figure 15.27. Then we can complete the alignment by the same steps to center the fringes and maximize their spacing.

In Practice

Ideally, we might want to put one of the adjustment mirrors at an image plane and the other at a pupil plane. Then if we could make our observations at image and pupil planes the alignment would be easier. However, in addition to the possible need for extra optics to implement this approach, it is usually a good design to avoid having a mirror at an image plane (field plane). Any dust or imperfection on the mirror will appear in focus in the image.

15.3.5.3 Pre-Lab Homework: Interferometer

Read Section 15.5.

How many turns of the adjustment screw are required to add one fringe across a beam of 1 cm diameter, if the adjustment screw is 3 cm from the pivot point of the mount and has 40 threads per inch?
Design the layout of your system, including distances and design of the beam expander. Assume that the mirrors will be on tip-tilt mounts that can be adjusted.
Find the index of refraction of air as a function of temperature. Determine how large a temperature change will be observable over an appropriate distance.

15.3.5.4 Lab Experiment Instructions: Interferometer

Assemble your system. Leave out the beam expander for the first alignment.
Align so that the two beams are coaxial.
Add the beam expander and adjust for best fringes.
Place a hot object under one of the beam paths and observe fringes.
Mount a camera at the output and collect images of fringe patterns.
Put a glass or plastic transparent slide in one path. See if you can observe imperfections in the slide.
Try an LED as a light source.

A photograph of this experiment is shown in Figure 15.28. Note the apertures used to reduce stray light, and the adjustable mirror mounts. Fringes recorded on the camera are shown.

The alignment of the interferometer is among the most challenging tasks among all of these experiments. Here are a few helpful tips.

- Put a lens after the recombining beamspliter to increase the size of the fringe pattern.
- Adjust for the largest fringes; if adjustment makes the fringes thinner, the reverse the direction of the adjustment.
- It may be easier to center the fringe pattern in one direction first and then center the other. When the pattern is nearly centered it is hard to decide which direction of motion is needed.
- Vibrations may make the fringes impossible to see. This experiment is best done on a vibration-isolating table.
- Expect some cross-talk between the directions on kinematic mounts.

15.3.5.5 Post-Lab Reflections: Interferometer

How well did your interferometer work? Did it meet your expectations?
What problems did you have using the camera?
Did the LED source work?
What problems might you have in making quantitative measurements?

15.4 Cleaning Optics

After some time, despite the best precautions, dirt on lens surfaces will absorb light, reducing the overall transmission. Perhaps even worse, some light will be scattered, reducing the contrast in an imaging system. Coatings make surfaces susceptible to damage during

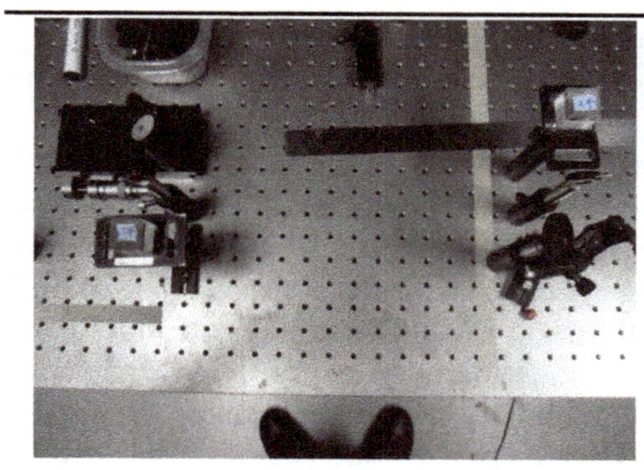

FIGURE 15.28
Interferometer. The layout on the left includes adjustable mirror mounts and apertures to minimize stray light. Fringes are shown on the right. (Left photo by Jane Lee, Northeastern University. Analysis and right image by Mohammadrasoul Taghavi.)

cleaning. The first rule is to avoid having and contamination come in contact with the surface so that less frequent cleaning is required. Dust and liquid are typical contaminants. One of the most common sources of contamination is oil from the skin deposited when the component is handled. It is also one of the hardest to clean. Cleaning fingerprints as soon as possible offers the best chance for their removal. Wearing clean gloves or finger covers and handling optics only by the edges will minimize this source of contamination. Dust can be reduced by working in a clean space and covering optics when not in use.

15.4.1 Preparation

There are a number of sources of information about cleaning optics. Read the instructions carefully. One source is https://www.thorlabs.com/newgrouppage9.cfm?objectgroup_id=9025 on cleaning lenses.

Ideally the manufacturer will provide cleaning instructions for every component, and if these are available, they are the best source for information. Coated lenses, front-surface mirrors, pellicles, many crystals, and gratings are among the commonly used components that require special cleaning instructions. For uncoated lenses and those with certain coatings, the generic instructions will be sufficient.

15.4.2 Practice

Most established optics labs have a "junk drawer," where old and often slightly damaged optics accumulate. If possible, practice cleaning optics on components from the junk drawer before attempting to clean optics that you plan to use. You can also test your technique on an old set of eyeglasses.

In practice you will want to use the least damaging cleaning techniques required for the component you are trying to clean, but this is an opportunity to try all of them.

Inspection: First examine the surface to see what needs to be done. If the problem involves one or two specks of dust only some dry air may be needed, but if there is a coating of dust or a fingerprint liquid solvents and perhaps some lens tissue will be required. Use a bright light, illuminate and view from different angles to see the extent of the problem.

Air: A few small specks of dust may be removed by blowing air across the surface. Cans of compressed air or a bulb may be used to blow air nearly parallel to the surface. It is important to avoid contaminants such as the propellant in a can of compressed air. If the laboratory has compressed air available, it should never be used for cleaning optics because the air will contain oil from the compressor. With care, clean and dry air can often be used on gratings and other sensitive components. You can test your skill on any piece of glass that has been in a dusty environment.

Immersion: Immersion in distilled water, possibly with a small amount of optical soap can remove dust that is not removed by air, and can possibly remove fingerprints. It is important that the water not contain any contaminants, so only distilled water should be used, and some components cannot be immersed. Try this with a pair of dirty eyeglasses.

Lens Tissue: If the above techniques are insufficient, more robust cleaning may be possible. Holding a piece of lens-cleaning tissue over the optic very close to the surface place a drop of solvent such as methanol on the paper and gently drag the paper across the surface. If the contamination is concentrated in one region, drag toward the nearest edge to move the contaminant across the minimum area. The paper may accumulate contaminants that you did not notice, so use a fresh sheet every time. Practice this on optics from the junk drawer.

15.4.3 The Real Test

When you begin a project, develop the habit of covering most of the components when not in use. Expensive components such as cameras will usually have covers that can be used. Paper cones can be placed over lenses, polarizers, and other components.

Inevitably you will sometimes need to clean optics. Contact the manufacturer or search for instructions for the particular component. Use the techniques discussed above in order, beginning with inspection to determine what level of cleaning may be needed. Examine the component after each step.

Damaged coatings will normally make a component unsuitable for use and it may be necessary to replace them. Expensive components can be returned to the manufacturer for recoating.

15.5 Laser Safety

Eq. 8.77: $d = \left(\frac{4}{\pi}\right)\left(\frac{\lambda}{d_0}\right) z_1$.

Lasers present a unique hazard for the user because of their ability to produce diffraction-limited beams with high power. Specifically, in a Gaussian beam of diameter, d_0, the diffraction limit at distance z_1 is given by Equation 8.77. so the angular diffraction limit is

$$\alpha = \left(\frac{4}{\pi}\right)\left(\frac{\lambda}{d_0}\right). \tag{15.13}$$

The radiance of a beam of power, P, is thus of order P/λ^2, which can be much larger than that of a thermal source.

Lasers can present hazards to the skin and especially to the eye. Depending on the wavelength, the light can be focused onto the retina to produce very dangerous levels of irradiance.

In this section, we will discuss some of the principles of laser safety and some guidelines. This is not meant to provide the user with a comprehensive knowledge of laser safety, nor to replace the training that is required in most organizations where lasers are used. The best source of information on the subject is the most recent edition of ANSI Standard Z–136–1 [136], which is based on extensive reviews of the biological effects of light on humans. Manufacturers worldwide label lasers according to this standard. Many other documents have been produced by different groups, but most are similar to this document. For example, the International Electro-technical Commission's IEC-60825 [106] is used in Europe and GB7247.1 [164] in China.

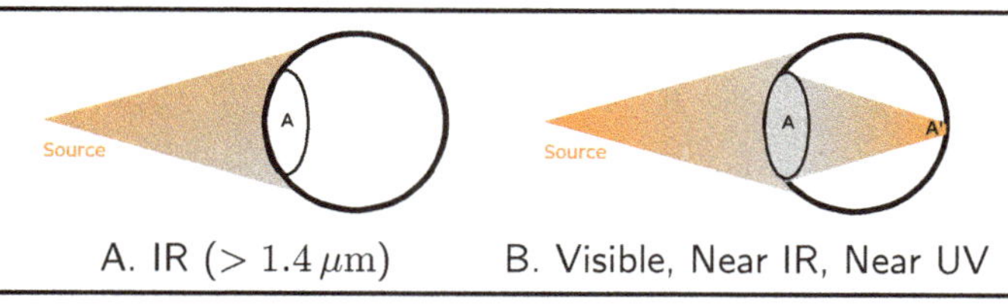

FIGURE 15.29
Eye safety. Infrared light beyond about $1.4\,\mu$m does not penetrate the eye. The eye hazard is primarily affected by the power per unit area, A. At visible and near-visible wavelengths, the eye focuses the light onto the retina so that the power on the eye's pupil (Area A) is concentrated on the retina in a much smaller area, A'.

The main hazards of laser light are photochemical and thermal, and the effects of both are dependent on wavelength and exposure time. We will first consider wavelength and then exposure time. We will conclude by discussing laser standards, classification, and protective eyeware.

15.5.1 Wavelength Issues

The effects of light on tissue include photochemical effects mostly at the shorter wavelength end of the optical spectrum and thermal effects. Differences in scattering and absorption across the spectrum can increase or decrease these effects. The most significant difference in safety is the result of focusing in the eye. As shown in Figure 15.29A, if light is strongly absorbed by the tissue, the potential damage is at the surface of the eye, and will increase with increasing power or energy per unit area. For most incoherent sources the irradiance is low. The most obvious exception is the sun, which is known to cause sunburn, skin cancers, and more. Other examples include arc lamps and ultraviolet emissions from welding, *etc*, which can damage the skin and the surface of the eye.

However, if the wavelength is transmitted through the eye (mostly water) it can focus on the retina as shown in Figure 15.29B. This effect is, of course, key to our vision. Still, most, but not all, incoherent sources are weak enough that no damage is done. However, a coherent wave falling on the surface of the eye can focus to a very small spot. Assuming a maximum iris aperture of 8 mm the diffraction-limited spot on the retina would be a few micrometers. Then all the power passing through the area A of the iris would be focused onto the retinal area, A' which is some 6 orders of magnitude smaller in area. The power is concentrated according to

$$I_{retina} = \frac{P}{A'} = I_{incident}\frac{A}{A'}, \tag{15.14}$$

and a laser

beam with irradiance of only $100\,\mu\text{W/cm}^2$ could result in $100\,\text{W/cm}^2$ or $1\,\text{MW/m}^2$ on the retina; approximately 1000 times the irradiance of the sun at the Earth's surface. The actual numbers are more complicated because the eye is not diffraction limited, and retinal tissue is different from other tissue, but the damage threshold is several orders of magnitude lower for visible and near–visible wavelengths than for others.

Safety standards allow several orders of magnitude higher exposure at wavelengths beyond $1.4\,\mu$m.

Fig. 12.9:

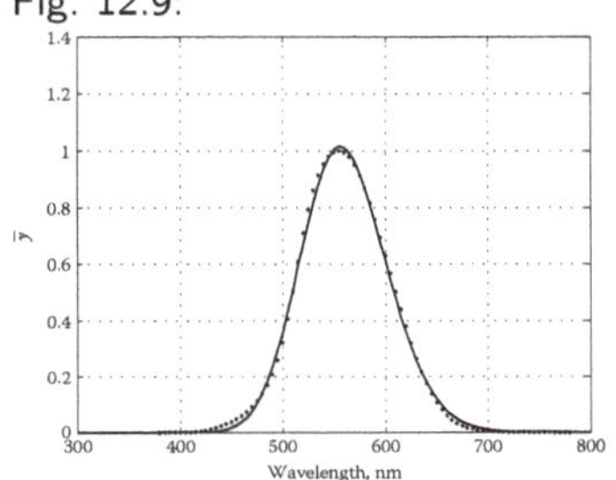

Which of the two situations in Figure 15.29 is most important depends primarily on the absorption of water, shown in Figure 15.30.

Lasers of wavelength greater than $1.4\,\mu$m are sometimes called "eye-safe" because absorption prevents the focusing of light onto the retina. This term is misleading because at high enough irradiance, damage can occur at the surface. Nevertheless, the safety standards allow several orders of magnitude higher irradiance at these wavelengths.

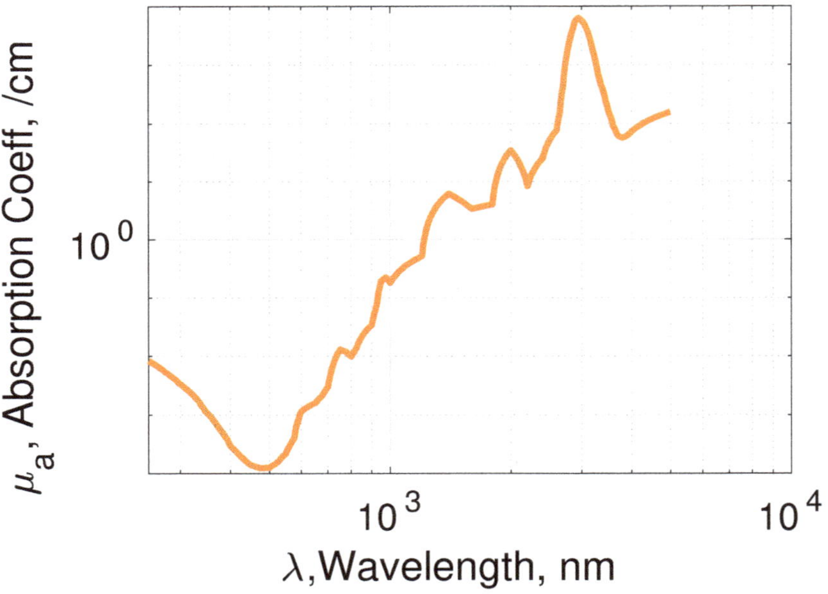

FIGURE 15.30
Absorption spectrum of water. Water absorbs light most weakly in and around the visible wavelength band. Visible, near-infrared and near-ultraviolet light can reach the retina, but longer wavelengths, especially beyond $1.4\,\mu\text{m}$, are blocked.

In Practice

Very long red wavelengths and very short infrared wavelengths can be seen on a diffuse viewing card and will look quite harmless, because the sensitivity of the eye is very low (see Figure 12.9). Nevertheless, these lasers can be quite powerful and do penetrate the eye to focus on the retina.

15.5.2 Temporal Effects

Some of the effects of light are cumulative so that a low power for a long time will produce the same result as a high power for a proportionally shorter time. The maximum permissible exposure is thus an amount of energy per unit area, often expressed in Joules per square centimeter. Other effects are self-limiting and the tissue can be exposed indefinitely to a maximum irradiance, in Watts per square centimeter. Thermal effects offer a good example, although similar temporal issues occur with photo-chemical effects.

Maximum permissible exposure is expressed in energy per area for very short exposures.

Pulses: When light is absorbed, the temperature rise depends on energy, increasing linearly in time with a constant amount of absorbed power as shown in Figure 15.31. If the irradiance is turned off, for example at the end of a short laser pulse, the tissue then cools by conduction and radiation. What is important is the maximum temperature, which depends only on the total energy absorbed.

For longer exposures to high PRF or CW lasers, the maximum permissible exposure is expressed in power per unit area, or irradiance.

CW or High PRF: Even under constant irradiance, the temperature will not increase indefinitely, because as the temperature rises above that of the surroundings, conductive and radiational cooling will occur, as shown in Figure 15.32. In the figure we consider a laser with a high pulse-repetition frequency (PRF). For each pulse we see the same effect as in Figure 15.31, except that the next pulse arrives before the temperature returns to its initial value. There is a continual increase in temperature up to a steady state value

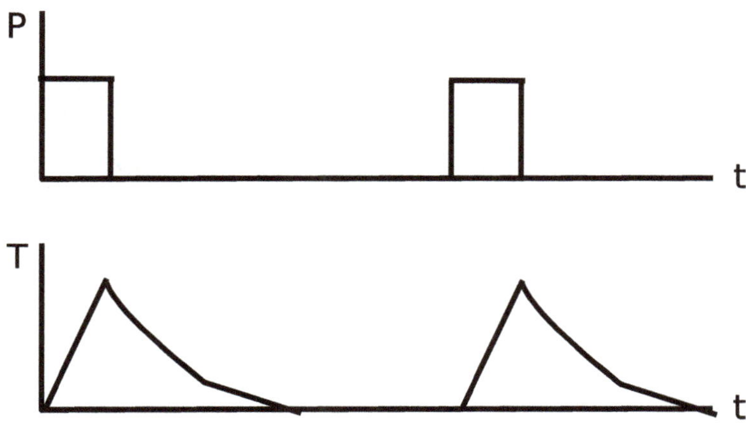

FIGURE 15.31
Pulsed laser thermal issues. Temperature rise is proportional to energy and so rises linearly during the pulse. The tissue cools after the end of the pulse.

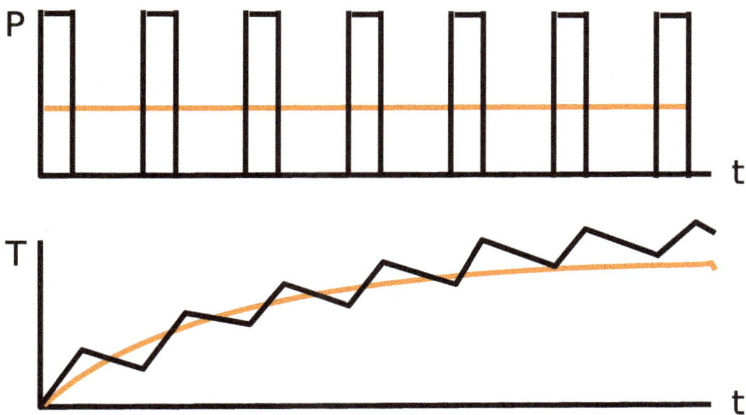

FIGURE 15.32
CW laser thermal issues. Absorbed power is ultimately balanced by cooling, leading to a constant temperature. Thus irradiance is the limit for CW (colored lines) or high PRF pulsed (black lines) laser safety.

where the total cooling during the pulse period is equal to the heating during one pulse. In this case the temperature approaches a limit based on the average power. The colored lines on the graphs are for a pulsed laser with a fast repetition rate, while the black ones are for a CW laser. In this case, the maximum permissible exposure is the irradiance that produces an acceptable steady–state temperature.

Maximum Permissible Exposure: We have discussed the temporal effects in terms of thermal damage, in the limiting cases of very short and very long exposures. There are also temporal differences associated with the photochemical effects, and we haven't

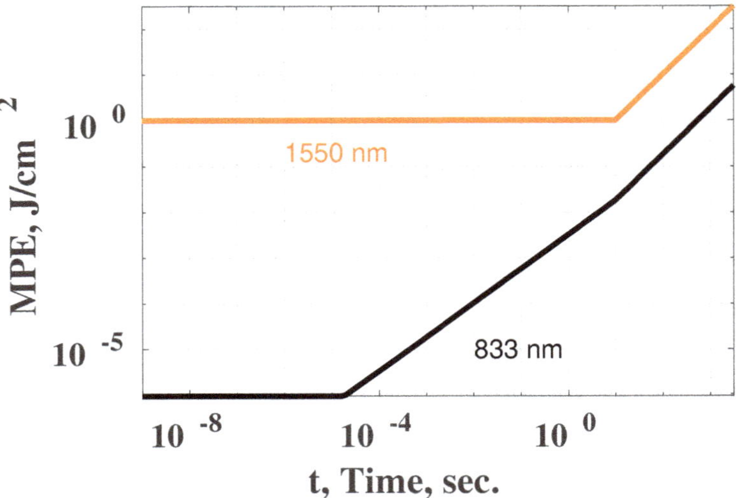

FIGURE 15.33
MPE. The maximum permissible exposure is shown in Joules per square centimeter. At short exposure times, the lines are horizontal indicating a constant energy density. At long times, the lines have slopes of one decade per decade, indicating constant power density. The exposure limits are significantly higher for the wavelength greater than $1.4\,\mu$m.

discussed what we mean by long and short exposures. The standards address all these issues and provide maximum permissible exposure (MPE) limits for the area-density of power and energy over all times from the shortest laser pulses to a full working day of about 30,000 seconds. Two examples are shown in Figure 15.33. For short times, the MPE is expressed in Joules per square centimeter, as discussed above. For longer times, the MPE is in Watts per square centimeter, so the plot is linear (one decade per decade on the logarithmic scale). The MPE for the longer wavelength is much higher than the shorter one, because the former wavelength does not penetrate the eye. This figure was plotted according to the limits prescribed in ANSI standard Z–136–1[136].

15.5.3 Laser Classification

So far, we've considered laser safety in terms of MPE limits as functions of wavelength and exposure time, and we can see that these are quite complicated functions. Considering the many issues associated with how lasers are used introduces more complexity. For example, if a laser is used outdoors, an observer might use binoculars or a telescope, which would increase the exposure. On the other hand a high-powered laser might be totally enclosed in a box and thus present no hazard during normal use. The laser user would be faced with a new challenge in determining safety with every laser in every situation. For this reason, a set of safety classes has been developed with different safety considerations applicable to each one. The overall classifications are presented simply and briefly in Table 15.2.

Different controls measures are required for different classes.

For each laser classification, the applicable standards include detailed descriptions of the potential hazards and rules for mitigating them. More strict rules may also be imposed by laboratories, institutions, employers, or government entities.

1: Class 1 lasers are considered safe for all exposures up to a full working day, and normally no controls are imposed. Regardless of the use of external optics these lasers will never exceed the MPE for any exposure time. Nevertheless, it is good practice to avoid pointing any laser toward a person. The surprise of sudden bright light could cause a fall

TABLE 15.2
Simplified Laser Classes. The four main classes of lasers are shown here. Different classes require different control measures to ensure safety.

Class	Simplified Description	Variations
1	Normally not Hazardous	1, 1M
2	Aversion Response is Protective (Visible Light Only)	2,2M
3	Potential Direct-Beam Eye Hazard	3R,3B,(3A)
4	Diffuse-Light Hazard, Fire Hazard, Air Contamination, Plasma	

or other injury, and the viewer could raise concerns as to whether the exposure was really safe.

A laser system can be considered Class 1 even if it contains a laser with a higher classification if the beam is completely enclosed and certain conditions are met with regard to the enclosure and interlocks. When such a system is opened, for example during repair, it is not considered Class 1.

2: In the visible spectrum, a viewer will see the laser light and blink, thereby limiting the exposure to less than 1/4 second. If the power is low enough, the energy received during this time will be less than the MPE. Such lasers are designated Class 2, and can usually be used without controls, in situations where a person can turn away from the beam upon exposure. Lasers outside the visible spectrum can never be Class 2 because the viewer will not respond with a blink.

3: Class 3 lasers can cause eye damage. In particular, some control measures are usually required for Class 3B lasers, such as warning signs and lights, access control and the use of protective eyeware. Some institutions and governments require registration of Class 3B lasers. A Class 3B laser should never be operated where anyone's eye could accidentally enter the beam path. Diffuse viewing of Class 3 (and 1 and 2) lasers is not an eye hazard. Class 3R is considered to present only a minimal hazard. The designation Class 3A is now obsolete but is similar to Class 3R. Note that Class 3 laser pointers are readily available despite the hazard.

4: A laser earns a Class 4 designation if it is hazardous to view a diffuse reflection, if it presents a burn hazard, or if it is capable of ionizing air to produce a contamination hazard. Usually these lasers are operated in a room with door interlocks so that the laser is turned off if the door is opened. Eye protection is essential. Special consideration may be needed to prevent fires from starting.

15.5.4 Protective Eyeware

For some lasers (usually Classes 3B and 4), protective eyeware is important. The goal of protective eyeware is simple; to reduce eye exposure to below the Class 1 MPE at the wavelength(s) in use. Eyeware comes in a variety of styles, often with protection on the sides, top, and bottom. Most eyeware is designed to provide protection while minimizing the disruption to the user's vision. The eyeware may, for example, provide a notch filter to block the specific wavelength while passing all others. In this case the view through the eyeware will be very natural. For example, simple glass or plastic lenses can provide protection from 10 μm wavelengths, while being nearly transparent to visible wavelengths. A person who wears glasses is afforded some protection from these wavelengths. For visible wavelengths, the eyeware can block a particular wavelength using thin film coatings as in Section 7.7. Because the coating needs to block light over a range of angles, the blocked bandwidth will be larger than the laser line, but still small enough that most of the visible spectrum is passed. However, it becomes expensive to have eyeware for every wavelength

Protective eyeware reduces the eye exposure below the Class 1 MPE.

Thin-Film Coatings: Sec. 7.7

present in a laboratory with many different lasers. Therefore, it is common to find eyeware to protect against a range of wavelengths. Argon laser users want protection from 514 nm, 488 nm, and other wavelengths emitted by their lasers. Likewise, near–infrared laser users may need protection from 750 nm and longer wavelengths. Some eyeware will protect against multiple combinations, with particularly high protection for wavelengths of more powerful lasers.

Eyeware performance is characterized by the OD at the wavelength(s) of interest.

Choosing the right eyeware for a particular laser is first a matter of reducing the exposure to Class 1. The protection is expressed as optical density (OD), which defines the transmission of the lens by

$$T = 10^{-OD}, \tag{15.15}$$

just as we defined ND in Equation 15.7. We use the term neutral density (ND) when the transmission is independent of wavelength, at least over a wide band such as the whole visible spectrum and optical density (OD) when it is a function of wavelength. Thus the minimum OD for a laser of power, P, is

$$OD_{min} = \log_{10} \frac{P_{Class1}}{P}. \tag{15.16}$$

Example

For example, at $\lambda = 640$ nm, the Class 1 limit according to the ANSI standard[136] is $P_{Class1} = 100\,\mu$W. If my laser is a Class 3B laser at $P = 100$ mW, then the requirement is $OD > 3$ at 640 nm. We would do well to choose $OD = 4$ or more to provide a margin of error.

In Practice

The amount of protection is normally indicated on the eyeware, with the OD at specific wavelengths or ranges of wavelengths. It is important to note the wavelength as well as the OD. Glass with $OD = 6$ at red wavelengths may do nothing to protect the user working with a green laser. Glasses with $OD = 6$ at the frequency–doubled Nd:YAG wavelength of 532 nm may not provide any protection at the fundamental wavelength of that laser, 1064 nm. Do not interpolate; Glasses with $OD = 6$ at 488 and 514 nm may have a much lower OD in between.

It is helpful to see the laser beam with the protective eyeware.

Outside the visible spectrum we may want to use as high OD as possible to protect from any laser we might use at a future time, and thus avoid having to buy higher OD later. However in the visible, we like to see the light for alignment, and too great an OD eliminates that possibility. It is useful to buy "alignment glasses" that provide sufficient OD, but not so much that the beam becomes invisible. In the example above, that required $OD > 3$ at 640 nm, we may want to avoid choosing anything above $OD = 5$, with which the power visible to the eye would be 1 μW.

15.5.5 Resources

The subject of laser safety is quite complicated, and rules vary somewhat across organizations. There are many complicated situations that affect laser safety. Scanning may make the computation of exposure time complicated. Outdoor use of high-powered lasers requires definition of a hazard area, beam size, the potential for the use of optical instruments, and more. There are potential hazards we have not discussed, commonly called non-beam hazards, such as electrical, chemical and cryogenic hazards, the use of compressed gasses, and more that are related to specific lasers.

Fortunately, there are many resources. In addition to the national and international standards we have discussed, there are rules by local governments, and individual workplaces. Occasionally standards are updated as new information becomes available to the organizations that write them. Almost every organization has a safety office that can provide expertise as needed. Many have administrative controls that require training and specific authorization to use lasers and require periodic inspections and "refresher" training. Some governments and employers require registration of lasers in certain classes. There are also vendors and consultants who will provide expertise for a fee.

Appendix

APPENDIX A: NOTATION AND DRAWINGS FOR GEOMETRIC OPTICS

Geometric optics necessarily involves a large number of variables, and many authors disagree on the notation. It is often a difficult task for the seasoned expert to keep in mind the meanings of various terms defined in different texts and papers. Presenting the notation as it is used can be helpful but can either become repetitive if it is discussed at each use, or can require the reader to search through many chapters to find the definitions. Although the chapters present each definition the first time it is used, this appendix summarizes them all in a single location. The reader may wish to refer to this appendix particularly during the study of the first five chapters of the text.

A.1 Direction

Whenever possible, we will draw figures so that light propagates from left to right. We will generally organize our sign conventions so that object and image distances are positive when light travels from the source to the imaging system and from the imaging system to the image.

A.2 Angles and the Plane of Incidence

Angles of incidence, reflection, and refraction are always measured from the normal to the surface. The incident ray and the normal define the plane of incidence. The reflected and refracted rays will also be in this plane.

A.3 Vertices

The first step in analyzing an optical system is to decide on the front and back vertices. Usually, for a simple lens, these are the intersections of the front and back surfaces of the glass, respectively with the axis. In a compound lens, they may be chosen as the intersections of the first and last glass surfaces with the axis. Alternatively, we may define the vertices in any way we like. For a camera lens as an example, it may be convenient to define the vertices as both being at a flat reference surface. We might use the location where the lens attaches to the camera body for example. In any case, however we choose the vertices, the mathematical results may be different, but the physical results must not. If, as an example, the front vertex is not at the first glass surface, we need to account for propagation of light from the vertex to the glass surface, probably using a translation matrix (Equation 3.5).

DOI: 10.1201/9781315157047-A

A.4 Notation Conventions

Most of the conventions are shown in Figures A.1 and A.2. We use capital letters to represent points and lowercase letters to represent corresponding distances. For example, F is the front focal point, located a distance f to the left of the front principal plane. In Figure A.2, f is negative. The principal planes are drawn with a thin solid line and a thick dashed one. The dashed part is on the left for the front principal plane and the right for the back one. A virtual image is represented by a dashed arrow. The aperture stop is represented by a block with diagonal stripes from top left to bottom right, and the field stop is represented similarly, but with the stripes from top right to bottom left. Most of the letters that will be used are shown in Table A.1.

A.4.1 Curvature

The radius of curvature of a mirror is positive if it is concave toward the source.

The radius of curvature of a dielectric interface is positive if it is convex toward the less-dense medium. There is great confusion about this choice in the literature. From an analytical perspective, it would be best to use a convention that is consistent with respect to the direction of the light, such as we do with mirrors. We have chosen to define the

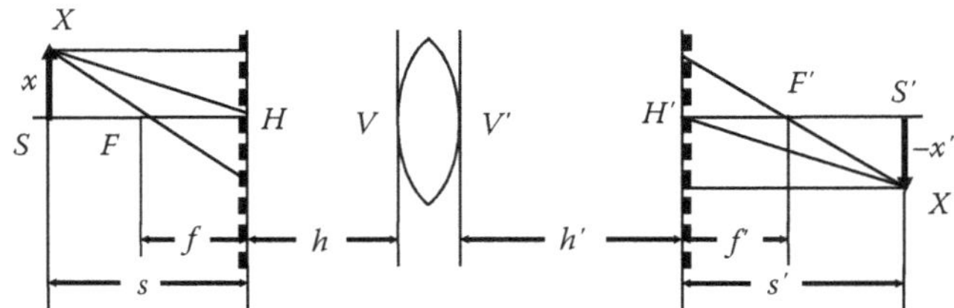

FIGURE A.1
Notation and conventions for drawings in imaging. Nearly every drawing of an imaging system has an axis. The object and image represented by thick vertical arrows. Actual light rays that could travel through the system are drawn with solid lines. The principal planes are drawn with a thin solid line and a thick dashed one. The dashed part is on the left for the front principal plane and the right for the back one. Capital letters represent points. For example, the front focal point is at F, and the front focal length is f. Every distance in this figure except the image height is positive.

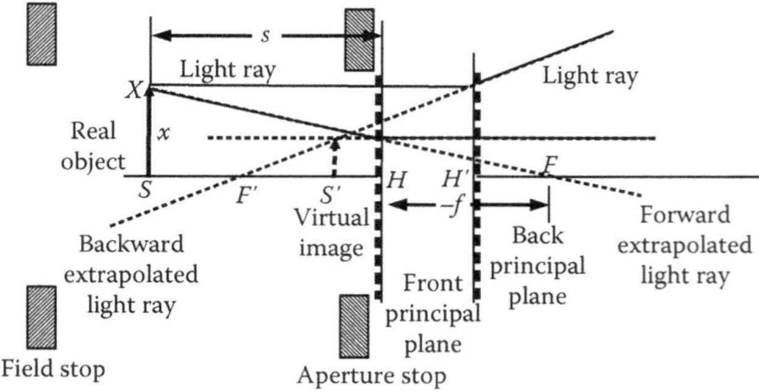

FIGURE A.2
Virtual rays and images. Actual light rays that could travel through the system are drawn with solid lines, but rays that are extrapolated in either direction are dashed. A virtual image is represented by a dashed arrow. The aperture stop is represented by a block with diagonal stripes from top left to bottom right, and the field stop is represented similarly, but with the stripes from top right to bottom left. Capital letters represent points. The front focal point is at F, and the front focal length is f, which is negative in this example.

TABLE A.1
Notation for Geometric Optics

Point Name	Notation	Length Name	Notation	Description
Center of curvature	R		r	Positive concave for mirror
Front vertex	V			
Back vertex	V'			
Front principal point	H		h	Left of V
Back principal point	H'		h'	Right of V
Front focal point	F	Front focal length	f	Left of H
Back focal point	F'	Back focal length	f'	Right of H'
Object point	S	Object distance	s	Left of H
Image point (base)	S'	Image distance	s'	Right of H'
Object point (tip)	X	Object height	x	Above S
Image point (tip)	X'	Image height	x'	Above S'

Capital letters represent points, and lowercase letters represent the corresponding distances. We may choose to define the front vertex, V, as the intersection of the first glass surface with the axis, and the back vertex, V' as the intersection of the last glass surface with the axis. Other definitions are possible and will result in different mathematical description, but the same physical system.

curvature as positive if it is convex toward the source. Therefore, a symmetric biconvex lens has radii of curvature with different signs for the two sides. An alternative convention, convenient from the perspective of the manufacturer, is to define a surface as positive if it is convex toward the medium with the lower index, normally air. A surface with "a radius of 10 cm" is then made with the same tool. This definition also has the advantage that surfaces with positive radii of curvature all have positive optical power. However, it makes the derivations of equations more complicated with the introduction of absolute values.

The radius of curvature of a Gaussian beam, ρ, is defined to be consistent with the definition of the distance from the waist, z (Chapter 9). If z and ρ are negative, then the wavefront is converging toward the waist, and if they are positive, it is diverging from the waist.

A.5 Solid and Dashed Lines

We use solid lines to indicate rays of light that pass through the optical system. The ray that passes through the front focal point leaves the back principal plane parallel to the axis. This ray starts as a diagonal from the object point, X, and is shown passing through F. However, after the ray intersects the front principal plane, it changes direction and continues parallel to the axis. The actual ray is a solid line, and the forward extension to show it passing through F is shown dashed. Likewise, after refraction, the extension backward is shown dashed back beyond the image. The ray that starts from X parallel to the axis leaves the second principal plane and passes through the back focal point, F'. Because the ray is refracted at the lens and does not actually pass through F', it is extended by a dashed line as needed. The image formed by the intersection of dashed lines is a virtual image and is thus itself shown with a dashed arrow.

A.6 Stops

The aperture stop and field stop are shown as thick boxes illustrating the size of their openings. The aperture stop, and any images of it (pupils), are shown with diagonal hatching from top left to bottom right. The field stop and its images (windows) use hatching in the orthogonal direction.

APPENDIX B: SOLID ANGLE

B.1 Solid Angle Defined

The concept of **solid angle** in three dimensions is similar to the concept of angle in two dimensions. We define an angle in radians as the circumference of that portion of a circle subtended by the angle, divided by the radius of the circle. The radian is strictly a "dimensionless parameter" in that it is a length divided by a length. Likewise, we define solid angle in three dimensions as the area of a portion of the surface of a sphere divided by the square of the radius. The unit of solid angle is called the **steradian** or sr. Like the radian, it is also a "dimensionless parameter." A semicircle corresponds to an angle of π radians and a full circle to 2π radians. In the same way, a hemisphere corresponds to a solid angle of 2π steradians and a full sphere to 4π steradians.

B.2 Integration over Solid Angle

Frequently, we need to integrate a function over solid angle. For example, we may want to integrate the intensity of a light source, I, over the numerical aperture of an imaging system to obtain total power, P, as in Section 12.1.1.2.

The integral over solid angle is accomplished as shown in Figure B.1. The increment of area on the unit sphere is found as follows. Thinking of the sphere in geographic coordinates, with the "North pole" being at the top (z),

$$\theta = 90° - latitude, \tag{B.1}$$

is zero at the North pole, 90° at the equator, and 180° at the South pole. The "longitude" is

$$longitude = \zeta. \tag{B.2}$$

The width of the strip of area shown is $d\theta$ in the north–south direction, and the length, in the east-west direction is $\sin\theta\, d\zeta$. The resulting integral is

$$P = \iint I\,(\theta, \phi) \sin\theta\, d\theta\, d\zeta. \tag{B.3}$$

If I can be considered constant and brought outside the integral, then

$$P = \Phi = I\Omega,$$

where the solid angle is given by

$$\Omega = \iint \sin\theta' d\theta' d\zeta. \tag{B.4}$$

DOI: 10.1201/9781315157047-B

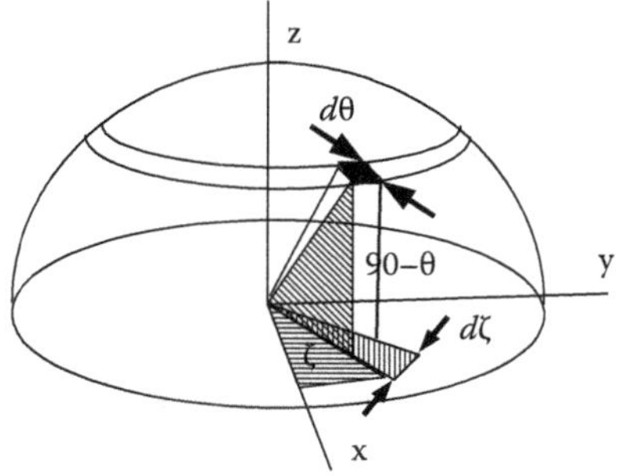

FIGURE B.1
Unit sphere. The solid angle is $\int\int \sin\theta\, d\theta\, d\zeta$. The integral over a hemisphere is 2π. The integral of any quantity, $X(\theta, \zeta)$, over solid angle is $\int\int X(\theta,\zeta) \sin\theta\, d\theta\, d\zeta$.

For cylindrically symmetric problems, the integral over ζ is from 0 to 2π, and the integral over θ is from 0 to Θ:

$$\Omega = \int_0^{2\pi}\int_0^{\Theta} \sin\theta' d\theta' d\zeta = 2\pi\left(1 - \cos\Theta\right), \qquad \text{(B.5)}$$

or, recalling the definition of numerical aperture in Equation 4.2,

$$\Omega = 2\pi\left(1 - \sqrt{1 - \left(\frac{NA}{n}\right)^2}\right) \qquad \text{(Cylindrical symmetry).} \qquad \text{(B.6)}$$

In the case of small numerical aperture, we can use Taylor's series for the square root, and obtain

$$\Omega \approx 2\pi\left\{1 - \left[1 - \frac{1}{2}\left(\frac{NA}{n}\right)^2\right]\right\},$$

or

$$\Omega \approx \pi\left(\frac{NA}{n}\right)^2 \qquad \text{(Small NA)} \qquad \text{(B.7)}$$

At the other extreme, to capture all the power, the integral is taken over ζ from 0 to 2π, and the integral over θ from 0 to $\pi/2$ for a hemisphere. From Equation B.6, $\Omega = 2\pi$. For a full sphere, the θ integral is taken to π and Equation B.4 produces $\Omega = 4\pi$. In fact, this integration can be realized physically. As we will see in Section 12.4.2, an integrating sphere can be used to collect all of the light and direct it to a detector for measurement.

To summarize the discussion of solid angle, the general expression is Equation B.4, and important special cases are Equation B.6, and its approximations, Equation B.7 as well as

$$\Omega = 2\pi \qquad \text{(Hemisphere)} \qquad \text{(B.8)}$$

and

$$\Omega = 4\pi \qquad \text{(Sphere).} \qquad \text{(B.9)}$$

APPENDIX C: MATRIX MATHEMATICS

For the reader who needs a review of the mathematics of matrices, we present a short "survival guide." The techniques needed to perform the calculations in matrix optics (Chapter 3), Jones and Mueller calculus (Chapter 6) and multilayer dielectric coatings (Section 7.7) will be covered. The reader interested in further study of these techniques can choose from a number of books on linear algebra. We will consider two-by-two matrices in this appendix because the matrices in these sections are all that size. Most of the operations described here are valid for larger matrices with the exception of calculating the determinant. We briefly use three-by-three matrices in our study of color in Section 12.3.3, and four-by-four in our discussion of Mueller matrices in Chapter 6, but we only use matrix multiplication and inversion in those sections.

C.1 Vectors and Matrices: Multiplication

A **vector** is normally represented as a matrix consisting of a single column. The vector notation we will use in the text is the bold character. The "vector V" is written as $\mathbf{V}$. For two dimensions,

$$\mathbf{V} = \begin{pmatrix} v_1 \\ v_2 \end{pmatrix}. \tag{C.1}$$

We use the calligraphic font, $\mathcal{M}$, for matrices. A **matrix** in two dimensions is given by four elements in two rows and two columns. If we wish to identify the matrix elements with subscripts, we place the row number first and the column number second. For example,

$$\mathcal{M} = \begin{pmatrix} m_{11} & m_{12} \\ m_{21} & m_{22} \end{pmatrix}. \tag{C.2}$$

Matrix multiplication is performed as follows:

1. Pick row m of the first matrix and column n of the second.
2. Multiply the elements of these two, element-by-element, and sum.
3. Place the sum in row m, column n of the answer.

The process is only defined if the number of columns in the first matrix is equal to the number of rows in the second so that the numbers of elements in the selected row and column are the same. The answer will be a matrix with as many rows as first matrix and as many columns as the second. We will consider two examples. Multiplying two-by-two

DOI: 10.1201/9781315157047-C

matrices,

$$\mathcal{Z} = \mathcal{X}\mathcal{Y}$$

$$z_{11} = x_{11}y_{11} + x_{12}y_{21} \qquad z_{12} = x_{11}y_{12} + x_{12}y_{22}$$

$$z_{21} = x_{21}y_{11} + x_{22}y_{21} \qquad z_{22} = x_{21}y_{12} + x_{22}y_{22}$$

$$\mathcal{Z} = \begin{pmatrix} x_{11}y_{11} + x_{12}y_{21} & x_{11}y_{12} + x_{12}y_{22} \\ x_{21}y_{11} + x_{22}y_{21} & x_{21}y_{12} + x_{22}y_{22} \end{pmatrix}. \tag{C.3}$$

Specifically, multiplication of a matrix and a vector results in a new vector,

$$\mathbf{B} = \mathcal{M}\mathbf{A}$$

$$\mathbf{B} = \begin{pmatrix} m_{11}a_1 + m_{12}a_2 \\ m_{21}a_1 + m_{22}a_2 \end{pmatrix} \tag{C.4}$$

C.2 Cascading Matrices

If the matrix $\mathcal{M}_{12}$ operates on the vector $\mathbf{V}_1$ to produce the vector $\mathbf{V}_2$, and then a new matrix $\mathcal{M}_{23}$ operates on the vector $\mathbf{V}_2$ to produce the vector $\mathbf{V}_3$, so that

$$\mathbf{V}_2 = \mathcal{M}_{12}\mathbf{V}_1 \qquad \mathbf{V}_3 = \mathcal{M}_{23}\mathbf{V}_2, \tag{C.5}$$

then

$$\mathbf{V}_3 = \mathcal{M}_{23}\mathcal{M}_{12}\mathbf{V}_1. \tag{C.6}$$

We can multiply the two matrices first, and say that

$$\mathbf{V}_3 = \mathcal{M}_{13}\mathbf{V}_1, \tag{C.7}$$

where

$$\mathcal{M}_{13} = \mathcal{M}_{23}\mathcal{M}_{12}. \tag{C.8}$$

Note as we progress from beginning to end, we multiply matrices from right to left.

C.3 Adjoint

We use the superscript † to represent the **Hermitian adjoint** or the transpose of the complex conjugate:

$$\mathcal{M} = \begin{pmatrix} m_{11} & m_{12} \\ m_{21} & m_{22} \end{pmatrix} \qquad \mathcal{M}^\dagger = \begin{pmatrix} m_{11}^* & m_{21}^* \\ m_{12}^* & m_{22}^* \end{pmatrix}, \tag{C.9}$$

or for a vector,

$$\mathbf{V} = \begin{pmatrix} v_1 \\ v_2 \end{pmatrix} \qquad \mathbf{V}^\dagger = \begin{pmatrix} v_1^* & v_2^* \end{pmatrix}. \tag{C.10}$$

We can show by carrying out the multiplications term-by-term that if

$$\mathbf{B} = \mathcal{M}\mathbf{A}$$

then

$$\mathbf{B}^\dagger = \mathbf{A}^\dagger \mathcal{M}^\dagger, \tag{C.11}$$

or if

$$\mathcal{Z} = \mathcal{X}\mathcal{Y}$$

then

$$\mathcal{Z}^\dagger = \mathcal{Y}^\dagger \mathcal{X}^\dagger. \tag{C.12}$$

In both cases, we take the Hermitian adjoint of the matrices to be multiplied, reverse the order, and multiply. The Hermitian adjoint of a column vector is a row vector.

C.4 Inner and Outer Products

The **inner product** or **dot product** of two vectors is a scalar:

$$\mathbf{V}^\dagger \mathbf{V} = v_1 v_1^* + v_2 v_2^*. \tag{C.13}$$

If the vector components represent field components, then the inner product represents the irradiance.

The **outer product** is defined as

$$\mathbf{V}\mathbf{V}^\dagger.$$

Following our rules for matrix multiplication in Section C.1,

$$\mathbf{V}\mathbf{V}^\dagger = \begin{pmatrix} v_1 v_1^* & v_1 v_2^* \\ v_2 v_1^* & v_2 v_2^* \end{pmatrix}. \tag{C.14}$$

C.5 Determinant

The determinant of a matrix is a special number associated with the matrix. For our purposes, using only two-by-two matrices,

$$\det \mathcal{M} = m_{11} + m_{22} - m_{12} - m_{21}. \tag{C.15}$$

The determinant of a matrix often has some physical significance. In Chapter 3, the determinant of the matrix is a constant, and this statement dictates the Abbe sine condition.

C.6 Inverse

The **inverse** of a matrix can be calculated provided that the determinant of the matrix is not zero. If

$$\mathbf{B} = \mathcal{M}\mathbf{A},$$

then

$$\mathbf{A} = \mathcal{M}^{-1}\mathbf{B}. \tag{C.16}$$

C.7 Unitary Matrices

A matrix, $\mathcal{M}$, is said to be **unitary** if

$$\mathcal{M}^\dagger \mathcal{M} = \mathcal{I} = \begin{pmatrix} 1 & 0 \\ 0 & 1 \end{pmatrix}. \tag{C.17}$$

The importance of unitary matrices is that they do not alter the inner product of a vector with itself. Thus, a unitary Jones matrix does not change the irradiance. A transformation of coordinates must be represented by a unitary matrix because the irradiance is independent of the coordinate system in which the field is expressed. A matrix for a device that does not change the irradiance, regardless of the input vector, is also unitary. In this case,

$$\mathbf{V}_{out}^\dagger \mathbf{V}_{out} = \mathbf{V}_{in}^\dagger \mathcal{M}^\dagger \mathcal{M} \mathbf{V}_{in} = \mathbf{V}_{in}^\dagger \mathbf{V}_{in}. \tag{C.18}$$

C.8 Eigenvectors

If the direction of a vector is unaltered when it is multiplied by a matrix, then it is called an **eigenvector** of the matrix. The vector $\mathbf{V}$ is an eigenvector of $\mathcal{M}$ if

$$\mathcal{M}\mathbf{V} = m\mathbf{V}, \tag{C.19}$$

where m is a scalar, called the **eigenvalue**. Eigenvectors are important in Chapter 6 because they are the vectors for which the function of a polarizing device is easy to describe. If we use a coordinate system consisting of the eigenvectors of a matrix, then the matrix is diagonal.

APPENDIX D: LIGHT PROPAGATION IN BIOLOGICAL TISSUE

Light provides spectroscopic information, is nonionizing, and is readily available at low cost. Optical techniques are highly desired in biology and medicine, where they provide

- Resolution less than 1 μm (subcellular)
- Depth of penetration (to centimeters with near-infrared light)
- Noninvasive imaging

Not all of these are possible in a single situation, however, and it is necessary to explore the compromises among them. Spurred by the development of lasers, detectors, and computers, combined with advances in chemistry, physics, and biology, the field of biomedical optical imaging has grown in recent years and is the subject of many books and countless papers. Here we present a brief overview of light propagation in tissue, primarily to help the reader understand some of the examples in the book.

When light is used to image inside tissue, it is scattered, absorbed, and refracted in complicated ways by the highly heterogeneous tissue as shown in Figure D.1. Some light is specularly reflected, according to the Fresnel equations of Section 6.4. Heterogeneities with sizes comparable to, or smaller than, the wavelength, will scatter any incident light over a large range of angles. In most tissue, the amount of directly transmitted, or **ballistic, light** that does not interact with these heterogeneities is so small that it usually cannot be measured. The **diffusely transmitted light** may be strong or weak depending on the tissue and wavelength. A thumb placed over a red LED on a telephone or similar device will transmit a considerable amount of the light. With a green light source, almost no light will be transmitted. With a white light source, the transmitted light will appear mostly red, as shorter wavelengths are absorbed and scattered. Near infrared light will penetrate further, up to several centimeters of tissue[39]. The reflected ballistic light that is scattered back from exactly one scattering point, is strong from points near the surface, but becomes increasingly weaker according to Beer's law, as depth, z, increases:

$$I(z) = I(0)\, e^{-\mu z}, \tag{D.1}$$

where μ is called the extinction coefficient. Most of the light scatters many times, before exiting in either direction as diffuse transmission or reflection, and the result is that the promise of high resolution that optics inherently provides is lost as a result of the scatter. Only the ballistic photons carry sufficient information to provide resolution near the diffraction limit, and they are too few to measure, with few exceptions. Quasi-ballistic or so-called **snake photons** that deviate only slightly from a straight line can provide information at greater depths with degraded but still useful resolution.

The ballistic and snake photons are few in number but they are sometimes sufficient to detect. Scattered from the tissue surface, they are easily detected, but as we image deeper they are first obscured by the "clutter" from other scatterers. The ballistic and snake photons may be selected either on the basis of their spatial behavior as in confocal

DOI: 10.1201/9781315157047-D

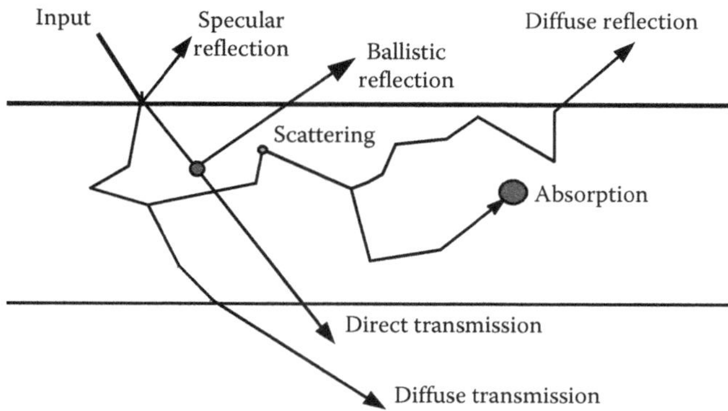

FIGURE D.1
Light in tissue. When light enters biological tissue, it can be reflected, refracted, transmitted, scattered, or absorbed. In thick tissue, the specular transmission is very small. The scattering is so strong that light may travel 6 or 10 times the ballistic transit time.

microscopy, or their temporal behavior, using short pulses, modulated sources, or the coherence gating properties of optical coherence tomography (OCT).

At greater depths, these photons are too few to be detected at all. Then we can only use the highly scattered photons in techniques such as diffusive optical tomography[257;23], and other techniques.

APPENDIX E: USEFUL MATRICES

In this appendix, we include some useful matrices.

E.1 ABCD Matrices

Table E.1 shows the fundamental ABCD matrices used in matrix optics (Chapter 3), and Table E.2 shows some useful ABCD matrices we have derived from the fundamental ones. In each case, the equation where the matrix is defined is shown.

TABLE E.1
Fundamental ABCD Matrices

Operation	Matrix	Equation
Translation	$\mathcal{T}_{12} = \begin{pmatrix} 1 & z_{12} \\ 0 & 1 \end{pmatrix}$	3.10
Refraction	$\mathcal{R} = \begin{pmatrix} 1 & 0 \\ \frac{n-n'}{n'r} & \frac{n}{n'} \end{pmatrix}$	3.17

E.2 Jones Matrices

Some fundamental and useful Jones matrices are shown in Table E.3.

TABLE E.2
Derived ABCD Matrices

Operation	Matrix	Equation
Thin lens	$\mathcal{L} = \begin{pmatrix} 1 & 0 \\ -\frac{P}{n'} & \frac{n}{n'} \end{pmatrix}$	3.25
Thin lens in air	$\mathcal{L} = \begin{pmatrix} 1 & 0 \\ -P & 1 \end{pmatrix} = \begin{pmatrix} 1 & 0 \\ -\frac{1}{f} & 1 \end{pmatrix}$	3.27
Imaging (conjugate planes)	$\mathcal{M}_{SS'} = \begin{pmatrix} m & 0 \\ -\frac{P}{n'} & \frac{n'}{n}\frac{1}{m} \end{pmatrix}$	3.50
Imaging in air	$\mathcal{M}_{SS'} = \begin{pmatrix} m & 0 \\ -\frac{1}{f} & \frac{1}{m} \end{pmatrix}$	
Telecentric: image to pupil	$\mathcal{M}_{FF'} = \begin{pmatrix} 0 & f_1 \\ -\frac{1}{f_1} & 0 \end{pmatrix}$	9.46
Telecentric: conjugate image planes	$\mathcal{M}_{SS'} = \begin{pmatrix} m & 0 \\ 0 & \frac{1}{m} \end{pmatrix}$	3.93

DOI: 10.1201/9781315157047-20

TABLE E.3
Some Jones Matrices

Operation	Matrix	Equation
Perfect $\hat{\mathbf{x}}$ polarizer	$\mathcal{P}_x = \begin{pmatrix} 1 & 0 \\ 0 & 0 \end{pmatrix}$	6.98
Realistic $\hat{\mathbf{x}}$ polarizer	$\mathcal{P}_x = \begin{pmatrix} \tau_{pass} & 0 \\ 0 & \tau_{block} \end{pmatrix}$	6.100
Perfect $\hat{\mathbf{y}}$ polarizer	$\mathcal{P}_y = \begin{pmatrix} 0 & 0 \\ 0 & 1 \end{pmatrix}$	6.99
Perfect polarizer at angle ζ	$\mathcal{P}_{\zeta_r} = \mathcal{R}(\zeta_r)\mathcal{P}_x\mathcal{R}^\dagger(\zeta_r)$	6.120
Arbitrary waveplate	$\mathcal{W} = \begin{pmatrix} e^{j\delta\phi/2} & 0 \\ 0 & e^{j\delta\phi/2} \end{pmatrix}$	6.108
Arbitrary waveplate (Alt.)	$\mathcal{W} = \begin{pmatrix} e^{j\delta\phi} & 0 \\ 0 & 1 \end{pmatrix}$	6.109
Quarter-wave plate	$\mathcal{Q} = \begin{pmatrix} 1 & 0 \\ 0 & j \end{pmatrix}$	6.111
Half-wave plate	$\mathcal{H} = \begin{pmatrix} 1 & 0 \\ 0 & -1 \end{pmatrix}$	6.112
Rotator	$\mathcal{R}(\zeta) = \begin{pmatrix} \cos\zeta & -\sin\zeta \\ \sin\zeta & \cos\zeta \end{pmatrix}$	6.113
Coordinate rotation	$\mathcal{J} = \mathcal{R}(\zeta_r)\mathcal{J}'\mathcal{R}^\dagger(\zeta_r)$	6.118
Mirror	$\mathcal{M} = \begin{pmatrix} 1 & 0 \\ 0 & -1 \end{pmatrix}$	6.135
Generic forward propagation	$\mathcal{J}_{forward} = \begin{pmatrix} j_{11} & j_{12} \\ j_{21} & j_{22} \end{pmatrix}$	6.136
Reverse propagation	$\mathcal{J}_{reverse} = \begin{pmatrix} j_{11} & -j_{21} \\ -j_{12} & j_{22} \end{pmatrix}$	6.136
Unitary matrix	$\mathcal{J}^\dagger\mathcal{J} = \mathcal{I}$	6.138

APPENDIX F: NUMERICAL CONSTANTS AND CONVERSION FACTORS

F.1 Fundamental Constants and Useful Numbers

Table F.1 contains some fundamental constants that are frequently used, and some numbers which will be useful to the optical engineer in some of the examples are considered in this book.

F.2 Conversion Factors

Important conversion factors are shown in Table F.2.

F.3 Wavelengths

Some laser wavelengths are included in Tables F.3 and F.4[112;153;12]. *The Laser Focus Buyer's Guide*[12], updated annually, is a good source of technical information about lasers and related products.

F.4 Indices of Refraction

Some typical indices of refraction are shown in Table F.5.

TABLE F.1
Some Useful Physical Constants

Quantity	Symbol	Value	Units
Speed of light	c	2.998×10^{8}	m/s
Planck's constant	h	6.626×10^{-34}	J s
Boltzmann's constant	k	1.381×10^{-23}	J/K
Electronic charge	e	1.602×10^{-19}	C
Thermal voltage at 273 K	kT/e	25.2×10^{-3}	V
Permeability constant	μ_0	1.26×10^{-6}	H/m
Permittivity constant	ϵ_0	8.85×10^{-12}	F/m
Avogadro's number	A_n	6.022045×10^{23}	/mol
Gas constant	R	1.98719	cal/mol/K
Distance from earth to moon		384, 403	km
Radius of moon		1, 737.4	km
Radius of Earth		6, 378.1	km

DOI: 10.1201/9781315157047-F

TABLE F.2
Conversion Factors

Quantity	Conversion
Energy/wavelength	J in eV $= \frac{1.24}{\lambda}$ (µm)
Length	2.54 cm/in.
	3.28 ft/m
	5280 ft/mi
	6080 ft/nm
	10^{10} Å/m
Mass	2.2 lb/kg

TABLE F.3
Some Laser Wavelength Ranges

Laser	Range	Units
Fiber-optical communication		
From	850	nm
To	1.8	µm
Dye lasers	190–4500	nm
Ti: sapphire	680–1070	nm
Carbon dioxide laser	9–11	µm
Hydrogen fluoride	2.6–3	µm
Deuterium fluoride	3.6–4	µm
Alexandrite	730–780	nm

TABLE F.4
Selected Laser Lines

Laser	Wavelength	Units
Argon dimer (UV)	125	nm
Argon ion (blue)	488.1	nm
Argon ion (green)	514.5	nm
Copper vapor	510.6	nm
Copper vapor	578.2	nm
Neodymium YAG laser	1.06	µm
Neodymium YAG laser	1.318	µm
Krypton chloride	222	nm
Krypton fluoride	248	nm
Xenon bromide	282	nm
Xenon chloride	308	nm
Xenon fluoride	351	nm
Nitrogen	337.1	nm
Helium-neon laser (red)	632.8	nm
Neodymium YaLo laser	1.35	µm
Ho:YAG	2.10	µm
Th:YAG	2.00	µm
Carbon dioxide laser P(20) line	10.59	µm

TABLE F.5
Indices of Refraction

Material	Index of Refraction
Pure water at 589.2 nm	1.33
Complex index of water at 420 nm	$1.33 + 4.68 \times 10^{-8}$ j
Sugar and water	Up to 1.49
Crown glass at 589.2 nm	1.52
Germanium at 10.59 μm	4.00
Zinc selenide at 10.59 μm	2.4
Biological materials: visible to near IR	
Extracellular fluid	1.35
Cytoplasm	1.37
Nucleus of cell	1.39
Mitochondria	1.42
Melanin	1.7
Elastin	1.536

F.5 Multiplier Prefixes

In SI units, it is customary to use a set of prefixes for multipliers that are powers of 10^3. These are given in Table F.6 with the standard symbols and some humorous examples. The reader is reminded to be cautious about the capitalization of the symbols.

F.6 Decibels

Decibels provide a convenient concept to describe gains and losses in electronic systems. The gain of an amplifier is described in terms of the power ratio of output to input and is expressed in dB (note lowercase "d," meaning "deci-," and capital "B" for "Bel") as

$$g = 10 \log (P_{OUT}/P_{IN}). \tag{F.1}$$

TABLE F.6
Multiplier Prefixes

Multiplier	Prefix	Symbol	Example
10^{-18}	atto-	a-	10^{-18} boys = 1 attoboy
10^{-15}	femto-	f-	10^{-15} bismols = 1 femtobismol
10^{-12}	pico-	p-	10^{-12} boos = 1 picoboo
10^{-9}	nano-	n-	10^{-9} goats = 1 nanogoat
10^{-6}	micro-	μ-	10^{-6} scopes = 1 microscope
10^{-3}	milli-	m-	10^{-3} cents = 1 millicent
10^{3}	Kilo-	k-	2×10^{3} mockingbirds = 2 kilomockingbirds
10^{6}	Mega-	M-	10^{6} phones = 1 Megaphone
10^{9}	Giga-	G-	10^{9} los = 1 Gigalo
10^{12}	Tera-	T-	10^{12} bulls = 1 Terabull
10^{15}	Peta-	P-	10^{15} lumas = 1 Petaluma
10^{18}	Exo-	E-	10^{18} skeletons = 1 Exoskeleton

In keeping with current usage, we are including only the powers of 10^3. Some of the humorous annotations are loosely borrowed from Standards for Inconsequential Trivia on p. 61 of *A Random Walk in Science*[243].

If the output and input impedances are the same, then the power gain is the square of the voltage (or current) gain, and

$$g = 10\log\left(\frac{v_{OUT}}{v_{IN}}\right)^2 = 20\log\left(v_{OUT}/v_{IN}\right). \tag{F.2}$$

Electrical signals are often expressed in dBm. This unit is not a measure of gain, but of signal level. It is the ratio of the power to 1 mW. Optical power is often expressed in dBm, particularly in the fiber-optics industry:

$$y = 10\log\left(P/10^{-3}\,\mathrm{W}\right). \tag{F.3}$$

APPENDIX G: SOLUTIONS TO CHAPTER PROBLEMS

Solutions to the chapter problems are presented in this appendix. The chapter problems are mostly developed from "real-world" examples encountered in the author's research activities. Therefore, unlike the problems in most textbooks, the reader's first task here is to determine what the question is. Many of the solutions involve computation, and in many cases there may be other solutions that produce the same or similar answers. In several cases, approximations are made, and the reader may arrive at slightly different answers with different approximations.

G.1 Chapter 1

Problem 1.1: Wavelength, Frequency, and Energy

Here are some "warmup" questions to build familiarity with the typical values of wavelength, frequency, and energy encountered in optics.

1.1a What is the frequency of an argon ion laser with a vacuum wavelength of 514 nm (green)?

Solution:

Use the equation,

$$f\lambda = c. \tag{G.1}$$

$$f = \frac{c}{\lambda} = \frac{c}{514 \times 10^{-9}\,\mathrm{m}} = 5.83 \times 10^{14}\,\mathrm{Hz}. \tag{G.2}$$

The frequency is 583 THz.

1.1b In a nonlinear crystal, we can generate sum and difference frequencies. If the inputs to such a crystal are the argon ion laser with a vacuum wavelength of 514 nm (green) and another argon ion laser with a vacuum wavelength of 488 nm (blue), what are the wavelengths of the sum and difference frequencies that are generated? Can we use glass optics in this experiment?

Solution:

Using Equation G.1,

$$f_{dif} = f_1 - f_2 \qquad \frac{c}{\lambda_{dif}} = \frac{c}{\lambda_1} - \frac{c}{\lambda_2} \tag{G.3}$$

$$\frac{1}{\lambda_{dif}} = \frac{1}{\lambda_1} - \frac{1}{\lambda_2} = \frac{1}{488\,\mathrm{nm}} - \frac{1}{514\,\mathrm{nm}} = \frac{1}{9650\,\mathrm{nm}}. \tag{G.4}$$

DOI: 10.1201/9781315157047-G

Solution to Problem 1.1b (Continued)

Likewise

$$\frac{1}{\lambda_{sum}} = \frac{1}{\lambda_1} + \frac{1}{\lambda_2} = \frac{1}{488\,\text{nm}} - \frac{1}{514\,\text{nm}} = \frac{1}{250\,\text{nm}}. \tag{G.5}$$

The difference and sum wavelengths, 9.645 µm and 250 nm, are infrared and ultraviolet, respectively. Neither of them is transmitted by glass (see Table 1.2).

1.1c A nonlinear crystal can also generate higher harmonics (multiples of the input laser's frequency). What range of input laser wavelengths can be used so that both the second and third harmonics will be in the visible spectrum?

Solution:

If the fundamental wavelength is λ_1, the wavelength of the nth harmonic is given by

$$\lambda_n = \frac{c}{f_n} = \frac{c}{nf_1} = \frac{\lambda_1}{n}. \tag{G.6}$$

We want

$$\frac{\lambda_1}{3} > 380\,\text{nm} \quad \text{and} \quad \frac{\lambda_1}{2} < 780\,\text{nm} \tag{G.7}$$

Thus, the short-wavelength limit is

$$\lambda_1 = 3 \times 380\,\text{nm} = 1140\,\text{nm}, \tag{G.8}$$

and the long-wavelength limit is

$$\lambda_1 = 2 \times 780\,\text{nm} = 1560\,\text{nm}, \tag{G.9}$$

Thus the usable band is 1.14–1.56 µm.

1.1d How many photons are emitted per second by a helium-neon laser (633 nm wavelength) with a power of 1 mW?

Solution:

The energy of a photon is

$$Q = h\nu = \frac{hc}{\lambda} = \frac{hc}{633 \times 10^{-9}\,\text{m}} = 3.14 \times 10^{-19}\,\text{J}. \tag{G.10}$$

The number of photons per second is

$$\frac{P}{Q} = \frac{10^{-3}\,\text{W}}{3.14 \times 10^{-19}\,\text{J}} = 3.19 \times 10^{15}/\text{s}. \tag{G.11}$$

1.1e It takes 4.18 J of energy to heat a cubic centimeter of water by 1 K. Suppose 100 mW of laser light is absorbed in a volume with a diameter of 2 µm and a length of 4 µm. Neglecting thermal diffusion, how long does it take to heat the water by 5 K?

Solution:

The energy required is

$$J = 5\,\text{K} \times 4.18\,\frac{\text{J}}{\text{K cm}^3} \times \frac{\pi(2 \times 10^{-4}\,\text{cm}/2)^2}{4 \times 10^{-4}\,\mu\text{m}} = 0.0016\,\text{J}. \tag{G.12}$$

From

$$J = Pt, \tag{G.13}$$

$$t = \frac{J}{P} = \frac{0.0016\,\text{J}}{100 \times 10^{-3}\,\text{W}} = 0.0164\,\text{s}, \tag{G.14}$$

or 16 ms.

1.1f A fluorescent material absorbs light at 488 nm and emits at 530 nm. What is the maximum possible efficiency, where efficiency means power out divided by power in.

Solution:

The energies of the absorbed and emitted photons are

$$Q_{in} = \frac{hc}{\lambda_{in}} \quad Q_{out} = \frac{hc}{\lambda_{out}} \tag{G.15}$$

The efficiency is

$$\frac{Q_{out}}{Q_{in}} = \frac{\lambda_{in}}{\lambda_{out}} = \frac{488\,\text{nm}}{530\,\text{nm}} = 0.92, \tag{G.16}$$

so the maximum possible efficiency is 92%.

Problem 1.2: Optical Path

1.2a We begin with an empty fish tank, 30 cm thick. Neglect the glass sides. We paint a scene on the back wall of the tank. How long does light take to travel from the scene on the back of the tank to the front wall?

Solution:

$$t = \frac{d}{c} = \frac{0.3\,\text{m}}{c} = 10^{-9}\,\text{s}. \tag{G.17}$$

A good rule of thumb is light travels a foot, or 30 cm, in a nanosecond.

1.2b Now we fill the tank with water and continue to neglect the glass. How far does the scene appear behind the front wall of the tank? Now how long does the light take to travel from back to front?

Solution:

See Figure 1.15. The apparent distance is shorter than the actual distance and is given by

$$d_{geom} = \frac{d}{n} = \frac{0.3\,\text{m}}{1.5} = 0.2\,\text{m}. \tag{G.18}$$

The transit time is increased to 1.5 ns,

$$t = \frac{nd}{c} = 1.5 \times 10^{-9}\,\text{s}, \tag{G.19}$$

so, although the distance appears shorter in water than it is in air, the transit time is longer than it would be in air.

Solution to Problem 1.2b (Continued)

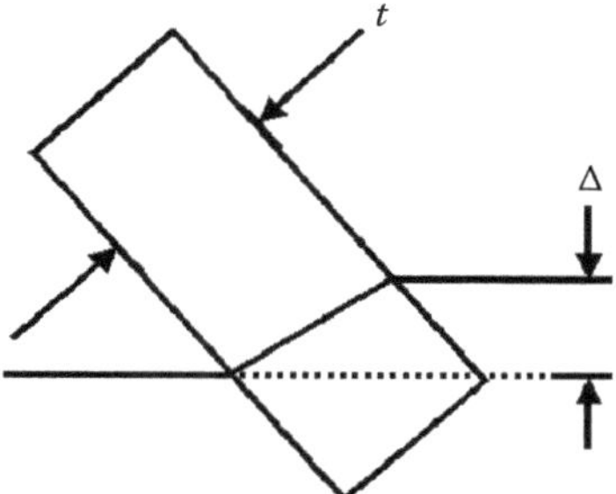

FIGURE P2.1
Layout for Problem 2.1.

1.2c What is the optical-path length difference experienced by light traveling through a glass coverslip 170 μm thick in comparison to an equivalent length of air?

Solution:

The path length is

$$nd = 1.5 \times 170\,\mu\text{m}. \tag{G.20}$$

The path length difference is thus

$$(n-1)\,d = 0.5 \times 170\,\mu\text{m} = 85\,\mu\text{m}. \tag{G.21}$$

G.2 Chapter 2

Problem 2.1: Dogleg

A beam of laser light is incident on a beam splitter of thickness, t, at an angle, θ, as shown in Figure P2.1. The index of refraction of the medium on each side is the same, n, and that of the beam splitter is n'. When I remove the beam splitter, the light falls on a detector. When I replace the beam splitter, the light follows a "dogleg" path, and I must move the detector in a direction transverse to the beam. How far?

2.1a Derive the general equation.

Solution:

$$\begin{aligned}\Delta &= t\sin\theta - t\tan\theta'\cos\theta = t\sin\theta - t\sin\theta'\frac{\cos\theta}{\cos\theta'}\\ &= t\sin\theta\left(1-\frac{n}{n'}\frac{\sqrt{1-\sin^2\theta}}{\sqrt{1-\sin^2\theta'}}\right) = t\sin\theta\left(1-\frac{n\sqrt{1-\sin^2\theta}}{n'\sqrt{1-\sin^2\theta(n/n')^2}}\right)\\ &= t\sin\theta\left(1-\sqrt{\frac{n^2-n^2\sin^2\theta}{(n')^2-n^2\sin^2\theta}}\right)\end{aligned} \tag{G.22}$$

The geometry is shown in Figure P2.2.

2.1b Check your result for $n' = n$ and for $n'/n \to \infty$. Calculate and plot vs. angle, for $t = 5\,\text{mm}$, $n = 1$, with $n' = 1.5, \ldots$

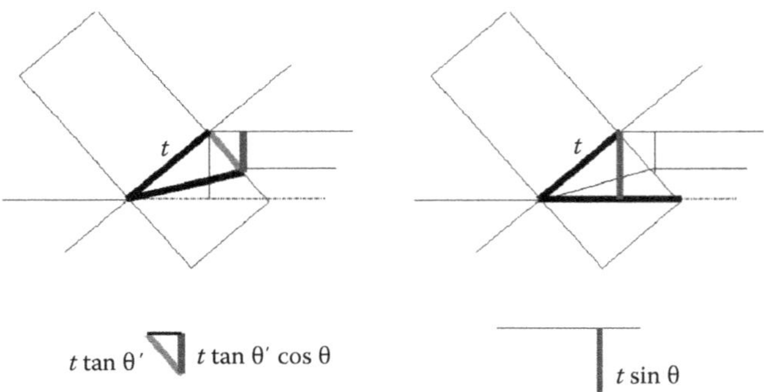

FIGURE P2.2
Solution for Problem 2.1, Part (a).

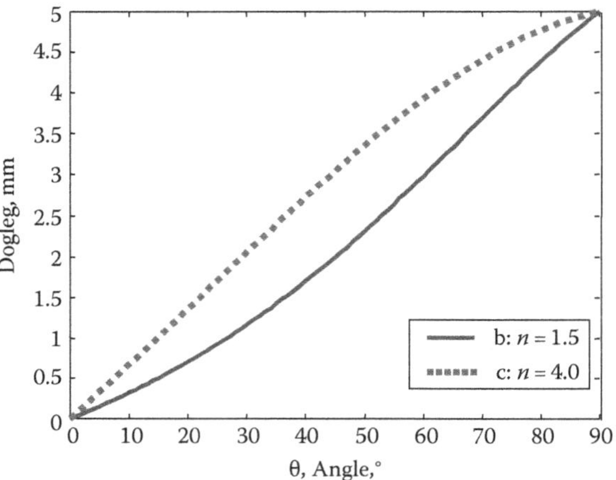

FIGURE P2.3
Solution for Problem 2.1, Parts (b) and (c).

2.1c and again with $n' = 4$.

Solution:

For

$$n' = n \quad \Delta = 0, \tag{G.23}$$

as is obviously expected.

$$n' \to \infty \quad \Delta \to t \sin\theta \tag{G.24}$$

because the ray will follow the normal. See the following MATLAB® code and the plot in Figure P2.3

```
nprime=1.5;
t=5; % units are mm
thetad=[0:90]; % units are degrees
sintheta=sin(thetad*pi/180);
%
% part b
Deltab=t*sintheta.*...
   (1-sqrt((n^2-n^2*sintheta.^2)./...
```

Solution to Problem 2.1c (Continued)

```
      (nprime^2-n^2*sintheta.^2)));
%
% part c
nprime=4;
Deltac=t*sintheta.*...
    (1-sqrt((n^2-n^2*sintheta.^2)./...
      (nprime^2-n^2*sintheta.^2)));
%
figure;
plot(thetad,Deltab,'-',thetad,Deltac,'--');
xlabel('\theta, Angle, Deg.');
ylabel('\Delta, Dogleg, mm');
title('HW1, Problem 1, Parts b and c');
legend('b: n=1.5','c: n=4.0',0);
```

Problem 2.2: Some Lenses

2.2a Consider the thin lens in Figure P2.4A. Let the focal length be $f = 10\,\text{cm}$. Plot the image distance as a function of object distance from 50 cm down to 5 cm.

Solution:

See the MATLAB code following Part (f) and the plot in Figure P2.5A.

2.2b Plot the magnification.

Solution:

See the MATLAB code following Part (f) and the plot in Figure P2.5B.

2.2c Consider the thin lens in Figure P2.4B. Let the focal length be $f = -10\,\text{cm}$. Plot the image distance as a function of object distance from 50 cm down to 5 cm.

Solution:

See the MATLAB code following Part (f) and the plot in Figure P2.5C.

2.2d Plot the magnification.

Solution:

See the MATLAB code following Part (f) and the plot in Figure P2.5D.

2.2e In Figure P2.4C, the lenses are both thin and have focal lengths of $f_1 = 20\,\text{cm}$ and $f_2 = 10\,\text{cm}$, and the separation is $d = 5\,\text{cm}$. For an object 40 cm in front of the first lens, where is the final image?

Solution:

The final image is 7.8 mm to the right of the second lens, as calculated in the MATLAB code following Part (f).

2.2f What is the magnification?

Solution:

The combined magnification is -0.22. The calculation is in the code below.

The following is a screen dump from MATLAB for all parts of this problem. Figure P2.5 shows the resulting plots.

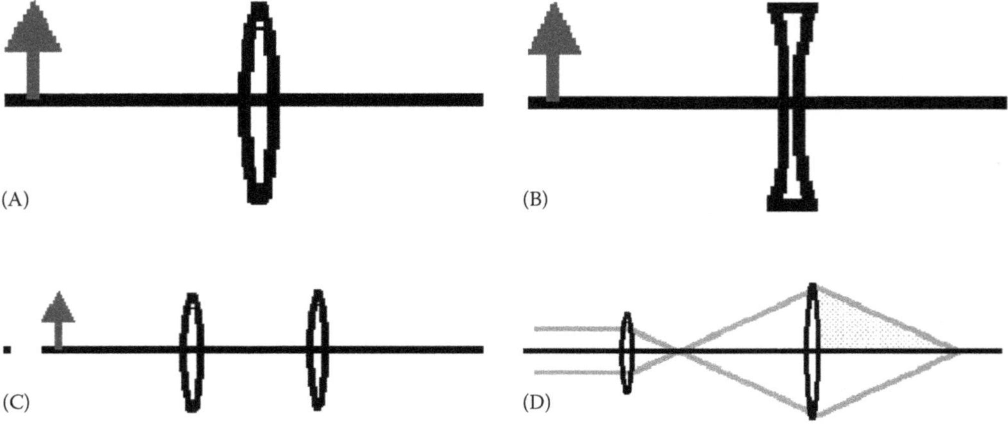

FIGURE P2.4
Sample lenses for Problem 2.2.

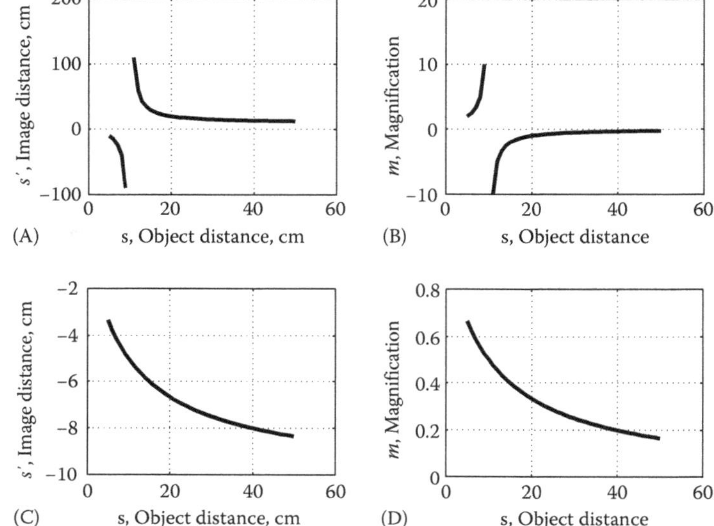

FIGURE P2.5
Plots for Problem 2.2. (A) Image distance for Part (a). (B) Magnification for Part (b). (C) Image distance for Part (c). (D) Magnification for Part (d).

Solution to Problem 2.2f (Continued)

```
echo on
% HW2P2 Script for Problem on lenses
%       in Optics for Engineers
% by Chuck DiMarzio
% Northeastern University
% July 2003, March 2011
f=10;
s=[5:1:50];
sprime=1./(1/f-1./s);
Warning: Divide by zero.

> In C:\document\working\10353\m10353b.m at line 13
```

```
fig1=figure;
plot(s,sprime);grid on;
%title('Part a');
xlabel('s, Object Distance, cm');
ylabel('s'', Image Distance, cm');
%
m=-sprime./s;
fig2=figure;plot(s,m);grid on;
%title('Part b');
xlabel('s, object distance');
ylabel('m, Magnification');
%
f=-10;
sprime=1./(1/f-1./s);
fig3=figure;;plot(s,sprime);grid on;
%title('Part c');
xlabel('s, Object Distance, cm');
ylabel('s'', Image Distance, cm');
%
m=-sprime./s;
title('Part d');
fig4=;plot(s,m);grid on;
xlabel('s, object distance');
ylabel('m, Magnification');
%
disp('Part e');
Part e
s1=40;
f1=20;
d=5;
f2=10;
s1prime=1/(1/f1-1/s1)
s1prime =
    40
s2=d-s1prime
s2 =
   -35
s2prime=1/(1/f2-1/s2)
s2prime =
    7.7778
disp('the image is this far to the right
 of the second lens')
the image is this far to the right of
 the second lens
disp(' ');
%
disp('Part f');
Part f
m1=-s1prime/s1
m1 =
    -1
m2=-s2prime/s2
m2 =
    0.2222
```

Solution to Problem 3.1a (Continued)

```
m=m1*m2
m =
   -0.2222
disp('m is the total magnification')
m is the total magnification
```

G.3 Chapter 3

Problem 3.1: Focusing a Laser Beam

The configuration in Figure P2.4D can be used to focus a laser beam to couple it into an optical fiber. We want the sine of the half-angle subtended by the last lens (shaded region) to equal the numerical aperture of the fiber, NA $= 0.4$. Let the second lens have a focal length of 16 mm and assume that the initial laser beam diameter is $D_1 = 1.4$ mm. We want the magnification of the second lens to be $-1/4$.

3.1a What is the focal length of the first lens?

Solution:

We know that

$$\frac{s_2'}{s_2} = -m_2 = \frac{1}{4} \tag{G.25}$$

The focus between the lenses is located to the left of the second lens by an amount s_2 given by

$$\frac{1}{s_2} + \frac{1}{s_2'} = \frac{1}{f_2} \tag{G.26}$$

$$\frac{1}{s_2} + \frac{1}{s_2/4} = \frac{5}{s_2} = \frac{1}{16} \tag{G.27}$$

$$s_2 = 5f_2 = 80 \qquad s_2' = 20 \tag{G.28}$$

where all distances are in millimeters.

The beam diameter at the second lens is related to the tangent of the angle in the shaded area, whose sine is the numerical aperture.

$$d_2 = 2\frac{NA}{\sqrt{1-NA^2}}s_2' = 17.5\,\text{mm}. \tag{G.29}$$

Thus, by similar triangles, noting that $s_1 \to \infty$,

$$\frac{d_1}{f_1} = \frac{d_2}{s_2} \tag{G.30}$$

$$f_1 = \frac{d_1}{d_2}s_2 = \frac{1.4}{17.5} \times 80 = 6.4\,\text{mm}. \tag{G.31}$$

3.1b What is the spacing between the two lenses, assuming thin lenses?

Solution:

$$f_1 + s_2 = 6.4 + 80 = 86.4\,\text{mm}. \tag{G.32}$$

3.1c Now, assume that the lenses are both 5 mm thick, the first is convex-plano and the second is biconvex. What is the spacing between the vertices of the lenses now? What is the distance to the image point from the second lens vertex? Where should the fiber be placed relative to the second lens? You may use an approximation and the fact that the lenses are glass.

Solution:

The principal planes of the first lens are at the left (convex) vertex and one third of the thickness (5/3 mm) to the right of that same vertex. Thus, the second principal plane of the first lens is 10/3 mm to the right of the plane surface. The principal planes of the biconvex lens are located toward the center of the lens by one third of the thickness from each vertex. Thus, for the distance between principal planes to be $z_{12} = 86.4\,\text{mm}$, the distance between vertices must be

$$w_{12} = 86.4 - \frac{10}{3} - \frac{5}{3} = 81.4\,\text{mm}. \tag{G.33}$$

Likewise, the distance to the fiber is

$$w_2' = s_2' - 5/3 = 18.33\,\text{mm}. \tag{G.34}$$

3.1d Use matrices and find the focal length of this pair of lenses and the location of the principal planes. Be sure to specify distances and directions from lens vertices.

Solution:

The matrix for the first lens, from front principal plane to back principal plane is

$$\begin{pmatrix} 1 & 0 \\ -\frac{1}{f_1} & 1 \end{pmatrix}. \tag{G.35}$$

Note that we could have worked from vertex to vertex assuming an index of refraction and computing the surface curvatures, but there is no real need to do this, and in the case of complicated lenses, which are often proprietary designs, you will not know these parameters.

The complete matrix is

$$\begin{pmatrix} 1 & 0 \\ -\frac{1}{f_2} & 1 \end{pmatrix} \begin{pmatrix} 1 & z_{12} \\ 0 & 1 \end{pmatrix} \begin{pmatrix} 1 & 0 \\ -\frac{1}{f_1} & 1 \end{pmatrix} \tag{G.36}$$

Note that we use z_{12} rather than w_{12}, because the lens matrices are from principal plane to principal plane.

Now, using MATLAB

```
% HW3P1  The focusing problem, part d: Principal Planes
% for homework in Optics for Engineers
% by Chuck DiMarzio
% Northeastern University
% revised 2 Nov 03, Mar 2011
%
d1=1.4;
na=0.4;
f2=16;
m2=-1/4;
s2=f2*(1+1/(-m2))
s2 =
    80
s2prime=-s2*m2
```

Solution to Problem 3.1d (Continued)

```
s2prime =
    20
tang=0.4/sqrt(1-0.4^2)
tang =
    0.4364
d2=2*s2prime*tang
d2 =
   17.4574
f1=s2*d1/d2
f1 =
    6.4156
disp('above solves a');
above solves a

z12=f1+s2
z12 =
   86.4156
w12=z12-5/3-10/3
w12 =
   81.4156
disp('above solves b,c');
above solves b,c

lens1=[1,0;-1/f1,1]
lens1 =
    1.0000         0
   -0.1559    1.0000
lens2=[1,0;-1/f2,1]
lens2 =
    1.0000         0
   -0.0625    1.0000
sep=[1,z12;0,1]
sep =
    1.0000   86.4156
         0    1.0000
m=lens2*sep*lens1
m =
  -12.4696   86.4156
    0.6235   -4.4010

f=-1/m(2,1)
f =
   -1.6039
d=(1-m(2,2))/m(2,1)
d =
    8.6626
dprime=(1-m(1,1))/m(2,1)
dprime =
   21.6039

disp(' Above solve d');
 Above solve d
```

Now, we need to interpret the results. The front principal plane of the combined lens is a distance D in front of the front principal plane of the first lens. Because the first lens is convex-plano, the result is that the principal plane is $D = 8.7\,\text{mm}$ to the left of the first vertex of the first lens. The back principal plane is D' to the right of the back principal plane of the second lens, $D' - 5/3\,\text{mm} = 19.9\,\text{mm}$ to the right of the second vertex of the second lens.

G.4 Chapter 4

Problem 4.1: Coaxial Lidar: Pupil and Image Conjugates (G)

Consider the coaxial lidar (laser radar) system shown in Figure P4.1. The laser beam is 1 in. in diameter, and is reflected from a 45° mirror having an elliptical shape so that its projection along the axis is a 2 in. circle. (This mirror, of course, causes an obscuration of light returning from the target to the detector. We will make good use of this feature.) The light is then reflected by a scanning mirror to a distant target. Some of the backscattered light is returned via the scanning mirror, reflected from the primary mirror to a secondary mirror, and then to a lens and detector. The focal lengths are 1 m for the primary, 12.5 cm for the secondary, and 5 mm for the detector lens. The diameters are, respectively, 8, 2, and 1 in. The detector, with a diameter of 100 μm, is at the focus of the detector lens, which is separated from the secondary by a distance of 150 cm. The primary and secondary are separated by the sum of their focal lengths.

4.1a Locate all the optical elements in object space, i.e., the space containing the target, and determine their size; the detector, detector lens, secondary lens, primary, obscuration caused by the transmitter folding mirror.

4.1b Identify the aperture stop and the field stop for a target at a long distance from the lidar.

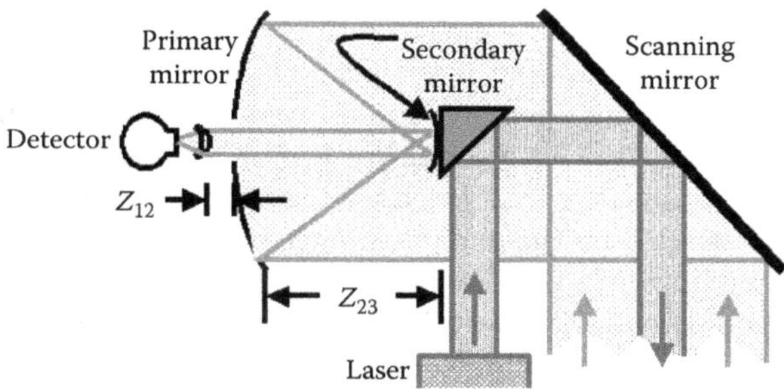

FIGURE P4.1
Coaxial lidar configuration.

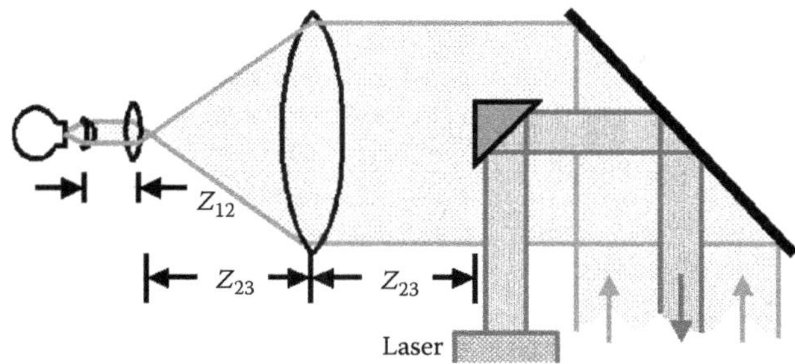

FIGURE P4.2
(Almost) straight-line layout of coaxial lidar.

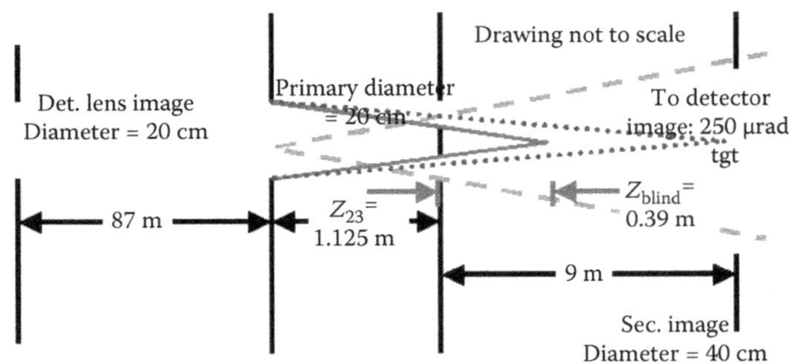

FIGURE P4.3
Coaxial lidar in object space.

4.1c As mentioned earlier, the folding mirror for the transmitter causes an obscuration of the received light. It turns out that this is a very useful feature. The scanning mirror surface will always contain some imperfections, scratches, and dirt, which will scatter light back to the receiver. Because it is very close, the effect will be strong and will prevent detection of weaker signals of interest. Now, if we place the scanning mirror close enough, the obscuration and the finite diameter of one or more lenses will block light scattered from the relatively small transmitter spot at the center of the mirror. Find the "blind distance," z_{blind}, inside which the scan mirror can be placed.

Solution:

We begin with the almost-straight-line layout in Figure P4.2. Here the mirrors are treated as lenses so that we can see the relationships between objects and images more clearly. The calculations are shown in the following MATLAB code, and the results are shown in Figure P4.3 **(a)** Locating the lens images is straightforward using the lens equation and magnification. **(b)** With a target at a long distance, the limiting aperture (entrance pupil) is the primary as shown by the dotted line. Having the primary as the limiting aperture is highly desirable as it is the biggest and likely most expensive element in the system. We would be wasting primary area if it were not the pupil. The primary is thus the aperture stop. From the center of the aperture stop, the cone subtended by the detector image is the smallest, so the detector image is the entrance window and the detector is the field stop. The field of view is shown by the dashed line. Having the detector act as the field stop is also highly desirable as we would otherwise be wasting detector area. Note that the secondary is much larger than it needs to be. **(c)** The solid line in the figure shows how the combination of the transmitter folding mirror and the primary serve to block light from a distance closer than z_{blind}.

```
echo on
% HW4P1 Coaxial laser radar
%
% for  problem in Optics For Engineers
% by Chuck DiMarzio, Northeastern University,
%August 2003
%
% units are meters
%
det0=100e-6; % detector diameter
f1=5e-2;
z12=150e-2;
f2=12.5e-2;
f3=1;
```

```
z23=f2+f3;
z34=z23;
% detector image through lenses 1 and 2
% is at back focus of lens 2
det2=det0*f2/f1
det2 =
  2.5000e-004
% detector image after lens 3 is at infinity
% thus diameter is infinite:  Specify angle instead
det3angle=det2/f3
det3angle =
  2.5000e-004

Ldet=2.54e-2; % one inch diameter detector lens
s2p=1/(1/f2-1/z12) % locate its image through lens 2
s2p =
    0.1364
Ldet2=abs(Ldet*s2p/z12) % find its size
Ldet2 =
    0.0023

s3=z23-s2p % locate it after lens 3
s3 =
    0.9886
s3p=1/(1/f3-1/s3)
s3p =
  -87.0000
Ldet3=abs(Ldet2*s3p/s3)
Ldet3 =
    0.2032

secondary=2*2.54e-2; % two inch diameter secondary
s3p=1/(1/f3-1/z23) % image through primary
s3p =
    9.0000
secondary3=secondary*s3p/z23  % find its size
secondary3 =
    0.4064

% finding the maximum blind distance
d3=8*.02454
d3 =
    0.1963
dobs=2*.0254
dobs =
    0.0508
% (zblind+z34)/zblind=(d3/2)/(dobs/2)
% =1+z34/zblind
%
zblind=z34/(d3/dobs-1)
zblind =
    0.3927
echo off;
```

G.5 Chapter 5

Problem 5.1: Spherical Aberrations

Here we consider how the focus moves as we use larger parts of an aperture. The lens has a focal length of 4 cm, and the object is located 100 cm in front of it. We may assume a thin lens with a diameter of 2 cm.

5.1a What is the f-number of this lens?

Solution:

$$\frac{f}{d} = \frac{4}{2} = 2. \tag{G.37}$$

The lens is f/2.

5.1b Plot the location of the focus of rays passing through the lens at a height, h, from zero to $d/2$, assuming the lens is biconvex.

5.1c Repeat assuming it is plano-convex.

5.1d Repeat assuming it is convex-plano.

5.1e Repeat for the optimal shape for minimizing spherical aberrations.

Solution:

The following MATLAB code produced Figure P5.1:

```
echo on;
% HW5p1 Spherical Aberrations Problem
% by Chuck DiMarzio for homework in Optics for Engineers
% Northeastern University August 2003
%
h=[0:.01:1];
n=1.5;
f=4; % cm
% 1/f = (n-1) (1/r1-1/r2)
% r2=-r1;
r1=(n-1)*2*f
r1 =
     4
r2=-r1;


s=100;
sp=1/(1/f-1/s)
sp =
    4.1667
p=(sp-s)/(sp+s)
p =
   -0.9200
q=(r2+r1)/(r2-r1)
q =
     0
Ls=h.^2/8/f^3/n/(n-1)*((n+2)/(n-1)*q^2+4*(n+1)*p*q...
   +(3*n+2)*(n-1)*p^2+n^3/(n-1));
sph_bicon=1./(1/sp+Ls);
```

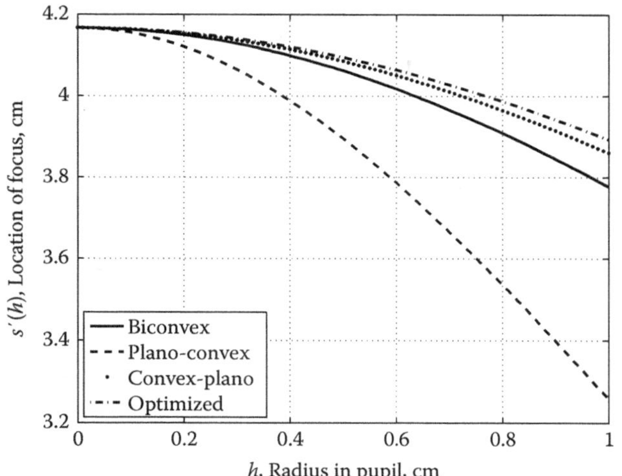

FIGURE P5.1
Plots for Problem 5.1.

```
r1=100000;
r2=-(n-1)*f;
q=(r2+r1)/(r2-r1)
q =
   -1.0000
Ls=h.^2/8/f^3/n/(n-1)*((n+2)/(n-1)*q^2+4*(n+1)*p*q...
   +(3*n+2)*(n-1)*p^2+n^3/(n-1));
sph_pc=1./(1/sp+Ls);

r2=100000;
r1=-(n-1)*f;
q=(r2+r1)/(r2-r1)
q =
    1.0000
Ls=h.^2/8/f^3/n/(n-1)*((n+2)/(n-1)*q^2+4*(n+1)*p*q...
   +(3*n+2)*(n-1)*p^2+n^3/(n-1));
sph_cp=1./(1/sp+Ls);

q=-2*(n^2-1)*p/(n+2)
q =
    0.6571
Ls=h.^2/8/f^3/n/(n-1)*((n+2)/(n-1)*q^2+4*(n+1)*p*q...
   +(3*n+2)*(n-1)*p^2+n^3/(n-1));
sph_opt=1./(1/sp+Ls);

fig1=figure;
plot(h,sph_bicon,'-',h,sph_pc,'--',h,sph_cp,':',
    h,sph_opt,'-.');
grid on;
xlabel('h, Radius in Pupil, cm');
ylabel('s''(h), Location of Focus, cm');
legend('Biconvex','Plano-Convex','Convex-Plano',...
  'Optimized',3);
echo off;
```

G.6 Chapter 6

Problem 6.1: Polarizer in an Attenuator

Polarizers can be useful for attenuating the level of light in an optical system. For example, consider two polarizers, with the second one aligned to pass vertically polarized light. The first can be rotated to any desired angle, with the angle defined to be zero when the two polarizers are parallel. The first parts of this problem can be solved quite simply without the use of Jones matrices.

6.1a What is the output polarization?

Solution:

Because the last polarizer axis is vertical, the output light will always be vertically polarized.

6.1b Plot the power transmission of the system as a function of angle, from 0° to 90°, assuming the incident light is unpolarized.

Solution:

The first polarizer will pass the light that is polarized parallel to its axis, which is half of the power in the unpolarized beam. The second polarizer will pass the vertical component, $\cos\theta$. The power will be proportional to the square of the field:

$$T_1 = 0.5\left(\cos\theta\right)^2 . \tag{G.38}$$

Matlab code for this and other parts is shown in 6.1f.

6.1c Plot the power transmission assuming vertically polarized incident light.

Solution:

If the incident light is vertically polarized, the first polarizer passes a component parallel to its axis, and the second does the same:

$$T_2 = \left(\cos\theta\ \cos\theta\right)^2 . \tag{G.39}$$

6.1d Plot the power transmission assuming horizontally polarized incident light.

Solution:

If the incident light is horizontally polarized, the relevant angle for the first polarizer is $90° - \theta$

$$T_3 = \left(\sin\theta\ \cos\theta\right)^2 . \tag{G.40}$$

6.1e Now, replace the first polarizer with a quarter-wave plate. I recommend you use Jones matrices for this part. In both 6.1e and 6.1f, the incident light is vertically polarized. Plot the power transmission as a function of angle from 0° to 90°.

Solution:

Now, we turn to the use of Jones matrices. For this part,

$$\mathcal{R} = \begin{pmatrix} \cos\theta & \sin\theta \\ -\sin\theta & \cos\theta \end{pmatrix} \tag{G.41}$$

$$\mathcal{Q} = \begin{pmatrix} e^{i\pi/4} & 0 \\ 0 & e^{-i\pi/4} \end{pmatrix} \tag{G.42}$$

$$\mathcal{P} = \begin{pmatrix} 1 & 0 \\ 0 & 0 \end{pmatrix}, \tag{G.43}$$

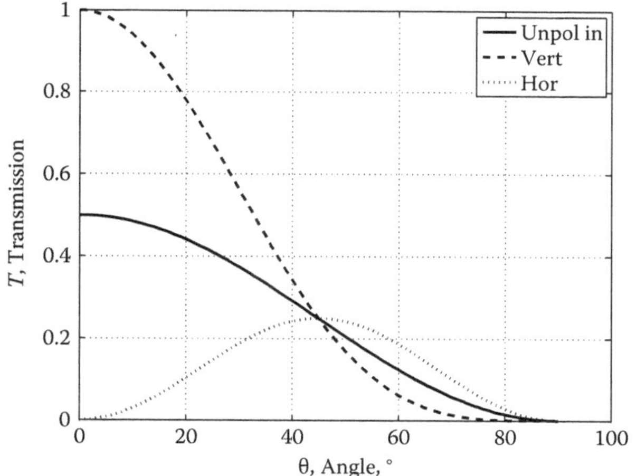

FIGURE P6.1
Transmissions of attenuators using rotating and fixed polarizers.

and the incident light is

$$E_0 = \begin{pmatrix} 1 \\ 0 \end{pmatrix}. \tag{G.44}$$

The field output is

$$E_q = \mathcal{PR}^*\mathcal{QR}E_0, \tag{G.45}$$

and the output power is proportional to the square of this. Note that this is probably not a good attenuator.

6.1f Repeat using a half-wave plate.

Solution:

This part is similar to the previous one, except that

$$E_q = \mathcal{PR}^*\mathcal{HR}E_0, \tag{G.46}$$

where

$$\mathcal{H} = \begin{pmatrix} e^{i\pi/4} & 0 \\ 0 & e^{-i\pi/4}. \end{pmatrix} \tag{G.47}$$

Plots for parts 6.1b through 6.1f are shown in Figures P6.1 and P6.2.

The MATLAB code for all parts is shown in the following:

```
% HW6p1 Problem on polarizers in an attenuator
% Optics for Engineers
% by Chuck DiMarzio
% Northeastern University, August 2003
thetad=[0:90];
theta=thetad*pi/180;
%
% For Unpolarized light in, we lose half in first polarizer.
% Field out of second is cos(theta) times input from first
%
```

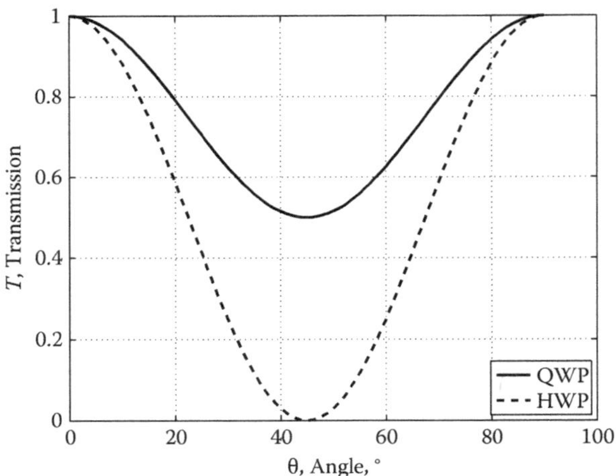

FIGURE P6.2
Transmissions of attenuators using rotating waveplate and fixed polarizer.

Solution to Problem 6.1f (Continued)

```
T1=0.5*(cos(theta)).^2;
%
% Vertical in
% Angle of incident polarization to first polarizer is
% theta.  Same for second.
%
T2=(cos(theta).*cos(theta)).^2;
%
% Horizontal in
% Angle of input polarization rel to first polarizer is
% 90 deg - theta
%
T3=(sin(theta).*cos(theta)).^2; fig1=figure;
plot(thetad,T1,'-',thetad,T2,'--',thetad,T3,':');grid on;
xlabel('\theta, Angle, deg.'); ylabel('T, Transmission');
legend('Unpol in','Vert','Hor');
%
%
% Set up the Jones matrices
e0=[1,0]';
p=[1,0;0,0];
e=exp(1);
qwp=[e^(i*pi/4),0;0,e^(-i*pi/4)];
hwp=[e^(i*pi/2),0;0,e^(-i*pi/2)];
for count=1:length(theta);
   theta1=theta(count);
r=[cos(theta1),sin(theta1);-sin(theta1),cos(theta1)];
eq=p*r'*qwp*r*e0;
Tq(count)=eq'*eq;
eh=p*r'*hwp*r*e0;
```

```
Th(count)=eh'*eh;
end;
fig2f=igure;
plot(thetad,Tq,'-',thetad,Th,'--');grid on;
xlabel('\theta, Angle, deg.');
ylabel('T, Transmission');
legend('QWP','HWP',4);
```

Problem 6.2: Electro-Optical Modulator

Suppose that an electro-optical modulator crystal is placed between crossed polarizers so that its principal axes are at 45° with respect to the axes of the polarizers. The crystal produces a phase change between fields polarized in the two principal directions of $\delta\phi = \pi V/V_\pi$, where V_π is some constant voltage and V is the applied voltage.

6.2a Write the Jones matrix for the crystal in the coordinate system of the polarizers.

Solution:

$$\mathcal{R}_{45}\mathcal{W}\mathcal{R}_{-45} = \frac{1}{\sqrt{2}}\begin{pmatrix}1 & -1\\ 1 & 1\end{pmatrix}\begin{pmatrix}e^{i\pi V/2V_\pi} & 0\\ 0 & e^{-i\pi V/2V_\pi}\end{pmatrix} \times \frac{1}{\sqrt{2}}\begin{pmatrix}1 & 1\\ -1 & 1\end{pmatrix} = \begin{pmatrix}\cos\pi V/2V_\pi & i\sin\pi V/2V_\pi\\ i\sin\pi V/2V_\pi & \cos\pi V/2V_\pi\end{pmatrix}. \qquad (G.48)$$

6.2b Calculate and plot the power transmission, T, through the whole system, polarizer–crystal–polarizer, as a function of applied voltage.

6.2c Next, add a quarter-wave plate with axes parallel to those of the crystal. Repeat the calculations of Part (b).

Solution:

See Figure P6.3 for Parts **(b)** and **(c)**.

6.2d What is the range of voltages over which the output power is linear to 1%?

Solution:

Notice that the signal varies like $\sin\pi\frac{V}{4V_\pi}$. Thus, the first two terms of a Taylor series would be

$$\sin x = x + \frac{x^3}{6}. \qquad (G.49)$$

The first term is the linear one, so the second term is the error:

$$\frac{x^3}{6} = 0.01 \qquad x = 0.39 \qquad V = \frac{0.39V_\pi}{4\pi} = 0.031V_\pi \qquad (G.50)$$

Problem 6.3: Laser Cavity

Consider a laser cavity with a gain medium having a power gain of 2% for a single pass. By this we mean that the output power will be equal to the input power multiplied by 1.02. The gain medium is a gas, kept in place by windows on the ends of the laser tube, at Brewster's angle. This is an infrared laser, and these windows are ZnSe, with an index of refraction of 2.4. To make the problem easier, assume the mirrors at the ends of the cavity are perfectly reflecting.

6.3a What is the Jones matrix describing a round trip through this cavity? What is the power gain for each polarization? Note that if the round-trip multiplier is greater than one, lasing can occur, and if not, it cannot.

Solution:

The transmission for P-polarized light is one, that for S-polarized light at Brewster's angle is

$$\frac{n^2-1}{n^2+1} = 0.704 \tag{G.51}$$

for each surface, or about 0.496 for two surfaces. Then for both plates,

$$\mathcal{B} = \begin{pmatrix} 1 & 0 \\ 0 & 0.496 \end{pmatrix}^2 \tag{G.52}$$

The gain medium is represented by the scalar multiplier $\sqrt{1.02}$. Putting these together, considering two passes through each Brewster plate,

$$\mathcal{M} = 1.02 \begin{pmatrix} 1 & 0 \\ 0 & 0.496 \end{pmatrix}^4 = \begin{pmatrix} 1.02 & 0 \\ 0 & 0.0617 \end{pmatrix} \tag{G.53}$$

Thus, only the P polarization will lase.

6.3b What is the tolerance on the angles of the Brewster plates to ensure that the laser will operate for light which is P-polarized with respect to the Brewster plates?

Solution:

Let us assume one plate is at the wrong angle. Then if the transmission for P polarization is less than $\sqrt{0.98}$, lasing will not occur. Therefore, the reflection coefficient for amplitude must be less than 0.02. This occurs between 66.2° and 68.3°, with the Brewster angle being at 67.4. Thus, there is a tolerance of about 1°.

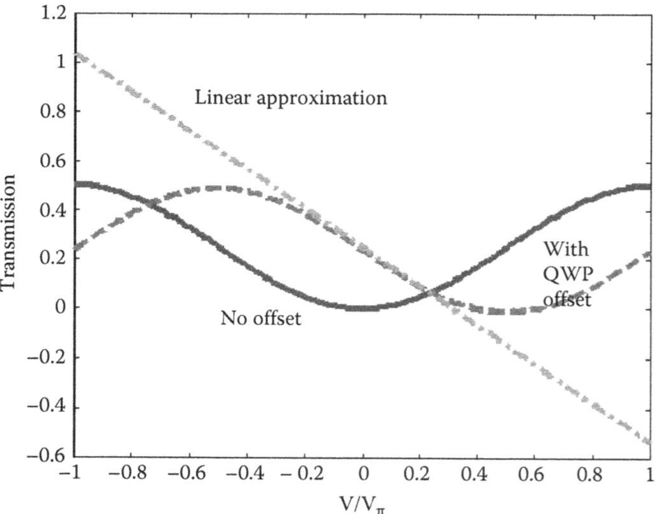

FIGURE P6.3
Transmission of electro-optical modulator as a function of relative applied voltage.

Solution to Problem 6.3c (Continued)

6.3c Now assume that because of sloppy manufacturing, the Brewster plates are rotated about the tube axis by 10° with respect to each other. What states of polarization are eigenvectors of the cavity? Will it lase?

Solution:

Now recalling

$$\mathcal{B} = \begin{pmatrix} 1 & 0 \\ 0 & 0.496 \end{pmatrix}^2 \tag{G.54}$$

the matrix is

$$1.02\mathcal{B}\mathcal{B}\mathcal{R}_{10}\mathcal{B}\mathcal{B}\mathcal{R}_{-10} \tag{G.55}$$

I calculated the eigenvectors as

$$\begin{pmatrix} 0.9994 \\ -0.346 \end{pmatrix} \qquad \begin{pmatrix} 0.1394 \\ 0.9902 \end{pmatrix}, \tag{G.56}$$

with eigenvalues

$$1.0014 \qquad 0.0629. \tag{G.57}$$

The first one will lase, and consists of linear polarization at −1.98°.

Problem 6.4: Depolarization in a Fiber (G)

Consider light passing through an optical fiber. We will use as a basis set the eigenvectors of the fiber. The fiber is birefringent, because of imperfections, bending, etc., but there are no polarization-dependent losses. In fact, to make things easier, assume there are no losses at all.

6.4a Write the Jones matrix for a fiber of length ℓ with birefringence Δn in a coordinate system in which the basis vectors are the eigenvectors of the fiber. Note that your result will be a function of wavelength. The reader is advised to keep in mind that the answer here is a simple one.

Solution:

$$\mathcal{J} = \begin{pmatrix} e^{i2\pi\Delta n\ell/(2\lambda)} & 0 \\ 0 & e^{-i2\pi\Delta n\ell}/(2\lambda) \end{pmatrix} \tag{G.58}$$

6.4b Suppose that the coherency matrix of the input light in this basis set is

$$\begin{pmatrix} a & b \\ b^* & c \end{pmatrix}. \tag{G.59}$$

What is the coherency matrix of the output as a function of wavelength?

Solution:

$$\begin{aligned} \mathcal{J}\mathcal{C}_{in}\mathcal{J}^\dagger &= \begin{pmatrix} e^{i2\pi n_x\ell/(2\lambda)} & 0 \\ 0 & e^{-i2\pi n_y\ell/(2\lambda)} \end{pmatrix} \begin{pmatrix} a & b \\ b^* & c \end{pmatrix} \\ &\times \begin{pmatrix} e^{-i2\pi n_x\ell/(2\lambda)} & 0 \\ 0 & e^{+i2\pi n_y\ell/(2\lambda)} \end{pmatrix} \\ &= \begin{pmatrix} a & be^{i2\pi\Delta n\ell/(2\lambda)} \\ b^*e^{-i2\pi\Delta n\ell}/(2\lambda) & c \end{pmatrix}. \end{aligned} \tag{G.60}$$

6.4c Assume now that the input light is composed of equal amounts of the two eigenvectors with equal phases, but has a linewidth $\Delta\lambda$. That is, there are an infinite number of waves present at all wavelengths from $\lambda - \Delta\lambda/2$ to $\lambda + \Delta\lambda/2$. Provided that $\Delta\lambda$ is much smaller than λ, write the coherency matrix in a way that expresses the phase as related to the center wavelength, λ, with a perturbation of $\Delta\lambda$.

Solution:

Let $\lambda = \lambda_0 + \lambda_1$. Then to first order, the exponent is

$$i2\pi\Delta n\ell/(2\lambda) \approx i2\pi\Delta n\ell/(2\lambda_0) + i2\pi\Delta n\ell\lambda_1/(2\lambda_0^2)$$

$$b_{out} = \frac{1}{\Delta\lambda}\int_{\lambda_0-\Delta\lambda/2}^{\lambda_0+\Delta\lambda/2} be^{i2\pi\Delta n\ell/(2\lambda)}d\lambda$$

$$= be^{i2\pi\Delta n\ell/(2\lambda_0)}\int_{-\Delta\lambda/2}^{\Delta\lambda/2} e^{i2\pi\Delta n\ell\lambda_1/(2\lambda_0^2)}d\lambda_1. \tag{G.61}$$

6.4d Now let us try to average the coherency matrix over wavelength. This would be a hard problem, so let us approach it in the following way: Show the value of b from the coherency matrix as a point in the complex plane for the center wavelength. Show what happens when the wavelength changes, and see what effect this will have on the average b. Next, write an inequality relating the length of the fiber, the line width, and the birefringence to ensure that depolarization is "small."

Solution:

The easiest way to do this is to show the integrand as moving around the unit circle in the complex plane. Then, the average must have a magnitude smaller than one, unless

$$i2\pi\Delta n\ell\lambda_1/(2\lambda_0^2) \ll 1. \tag{G.62}$$

Alternatively, the integral is not hard to do:

$$\frac{(2\lambda_0^2)}{i2\pi\Delta n\ell}e^{i2\pi\Delta n\ell\Delta\lambda/(2\lambda_0^2)} - e^{-i2\pi\Delta n\ell\Delta\lambda/(2\lambda_0^2)}$$

$$= 2\frac{\sin 2\pi\Delta n\ell\Delta\lambda/(2\lambda_0^2)}{i\frac{2\pi\Delta n\ell}{(2\lambda_0^2)}} \tag{G.63}$$

Note that this has its first zero at

$$2\pi\Delta n\ell\Delta\lambda/(2\lambda_0^2) = \pi \tag{G.64}$$

$$\frac{\ell}{\lambda_0} = \frac{\lambda_0}{\Delta\lambda}\frac{1}{\Delta n} \tag{G.65}$$

That is, the first zero is where the length in wavelengths is equal to the reciprocal of the line width in wavelengths, divided by the birefringence.

Thus, we can maintain polarization by keeping

$$\frac{\ell}{\lambda_0} \ll \frac{\lambda_0}{\Delta\lambda}\frac{1}{\Delta n} \tag{G.66}$$

G.7 Chapter 7

Problem 7.1: Laser Tuning

Suppose that I place the back mirror of a laser on a piezoelectric transducer that moves the mirror axially through a distance of 10 μm with an applied voltage of 1 kV, and that the relationship between position and voltage is linear. The laser has a nominal wavelength of 633 nm. The cavity length is 20 cm.

7.1a What voltage is required to sweep the cavity through one free spectral range?

Solution:

The easy way to handle this is to note that we have to change the cavity length by half a wavelength to move the $q-1$ mode to where the q mode was. Thus,

$$V_{FSR} = 1000\,\text{V}\frac{633/2}{10,000} = 31.65\,\text{V} \tag{G.67}$$

7.1b What is the free spectral range in frequency units?

Solution:

$$\frac{c}{2\ell} = \frac{3\times 10^8\,\text{m/s}}{2\times 0.2\,\text{m}} = 7.5\times 10^8\,\text{Hz}. \tag{G.68}$$

and the answer is 750 MHz.

7.1c What voltage is required to tune the laser frequency by 50 MHz?

Solution:

$$V_{50} = V_{FSR}\frac{50\,\text{MHz}}{FSR} = 31.65\,\text{V}\times\frac{50\times 10^6\,\text{Hz}}{750\times 10^6\,\text{Hz}} = 2.1\,\text{V}. \tag{G.69}$$

Problem 7.2: Interference Filters

Design a pair of layers for a high-reflection stack for 633 nm light, using $n = 1.35$ and $n = 2.3$. Assume light coming from air to glass. This problem is very similar to the one solved in the text. Try to solve it without looking at that solution.

7.2a Write the matrix for the layer.

7.2b Plot the reflection for a stack with five pairs of layers. Plot from 400 to 1600 nm.

7.2c Repeat for nine pairs.

```
% hw7p2.m  Interference filters homework problem
%
% by Chuck DiMarzio, Northeastern University
%  August 2003
%
count=0;
lambda_choice=488e-9;
lambdalist=[400:10:1600]*1e-9;
for lambda=lambdalist;
   count=count+1;
n0=1;nt=1.5;
n1=1.35;n2=2.3;
s1=lambda_choice/4/n1;
```

```
s2=lambda_choice/4/n2;
term1=2*pi*n1*s1/lambda;
term2=2*pi*n2*s2/lambda;
m1=[cos(term1),-i/n1*sin(term1);-i*n1*sin(term1),cos(term1)];
m2=[cos(term2),-i/n2*sin(term2);-i*n2*sin(term2),cos(term2)];
m=(m1*m2)^5;
rho5(count)=(n0*m(1,1)+n0*nt*m(1,2)-m(2,1)-nt*m(2,2))/...
   (n0*m(1,1)+n0*nt*m(1,2)+m(2,1)+nt*m(2,2));
end;
count=0;
for lambda=lambdalist;
   count=count+1;
n0=1;nt=1.5;
n1=1.35;n2=2.3;
s1=lambda_choice/4/n1;
s2=lambda_choice/4/n2;
term1=2*pi*n1*s1/lambda;
term2=2*pi*n2*s2/lambda;
m1=[cos(term1),-i/n1*sin(term1);-i*n1*sin(term1),
  cos(term1)];
m2=[cos(term2),-i/n2*sin(term2);-i*n2*sin(term2),
  cos(term2)];
m=(m1*m2)^9;
rho9(count)=(n0*m(1,1)+n0*nt*m(1,2)-m(2,1)-nt*m(2,2))
  /...
   (n0*m(1,1)+n0*nt*m(1,2)+m(2,1)+nt*m(2,2));
end;
fig1=figure;plot(lambdalist*1e9,abs(rho5),'-',...
                  lambdalist*1e9,abs(rho9),'--');
grid on;
xlabel('Wavelength, nm');ylabel('\rho, Reflectivity')
```

Results are shown in Figure P7.1.

Problem 7.3: Quadrature Interference (G)

Consider a coherent laser radar in which the signal beam returning from the target is linearly polarized. We place a quarter-wave plate in the local oscillator arm so that this beam becomes circularly polarized. Now, after the recombining beam splitter, we have a polarizing beam splitter that separates the beam into two components at $\pm 45°$ with respect to the signal polarization. A separate detector is used to detect each polarization.

7.3a Show that we can use the information from these two detectors to determine the magnitude and phase of the signal.

Solution:

Use coordinates based on the detectors. The recombined signal is

$$\frac{s}{\sqrt{2}}\begin{pmatrix}1\\1\end{pmatrix}+\frac{r}{\sqrt{2}}\begin{pmatrix}1\\i\end{pmatrix}, \tag{G.70}$$

where

- s is the signal field
- r is the reference field

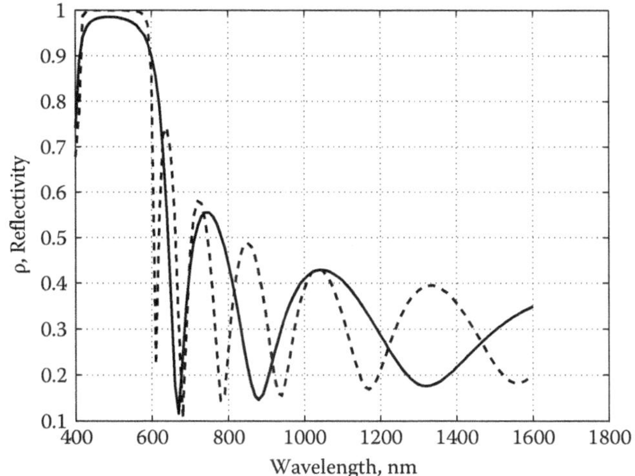

FIGURE P7.1
Reflection of interference filter. The lower (solid) curve is for five pairs of layers and the upper (dashed) for nine.

Detectors see

$$C_0 = \left| \begin{pmatrix} 1 & 0 \\ 0 & 0 \end{pmatrix} \frac{1}{\sqrt{2}} \begin{pmatrix} s+r \\ s+ir \end{pmatrix} \right|^2 = \frac{1}{2}\left[|s|^2 + |r|^2 + s^*r + sr^*\right] \tag{G.71}$$

$$C_1 = \left| \begin{pmatrix} 0 & 0 \\ 0 & 1 \end{pmatrix} \frac{1}{\sqrt{2}} \begin{pmatrix} s+r \\ s+ir \end{pmatrix} \right|^2 = \frac{1}{2}\left[|s|^2 + |r|^2 + s^*ir + s(-i)r^*\right] \tag{G.72}$$

Keeping only the AC terms,

$$C_{0AC} + iC_{1AC} = sr^* + s^*r. \tag{G.73}$$

7.3b What happens if the sensitivities of the two detectors are unequal. Explain in the frequency domain, as quantitatively as possible.

Solution:

Follow the same line as in Part (a), but include multiplicative constants, a_i.

$$C_0 = \left| \begin{pmatrix} 1 & 0 \\ 0 & 0 \end{pmatrix} \frac{1}{\sqrt{2}} \begin{pmatrix} s+r \\ s+ir \end{pmatrix} \right|^2 = \frac{1}{2}\left[|s|^2 + |r|^2 + s^*r + sr^*\right] \tag{G.74}$$

$$\begin{aligned} C_1 &= a_0 \left| \begin{pmatrix} 0 & 0 \\ 0 & 1 \end{pmatrix} \frac{1}{\sqrt{2}} \begin{pmatrix} s+r \\ s+ir \end{pmatrix} \right|^2 \\ &= a_1 \frac{1}{2}\left[|s|^2 + |r|^2 + s^*ir + s(-i)r^*\right] \end{aligned} \tag{G.75}$$

Keeping only the AC terms,

$$C_{0AC} + iC_{1AC} = (a_0 + a_1)\, sr * + (a_0 - a_1)\, s * r \tag{G.76}$$

Now the second term, representing the conjugate of the signal, or the negative frequency part, which was zero in Part (a), is nonzero in Part (b). In the worst case, if the coefficients are very different, we get a constant times $s + s*$, and the direction of the Doppler signal is again ambiguous.

7.3c Suppose now that both beams are linearly polarized, but at 45° with respect to each other. I pass the combination through a quarter-wave plate with one axis parallel to the reference polarization. I then pass the result to a polarizing beam splitter with its axes at 45° to those of the waveplate and detect the two outputs of this beam splitter with separate detectors. Write the mixing terms in the detected signals. Is it possible to determine the phase and amplitude?

Solution:

The signal after the quarter-wave plate is

$$\begin{pmatrix} 1 & 0 \\ 0 & i \end{pmatrix} \begin{pmatrix} \frac{s}{\sqrt{2}} + r \\ \frac{s}{\sqrt{2}} \end{pmatrix} = \begin{pmatrix} \frac{s}{\sqrt{2}} + r \\ \frac{is}{\sqrt{2}} \end{pmatrix} \tag{G.77}$$

Now we need to rotate 45° to get into the coordinate system of the detectors;

$$\frac{1}{\sqrt{2}} \begin{pmatrix} 1 & -1 \\ 1 & 1 \end{pmatrix} \begin{pmatrix} \frac{s}{\sqrt{2}} + r \\ \frac{is}{\sqrt{2}} \end{pmatrix} = \frac{1}{\sqrt{2}} \begin{pmatrix} s + 2r - is \\ s + 2r + is \end{pmatrix}. \tag{G.78}$$

Passing this through the polarizers and taking the magnitude squared in each case,

$$C_0 = \frac{1}{4} \left| s - is + 2r \right|^2 \tag{G.79}$$

$$C_1 = \frac{1}{4} \left| s + is + 2r \right|^2 . \tag{G.80}$$

These are the same signals as before except for multiplication of the signal by $(1 - i)$. That is, we have two components of the signal in quadrature as in the previous case.

G.8 Chapter 8

Problem 8.1: Resolution

A car approaches along a long straight road. Initially, it is so far away that the two headlights appear as one.

8.1a Using the Rayleigh criterion for resolution, at what distance do we begin to see that they are two separate lights? Use reasonable numbers for the spacing of the headlights and the diameter of the pupil of the eye.

Solution:

$$d = 1.22 \frac{\lambda}{D} z, \tag{G.81}$$

where

- d is the spacing of the lights
- D is the aperture diameter of the pupil

$$z = \frac{Dd}{1.22\lambda} = \frac{1.5\,\text{m} \times 0.008\,\text{m}}{1.22500 \times 10^{-9}\,\text{m}} = 1.97 \times 10^4\,\text{m}, \tag{G.82}$$

or about 20 km.

8.1b Now, consider the fact that we have two eyes, separated by a distance of a few centimeters. Does this change your answer? Why or why not?

Solution:

The headlights image separately on each retina, so the eyes have no mechanism for comparing two different wavefronts across the two eyes. In order to improve the resolution, the signals received by the two eyes would have to be processed coherently in some way. Once the retinas have detected the signals, this is no longer possible.

G.9 Chapter 9

Problem 9.1: Gaussian Beams

A helium-neon laser (633 nm) produces a collimated output beam with a diameter of 0.7 mm.

9.1a If the laser is 20 cm long, what are the radii of curvature of the two mirrors?

Solution:

The output being collimated dictates that the wavefronts are plane there, which means that the front mirror is a plane mirror. For the rear mirror;

$$b = \frac{\pi d_0^2}{4\lambda} = \frac{\pi(7 \times 10^{-4}\,\text{m})^2}{4 \times 10^{-9}\,\text{m}} = 0.61\,\text{m} \tag{G.83}$$

$$\rho = z + \frac{b^2}{z} = 0.2\,\text{m} + \frac{(0.61\,\text{m})^2}{0.2} = 2.1\,\text{m} \tag{G.84}$$

9.1b How big is the beam at the back mirror?

Solution:

$$d = d_0\sqrt{1 + \left(\frac{z}{b}\right)^2} = 0.7\,\text{mm} \times \sqrt{1 + \left(\frac{20}{61}\right)^2} = 0.74\,\text{mm} \tag{G.85}$$

9.1c I now wish to launch the beam into a fiber whose core is best matched to a Gaussian beam with a diameter of 5 µm. Figure out how to do this, using one or more lenses in a reasonable space.

Solution:

There are a number of good solutions. The easiest is to start from the waist of the beam entering the fiber. Find where the beam is 0.7 mm in diameter.

$$d = d_0\sqrt{1 + \left(\frac{z}{b}\right)^2} \tag{G.86}$$

$$\left(\frac{z}{b}\right) = \sqrt{\left(\frac{d}{d_0}\right)^2 - 1} \tag{G.87}$$

$$z = \pm b\sqrt{\left(\frac{d}{d_0}\right)^2 - 1} = \pm\frac{\pi d_0^2}{4\lambda}\sqrt{\left(\frac{d}{d_0}\right)^2 - 1}. \tag{G.88}$$

With all distances in micrometers,

$$\begin{aligned} z &= \pm\frac{\pi 5^2}{4\times 0.633}\sqrt{\left(\frac{700}{5}\right)^2 - 1} \\ &\approx \pm\frac{\pi 5^2}{4\times 0.633}\frac{700}{5} = \pm 4.4\times 10^3. \end{aligned} \tag{G.89}$$

We take the negative square root because we are to the left of the waist, so the waist is at −4.4 mm. We place a lens at this location and put the laser immediately in front of it. The focal length of the lens must be chosen so that the out radius of curvature is ρ given by

$$\rho = z + \frac{b^2}{z} \tag{G.90}$$

with

$$b = \left(\frac{\pi d_0^2}{4\lambda}\right) = \left(\frac{\pi 5^2}{4\times 0.633}\right) = 31\ \mu\text{m} \tag{G.91}$$

$$\rho = -4.4\,\text{mm} + \frac{0.031\ \text{mm}}{-4.4\ \text{mm}} \approx -4.4\,\text{mm} \tag{G.92}$$

Thus, the lens must have a focal length of 4.4 mm.

Problem 9.2: Diameter of a Laser Amplifier

I wish to build an amplifier tube 6 m long to amplify a carbon dioxide laser beam at a wavelength of 10.59 μm. I can change the beam diameter and radius of curvature entering the amplifier, and I can place lenses at the output so I do not care about the curvature of the wavefront at the output. I need to keep the beam diameter over the entire length as small as possible to keep the cost of the amplifier low. What is the waist diameter? Where is it located? What is the largest size of the Gaussian beam along the tube?

Solution:

To minimize the beam diameter along the tube as much as possible, we need to place the waist at the center, or 3 m from the end. Then, if the waist diameter is d_0, the largest diameter, which occurs at the tube ends, is

$$d = d_0\sqrt{1 + \left(\frac{z}{b}\right)^2}. \tag{G.93}$$

To minimize d, it is sufficient to minimize d^2.

$$d^2 = d_0^2 + d_0^2\left(\frac{4\lambda z}{\pi d_0^2}\right)^2. \tag{G.94}$$

$$d^2 = d_0^2 + \frac{16\lambda^2 z^2}{\pi^2 d_0^2}. \tag{G.95}$$

$$\frac{d\left(d^2\right)}{d\left(d_0^2\right)} = 1 - \frac{16\lambda^2 z^2}{\pi^2 d_0^4} = 0. \tag{G.96}$$

The solution is

$$d_0^2 = \frac{4\lambda z}{\pi} = 0 \tag{G.97}$$

$$d_0 = 6.4\,\text{mm}. \tag{G.98}$$

Now,

$$b = \frac{\pi d_0^2}{4\lambda}. \tag{G.99}$$

Not surprisingly, this is 3 m. We want the Rayleigh range to be half the tube length. Now at the ends of the tube,

$$d = d_0\sqrt{1 + \left(\frac{z}{b}\right)^2} = d_0\sqrt{2} = 9\,\text{mm}. \tag{G.100}$$

We would probably make the laser amplifier tube at least 1.5 times as large, because more severe truncation of the Gaussian beam would change the diffraction behavior.

G.10 Chapter 10

Problem 10.1: Köhler Illumination (G)

A common way of illuminating an object for a microscope is to focus the filament of a tungsten lamp on the pupil of the system (which means focusing it in the back focal plane of the objective lens). The image plane will then contain the two-dimensional (2-D) Fourier transform of the filament. The goal is to make illumination very uniform. Start with the following code:

```
\% Koehler Illumination
\%
\% By Chuck DiMarzio, Northeastern University,
%April 2002
\%
\% coherent source
\%
[x,y]=meshgrid(1:256,1:256);
coh=zeros(size(x));
coh(60:69,50:200)=1;
coh(80:89,50:200)=1;
coh(100:109,50:200)=1;
coh(120:129,50:200)=1;
coh(140:149,50:200)=1;
coh(160:169,50:200)=1;
coh(180:189,50:200)=1;
fig1=figure;
subplot(2,2,1);imagesc(coh);colormap(flipud(bone));
  colorbar;
title('Coherent Source');
```

Solution to Problem 10.1c (Continued)

10.1a Now, take the 2-D Fourier transform and image the magnitude of the result. A couple of hints are in order here. Try "help fft2," and "help fftshift." If you have not used subplots, try "help subplot" to see how to display all the answers in one figure.

10.1b The aforementioned result is not very uniform, but the prediction was that it would be. See if you can figure out why it does not work. You have two more subplots to use. On the third one, show the corrected source.

10.1c On the fourth one, show the correct image (Fourier transform).

There are different degrees of correctness possible in Part (b), which will affect the results somewhat. You only need to develop a model that is "correct enough" to show the important features.

Solution:

```
% HW10P1   Koehler Illumination
%
% By Chuck DiMarzio, Northeastern University, April 2002
%
% coherent source
%
[x,y]=meshgrid(1:256,1:256);
coh=zeros(size(x));
coh(60:69,50:200)=1;
coh(80:89,50:200)=1;
coh(100:109,50:200)=1;
coh(120:129,50:200)=1;
coh(140:149,50:200)=1;
coh(160:169,50:200)=1;
coh(180:189,50:200)=1;
fig1=figure;
subplot(2,2,1);imagesc(coh);colormap(flipud(bone));colorbar;
title('Coherent Source');
imgcoh=abs(fftshift(fft2(coh))).^2;
subplot(2,2,2);imagesc(imgcoh);colormap(flipud(bone));
    colorbar;
title('Coherent Image');
%
% incoherent source
inc=coh.*exp(i*2*pi*rand(size(coh)));
subplot(2,2,3);imagesc(real(inc));colormap(flipud(bone));
    colorbar;
title('Incoherent Source');
imginc=abs(fftshift(fft2(inc))).^2;
subplot(2,2,4);imagesc(imginc);colormap(flipud(bone));
    colorbar;
title(''Incoherent Image');
```

Results are shown in Figure P10.1. The essential point here is this: Each point in the pupil plane illuminates the whole image plane. If the light is coherent, these contributions interfere in just the right places to produce the Fourier transform of the pupil function. In the case of incoherent light, the interference effects are random and tend to cancel,

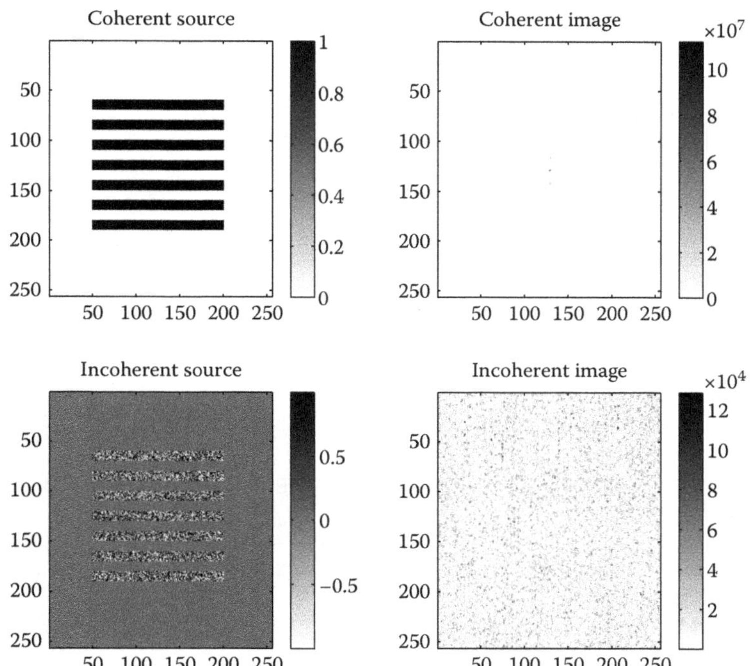

FIGURE P10.1
Results for Problem 10.1.

producing more-or-less uniform illumination. The more points contributing, the smaller will be the fluctuations.

Note that in the picture, I used "color map(flipud(bone))" to produce something that would print well in black and white.

G.11 Chapter 11

Problem 11.1: Point-Spread Function

Generate bar charts with one bar pair (light and dark) per 8 μm. The bars are in the x direction for one and y for the other. Then generate an elliptical Gaussian point-spread function with diameters of $d_x = 4\,\mu\text{m}$ and $d_y = 2\,\mu\text{m}$. Show the images of the bar charts viewed through a system having this point-spread function.
Solution:

The code follows, and the images are in Figure P11.1. The width of the point-spread function is larger than one of the bars in the x direction, and the loss of contrast is severe. The result is less extreme in the y direction.

```
%%% hw11p1.m Bar chart and psf.
% Define the image axes;
dx=0.1e-6;
xaxis=[0:511]*dx;
[x,y]=meshgrid(xaxis,xaxis);
barwidth=8e-6; % Make the bar chart
xbarchart=double(sin(2*pi*x/barwidth)>0);
ybarchart=double(sin(2*pi*y/barwidth)>0);
xaxispsf=([0:127]-64)*dx;  % Different axis for psf,
    smaller
```

Solution to Problem 11.1 (Continued)

```
[xx,yy]=meshgrid(xaxispsf,xaxispsf);
wx=4e-6;wy=2e-6;  %  Make the point--spread function
psf=exp(-2*(xx/wx).^2-2*(yy/wy).^2);
ximage=conv2(xbarchart,psf,'same'); % Generate the
    images
yimage=conv2(ybarchart,psf,'same');

fig1=figure;imagesc(xaxis,xaxis,xbarchart);axis image;
colormap(flipud(gray));
xlabel('x, \mu m');ylabel('x, \mu m');

fig2=figure;imagesc(xaxis,xaxis,ybarchart);axis image;
colormap(flipud(gray));
xlabel('x, \mu m');ylabel('x, \mu m');

fig3=figure;imagesc(xaxis,xaxis,ximage);axis image;
colormap(flipud(gray));

fig4=figure;imagesc(xaxis,xaxis,yimage);axis image;
colormap(flipud(gray));
xlabel('x, \mu m');ylabel('x, \mu m');

fig5=figure;imagesc(xaxispsf,xaxispsf,psf);axis image;
colormap(flipud(gray));
xlabel('x, \mu m');ylabel('x, \mu m')
```

G.12 Chapter 12

Problem 12.1: Radiation from a Human

For this problem, suppose that the surface area of a human being is about 1 m^2.

12.1a What is the power radiated in an infrared wavelength band from 10.0 to 10.1 μm? How many photons per second is this?

12.1b What is the power radiated in a green wavelength band from 500 to 600 nm? In what time is the average number of photons equal to one?

Solution:

We need to integrate the Planck equation, Equation 12.81, between the indicated limits. The body mass function used in Chapter 12 is useful for doing this integration numerically.

For the infrared band, 10.0 to 10.1 μm, the integral is about $P = 4\,\mathrm{W/m^2}$ or 4 W for the whole body.

The number of photons per second is

$$\frac{P}{h\nu} = 2.1 \times 10^{20}. \tag{G.101}$$

For the visible wavelengths from 500 to 600 nm, the power is 1.7×10^{-26} W.

The photon energy is $J_1 = hc/\lambda = 3.6 \times 10^{-19}$ J, taking the central wavelength of 550 nm. The time is

$$t = \frac{J_1}{P} = 2 \times 10^7\,\mathrm{s}, \tag{G.102}$$

or about 240 days.

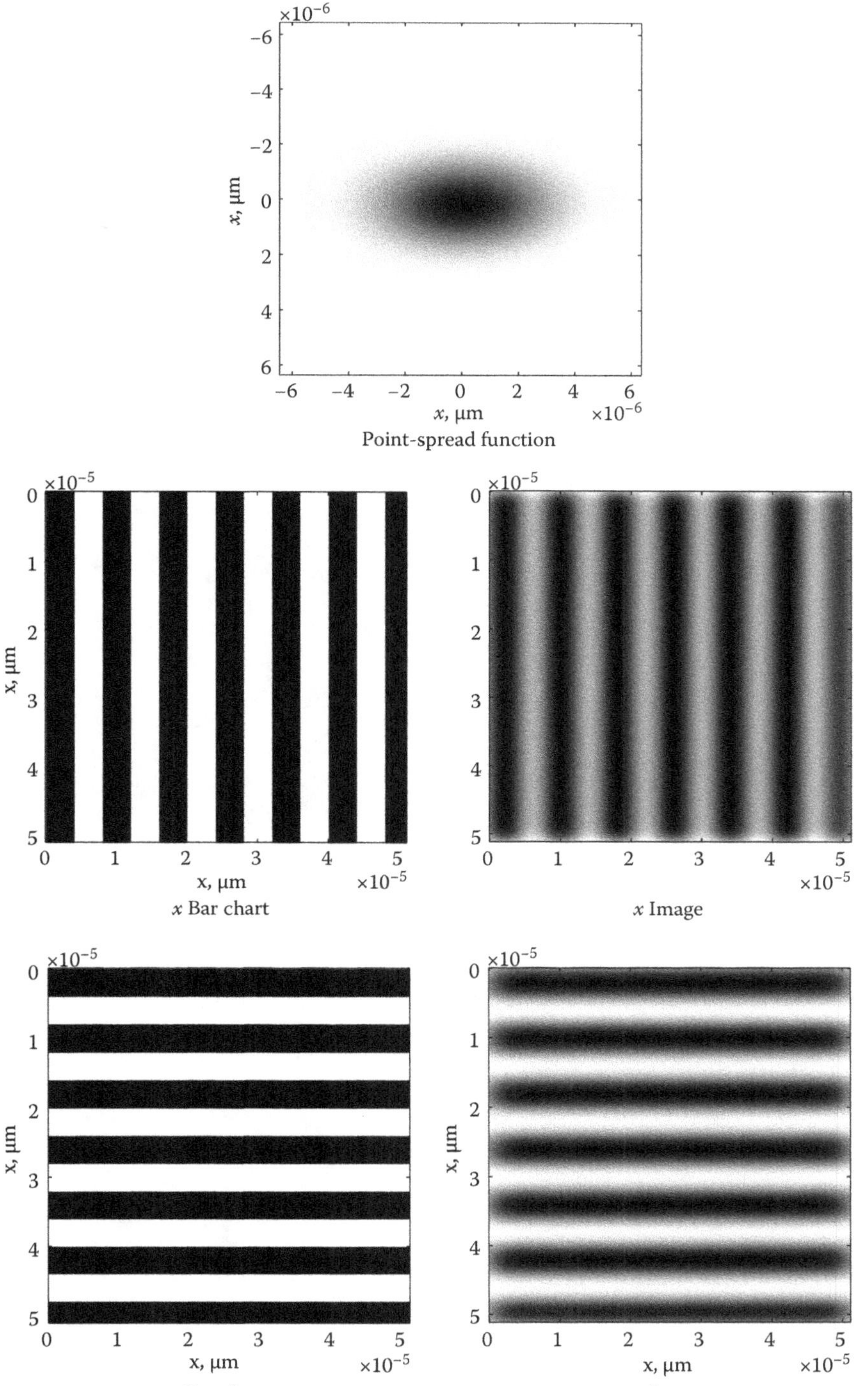

FIGURE P11.1
Bar charts and images.

Bibliography

[1] Comptes rendus de la 16^e cgpm (1979).

[2] Exodus 38:8.

[3] (1931). *Commission internationale de l'Eclairage proceedings.* Cambridge: Cambridge University Press.

[4] (1932). *Commission Internationale de l'Eclairage Proceedings.* Cambridge, UK: Cambridge University Press.

[5] (1963). *Introduction to Statistical Optics.* Reading, MA: Addison-Wesley.

[6] (1989). *Oxford English Dictionary* (2nd ed.). Oxford University Press. OED Online.

[7] (1995). Cie technical report 13.3-1995: Method of measuring and specifying colour rendering of light sources (3rd ed.) Commission internationale de l'Eclairage, Vienna.

[8] (2003). Astm standard g173-03, 2003, standard tables for reference solar spectral irradiance at air mass 1.5: Direct normal and hemispherical for a 37 degree tilted surface. ASTM International, West Conshohocken, PA.

[9] (2004). *Introduction to Statistical Optics.* Mineola, NY: Dover.

[10] (2006). Imaging intracellular fluorescent proteins at nanometer resolution. *Science 313*(5793), 1642–1645.

[11] (2007). Cie cie technical report 177:2007 colour rendering of white led light sources. Commission internationale de l'Eclairage, Vienna.

[12] (2010). *Laser Focus Buyers' Guide.* Nashua, NH: Penwell.

[13] Abramowitz, M. and I. A. Stegun (1972, December). *Handbook of Mathematical Functions, Tenth Printing.* U.S. Government Printing Office.

[14] Alexandrini, H. (1900). Opera quae supersunt omnia. In L. Nix and W. Schmidt (Eds.), *Mechanicaet catoptrica.* Leipzig: Teuber.

[15] Alferov, Z. and R. Kazarinov (1963). Semiconductor laser with electrical pumping. *USSR Patent 181737.*

[16] Amirault, C. T. and C. A. DiMarzio (1985, May). Precision pointing using a dual-wedge scanner. *Applied Optics 24.*

[17] Azzam, R. and N. Bashara (1987). *Ellipsometry and Polarized Light.* Elsevier Science Ltd.

[18] Balanis, C. (1989). *Advanced Engineering Electromagnetics.* New York: Wiley.

[19] Banerjee, P. P. (Ed.) (2003). *Nonlinear Optics, Theory, Numerical Modeling, and Application.* Optical Science and Engineering. Boca Raton, FL: CRC Press.

[20] Bayer, B. E. (1976). Color imaging array. *US Patent, 3,971,065.*

[21] Bloembergen, N. (1997). *Nonlinear Optics* (4th ed.). Singapore: World Scientfic Publishing Co.

[22] Bloom, D. and G. Bjorklund (1977). Conjugate wave-front generation and image reconstruction by four-wave mixing. *Applied Physics Letters 31*, 592.

[23] Boas, D., D. Brooks, E. Miller, C. DiMarzio, M. Kilmer, R. Gaudette, and Q. Zhang (2001). Imaging the body with diffuse optical tomography. *IEEE Signal Processing Magazine 18*(6), 57–75.

[24] Boas, D. A., C. Pitris, and N. Ramanujam (Eds.) (2011). *Handbook of Biomedical Optics.* Boca Raton, FL: CRC Press.

[25] Bohren, C. and D. Huffman (1983). *Absorption and Scattering of Light by Small Particles.* New York: John Wiley & Sons.

[26] Bohren, C. and J. Sardie (1981). Utilization of solar radiation by polar animals: an optical model for pelts; an alternative explanation. *Applied Optics 20*(11), 1894–1896.

[27] Borel, P. (1655). De Vero Telescopii Inventore et Centuria Observationum Microscopicarum.

[28] Born, M. and E. Wolf (1980). *Principles of Optics.* New York: Pergamon.

[29] Bouma, B. and G. Tearney (2002). *Handbook of Optical Coherence Tomography.* Informa Healthcare.

[30] Boyd, R. L. F. (Ed.) (1992). *Astronomical Photometry.* Springer.

[31] Boyd, R. W. (2003). *Nonlinear Optics* (2nd ed.). Amsterdam: Academic Press.

[32] Boyle, W. and G. Smith (1970). Charge coupled semiconductor devices. *Bell Sys. Tech. J 49*, 587.

[33] Boyle, W. and G. Smith (1974). Information Storage Devices. US Patent 3,858,232.

[34] Bridges, W. (1964). Laser oscillation in singly ionized argon in the visible spectrum. *Applied Physics Letters 4*, 128–130.

[35] Bridges, W. B. (1968). Bridges Ionized Noble Gas Laser. US Patent 3,395,364.

[36] Bringi, V. N. and V. Chandrasekar (2001). *Polarimetric Doppler Weather Radar: Principles and Applications.* Cambridge University Press.

[37] Buchwald, J. (1989). *The Rise of the Wave Theory of Light: Optical Theory and Experiment in the Early Nineteenth Century.* University of Chicago Press.

[38] Buzhan, P., B. Dolgoshein, L. Filatov, A. Ilyin, V. Kantzerov, V. Kaplin, A. Karakash, F. Kayumov, S. Klemin, E. Popova, et al. (2003). Silicon photomultiplier and its possible applications. *Nuclear Instruments and Methods in Physics Research Section A: Accelerators, Spectrometers, Detectors and Associated Equipment 504*(1-3), 48–52.

[39] Cheong, W. F., S. A. Prahl, and A. J. Welch (1990). A review of the optical properties of biological tissues. *IEEE Journal of Quantum Electronics 26*(12), 2166–2185.

[40] Chu, S. and C. Wieman (1989). Laser cooling and trapping of atoms introduction to feature issue. *Journal of the Optical Society of America B Optical Physics 6*, 2020.

[41] Clay, R. S. and T. H. Court (1975). *The History of the Microscope: Compiled From Original Instruments and Documents, Up To the Introduction of the Achromatic Microscope.* Holland Press.

[42] Colladon, D. (1842). Sur les réflexions d'un rayon de lumiere al'interieur d'une veine liquide parabolique. *CR Acad. Sci 15*, 800–802.

[43] Collins, Jr., S. A. (1964). Analysis of optical resonators involving focusing elements. *Appl. Opt. 3*(11), 1263–1275.

[44] committeee ISO/TC42, T. (2017). *Photography — Electronic still picture imaging — Resolution and spatial frequency responses: ISO Standard 12233:2017.* International Standards Organization.

[45] Cova, S., M. Ghioni, A. Lacaita, C. Samori, and F. Zappa (1996). Avalanche photodiodes and quenching circuits for single-photon detection. *Applied Optics 35*(12), 1956–1976.

[46] Creath, K. (2011). Dynamic phase imaging for in vitro process monitoring and cell tracking. In *Engineering in Medicine and Biology Society, EMBC, 2011 Annual International Conference of the IEEE*, pp. 5977–5980. IEEE.

[47] Crew, H., C. Huygens, T. Young, A. Fresnel, and F. Arago (1900). *The Wave Theory of Light: Memoirs of Huygens, Young and Fresnel.* American Book Company.

[48] Davis, B. (1993). Making Light of Stamp Collecting. *Optics and Photonics News 4*(4), 44–45.

[49] Den Dekker, A. J. and A. Van den Bos (1997). Resolution: a survey. *Journal of the Optical Society of America A 14*(3), 547–557.

[50] Denk, W., J. Strickler, and W. Webb (1990). Two-photon laser scanning fluorescence microscopy. *Science 248*, 73.

[51] Descartes, R. (1637). *La Dioptrique.*

[52] Deschamps, G. A. (1971). Gaussian Beams as a bundle of complex rays. *Electron. Lett 7*, 648–685.

[53] Dijksterhuis, F. J. (2004). *Lenses and Waves: Christiaan Huygens and the Mathematical Science of Optics in the Seventeenth Century.* Dordrecht, The Neatherlands: Kluwer.

[54] Dimarzio, C., C. Harris, J. W. Bilbro, E. A. Weaver, D. C. Burnham, and J. N. Hallock (1979, September). Pulsed Laser Doppler Measurements of Wind Shear. *Bulletin of the American Meteorological Society 60*, 1061–1068.

[55] DiMarzio, C. A. and S. C. Lindberg (1992, July). Signal-to-noise equations for heterodyne laser radar. *Applied Optics 31*(21), 4240–4246.

[56] Doppler, C. (1842). *Über das farbige Licht der Doppelsterne und einiger anderer Gestirne des Himmels.* Borrosch & André.

[57] Duncan, M., J. Reintjes, and T. Manuccia (1982). Scanning coherent anti-Stokes Raman microscope. *Optics Letters 7*(8), 350–352.

[58] Dunn, A. and R. Richards-Kortum (1996). Three-dimensional computation of light scattering from cells. *IEEE Journal on Selected Topics in Quantum Electronics 2*(4), 898 – 905.

[59] Dwyer, P. J., C. A. DiMarzio, and M. Rajadhyaksha (2007, April). Confocal theta line-scanning microscope for imaging human tissues. *Applied Optics 46*, 1843–1851.

[60] Einstein, A. (1905). Über einen die erzeugung und verwandlung des lichtes betreffenden heuristischen gesichtspunkt. *Annalen der Physik 322*(6), 132–148.

[61] Einstein, A. (1906). Zur Theorie der Lichterzeugung und Lichtabsorption. *Annalen der Physik 20*, 199–206.

[62] Einstein, A. (1914). Method for the determination of the statistical values of observations concerning quantities subject to irregular fluctuations. *Arch. Sci. Phys. et Natur. 37*(4), 254–56.

[63] Einstein, A. (1917). On the quantum theory of radiation. *Phys. Z 18*, 121.

[64] Euclid (1895). Optica. In J. L. Heiberg (Ed.), *Euclidis Opera Omnia*, Volume 7. Leipzig: Teuber.

[65] Dereniak E. L. and D. G. Crowe (1984). *Optical Radiation Detectors.* Pure and Applied Optics. New York: Wiley.

[66] Fairman, H., M. Brill, and H. Hemmendinger (1998). How the CIE 1931 color-matching functions were derived from Wright-Guild data. *Color Research & Application 22*(1), 11–23.

[67] Feynman, R. P., R. B. Leighton, and M. Sands (1963). *Feynman Lectures on Physics*, Volume 1. Reading, MA: Addison-Wesley. Chapter 26.

[68] Fine, S. and W. Hansen (1971/10/). Optical second harmonic generation in biological systems. *Applied Optics 10*(10), 2350 – 3.

[69] Fleck, J., J. Morris, and M. Feit (1976). Time-dependent propagation of high energy laser beams through the atmosphere. *Applied Physics A: Materials Science & Processing 10*(2), 129–160.

[70] Ford, B. J. (2001). The Royal Society and the microscope. *Notes and Records 55*(1), 29.

[71] Franken, P. A., A. E. Hill, C. W. Peters, and G. Weinreich (1961, Aug). Generation of optical harmonics. *Phys. Rev. Lett. 7*(4), 118–119.

[72] Freund, I. and M. Deutsch (1986). Second-harmonic microscopy of biological tissue. *Optics Letters 11*(2), 94.

[73] Gabor, D. (1948). A new microscopic principle. *Nature 161*(4098), 777–778.

[74] Gauss, C. F. (1841). *Dioptrische Untersuchungen.* Dieterich.

[75] Geary, J. M. (2002). *Introduction to Lens Design, With Practical ZEMAX Examples.* Richmond, VA: Willmann-Bell.

[76] Geusic, J., H. Marcos, and L. Van Uitert (1964). Laser Oscillations in Nd-Doped Yttrium Aluminum, Yttrium Gallium and Gadolinium Garnets. *Applied Physics Letters 4*(10), 182–184.

[77] Ghiglia, D. and M. Pritt (1998). *Two-Dimensional Phase Unwrapping: Theory, Algorithms, and Software.* New York: John Wiley and Sons.

[78] Giordmaine, J. A. and R. C. Miller (1965, Jun). Tunable coherent parametric oscillation in linbo3 at optical frequencies. *Phys. Rev. Lett. 14*(24), 973–976.

[79] Goebel, D. (1967). Generalized integrating-sphere theory. *Applied Optics 6*(1), 125–128.

[80] Goldstein, D. (2003). *Polarized Light* (2nd ed.). Lasers, Optics & Optoelectronics. New York: Marcel Dekker.

[81] Goodman, J. W. (1985). *Statistical Optics*. New York: Wiley.

[82] Goodman, J. W. (2005). *Introduction to Fourier Optics* (3rd ed.). Englewood, CO: Roberts & Company. 491 Pages.

[83] Goodwin, E. P. and J. C. Wyant (2006). *Field Guide to Interferometric Optical Testing*. Bellingham, WA: SPIE Press.

[84] Goppert-Mayer, M. (1931). Ueber Elementarakte mit zwei Quantenspruengen. *Ann. Phys 9*, 273–295.

[85] Gordon, J., H. Zeiger, and C. Townes (1955). Molecular Microwave Oscillator and New Hyperfine Structure in the Microwave Spectrum of NH3, n Phys. Rev.. 95, Jul 1954, pp. 282–284. *The Maser–New Type of Microwave Amplifier, Frequency Standard, and Spectrometer, Phys. Rev 99*, 1264–1274.

[86] Greivenkamp, J. and J. Bruning (1992). Phase shifting interferometry. *Optical Shop Testing*, 501–599.

[87] Grojean, R., J. Sousa, and M. Henry (1980). Utilization of solar radiation by polar animals: an optical model for pelts. *Appl. Opt 19*, 339–346.

[88] Gustafsson, M. G. (2000). Surpassing the lateral resolution limit by a factor of two using structured illumination microscopy. *Journal of Microscopy 198*(2), 82–87.

[89] Hambley, A. R. (2000). *Electronics* (2nd ed.). Prentice-Hall.

[90] Hardy, A. C. and F. H. Perrin (1929). *The Principles of Optics*. McGraw–Hill.

[91] Hariharan, P. (1985). *Optical Interferometry*. Academic Press.

[92] Hecht, J. (2004). *City of Light: The Story of Fiber Optics*. Oxford University Press, USA.

[93] Hecht, J. (2005). Beam: the race to make the laser. *Optics and Photonics News 16*(7), 24–29.

[94] Hellwarth, R. (1977). Generation of time-reversed wave fronts by nonlinear refraction. *Journal of the Optical Society of America 67*(1), 1–3.

[95] Hendley, C. and S. Hecht (1949). The colors of natural objects and terrains, and their relation to visual color deficiency. *J. Opt. Soc. Am 39*, 870–873.

[96] Hoffmann, B. (1959). *The Strange Story of the Quantum: An Account for the General Reader of the Growth of the Ideas Underlying Our Present Atomic Knowledge* (2nd ed.). New York: Dover.

[97] Hogenboom, D. O. and C. A. DiMarzio (1998, May). Quadrature detection of a doppler lidar signal. *Applied Optics 37*(13), 2569–2572.

[98] Hooke, R. (1665). *Micrographia: or Some Physiological Descriptions of Minute Bodies, Made by Magnifying Glasses with Observations and Inquiries Thereupon*. James Allestry.

[99] Hopkins, H. (1951). The concept of partial coherence in optics. *Proceedings of the Royal Society of London. Series A, Mathematical and Physical Sciences*, 263–277.

[100] Houston, W. (1926). The Fine Structure and the Wave-Length of the Balmer Lines. *The Astrophysical Journal 64*, 81.

[101] Huang, D., E. A. Swanson, C. P. Lin, J. S. Schuman, W. G. Stinson, W. Chang, M. R. Hee, T. Flotte, K. Gregory, C. A. Puliafito, and J. G. Fujimoto (1991, November). Optical coherence tomography. *Science 254*, 1178–1181.

[102] Huygens, C. (1913). *Abhandlung ueber das Licht. Worin die Ursachen der Vorgaenge bei seiner Zurueckwerfung und Brechung und besonders bei der eigenthuemlichen Brechung des islaendischen Spathes dargelegt sind (1678)*. Ostwald's Klassiker der exakten Wissenschaften, Leipzig: Engelmann, c1913, 3.Auflage.

[103] Warger II, W. C. and C. A. DiMarzio (2009). Computational signal-to-noise ratio analysis for optical quadrature microscopy. *Opt. Express 17*(4), 2400–2422.

[104] Ikeda, T., G. Popescu, R. R. Dasari, and M. S. Feld (2005). Hilbert phase microscopy for investigating fast dynamics in transparent systems. *Optics Letters 30*(10), 1165–1167.

[105] Iliffe, R. and J. Young (2004). REPORT. Newton on the Net: first and prospective fruits of a Royal Society grant. *Notes and Records of the Royal Society of London 58*(1), 83–88.

[106] International Electrotechnical Commission (2014). *IEC Standard 60825 (2014) Safety of Laser Products.* International Electrotechnical Commission.

[107] Ishimaru, A. (1997). *Wave Propagation and Scattering in Random Media.* IEEE Press.

[108] Jackson, J. D. (1975). *Classical Electrodynamics* (2nd ed.). New York: Wiley.

[109] Jacquez, J. and H. Kuppenheim (1955). Theory of the integrating sphere. *J. Opt. Soc. Am 45*(6), 460–470.

[110] Javan, A. (1976, December 21). Gas detector. US Patent 3,998,557.

[111] Jeans, J. H. (1905). On the partition of energy between matter and the ether. *Phil. Mag 10*, 91–98.

[112] Jelalian, A. V. (1991). *Laser Radar Systems.* Norwood, MA: Artech House.

[113] Jensen, S. and R. Hellwarth (1978). Observation of the time-reversed replica of a monochromatic optical wave. *Applied Physics Letters 32*, 166.

[114] Johnson, J. (1928). Thermal agitation of electricity in conductors. *Physical Review 32*(1), 97–109.

[115] Jones, R. C. (1941). New calculus for the treatment of optical systems. *J. Opt. Soc. Am. 31*, 488–493.

[116] Joyce, W. B. and A. Joyce (1976). Descartes, newton, and snell's law. *J. Opt. Soc. Am. 66*(1), 1–8.

[117] Kennedy, J. F. (1961, 16 Sept). Transcript of kennedy remarks on space challenge. *New York Times*, 16.

[118] Khintchine, A. (1934). Correlation theory of stationary stochastic processes. *Math. Ann. 109*, 604–15.

[119] Kingston, R. H. (1978). *Detection of Optical and Infrared Radiation*, Volume 10 of *Springer Series in Optical Sciences.* New York: Springer Verlag.

[120] Klar, T., E. Engel, and S. Hell (2001, Dec). Breaking Abbe's diffraction resolution limit in fluorescence microscopy with stimulated emission depletion beams of various shapes. *Phys Rev E Stat Nonlin Soft Matter Phys 64*, 066613.

[121] Klein, C. A. and T. A. Dorschner (1989). Power handling capability of faraday rotation isolators for co2 laser radars. *Appl. Opt. 28*(5), 904–914.

[122] Knüttel, A., S. Bonev, and W. Knaak (2004). New method for evaluation of in vivo scattering and refractive index properties obtained with optical coherence tomography. *Journal of Biomedical Optics 9*(2), 265–273.

[123] Kobayashi, E., I. Sakuma, K. Konishi, M. Hashizume, and T. Dohi (2004). A robotic wide-angle view endoscope using wedge prisms. *Surgical Endoscopy 18*(9), 1396–1398.

[124] Koechner, W. (1970). Thermal lensing in a Nd: YAG laser rod. *Applied Optics 9*(11), 2548–2553.

[125] Kogelnik, H. and T. Li (1966). Lasers and resonators. *Appl. Opt 5*, 1550–1567.

[126] Kowalski, G. J. (1996). Numerical simulation of transient thermoreflectance of thin films in the picosecond regime. In A. M. Khounsary (Ed.), *Proc. SPIE, High Heat Flux Engineering III*, Volume 2855, pp. 138–146. SPIE.

[127] Krane, K. S. (1996). *Modern Physics* (2nd ed.). New York: Wiley.

[128] Kroemer, H. (1963). A proposed class of hetero-junction injection lasers. *Proceedings of the IEEE 51*(12), 1782–1783.

[129] Kroemer, H. (1967). Solid state radiation emitters. US Patent 3,309,553.

[130] Laikin, M. (Ed.) (2006). *Lens Design* (4th ed.). Optical Science and Engineering. Boca Raton, FL: CRC Press.

[131] Lakowicz, J. R. (2006). *Principles of Fluorescence Spectroscopy* (3rd ed.). Springer.

[132] Lamm, H. (1930). Biegsame optische gerate. *Zeitschrift Instrumentenkunde 50*, 579–581.

[133] Land, E. (1939). Laminated light polarizer. US Patent 2,168,221.

[134] Land, E. H. (1951). Some aspects of the development of sheet polarizers. *J. Opt. Soc. Am. 41*(12), 957–962.

[135] Langley, S. P. (1900). *Annals of the Astrophysical Observatory of the Smithsonian Institution Volume I.* Smithsonian Institution.

[136] Laser Institute of America (2007). *ANSI Standard Z136.1 (2007) American National Standard for the Safe Use of Lasers.* Laser Institute of America.

[137] Lavigne, D. and N. Øritsland (1974a). Black polar bears. *Nature 251*, 218–219.

[138] Lavigne, D. and N. Øritsland (1974b). Ultraviolet photography: a new application for remote sensing of mammals. *Canadian Journal of Zoology 52*(7), 939–941.

[139] Lebedev, P. (1901). Untersuchungen
“uber die Druckkr
“afte des Lichtes. *Ann. Phys 6*(433), 3.

[140] Leeuwenhoek, A. V. (1977). *The Selected Works of Antony van Leeuwenhoek Containing His Microscopical Discoveries in Many of the Works of Nature.* Arno Press.

[141] Lindberg, D. (1971). Lines of Influence in Thirteenth-Century Optics: Bacon, Witelo, and Pecham. *Speculum: A Journal of Mediaeval Studies 46*(1), 66–83.

[142] Lindberg, D. C. (1967). Alhazen's theory of vision and its reception in the west. *Isis 58*(3), 321–341.

[143] Lorentz, H. A. (1895). *Versuch einer Theorie der elektrischen und optischen Erscheinungen in bewegten Körpern.* Leiden: E. J. Brill.

[144] Lorentz, H. A. (2005). *Versuch einer Theorie der elektrischen und optischen Erscheinungen in bewegten Körpern (Reprint).* Elibron.

[145] Mach, L. (1892). Uber einen interferenzrefraktor. *Z. Instrumentenkunde 12*, 89–93.

[146] Mahoney, M. S. (1994). *The Mathematical Career of Pierre de Fermat, 1601-1665* (2nd ed.). Princeton University Press.

[147] Malacara, D. and B. J. Thompson (2001). *Handbook of Optical Engineering.* Lasers, Optics & Optoelectronics. New York: Marcel Dekker.

[148] Malus, E.-L. (1810). *Théorie de la Double Réfraction de la Lumiére dans les Substances Cristallisées.* Baudoin.

[149] Marchand, E. (1970). Ray tracing in gradient-index media. *J. Opt. Soc. Am 60*, 1–2.

[150] Marchand, E. (1972). Ray tracing in cylindrical gradient-index media. *Applied Optics 11*(5), 1104–1106.

[151] Marchand, E. W. (1978). *Gradient Index Optics.* New York: Academic Press.

[152] Maxwell, J. C. (1873). *A Treatise on Electricity and Magnetism.* MacMillan.

[153] Measures, R. M. (1984). *Remote Sensing: Fundamentals and Applications.* New York: Wiley.

[154] Michelson, A. (1927). Measurement of the velocity of light between Mount Wilson and Mount San Antonio. *The Astrophysical Journal 65*(1), 1–22.

[155] Michelson, A. A., H. A. Lorentz, D. C. Miller, R. J. Kennedy, E. R. Hedrick, and P. S. Epstein (1928, December). Conference on the Michelson-Morley Experiment Held at Mount Wilson, February, 1927. *Astrophysical Journal 68*, 341–+.

[156] Michelson, A. A. and E. W. Morley (1887). On the relative motion of the Earth and the luminiferous ether. *American Journal of Science 34*, 333–345.

[157] Mie, G. (1908). Beitr
“age zur Optik tr
“uber Medien, speziell kolloidaler Metall
“osungen. *Annalen der Physik 330*(3), 377–445.

[158] Millikan, R. A. (1916, Mar). A direct photoelectric determination of Planck's h. *Phys. Rev. 7*(3), 355–388.

[159] Minsky, M. (1988). Memoir on inventing the confocal microscope. *Scanning 10*, 128–138.

[160] Minsky, M. M. (1955, October). Confocal scanning microscope. *U. S. Patent Number 3,013,467.*

[161] Moore, D. (1975). Ray tracing in gradient-index media. *J. Opt. Soc. Am 65*, 451–455.

[162] Motamedi, M., A. Welch, W. Cheong, S. Ghaffari, and O. Tan (1988). Thermal lensing in biologic medium. *IEEE Journal of Quantum Electronics 24*(4), 693–696.

[163] Mueller, H. (1948). The foundations of optics (abstract). *Journal of the Optical Society of America 38*, 661.

[164] National Standardsof People's Republic of China (2012). *Safety of Laser Products.* National Standardsof People's Republic of China.

[165] Newton, I. (1704). *Opticks: Or, A treatise of the Reflections, Refractions, Inflexions and Colours of Light. Also Two Treatises of the Species and Magnitude of Curvilinear Figures.* The Royal Society.

[166] Niklaus, F., A. Decharat, C. Jansson, and G. Stemme (2008). Performance model for uncooled infrared bolometer arrays and performance predictions of bolometers operating at atmospheric pressure. *Infrared Physics & Technology 51*(3), 168–177.

[167] Nomarski, C. (1953). Dispositif interferentiel a polarisation pour l'etudedes objets transparents ou opaques appartenant a la classe des objets de phase. *French Patent* (1,059,124).

[168] Nomarski, G. and A. Weill (1954). Sur l'observation des figures de croissance des cristaux par les méthodes interferentielles à deux ondes. *Bull. Soc. Fr. Mineral 104*, 292–297.

[169] Ober, R. J., E. S. Ward, and J. Chao (2020). *Quantitative Bioimaging: An Introduction to Biology, Instrumentation, Experiments, and Data Analysis for Scientists and Engineers.* CRC Press.

[170] Osmundsen, J. A. (1960, 8 July). Light amplification claimed by scientist. *New York Times*, 1.

[171] Parimi, P., W. Lu, P. Vodo, and S. Sridhar (2003). Photonic crystals: Imaging by flat lens using negative refraction. *Nature 426*(6965), 404.

[172] Patel, C. (1964). Continuous-wave laser action on vibrational rotational transitions of carbon dioxide. *Phys Rev 136*, A1187–A1193.

[173] Pendry, J. (2000). Negative refraction makes a perfect lens. *Physical Review Letters 85*(18), 3966–3969.

[174] Pepper, D. (1986). Applications of optical phase conjugation. *Scientific American 254*, 74–83.

[175] Peticolas, W. L., J. P. Goldsborough, and K. E. Rieckhoff (1963, Jan). Double photon excitation in organic crystals. *Phys. Rev. Lett. 10*(2), 43–45.

[176] Pickering, J., S. Prahl, N. Van Wieringen, J. Beek, H. Sterenborg, and M. Van Gemert (1993). Double-integrating-sphere system for measuring the optical properties of tissue. *Applied Optics 32*(4), 399–410.

[177] Planck, M. (1901). Über die elementarquanta der materie und der elektricität. *Annalen der Physik 309*(3), 564–566.

[178] Planck, M. (1909). Zur theorie der wärmestrahlung. *Annalen der Physik 336*(4), 758–768.

[179] Ploetz, E., S. Laimgkuber, S. Berner, W. Zinth, and P. Gilch (2007/05/). Femtosecond stimulated raman microscopy. *Applied Physics B (Lasers and Optics) B87*(3), 389–393.

[180] Pohl, D. W., W. Denk, and M. Lanz (1984). Optical stethoscopy: Image recording with resolution $\lambda/20$. *Applied Physics Letters 44*(7), 651–653.

[181] Poincaré, H. (1892). *Traité de la Lumiére.*

[182] Popescu, G., L. P. Deflores, J. C. Vaughan, K. Badizadegan, H. Iwai, R. R. Dasari, and M. S. Feld (2004). Fourier phase microscopy for investigation of biological structures and dynamics. *Optics Letters 29*(21), 2503–2505.

[183] Prahl, S., M. van Gemert, and A. Welch (1993). Determining the optical properties of turbid mediaby using the adding–doubling method. *Applied Optics 32*(4), 559–568.

[184] Preciado, J., B. Rubinsky, D. Otten, B. Nelson, M. Martin, and R. Greif (2002). Radiative properties of polar bear hair.

[185] Raman, C. and K. Krishnan (1928). A new type of secondary radiation. *Nature 121*(3048), 501.

[186] Rayleigh, L. (1871). On the light from the sky, its polarization and colour. *Philos. Mag 41*(107), 274.

[187] Rayleigh, L. (1879). Investigations in optics with special reference to the spectroscope, 1: resolving, or separating power of optical instruments. *Philosophical Magazine. Reprinted in: Scientific Papers of Lord Rayleigh 1*(8).

[188] Rayleigh, L. (1881). On the electromagnetic theory of light. *Philos. Mag 12*, 81–101.

[189] Risnero, F. (Ed.) (1572). *Opticae thesaurus Alhazeni Arabis libri septem, nunc primum.* Basel.

[190] Rogalski, A. (Ed.) (2006). *Infrared Detectors* (2nd ed.). Optics, Lasers, and Photonics. Boca Raton, FL: CRC Press.

[191] Ronchi, V. (1923). Le frangie di combinazione nello studio delle superficie e dei sistemi ottici. *Riv. Ottica Mecc. Precis 2*, 9–35.

[192] Ronchi, V. (1956). *Histoire de la lumière.* Colin.

[193] Ronchi, V. (1970). *The Nature of Light.* Harvard University Press.

[194] Ronchi, V. and E. Rosen (1991). *Optics, the Science of Vision.* Dover Publications.

[195] Rosell, F. (1960). Prism scanner. *Journal of the Optical Society of America 50*(6).

[196] Rosen, E. (1956, Jan). The invention of eyeglasses. I. *J Hist Med Allied Sci 11*, 13–46.

[197] Rothwell, E. J. and M. J. Cloud (2001). *Electromagnetics* (2nd ed.). Boca Raton, FL: CRC Press.

[198] Roychoudhuri, C. and K. R. Lefebvre (1995). Van cittert-zernike theorem for introductory optics course using the concept of fringe visibility. Volume 2525, pp. 148–160. SPIE.

[199] Rust, M. J., M. Bates, and X. Zhuang (2006, Oct). Sub-diffraction-limit imaging by stochastic optical reconstruction microscopy (STORM). *Nat Methods 3*(10), 793–795.

[200] Schawlow, A. and C. Townes (1958). Infrared and optical masers. *Physical Review 112*(6), 1940–1949.

[201] Selloy, S. M. (1967). Index of refraction of air. In *Handbook of Chemistry and Physics. 48th edition*, pp. E-160. The Chemical Rubber Co.

[202] Shalaev, V. (2007). Optical negative-index metamaterials. *Nature 1*(1).

[203] Shannon, C. (1949, jan.). Communication in the presence of noise. *Proceedings of the IRE 37*(1), 10–21.

[204] Shannon, R. R. (1997). *The Art and Science of Optical Design.* Cambridge, UK: Cambridge University Press.

[205] Shapiro, I. (1973). On the History of the Discovery of the Maxwell Equations. *Physics-Uspekhi 15*(5), 651–659.

[206] Shelby, R., D. Smith, and S. Schultz (2001). Experimental verification of a negative index of refraction. *Science 292*(5514), 77.

[207] Sheldon, S., L. Knight, and J. Thorne (1982). Laser-induced thermal lens effect: a new theoretical model. *Applied Optics 21*(9), 1663–1669.

[208] Shen, L. and J. Kong (1987). *Applied Electromagnetism.* Brooks/Cole.

[209] Shurcliff, W. A. (1962). *Polarized Light, Production and Use.* Cambridge, MA: Harvard University Press.

[210] Siegman, A. E. (1976). *Lasers.* Sausalito, CA: University Science Books.

[211] Siegman, A. E. (2000). Laser beams and resonators: the 1960s. *Selected Topics in Quantum Electronics, IEEE Journal of 6*(6), 1380–1388.

[212] Singer, C. (1914). Notes on the early history of microscopy. *Proceedings of the Royal Society of Medicine 7*(Sect Hist Med), 247.

[213] Smith, D. and N. Kroll (2000). Negative refractive index in left-handed materials. *Physical Review Letters 85*(14), 2933–2936.

[214] Smith, G. H. (1998). *Practical Computer-Aided Lens Design.* Richmond, VA: Willmann-Bell.

[215] Smith, W. J. (1992). *Modern Lens Design.* New York: McGraw Hill.

[216] Sparrow, C. (1916). On spectroscopic resolving power. *The Astrophysical Journal 44*, 76.

[217] Stachel, J. (Ed.) (2005). *Einstein's Miraculous Year; Five Papers that Changed the Face of Phyiscs.* Princeton, NJ: Princeton University Press.

[218] Stokes, G. G. (1852). On the composition and resolution of streams of polarized light from different sources. *Trans. Cambridge Phil. Soc. 9*, 399–416. (reprinted in Mathematical and Physical Papers. London, U.K.: Cambridge University Press, 1901, vol. 3, pp. 233–250).

[219] Stone, J. A. and J. H. Zimmerman (2001). Index of refraction of air. *National Institute of Science and Technology, Gaithersburg, Maryland 12.*

[220] Stratton, J. (2007). *Electromagnetic Theory.* Piscataway, NJ: Wiley-IEEE Press.

[221] Sutherland, R. L. (Ed.) (2003). *Handbook of Nonlinear Optics.* Optical Science and Engineering. Boca Raton, FL: CRC Press.

[222] Tao, X., H. Cho, and F. Janabi-Sharifi (2010). Optical design of a variable view imaging system with the combination of a telecentric scanner and double wedge prisms. *Applied Optics 49*(2), 239–246.

[223] Teuchner, K., J. Ehlert, W. Freyer, D. Leupold, P. Altmeyer, M. Stuecker, and K. Hoffmann (2000). Fluorescence studies of melanin by stepwise two-photon femtosecond laser excitation. *Journal of Fluorescence 10*(3), 275 – 281.

[224] Thompson, A. and B. N. Taylor (2008, March). *Guide for the Use of the International System of Units (SI),* Volume SP811 (2008 Edition). Gaithersburg, MD 20899: National Institute of Standards and Technology.

[225] Thorlabs (2010). *Thorlabs Catalog* (20th ed.). Newton, NJ: Thorlabs.

[226] Tributsch, H., H. Goslowsky, U. Küppers, and H. Wetzel (1990). Light collection and solar sensing through the polar bear pelt. *Solar Energy Materials 21*(2-3), 219–236.

[227] Tzannes, A. and J. Mooney (1995). Measurement of the modulation transfer function of infrared cameras. *Optical Engineering 34*, 1808–1817.

[228] United States Armed Forces Support Center (1959, May). *Military Standard: Photographic Lenses.* Superintendent of Documents, U.S. Govenment Printing Office: United States Armed Forces Support Center.

[229] Unser, M. (2000). Sampling–50 years after shannon. In *Proceedings of the IEEE,* pp. 569–587.

[230] Valero, E., J. Nieves, S. Nascimento, K. Amano, and D. Foster (2007). Recovering spectral data from natural scenes with an RGB digital camera and colored filters. *Color Research and Application 32*(5), 352–359.

[231] van Cittert, P. H. (1934). Die Wahrscheinliche Schwingungsverteilung in Einer von Einer Lichtquelle Direkt Oder Mittels Einer Linse Beleuchteten Ebene. *Physica 1*, 201–210.

[232] van der Hulst, H. C. (1957). *Light Scattering by Small Particles.* New York: John Wiley & Sons.

[233] Verdet, É. (1870). *Leçons d'optique physique.* Imprimerie impériale.

[234] Vernon, R. and B. Huggins (1980). Extension of the Jones matrix formalism to reflection problems and magnetic materials. *Journal of the Optical Society of America 70*(11), 1364–1370.

[235] Veselago, V. (1968). The Electrodynamics of Substances with Simultaneously Negative Values of ϵ and μ. *Physics-Uspekhi 10*(4), 509–514.

[236] Vincelette, R., A. Welch, R. Thomas, B. Rockwell, and D. Lund (2008). Thermal lensing in ocular media exposed to continuous-wave near-infrared radiation: the 1150–1350-nm region. *Journal of Biomedical Optics 13*, 054005.

[237] Vo-Dinh, T. (Ed.) (2003). *Biomedical Photonics Handbook.* Boca Raton, FL: CRC Press.

[238] Wachtler, T., T. Lee, and T. Sejnowski (2001). Chromatic structure of natural scenes. *Journal of the Optical Society of America A 18*(1), 65–77.

[239] Wadsworth, F. (1903). XXXIV. On the effect of absorption on the resolving power of prism trains, and on methods of mechanically compensating this effect. *Philosophical Magazine Series 6 5*(27), 355–374.

[240] Walsberg, G. (1983). Coat color and solar heat gain in animals. *Bioscience 33*(2), 88–91.

[241] Warger, II, W. and C. DiMarzio (2007). Dual-wedge scanning confocal reflectance microscope. *Optics Letters 32*(15), 2140–2142.

[242] Warger, II, W. C. and C. A. DiMarzio (2008, December). Compact confocal microscope, based on a dual-wedge scanner. United States Patent 7,471,450.

[243] Weber, R. (1973). *A Random Walk in Science.* Taylor & Francis.

[244] Westfall, R. S. (1985). Science and patronage: Galileo and the telescope. *Isis 76*(1), 11–30.

[245] Whalen, R. and G. Kowalski (1999//). Microscale heat transfer in thermally stimulated nonlinear optical materials. In *Proceedings of the SPIE - The International Society for Optical Engineering*, Volume 3798, USA, pp. 85–92.

[246] Whittaker, E. (1960). *A History of the Theories of Aether and Electricity: The Classical Theories. 2. The Modern Theories, 1900-1926.* Harper.

[247] Whittaker, J. M. (1929). The fourier theory of the cardinal functions. In *Proc. Math. Soc. Edinburgh*, Volume 1, pp. 140.

[248] Whittaker, J. M. (1935). *Interpolatory Function Theory*, Chapter IV. Number 33. Cambridge University Press.

[249] Wien, W. (1893, Feb). Eine neue beziehung der strahlung schwarzer körper zum zweiten haupsatz. der wärmetheorie. *Sitzb. Akad. Wiss. Berlin*, 55–62.

[250] Wiener, N. (1930). Generalized harmonic analysis. *Acta Math. 55*, 117–258.

[251] Wolfe, W. and G. Zissis (1978). *The Infrared Handbook.* Office of Naval Research, Department of the Navy.

[252] Worrall, J. (1994). How to remain (reasonably) optimistic: Scientific realism and the "luminiferous ether". *PSA: Proceedings of the Biennial Meeting of the Philosophy of Science Association 1994*, 334–342.

[253] Yaglom, A. (1987, Oct). Einstein's 1914 paper on the theory of irregularly fluctuating series of observations. *ASSP Magazine, IEEE 4*(4), 7–11.

[254] Yariv, A. (1985). *Quantum Electronics* (3rd ed.). John Wiley & Sons.

[255] Yariv, A. and D. Pepper (1977). Amplified reflection, phase conjugation, and oscillation in degenerate four-wave mixing. *Optics Letters 1*(1), 16.

[256] Yee, K. (1966). Numerical solution of initial boundary value problems involving maxwell's equation in isotropic media. *IEEE Transactions on Antennas and Propagation 40*, 1068–1075.

[257] Yodh, A. and B. Chance (1995). Spectroscopy and imaging with diffusing light. *Physics Today 48*(3), 34–41.

[258] Young, T. (1845). *A Course of Lectures on Natural Philosophy and the Mechanical Arts.* London: Taylor and Walton.

[259] Zehnder, L. (1891). Ein neuer interferenzrefraktor. *Z. Instrumentenkunde 11*, 275–285.

[260] Zernike, F. (1935). Das Phasenkontrastverfahren bei der mikroskopischen Beobachtung. *Z. Tech. Phys 16*, 454–457.

[261] Zernike, F. (1938). The concept of degree of coherence and its application to optical problems. *Physica 5*(8), 785–795.

Index

For Product Safety Concerns and Information please contact our EU
representative GPSR@taylorandfrancis.com
Taylor & Francis Verlag GmbH, Kaufingerstraße 24, 80331 München, Germany

www.ingramcontent.com/pod-product-compliance
Lightning Source LLC
Chambersburg PA
CBHW082101220326
41598CB00066BA/4549

* 9 7 8 1 4 8 2 2 6 3 2 3 7 *